International Handbook on Information Systems

Series Editors

Peter Bernus, Jacek Błażewicz, Günter Schmidt, Michael Shaw

Titles in the Series

Jacek Błażewicz
Klaus H. Ecker
Erwin Pesch
Günter Schmidt
Jan Węglarz

Handbook
on Scheduling

From Theory to Applications

With 144 Figures and 28 Tables

  Springer

Professor Dr. Jacek Błażewicz
Poznań University of Technology
Institute of Computing Science
Piotrowo 3a
60-965 Poznań
Poland
jblazewicz@cs.put.poznan.pl

Professor Dr. Klaus H. Ecker
Ohio University
School of Electrical Engineering
and Computer Science
Stocker Center 339
Athens OH 45701
USA
ecker@ohio.edu

Professor Dr. Günter Schmidt
University of Saarland
Information and Technology
Management
Bau A54, Im Stadtwald
66123 Saarbrücken
Germany
gs@itm.uni-sb.de

Professor Dr. Jan Węglarz
Poznań University of Technology
Institute of Computing Science
Piotrowo 3a
60-965 Poznań
Poland
jan.weglarz@cs.put.poznan.pl

Professor Dr. Erwin Pesch
University of Siegen
FB 5 - Management Information
Sciences
Hölderlinstraße 3
57068 Siegen
Germany
erwin.pesch@uni-siegen.de

Library of Congress Control Number: 2007925715

ISBN 978-3-540-28046-0 Springer Berlin Heidelberg New York

Springer is a part of Springer Science+Business Media

springer.com

© Springer-Verlag Berlin Heidelberg 2007

Production: LE-TEX Jelonek, Schmidt & Vöckler GbR, Leipzig
Cover-design: WMX Design GmbH, Heidelberg

SPIN 11532057   42/3180YL - 5 4 3 2 1 0   Printed on acid-free paper

FOREWORD

This handbook is in a sense a continuation of *Scheduling Computer and Manufacturing Processes* [1], two editions of which have received kind acceptance of a wide readership. As the previous volume, it is the result of a long lasting German-Polish collaboration. However, due to important reasons, it has a new form. Namely, following the suggestions of the Publisher, we decided to prepare a handbook filling out a gap on the market in the area. The gap concerns a unified approach to the most important scheduling models and methods with the special emphasis put on their relevance to practical situations. Thus, in comparison with [1], the contents has been changed significantly. This concerns not only corrections we have introduced, following the suggestions made by many readers (we are very grateful to all of them) and taking into account our own experience, but first of all this means that important new material has been added. It is characterized in Chapter 1, and, generally speaking, covers a transition from theory to applications in a wide spectrum of scheduling problems. Independently of this, in all chapters new results have been reported and new illustrative material, including real-world problems, has been given.

We very much hope that in this way the handbook will be of interest to a much wider readership than the former volume, the fact which has been underlined in the title.

During the preparation of the manuscript many colleagues have discussed with us different topics presented in the book. We are not able to list all of them but we would like to express our special gratitude towards Oliver Braun, Nadia Brauner, Joachim Breit, Edmund Burke, Ulrich Dorndorf, Maciej Drozdowski, Gerd Finke, Graham Kendall, Tamas Kis, Misha Kovalyov, Wieslaw Kubiak, Chung-Yee Lee, Ceyda Oguz, Toan Phan-Huy, Malgorzata Sterna, Denis Trystram and Dominique deWerra. Special thanks are due to Adam Janiak who corrected and updated Section 12.3.

References

[1]   J. Błażewicz, K. Ecker, E.Pesch, G. Schmidt, J. Węglarz, *Scheduling Computer and Manufacturing Processes*, Springer Verlag, Berlin, 2001 (2-nd edition).

Contents

1 Introduction

Scheduling problems can be understood in general as the problems of allocating resources over time to perform a set of tasks being parts of some processes, among which computational and manufacturing ones are most important. Tasks individually compete for resources which can be of a very different nature, e.g. manpower, money, processors (machines), energy, tools. The same is true for task characteristics, e.g. ready times, due dates, relative urgency weights, functions describing task processing in relation to allotted resources. Moreover, a structure of a set of tasks, reflecting relations among them, can be defined in different ways. In addition, different criteria which measure the quality of the performance of a set of tasks can be taken into account.

It is easy to imagine that scheduling problems understood so generally appear almost everywhere in real-world situations. Of course, there are many aspects concerning approaches for modeling and solving these problems which are of general methodological importance. On the other hand, however, some classes of scheduling problems have their own specificity which should be taken into account. Since it is rather impossible to treat all these classes with the same attention in a framework of one book, some constraints must be put on the subject area considered. In the case of this handbook these constraints are as follows.

First of all we focus on the problems motivated by applications from industry and service operations management as well as from case studies of real - life problems. Among others there is a detailed description of optimization procedures for acrylic-glass production and the production of helicopter parts in a flexible manufacturing system. We will describe the backbone of an efficient decision support system for airport gate scheduling as well as a flexible flow shop scheduling system in order to manufacture screws and car power brakes.

Second, we deal with deterministic scheduling problems (cf. [Bak74, BCSW86, Bru04, CCLL95, CMM67, Cof76, Eck77, Fre82, GK87, Len77, Leu04, LLR+93, Pin01, Pin05, Rin76, TGS94, TSS94]), i.e. those in which no variable with a non-deterministic (e.g. probabilistic) description appears. Let us stress that this assumption does not necessarily mean that we deal only with static problems in which all characteristics of a set of tasks and a set of resources are known in advance. We consider also dynamic problems in which some parameters such as task ready times are unknown in advance, and we do not assume any a priori knowledge about them; this approach is even more realistic in many practical situations.

Third, we consider problems in which a set of resources always contains processors (machines). This means that we take into account the specificity of these particular resources in modeling and solving corresponding scheduling problems, but it does not mean that all presented methods and approaches are restricted to this specificity only. The main reason for which we differentiate processors (we even do not call them "resources" for the convenience of a reader) is that we like to expose especially two broad (and not exclusive) areas of practical applications of the considered problems, namely computer and manufacturing systems.

After the explanation of the handbook's title, we can pass to the description of some of its deeper specificities. They can be meant as compromise we accepted in multi-objective decision situations we had to deal with before and during the preparation of the text. At the beginning, we found a compromise between algorithmic (rather quantitative) and knowledge-based (rather qualitative) approaches to scheduling. We decided to present in the first chapters the algorithmic approach, and at the end to show how it can be integrated with the approach coming from the area of Artificial Intelligence, in order to create a pretty general and efficient tool for solving a broad class of practical problems. In this way we also hopefully found a compromise between rather more computer and rather more manufacturing oriented audience.

The second compromise was concerned with the way of presenting algorithms: formal or descriptive. Basically we decided to adopt a Pascal-like notation, although we allowed for few exceptions in cases where such a presentation would be not efficient.

Next we agreed that the presented material should realize a reasonable compromise between the needs of readers coming from different areas, starting from those who are beginners in the field of scheduling and related topics, and ending with specialists in the area. Thus we included some preliminaries concerning basic notions from discrete mathematics (problems, algorithms and methods), as well as, besides the most recent, also the most important classical results.

Summing up the above compromises, we think that the handbook can be addressed to a quite broad audience, including practitioners and researchers interested in scheduling, and also to graduate or advanced undergraduate students in computer science/engineering, operations research, industrial engineering, management science, business administration, information systems, and applied mathematics curricula.

Finally, we present briefly the outline of the handbook.

In **Chapter 2** basic definitions and concepts used throughout the book are introduced. One of the main issues studied here is the complexity analysis of combinatorial problems. As a unified framework for this presentation the concept of a combinatorial search problem is used. Such notions as: decision and optimization problems, their encoding schemes, input length, complexity classes of problems, are discussed and illustrated by several examples. Since the majority of scheduling problems are computationally hard, two general approaches dealing with such problems are briefly discussed: enumerative and heuristic. First, general enumerative approaches, i.e. dynamic programming and branch and bound are shortly presented. Second, heuristic algorithms are introduced and the ways of analysis of their accuracy in the worst case and on the average are described. Then, we introduce the ideas of general local search metaheuristics known under names: simulated annealing, tabu search, and ejection chains as well as genetic algorithms. In contrast with previous approaches (e.g. hill-climbing) to deal with combinatorially explosive search spaces about which little knowledge is known a priori, the above mentioned metaheuristics, in combination with problem specific knowledge, are able to escape a local optimum. Basically, the only thing that matters, is the definition of a neighborhood structure in order to step from a current solution to a next, probably better one. The neighborhood structure has to be defined within the setting of the specific problem and with respect to the accumulated knowledge of the previous search process. Furthermore, frequently some random elements guarantee a satisfactory search diversification over the solution space. During the last years the metaheuristics turned out to be very successful in the scheduling area; specific applications will be described in Chapters 8,9,10 and 14.

The chapter is complemented by a presentation of the notions from sets and relations, as well as graphs and networks, which will be used in the later chapters.

In **Chapter 3** definitions, assumptions and motivations for deterministic scheduling problems are introduced. We start with the set of tasks, the set of processors (machines) and the set of resources, and with two basic models in deterministic scheduling, i.e. parallel processors and dedicated processors. Then, schedules and their performance measures (optimality criteria) are described. After this introduction, possible ways of analyzing scheduling problems are described with a special emphasis put to solution strategies of computationally hard problems. Finally, motivations for the use of the deterministic scheduling model as well as an interpretation of results, are discussed. Two major areas of applications, i.e. computer and manufacturing systems are especially taken into account. These considerations are complemented by a description of a classification scheme which enables one to present deterministic scheduling problems in a short and elegant way.

Chapter 4 deals with single-processor scheduling. The results given here are mainly concerned with polynomial time optimization algorithms. Their presentation is divided into several sections taking into account especially the following optimality criteria: schedule length, mean (and mean weighted) flow time, and due date involving criteria, such as maximum lateness, number of tardy tasks, mean (and mean weighted) tardiness and a combination of earliness and lateness. In each case polynomial time optimization algorithms are presented, taking into account problem parameters such as the type of precedence constraints, possibility of task preemption, processing and arrival times of tasks, etc. These basic results are complemented by some more advanced ones which take into account change-over cost, also called lot size scheduling, and more general cost functions. Let us stress that in this chapter we present a more detailed classification of subcases as compared to the following chapters. This follows from the fact that the algorithms for the single-processor model are useful also in more complicated cases, whenever a proper decomposition of the latter is carried out. On the other hand, its relative easiness makes it possible to solve optimally in polynomial time many more subcases than in the case of multiple processors.

Chapter 5 carries on an analysis of scheduling problems where multiple parallel processors are involved. As in Chapter 4, a presentation of the results is divided into several subsections depending mainly on the criterion considered and then on problem parameters. Three main criteria are analyzed: schedule length, mean flow time and maximum lateness. A further division of the presented results takes in particular into account the type of processors considered, i.e. identical, uniform or unrelated processors, and then parameters of a set of tasks. Here, scheduling problems are more complicated than in Chapter 4, so not as many optimization polynomial time algorithms are available as before. Hence, more attention is paid to the presentation of polynomial time heuristic algorithms with guaranteed accuracy, as well as to the description of some enumerative algorithms.

In **Chapter 6** new scheduling problems arising in the context of rapidly developing manufacturing as well as parallel computer systems, are considered. When formulating scheduling problems in such systems, one must take into account the fact that some tasks have to be processed on more than one processor at a time. On the other hand, communication issues must be also taken into account in systems where tasks (program modules) are assigned to different processors and exchange information between each other. In the chapter three models are discussed in a sequel. The first model assumes that each so-called multiprocessor task may require more than one processor at a time and communication times are implicitly included in tasks' processing times. The second model assumes that uniprocessor tasks, each assigned to one processor, communicate explicitly via directed links of the task graph. More precise approaches distinguish between coarse grain and fine grain parallelism and discuss their impact on communication delays. Furthermore, task duplication often leads to shorter schedules; this is in particular the case if the communication times are large compared to the processing times. The last model is a combination of the first two models and involves the so called divisible tasks.

Chapter 7 deals with another type of scheduling problems where the tasks are periodic in the sense that they are processed repeatedly and with given frequencies. Particularly in real-time systems designed for controlling some technical facility we are confronted with problems where sets of periodic tasks are to be processed on a single processor or on a distributed or parallel processor system. The chapter starts with a short introduction to real-time systems and discusses characteristic properties and general functional requirements of such systems. Then strategies for scheduling sets of periodic tasks on a single processor and on a multiprocessor system are presented, and the classical results for the rate monotonic and earliest deadline scheduling strategies are discussed from their properties and performance points of view. Another important issue regards runtime problems that appear if tasks use of non-preemptable (non-withdrawable) resources. Finally, several variants of the periodic task model allowing higher flexibility as compared to the simple periodic task model are presented.

In **Chapter 8** flow shop scheduling problems are described, i.e. scheduling a set of jobs (composed of tasks) in shops with a product machine layout. Thus, the jobs have the same manufacturing order. Recent local search heuristics as well as heuristics relying on the two-machine flow shop scheduling problem - which can easily be solved optimally - are considered. Some special flow shops are introduced, e.g. permutation and no-wait ones. The hybrid or flexible flow-shop problem is a generalization of the flow shop in such a way that every job can be processed by one among several machines on each machine stage. In recent years a number of effective exact methods have been developed. A major reason for this progress is the development of new job and machine based lower bounds as well as the rapidly increasing importance of constraint programming. We provide a comprehensive and uniform overview on exact solution methods for flexible flowshops with branching, bounding and propagation of constraints, under two different objective functions: minimizing the makespan of a schedule and the mean flow time. For some simple cases we present heuristics with known worst case performance and then describe a branch and bound algorithm for the general case.

In **Chapter 9** we consider the open shop problem where jobs without any precedence constraints are supposed to be scheduled. Only few exact solution methods are available and we motivate our presentation with a description of optimal results for small open shop scheduling problems. We continue describing a branch-and-bound algorithm for solving this problem which performs better than any other existing algorithm. The key to the efficiency of the algorithm lies in the following approach: instead of analyzing and improving the search strategies for finding solutions, the focus is on constraint propagation based methods for reducing the search space. For the first time, many problem instances are solved to optimality in a short amount of computation time.

In **Chapter 10** job shop scheduling problems are investigated. This is the most general form of manufacturing a set of different jobs on a set of machines where each job is characterized by its specific machine order reflecting the jobs production process. The most successful branch and bound ideas are described and we will see that their branching structure is reflected in the neighborhood definitions of many local search methods. In particular tabu search, ejection chains, genetic algorithms as well as the propagation of constraints - this is closely related to the generation of valid inequalities - turned out to become the currently most powerful solution approaches. Moreover, we introduce priority rule based scheduling and describe a well-known opportunistic approach: the shifting bottleneck procedure.

Chapter 11 deals with scheduling problems where the availabilities of processors to process tasks are limited. In the preceding chapters the basic model assumes that all machines are continuously available for processing throughout the planning horizon. This assumption might be justified in some cases but it does not apply if certain maintenance requirements, breakdowns or other constraints that cause the machines not to be available for processing have to be considered. In this chapter we generalize the basic model in this direction and discuss results related to one machine, parallel machines, and shop scheduling problems where machines are not continuously available for processing.

Chapter 12 deals with resource constrained scheduling. In the first two sections it is assumed that tasks require for their processing processors and certain fixed amounts of discrete resources. The first section presents the classical model where schedule length is to be minimized. In this context several polynomial time optimization algorithms are described. In the next section this model is generalized to cover also the case of multiprocessor tasks. Two algorithms are presented that minimize schedule length for preemptable tasks under different assumptions concerning resource requirements. The last section deals with problems in which additional resources are continuous, i.e. continuously-divisible. We study three classes of scheduling problems of that type. The first one contains problems with parallel processors and tasks described by continuous functions relating their processing speeds to the resource amount allotted at a time. The next two classes are concerned with single processor problems where task processing times or ready times, respectively, are continuous functions of the allotted resource amount.

Constraint propagation is the central topic of **Chapter 13**. It is an elementary method for reducing the search space of combinatorial search and optimization problems which has become more and more important in the last decades. The basic idea of constraint propagation is to detect and remove inconsistent variable assignments that cannot participate in any feasible solution through the repeated analysis and evaluation of the variables, domains and constraints describing a specific problem instance. We describe efficient constraint propagation methods also known as consistency tests for the disjunctive scheduling problem (DSP) applications of which will be introduced in machine scheduling chapters 8 to 10. We will further present and analyze both new and classical consistency tests involving a higher number of variables. They still can be implemented efficiently in a polynomial time. Further, the concepts of energetic reasoning and shaving are analyzed and discussed.

The other contribution is a classification of the consistency tests derived according to the domain reduction achieved. The particular strength of using consistency tests is based on their repeated application, so that the knowledge derived is propagated, i.e. reused for acquiring additional knowledge. The deduction of this knowledge can be described as the computation of a fixed point. Since this fixed point depends upon the order of the application of the tests, we first derive a necessary condition for its uniqueness. We then develop a concept of dominance which enables the comparison of different consistency tests as well as a method for proving dominance.

Chapter 14 is devoted to problems which perhaps closer reflect some specific features of scheduling in flexible manufacturing systems than other chapters do. Dynamic job shops are considered, i.e. such in which some events, particularly job arrivals, occur at unknown times. A heuristic for a static problem with mean tardiness as a criterion is described. It solves the problem at each time when necessary, and the solution is implemented on a rolling horizon basis. The next section deals with simultaneous assignment of machines and vehicles to jobs. This model is motivated by the production of helicopter parts in somefactory First we solve in polynomial time the problem of finding a feasible vehicle schedule for a given assignment of tasks to machines, and then present a dynamic programming algorithm for the general case. In the last section we are modeling manufacturing of acrylic-glass as a batch scheduling problem on parallel processing units under resource constraints. This section introduces the real world problem and reveals potential applications of some of the material in the previous chapters. In particular, a local search heuristic is described for constructing production sequences.

Chapter 15 serves two purposes. On one hand, we want to introduce a quite general solution approach for scheduling problems as they appear not only in manufacturing environments. On the other hand, we also want to survey results from interactive and knowledge-based scheduling which were not covered in this handbook so far. To combine both directions we introduce some broader aspects like computer integrated manufacturing and object-oriented modeling. The common goal of this chapter is to combine results from different areas to treat scheduling problems in order to answer quite practical questions. To achieve this we first concentrate on the relationship between the ideas of computer integrated manufacturing and the requirements concerning solutions of the scheduling problems. We present an object-oriented reference model which is used for the implementation of the solution approach. Based on this we discuss the outline of an intelligent production scheduling system using open loop interactive and closed loop knowledge-based problem solving. For reactive scheduling we suggest to use concepts from soft computing. Finally, we make some proposals concerning the integration of solution approaches discussed in the preceding chapters with the ideas developed in this chapter. We use an example to clarify the approach of integrated problem solving and discuss the impact for computer integrated manufacturing.

References

Bak74   K. Baker, *Introduction to Sequencing and Scheduling*, J. Wiley, New York, 1974.

BCSW86   J. Błażewicz, , W. Cellary, R. Słowiński , J. Węglarz, *Scheduling under Resource Constraints: Deterministic Models*, J. C. Baltzer, Basel, 1986.

Bru04   P. Brucker, *Scheduling Algorithms*, Springer, 4. edition, Berlin, 2004.

CCLL95   P. Chretienne, E. G. Coffman, J. K. Lenstra, Z. Liu (eds.), *Scheduling Theory and its Applications*, Wiley, New York, 1995.

CMM67   R. W. Conway, W. L. Maxwell, L. W. Miller, *Theory of Scheduling*, Addison-Wesley, Reading, Mass. 1967.

Cof76   E. G. Coffman, Jr. (ed.), *Scheduling in Computer and Job Shop Systems*, J. Wiley, New York, 1976.

Eck77   K. Ecker, *Organisation von parallelen Prozessen*, BI-Wissenschaftsverlag, Mannheim, 1977.

Fre82   S. French, *Sequencing and Scheduling: An Introduction to the Mathematics of the Job-Shop*, Horwood, Chichester, 1982.

GK87   S. K. Gupta, J. Kyparisis, Single machine scheduling research, *OMEGA Internat. J. Management Sci.* 15, 1987, 207-227.

Len77   J. K. Lenstra, *Sequencing by Enumerative Methods*, Mathematical Centre Tract 69, Amsterdam, 1977.

Leu04   J.Y.-T. Leung (ed.), *Handbook of Scheduling: Algorithms, Models and Performance Analysis*, Chapman & Hall, Boca Raton, 2004.

LLR+93   E. L. Lawler, J. K. Lenstra, A. H. G. Rinnooy Kan, D. B. Shmoys, Sequencing and scheduling: Algorithms and complexity, in: S. C. Graves, A. H. G. Rinnooy Kan, P. H. Zipkin (eds.), *Handbook in Operations Research and Management Science, Vol. 4: Logistics of Production and Inventory*, Elsevier, Amsterdam, 1993.

Pin01   M. Pinedo, *Scheduling: Theory, Algorithms, and Systems*, Prentice Hall, 2.edition, Englewood Cliffs, N. J., 2001.

Pin05   M. Pinedo, *Planning and Scheduling in Manufacturing and Services*, Springer, New York, 2005.

Rin76   A. H. G. Rinnooy Kan, *Machine Scheduling Problems; Classification, Complexity and Computations,* Martinus Nijhoff, The Hague, 1976.

TGS94   V. S. Tanaev, V. S. Gordon and Y. M. Shafransky, *Scheduling Theory. Single-Stage Systems*, Kluwer, Dordrecht, 1994.

TSS94   V. S. Tanaev, Y. N. Sotskov and V. A. Strusevich, *Scheduling Theory. Multi-Stage Systems*, Kluwer, Dordrecht, 1994.

2 Basics

In this chapter we provide the reader with basic notions used throughout the book. After a short introduction into sets and relations, decision problems, optimization problems and the encoding of problem instances are discussed. The way algorithms will be represented and problem membership of complexity classes are other essential issues which will be discussed. Afterwards graphs, especially certain types such as precedence graphs and networks that are important for scheduling problems, are presented. The last two sections deal with algorithmic methods used in scheduling such as enumerative algorithms (e. g. dynamic programming and branch and bound) and heuristic approaches (e. g. tabu search, simulated annealing, ejection chains, and genetic algorithms).

2.1 Sets and Relations

Sets are understood to be any collection of distinguishable objects, such as the set $\{1, 2, \cdots\}$ of natural numbers, denoted by $I\!N$, the set $I\!N_0$ of non-negative integers, the set of real numbers, $I\!R$, or the set of non-negative reals $I\!R_{\geq 0}$. Given real numbers a and b, $a \leq b$, then $[a, b]$ denotes the *closed interval* from a to b, i.e. the set of reals $\{x \mid a \leq x \leq b\}$. *Open intervals* $((a, b) := \{x \mid a < x < b\})$ and *half open intervals* are defined similarly.

In scheduling theory we are normally concerned with finite sets; so, unless infinity is stated explicitly, the sets are assumed to be finite.

For set S, $|S|$ denotes its *cardinality*. The *power set* of S (i.e. the set of all subsets of S) is denoted by $\mathcal{P}(S)$. For an integer k, $0 \leq k \leq |S|$, the set of all subsets of cardinality k is denoted by $\mathcal{P}_k(S)$.

The *Cartesian product* $S_1 \times \cdots \times S_k$ of sets $S_1, \cdots, S_k$ is the set of all tuples of the form $(s_1, s_2, \cdots, s_k)$ where $s_i \in S_i$, $i = 1, \cdots, k$, i.e. $S_1 \times \cdots \times S_k = \{(s_1, \cdots, s_k) \mid s_i \in S_i, i = 1, \cdots, k\}$. The k-fold Cartesian product $S \times \cdots \times S$ is denoted by S^k.

Given sets $S_1, \cdots, S_k$, a subset Q of $S_1 \times \cdots \times S_k$ is called a *relation over* $S_1, \cdots, S_k$. In the case $k = 2$, Q is called a *binary relation*. For a binary relation Q over S_1 and S_2, the sets S_1 and S_2 are called *domain* and *range*, respectively. If Q is a relation over $S_1, \cdots, S_k$, with $S_1 = \cdots = S_k = S$, then we simply say: Q is a (*k-ary*) *relation over* S. For example, the set of edges of a directed graph (see Section 2.3) is a binary relation over the vertices of the graph.

Let S be a set, and Q be a binary relation over S. Then, $Q^{-1} = \{(a, b) \mid (b, a) \in Q\}$ is the *inverse* to Q. Relation Q is *symmetric* if $(a, b) \in Q$ implies $(b, a) \in Q$. Q is *antisymmetric* if for $a \ne b$, $(a, b) \in Q$ implies $(b, a) \notin Q$. Q is *reflexive* if $(a, a) \in Q$ for all $a \in S$. Q is *irreflexive* if $(a, a) \notin Q$ for all $a \in S$. Q is *transitive* if for all $a, b, c \in S$, $(a, b) \in Q$ and $(b, c) \in Q$ implies $(a, c) \in Q$.

A binary relation over S is called a *partial order* (partially ordered set, *poset*) if it is reflexive, antisymmetric and transitive. A binary relation over S is called an *equivalence* relation (over S) if it is reflexive, symmetric, and transitive.

Given set $\mathcal{J}$ of n closed intervals of reals, $\mathcal{J} = \{I_i \mid I_i = [a_i, b_i], a_i \le b_i, i = 1, \cdots, n\}$, a partial order $\le_I$ on $\mathcal{J}$ can be defined by

$$I_i \le_I I_j \iff (I_i = I_j) \text{ or } (b_i \le a_j), \ i, j \in \{1, \cdots, n\}.$$

A poset is called *interval order* if there exists a set of intervals whose partial order $\le_I$ represents the poset.

Let $l = (n_1, \cdots, n_k)$ and $l' = (n'_1, \cdots, n'_{k'})$ be sequences of integers, and $k, k' \ge 0$. If $k = 0$ then l is the empty sequence. We say that l is *lexicographically smaller* than l', written $l \lessdot l'$, if

(*i*) the two sequences agree up to some index j, but $n_{j+1} < n'_{j+1}$ (i.e. there exists j, $0 \le j \le k$, such that for all i, $1 \le i \le j$, $n_i = n'_i$ and $n_{j+1} < n'_{j+1}$), or if

(*ii*) sequence l is shorter, and the two sequences agree up to the length of l (i.e. $k < k'$ and $n_i = n'_i$ for all i, $1 \le i \le k$).

If Q is a binary relation over set S, then $Q^2 = Q \circ Q$ is the *relational product* of Q and Q. Generally, we write Q^0 for $\{(a, a) \mid a \in S\}$, $Q^1 = Q$, and $Q^{i+1} = Q^i \circ Q$ for $i > 1$. The union $Q^* = \cup\{Q^i \mid i \ge 0\}$ is called the *transitive closure* of Q.

A *function* from $\mathcal{A}$ to $\mathcal{B}$ ($\mathcal{A} \to \mathcal{B}$; $\mathcal{A}$ and $\mathcal{B}$ are not necessarily finite) is a relation F over $\mathcal{A}$ and $\mathcal{B}$ such that for each $a \in \mathcal{A}$ there exists just one $b \in \mathcal{B}$ for which $(a, b) \in F$; instead of $(a, b) \in F$ we usually write $F(a) = b$. Set $\mathcal{A}$ is called the *domain* of F and set $\{b \mid b \in \mathcal{B}, \exists a \in \mathcal{A}, (a, b) \in F\}$ is called the *range* of F. F is called *surjective*, or *onto* $\mathcal{B}$ if for each element $b \in \mathcal{B}$ there is at least one element $a \in \mathcal{A}$ such that $F(a) = b$. Function F is said to be *injective*, or *one-one* if for each pair of elements, $a_1, a_2 \in \mathcal{A}$, $F(a_1) = F(a_2)$ implies $a_1 = a_2$. A function that is both surjective and injective is called *bijective*. A bijective function $F: \mathcal{A} \to \mathcal{A}$ is called a *permutation* of $\mathcal{A}$. Though we are able to represent functions in special cases by means of tables we usually specify functions in a more or less abbreviated way that specifies how the function values are to be determined. For example, for $n \in I\!N$, the factorial function $n!$ denotes the set of pairs $\{(n, m) \mid n \in I\!N, \; m = n \cdot (n - 1) \cdots 3 \cdot 2\}$. Other examples of functions are polynomials, exponential functions and logarithms.

We will say that function $f : I\!N \to I\!R$ is *of order g*, written $O(g)$, if there exist constants c and $k_0 \in I\!N$ such that $f(k) \leq cg(k)$ for all $k \geq k_0$.

2.2 Problems, Algorithms, Complexity

2.2.1 Problems and Their Encoding

In general, the scheduling problems we consider belong to a broader class of combinatorial search problems. A *combinatorial search problem* Π is a set of pairs (I, A), where I is called an *instance* of a problem, i.e. a finite set of *parameters* (understood generally, e.g. numbers, sets, functions, graphs) with specified values, and A is an *answer* (*solution*) to the instance. As an example of a search problem let us consider *merging* two sorted sequences of real numbers. Any instance of this problem consists of two finite sequences of reals e and f sorted in non-decreasing order. The answer is the sequence g consisting of all the elements of e and f arranged in non-decreasing order.

Let us note that among search problems one may also distinguish two subclasses: optimization and decision problems. An *optimization problem* is defined in such a way that an answer to its instance specifies a solution for which a value of a certain objective function is at its optimum (an *optimal solution*). On the other hand, an answer to an instance of a *decision problem* may take only two values, either "yes" or "no". It is not hard to see, that for any optimization problem, there always exists a decision counterpart, in which we ask (in the case of minimization) if there exists a solution with the value of the objective function less than or equal to some additionally given threshold value y. (If in the basic problem the objective function has to be maximized, we ask if there exists a solution with the value of the objective function $\geq y$.) The following example clarifies these notions.

Example 2.2.1 Let us consider an optimization *knapsack problem*.

Knapsack

Instance: A finite set of elements $\mathcal{A} = \{a_1, a_2, \cdots, a_n\}$, each of which has an integer weight $w(a_i)$ and value $v(a_i)$, and an integer capacity b of a knapsack.

Answer: Subset $\mathcal{A}' \subseteq \mathcal{A}$ for which $\sum\limits_{a_i \in \mathcal{A}'} v(a_i)$ is at its maximum, subject to the constraint $\sum\limits_{a_i \in \mathcal{A}'} w(a_i) \leq b$ (i.e. the total value of chosen elements is at its maximum and the sum of weights of these elements does not exceed knapsack capacity b).

The corresponding decision problem is denoted as follows. (To distinguish optimization problems from decision problems the latter will be denoted using capital letters.)

KNAPSACK

Instance: A finite set of elements $\mathcal{A} = \{a_1, a_2, \cdots, a_n\}$, each of which has an integer weight $w(a_i)$ and value $v(a_i)$, an integer knapsack capacity b and threshold value y.

Answer: "Yes" if there exists subset $\mathcal{A}' \subseteq \mathcal{A}$ such that

$$\sum_{a_i \in \mathcal{A}'} v(a_i) \geq y \text{ and } \sum_{a_i \in \mathcal{A}'} w(a_i) \leq b.$$

Otherwise "No".   □

When considering search problems, especially in the context of their solution by computer algorithms, one of the most important issues that arises is a question of data structures used to encode problems. Usually to encode instance I of problem Π (that is particular values of parameters of problem Π) one uses a finite *string* of symbols $x(I)$. These symbols belong to a predefined finite set Σ (usually called an *alphabet*) and the way of coding instances is given as a set of encoding rules (called *encoding scheme e*). By *input length* (*input size*) $|I|$ of instance I we mean here the length of string $x(I)$. Let us note that the requirement that an instance of a problem is encoded by a finite string of symbols is the only constraint imposed on the class of search problems which we consider here. However, it is rather a theoretical constraint, since we will try to characterize algorithms and problems from the viewpoint of the application of real computers.

Now the encoding scheme and its underlying data structure is defined in a more precise way. For representation of mathematical objects we use set Σ that contains the usual characters, i.e. capital and small Arabic letters, capital and small Greek letters, digits $(0, \cdots, 9)$, symbols for mathematical operations such as $+$, $-$, $\times$, $/$, and various types of parentheses and separators. The class of mathematical objects, $\mathcal{A}$, is then mapped to the set Σ^* of words over the alphabet Σ by means of a function $\rho\colon \mathcal{A} \to \Sigma^*$, where Σ^* denotes the set of all finite strings (words) made up of symbols belonging to Σ. Each mathematical object $A \in \mathcal{A}$ is represented as a *structured string* in the following sense: Integers are represented by their decimal representation. A square matrix of dimension n with integer elements will be represented as a finite list whose first component represents matrix dimension n, and the following n^2 components represent the integer matrix elements in some specific order. For example, the list is a structured string of the form $(n, a(1, 1), \cdots, a(1, n), a(2, 1), \cdots, a(2, n), \cdots, a(n, n))$ where n and all the

$a(i,j)$ are structured strings representing integers. The length of encoding (i.e. the complexity of storing) an integer k would then be of order $\log k$, and that of a matrix would be of order $n^2 \log k$ where k is an upper bound for the absolute value of each matrix element. Real numbers will be represented either in decimal notation (e.g. 3.14159) or in half-logarithmic representation using mantissa and exponent (e.g. $0.314159 \cdot 10^1$). Functions may be represented by tables which specify the function (range) value for each domain value. Representations of more complicated objects (e.g. graphs) will be introduced later, together with the definition of these types of objects.

As an example let us consider encoding of a particular instance of the knapsack problem defined in Example 2.2.1. Let the number n of elements be equal to 6 and let an encoding scheme define values of parameters in the following order: n, weights of elements, values of elements, knapsack's capacity b. A string coding an exemplary instance is: $6, 4, 2, 12, 15, 3, 7, 1, 4, 8, 12, 5, 7, 28$.

The above remarks do not exclude the usage of any other *reasonable encoding scheme* which does not cause an exponential growth of the input length as compared with other encoding schemes. For this reason one has to exclude unary encoding in which each integer k is represented as a string of k1's. We see that the length of encoding this integer would be k which is exponentially larger, as compared to the above decimal encoding.

In practice, it is worthwhile to express the input length of an instance as a function depending on the number of elements of some set whose cardinality is dominating for that instance. For the knapsack problem defined in Example 2.2.1 this would be the number of elements n, for the merging problem - the total number of elements in the two sequences, for the scheduling problem - the number of tasks. This assumption, usually made, in most cases reduces practically to the assumption that a computer word is large enough to contain any of the binary encoded numbers comprising an instance. However, in some problems, for example those in which graphs are involved, taking as input size the number of nodes may appear too great a simplification since the number of edges in a graph may be equal to $n(n-1)/2$. Nevertheless, in practice one often makes this simplification to unify computational results. Let us note that this simplification causes no exponential growth of input length.

2.2.2 Algorithms

Let us now pass to the notion of an algorithm and its complexity function. An algorithm is any procedure for solving a problem (i.e. for giving an answer). We will say that an algorithm *solves search problem* Π, if it finds a solution for any instance I of Π. In order to keep the representation of algorithms easily understandable we follow a structural approach that uses language concepts known from structural programming, such as case statements, or loops of various kinds. Like functions or procedures, algorithms may also be called in an algorithm. Parameters may be used to import data to or export data from the algorithm. Besides these, we also use mathematical notations such as set-theoretic notations.

In general, an algorithm consists of two parts: a *head* and a *method*. The head starts with the keyword **Algorithm,** followed by an identifying number and, optionally, a descriptor (a name or a description of the purpose of the algorithm) and a reference to the author(s) of the algorithm. Input and output parameters are omitted in cases where they are clear from the context. In other cases, they are specified as a parameter list. In even more complex cases, two fields, *Input (Instance):* and *Output (Answer):* are used to describe parameters, and a field *Method:* is used to describe the main idea of the algorithm. The *method* part is a block of instructions. As in PASCAL, a block is embraced by **begin** and **end**. Each block is considered as a sequence of instructions. An instruction itself may again be a block, an assignment-, an else-, or a case- operation, or a loop (**for, while, repeat** ··· **until,** or a general **loop**), a **call** of another algorithm, or an exit instruction to terminate a loop instruction (**exit loop,** etc.) or the algorithm or procedure (just **exit**). The right hand side of an assignment operation may be any mathematical expression, or a function call. Case statements partition actions of the algorithm into several branches, depending on the value of a control variable. Loop statements may contain formulations such as: "**for all** $a \in \mathcal{M}$ **do** ···" or " **while** $\mathcal{M} \neq \varnothing$ **do** ···". If a loop is preempted by an exit statement the algorithm jumps to the first statement after the loop. Comments are started with two minus signs and are finished at the end of the line. If a comment needs more than one line, each comment line starts with '−−'.

Algorithms should reflect the main idea of the method. Details like output layouts are omitted. Names for procedures, functions, variables etc. are chosen so that they reflect the semantics behind them. As an example let us consider an algorithm solving the problem of merging two sequences as defined at the beginning of this section.

Algorithm 2.2.2 *merge.*
Input: Two sequences of reals, $e = (e[1], \cdots, e[n])$ and $f = (f[1], \cdots, f[m])$, both sorted in non-decreasing order.
Output: Sequence $g = (g[1], \cdots, g[n+m])$ in which all elements are arranged in non-decreasing order.

```
begin
i := 1; j := 1; k := 1;    -- initialization of counters
while (i ≤ n) and (j ≤ m) do
        -- the while loop merges elements of sequences e and f into g;
        -- the loop is executed until all elements of one of the sequences are merged
    begin
    if e[i] < f[j]
    then begin g[k] := e[i]; i := i+1; end
    else begin g[k] := f[j]; j := j+1; end;
```

```
    k := k+1;
    end;
if  i ≤ n          -- not all elements of sequence e have been merged
then for  l := i to  n do  g[k+l−i] := e[l]
else
    if  j ≤ m       -- not all elements of sequence f have been merged
    then for  l := j to  m do  g[k+l−j] := f[l];
end;
```

The above algorithm returns as an answer sequence g of all the elements of e and f, sorted in non-decreasing order of the values of all the elements.

As another example, consider the search problem of sorting in non-decreasing order a sequence $e = (e[1], \cdots, e[n])$ of $n = 2^k$ reals (i.e. n is a power of 2). The algorithm *sort* (Algorithm 2.2.4) uses two other algorithms that operate on sequences: $msort(i,j)$ and $merge1(i,j,k)$. If the two parameters of *msort*, i and j, obey $1 \leq i < j \leq n$, then $msort(i, j)$ sorts the elements of the subsequence $(e[i], \cdots, e[j])$ of e non-decreasingly. Algorithm *merge1* is similar to *merge* (Algorithm 2.2.2): $merge1(i, j, k)$ $(1 \leq i \leq j < k \leq n)$ takes the elements from the two adjacent and already sorted subsequences $(e[i], \cdots, e[j])$ and $(e[j+1], \cdots, e[k])$ of e, and merges their elements into $(e[i], \cdots, e[k])$.

Algorithm 2.2.3 *msort(i,j)*.
```
begin
case  (i,j) of     -- depending on relative values of i and j,
                   -- three subcases are considered
    i = j:  exit;      -- terminate msort
    i = j−1:  if  e[i] > e[j] then  Exchange  e[i]  and  e[j];
    i < j−1:
      begin
      call  msort(i,⌊(j+i)/2⌋); [1]
          -- sorts elements of subsequence (e[i], ⋯, e[⌊(j+i)/2⌋])
      call  msort(⌊(j+i)/2⌋+1,j);
          -- sorts elements of subsequence (e[⌊(j+i)/2⌋+1], ⋯, e[j])
      call  merge1(i,⌊(j+i)/2⌋,j);
          -- merges sorted subsequences into sequence (e[i], ⋯, e[j])
      end;
    end;
end;
```

Algorithm 2.2.4 *sort*.
```
begin
```

---

[1] $\lfloor x \rfloor$ denotes the largest number less than or equal to x.

```
read(n);
read((e[1],···,e[n]));
call  msort(1,n);
end;
```

Notice that in the case of an optimization problem one may also consider an *approximate* (*sub-optimal*) *solution* that is *feasible* (i.e. fulfills all the conditions specified in the description of the problem) but does not extremize the objective function. It follows that one can also consider *heuristic* (*sub-optimal*) algorithms which tend toward but do not guarantee the finding of optimal solutions for any instance of an optimization problem. An algorithm which always finds an optimal solution will be called an *optimization* or *exact* algorithm.

2.2.3   Complexity

Let us turn now to the analysis of the computational complexity of algorithms. By the *time complexity function* of algorithm A solving problem Π we understand the function that maps each input length of an instance I of Π into a maximal number of elementary steps (or time units) of a computer, which are needed to solve an instance of that size by algorithm A.

It is obvious that this function will not be well defined unless the encoding scheme and the model of computation (computer model) are precisely defined. It appears, however, that the choice of a particular reasonable encoding scheme and a particular realistic computer model has no influence on the distinction between polynomial- and exponential time algorithms which are the two main types of algorithms from the computational complexity point of view [AHU74]. This is because all realistic models of computers [2] are equivalent in the sense that if a problem is solved by some computer model in time bounded from above by a polynomial in the input length (i.e. in polynomial time), then any other computer model will solve that problem in time bounded from above by a polynomial (perhaps of different degree) in the input length [AHU74]. Thus, to simplify the computation of the complexity of polynomial algorithms, we assume that, if not stated otherwise, the operation of writing a number as well as addition, subtraction and comparison of two numbers are elementary operations of a computer that need the same amount of time, if the length of a binary encoded number is bounded from above by a polynomial in the computation time of the whole algorithm. Otherwise, a logarithmic cost criterion is assumed. Now, we define the two types of algorithms.

---

[2] By "realistic" we mean here such computer models which in unit time may perform a number of elementary steps bounded from above by a polynomial in the input length. This condition is fulfilled for example by the one-tape Turing machine, the k-tape Turing machine, or the random access machine (RAM) under logarithmic cost of performing a single operation.

A *polynomial time (polynomial) algorithm* is one whose time complexity function is $O(p(k))$, where p is some polynomial and k is the input length of an instance. Each algorithm whose time complexity function cannot be bounded in that way will be called an *exponential time algorithm*.

Let us consider two algorithms with time complexity functions k and 3^k, respectively. Let us assume moreover that an elementary step lasts 1 μs and that the input length of the instance solved by the algorithms is $k = 60$. Then one may calculate that the first algorithm solves the problem in 60 μs while the second needs $1.3 \cdot 10^{13}$ centuries. This example illustrates the fact that indeed the difference between polynomial- and exponential time algorithms is large and justifies definition of the first algorithm as a "good" one and the second as a "bad" one [Edm65].

If we analyze time complexity of Algorithm 2.2.2, we see that the number of instructions being performed during execution of the algorithm is bounded by $c_1(n+m) + c_2$, where c_1 and c_2 are suitably chosen constants, i.e. the number of steps depends linearly on the total number of elements to be merged.

Now we estimate the time complexity of Algorithm 2.2.4. The first two *read* instructions together take $O(n)$ steps, where reading one element is assumed to take constant ($O(1)$) time. During execution of *msort*$(1, n)$, the sequence of elements is divided into two subsequences, each of length $n/2$; *msort* is applied recursively on the subsequences which will thus be sorted. Then, procedure *merge*1 is applied, which combines the two sorted subsequences into one sorted sequence. Now let $T(m)$ be the number of steps *msort* performs to sort m elements. Then, each call of *msort* within *msort* involves sorting of $m/2$ elements, so it takes $T(m/2)$ time. The call of *merge*1 can be performed in a number of steps proportional to $m/2 + m/2 = m$, as can easily be seen. Hence, we get the recursion

$$T(m) = 2T(m/2) + cm,$$

where c is some constant. One can easily verify that there is a constant c' such that $T(m) = c'm\log m$ solves the recursion[3]. Taking all steps of Algorithm 2.2.4 together we get the time complexity $O(\log n) + O(n) + O(n\log n) = O(n\log n)$.

Unfortunately, it is not always true that we can solve problems by algorithms of linear or polynomial time complexity. In many cases only exponential algorithms are available. We will take now a closer look to inherent complexity of some classes of search problems to explain the reasons why polynomial algorithms are unlikely to exist for these problems.

As we said before, there exist two broad subclasses of search problems: decision and optimization problems. From the computational point of view both classes may be analyzed much in the same way (strictly speaking when their computational hardness is analyzed). This is because a decision problem is computationally not harder than the corresponding optimization problem. That means that if one is able to solve an optimization problem in an "efficient" way (i.e. in

---

[3] We may take any fixed base for the logarithm, e.g. 2 or 10.

polynomial time), then it will also be possible to solve a corresponding decision problem efficiently (just by comparing an optimal value of the objective function [4] to a given constant y). On the other hand, if the decision problem is computationally "hard", then the corresponding optimization problem will also be "hard" [5].

Now, we can turn to the definition of the most important complexity classes of search problems. Basic definitions will be given for the case of decision problems since their formulation permits an easier treatment of the subject. One should, however, remember the above dependencies between decision and optimization problems. We will also point out the most important implications. In order to be independent of a particular type of a computer we have to use an abstract model of computation. From among several possibilities, we choose the *deterministic Turing machine (DTM)* for this purpose. Despite the fact that this choice was somehow arbitrary, our considerations are still *general* because all the realistic models of computations are polynomially related.

Class **P** consists of all decision problems that may be solved by the deterministic Turing machine in time bounded from above by a polynomial in the input length. Let us note that the corresponding (broader) class of all search problems solvable in polynomial time, is denoted by **FP** [Joh90a]. We see that both, the problem of merging two sequences and that of sorting a sequence belong to that class. In fact, class **FP** contains all the search problems which can be solved efficiently by the existing computers.

It is worth noting that there exists a large class of decision problems for which no polynomial time algorithms are known, for which, however, one can verify a positive answer in polynomial time, provided there is some additional information. If we consider for example an instance of the KNAPSACK problem defined in Example 2.2.1 and a subset $\mathcal{A}_1 \subseteq \mathcal{A}$ defining additional information, we may easily check in polynomial time whether or not the answer is "yes" in the case of this subset. This feature of polynomial time verifiability rather than solvability is captured by a *non-deterministic Turing machine (NDTM)* [GJ79].

We may now define *class NP* of decision problems as consisting of all decision problems which may be solved in polynomial time by an NDTM.

It follows that $P \subseteq NP$. In order to define the most interesting class of decision problems, i.e. the class of NP-complete problems, one has to introduce the definition of a polynomial transformation. A *polynomial transformation* from problem Π_2 to problem Π_1 (denoted by $\Pi_2 \propto \Pi_1$) is a function f mapping the set of all instances of Π_2 into the set of instances of Π_1, that satisfies the following two conditions:

---

[4] Strictly speaking, it is assumed that the objective function may be calculated in polynomial time.

[5] Many decision problems and corresponding optimization problems are linked even more strictly, since it is possible to prove that a decision problem is not easier than the corresponding optimization problem [GJ79].

1. for each instance I_2 of Π_2 the answer is "yes" if and only if the answer for $f(I_2)$ of Π_1 is also "yes",

2. f is computable in polynomial time (depending on problem size $|I_2|$) by a DTM.

We say that decision problem Π_1 is *NP-complete* if $\Pi_1 \in \textbf{NP}$ and for any other problem $\Pi_2 \in \textbf{NP}$, $\Pi_2 \propto \Pi_1$ [Coo71].

It follows from the above that if there existed a polynomial time algorithm for some NP-complete problem, then any problem from that class (and also from the **NP** class of decision problems) would be solvable by a polynomial time algorithm. Since NP-complete problems include classical hard problems (as for example HAMILTONIAN CIRCUIT, TRAVELING SALESMAN, SATISFIABILITY, INTEGER PROGRAMMING) for which, despite many attempts, no one has yet been able to find polynomial time algorithms, probably all these problems may only be solved by the use of exponential time algorithms. This would mean that **P** is a proper subclass of **NP** and the classes **P** and NP-complete problems are disjoint.

Another consequence of the above definitions is that, to prove the NP-completeness of a given problem Π, it is sufficient to transform polynomially a known NP-complete problem to Π. SATISFIABILITY was the first decision problem proved to be NP-complete [Coo71]. The current list of NP-complete problems contains several thousands, from different areas. Although the choice of an NP-complete problem which we use to transform into a given problem in order to prove the NP-completeness of the latter, is theoretically arbitrary, it has an important influence on the way a polynomial transformation is constructed [Kar72]. Thus, these proofs require a good knowledge of NP-complete problems, especially characteristic ones in particular areas.

As was mentioned, decision problems are not computationally harder than the corresponding optimization ones. Thus, to prove that some optimization problem is computationally hard, one has to prove that the corresponding decision problem is NP-complete. In this case, the optimization problem belongs to the class of *NP-hard problems*, which includes computationally hard search problems. On the other hand, to prove that some optimization problem is easy, it is sufficient to construct an optimization polynomial time algorithm. The order of performing these two steps follows mainly from the intuition of the researcher, which however, is guided by several hints. In this book, by "open problems" from the computational complexity point of view we understand those problems which neither have been proved to be NP-complete nor solvable in polynomial time.

Despite the fact that all NP-complete problems are computationally hard, some of them may be solved quite efficiently in practice (as for example the KNAPSACK problem). This is because the time complexity functions of algorithms that solve these problems are bounded from above by polynomials in two variables: the input length $|I|$ and the maximal number $\max(I)$ appearing in an instance I. Since in practice $\max(I)$ is usually not very large, these algorithms have good computational properties. However, such algorithms, called *pseudopolynomial*, are not really of polynomial time complexity since in reasonable encoding schemes all numbers are encoded binary (or in another integer base greater than 2). Thus, the length of a string used to encode $\max(I)$ is $\log \max(I)$ and the time complexity function of a polynomial time algorithm would be $O(p(|I|, \log \max(I)))$ and not $O(p(|I|, \max(I)))$, for some polynomial p. It is also obvious that pseudopolynomial algorithms may perhaps be constructed for *number problems*, i.e. those problems Π for which there does not exist a polynomial p such that $\max(I) \leq p(|I|)$ for each instance I of Π. The KNAPSACK problem as well as TRAVELING SALESMAN and INTEGER PROGRAMMING belong to number problems; HAMILTONIAN CIRCUIT and SATISFIABILITY do not. However, there might be number problems for which pseudopolynomial algorithms cannot be constructed [GJ78].

The above reasoning leads us to a deeper characterization of a class of NP-complete problems by distinguishing problems which are NP-complete in the strong sense [GJ78, GJ79].

For a given decision problem Π and an arbitrary polynomial p, let Π_p denote the subproblem of Π which is created by restricting Π to those instances for which $\max(I) \leq p(|I|)$. Thus Π_p is not a number problem.

Decision problem Π is *NP-complete in the strong sense (strongly NP-complete)* if $\Pi \in NP$ and there exists a polynomial p defined for integers for which Π_p is NP-complete.

It follows that if Π is NP-complete and it is not a number problem, then it is NP-complete in the strong sense. Moreover, if Π is NP-complete in the strong sense, then the existence of a pseudopolynomial algorithm for Π would be equivalent to the existence of polynomial algorithms for all NP-complete problems, and thus would be equivalent to the equality $P = NP$. It has been shown that TRAVELING SALESMAN and 3-PARTITION are examples of number problems that are NP-complete in the strong sense [GJ79, Pap94].

From the above definition it follows that to prove NP-completeness in the strong sense for some decision problem Π, one has to find a polynomial p for which Π_p is NP-complete, which is usually not an easy way. To make this proof easier one may use the concept of pseudopolynomial transformation [GJ78].

To end this section, let us stress once more that the membership of a given search problem in class FP or in the class of NP-hard problems does not depend on the chosen encoding scheme if this scheme is reasonable as defined earlier. The differences in input lengths for a given instance that follow from particular encoding schemes have only influence on the complexity of the polynomial (if the problem belongs to class FP) or on the complexity of the exponential algorithm (if the problem is NP-hard). On the other hand, if numbers are written unary, then pseudopolynomial algorithms would become polynomial because of the artificial increase in input lengths. However, problems NP-hard in the strong sense would remain NP-hard even in the case of such an encoding scheme. Thus, they are also called *unary NP-hard* [LRKB77].

2.3 Graphs and Networks

2.3.1 Basic Notions

A *graph* is a pair $G = (\mathcal{V}, \mathcal{E})$ where $\mathcal{V}$ is the set of *vertices* or *nodes*, and $\mathcal{E}$ is the set of *edges*. If $\mathcal{E}$ is a binary relation over $\mathcal{V}$, then G is called a *directed graph* (or *digraph*). If $\mathcal{E}$ is a set of two-element subsets of $\mathcal{V}$, i.e. $\mathcal{E} \subseteq P_2(\mathcal{V})$, then G is an *undirected* graph.

A graph $G' = (\mathcal{V}', \mathcal{E}')$ is a *subgraph* of $G = (\mathcal{V}, \mathcal{E})$ (denoted by $G' \subseteq G$), if $\mathcal{V}' \subseteq \mathcal{V}$, and $\mathcal{E}'$ is the set of all edges of $\mathcal{E}$ that connect vertices of $\mathcal{V}'$.

Let $G_1 = (\mathcal{V}_1, \mathcal{E}_1)$ and $G_2 = (\mathcal{V}_2, \mathcal{E}_2)$ be graphs whose vertex sets $\mathcal{V}_1$ and $\mathcal{V}_2$ are not necessarily disjoint. Then $G_1 \cup G_2 = (\mathcal{V}_1 \cup \mathcal{V}_2, \mathcal{E}_1 \cup \mathcal{E}_2)$ is the *union* graph of G_1 and G_2, and $G_1 \cap G_2 = (\mathcal{V}_1 \cap \mathcal{V}_2, \mathcal{E}_1 \cap \mathcal{E}_2)$ is the *intersection* graph of G_1 and G_2.

Digraphs G_1 and G_2 are *isomorphic* if there is a bijective mapping $\chi \colon \mathcal{V}_1 \to \mathcal{V}_2$ such that $(v_1, v_2) \in \mathcal{E}_1$ if and only if $(\chi(v_1), \chi(v_2)) \in \mathcal{E}_2$.

A (undirected) *path* in a graph or in a digraph $G = (\mathcal{V}, \mathcal{E})$ is a sequence $i_1, \cdots, i_r$ of distinct nodes of $\mathcal{V}$ satisfying the property that either $(i_k, i_{k+1}) \in \mathcal{E}$ or $(i_{k+1}, i_k) \in \mathcal{E}$ for each $k = 1, \cdots, r-1$. A *directed path* is defined similarly, except that $(i_k, i_{k+1}) \in \mathcal{E}$ for each $k = 1, \cdots, r-1$. A (undirected) *cycle* is a path together with an edge (i_r, i_1) or (i_1, i_r). A *directed cycle* is a directed path together with the edge (i_r, i_1). We will call a graph (digraph) G *acyclic* if it contains no (directed) cycle.

Two vertices i and j of G are said to be *connected* if there is at least one undirected path between i and j. G is *connected* if all pairs of vertices are connected; otherwise it is *disconnected*.

Let v and w be vertices of the digraph $G = (\mathcal{V}, \mathcal{E})$. If there is a directed path from v to w, then w is called *successor* of v, and v is called *predecessor* of w. If $(v, w) \in \mathcal{E}$, then vertex w is called *immediate successor* of v, and v is called *immediate predecessor* of w. The set of immediate successors of vertex v is denoted by isucc(v); the sets succ(v), ipred(v), and pred(v) are defined similarly. The cardinality of ipred(v) is called *in-degree* of vertex v, whereas *out-degree* is the cardinality of isucc(v). A vertex v that has no immediate predecessor is called *initial vertex* (i.e. ipred(v) = $\varnothing$); a vertex v having no immediate successors is called *final* (i.e. isucc(v) = $\varnothing$).

Directed or undirected graphs can be represented by means of their adjacency matrix. If $V = \{v_1, \cdots, v_n\}$, the *adjacency matrix* is a binary (n, n)-matrix A. In case of a directed graph, $A(i, j) = 1$ if there is an edge from v_i to v_j, and $A(i, j) = 0$ otherwise. In case of an undirected graph, $A(i, j) = 1$ if there is an edge between v_i and v_j, and $A(i, j) = 0$ otherwise. The complexity of storage (space complexity) is $O(n^2)$. If the adjacency matrix is sparse, as e.g. in case of trees, there are better ways of representation, usually based on *linked lists*. For details we refer to [AHU74].

In many situations, it is appropriate to use a generalization of graphs called hypergraphs. Following [Ber73] a finite *hypergraph* is a pair $H = (V, \mathcal{H})$ where V is a finite set of vertices, and $\mathcal{H} \subseteq \mathcal{P}(V)$ is a set of subsets of V. The elements of $\mathcal{H}$ are referred to as *hyperedges*. Hypergraphs can be represented as bipartite graphs (see below): Let G_H be the graph whose vertex set is $V \cup \mathcal{H}$, and the set of edges is defined as $\{\{v, h\} \mid h \in \mathcal{H}, \text{ and } v \in h\}$.

2.3.2   Special Classes of Digraphs

A digraph $G = (V, \mathcal{E})$ is called *bipartite* if its vertex set V can be partitioned into two subsets V_1 and V_2 such that for each edge $(i, j) \in \mathcal{E}$, $i \in V_1$ and $j \in V_2$.

If a digraph $G = (V, \mathcal{E})$ contains no directed cycle and no transitive edges (i.e. pairs (u, w) of vertices for which there exists a directed path from u to w), it will be called a *precedence graph*. A corresponding binary relation will be called a *precedence relation* $\prec$ over set V. A precedence graph $G = (V, \mathcal{E})$ (we also write $(V, \prec)$, where $\prec$ is the corresponding precedence relation) can always be enlarged to a partially ordered set (poset, see Section 2.1) $\prec^*$ by adding transitive edges and all reflexive pairs (v, v) ($v \in V$) to $\mathcal{E}$. On the other hand, given a poset (V, Q), where Q is a partial order over set V, we can always construct a precedence graph $(V, \mathcal{E})$ in the following way: $\mathcal{E}$ is obtained by taking those pairs of elements (u, w), $u \neq w$, for which no sequence $v_1, \cdots, v_k$ of elements with $(u, v_1) \in Q$, $(v_i, v_{i+1}) \in Q$ for $i = 1, \cdots, k-1$, and $(v_k, w) \in Q$ can be found. It can be constructed from a given poset in $O(|V|^{2.8})$ time [AHU74].

A digraph $G = (V, \mathcal{E})$ is called a *chain* if in the corresponding poset (V, Q) for any two vertices v and $v' \in V$, $v \neq v'$, either $(v, v') \in Q$ or $(v', v) \in Q$ (such a poset is usually called a *linear order*). An *anti-chain* is a (directed) graph $(\mathcal{V}, \mathcal{E})$ where $\mathcal{E} = \emptyset$.

An *out-tree* is a precedence graph where exactly one vertex has in-degree 0, and all the other vertices have in-degree 1. If $G = (\mathcal{V}, \mathcal{E})$ is an out-tree, then graph $G' = (\mathcal{V}, \mathcal{E}^{-1})$ is called an *in-tree*. An *out-forest* (*in-forest*) is a disjoint union of out-trees (in-trees), respectively. An *opposing forest* is a disjoint union of in-trees and out-trees.

A precedence graph $(\{a, b, c, d\}, \prec)$ has *N-structure* if $a \prec c$, $b \prec c$, $b \prec d$, $a \nprec d$, $d \nprec a$, $a \nprec b$, $b \nprec a$, $c \nprec d$, and $d \nprec c$ (see also Figure 2.3.1). A precedence graph P is *N-free* if it contains no subset isomorphic to an *N*-structure.

To define another interesting class of graphs let us consider a finite set $\mathcal{V}$ and a collection $(I_v)_{v \in \mathcal{V}}$ of intervals I_v on the reals. This collection defines a partial order $\prec$ on $\mathcal{V}$ as follows:

$$v \prec w \iff I_v \text{ is entirely before } I_w.$$

Such a partial order is called an *interval order*. Without loss of generality, we may assume that the intervals have the form $[n_1, n_2]$ with n_1 and n_2 integral. It can be shown that $\prec$ is an interval order if and only if the transitive closure of this order does not contain $2K_2$ (see Figure 2.3.2) as an induced subgraph [Fis70].

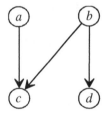

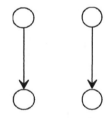

Figure 2.3.1 *N-structured precedence graph.*   **Figure 2.3.2** *Graph* $2K_2$.

Finally we introduce a class of precedence graphs that has been considered frequently in literature. Let $S = (\mathcal{V}, \prec)$ be a precedence graph, and let for each $v \in \mathcal{V}$, $P_v = (\mathcal{V}_v, \prec_v)$ be a precedence graph, where all the sets $\mathcal{V}_v$ ($v \in \mathcal{V}$) and $\mathcal{V}$ are pair-wise disjoint. Let $\mathcal{U} = \bigcup_{v \in \mathcal{V}} \mathcal{V}_v$. Define $(\mathcal{U}, \prec_{\mathcal{U}})$ as the following precedence graph: for $p, q \in \mathcal{U}$, $p \prec_{\mathcal{U}} q$ if either there are $v, v' \in \mathcal{V}$ with $v \prec v'$ such that p is a final vertex in $(\mathcal{V}_v, \prec_v)$ and q is an initial vertex in $(\mathcal{V}_{v'}, \prec_{v'})$, or there is $v \in \mathcal{V}$ with $p, q \in \mathcal{V}_v$, and $p \prec_v q$. Then $(\mathcal{U}, \prec_{\mathcal{U}})$ is called the *lexicographic sum* of $(P_v)_{v \in \mathcal{V}}$ over S. Notice that each vertex v of the digraph $S = (\mathcal{V}, \prec)$ is replaced by the digraph $(\mathcal{V}_v, \prec_v)$, and if vertex v is connected to v' in S (i.e. $v \prec$

v'), then each final vertex of $(\mathcal{V}_v, \prec_v)$ is connected to each initial vertex of $(\mathcal{V}_{v'}, \prec_{v'})$.

We need two special cases of lexicographic sums: If $S = (\mathcal{V}, \prec)$ is a chain, the lexicographic sum of $(P_v)_{v \in \mathcal{V}}$ over S is called a *linear sum*. If S is an anti-chain (i.e. $v_1 \prec v_2 \Rightarrow v_1 = v_2$), then the lexicographic sum of $(P_v)_{v \in \mathcal{V}}$ over S is called *disjoint sum*. A *series-parallel* precedence graph is a precedence graph that can be constructed from one-vertex precedence graphs by repeated application of the operations linear sum and disjoint sum. Opposing forests are examples of series-parallel digraphs. Another example is shown in Figure 2.3.3.

Without proof we mention some properties of series-parallel graphs. A precedence graph $G = (\mathcal{V}, \mathcal{E})$ is series-parallel if and only if it is N-free. The question if a digraph is series-parallel can be decided in $O(|\mathcal{V}| + |\mathcal{E}|)$ time [VTL82].

The structure of a series-parallel graph as it is obtained by successive applications of linear sum and disjoint sum operations can be displayed by a *decomposition tree*. Figure 2.3.4 shows a decomposition tree for the series-parallel graph of Figure 2.3.3. Each leaf of the decomposition tree is identified with a vertex of the series-parallel graph. An *S-node* represents an application of linear sum (series composition) to the sub-graphs identified with its children; the ordering of these children is important: we adopt the convention that left precedes right. A *P-node* represents an application of the operation of disjoint sum (parallel composition) to the subgraphs identified with its children; the ordering of these children is of no relevance for the disjoint sum. The series or parallel relationship of any pair of vertices can be determined by finding their least common ancestor in the decomposition tree.

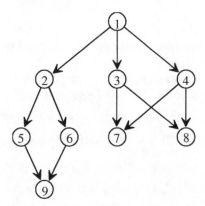

Figure 2.3.3 *Example of a series-parallel digraph.*

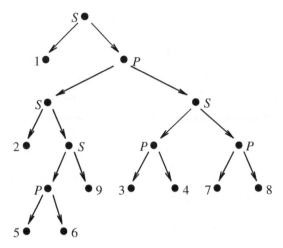

Figure 2.3.4 *Decomposition tree of the digraph of Figure 2.3.3.*

2.3.3 Networks

In this section the problem of finding a maximum flow in a network is considered. We will analyze the subject rather thoroughly because of its importance for many scheduling problems.

By a *network* we will mean a directed graph $G = (\mathcal{V}, \mathcal{E})$ without loops and parallel edges, where each edge $e \in \mathcal{E}$ is assigned a *capacity* $c(e) \in I\!R_{\geq 0}$, and sometimes a cost of a unit flow. Usually in the network two vertices s and t, called a *source* and a *sink*, respectively, are specified.

A real-valued *flow* function ρ is to be assigned to each edge such that the following conditions hold for some $F \in I\!R_{\geq 0}$:

$$0 \leq \rho(e) \leq c(e) \text{ for each } e \in \mathcal{E}, \tag{2.3.1}$$

$$\sum_{e \in IN(v)} \rho(e) - \sum_{e \in OUT(v)} \rho(e) = \begin{cases} -F & \text{for } v = s \\ 0 & \text{for } v \in \mathcal{V} - \{s, t\} \\ F & \text{for } v = t, \end{cases} \tag{2.3.2}$$

where $IN(v)$ and $OUT(v)$ are the sets of edges *incoming* to vertex v and *outgoing* from vertex v, respectively. The *total flow* (the *value of flow*) F of ρ is defined by

$$F := \sum_{e \in IN(t)} \rho(e) - \sum_{e \in OUT(t)} \rho(e). \tag{2.3.3}$$

Given a network, in the *maximum flow problem* we want to find a flow function ρ which obeys the above conditions and for which total flow F is at its maximum.

Now, some important notions will be defined and their properties will be discussed. Let S be a subset of the set of vertices $\mathcal{V}$ such that $s \in S$ and $t \notin S$, and let $\bar{S}$ be the complement of S, i.e. $\bar{S} = \mathcal{V} - S$. Let $(S, \bar{S})$ denote a set of edges of network G, each of which has its starting vertex in S and its target vertex in $\bar{S}$. Set $(\bar{S}, S)$ is defined in a similar way. Given some subset $S \subseteq \mathcal{V}$, either set, $(S, \bar{S})$ and $(\bar{S}, S)$, will be called *cut* defined by S.

Following definition (2.3.3) we see that the value of flow is measured at the sink of the network. It is however, possible to measure this value at any cut [Eve79, FF62].

Lemma 2.3.1 *For each subset of vertices $S \subseteq \mathcal{V}$, we have*

$$F = \sum_{e \in (S,\bar{S})} \rho(e) - \sum_{e \in (\bar{S},S)} \rho(e).$$
(2.3.4)

$\square$

Let us denote by $c(S)$ the *capacity of a cut* defined by S,

$$c(S) = \sum_{e \in (S,\bar{S})} c(e).$$
(2.3.5)

It is possible to prove the following lemma, which specifies a relation between the value of a flow and the capacity of any cut [FF62].

Lemma 2.3.2 *For any flow function ρ having the value F and for any cut defined by S we have*

$$F \leq c(S).$$
(2.3.6)

$\square$

From the above lemma we get immediately the following corollary that specifies a relation between maximum flow and a cut of minimum capacity.

Corollary 2.3.3 *If $F = c(S)$, then F is at its maximum, and S defines a cut of minimum capacity.* $\square$

Let us now define, for a given flow ρ, an *augmenting path* as a path from s to t, (not necessarily directed), which can be used to increase the value of the flow. If an edge e belonging to that path is directed from s to t, then $\rho(e) < c(e)$, otherwise no increase in the flow value on that path would be possible. On the other hand, if such an edge e is directed from t to s, then $\rho(e) > 0$ must be satisfied in order to be able to increase the flow value F by decreasing $\rho(e)$.

Example 2.3.4 As an example let us consider the network given in Figure 2.3.5(a). Each edge of this network is assigned two numbers, $c(e)$ and $\rho(e)$. It is easy to check that flow ρ in this network obeys conditions (2.3.1) and (2.3.2) and its value is equal to 3. An augmenting path is shown in Figure 2.3.5(b). The flow on edge $(5, 4)$ can be decreased by one unit. All the other edge flows on that path can be increased by one unit. The resulting network with a new flow is shown in Figure 2.3.5(c).   □

(a) $c(e) / \rho(e)$

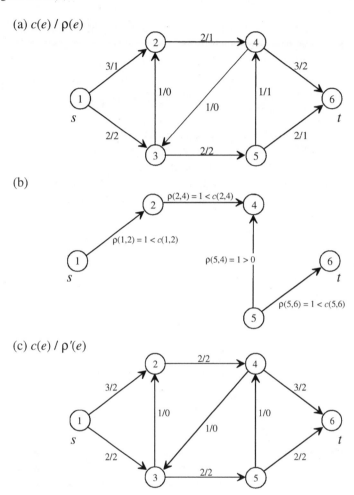

(b)

(c) $c(e) / \rho'(e)$

Figure 2.3.5 *A network for Example 2.3.4:*
(a) a flow $\rho(e)$ is assigned to each edge,
(b) an augmenting path,
(c) a new flow $\rho'(e)$.

The first method proposed for the construction of a flow of a maximum value was given by Ford and Fulkerson [FF62]. This method consists in finding an augmenting path in a network and increasing the flow value along this path until no such path remains in the network. Convergence of such a general method could be proved for integer capacities only. A corresponding algorithm is of pseudopolynomial complexity [FF62, Eve79].

An important improvement of the above algorithm was made by Edmonds and Karp [EK72]. They showed that if the shortest augmenting path is chosen at every step, then the complexity of the algorithm reduces to $O(|\mathcal{V}|^3|\mathcal{E}|)$, no matter what are the edge capacities. Further improvements in algorithmic efficiency of network flow algorithm were made by Dinic [Din70] and Karzanov [Kar74], whose algorithms' running times are $O(|\mathcal{V}|^2|\mathcal{E}|)$ and $O(|\mathcal{V}|^3)$, respectively. An algorithm proposed by Cherkassky [Che77] allows for solving the max-flow problem in time $O(|\mathcal{V}|^2|\mathcal{E}|^{1/2})$.

Below, Dinic's algorithm will be described, since despite its relatively high worst case complexity function, its average running time is low [Che80], and the idea behind it is quite simple. It uses the notion of a *layered network* which contains all the shortest paths in a network. This allows for a parallel increase of flows in all such paths, which is the main reason of the efficiency of the algorithm.

In order to present this algorithm, the notion of *usefulness* of an edge for a given flow is introduced. We say that edge e having flow $\rho(e)$ is *useful* from u to v, if one of the following conditions is fulfilled:

1) if the edge is directed from u to v then $\rho(e) < c(e)$;

2) otherwise, $\rho(e) > 0$.

For a given network $G = (\mathcal{V}, \mathcal{E})$ and flow ρ, the following algorithm determines a corresponding layered network.

Algorithm 2.3.5 *Construction of a layered network for a given network $G = (\mathcal{V}, \mathcal{E})$ and flow function ρ* [Din70].

begin
Set $\mathcal{V}_0 := \{s\}$; $\mathcal{T}:= \{\varnothing\}$; $i := 0$;
while $t \notin \mathcal{T}$ **do**
  begin
  Construct subset $\mathcal{T}:= \{v \mid v \notin \mathcal{V}_j$ for $j \leq i$ and there exists a useful edge
      from any of the vertices of $\mathcal{V}_i$ to $v\}$;
      $--$ subset $\mathcal{T}$ contains vertices comprising a new layer of the layered network
  $\mathcal{V}_{i+1} := \mathcal{T}$; $--$ a new layer of the network has been constructed
  $i := i+1$;
  if $\mathcal{T} = \varnothing$ **then exit**;
      $--$ no layered network exists, the flow value F is at its maximum
  end;
$l := i$; $\mathcal{V}_l := \{t\}$;

```
for j := 1 to l do
  begin
```
$\mathcal{E}_j := \{e \mid e \text{ is a useful edge from a vertex belonging to layer } \mathcal{V}_{j-1} \text{ to a vertex belonging to layer } \mathcal{V}_j\};$
```
    for all e ∈ E_j do
```
if $e = (u,v)$ **and** $u \in \mathcal{V}_{j-1}$ **and** $v \in \mathcal{V}_j$
then $\tilde{c}(e) := c(e) - \rho(e)$
else
    if $e = (v,u)$ **and** $u \in \mathcal{V}_{j-1}$ **and** $v \in \mathcal{V}_j$
    then
      begin
      $\tilde{c}(e) := \rho(e);$
      Change the orientation of the edge, so that $e = (u,v);$
      end;
  end;   -- a layered network with new edges and capacities has been constructed
end;

In such a layered network a new flow function $\tilde{\rho}$ with $\tilde{\rho} = 0$ for each edge e is assumed. Then a maximal flow is searched for, i.e. one such that for each path $v_0 (= s)$, v_1, $v_2, \cdots, v_{l-1}$, $v_l (= t)$, where $e_j = (v_{j-1}, v_j) \in \mathcal{E}_j$ and $v_j \in \mathcal{V}_j$, $j = 1$, $2, \cdots, l$, there exists at least one edge e such that $\tilde{\rho}(e_j) = \tilde{c}(e_j)$.

Let us note, that such a maximal flow may not be of maximum value. This fact is illustrated in Figure 2.3.6 where all capacities $\tilde{c}(e) = 1$. The flow depicted in this figure is maximal and its value $F = 1$. It is not hard, however, to construct a flow of value $F = 2$.

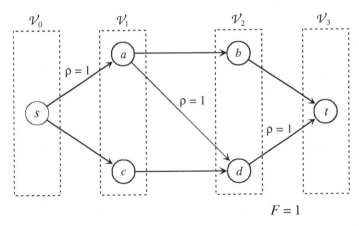

Figure 2.3.6 *An example of a maximal flow which is not of maximum value.*

The construction of a maximal flow for a given layered network is shown below. It consists in finding augmenting paths by means of a *labeling procedure*. For this purpose a depth first search label algorithm is used, that labels all the nodes of the layered network, i.e. assigns to node u, if any, a label $lab(e)$ that corresponds to edge $e = (v, u)$ in a layered network. The algorithm uses for each node v a list isucc(v) of all immediate successors of v (i.e. all nodes u for which an arc (v, u) exists in the layered network). Let us note that, if v belongs to layer $\mathcal{V}_j$, then $u \in$ isucc(v) belongs to layer $\mathcal{V}_{i+1}$, and edge $(v, u) \in \mathcal{E}_j$. The algorithm uses recursively an algorithm $label(v)$ that labels nodes being successors of v. Boolean variable $new(v)$ is used to check whether or not a given node has been visited and consequently labeled. The algorithms are as follows.

Algorithm 2.3.6 *label(v).*

```
begin
new(v) := false;          -- node v has been visited and labeled
for all u ∈ isucc(v) do
if new(u) then
   begin
```
$$\textbf{if } e = (v,u) \in \bigcup_{j=1}^{l} \mathcal{E}_j \textbf{ then } lab(u) := e;$$
```
   call label(u);
      end;          -- all successors of node v have been labeled
end;
```

Algorithm 2.3.7 *label.*

```
begin
lab(s) := 0;     -- a source of layered network has been labeled
for all v ∈ V do  new(v) := true;     -- initialization
call label(s);
end;      -- all successors of s in the layered network are now visited and labeled
```

Using the above algorithms as subroutines the following algorithm constructs a maximal flow in the layered network. The algorithm will stop whenever no augmenting path exists; in this case the flow is maximal [Din70] (see also [Eve79]).

Algorithm 2.3.8 *Construction of a maximal flow in a layered network* [Din70].

```
begin
```
$$\textbf{for all } e \in \bigcup_{j=1}^{l} \mathcal{E}_j \textbf{ do}$$
```
   begin
   ρ₁(e) := p̃(e) := 0;
   c₁(e) := c̃(e);
```

end;   -- initialization phase
loop
    call *label;*   -- all nodes, if any, have been labeled
    if node t is not labeled then **exit;**
        -- no augmenting path exists
        -- a maximal flow in a layered network has been constructed
    Find an augmenting path ap starting from node t backward and using labels;
    $\Delta := \min\{c_1(e) \mid e \in ap\};$
    for all $e \in ap$ **do**
        begin
        $\rho_1(e) := \Delta;$
        $\tilde{\rho}(e) := \tilde{\rho}(e) + \rho_1(e);$
        $c_1(e) := c_1(e) - \Delta;$
        end;   -- the value of a flow is increased along an augmenting path
    for all e **with** $c_1(e) = 0$ **do** Delete e from the layered network;
    repeat
        Delete all nodes which have either no incoming or no outgoing edges;
        Delete all edges incident with such nodes;
    until all such edges and nodes are deleted;
    for all $e \in \bigcup\limits_{j=1}^{l} \mathcal{E}_j$ **do** $\rho_1(e) := 0;$
end loop;
end;

The flow constructed by the above algorithm is used to obtain a new flow in the original network. Next, a new layered network is created and the above procedure is repeated until no new layered network can be constructed. The obtained flow has a maximum value. This is summarized in the next algorithm.

Algorithm 2.3.9 *Construction of a flow of maximum value* [Din70].

begin
$\rho(e) := 0$ for all $e \in \mathcal{E};$
loop
    call Algorithm 2.3.5;
        -- a new layered network is constructed for a flow function ρ
        -- if no layered network exists, then the flow has maximum value
    call Algorithm 2.3.8;   -- a new maximal flow $\tilde{\rho}$ is constructed
    for all $e \in \mathcal{E}$ **do**
        begin
        if $u \in \mathcal{V}_{j-1}$ **and** $v \in \mathcal{V}_j$ **and** $e = (u, v) \in \mathcal{E}$
        then $\rho(e) := \rho(e) + \tilde{\rho}(e);$
            -- the value of the flow increases if edge e has the same direction

```
                 -- in the original and in the layered network
        if  u ∈ 𝒱_{j-1}  and  v ∈ 𝒱_j  and  e = (v, u) ∈ 𝓔
        then  ρ(e) := ρ(e) − ρ̃(e);
                 -- the value of the flow decreases if edge e has opposite directions
                 -- in the original and in the layered network
        end;
      -- the flow in the original network is augmented using the
      -- constructed maximal flow values
end loop;
end;
```

To analyze the complexity of the above approach let us call one loop of Algorithm 2.3.9 a *phase*. We see that one phase consists of finding a layered network, constructing a maximal flow $\tilde{\rho}$ in the latter and improving the flow in the original network. It can be proved [Din70, Eve79] that the number of phases is bounded from above by $O(|\mathcal{V}|)$. The most complex part of each phase is to find a maximal flow in a layered network. Since in Algorithm 2.3.8 a depth first search procedure has been used for visiting a network, the complexity of one phase is $O(|\mathcal{V}||\mathcal{E}|)$. The overall complexity of Dinic's approach is thus $O(|\mathcal{V}|^2|\mathcal{E}|)$.

Further generalizations of the subject include networks with lower bounds on edge flows, networks with linear total cost function of the flow where a flow of maximum value and of minimum total cost is looked for, and a transportation problem being a special case of the latter. All these problems can be solved in time bounded from above by a polynomial in the number of nodes and edges of the network. We refer the reader to [AMO93] or [Law76] where a detailed analysis of the subject is presented.

2.4 Enumerative Methods

In this section we describe very briefly two general methods of solving many combinatorial problems [6], namely the method of dynamic programming and the method of branch and bound. Few remarks should be made at the beginning, concerning the scope of this presentation. First, we will not go into details, since both methods are broadly treated in literature, including basic scheduling books [Bak74, Len77, Rin76], and our presentation should only fulfill the needs of this book. In particular, we will not perform a comparative study of the methods - the interested reader is referred to [Cof76]. We will also not present examples, since they will be given in the later chapters.

---

[6] Dynamic programming can also be used in a wider context (see e.g. [Den82, How69, DL79]).

Before passing to the description of the methods let us mention that they are of *implicit enumeration* variety, because they consider certain solutions only indirectly, without actually evaluating them explicitly.

2.4.1   Dynamic Programming

Fundamentals of *dynamic programming* were elaborated by Bellman in the 1950's and presented in [Bel57, BD62]. The name "Dynamic Programming" is slightly misleading, but generally accepted. A better description would be "recursive" or "multistage" optimization, since it interprets optimization problems as *multistage decision processes.* It means that the problem is divided into a number of stages, and at each stage a decision is required which impacts on the decisions to be made in later stages. Now, Bellman's principle of optimality is applied to draw up a recursive equation which describes the optimal criterion value at a given stage in terms of the previously obtained one. This principle can be formulated as follows: Starting from any current stage, an optimal policy for the rest of the process, i.e. for subsequent stages, is independent of the policy adopted in the previous stages. Of course, not all optimization problems can be presented as multistage decision processes for which the above principle is true. However, the class of problems for which it works is quite large. For example, it contains problems with an additive optimality criterion, but also other problems as we will show in Sections 5.1.1 and 10.4.3.

If dynamic programming is applied to a combinatorial problem, then in order to calculate the optimal criterion value for any subset of size k, we first have to know the optimal value for each subset of size $k-1$. Thus, if our problem is characterized by a set of n elements, the number of subsets considered is 2^n. It means that dynamic programming algorithms are of exponential computational complexity. However, for problems which are NP-hard (but not in the strong sense) it is often possible to construct pseudopolynomial dynamic programming algorithms which are of practical value for reasonable instance sizes.

2.4.2   Branch and Bound

Suppose that given a finite [7] set S of feasible solutions and a criterion $\gamma: S \to \mathbb{R}$, we want to find $S^* \in S$ such that $\gamma(S^*) = \min_{S \in S}\{\gamma(S)\}$.

Branch and bound finds S^* by implicit enumeration of all $S \in S$ through examination of increasingly smaller subsets of S. These subsets can be treated as sets of solutions of corresponding sub-problems of the original problem. This

---

[7] In general, $|S|$ can be infinite (see, e.g. [Mit70]).

way of thinking is especially motivated if the considered problems have a clear practical interpretation, and we will adopt this interpretation in the book.

As its name implies, the branch and bound method consists of two fundamental procedures: branching and bounding. *Branching* is the procedure of partitioning a large problem into two or more sub-problems usually mutually exclusive[8]. Furthermore, the sub-problems can be partitioned in a similar way, etc. *Bounding* calculates a *lower bound* on the optimal solution value for each sub-problem generated in the branching process. Note that the branching procedure can be conveniently represented as a *search* (or *branching*) *tree*. At level 0, this tree consists of a single node representing the original problem, and at further levels it consists of nodes representing particular sub-problems of the problem at the previous level. Edges are introduced from each problem node to each of its sub-problems nodes. A list of unprocessed nodes (also called active nodes) corresponding to sub-problems that have not been eliminated and whose own sub-problems have not yet been generated, is maintained.

Suppose that at some stage of the branch and bound process a (complete) solution S of criterion value $\gamma(S)$ has been obtained. Suppose also that a node encountered in the process has an associated lower bound $LB > \gamma(S)$. Then the node needs not be considered any further in the search for S^*, since the resulting solution can never have a value less than $\gamma(S)$. When such a node is found, it is eliminated, and its branch is said to be *fathomed*, since we do not continue the bounding process from it. The solution used for checking if a branch is fathomed is sometimes called a *trial* solution. At the beginning it may be found using a special heuristic procedure, or it can be obtained in the course of the tree search, e.g. by pursuing the tree directly to the bottom as rapidly as possible. At any later stage the best solution found so far can be chosen as a trial one. The value $\gamma(S)$ for a trial solution S is often called an *upper bound*. Let us mention that a node can be eliminated not only on the basis of lower bounds but also by means of so-called elimination criteria provided by dominance properties or feasibility conditions developed for a given problem.

The choice of a node from the set of generated nodes which have so far neither been eliminated nor led to branching is due to the chosen *search strategy*. Two search strategies are used most frequently: jumptracking and backtracking. *Jumptracking* implements a *frontier search* where a node with a minimal lower bound is selected for examination, while *backtracking* implements a *depth first search* where the descendant nodes of a parent node are examined either in an arbitrary order or in order of non-decreasing lower bounds. Thus, in the jumptracking strategy the branching process jumps from one branch of the tree to another, whereas in the backtracking strategy it first proceeds directly to the bottom along some path to find a trial solution and then retraces that path upward up to the first level with active nodes, and so on. It is easy to notice that jumptracking tends to construct a fairly large list of active nodes, while backtracking maintains

---

[8] If this is not the case, we speak rather about a division of S instead of its partition.

relatively few nodes on the list at any time. However, an advantage of jumptrack-ing is the quality of its trial solutions which are usually much closer to optimum than the trial solutions generated by backtracking, especially at early stages. Deeper comparative discussion of characteristics of the search strategies can be found in [Agi66, LW66].

Summing up the above considerations we can say that in order to implement the scheme of the branch and bound method, i.e. in order to construct a branch and bound algorithm for a given problem, one must decide about

(*i*) the branching procedure and the search strategy,

(*ii*) the bounding procedure or elimination criteria.

Making the above decisions one should explore the problem specificity and observe the compromise between the length of the branching process and time overhead concerned with computing lower bounds or trial solutions. However, the actual computational behavior of branch and bound algorithms remains un-predictable and large computational experiments are necessary to recognize their quality. It is obvious that the computational complexity function of a branch and bound algorithm is exponential in problem size when we search for an optimal solution. However, the approach is often used for finding suboptimal solutions, and then we can obtain polynomial time complexity by stopping the branching process at a certain stage or after a certain time period elapsed.

2.5 Heuristic and Approximation Algorithms

As already mentioned, scheduling problems belong to a broad class of combina-torial optimization problems (cf. Section 2.2.1). To solve these problems one tends to use optimization algorithms which for sure always find optimal solu-tions. However, not for all optimization problems, polynomial time optimization algorithms can be constructed. This is because some of the problems are NP-hard. In such cases one often uses *heuristic (suboptimal) algorithms* which tend toward but do not guarantee the finding of optimal solutions for any instance of an optimization problem. Of course, the necessary condition for these algorithms to be applicable in practice is that their worst-case complexity function is bounded from above by a low-order polynomial in the input length. A sufficient condition follows from an evaluation of the distance between the solution value they produce and the value of an optimal solution. This evaluation may concern the worst case or a mean behavior.

2.5.1 Approximation Algorithms

We will call heuristic algorithms with analytically evaluated accuracy *approxi-mation algorithms*. To be more precise, we give here some definitions, starting with the worst case analysis [GJ79].

If Π is a minimization (maximization) problem, and I is any instance of it, we may define the ratio $R_A(I)$ for an approximation algorithm A as

$$R_A(I) = \frac{A(I)}{OPT(I)} \qquad \left(R_A(I) = \frac{OPT(I)}{A(I)} \right),$$

where $A(I)$ is the value of the solution constructed by algorithm A for instance I, and $OPT(I)$ is the value of an optimal solution for I. The *absolute performance ratio* R_A for an approximation algorithm A for problem Π is then given as

$$R_A = \inf\{r \geq 1 \mid R_A(I) \leq r \text{ for all instances of } \Pi\}.$$

The *asymptotic performance ratio* R_A^∞ for A is given as

$$R_A^\infty = \inf\{r \geq 1 \mid \text{for some positive integer } K, R_A(I) \leq r \text{ for}$$
$$\text{all instances of } \Pi \text{ satisfying } OPT(I) \geq K\}.$$

The above formulas define a measure of the "goodness" of approximation algorithms. The closer R_A^∞ is to 1, the better algorithm A performs. However, for some combinatorial problems it can be proved that there is no hope of finding an approximation algorithm of a specified accuracy, i.e. this question is as hard as finding a polynomial time algorithm for any NP-complete problem.

Analysis of the worst-case behavior of an approximation algorithm may be complemented by an analysis of its mean behavior. This can be done in two ways. The first consists in assuming that the parameters of instances of the considered problem Π are drawn from a certain distribution D and then one analyzes the *mean performance* of algorithm A.

In such an analysis it is usually assumed that all parameter values are realizations of independent probabilistic variables of the same distribution function. Then, for an instance I_n of the considered optimization problem (n being a number of generated parameters) a probabilistic value analysis is performed. The result is an asymptotic value $OPT(I_n)$ expressed in terms of problem parameters. Then, algorithm A is probabilistically evaluated by comparing solution values $A(I_n)$ it produces ($A(I_n)$ being independent probabilistic variables) with $OPT(I_n)$ [Rin87]. The two evaluation criteria used are absolute error and relative error. The *absolute error* is defined as a difference between the approximate and optimal solution values

$$a_n = A(I_n) - OPT(I_n).$$

On the other hand, the *relative error* is defined as the ratio of the absolute error and the optimal solution value

$$b_n = \frac{A(I_n) - OPT(I_n)}{OPT(I_n)}.$$

Usually, one evaluates the convergence of both errors to zero. Three types of convergence are distinguished. The strongest, i.e. *almost sure convergence* for a sequence of probabilistic variables y_n which converge to constant c is defined as

$$Pr\{\lim_{n\to\infty} y_n = c \} = 1.$$

The latter implies a weaker *convergence in probability*, which means that for every $\varepsilon > 0$,

$$\lim_{n\to\infty} Pr\{|y_n - c| > \varepsilon\} = 0.$$

The above convergence implies the first one if the following additional condition holds for every $\varepsilon > 0$:

$$\sum_{j=1}^{\infty} Pr\{|y_n - c| > \varepsilon\} < \infty.$$

Finally, the third type of convergence, *convergence in expectation* holds if

$$\lim_{n\to\infty} |E(y_n) - c| = 0,$$

where $E(y_n)$ is the mean value of y_n.

It follows from the above definitions, that an approximation algorithm A is the best from the probabilistic analysis point of view if its absolute error almost surely converges to 0. Algorithm A is then called *asymptotically optimal*.

At this point one should also mention an analysis of the *rate of convergence* of the errors of approximation algorithms which may be different for algorithms whose absolute or relative errors are the same. Of course, the higher the rate, the better the performance of the algorithm.

It is rather obvious that the mean performance can be much better than the worst case behavior, thus justifying the use of a given approximation algorithm. A main obstacle is the difficulty of proofs of the mean performance for realistic distribution functions. Thus, the second way of evaluating the mean behavior of heuristic algorithms are computational experiments, which is still used very often. In the latter approach the values of the given criterion, constructed by the given heuristic algorithm and by an optimization algorithm are compared. This comparison should be made for a representative sample of instances. There are some practical problems which follow from the above statement and they are discussed in [SVW80].

2.5.2 Local Search Heuristics

In recent years more generally applicable heuristic algorithms for combinatorial optimization problems became known under the name *local search*. Primarily, they are designed as universal global optimization methods operating on a high-level solution space in order to guide heuristically lower-level local decision rules' performance to their best outcome. Hence, local search heuristics are often called *meta-heuristics* or strategies with knowledge-engineering and learning capabilities reducing uncertainty while knowledge of the problem setting is exploited and acquired in order to improve and accelerate the optimization process. The desire to achieve a certain outcome may be considered as the basic guide to appropriate knowledge modification and inference as a process of transforming some input information into the desired goal dependent knowledge.

Hence, in order to be able to transform knowledge, one needs to perform inference and to have memory which supplies the background knowledge needed to perform the inference and records the results of the inference for future use. Obviously, an important issue is the extent to which problem-specific knowledge must be used in the construction of *learning* algorithms (in other words the power and quality of inferencing rules) capable to provide significant performance improvements. Very general methods having a wide range of applicability in general are weak with respect to their performance. Problem specific methods achieve a highly efficient learning but with little use in other problem domains. Local search strategies are falling somewhat in between these two extremes, where genetic algorithms or neural networks tend to belong to the former category while tabu search or simulated annealing etc. are counted as examples of the second category. Anyway, these methods can be viewed as tools for searching a space of legal alternatives in order to find a best solution within reasonable time limitations. What is required are techniques for rapid location of high-quality solutions in large-size and complex search spaces and without any guarantee of optimality. When sufficient knowledge about such search spaces is available a priori, one can often exploit that knowledge (inference) in order to introduce problem-specific search strategies capable of supporting to find rapidly solutions of higher quality. Without such an a priori knowledge, or in cases where close to optimum solutions are indispensable, information about the problem has to be accumulated dynamically during the search process. Likewise obtained long-term as well as short-term memorized knowledge constitutes one of the basic parts in order to control the search process and in order to avoid getting stuck in a locally optimal solution. Previous approaches dealing with combinatorially explosive search spaces about which little knowledge is known a priori are unable to learn how to escape a local optimum. For instance, consider a random search. This can be effective if the search space is reasonably dense with acceptable solutions, such that the probability to find one is high. However, in most cases finding an acceptable solution within a reasonable amount of time is impossible because random search is not using any knowledge generated during the search process in order to improve its performance. Consider hill-climbing in which better solutions are found by exploring solutions "close" to a current and best one found so far. Hill-climbing techniques work well within a search space with relatively "few" hills. Iterated hill-climbing from randomly selected solutions can frequently improve the performance, however, any global information assessed during the search will not be exploited. Statistical sampling techniques are typical alternative approaches which emphasize the accumulation and exploitation of more global information. Generally speaking they operate by iteratively dividing the search space into regions to be sampled. Regions unlikely to produce acceptable solutions are discarded while the remaining ones will be subdivided for further sampling. If the number of useful sub-regions is small this search process can be effective. However, in case that the amount of a priori search space knowledge is pretty small, as is the case for many applications in business and engineering, this strategy frequently is not satisfactory.

Combining hill-climbing as well as random sampling in a creative way and introducing concepts of learning and memory can overcome the above mentioned deficiencies. The obtained strategies dubbed "local search based learning" are known, for instance, under the names tabu search and genetic algorithms. They provide general problem solving strategies incorporating and exploiting problem-specific knowledge capable even to explore search spaces containing an exponentially growing number of local optima with respect to the problem defining parameters.

A brief outline of what follows is to introduce the reader into extensions of the hill-climbing concept which are simulated annealing, tabu search, ejection chains, and genetic algorithms. Let us mention that they are particular specifications of the above mentioned knowledge engineering and learning concept reviewed in [Hol75, Mic97, Jon90]. Tabu search develops to become the most popular and successful general problem solving strategy. Hence, attention is drawn to a couple of tabu search issues more recently developed. e.g. ejection chains. Parts of this section can also be found embedded within a problem related setting in [CKP95, PG97].

To be more specific consider the minimization problem min $\{\gamma(x) \mid x \in S\}$ where γ is the objective function, i.e. the desired goal, and S is the search space, i.e. the set of feasible solutions of the problem. One of the most intuitive solution approaches to this optimization problem is to start with a known feasible solution and slightly perturb it while decreasing the value of the objective function. In order to realize the concept of slight perturbation let us associate with every x a subset $\mathcal{N}(x)$ of S, called *neighborhood* of x. The solutions in $\mathcal{N}(x)$, or neighbors of x, are viewed as perturbations of x. They are considered to be "close" to x. Now the idea of a simple local search algorithm is to start with some initial solution and move from one neighbor to another neighbor as long as possible while decreasing the objective value. This local search approach can be seen as the basic principle underlying many classical optimization methods, like the gradient method for continuous non-linear optimization or the simplex method for linear programming. Some of the important issues that have to be dealt with when implementing a local search procedure are how to pick the initial solution, how to define neighborhoods and how to select a neighbor of a given solution. In many cases of interest, finding an initial solution creates no difficulty. But obviously, the choice of this starting solution may greatly influence the quality of the final outcome. Therefore, local search algorithms may be run several times on the same problem instance, using different (e.g. randomly generated) initial solutions. Whether or not the procedure will be able to significantly ameliorate a poor solution often depends on the size of the neighborhoods. The choice of neighborhoods for a given problem is conditioned by a trade-off between quality of the solution and complexity of the algorithm, and is generally to be resolved by experiments. Another crucial issue in the design of a local search algorithm is the selection of a neighbor which improves the value of the objective function. Should the first neighbor found improving upon the current solution be picked, the best one, or still some other candidate? This question is rarely to be answered through theoretical considerations. In particular, the effect of the selection criterion on the quality of the final solution, or on the number of iterations of the procedure is often hard to predict (although, in some cases, the number of neighbors can rule out an exhaustive search of the neighborhood, and hence, the selection of the best neighbor). Here again experiments with various strategies are required in order to make a decision. The attractiveness of local search procedures stems from their wide applicability and (usually) low empirical complexity (see [JPY88] and [Yan90] for more information on the theoretical complexity of local search). Indeed, local search can be used for highly intricate problems, for which analytical models would involve astronomical numbers of variables and constraints, or about which little problem-specific knowledge is available. All that is needed here is a reasonable definition of neighborhoods, and an efficient way of searching them. When these conditions are satisfied, local search can be implemented to quickly produce good solutions for large instances of the problem. These features of local search explain that the approach has been applied to a wide diversity of situations, see [PV95, GLTW93, Ree93, AL97]. In the scheduling area we would like to emphasize on two excellent surveys, [AGP95] as well as [VAL96].

Nevertheless, local search in its most simple form, the *hill-climbing*, stops as soon as it encounters a local optimum, i.e., a solution x such that $\gamma(x) \leq \gamma(y)$ for all y in $\mathcal{N}(x)$. In general, such a local optimum is not a global optimum. Even worse, there is usually no guarantee that the value of the objective function at an arbitrary local optimum comes close to the optimal value. This inherent shortcoming of local search can be palliated in some cases by the use of multiple restarts. But, because NP-hard problems often possess many local optima, even this remedy may not be potent enough to yield satisfactory solutions. In view of this difficulty, several extensions of local search have been proposed, which offer the possibility to escape local optima by accepting occasional deteriorations of the objective function. In what follows we discuss successful approaches based on related ideas, namely *simulated annealing* and *tabu search*. Another interesting extension of local search works with a population of feasible solutions (instead of a single one) and tries to detect properties which distinguish good from bad solutions. These properties are then used to construct a new population which hopefully contains a better solution than the previous one. This technique is known under the name *genetic algorithm*.

Simulated Annealing

Simulated annealing was proposed as a framework for the solution of combinatorial optimization problems by Kirkpatrick, Gelatt and Vecchi and, independently, by Cerny, cf. [KGV83, Cer85]. It is based on a procedure originally devised by Metropolis et al. in [MRR+53] to simulate the annealing (or slow cooling) of solids, after they have been heated to their melting point. In simulated annealing procedures, the sequence of solutions does not roll monotonically down towards a local optimum, as was the case with local search. Rather, the solutions trace an up-and-down random walk through the feasible set S, and this walk is loosely guided in a "favorable" direction. To be more specific, we describe the kth iteration of a typical simulated annealing procedure, starting from a current solution x. First, a neighbor of x, say $y \in \mathcal{N}(x)$, is selected (usually, but not necessarily, at random). Then, based on the amplitude of $\Delta := \gamma(x) - \gamma(y)$, a transition from x to y (i.e., an update of x by y) is either accepted or rejected. This decision is made non-deterministically: the transition is accepted with probability $ap_k(\Delta)$, where ap_k is a probability distribution depending on the iteration count k. The intuitive justification for this rule is as follows. In order to avoid getting trapped early in a local optimum, transitions implying a deterioration of the objective function (i.e., with $\Delta < 0$) should be occasionally accepted, but the probability of acceptance should nevertheless increase with Δ. Moreover, the probability distributions are chosen so that $ap_{k+1}(\Delta) \leq ap_k(\Delta)$. In this way, escaping local optima is relatively easy during the first iterations, and the procedure explores the set S freely. But, as the iteration count increases, only improving transitions tend to be accepted, and the solution path is likely to terminate in a local optimum. The procedure stops if the value of the objective function remains constant in L (a termination parameter) consecutive iterations, or if the number of iterations becomes too large. In most implementations, and by analogy with the original procedure of Metropolis et al. [MRR+53], the probability distributions ap_k take the form:

$$ap_k (\Delta) = \begin{cases} 1 & \text{if } \Delta \geq 0 \\ e^{c_k \Delta} & \text{if } \Delta < 0, \end{cases}$$

where $c_{k+1} \geq c_k \geq 0$ for all k, and $c_k \to \infty$ when $k \to \infty$. A popular choice for the parameter c_k is to hold it constant for a number $L(k)$ of consecutive iterations, and then to increase it by a constant factor: $c_{k+1} = \alpha^{k+1} c_0$. Here, c_0 is a small positive number, and α is slightly larger than 1. The number $L(k)$ of solutions visited for each value of c_k is based on the requirement to achieve a quasi equilibrium state. Intuitively this is reached if a fixed number of transitions is accepted. Thus, as the acceptance probability approaches 0 we would expect $L(k) \to \infty$. Therefore $L(k)$ is supposed to be bounded by some constant B to avoid long chains of trials for large values of c_k. It is clear that the choice of the termination parameter and of the distributions ap_k $(k = 1, 2, \cdots)$ (the so-called *cooling schedule*) strongly influences the performance of the procedure. If the cooling is too rapid (e.g. if B is small and α is large), then simulated annealing tends to behave like local search, and gets trapped in local optima of poor quality. If the cooling is too slow, then the running time becomes prohibitive. Starting from an initial solution x_{start} and parameters c_0 and α a generic simulated annealing algorithm can be presented as follows.

Algorithm 2.5.1 *Simulated annealing* [LA87, AK89].
begin
Initialize $(x_{\text{start}}, c_0, \alpha)$;
$k := 0$;
$x := x_{\text{start}}$;
repeat
   Define $L(k)$ or B;
    for $t := 1$ to $L(k)$ **do**
    begin
       Generate a neighbor $y \in \mathcal{N}(x)$;
       $\Delta := \gamma(x) - \gamma(y)$;
       $ap_k(\Delta) := e^{c_k \Delta}$;
       if random$[0,1] \leq ap_k(\Delta)$ **then** $x := y$
    end;
   $c_{k+1} := \alpha c_k$;
   $k := k+1$;
until some stopping criterion is met
end;

Under some reasonable assumptions on the cooling schedule, theoretical results can be established concerning convergence to a global optimum or the complexity of the procedure (see [LA87, AK89]). In practice, determining appropriate values for the parameters is a part of the fine tuning of the implementation, and still relies on experiments. We refer to the extensive computational studies in [JAMS89, JAMS91] for the wealth of details on this topic. If the number of iterations during the search process is large, the repeated computation of the acceptance probabilities becomes a time consuming factor. Hence, *threshold accepting* as a deterministic variant of the simulated annealing has been introduced in [DS90]. The idea is not to accept transitions with a certain probability that changes over time but to accept a new solution if the amplitude $-\Delta$ falls below a certain threshold which is lowered over time. Simulated annealing has been applied to several types of combinatorial optimization problems, with various degrees of success (see [LA87, AK89, and JAMS89, JAMS91] as well as the bibliography [CEG88]).

As a general rule, one may say that simulated annealing is a reliable procedure to use in situations where theoretical knowledge is scarce or appears difficult to apply algorithmically. Even for the solution of complex problems, simulated annealing is relatively easy to implement, and usually outperforms a hill-climbing procedure with multiple starts.

Tabu Search

Tabu search is a general framework, which was originally proposed by Glover, and subsequently expanded in a series of papers [GL97, Glo77, Glo86, Glo89, Glo90a, Glo90b, GM86, WH89]. One of the central ideas in this proposal is to guide deterministically the local search process out of local optima (in contrast with the non-deterministic approach of simulated annealing). This can be done using different criteria, which ensure that the loss incurred in the value of the objective function in such an "escaping" step (a move) is not too important, or is somehow compensated for.

A straightforward criterion for leaving local optima is to replace the improvement step in the local search procedure by a "least deteriorating" step. One version of this principle was proposed by Hansen under the name steepest descent mildest ascent (see [HJ90], as well as [Glo89]). In its simplest form, the resulting procedure replaces the current solution x by a solution $y \in \mathcal{N}(x)$ which maximizes $\Delta := \gamma(x) - \gamma(y)$. If during L (a termination parameter) iterations no improvements are found, the procedure stops. Notice that Δ may be negative, thus resulting in a deterioration of the objective function. Now, the major defect of this simple procedure is readily apparent. If Δ is negative in some transition from x to y, then there will be a tendency in the next iteration of the procedure to reverse the transition, and go back to the local optimum x (since x improves on y). Such a reversal would cause the procedure to oscillate endlessly between x and y. Therefore, throughout the search a (dynamic) list of forbidden transitions, called *tabu list* (hence the name of the procedure) is maintained. The purpose of this list is not to rule out cycling completely (this would in general result in heavy bookkeeping and loss of flexibility), but at least to make it improbable. In the framework of the steepest descent mildest ascent procedure, we may for instance implement this idea by placing solution x in a tabu list TL after every transition away from x. In effect, this amounts to deleting x from S. But, for reasons of flexibility, a solution would only remain in the tabu list for a limited number of iterations, and then should be freed again. To be more specific the transition to the neighbor solution, i.e. a move, may be described by one or more attributes. These attributes (when properly chosen) can become the foundation for creating a so-called attribute based memory. For example, in a 0–1 integer programming context the attributes may be the set of all possible value assignments (or changes in such assignments) for the binary variables. Then two attributes which denote that a certain binary variable is set to 1 or 0, may be called complementary to each other. A move may be considered as the assignment of the compliment attribute to the binary variable. That is, the complement of a move cancels the effect of the considered move. If a move and its complement are performed, the same solution is reached as without having performed both moves. Moves eventually leading to a previously visited solution may be stored in the tabu list and are hence forbidden or tabu. The tabu list may be derived from the running list (RL), which is an ordered list of all moves (or their attributes) performed throughout the search. That is, RL represents the trajectory of solutions encountered. Whenever the length of RL is limited the attribute based memory of tabu search based on exploring RL is structured to provide a short term memory function. Now, each iteration consist of two parts: The guiding or tabu process and the application process. The tabu process updates the tabu list hereby requiring the actual RL; the application process chooses the best move that is not tabu and updates RL. For faster computation or storage reduction both processes are often combined. The application process is a specification on, e.g., the neighborhood definition and has to be defined by the user. The tabu navigation method is a rather simple approach requiring one parameter l called *tabu list length*. The tabu navigation method disallows choosing any complement of the l most recent moves of the running list in order to establish the next move. Hence, the tabu list consists of a (complementary) copy of the last part of RL. Older moves are disregarded. The tabu status derived from the l most recent moves forces the algorithm to go l moves away from any explored solution before the first step backwards is allowed. Obviously, this approach may disallow more moves than necessary to avoid returning to a yet visited solution. This encourages the intention to keep l as small as possible without disregarding the principle aim of never exploring a solution twice. Consequently, if l is too small the algorithm probably will return to a local optimum just left. If a solution is revisited the same sequence of moves may be repeated consecutively until the algorithm eventually stops, i.e. the search process is cycling. Thus danger of cycling favors large values for l. An adequate value for l has to be adopted with respect to the problem structure, the cardinality of the considered problem instances (especially problem size), the objective, etc. The parameter l is usually fixed but could also be randomly or systematically varied after a certain number of iterations. The fact that the tabu navigation method disallows moves which are not necessarily tabu led to the development of a so called aspiration level criterion which may override the tabu status of a move. The basic form of the aspiration level criterion is to choose a move in spite of its tabu status if it leads to an objective function value better than the best obtained in all preceding iterations. Another possible implementation would be to create a tabu list $TL(y)$ for every solution y within the solution space S. After a transition from x to y, x would be placed in the list $TL(y)$, meaning that further transitions from y to x are forbidden (in effect, this amounts to deleting x from $\mathcal{N}(y)$). Here again, x should be dropped from $TL(y)$ after a number of transitions. For still other possible definitions of tabu lists, see e.g. [Glo86, Glo89, GG89, HJ90, HW90]. Tabu search encompasses many features beyond the possibility to avoid the trap of local optimality and the use of tabu lists. Even though we cannot discuss them all in the limited framework of this survey, we would like to mention two of them, which provide interesting links with artificial intelligence and with genetic algorithms. In order to guide the search, Glover suggests recording some of the salient characteristics of the best solutions found in some phase of the procedure (e.g., fixed values of the variables in all, or in a majority of those solutions, recurring relations between the values of the variables, etc.). In a subsequent phase, tabu search can then be restricted to the subset of feasible solutions presenting these characteristics. This enforces what Glover calls a "regional intensification" of the search in promising "regions" of the feasible set. An opposite idea may also be used to "diversify" the search. Namely, if all solutions discovered in an initial phase of the search procedure share some common features, this may indicate that other regions of the solution space have not been sufficiently explored. Identifying these unexplored regions may be helpful in providing new starting solutions for the search. Both ideas, of search intensification or diversification, require the capability of recognizing recurrent patterns within subsets of solutions. In many applications the aforementioned simple tabu search strategies are already very successful, cf. [GLTW93, PV95, OK96]. A brief outline of the tabu search algorithm can be presented as follows.

Algorithm 2.5.2 *Tabu search* [Glo89, Glo90a, Glo90b].
begin
Initialize $(x$, tabu list TL, running list RL, aspiration function $A(\Delta, k))$;
$x_{best} := x$;
$k := 1$;
Specify the tabu list length l_k at iteration k;
$RL := \varnothing$;
$TL := \varnothing$;
$\alpha := \infty$;
repeat
  repeat
    Generate neighbor $y \in \mathcal{N}(x)$;
    $\Delta := \gamma(x) - \gamma(y)$;
    Calculate the aspiration value $A(\Delta, k)$;
  until $A(\Delta, k) < \alpha$ **or** $\Delta = \max\{ \gamma(x) - \gamma(y) \mid y$ is not tabu$\}$;
  Update RL, i.e. $RL := RL \cup \{$some attributes of $y\}$;
  $TL := \{$the last l_k non-complimentary entries of $RL\}$;
  if $A(\Delta, k) < \alpha$ **then** $\alpha := A(\Delta, k)$;
  $x := y$;

if $\gamma(y) < \gamma(x_{best})$ **then** $x_{best} := y;$
  $k := k+1;$
until some stopping criterion is met
end;

As mentioned above, tabu search may be applied in a more advanced way to incorporate different means for solid theoretical foundations. Other concepts have been developed like the reverse elimination method or the reactive tabu search incorporating a memory employing simple reactive mechanisms that are activated when repetitions of solutions are discovered throughout the search, see e.g. [GL97].

Ejection Chains

Variable depth methods, whose terminology was popularized by Papadimitriou and Steiglitz [PS82], have had an important role in heuristic procedures for optimization problems. The origins of such methods go back to prototypes in network and graph theory methods of the 1950s and 1960s. A class of these procedures called *ejection chain* methods has proved highly effective in a variety of applications, see [LK73] which is a special instance of an ejection chain on the TSP, and [Glo91, Glo96, DP94, Pes94, PG97, Reg98].

Ejection chain methods extend ideas exemplified by certain types of shortest path and alternating path constructions. The basic moves for a transition from one solution to another are compound moves composed of a sequence of paired steps. The first component of each paired step in an ejection chain approach introduces a change that creates a dislocation (i.e., an inducement for further change), while the second component creates a change designed to restore the system. The dislocation of the first component may involve a form of unfeasibility, or may be heuristically defined to create conditions that can be usefully exploited by the second component. Typically, the restoration of the second component may not be complete, and hence in general it is necessary to link the paired steps into a chain that ultimately achieves a desired outcome. The ejection terminology comes from the typical graph theory setting where each of the paired steps begins by introducing an element (such as a node, edge or path) that disrupts the graph's preferred structure, and then is followed by ejecting a corresponding element, in a way that recovers a critical portion of the structure. A chain of such steps is controlled to assure the preferred structure eventually will be fully recovered (and preferably, fully recovered at various intermediate stages by means of trial solutions). The candidate element to be ejected in such instances may not be unique, but normally comes from a limited set of alternatives. The alternating path construction [Ber62] gives a simple illustration. Here, the preferred graph structure requires a degree constraint to be satisfied at each node (bounding the number of edges allowed to enter the node). The first component of a paired step introduces an edge that violates such a degree constraint, causing too many edges to enter a particular node, and thus is followed by a second component that ejects one of the current edges at the node so that the indicated constraint may again be satisfied. The restoration may be incomplete, since the ejected edge may leave another node with too few edges, and thus the chain is induced to continue. A construction called a reference structure becomes highly useful for controlling such a process, in order to restore imbalances at each step by means of special trial solution moves, see [Glo91, Glo96, PG97]. Loosely speaking, a reference structure is a representation of a (sometimes several) feasible solution such that, however, a very small number of constraints may be violated. Finding a feasible solution from a reference structure must be a trivial task which should be performable in constant time. Ejection chain processes of course are not limited to graph constructions. For example, they can be based on successively triggered changes in values of variables, as illustrated by a linked sequence of zero-one exchanges in multiple choice integer programming applications or by linked "bound escalations" in more general integer programs. The approach can readily be embedded in a complete tabu search implementation, or in a genetic algorithm or simulated annealing implementation. Such a method can also be used as a stand-alone heuristic, which terminates when it is unable to find an improved solution at the conclusion of any of its constructive passes. (This follows the customary format of a variable depth procedure.) As our construction proceeds, we therefore note the trial solutions (e.g. feasible tours in case of a TSP) that would result by applying these feasibility-recovering transformations after each step, keeping track of the best. At the conclusion of the construction we simply select this best trial solution to replace the current solution, provided it yields an improvement. In this process, the moves at each level cannot be obtained by a collection of independent and non-intersecting moves of previous levels. The list of forbidden (tabu) moves grows dynamically during variable depth search iteration and is reset at the beginning of the next iteration. In the subsequent algorithmic description we designate the lists of variables (in the basis of a corresponding LP solution) locked in and out of the solution by the names tabu-to-drop and tabu-to-add, where the former contains variables added by the current construction (hence which must be prevented from being dropped) and the latter contains variables dropped by the current construction (hence which must be prevented from being added). The resulting ejection chain procedure is shown in Algorithm 2.5.3. We denote the cost of a solution x by $\gamma(x)$. The cost difference of a solution x' and x, i.e. $\gamma(x) - \gamma(x')$, where x' results from x by replacing variable i by variable j will be defined by γ_{ij}. The reference structure that results by performing d ejection steps, is denoted by $x(d)$, where d is the "depth" of the ejection chain (hence $x = x(0)$ for a given starting solution x).

Algorithm 2.5.3 *Ejection chain* [PG97].
begin
Start with an initial solution x_{start};

$x := x_{start};$ $x^* := x_{start};$
Let s be any variable in x; $--$ s is the root
$k^* := s;$
repeat
   $d := 0;$ $--$ d is the current search depth
   while there are variables in $x(d)$ that are not tabu-to-drop
           and variables outside of $x(d)$ that are not tabu-to-add **do**
      begin
      $i := k^*;$
      $d := d+1;$
      Find the best component move that maintains the reference structure,
          where this 'best' is given by the variable pair i, j for which the gain
          $\gamma_{i^*j^*} = \max\{\gamma_{ij} \mid j$ is not a variable in $x(d-1)$ and i is a variable
                    in $x(d-1); j$ is not tabu-to-add; i is not tabu-to-drop$\};$
      Perform this move, i.e. introduce variable j^* and remove variable i^*
          thus obtaining $x(d)$ as a new reference structure at search depth $d;$
      j^* becomes tabu-to-drop and i^* becomes tabu-to-add;
      end;
   Let d^* denote the search depth at which the best solution $x^*(d^*)$ with
       $\gamma(x^*(d^*)) = \min\{\gamma(x^*(d)) \mid 0 < d \leq n\}$ has been found;
   if $d^* > 0$ **then** $x^* := x^*(d^*);$ $x := x^*;$
until $d^* = 0;$
end;

The above procedure describes in its inner **repeat** . . . **until** loop one iteration of an ejection chain search. The **while** . . . **do** describes one component move. Starting with an initially best solution $x^*(0)$, the procedure executes a construction that maintains the reference structure for a certain number of
component moves. The new currently best trial solution $x^*(d^*)$, encountered at
depth d^*, becomes the starting point for the next ejection chain iteration. The iterations are repeated as long as an improvement is possible. The maximum depth
of the construction is reached if all variables in the current solution x are set tabu-
to-drop. The step leading from a solution x to a new solution consists of a varying number d^* of component moves, hence motivating the "variable depth" terminology. A continuously growing tabu list avoids cycling of the search procedure. As an extension of the algorithm (not shown here), the whole **repeat**
. . . **until** part could easily be embedded in yet another control loop leading
to a multi-level (parallel) search algorithm, see [Glo96].

Genetic Algorithms

As the name suggests, *genetic algorithms* are motivated by the theory of evolution; they date back to the early work described in [Rec73, Hol75, Sch77], see also [Gol89] and [Mic97]. They have been designed as general search strategies and optimization methods working on populations of feasible solutions. Working with populations permits to identify and explore properties which good solutions have in common (this is similar to the regional intensification idea mentioned in our discussion of tabu search). Solutions are encoded as strings consisting of elements chosen from a finite alphabet. Roughly speaking, a genetic algorithm aims at producing near-optimal solutions by letting a set of strings, representing random solutions, undergo a sequence of unary and binary transformations governed by a selection scheme biased towards high-quality solutions. Therefore, the quality or *fitness* value of an individual in the population, i.e. a string, has to be defined. Usually it is the value of the objective function or some scaled version of it. The transformations on the individuals of a population constitute the *recombination* steps of a genetic algorithm and are performed by three simple operators. The effect of the operators is that implicitly good properties are identified and combined into a new population which hopefully has the property that the value of the best individual (representing the best solution in the population) and the average value of the individuals are better than in previous populations. The process is then repeated until some stopping criteria are met. It can be shown that the process converges to an optimal solution with probability one (cf. [EAH91]). The three basic operators of a classical genetic algorithm when a new population is constructed are *reproduction, crossover* and *mutation*.

Via reproduction a new temporary population is generated where each member is a replica of a member of the old population. A copy of an individual is produced with probability proportional to its fitness value, i.e. better strings probably get more copies. The intended effect of this operation is to improve the quality of the population as a whole. However, no genuinely new solutions and hence no new information are created in the process. The generation of such new strings is handled by the crossover operator.

In order to apply the crossover operator the population is randomly partitioned into pairs. Next, for each pair, the crossover operator is applied with a certain probability by randomly choosing a position in the string and exchanging the tails (defined as the substring starting at the chosen position) of the two strings (this is the simplest version of a crossover). The effect of the crossover is that certain properties of the individuals are combined to new ones or other properties are destroyed. The construction of a crossover operator should also take into consideration that fitness values of offspring are not too far from those of their parents, and that offspring should be genetically closely related to their parents.

The mutation operator which makes random changes to single elements of the string only plays a secondary role in genetic algorithms. Mutation serves to maintain diversity in the population (see the previous section on tabu search).

Besides unary and binary recombination operators, one may also introduce operators of higher arities such as consensus operators, that fix variable values common to most solutions represented in the current population. Selection of individuals during the reproduction step can be realized in a number of ways: one could adopt the scenario of [Gol89] or use deterministic ranking. Further it matters whether the newly recombined offspring compete with the parent solutions or simply replace them.

The traditional genetic algorithm, based on a binary string representation of solutions, is often unsuitable for combinatorial optimization problems because it is very difficult to represent a solution in such a way that sub-strings have a meaningful interpretation. Nevertheless, the number of publications on genetic algorithm applications to sequencing and scheduling problems exploded.

Problems from combinatorial optimization are well within the scope of genetic algorithms and early attempts closely followed the scheme of what Goldberg [Gol89] calls a *simple genetic algorithm*. Compared to standard heuristics, genetic algorithms are not well suited for fine-tuning structures which are very close to optimal solutions. Therefore, it is essential, if a competitive genetic algorithm is desired, to compensate for this drawback by incorporating (local search) improvement operators into the basic scheme. The resulting algorithm has then been called *genetic local search heuristic* or *genetic enumeration* (cf. [Joh90b, UAB+91, Pes94, DP95]). Each individual of the population is then replaced by a locally improved one or an individual representing a locally optimal solution, i.e. an improvement procedure is applied to each individual either partially (to a certain number of iterations, [KP94]) or completely. Some type of improvement heuristic may also be incorporated into the crossover operator, cf. [KP94].

Putting things into a more general framework, a solution of a combinatorial optimization problem may be considered as resolution of a sequence of local decisions (such as priority rules or even more complicated ones). In an enumeration tree of all possible decision sequences the solutions of the problem are represented as a path corresponding to the different decisions from the root of the tree to a leaf (hence the name genetic enumeration). While a branch and bound algorithm learns to find those decisions leading to an optimal solution (with respect to the space of all decision sequences) genetics can guide the search process in order to learn to find the most promising decision combinations within a reasonable amount of time, see [Pes94, DP95]. Hence, instead of (implicitly) enumerating all decision sequences a rudimentary search tree will be established. Only a polynomial number of branches can be considered where population genetics drives the search process into those regions which more likely contain optimal solutions. The scheme of a genetic enumeration algorithm is subsequently described; it requires further refinement in order to design a successful genetic algorithm.

Algorithm 2.5.4 *Genetic enumeration* [DP95, Pes94].
begin
Initialization: Construct an initial population of individuals each of which is a
    string of local decision rules;
Assessment / Improvement: Assess each individual in the current population
    introducing problem specific knowledge by special purpose heuristics (such as
    local search) which are guided by the sequence of local decisions;
if special purpose heuristics lead to a new string of local decision rules
then
    replace each individual by the new one, for instance, a locally optimal one;
repeat
    Recombination: Extend the current population by adding individuals obtained
        by unary and binary transformations (crossover, mutation) on one or two
        individuals in the current population;
    Assessment / Improvement: Assess each individual in the current population
        introducing problem specific knowledge by special purpose heuristics
        (such as local search) which are guided by the sequence of local decisions;
    if special purpose heuristics lead to a new string of local decision rules
    then
        replace each individual by the new one, for instance, a locally optimal one;
until some stopping criterion is met
end;

It is an easy exercise to recognize that the simple genetic algorithm as well as genetic local search fits into the provided framework.

For a successful genetic algorithm in combinatorial optimization a genetic meta-strategy is indispensable in order to guide the operation of good special purpose heuristics and to incorporate problem-specific knowledge. An older concept of a population based search technique which dates back in its origins beyond the early days of genetic algorithms is introduced in [Glo95] and called *scatter search*. The idea is to solve 0–1 programming problems departing from a solution of a linear programming relaxation. A set of reference points is created by perturbing the values of the variables in this solution. Then new points are defined as selected convex combinations of reference points that constitute good solutions obtained from previous solution efforts. Non-integer values of these points are rounded and then heuristically converted into candidate solutions for the integer programming problem. The idea parallels and extends the idea basic to the genetic algorithm design, namely, combining parent solutions in some way in order to obtain new offspring solutions. One of the issues that differentiates scatter search from the early genetic algorithm paradigm is the fact that the former creates new points strategically rather than randomly. Scatter search does not prespecify the number of points it will generate to retain. This can be adaptively established by considering the solution quality during the generation process. The "data perturbation idea" meanwhile has gained considerable attention within the

GA community. In [LB95] it is transferred as a tool for solving resource con-strained project scheduling problems with different objective functions. The ba-sic idea of their approach may be referred to as a "data perturbation" methodol-ogy which makes use of so-called problem space based neighborhoods. Given a well-known concept for deriving feasible solutions (e.g. a priority rule), a search approach is employed on account of the problem data and respective perturba-tions. By modifying (i.e. introducing some noise or perturbation) the problem data used for the priority values of activities, further solutions within a certain neighborhood of the original data are generated.

The ideas mentioned above are paving the way in order to do some steps into the direction of machine learning. This is in particular true if learning is consid-ered to be a right combination of employing inference on memory. Thus, local search in terms of tabu search and genetic algorithms emphasize such a unified approach in all successful applications. This probably resembles most the human way of thinking and learning.

References

Agi66   N. Agin, Optimum seeking with branch and bound, *Management Sci.* 13, 1966, B176-185.

AGP95   E. J. Anderson, C. A. Glass, C. N. Potts, Local search in combinatorial optimi-zation: applications in machine scheduling, Working paper, University of Southampton, 1995.

AHU74   A. V. Aho, J. E. Hopcroft, J. D. Ullman, *The Design and Analysis of Computer Algorithms*, Addison-Wesley, Reading, Mass., 1974.

AK89   E. H. L. Aarts, J. Korst, *Simulated Annealing and Boltzmann Machines,* J. Wiley, Chichester, 1989.

AL97   E. H. L. Aarts, J. K. Lenstra (eds.), *Local Search in Combinatorial Optimiza-tion*, Wiley, New York, 1997.

AMO93   R. K. Ahuja, T. L. Magnanti, J. B. Orlin, *Network Flows*, Prentice Hall, Engle-wood Cliffs, N.J., 1993.

Bak74   K. Baker, *Introduction to Sequencing and Scheduling*, J. Wiley, New York, 1974.

BD62   R. Bellman, S. E. Dreyfus, *Applied Dynamic Programming*, Princeton Univer-sity Press, Princeton, N.J., 1962.

Bel57   R. Bellman, *Dynamic Programming*, Princeton University Press, Princeton, N.J., 1957.

Ber62   C. Berge, *Theory of Graphs and its Applications*, Methuen, London, 1962.

Ber73   C. Berge, *Graphs and Hypergraphs*, North Holland, Amsterdam, 1973.

CEG88   N. E. Collins, R. W. Eglese, B. L. Golden, Simulated annealing - an annotated bibliography, *American J. Math. Management Sci.* 8, 1988, 209-307.

Cer85   V. Cerny, Thermodynamical approach to the traveling salesman problem; an efficient simulation algorithm, *J. Optimization Theory and Applications* 45, 1985, 41-51.

Che77   B. V. Cherkasskij, Algoritm postrojenija maksimalnogo potoka w sieti so sloznostju $0(V^2E^{1/2})$ operacij, *Matematiczeskije Metody Reszenija Ekonomiczeskich Problem* 7, 1977, 117-125.

Che80   T.-Y. Cheung, Computational comparison of eight methods for the maximum network flow problem, *ACM Trans. Math. Software* 6, 1980, 1-16.

CHW87   M. Chams, A. Hertz, D. de Werra, Some experiments with simulated annealing for colouring graphs, *European J. Oper. Res.* 32, 1987, 260-266.

CKP95   Y. Crama, A. Kolen, E. Pesch, Local search in combinatorial optimization, *Lecture Notes in Computer Science* 931, 1995, 157-174.

Cof76   E. G. Coffman, Jr. (ed.), *Scheduling in Computer and Job Shop Systems*, J. Wiley, New York, 1976.

Coo71   S. A. Cook, The complexity of theorem proving procedures, *Proc. 3rd ACM Symposium on Theory of Computing*, 1971, 151-158.

Den82   E. V. Denardo, *Dynamic Programming: Models and Applications*, Prentice-Hall, Englewood Cliffs, N.J., 1982.

Din70   E. A. Dinic, Algoritm reszenija zadaczi o maksimalnom potokie w sieti so stepennoj ocenkoj, *Dokl. Akad. Nauk SSSR* 194, 1970, 1277-1280.

DL79   S. E. Dreyfus, A. M. Law, *The Art and Theory of Dynamic Programming*, Academic Press, New York, 1979.

DP94   U. Dorndorf, E. Pesch, Fast clustering algorithms, *ORSA J. Comput.* 6, 1994, 141-153.

DP95   U. Dorndorf, E. Pesch, Evolution based learning in a job shop scheduling environment, *Computers and Oper. Res.* 22,1995, 25-40.

DS90   G. Dueck, T. Scheuer, Threshold accepting: a general purpose optimization algorithm appearing superior to simulated annealing, *J. Comp. Physics* 90, 1990, 161-175.

EAH91   A. E. Eiben, E. H. L. Aarts, K. H. van Hee, Global convergence of genetic algorithms: A Markov Chain analysis, *Lecture Notes in Computer Science* 496, 1991, 4-9.

Edm65   J. Edmonds, Paths, trees and flowers, *Canadian J. Math.* 17, 1965, 449-467.

EK72   J. Edmonds, R. M. Karp, Theoretical improvement in algorithmic efficiency for network flow problem, *J. Assoc. Comput. Mach.* 19, 1972, 248-264.

Eve79   S. Even, *Graph Algorithms*, Computer Science Press Inc., New York, 1979.

FF62   L. R. Ford, Jr., D. R. Fulkerson, *Flows in Networks*, Princeton University Press, Princeton, N.J., 1962.

Fis70   P.C.Fishburn, Intransitive indifference in preference theory: A survey, *Oper. Res.* 18 (1970) 207-228

GG89   F. Glover, H. J. Greenberg, New approaches for heuristic search: A bilateral linkage with artificial intelligence, *European J. Oper. Res.* 13, 1989, 563-573.

GJ78   M. R. Garey, D. S. Johnson, Strong NP-completeness results: motivation, examples, and implications, *J. Assoc. Comput. Mach.* 25, 1978, 499-508.

GJ79   M. R. Garey, D. S. Johnson, *Computers and Intractability: A Guide to the Theory of NP-Completeness*, W. H. Freeman, San Francisco, 1979.

GL97   F. Glover, M. Laguna, *Tabu Search*, Kluwer Academic Publ., Boston, 1997.

Glo77   F. Glover, Heuristic for integer programming using surrogate constraints, *Decision Sciences* 8, 1977, 156-160.

Glo86   F. Glover, Future paths for integer programming and links to artificial intelligence, *Computers and Oper. Res.* 13,1986, 533-549.

Glo89   F. Glover, Tabu-search - Part I, *ORSA J. Comput.* 1, 1989, 190-206.

Glo90a   F. Glover, Tabu Search - Part II, *ORSA J. Comput.* 2, 1990, 4-32.

Glo90b   F. Glover, Tabu search: a tutorial, *Interfaces* 20(4), 1990, 74-94.

Glo91   F. Glover, Multilevel tabu search and embedded search neighborhoods for the traveling salesman problem, Working paper, University of Colorado, Boulder, 1991.

Glo96   F. Glover, Ejection chains, reference structures and alternating path methods for traveling salesman problems, *Discrete Appl. Math.* 65, 1996, 223-253.

Glo95   F. Glover, Scatter search and star-paths: Beyond the genetic metaphor, *OR Spektrum* 17, 1995, 125-137.

GLTW93   F. Glover, M. Laguna, E. Taillard, D. de Werra (eds.), *Tabu Search*, Annals of Operations Research 41, Baltzer, Basel, 1993.

GM86   F. Glover, C. McMillan, The general employee scheduling problem: An integration of MS and AI, *Computers and Oper. Res.* 13, 1986, 563-573.

Gol89   D. E. Goldberg, *Genetic Algorithms in Search, Optimization and Machine Learning*, Addison-Wesley, Reading, Mass., 1989.

HJ90   P. Hansen, B. Jaumard, Algorithms for the maximum satisfiability problem, *Computing* 44, 1990, 279-303.

Hol75   J. H. Holland, *Adaptation in Natural and Artificial Systems*, The University of Michigan Press, Ann Arbor, 1975.

How69   R. A. Howard, *Dynamic Programming and Markov Processes*, MIT Press, Cambridge, Mass., 1969.

HW90   A. Hertz, D. de Werra, The tabu search metaheuristic: How we use it, *Ann. Math. Artif. Intell.* 1, 1990, 111-121.

JAMS89   D. S. Johnson, C. R. Aragon, L. A. McGeoch, C. Schevon, Optimization by simulated annealing: An experimental evaluation; Part I, Graph partitioning, *Oper. Res.* 37, 1989, 865-892.

JAMS91   D. S. Johnson, C. R. Aragon, L. A. McGeoch, C. Schevon, Optimization by simulated annealing: An experimental evaluation; Part II, Graph coloring and number partitioning, *Oper. Res.* 39, 1991, 378-406.

Joh90a   D. S. Johnson, A Catalog of Complexity Classes, in: J. van Leeuwen (ed.), *Handbook of Theoretical Computer Science*, Elsevier, New York, 1990, Ch.2.

Joh90b   D. S. Johnson, Local optimization and the traveling salesman problem, *Lecture Notes in Computer Science* 443, 1990, 446-461.

Jon90   K. de Jong, Genetic-algorithm-based learning, in: Y. Kodratoff, R. Michalski (eds.) *Machine Learning*, Vol. III, Morgan Kaufmann, San Mateo, 1990, 611-638.

JPY88   D. S. Johnson, C. H. Papadimitriou, M. Yannakakis, How easy is local search? *J. Computer System Sci.* 37, 1988,79-100.

Kar72   R. M. Karp, Reducibility among combinatorial problems, in: R. E. Miller, J. W. Thatcher (eds.), *Complexity of Computer Computation*, Plenum Press, New York, 1972, 85-104.

Kar74   A. W. Karzanov, Nachozdenije maksimalnogo potoka w sieti metodom predpotokow, *Dokl. Akad. Nauk SSSR* 215, 1974, 434-437.

KGV83   S. Kirkpatrick, C. D. Gelatt Jr., M. P. Vecchi, Optimization by simulated annealing, *Science* 220, 1983, 671-680.

KP94   A. Kolen, E. Pesch, Genetic local search in combinatorial optimization, *Discrete Appl. Math.* 48, 1994, 273-284.

Kub87   M. Kubale, The complexity of scheduling independent two-processor tasks on dedicated processors, *Inform. Process. Lett.* 24, 1987, 141-147.

LA87   P. J. M. van Laarhoven, E. H. L. Aarts, *Simulated Annealing: Theory and Applications*, Reider, Dordrecht, 1987.

Law76   E. L. Lawler, *Combinatorial Optimization: Networks and Matroids*, Holt, Rinehart and Winston, New York, 1976.

LB95   V. J. Leon, R. Balakrishnan, Strength and adaptability of problem-space based neigborhoods for resource constrained scheduling, *OR Spektrum* 17, 1995, 173-182.

Len77   J. K. Lenstra, *Sequencing by Enumerative Methods*, Mathematical Centre Tracts 69, Amsterdam, 1977.

LK73   S. Lin, B. W. Kernighan, An effective heuristic algorithm for the traveling salesman problem, *Oper. Res.* 21, 1973, 498-516.

LRKB77   J. K. Lenstra, A. H. G. Rinnooy Kan, P. Brucker, Complexity of machine scheduling problems, *Ann. Discrete Math.* 1, 1977, 343-362.

LW66   E. L. Lawler, D. E. Wood, Branch and bound methods: a survey, *Oper. Res.* 14, 1966, 699-719.

Mic97   Z. Michalewicz, *Genetic Algorithms + Data Structures = Evolution Programs*, Springer, Berlin, 1997.

Mit70   L. G. Mitten, Branch-and-bound methods: general formulation and properties, *Oper. Res.* 18, 1970, 24-34.

MRR+53   M. Metropolis, A. Rosenbluth, M. Rosenbluth, A. Teller, E. Teller, Equation of state calculations by fast computing machines, *J. Chemical Physics* 21, 1953, 1087-1092.

OK96   I. H. Osman, J. P. Kelly, *Meta-Heuristics: Theory and Applications*, Kluwer, Dordrecht, 1996.

Pap94   C. H. Papadimitriou, *Computational Complexity*, Addison-Wesley, Reading, Mass., 1994.

Pes94   E. Pesch, *Learning in Automated Manufacturing,* Physica, Heidelberg, 1994.

PG97   E. Pesch, F. Glover, TSP ejection chains, *Discrete Appl. Math.* 76, 1997, 165-181.

PS82   C. H. Papadimitriou, K. Steiglitz, *Combinatorial Optimization: Algorithms and Complexity*, Prentice-Hall, Englewood Cliffs, N.J., 1982.

PV95   E. Pesch, S. Voß (eds.), *Applied Local Search*, OR Spektrum 17, 1995.

Rec73   I. Rechenberg, *Optimierung technischer Systeme nach Prinzipien der biologischen Evolution*, Problemata, Frommann-Holzboog, 1973.

Ree93   C. Reeves (ed.), *Modern Heuristic Techniques for Combinatorial Problems*, Blackwell Scientific Publishing, 1993.

Reg98   C. Rego, A subpath ejection method for the vehicle routing problem, *Management Sci.* 44, 1998, 1447-1459.

Rin76   A. H. G. Rinnooy Kan, *Machine Scheduling Problems: Classification, Complexity and Computations*, Martinus Nijhoff, The Hague, 1976.

Rin87   A. H. G. Rinnooy Kan, Probabilistic analysis of approximation algorithms, *Ann. Discrete Math.* 31, 1987, 365-384.

Sch77   H.-P. Schwefel, *Numerische Optimierung von Computer-Modellen mittels der Evolutionsstrategie*, Birkhäuser, Basel, 1977.

SVW80   E. A. Silver, R. V. Vidal, D. de Werra, A tutorial on heuristic methods, *European J. Oper. Res.* 5, 1980, 153-162.

UAB+91   N. L. J. Ulder, E. H. L. Aarts, H.-J. Bandelt, P. J. M. van Laarhoven, E. Pesch, Genetic local search algorithms for the traveling salesman problem, *Lecture Notes in Computer Science* 496, 1991, 109-116.

VAL96   R. J. M. Vaessens, E. H. L. Aarts, J. K. Lenstra, Job shop scheduling by local search, *ORSA J. Comput.* 13, 1996, 302-317.

VTL82   J. Valdes, R. E. Tarjan, E. L. Lawler, The recognition of series parallel digraphs, *SIAM J. Comput.* 11, 1982, 298-313.

WH89   D. de Werra, A. Hertz, Tabu search techniques: a tutorial and an application to neural networks, *OR Spektrum* 11, 1989, 131-141.

Yan90   M. Yannakakis, The analysis of local search problems and their heuristics, *Lecture Notes in Computer Science* 415, 1990, 298-311.

3 Definition, Analysis and Classification of Scheduling Problems

Throughout this book we are concerned with scheduling computer and manufacturing processes. Despite the fact that we deal with two different areas of applications, the same model could be applied. This is because the above processes consist of complex activities to be scheduled, which can be modeled by means of tasks (or jobs), relations among them, processors, sometimes additional resources (and their operational functions), and parameters describing all these items in greater detail. The purpose of the modeling is to find optimal or sub-optimal schedules in the sense of a given criterion, by applying best suited algorithms. These schedules are then used for the original setting to carry out the various activities. In this chapter we introduce basic notions used for such a modeling of computer and manufacturing processes.

3.1 Definition of Scheduling Problems

In general, scheduling problems considered in this book are characterized by three sets: set $T = \{T_1, T_2, \cdots, T_n\}$ of n *tasks*, set $P = \{P_1, P_2, \cdots, P_m\}$ of m *processors* (*machines*) and set $R = \{R_1, R_2, \cdots, R_s\}$ of s types of *additional resources* R. Scheduling, generally speaking, means to assign processors from P and (possibly) resources from R to tasks from T in order to complete all tasks under the imposed constraints. There are two general constraints in classical scheduling theory. Each task is to be processed by at most one processor at a time (plus possibly specified amounts of additional resources) and each processor is capable of processing at most one task at a time. In Chapters 6 and 12 we will show some new applications in which the first constraint will be relaxed.

We will now characterize the processors. They may be either *parallel*, i.e. performing the same functions, or *dedicated* i.e. specialized for the execution of certain tasks. Three types of parallel processors are distinguished depending on their speeds. If all processors from set P have equal task processing speeds, then we call them *identical*. If the processors differ in their speeds, but the *speed* b_i of each processor is constant and does not depend on the task in T, then they are called *uniform*. Finally, if the speeds of the processors depend on the particular task processed, then they are called *unrelated*.

In case of dedicated processors there are three models of processing sets of tasks: *flow shop, open shop* and *job shop*. To describe these models more precisely, we assume that tasks form n subsets[1] (*chains* in case of flow- and job shops), each subset called a *job*. That is, job J_j is divided into n_j tasks, T_{1j}, $T_{2j}, \cdots, T_{n_j j}$, and two adjacent tasks are to be performed on different processors. A set of jobs will be denoted by $\mathcal{J}$. In an open shop the number of tasks is the same for each job and is equal to m, i.e. $n_j = m$, $j = 1, 2, \cdots, n$. Moreover, T_{1j} should be processed on P_1, T_{2j} on P_2, and so on. A similar situation is found in flow shop, but, in addition, the processing of $T_{i-1 j}$ should precede that of T_{ij} for all $i = 1, \cdots, n_j$ and for all $j = 1, 2, \cdots, n$. In a general job shop system the number n_j is arbitrary. Usually in such systems it is assumed that buffers between processors have unlimited capacity and a job after completion on one processor may wait before its processing starts on the next one. If, however, buffers are of zero capacity, jobs cannot wait between two consecutive processors, thus, a *no-wait property* is assumed.

In general, task $T_j \in \mathcal{T}$ is characterized by the following data.

1. *Vector of processing times* $\boldsymbol{p}_j = [p_{1j}, p_{2j}, \cdots, p_{mj}]^\mathrm{T}$, where p_{ij} is the time needed by processor P_i to process T_j. In case of identical processors we have $p_{ij} = p_j$, $i = 1, 2, \cdots, m$. If the processors in $\mathcal{P}$ are uniform, then $p_{ij} = p_j/b_i$, $i = 1, 2, \cdots, m$, where p_j is the *standard processing time* (usually measured on the slowest processor) and b_i is the *processing speed factor* of processor P_i. In case of shop scheduling, the vector of processing times describes the processing requirements of particular tasks comprising one job; that is, for job J_j we have $\boldsymbol{p}_j = [p_{1j}, p_{2j}, \cdots, p_{n_j j}]^\mathrm{T}$, where p_{ij} denotes the processing time of T_{ij} on the corresponding processor.

2. *Arrival time* (or *ready time*) r_j, which is the time at which task T_j is ready for processing. If the arrival times are the same for all tasks from $\mathcal{T}$, then it is assumed that $r_j = 0$ for all j.

3. *Due date* d_j, which specifies a time limit by which T_j should be completed; usually, penalty functions are defined in accordance with due dates.

4. *Deadline* $\tilde{d}_j$, which is a "hard" real time limit by which T_j must be completed.

5. *Weight (priority)* w_j, which expresses the relative urgency of T_j.

6. *Resource request* (if any), as defined in Chapter 12.

---

[1] Thus, the number of tasks in $\mathcal{T}$ is assumed to be $\geq n$.

Unless stated otherwise we assume that all these parameters, p_j, r_j, d_j, $\tilde{d}_j$, and w_j, are integers. In fact, this assumption is not very restrictive, since it is equivalent to permitting arbitrary rational values. We assume moreover, that tasks are assigned all required resources whenever they start or resume their processing and that they release all the assigned resources whenever they are completed or preempted. These assumptions imply that deadlock cannot occur.

Next, some definitions concerning task preemptions and precedence constraints among tasks are given. A schedule is called *preemptive* if each task may be preempted at any time and restarted later at no cost, perhaps on another processor. If preemption of all the tasks is not allowed we will call the schedule *non-preemptive*.

In set $\mathcal{T}$ *precedence constraints* among tasks may be defined. $T_i \prec T_j$ means that the processing of T_i must be completed before T_j can be started. In other words, in set $\mathcal{T}$ a precedence relation $\prec$ is defined. The tasks in set $\mathcal{T}$ are called *dependent* if the order of execution of at least two tasks in $\mathcal{T}$ is restricted by this relation. Otherwise, the tasks are called *independent*. A task set with precedence relation is usually represented as a directed graph (a digraph) in which nodes correspond to tasks and arcs to precedence constraints (a *task-on-node graph*). It is assumed that no transitive arcs exist in precedence graphs. An example of a set of dependent tasks is shown in Figure 3.1.1(a) (nodes are denoted by T_j/p_j). Several special types of precedence graphs have already been described in Section 2.3.2. Let us notice that in the case of dedicated processors (except in open shop systems) tasks that constitute a job are always dependent, but the jobs themselves can be either independent or dependent. There is another way of representing task dependencies which is useful in certain circumstances. In this so-called *activity network*, precedence constraints are represented as a *task-on-arc graph*, where arcs represent tasks and nodes time events. Let us mention here a special graph of this type called *uniconnected activity network* (*uan*), which is defined as a graph in which any two nodes are connected by a directed path in one direction only. Thus, all nodes are uniquely ordered. For every precedence graph one can construct a corresponding activity network (and vice versa), perhaps using dummy tasks of zero length. The corresponding activity network for the precedence graph from Figure 3.1.1(a), is shown in Figure 3.1.1(b). Note that we will show in Section 5.1.1. the equivalence of the uniconnected activity network and the interval order task-on-node representation (cf. also [BK02]).

Task T_j will be called *available* at time t if $r_j \leq t$ and all its predecessors (with respect to the precedence constraints) have been completed by time t.

Now we will give the definitions concerning schedules and optimality criteria. A *schedule* is an assignment of processors from set $\mathcal{P}$ (and possibly resources from set $\mathcal{R}$) to tasks from set $\mathcal{T}$ in time such that the following conditions are satisfied:

– at every moment each processor is assigned to at most one task and each task is processed by at most one processor[2],
– task T_j is processed in time interval $[r_j, \infty)$,
– all tasks are completed,
– if tasks T_i, T_j are in relation $T_i \prec T_j$, the processing of T_j is not started before T_i is completed,
– in the case of non-preemptive scheduling no task is preempted (then the schedule is called *non-preemptive*), otherwise the number of preemptions of each task is finite[3] (then the schedule is called *preemptive*),
– resource constraints, if any, are satisfied.

To represent schedules we will use the so-called *Gantt charts*. An example schedule for the task set of Figure 3.1.1 on three parallel, identical processors is shown in Figure 3.1.2. The following parameters can be calculated for each task $T_j, j = 1, 2, \cdots, n$, processed in a given schedule:

completion time C_j,

flow time $F_j = C_j - r_j$, being the sum of waiting and processing times;

lateness $L_j = C_j - d_j$,

tardiness $D_j = \max\{C_j - d_j, 0\}$;

earliness $E_j = \max\{d_j - C_j, 0\}$.

For the schedule given in Figure 3.1.2 one can easily calculate the two first parameters. In vector notation these are $C = [3, 4, 5, 6, 1, 8, 8, 8]$ and $F = C$. The other two parameters could be calculated, if due dates would be defined. Suppose that due dates are given by the vector $d = [5, 4, 5, 3, 7, 6, 9, 12]$. Then the latenesses, tardinesses and earliness for the tasks in the schedule are: $L = [-2, 0, 0, 3, -6, 2, -1, -4]$, $D = [0, 0, 0, 3, 0, 2, 0, 0]$, $E = [2, 0, 0, 0, 6, 0, 1, 4]$.

To evaluate schedules we will use three main *performance measures* or *optimality criteria*:

Schedule length (makespan) $C_{max} = \max\{C_j\}$,

mean flow time $\bar{F} = \dfrac{1}{n}\sum\limits_{j=1}^{n} F_j$,

or *mean weighted flow time* $\bar{F}_w = \sum\limits_{j=1}^{n} w_j F_j / \sum\limits_{j=1}^{n} w_j$,

maximum lateness $L_{max} = \max\{L_j\}$.

---

[2] As we mentioned, this assumption can be relaxed.
[3] This condition is imposed by practical considerations only.

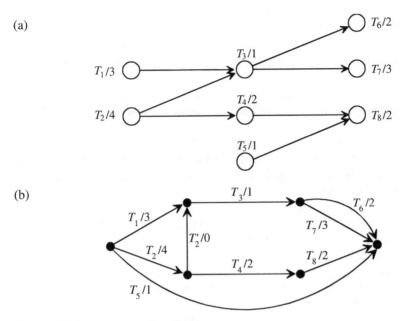

Figure 3.1.1 *An example task set*
(a) task-on-node representation
(b) task-on-arc representation (dummy tasks are primed).

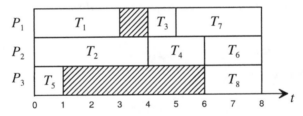

Figure 3.1.2 *A schedule for the task set given in Figure* 3.1.1.

In some applications, other related criteria may be used, as for example: *mean tardiness* $\bar{D} = \frac{1}{n} \sum\limits_{j=1}^{n} D_j$, *mean weighted tardiness* $\bar{D}_w = \sum\limits_{j=1}^{n} w_j D_j / \sum\limits_{j=1}^{n} w_j$, *mean earliness* $\bar{E} = \frac{1}{n} \sum\limits_{j=1}^{n} E_j$, *mean weighted earliness* $\bar{E}_w = \sum\limits_{j=1}^{n} w_j E_j / \sum\limits_{j=1}^{n} w_j$, *number of tardy tasks* $U = \sum\limits_{j=1}^{n} U_j$, where $U_j = 1$ if $C_j > d_j$, and 0 otherwise, or *weighted number of tardy tasks* $U_w = \sum\limits_{j=1}^{n} w_j U_j$.

Again, let us calculate values of particular criteria for the schedule in Figure 3.1.2. They are: schedule length $C_{max} = 8$, mean flow time $\bar{F} = 43/8$, maximum lateness $L_{max} = 3$, mean tardiness $\bar{D} = 5/8$, mean earliness $\bar{E} = 13/8$, and number of tardy jobs $U = 2$. The other criteria can be evaluated if weights of tasks are specified.

A schedule for which the value of a particular performance measure γ is at its minimum will be called *optimal*, and the corresponding value of γ will be denoted by γ^*.

We may now define the *scheduling problem* Π as a set of parameters described in this subsection [4] not all of which have numerical values, together with an optimality criterion. An *instance I* of problem Π is obtained by specifying particular values for all the problem parameters.

We see that scheduling problems are in general of optimization nature (cf. Section 2.2.1). However, some of them are originally formulated in decision version. An example is scheduling to meet deadlines, i.e. the problem of finding, given a set of deadlines, a schedule with no late task. However, both cases are analyzed in the same way when complexity issues are considered.

A *scheduling algorithm* is an algorithm which constructs a schedule for a given problem Π. In general, we are interested in optimization algorithms, but because of the inherent complexity of many problems of that type, approximation or heuristic algorithms will be discussed (cf. Sections 2.2.2 and 2.5).

Scheduling problems, as defined above, may be analyzed much in the same way as discussed in Chapter 2. However, their specificity raises some more detailed questions which will be discussed in the next section.

3.2 Analysis of Scheduling Problems and Algorithms

Deterministic scheduling problems are a part of a much broader class of combinatorial optimization problems. Thus, the general approach to the analysis of these problems can follow similar lines, but one should take into account their peculiarities. It is rather obvious that very often the time we can devote to solving particular scheduling problems is seriously limited so that only low order polynomial time algorithms may be used. Thus, the examination of the complexity of these problems should be the basis of any further analysis.

It has been known for some time [Coo71, Kar72] (cf. Section 2.2) that there exists a large class of combinatorial optimization problems for which most probably no *efficient optimization* algorithms exist. These are the problems whose decision counterparts (i.e. problems formulated as questions with "yes" or

---

[4] Parameters are understood generally, including e.g. relation $\prec$.

"no" answers) are NP-complete. The optimization problems are called NP-hard in this case. We refer the reader to [GJ79] and to Section 2.2 for a comprehensive treatment of the NP-completeness theory, and in the following we assume knowledge of its basic concepts like NP-completeness, NP-hardness, polynomial time transformation, etc. It follows that the complexity analysis answers the question whether or not an analyzed scheduling problem may be solved (i.e. an optimal schedule found) in time bounded from above by a polynomial in the input length of the problem (i.e. in polynomial time). If the answer is positive, then an optimization polynomial time algorithm must have been found. Its usefulness depends on the order of its worst-case complexity function and on the particular application. Sometimes, when the worst-case complexity function is not low enough, although still polynomial, a mean complexity function of the algorithm may be sufficient. This issue is discussed in detail in [AHU74]. On the other hand, if the answer is negative, i.e. when the decision version of the analyzed problem is NP-complete, then there are several other ways of further analysis.

First, one may try to relax some constraints imposed on the original problem and then solve the relaxed problem. The solution of the latter may be a good approximation to the solution of the original problem. In the case of scheduling problems such a relaxation may consist of

- allowing preemptions, even if the original problem dealt with non-preemptive schedules,

- assuming unit-length tasks, when arbitrary-length tasks were considered in the original problem,

- assuming certain types of precedence graphs, e.g. trees or chains, when arbitrary graphs were considered in the original problem, etc.

Considering computer applications, especially the first relaxation can be justified in the case when parallel processors share a common primary memory. Moreover, such a relaxation is also advantageous from the viewpoint of certain optimality criteria.

Second, when trying to solve NP-hard scheduling problems one often uses approximation algorithms which tend to find an optimal schedule but do not always succeed. Of course, the necessary condition for these algorithms to be applicable in practice is that their worst-case complexity function is bounded from above by a low-order polynomial in the input length. Their sufficiency follows from an evaluation of the difference between the value of a solution they produce and the value of an optimal solution. This evaluation may concern the worst case or a mean behavior. To be more precise, we use here notions that have been introduced in Section 2.5, i.e. absolute performance ratio R_A and asymptotic performance ratio R_A^∞ of an approximation algorithm A.

These notions define a measure of "goodness" of approximation algorithms; the closer R_A^∞ is to 1, the better algorithm A performs. However, for some combinatorial problems it can be proved that there is no hope of finding an approximation algorithm of a certain accuracy, i.e. this question is as hard as finding a polynomial time algorithm for any NP-complete problem.

Analysis of the worst-case behavior of an approximation algorithm may be complemented by an analysis of its mean behavior. This can be done in two ways. The first consists in assuming that the parameters of instances of the considered problem Π are drawn from a certain distribution, and then the *mean performance* of algorithm A is analyzed. One may distinguish between the *absolute error* of an approximation algorithm, which is the difference between the approximate and optimal values and the *relative error*, which is the ratio of these two (cf. Section 2.5). Asymptotic optimality results in the stronger (absolute) sense are quite rare. On the other hand, asymptotic optimality in the relative sense is often easier to establish. It is rather obvious that the mean performance can be much better than the worst case behavior, thus justifying the use of a given approximation algorithm. A main obstacle is the difficulty of proofs of the mean performance for realistic distribution functions. Thus, the second way of evaluating the mean behavior of approximation algorithms, consisting of experimental studies, is still used very often. In the latter approach, one compares solutions, in the sense of the values of an optimality criterion, constructed by a given approximation algorithm and by an optimization algorithm. This comparison should be made for a large, representative sample of instances.

In this context let us mention the most often used approximation scheduling algorithm which is the so-called *list scheduling algorithm* (which is in fact a general approach). In this algorithm a certain list of tasks is given and at each step the first available processor is selected to process the first available task on the list. The accuracy of a particular list scheduling algorithm depends on the given optimality criterion and the way the list has been constructed.

The third and last way of dealing with hard scheduling problems is to use exact enumerative algorithms whose worst-case complexity function is exponential in the input length. However, sometimes, when the analyzed problem is not NP-hard in the strong sense, it is possible to solve it by a pseudopolynomial optimization algorithm whose worst-case complexity function is bounded from above by a polynomial in the input length and in the maximum number appearing in the instance of the problem. For reasonably small numbers such an algorithm may behave quite well in practice and it can be used even in computer applications. On the other hand, "pure" exponential algorithms have probably to be excluded from this application, but they may be used sometimes for other scheduling problems which can be solved by off-line algorithms.

The above discussion is summarized in a schematic way in Figure 3.2.1. In the following chapters we will use the above scheme when analyzing scheduling problems.

3.3 Motivations for Deterministic Scheduling Problems

In this section, an interpretation of the assumptions and results in deterministic scheduling theory which motivate and justify the use of this model, is presented. We will underline especially computer applications, but we will also refer to manufacturing systems, even if the practical interpretation of the model is not for this application area. In a manufacturing environment deterministic scheduling is also known as *predictive*. Its complement is *reactive scheduling*, which can also be regarded as deterministic scheduling with a shorter planning horizon.

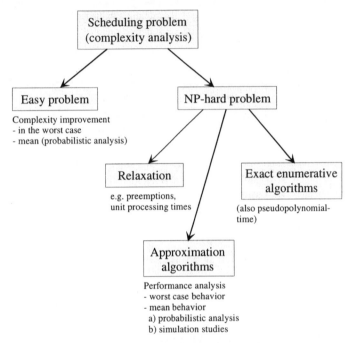

Figure 3.2.1 *An analysis of a scheduling problem - schematic view.*

Let us begin with an analysis of processors (machines). *Parallel processors* may be interpreted as central processors which are able to process every task (i.e. every program). *Uniform processors* differ from each other by their speeds, but they do not prefer any type of tasks. *Unrelated processors*, on the contrary, are specialized in the sense that they prefer certain types of tasks, for example numerical computations, logical programs, or simulation procedures. The processors may have different instruction sets, but they are still of comparable processing capacity so they can process tasks of any type, only processing times may be different. In manufacturing systems, pools of machines exist where all the machines have the same capability (except possibly speed) to process tasks.

Completely different from the above are *dedicated* processors (dedicated machines) which may process only certain types of tasks. The interpretation of this model for manufacturing systems is straightforward but it can also be applied to computer systems. As an example let us consider a computer system consisting of an input processor, a central processor and an output processor. It is not difficult to see that such a system corresponds to a flow shop with $m = 3$. On the other hand, a situation in which each task is to be processed by an input/output processor, then by a central processor and at the end again by the input/output processor, can easily be modeled by a job shop system with $m = 2$. As far as an open shop is concerned, there is no obvious computer interpretation. But this case, like the other shop scheduling problems, has great significance in other applications, especially in an industrial environment.

By an *additional resource* we understand in this book a "facility" besides processors the tasks to be performed compete for. The competition aspect in this definition should be stressed, since "facilities" dedicated to only one task will not be treated as resources in this book. In computer systems, for example, messages sent from one task to another specified task will not be considered as resources. In manufacturing environments tools, material, transport facilities, etc. can be treated as additional resources.

Let us now consider the assumptions associated with the task set. As mentioned in Section 3.1, in deterministic scheduling theory a priori knowledge of ready times and processing times of tasks is usually assumed. As opposed to other practical applications, the question of a priori knowledge of these parameters in computer systems needs a thorough comment.

Ready times are obviously known in systems working in an off-line mode and in control systems in which measurement samples are taken from sensing devices at fixed time moments.

As far as *processing times* are concerned, they are usually not known a priori in computer systems. Despite this fact the solution of a deterministic scheduling problem may also have an important interpretation in these systems. First, when scheduling tasks to meet deadlines, the only approach (when the task processing times are not known) is to solve the problem with assumed upper bounds on the processing times. Such a bound for a given task may be implied by the worst case complexity function of an algorithm connected with that task. Then, if all deadlines are met with respect to the upper bounds, no deadline will be exceeded for the real task processing times [5]. This approach is often used in a broad class of computer control systems working in a hard real time environment, where a certain set of control programs must be processed before taking the next sample from the same sensing device.

---

[5] However, one has to take into account list scheduling anomalies which will be explained in Section 5.1.

Second, instead of exact values of processing times one can take their mean values and, using the procedure described by Coffman and Denning in [CD73], calculate an optimistic estimate of the mean value of the schedule length.

Third, one can measure the processing times of tasks *after* processing a task set scheduled according to a certain algorithm A. Taking these values as an input in the deterministic scheduling problem, one may construct an optimal schedule and compare it with the one produced by algorithm A, thus evaluating the latter.

Apart from the above, optimization algorithms for deterministic scheduling problems give some indications for the construction of heuristics under weaker assumptions than those made in stochastic scheduling problems, cf. [BCSW86].

The existence of *precedence constraints* in computer systems also requires an explanation. In the simplest case the results of certain programs may be the input data for others. Moreover, precedence constraints may also concern parts of the same program. A conventional, serially written program, may be analyzed by a special procedure looking for parallel parts in it (see for example [RG69, Rus69], or [Vol70]). These parts may also be defined by the programmer who can use special programming languages supporting parallel concepts. Apart from this, a solution of certain reliability problems in operating systems, as for example the *determinacy problem* (see [ACM70, Bae74, Ber66]), requires an introduction of additional precedence constraints.

We will now discuss particular *optimality criteria* for scheduling problems from their practical significance point of view. Minimizing *schedule length* is important from the viewpoint of the owner of a set of processors (machines), since it leads to both, the maximization of the processor utilization factor (within schedule length C_{max}), and the minimization of the maximum in-process time of the scheduled set of tasks. This criterion may also be of importance in a computer control system in which a task set arrives periodically and is to be processed in the shortest time.

The *mean flow time* criterion is important from the user's viewpoint since its minimization yields a minimization of the mean response time and the mean in-process time of the scheduled task set.

Due date involving criteria are of great importance in manufacturing systems, especially in those that produce to specific customer orders. Moreover, the *maximum lateness* criterion is of great significance in computer control systems working in the hard real time environment since its minimization leads to the construction of a schedule with no task late whenever such schedules exist (i.e. when $L_{max}^{*} \leq 0$ for an optimal schedule).

The criteria mentioned above are basic in the sense that they require specific approaches to the construction of schedules.

3.4 Classification of Deterministic Scheduling Problems

The great variety of scheduling problems we have seen from the preceding section motivates the introduction of a systematic notation that could serve as a basis for a classification scheme. Such a notation of problem types would greatly facilitate the presentation and discussion of scheduling problems. A notation proposed by Graham et al. [GLL+79] and Błażewicz et al. [BLRK83] will be presented next and then used throughout the book.

The notation is composed of three fields $\alpha \mid \beta \mid \gamma$. They have the following meaning: The first field $\alpha = \alpha_1, \alpha_2$ describes the processor environment. Parameter $\alpha_1 \in \{\varnothing, P, Q, R, O, F, J\}$ characterizes the type of processor used:

$\alpha_1 = \varnothing$: single processor [6],

$\alpha_1 = P$: identical processors,

$\alpha_1 = Q$: uniform processors,

$\alpha_1 = R$: unrelated processors,

$\alpha_1 = O$: dedicated processors: open shop system,

$\alpha_1 = F$: dedicated processors: flow shop system,

$\alpha_1 = J$: dedicated processors: job shop system.

Parameter $\alpha_2 \in \{\varnothing, k\}$ denotes the number of processors in the problem:

$\alpha_2 = \varnothing$: the number of processors is assumed to be variable,

$\alpha_2 = k$: the number of processors is equal to k (k is a positive integer).

The second field $\beta = \beta_1, \beta_2, \beta_3, \beta_4, \beta_5, \beta_6, \beta_7, \beta_8$ describes task and resource characteristics. Parameter $\beta_1 \in \{\varnothing, pmtn\}$ indicates the possibility of task preemption:

$\beta_1 = \varnothing$: no preemption is allowed,

$\beta_1 = pmtn$: preemptions are allowed.

Parameter $\beta_2 \in \{\varnothing, res\}$ characterizes additional resources:

$\beta_2 = \varnothing$: no additional resources exist,

$\beta_2 = res$: there are specified resource constraints; they will be described in detail in Chapter 12.

Parameter $\beta_3 \in \{\varnothing, prec, uan, tree, chains\}$ reflects the precedence constraints:

$\beta_3 = \varnothing, prec, uan, tree, chains$: denotes respectively independent tasks, general precedence constraints, uniconnected activity networks, precedence constraints forming a tree or a set of chains.

Parameter $\beta_4 \in \{\varnothing, r_j\}$ describes ready times:

---

[6] In this notation $\varnothing$ denotes an empty symbol which will be omitted in presenting problems.

$\beta_4 = \emptyset$: all ready times are zero,

$\beta_4 = r_j$: ready times differ per task.

Parameter $\beta_5 \in \{\emptyset, p_j = p, \underline{p} \leq p_j \leq \overline{p}\}$ describes task processing times:

$\beta_5 = \emptyset$: tasks have arbitrary processing times,

$\beta_5 = (p_j = p)$: all tasks have processing times equal to p units,

$\beta_5 = (\underline{p} \leq p_j \leq \overline{p})$: no p_j is less than $\underline{p}$ or greater than $\overline{p}$.

Parameter $\beta_6 \in \{\emptyset, \tilde{d}\}$ describes deadlines:

$\beta_6 = \emptyset$: no deadlines are assumed in the system (however, due dates may be defined if a due date involving criterion is used to evaluate schedules),

$\beta_6 = \tilde{d}$: deadlines are imposed on the performance of a task set.

Parameter $\beta_7 \in \{\emptyset, n_j \leq k\}$ describes the maximal number of tasks constituting a job in case of job shop systems:

$\beta_7 = \emptyset$: the above number is arbitrary or the scheduling problem is not a job shop problem,

$\beta_7 = (n_j \leq k)$: the number of tasks for each job is not greater than k.

Parameter $\beta_8 \in \{\emptyset, no\text{-}wait\}$ describes a no-wait property in the case of scheduling on dedicated processors:

$\beta_8 = \emptyset$: buffers of unlimited capacity are assumed,

$\beta_8 = no\text{-}wait$: buffers among processors are of zero capacity and a job after finishing its processing on one processor must immediately start on the consecutive processor.

The third field, γ, denotes an optimality criterion (performance measure), i.e. $\gamma \in \{C_{max}, \Sigma C_j, \Sigma w_j C_j, L_{max}, \Sigma D_j, \Sigma w_j D_j, \Sigma E_j, \Sigma w_j E_j, \Sigma U_j, \Sigma w_j U_j, -\}$, where $\Sigma C_j = \overline{F}$, $\Sigma w_j C_j = \overline{F}_w$, $\Sigma D_j = \overline{D}$, $\Sigma w_j D_j = \overline{D}_w$, $\Sigma E_j = \overline{E}$, $\Sigma w_j E_j = \overline{E}_w$, $\Sigma U_j = U$, $\Sigma w_j U_j = U_w$ and "$-$" means testing for feasibility whenever scheduling to meet deadlines is considered.

The use of this notation is illustrated by Example 3.4.1.

Example 3.4.1

(a) Problem $P \| C_{max}$ reads as follows: *Scheduling of non-preemptable and independent tasks of arbitrary processing times (lengths), arriving to the system at time 0, on parallel, identical processors in order to minimize schedule length.*

(b) $O3 | pmtn, r_j | \Sigma C_j$ stands for: *Preemptive scheduling of arbitrary length tasks arriving at different time moments in the three machine open shop, where the objective is to minimize mean flow time.*   □

At this point it is worth mentioning that scheduling problems are closely related in the sense of polynomial transformation[7]. Some basic polynomial transformations between scheduling problems are shown in Figure 3.4.1. For each graph in the figure, the presented problems differ only by one parameter (e.g. by type and number of processors, as in Figure 3.4.1(a)) and the arrows indicate the direction of the polynomial transformation. These simple transformations are very useful in many situations when analyzing new scheduling problems. Thus, many of the results presented in this book can immediately be extended to cover a broader class of scheduling problems.

(a)

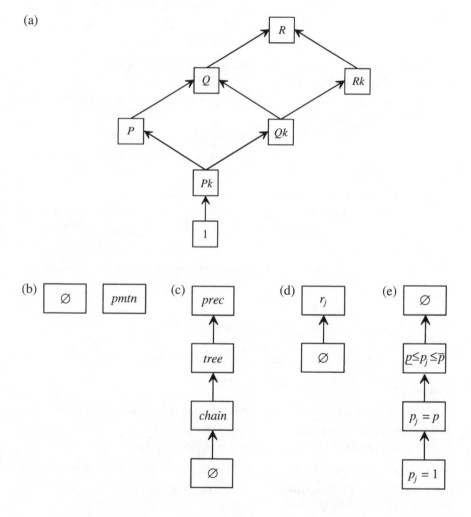

(b)   (c)   (d)   (e)

---

[7] This term has been explained in Section 2.2.

(f)

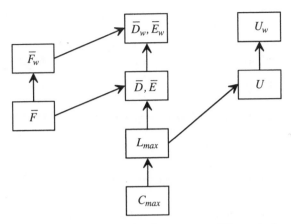

Figure 3.4.1 *Graphs showing interrelations among different values of particular parameters*
(a) *processor environment*
(b) *possibility of preemption*
(c) *precedence constraints*
(d) *ready times*
(e) *processing times*
(f) *optimality criteria.*

References

ACM70   ACM Record of the project MAC conference on concurrent system and parallel computation, Wood's Hole, Mass, 1970.

AHU74   A. V. Aho, J. E. Hopcroft, J. D. Ullman, *The Design and Analysis of Computer Algorithms*, Addison-Wesley, Reading, Mass., 1974.

Bae74   J. L. Baer, Optimal scheduling on two processors of different speeds, in: E. Gelenbe, R. Mahl (eds.), *Computer Architecture and Networks*, North Holland, Amsterdam, 1974.

BCSW86   J. Błażewicz, W. Cellary, R. Słowiński, J. Węglarz, *Scheduling under Resource Constraints: Deterministic Models*, J. C. Baltzer, Basel, 1986.

Ber66   A. J. Bernstein, Analysis of programs for parallel programming, *IEEE Trans. Comput.* EC-15, 1966, 757-762.

BK02   J.Blazewicz, D.Kobler, Review of properties of different precedence graphs for scheduling problems, *European J. of Oper. Res.* 142 (2002) 435-443

BLRK83   J. Błażewicz, J. K. Lenstra, A. H. G. Rinnooy Kan, Scheduling subject to resource constraints: classification and complexity, *Discrete Appl. Math.* 5, 1983, 11-24.

CD73   E. G. Coffman, Jr., P. J. Denning, *Operating Systems Theory*, Prentice-Hall, Englewood Cliffs, N.J., 1973.

Coo71   S. A. Cook, The complexity of theorem proving procedures, *Proc. 3rd ACM Symposium on Theory of Computing*, 1971, 151-158.

GJ79   M. R. Garey, D. S. Johnson, *Computers and Intractability: A Guide to the Theory of NP-Completeness*, W. H. Freeman, San Francisco, 1979.

GLL+79   R. L. Graham, E. L. Lawler, J. K. Lenstra, A. H. G. Rinnooy Kan, Optimization and approximation in deterministic sequencing and scheduling theory: a survey, *Ann. Discrete Math.* 5, 1979, 287-326.

Kar72   R. M. Karp, Reducibility among combinatorial problems, in: R. E. Miller, J. W. Thatcher (eds.), *Complexity of Computer Computations*, Plenum Press, New York, 1972, 85-104.

RG69   C. V. Ramamoorthy, M. J. Gonzalez, A survey of techniques for recognizing parallel processable streams in computer programs, *AFIPS Conference Proceedings, Fall Joint Computer Conference*, 1969, 1-15.

Rus69   E. C. Russel, Automatic program analysis, Ph.D. thesis, Dept. of Engineering, University of California, Los Angeles, 1969.

Vol70   S. Volansky, Graph model analysis and implementation of computational sequences, Ph.D. thesis, Rep. No.UCLA-ENG-7048, School of Engineering Applied Sciences, University of California, Los Angeles, 1970.

4 Scheduling on One Processor

Single machine scheduling (SMS) problems seem to have received substantial attention because of several reasons. These types of problems are important both because of their own intrinsic value, as well as their role as building blocks for more generalized and complex problems. In a multi-processor environment single processor schedules may be used in bottlenecks, or to organize task assignment to an expensive processor; sometimes an entire production line may be treated as a single processor for scheduling purposes. Also, compared to multiple processor scheduling, SMS problems are mathematically more tractable. Hence, more problem classes can be solved in polynomial time, and a larger variety of model parameters, such as various types of cost functions, or an introduction of change-over cost, can be analyzed. Single processor problems are thus of rather fundamental character and allow for some insight and development of ideas when treating more general scheduling problems.

The relative simplicity of the single-processor scheduling on one hand, and its fundamental character also for multiprocessor scheduling problems on the other hand, motivate to discuss the single processor case to a wider extent. In the next five sections we will study scheduling problems on one processor with the objective to minimize the following criteria: schedule length, mean (and mean weighted) flow time, due date involving criteria such as different lateness or tardiness functions, change-over cost and different maximum and mean cost functions.

4.1 Minimizing Schedule Length

One of the simplest type of scheduling problems considered here is the problem $1 \mid prec \mid C_{max}$, i.e. one in which all tasks are assumed to be non-preemptable, ordered by some precedence relation, and available at time $t = 0$. It is trivial to observe that in whatever order in accordance with the precedence relation the tasks are assigned to the processor, the schedule length is $C_{max} = \sum_{j=1}^{n} p_j$. If each task has a given release time (ready time), an optimal schedule can easily be obtained by a polynomial time algorithm where tasks are scheduled in the order of non-decreasing release times. Similarly, if each task has a given deadline, the earliest deadline scheduling rule would produce an optimal solution provided there exists a schedule that meets all the deadlines. Thus in fact, problems $1 \mid r_j \mid C_{max}$ and $1 \mid \mid L_{max}$ are equivalent as far as their complexities and solution techniques are concerned. The situation becomes considerably more complex from the algorithmic complexity point of view if both, release times and deadlines restrict task processing.

In the following section, for each task there is specified a release time and a deadline by which the task is to be completed. The aim is then to find a schedule that meets all the given deadlines and, in addition, minimizes C_{max}.

4.1.1 Scheduling with Release Times and Deadlines

Problem $1 \mid r_j, \tilde{d}_j \mid C_{max}$

In case of problem $1 \mid r_j, \tilde{d}_j \mid C_{max}$, i.e. if the tasks are allowed to have unequal processing times, a transformation from the 3-PARTITION problem [1] shows that the problem is NP-hard in the strong sense, even for integer release times and deadlines [LRKB77]. Only if all tasks have unit processing times, an optimization algorithm of polynomial time complexity is available.

The general problem can be solved by applying a branch and bound algorithm. Bratley et al. [BFR71] proposed an algorithm which is shortly described below.

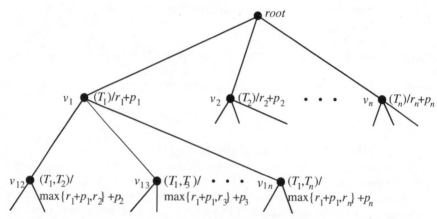

Figure 4.1.1   *Search tree in the branch and bound algorithm of Bratley et al.* [BFR71].

All possible task schedules are implicitly enumerated by a search tree construction, as shown in Figure 4.1.1. From the root node of the tree we branch to n new

---

[1] The 3-PARTITION problem is defined as follows (see [GJ79]).

   Instance: A finite set $\mathcal{A}$ of $3m$ elements, a bound $B \in \mathbb{N}$, and a "size" $s(a) \in \mathbb{N}$ for each $a \in \mathcal{A}$, such that each $s(a)$ satisfies $B/4 < s(a) < B/2$ and such that $\sum_{a \in \mathcal{A}} s(a) = mB$.

   Answer: "Yes" if $\mathcal{A}$ can be partitioned into m disjoint sets $S_1, S_2, \cdots, S_m$ such that, for $1 \le i \le m$, $\sum_{a \in S_i} s(a) = B$. Otherwise "No".

nodes at the first level of descendant nodes. The ith of these nodes, v_i, represents the assignment of task T_i to be the first in the schedule, $i = 1, \cdots, n$. Associated with each node is the completion time of the corresponding task, i.e. $r_i + p_i$ for node v_i. Next we branch from each node on the first level to $n-1$ nodes on the second level. Each of these represents the assignment of one of the $n-1$ unassigned tasks to be the second in the schedule. Again, the completion time is associated with each of the second level nodes. If v_{ij} is the successor node of v_i to which task T_j is assigned, the associated completion time would be $\max\{r_i + p_i, r_j\} + p_j$. This value represents the completion time of the partial schedule (T_i, T_j). Continuing that way, on level k, $1 \le k \le n$, there are $n-k+1$ new nodes generated from each node of the preceding level. It is evident that all the $n!$ possible different schedules will be enumerated that way.

The order in which the nodes of the tree are examined is based on a backtracking search strategy. However, the algorithm uses two criteria to reduce the number of search steps.

(i) *Exceeding deadlines*. Consider node v at level $k-1$, and its $n-k+1$ immediate successors on level k of the tree. If the completion time associated with at least one of these nodes exceeds the deadline of the task added at level k, then all $n-k+1$ nodes may be excluded from further consideration. This follows from the fact that if any of these tasks exceeds its deadline at level k (i.e. this task is at k^{th} position in the schedule), it will certainly exceed its deadline if scheduled later. Since all the successors of node v represent orderings in which the task in question is scheduled later, they may be omitted.

(ii) *Problem decomposition*. Consider level k of the search tree and suppose we generate a node on that level for task T_i. This is equivalent to assigning task T_i in position k of the schedule. If the completion time C_i of T_i in this position is less than or equal to the smallest release time r_{min} among the yet unscheduled tasks, then the problem decomposes at level k, and there is no need to enter another branch of the search tree, i.e. one doesn't need to backtrack beyond level k. The reason for this strong exclusion feature is that the best schedule for the remaining $n-k$ tasks may not be started prior to the smallest release time among these tasks, and hence not earlier than the completion time C_i of the first k tasks.

Example 4.1.1 To demonstrate the idea of the branch and bound algorithm described above consider the following sample problem of four tasks and vectors describing respectively task release times, processing times, and deadlines, $r = [4, 1, 1, 0]$, $p = [2, 1, 2, 2]$, and $\tilde{d} = [7, 5, 6, 4]$. The branch and bound algorithm would scan the nodes of the search tree shown in Figure 4.1.2 in some order that depends on the implementation of the algorithm. At each node the above criteria (i) and (ii) are checked. We see that schedules (T_4, T_2, T_3, T_1) and (T_4, T_3, T_2, T_1), when started at time 0, obey all release times and deadlines. When a schedule is obtained, its optimality must be checked (a criterion for doing this is given in Lemma 4.1.2). □

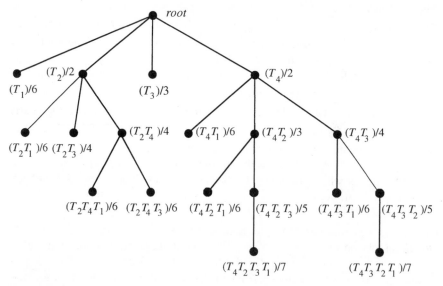

Figure 4.1.2 *Complete search tree of the sample problem of Example* 4.1.1.

To recognize an optimal solution we focus our attention on certain groups of tasks in a given feasible schedule. A *block* is a group of tasks such that the first task starts at its release time and all the following tasks to the end of the schedule are processed without idle times. Thus the length of a block is the sum of processing times of the tasks in the block. If a block has the property that the release times of all the tasks in the block are greater than or equal to the release time of the first task in the block (in that case we will say that "the block satisfies the *release time property*"), then the schedule found for this block is clearly optimal.

A block satisfying the release time property may be found by scanning the given schedule, starting from the last task and attempting to find a group of tasks of the described property. In particular, if T_{α_n} is the last task in the schedule and $C_{max} = r_{\alpha_n} + p_{\alpha_n}$, then $\{T_{\alpha_n}\}$ is a block satisfying the release time property. Another example is a schedule $(T_{\alpha_1}, \cdots, T_{\alpha_n})$ whose length is $\min_{j}\{r_j\} + \Sigma p_j$. In this case the block consists of all the tasks to be performed.

The following lemma can be used to prove optimality of a schedule.

Lemma 4.1.2 *If a schedule for problem* $1\,|\,r_j, \tilde{d}_j\,|\,C_{max}$ *satisfies the release time property then it is optimal.*

Proof. The lemma follows immediately from the block definition and the release time property.   □

The condition of release time property is sufficient but not necessary, as can be seen from simple examples. In this case this lemma cannot be used to prove optimality of a schedule constructed in the branch and bound procedure Then the completion time C of the schedule can still be used for bounding further solutions. This can be done by reducing all deadlines $\tilde{d}_j$ to be at most $C-1$, which ensures that if other feasible schedules exist, only those that are better than the solution at hand, are generated.

If task preemption is allowed, the problem $1\,|\,pmtn, r_j, \tilde{d}_j\,|\,C_{max}$ can be formulated as a maximum flow problem and can thus be solved in polynomial time [BFR71].

Problem $1\,|\,r_j, p_j = 1, \tilde{d}_j\,|\,C_{max}$

As already mentioned, if all release times are zero, the earliest deadline algorithm would be exact. Now, in the case of unequal and non-integer release times, it may happen that task T_i, though available for processing, must give preference to another task T_j with larger release time, because of $\tilde{d}_j < \tilde{d}_i$. Hence, in such a situation some idle interval should be introduced in the schedule in order to gain feasibility. These idle intervals are called *forbidden regions* [GJST81]. A forbidden region is an interval (f_1, f_2) of time (open both on the left and right) during which no task is allowed to start if the schedule is to be feasible. Notice that we do not forbid execution of a task during (f_1, f_2) that had been started at time f_1 or earlier. Algorithm 4.1.3 shows how forbidden regions are used (a technique how forbidden regions can be found is described in Algorithm 4.1.5). Let us assume for the moment that we have found a finite set of forbidden regions $F_1, \cdots, F_m$.

The following algorithm represents a basic way of how a feasible schedule can be generated. The algorithm schedules n unit time tasks, all of which must be completed by some time $\tilde{d}$. Release times are of no concern, but no task is allowed to start within one of the given forbidden regions $F_1, \cdots, F_m$. The algorithm finds the latest possible time by which the first task must start if all of them are to be completed by time $\tilde{d}$, without starting any task in a forbidden region.

Algorithm 4.1.3 *Backscheduling of a set of unit time tasks $\{T_1, \cdots, T_n\}$ with no release times and common deadline $\tilde{d}$, considering a set of forbidden regions* [GJST 81].

```
begin
Order the tasks arbitrarily as T_1, ··· , T_n;
for i := n downto 1 do
    Start T_i at the latest time s_i ≤ s_{i+1} − 1 (or d̃ − 1, if i = n) which does not fall into
        a forbidden region;
end;
```

Lemma 4.1.4 *The starting time s_1 found for T_1 by Algorithm 4.1.3 is such that, if all the given tasks (including T_1) were to start at times strictly greater than s_1, with none of them starting in one of the given forbidden regions, then at least one task would not be completed by time $\tilde{d}$.*

Proof. Consider a schedule found by Algorithm 4.1.3. Let $h_0 = s_1$, and let $h_1, \cdots, h_j$ be the starting times of the idle periods (if any) in the schedule, and let $h_{j+1} = \tilde{d}$ (see Figure 4.1.3). Notice that whenever (t_1, t_2) is an idle period, it must be the case that $(t_1-1, t_2-1]$ is part of some forbidden region, for otherwise Algorithm 4.1.3 would have scheduled some task to overlap or finish during $(t_1, t_2]$. Now consider any interval $(h_i, h_{i+1}]$, $0 \le i \le j$. By definition of the times h_i, the tasks that are finished in the interval are scheduled with no idle periods separating them and with the rightmost one finishing at time h_{i+1}. It follows that Algorithm 4.1.3 processes the maximum possible number of tasks in each interval $(h_i, h_{i+1}]$. Any other schedule that started all the tasks later than time s_1 and finished them all by time $\tilde{d}$ would have to exceed this maximum number of tasks in some interval $(h_i, h_{i+1}]$, $1 \le i \le j$, which is a contradiction.   □

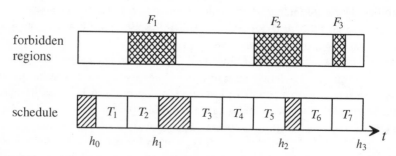

Figure 4.1.3 *A schedule with forbidden regions and idle periods.*

We will use Algorithm 4.1.3 as follows. Consider any two tasks T_i and T_j such that $\tilde{d}_i \le \tilde{d}_j$. We focus our interest on the interval $[r_i, \tilde{d}_j]$, and assume that we have already found a set of forbidden regions in this interval. We then apply Algorithm 4.1.3, with $\tilde{d} = \tilde{d}_j$ and with these forbidden regions, to the set of all tasks T_k satisfying $r_i \le r_k \le \tilde{d}_k \le \tilde{d}_j$. Let s be the latest possible start time found by Algorithm 4.1.3 in this case. There are two possibilities which are of interest. If $s < r_i$, then we know from Lemma 4.1.4 that there can be no feasible schedule since all these tasks must be completed by time $\tilde{d}$, none of them can be started before r_i, but at least one must be started by time $s < r_i$ if all are to be completed by time $\tilde{d}$. If $r_i \le s < r_i+1$, then we know that $(s-1, r_i)$ can be declared to be a forbidden region, since any task started in that region would not belong to our set (its release time is less than r_i) and it would force the first task of our set to be started later than s, thus preventing these tasks from being completed by $\tilde{d}$.

The algorithm presented next essentially applies Algorithm 4.1.3 to all such pairs of release times and deadlines in such a manner as to find forbidden regions from right to left. This is done by "considering the release times" in order from largest to smallest. To process a release time r_i, for each deadline $\tilde{d}_j \geq \tilde{d}_i$ the number of tasks is determined which cannot start before r_i and which must be completed by $\tilde{d}_j$. Then Algorithm 4.1.3 is used (with $\tilde{d} = \tilde{d}_j$) to determine the latest time at which the earliest such task can start. This time is called the *critical time* e_j for deadline $\tilde{d}_j$ (with respect to r_i). Letting e denote the minimum of all these critical times with respect to r_i, failure is declared in case of $e < r_i$, or $(e-1, r_i)$ is declared to be a forbidden region if $r_i \leq e$. Notice that by processing release times from largest to smallest, all forbidden regions to the right of r_i will have been found by the time that r_i is processed.

Once the forbidden regions are found in this way, we schedule the full set of tasks starting from time 0, using the *earliest deadline rule*. This proceeds by initially setting t to the least non-negative time not in a forbidden region and then assigning start time t to a task with lowest deadline among those ready at t. At each subsequent step, we first update t to the least time which is greater than or equal to the finishing time of the last scheduled task, and greater than or equal to the earliest ready time of an unscheduled task, and which does not fall into a forbidden region. Then we assign start time t to a task with lowest deadline among those ready (but not previously scheduled) at t.

Algorithm 4.1.5 *for problem* $1 \mid r_j, p_j = 1, \tilde{d}_j \mid C_{max}$ [GJST81].
```
begin
Order tasks so that r₁ ≤ r₂ ≤ ··· ≤ rₙ;
F := ∅;     -- the set of forbidden intervals is initially empty
for i := n downto 1 do
   begin
   for each task Tⱼ with d̃ⱼ ≥ d̃ᵢ do
      begin
      if eⱼ is undefined then eⱼ := d̃ⱼ-1 else eⱼ := eⱼ-1;
      while eⱼ ∈ F for some forbidden region F = (f₁,f₂) ∈ F, do eⱼ := f₁;
      end;
   if i = 1 or rᵢ₋₁ < rᵢ
   then
      begin
      e := min{eⱼ | eⱼ is defined};
      if e < rᵢ then begin write('No feasible schedule'); exit; end;
      if rᵢ ≤ e < rᵢ+1 then F := F ∪ (e-1,rᵢ);
      end;
```

```
   end;
t := 0;
while  T ≠ ∅  do
   begin
   if  r_i > t  for all  T_i ∈ T  then  t := min {r_i};
                                            T_i ∈ T
   while  t ∈ F for some forbidden region  F = (f_1, f_2) ∈ F,  do  t := f_2;
   Assign  T_k ∈ {T_i | T_i ∈ T  such that  d̃_k = min{d̃_i} and r_k ≤ t} next to the
      processor;
   t := t + 1;
   end;
end;
```

The following facts concerning Algorithm 4.1.5 can easily be proved [GJST81].

(*i*)   If the algorithm exits with failure, then there is no feasible schedule.

(*ii*) If the algorithm does not declare failure, then it finds a feasible schedule;
        this schedule has minimum makespan among all feasible schedules.

(*iii*) The time complexity of the algorithm is $O(n^2)$.

In [GJST81], there is also presented an improved version of Algorithm 4.1.5
which runs in time $O(n\log n)$.

Problem $1 \mid prec, r_j, \tilde{d}_j \mid C_{max}$

Problem $1 \mid prec, r_j, \tilde{d}_j \mid C_{max}$ is NP-hard in the strong sense because problem
$1 \mid r_j, \tilde{d}_j \mid C_{max}$ already is. However, if all tasks have unit processing times (i.e.
problem $1 \mid prec, r_j, p_j = 1, \tilde{d}_j \mid C_{max}$) we can replace the problem by one of type
$1 \mid r_j, p_j = 1, \tilde{d}_j \mid C_{max}$, which can then be solved optimally in polynomial time.
We will describe this approach below.

Given schedule S, let s_i be the starting time of task T_i, $i = 1, \cdots, n$. A schedule
is called *normal* if, for any two tasks T_i and T_j, $s_i < s_j$ implies that $\tilde{d}_i \leq \tilde{d}_j$ or $r_j >$
s_i. Release times and deadlines are called *consistent with the precedence relation*
if $T_i \prec T_j$ implies that $r_i + 1 \leq r_j$ and $\tilde{d}_i \leq \tilde{d}_j - 1$. The following lemma proves that
the precedence constraints are not of essential relevance if there is only one
processor.

Lemma 4.1.6 *If the release times and deadlines are consistent with the prece-*
dence relation, then any normal one-processor schedule that satisfies the release
times and deadlines must also obey the precedence relation.

Proof. Consider a normal schedule, and suppose that $T_i \prec T_j$ but $s_i > s_j$. By the
consistency assumption we have $r_i < r_j$ and $\tilde{d}_i < \tilde{d}_j$. However, these, together with

$r_j \leq s_j$, cause a violation of the assumption that the schedule is normal, a contradiction from which the result follows.   □

Release times and deadlines can be made consistent with the precedence relation $\prec$ if release times are redefined by

$$r'_{\alpha_j} = \max \left(\{r_{\alpha_j}\} \cup \{r'_{\alpha_i}+1 \mid T_{\alpha_i} \prec T_{\alpha_j}\} \right),$$

and deadlines are redefined by

$$\tilde{d}'_{\alpha_j} = \min \left(\{\tilde{d}_{\alpha_j}\} \cup \{\tilde{d}'_{\alpha_i}-1 \mid T_{\alpha_j} \prec T_{\alpha_i}\} \right).$$

These changes obviously do not alter the feasibility of any schedule. Furthermore, it follows from Lemma 4.1.6 that a precedence relation is essentially irrelevant when scheduling on one processor. Having arrived at a problem of type $1 \mid r_j, p_j = 1, \tilde{d}_j \mid C_{max}$ we can apply Algorithm 4.1.5 .

4.1.2   Scheduling with Release Times and Delivery Times

In this type of problems, task T_j is available for processing at time r_j, needs processing time p_j, and, finally, has to spend some "delivery" time q_j in the system after its processing. We will generalize the notation introduced in Section 3.4 and write $1 \mid r_j, delivery\ times \mid C_{max}$ for this type of problems. The aim is to find a schedule for tasks $T_1, \cdots, T_n$ such that the final completion time is minimal.

One may think of a production process consisting of two stages where the first stage is processed on a single processor, and in the second stage some finishing operations are performed which are not restricted by the bottleneck processor. We will see in Section 4.3.1 that maximum lateness problems are very closely related to the problem considered here. Numerous authors, e.g. [BS74, BFR73, FTM71, Pot80a, Pot80b, LLRK76, and Car82], studied this type of scheduling problems. Garey and Johnson [GJ79] proved the problem to be NP-hard in the strong sense.

Problem $1 \mid r_j , delivery\ times \mid C_{max}$

Schrage [Sch71] presented a heuristic algorithm which follows the idea that a task of maximal delivery time among those of earliest release time is chosen. The algorithm can be implemented with time complexity $O(n\log n)$.

Algorithm 4.1.7 *Schrage's algorithm for* $1 \mid r_j, delivery\ times \mid C_{max}$ [Car82].

```
begin
t := min {r_j};   C_max := t;
    T_j ∈ T
while  T ≠ ∅ do
```

```
begin
```
$\mathcal{T}' := \{T_j \mid T_j \in \mathcal{T}, \text{ and } r_j \leq t\};$

Choose $T_j \in \mathcal{T}'$ such that $p_j = \max\limits_{T_k \in \mathcal{T}'} \{p_k \mid q_k = \max\limits_{T_l \in \mathcal{T}'} \{q_l\}\};$

Schedule T_j at time t;

$\mathcal{T} := \mathcal{T} - \{T_j\};$

$C_{max} := \max\{C_{max}, t+p_j+q_j\};$

$t := \max\{t+p_j, \min\limits_{T_l \in \mathcal{T}} \{r_l\}\};$

```
   end;
end;
```

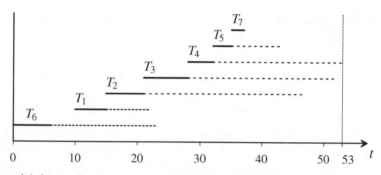

Figure 4.1.4   *A schedule generated by Schrage's algorithm for Example* 4.1.8.

Example 4.1.8 [Car82] Consider seven tasks with release times $r = [10, 13, 11, 20, 30, 0, 30]$, processing times $p = [5, 6, 7, 4, 3, 6, 2]$, and delivery times $q = [7, 26, 24, 21, 8, 17, 0]$. Schrage's algorithm determines the schedule $(T_6, T_1, T_2, T_3, T_4, T_5, T_7)$ of length 53, which is shown in Figure 4.1.4. Execution on the single processor is represented by solid lines, and delivery times are represented by dashed lines. An optimal schedule, however, would be $(T_6, T_3, T_2, T_4, T_1, T_5, T_7)$, and its total length is 50.   □

Carlier [Car82] improved the performance of Schrage's algorithm. Furthermore, he presented a branch and bound algorithm for the problem.

Problem $1 \mid pmtn, r_j, delivery\ times \mid C_{max}$

If task execution is allowed to be preempted, an optimal schedule can be constructed in $O(n\log n)$ time. We simply modify the **while**-loop in Schrage's algorithm such that processing of a task is preempted as soon as a task with a higher priority becomes available ("preemptive version" of Schrage's algorithm). The following result is mentioned without a proof (cf. [Car82]).

Theorem 4.1.9 *The preemptive version of Schrage's algorithm generates optimal preemptive schedules in* $O(n \log n)$ *time. The number of preemptions is not greater than* $n - 1$.   □

4.2 Minimizing Mean Weighted Flow Time

This section deals with scheduling problems subject to minimizing $\Sigma w_j C_j$. The problem $1 \| \Sigma w_j C_j$ can be optimally solved by scheduling the tasks in order of non-decreasing ratios p_j / w_j of processing times and weights. In the special case $1 \| \Sigma C_j$ (all weights are equal to 1), this reduces to the *shortest processing time* (*SPT*) rule.

The problem of minimizing the sum of weighted completion times subject to release dates is strongly NP-hard, even if all weights are 1 [LRKB77]. In the preemptive case, $1 \mid pmtn, r_j \mid \Sigma C_j$ can be solved optimally by a simple extension of the SPT rule [Smi56], whereas $1 \mid pmtn, r_j \mid \Sigma w_j C_j$ turns again out to be strongly NP-hard [LLL+84].

If deadlines are introduced, the situation is similar: $1 \mid \tilde{d}_j \mid \Sigma C_j$ can be solved in polynomial time, but the weighted case $1 \mid \tilde{d}_j \mid \Sigma w_j C_j$ is strongly NP-hard. Several elimination criteria and branch and bound algorithms have been proposed for this problem.

If the order of task execution is restricted by arbitrary precedence constraints, the problem $1 \mid prec \mid \Sigma w_j C_j$ becomes NP-hard [LRK78]. This remains true, even if all processing times p_j are 1 or all weights w_j are 1. For special classes of precedence constraints, however, polynomial time optimization algorithms are known.

Problem $1 \| \Sigma w_j C_j$

Suppose each task $T_j \in \mathcal{T}$ has a specified processing time p_j and weight w_j; the problem of determining a schedule with minimal weighted sum of task completion times, i.e. for which $\Sigma w_j C_j$ is minimal, can be optimally solved by means of Smith's "ratio rule" [Smi56], also known as Smith's *weighted shortest processing time* (*WSPT*) rule: Any schedule is optimal that puts the tasks in order of non-decreasing ratios p_j / w_j. In the special case that all tasks have equal weights, any schedule is optimal which places the tasks according to *SPT* rule, i.e. in non-decreasing order of processing times.

In order to prove the optimality of the WSPT rule for $1 \| \Sigma w_j C_j$, we present a far more general result due to Lawler [Law83] that includes $1 \| \Sigma w_j C_j$ as a special case: Given a set T of n tasks and a real-valued function γ which assigns value $\gamma(\pi)$ to each permutation π of tasks, find a permutation π^* such that

$$\gamma(\pi^*) = \min\{\gamma(\pi) \mid \pi \text{ is a permutation of task set } T\}.$$

If we know nothing about the structure of function γ, there is clearly nothing to be done except evaluating $\gamma(\pi)$ for each of the $n!$ possible different permutations of the task set. But for a given function γ we can sometimes find a transitive and complete relation $\precsim$ on the set of tasks with the property that for any two tasks T_i, T_k, and for any permutation of the form $\alpha T_i T_k \delta$ we have

$$T_i \precsim T_k \Rightarrow \gamma(\alpha T_i T_k \delta) \le \gamma(\alpha T_k T_i \delta). \tag{4.2.1}$$

If such a relation exists for a given function γ, we say: "γ *admits the relation* $\precsim$", or: "$\precsim$ *is a task interchange relation for* γ". This means that whenever T_i and T_k occur as adjacent tasks with T_k before T_i in a schedule, we are at least as well off to interchange their order. This relation is also referred to as the *adjacent pairwise interchange property*. Hence we have the following theorem:

Theorem 4.2.1 *If γ admits a task interchange relation $\precsim$, then an optimal permutation π^* can be found by ordering the tasks according to $\precsim$.* □

Consider, for example, *Smith's WSPT rule*,

$$T_i \precsim T_k \Leftrightarrow p_i/w_i \le p_k/w_k. \tag{4.2.2}$$

If the last task in the subsequence α in (4.2.1) finishes at time t, the cost $\Sigma w_j C_j$ of $\alpha T_i T_k \delta$ will be $w_i(t + p_i) + w_k(t + p_i + p_k) + C$ where C considers all the costs of tasks in the subsequences α and δ. If T_i and T_k are interchanged, the cost of $\alpha T_k T_i \delta$ will be $w_k(t + p_k) + w_i(t + p_k + p_i) + C$. Clearly, because of (4.2.2), the first sequence is of smaller cost than the second. As a consequence, the function $\Sigma w_j C_j$ admits Smith's *WSPT* rule, hence, by Theorem 4.2.1, this rule solves $1 \Vert \Sigma w_j C_j$ optimally.

Example 4.2.2 Let $T = \{T_1, \cdots, T_{10}\}$, with processing times and weights given by vectors $p = [16, 12, 19, 4, 7, 11, 12, 10, 6, 8]$ and $w = [2, 4, 3, 2, 5, 5, 1, 3, 6, 2]$. The optimal schedule is obtained by sorting the tasks in order of non-decreasing values of p_j/w_j, i.e. we get the task list $(T_9, T_5, T_4, T_6, T_2, T_8, T_{10}, T_3, T_1, T_7)$. The weighted sum of completion times is $6 \cdot 6 + 13 \cdot 5 + 17 \cdot 2 + 28 \cdot 5 + 40 \cdot 4 + 50 \cdot 3 + 58 \cdot 2 + 79 \cdot 3 + 95 \cdot 2 + 105 \cdot 1 = 1233$. Note that interchanging any two tasks in the schedule causes an increase of $\Sigma w_j C_j$. □

Problem $1 \mid r_j \mid \Sigma w_j C_j$

If the task ready times are not identical, the problem has been proved to be NP-hard even in the case that all weights are 1 [LRKB77]. We will present two heuristic algorithms for scheduling tasks with equal weights, where each rule specifies priority criteria for adding a task to an existing partial schedule, $S_{\mathcal{U}}$ of already scheduled tasks $\mathcal{U} \subseteq \mathcal{T}$, starting with $\mathcal{U} = \emptyset$.

Suppose that the schedule is constructed by adding one task at a time, starting from the empty schedule. At any point, we have a partial schedule $S_{\mathcal{U}}$ of task set $\mathcal{U} \subseteq \mathcal{T}$, $S_{\mathcal{U}} = (T_{\alpha_1}, \cdots, T_{\alpha_{|\mathcal{U}|}})$. The earliest start time of task T_j, s_j, and its completion time, C_j, satisfy

$$
s_j \begin{cases}
= r_j & \text{if } j = \alpha_1 \\
= \max\{r_j, C_{\alpha_{j-1}}\} & \text{if } j = \alpha_j, j \neq 1, \\
\geq \max\{r_j, C_{\alpha_{|\mathcal{U}|}}\} & \text{if } T_j \in \mathcal{T} - \mathcal{U};
\end{cases}
\tag{4.2.3}
$$

$$
C_j = s_j + p_j.
\tag{4.2.4}
$$

The two heuristics are as follows.

A. *The earliest completion time (ECT) rule*: Select task T_j with $C_j = \min\{C_i \mid T_i \in \mathcal{T} - \mathcal{U}\}$. Break ties by choosing T_j with $s_j = \min_i\{s_i\}$, and further ties by choosing T_j with minimum index j. Update s_j and C_j using (4.2.3) and (4.2.4).

B. *The earliest start time (EST) rule*: Select task T_j with $s_j = \min\{s_i \mid T_i \in \mathcal{T} - \mathcal{U}\}$. Break ties by choosing T_j with $C_j = \min_i\{C_i\}$, and further ties by choosing T_j with minimum index j. Update s_j and C_j using (4.2.3) and (4.2.4).

For these two heuristics, no accuracy bounds are known. The main difficulty arises from the fact that, since $r_j \geq 0$, idle times may be inserted in the optimal schedule. Consider the following example.

Example 4.2.3 Let $\mathcal{T} = \{T_1, \cdots, T_5\}$ with processing times $p = [3, 18, 17, 21, 25]$ and ready times $r = [35, 22, 34, 37, 66]$. The *ECT* rule results in the schedule $(T_1, T_3, T_2, T_4, T_5)$. The final values of the earliest start time s_j and the completion times C_j are given by the vectors $s = [35, 55, 38, 73, 94]$ and $C = [38, 73, 55, 94, 119]$, respectively, and the sum of completion times is 379. An optimal schedule, however, would be $(T_2, T_1, T_3, T_5, T_4)$ whose sum of completion times is 330. $\square$

An enumerative algorithm for solving the problem with equal weights optimally was presented by Dessouky and Deogun [DD81]. This is a branch and bound algorithm using a search tree in which a node at level k represents a partial schedule. If $S_{\mathcal{U}}$ is such a partial schedule for a subset $\mathcal{U}$ of k tasks, then let $C^*_{S_{\mathcal{U}}}$ denote the minimal total completion time of any schedule starting with $S_{\mathcal{U}}$. For each node at level k, if $S_{\mathcal{U}} = (T_{\alpha_1}, \cdots, T_{\alpha_k})$ is the corresponding partial schedule, a lower bound $\underline{C}_{S_{\mathcal{U}}}$ and an upper bound $\bar{C}_{S_{\mathcal{U}}}$ on $C^*_{S_{\mathcal{U}}}$ are computed. A successor node at level $k+1$ is obtained by selecting a task $T_i \in \mathcal{T} - \mathcal{U}$ and adding it to $S_{\mathcal{U}}$ in position $k+1$ to form partial schedule $(T_{\alpha_1}, \cdots, T_{\alpha_k}, T_i)$.

At any iteration, the branch and bound search chooses for branching a node that has currently the lowest lower bound $\underline{C}_{S_{\mathcal{U}}}$. Among the nodes generated from the same parent node, dominance is tested. A partial schedule $S_i = (T_{\alpha_1}, \cdots, T_{\alpha_k}, T_i)$ is *dominated* if another partial schedule $S_j = (T_{\alpha_1}, \cdots, T_{\alpha_k}, T_j)$ exists, and $C^*_{S_i} \geq C^*_{S_j}$. A node whose partial schedule has been found dominated by that of another node is eliminated from further consideration.

The crucial steps are indeed those where lower and upper bounds for the total completion time are estimated. For this, a number of tests are available (see [DD81]).

An extension of this branch and bound algorithm to the case of unequal weights is presented in [BR82].

The case in which the tasks have unit processing times can be solved in polynomial time [LRK80]. The preemptive case, $1 \mid pmtn, r_j \mid \Sigma C_j$, can be solved optimally by a simple modification of Smith's *WSPT* rule [Smi56], whereas $1 \mid pmtn, r_j \mid \Sigma w_j C_j$ turns out to be strongly NP-hard [LLL+84].

Problem $1 \mid \tilde{d}_j \mid \Sigma w_j C_j$

Each task T_j becomes available for processing at time zero, has processing time p_j, a deadline $\tilde{d}_j$ by which it must be completed (i.e. $C_j \leq \tilde{d}_j, j = 1, \cdots, n$), and has a positive weight w_j. The tasks are to be processed without preemption. The objective is to find a schedule of the tasks which minimizes the sum of weighted completion times $\Sigma w_j C_j$, subject to meeting all deadlines. This problem was first studied by Smith, who found a simple solution procedure for both, situations with no deadlines, and those with deadlines, but with equal weights. Emmons [Emm75] showed that Smith's procedure does not extend to the case of unequal weights, and from Lenstra [Len77] we know that problem $1 \mid \tilde{d}_j \mid \Sigma w_j C_j$ is NP-hard. Burns [Bur76] constructed a pairwise interchange heuristic for this problem that was improved by Miyazaki [Miy81]. Bansal [Ban80] developed an optimization algorithm based on a branch and bound approach and dominance criteria, and used Smith's *WSPT* rule to calculate lower bounds. Potts and van Wassenhove [PW83] presented a branch and bound algorithm based on a Lagrangian relaxation of the problem and found additional dominance criteria. Similar improvements have been presented by Kalra and Khurana [KK83], Posner [Pos85] and Bagchi and Ahmadi [BA87]. The latter used a task-splitting procedure to compute lower bounds for the weighted sum of completion times.

In the following we will assume that at least one feasible schedule exists for the given problem; this is easily checked by ordering the tasks in non-decreasing order of deadlines. If any of the tasks in this sequence is completed after its deadline, then no feasible schedule exists. It can be shown that if tasks have agreeable deadlines, i.e. $p_j/w_j \leq p_k/w_k$ implies $\tilde{d}_j \leq \tilde{d}_k$ for all tasks T_j and T_k, then an optimal solution is obtained by ordering the tasks in non-decreasing order of their deadlines.

Another interesting heuristic algorithm for $1|\tilde{d}_j|\Sigma w_j C_j$ is *Smith's backward scheduling rule* [Smi56]. Provided there exists a schedule in which all tasks meet their deadlines, the algorithm chooses one task with largest ratio p_j/w_j among all tasks T_j with $\tilde{d}_j \geq p_1 + \cdots + p_n$, and schedules the selected task last. It then continues by choosing an element of ratio among the remaining $n-1$ tasks and placing it in front of the already scheduled tasks, etc.

Algorithm 4.2.4 *Smith's backward scheduling rule for* $1|\tilde{d}_j|\Sigma w_j C_j$ [Smi56].

begin

$p := \sum\limits_{i=1}^{n} p_i$;

while $\mathcal{T} \neq \emptyset$ **do**

  begin

    $\mathcal{T}_p := \{T_j | T_j \in \mathcal{T}, \tilde{d}_j \geq p\}$;

    Choose task $T_j \in \mathcal{T}_p$ such that p_j/w_j is maximal;

    Schedule T_j in position n;

    $n := n-1$;

    $\mathcal{T} := \mathcal{T} - \{T_j\}$;

    $p := p - p_j$;

  end;

end;

This algorithm can be implemented to run in $O(n\log n)$ time. We also know that the algorithm is exact in the following cases (cf. [PW83]):

(*i*)   unit processing times, i.e. for the problem $1|p_j=1, \tilde{d}_j|\Sigma w_j C_j$,

(*ii*) unit weights, i.e. for problem $1|\tilde{d}_j|\Sigma C_j$,

(*iii*) agreeable weights, i.e. for problems where $p_i \leq p_j$ implies $w_i \geq w_j$ for all i and $j \in \{1, \cdots, n\}$.

However, in case of arbitrary weights, simple examples show that this algorithm is not exact.

We will present a branch and bound algorithm for $1|\tilde{d}_j|\Sigma w_j C_j$. In order to reduce the search for an optimal solution, dominance conditions are useful.

Dominance theorems usually specify that if certain conditions are satisfied, then task T_i precedes task T_j in at least one optimal schedule. When such conditions are satisfied, we say that task T_i is a *predecessor* of task T_j, and T_j is *successor* of T_i. In that way, dominance theorems result in a set of precedence constraints between pairs of tasks. It is clear that any enumerative algorithm can restrict its search to schedules obeying these precedence constraints. Hence, if many precedence constraints are found, the number of schedules to be investigated can be considerably reduced. Following [PW83], we formulate without proof three examples of such constraints.

Lemma 4.2.5 *Let* $\mathcal{T}' \subseteq \mathcal{T}$ *be a subset of tasks chosen such that for any* $T_i \in \mathcal{T} - \mathcal{T}'$ *and for any* $T_j, T_k \in \mathcal{T}'$ *with* $\tilde{d}_j \leq \tilde{d}_k$, $p_i/w_i \leq p_j/w_j \leq p_k/w_k$ *holds. Then for any pair of tasks* $T_i \in \mathcal{T}$ *and* $T_j \in \mathcal{T}'$ *with* $\tilde{d}_i \leq \tilde{d}_j$, *there exists an optimal schedule in which task* T_i *appears before task* T_j. $\qquad\square$

For the next lemma, let $\mathcal{A}_i$ denote the set of tasks which, according to the precedence condition of Lemma 4.2.5, are successors of task T_i $(i = 1, \cdots, n)$.

Lemma 4.2.6 *If* $p_i \leq p_j$, $w_i \geq w_j$ *and* $\min\{\tilde{d}_i, p_j + \sum_{T_k \in \mathcal{T} - \mathcal{A}_j} p_k\} \leq \tilde{d}_j$, *then there exists an optimal schedule in which task* T_i *is processed before task* T_j. $\qquad\square$

Lemma 4.2.7 *If the tasks are renumbered so that* $\tilde{d}_1 \leq \cdots \leq \tilde{d}_n$, *and if* $p_j + \sum_{k=1}^{i} p_k > \tilde{d}_i$ *for some* j *with* $1 \leq i \leq j \leq n$, *then tasks* $T_1, \cdots, T_i$ *are scheduled before task* T_j *in any feasible schedule.* $\qquad\square$

Obviously, each deadline that exceeds the total processing time $p = \Sigma p_i$ can be replaced by p without any changes of the resulting schedule. In addition, after some precedence conditions have been derived, the deadline of each task T_i can be reset to $\tilde{d}_i = \min\{\tilde{d}_i, p - \sum_{T_k \in \mathcal{A}_i} p_k\}$ $(i = 1, \cdots, n)$ where $\mathcal{A}_i$ is the set of successors of task T_i. Furthermore, the deadline of any task T_i which is predecessor of another task T_j is reset using $\tilde{d}_i = \min\{\tilde{d}_i, \tilde{d}_j - p_j - \sum_{T_k \in \mathcal{A}_i \cap \mathcal{B}_j} p_k\}$, where $\mathcal{B}_j$ is the set of predecessors of task T_j.

Reducing deadlines that way may induce additional precedence conditions between tasks. Lemmas 4.2.6 and 4.2.7 are applied repeatedly until no additional precedences can be found. It is indeed our aim to find as many precedences as possible because they allow to reduce the deadlines, and thus decrease the number of potential schedules in the branch and bound algorithm.

Scheduling a set of tasks according to Smith's backward scheduling rule allows to partition the task set $\mathcal{T}$ into *blocks* $\mathcal{T}_1, \cdots, \mathcal{T}_k$. Assume that the tasks have been renumbered so that the schedule generated by Algorithm 4.2.4 is $(T_1, \cdots, T_n)$. A task $T_{l'}$ is called *final* if $\tilde{d}_i \leq C_{l'}$ for $i = 1, \cdots, l'$ (implying $C_{l'} = \tilde{d}_{l'}$). The reasoning behind this definition is that tasks $T_1, \cdots, T_{l'}$ must be scheduled before all other tasks in any feasible schedule. A set of tasks $\mathcal{T}_i = \{T_{\alpha_i}, \cdots, T_{\beta_i}\}$ forms a *block* if the following conditions are satisfied:

(i) $\alpha_i = 1$, or task $T_{\alpha_i - 1}$ is final,

(ii) task T_i is not final for $i = \alpha_i, \cdots, \beta_i - 1$,

(iii) T_{β_i} is final.

If the deadlines force tasks $T_1, \cdots, T_{\beta_i}$ to be scheduled before all other tasks, then the previous deadline adjustment procedures will ensure that $\tilde{d}_j \leq C_{\beta_i}$ for $j = 1, \cdots, \beta_i$, and $C_{\beta_i} = \tilde{d}_{\beta_i}$, thus, T_{β_i} will be the last task in a block.

The following theorem gives a sufficient condition for a schedule generated by Smith's backward scheduling rule to be exact.

Theorem 4.2.8 *A schedule generated by Smith's backward scheduling rule is optimal if there is a block partition (in the above sense) of the given task set, the tasks within each block being scheduled in non-decreasing order of p_j/w_j.*

Proof. Suppose that the construction of blocks results in k blocks $\mathcal{T}_1, \cdots, \mathcal{T}_k$. It is clear that all tasks in block $\mathcal{T}_j$ must precede all tasks in block $\mathcal{T}_{j+1}$, $j = 1, \cdots, k-1$ in any feasible schedule. Therefore, the problem decomposes into sub-problems each of which involves scheduling tasks within a block. From Smith's backward scheduling rule we know that if the tasks within a block are scheduled in non-decreasing order of p_j/w_j, then the schedule is optimal. $\square$

Example 4.2.9 Let $\mathcal{T} = \{T_1, \cdots, T_8\}$ with processing times, deadlines and weights as follows: $p = [4, 3, 8, 2, 4, 7, 5, 4]$, $\tilde{d} = [13, 8, 38, 14, 9, 40, 25, 22]$, $w = [2, 6, 3, 3, 4, 2, 9, 2]$. A feasible schedule is $(T_2, T_5, T_1, T_4, T_8, T_7, T_3, T_6)$. Applying Lemmas 4.2.5-4.2.7 allows to reduce the deadlines to $\tilde{d} = [13, 8, 37, 13, 9, 37, 22, 22]$, and Algorithm 4.2.4 defines the heuristic schedule $(T_2, T_4, T_5, T_1, T_7, T_8, T_3, T_6)$. We see that tasks T_1 and T_6 are final, hence both, the first four and the last four tasks define a partial schedule. As within the partial schedules the values of p_j/w_j are non-decreasing, both partial schedules are optimal; hence the total schedule is optimal. $\square$

In general we will not be able to partition the given set of tasks into blocks. Algorithm 4.2.4 will then produce a schedule S that is not necessarily optimal. However, as this schedule may be considered as an approximate solution, its value $\gamma_S = \Sigma w_j C_j$ serves as an upper bound for the value of an optimal schedule.

A branch and bound method can now be applied in the following way: a node at level l of the search tree corresponds to a final partial schedule in which tasks are scheduled in the last l positions. The value of the partial schedule represents a lower bound for the schedule that can be obtained by descending from that node. Hence, if the lower bound is greater than or equal to any upper bound γ_S, the node can be discarded from further consideration.

An interesting modification of the problem is to allow tasks to be tardy up to a given *maximum allowable tardiness* $D \geq 0$, i.e. the objective is to minimize $\Sigma w_j C_j$ subject to $C_j - \tilde{d}_j \leq D$ for $j = 1, \cdots, n$. This problem is called *constrained weighted completion time (CWCT) problem* [CS86]. This has been shown to be NP-hard by Lenstra et al. [LRKB77]. From Chand and Schneeberger [CS86] we know that the CWCT problem can be solved optimally, e.g. in the case that the weight w_j of each task is a non-increasing function of the processing time p_j. Furthermore they discussed a worst-case analysis of the *WSPT* heuristic and showed that the accuracy performance ratio can become arbitrarily large in the worst case.

The case where tasks have unit processing times and both, release times and deadlines, is solvable as a linear assignment problem in $O(n^3)$ time [LRK80]. As can easily be shown there is no advantage to preempt task execution, as any solution that is optimal for $1 \mid \tilde{d}_j \mid \Sigma w_j C_j$ is also optimal for $1 \mid pmtn, \tilde{d}_j \mid \Sigma w_j C_j$. Consequently $1 \mid pmtn, \tilde{d}_j \mid \Sigma C_j$ can be solved in polynomial time, and the problems $1 \mid pmtn, \tilde{d}_j \mid \Sigma w_j C_j$ and $1 \mid pmtn, r_j, \tilde{d}_j \mid \Sigma w_j C_j$ are NP-hard.

Problem $1 \mid prec \mid \Sigma w_j C_j$

For general precedence constraints, Lawler [Law78] and Lenstra and Rinnooy Kan [LRK78] showed that the problem is NP-hard. Sidney [Sid75] presented a decomposition approach which produces an optimal schedule. Among others, Potts [Pot85] presented an especially interesting branch and bound algorithm where lower bounds are derived using a Lagrangian relaxation technique in which the multipliers are determined by the cost reduction method. Optimization scheduling algorithms running in polynomial time have been presented for tree-like precedences [Hor72, AH73], for series-parallel precedences [Sid75, IIN81], and for more general precedence relations [BM83, MS89].

Following [Sid75], a subset $\mathcal{U} \subset \mathcal{T}$ is said to have *precedence* over subset $\mathcal{V} \subset \mathcal{T}$ if there exist tasks $T_i \in \mathcal{U}$ and $T_j \in \mathcal{V}$ such that $T_j \in \text{succ}(T_i)$. If this is the case we will write $\mathcal{U} \to \mathcal{V}$. A set $\mathcal{U} \subset \mathcal{T}$ is said to be *initial* in $(\mathcal{T}, \prec)$ if $(\mathcal{T} - \mathcal{U}) \nrightarrow \mathcal{U}$, i.e. if $(\mathcal{T} - \mathcal{U}) \to \mathcal{U}$ is not true. In effect, no task from $\mathcal{T} - \mathcal{U}$ has a successor in $\mathcal{U}$, or, in other words, for each task in $\mathcal{U}$, all its predecessors are in $\mathcal{U}$, too. Obviously, there exists a feasible task order in which the elements of set $\mathcal{U}$ are arranged before that of $\mathcal{T} - \mathcal{U}$.

For a non-empty set $\mathcal{U} \subset \mathcal{T}$, define $p(\mathcal{U}) = \sum_{T_i \in \mathcal{U}} p_i$, $w(\mathcal{U}) = \sum_{T_i \in \mathcal{U}} w_i$, and $\rho(\mathcal{U}) = p(\mathcal{U})/w(\mathcal{U})$. We are interested in initial task sets that have some minimality property. Set $\mathcal{U} \subset \mathcal{T}$ is said to be ρ^*-*minimal* for $(\mathcal{T}, \prec)$ if

(*i*) $\mathcal{U}$ is initial in $(\mathcal{T}, \prec)$,

(*ii*) $\rho(\mathcal{U}) \le \rho(\mathcal{V})$ for any $\mathcal{V}$ which is initial in $(\mathcal{T}, \prec)$, and

(*iii*) $\rho(\mathcal{U}) < \rho(\mathcal{V})$ for each proper initial subset $\mathcal{V} \subset \mathcal{U}$.

With these notations we are able to formulate the following algorithm.

Algorithm 4.2.10 *Sidney's decomposition algorithm for* $1 \mid prec \mid \Sigma w_j C_j$ [Sid75, IIN81].

```
begin
while T ≠ ∅ do
   begin
   Determine task set U that is ρ*-minimal for (T, ≺);
   Schedule the members of task set U optimally;
   T := T - U;
   end;
end;
```

From Sidney [Sid75] we know that a schedule is optimal if and only if it can be generated by this algorithm. Instead of proving this fact, we give an intuitive explanation why the algorithm works. Observe that at each step of the iteration, the next subset added to the current schedule is an available subset (i.e. an initial subset) that minimizes $\rho(\mathcal{U}) = p(\mathcal{U})/w(\mathcal{U})$. Thus, subsets containing tasks with small processing times will be favored, which is consistent with the fact that such tasks delay future tasks by relatively little amounts of time. Also, subsets containing tasks with high deferral rates are favored, as we would expect from the fact that it is costly to delay such tasks.

For implementing the first instruction of the **while**-loop, Ichimori et al. [IIN81] gave an algorithm of time complexity $O(n^4)$. Consequently, because of the NP-hardness of the problem, the second step of Algorithm 4.2.10 must be of exponential time complexity. Only for special types of precedence graphs such as series parallel graphs, the second step of the Algorithm 4.2.10 can be implemented to run in time polynomially bounded in the number of tasks.

Note that in the special case for which there are no precedence constraints (i.e. $\prec$ is empty), Algorithm 4.2.10 reduces to the Smith's ratio rule introduced in (4.2.2).

Example 4.2.11 [Sid75] Let $\mathcal{T} = \{T_1, \cdots, T_7\}$, and let the processing times and weights be given by the vectors $p = [5, 8, 3, 5, 3, 7, 6]$ and $w = [1, 1, 1, 1, 1, 1, 1]$. The precedence constraints are shown in Figure 4.2.1. The subset $\mathcal{U} = \{T_1, T_3\}$ is initial, and $p(\mathcal{U}) = 8$, $w(\mathcal{U}) = 2$, $\rho(\mathcal{U}) = 4$. It is easy to verify that there is no other initial subset $\mathcal{V}$ with the property $\rho(\mathcal{V}) < \rho(\mathcal{U})$. Furthermore, the only proper subset of $\mathcal{U}$ that is initial in $(\mathcal{T}, \prec)$ is $\{T_1\}$, with $\rho(T_1) = p_1 = 5 > \rho(\mathcal{U})$. Hence $\mathcal{U}$ is ρ^*-minimal.

If $\mathcal{U}$ is the first subset selected in the **while**-loop of Algorithm 4.2.10, the schedule will start with tasks T_1 and T_3, and the algorithm proceeds with task set $\{T_2, T_4, T_5, T_6, T_7\}$. Next the ρ^*-minimal subset $\{T_2, T_4, T_5\}$ could be chosen, which gives the partial schedule $(T_1, T_3, T_2, T_5, T_4)$, etc.   □

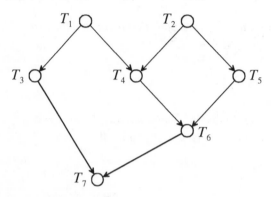

Figure 4.2.1 *A precedence graph for Example* 4.2.11.

A series of branch and bound algorithms have been developed for problem $1 \mid prec \mid \Sigma w_j C_j$ during the last decades. A more recent algorithm was presented by Potts [Pot85] where lower bounds are derived using a Lagrangian relaxation technique in which the multipliers are determined by a cost reduction method. A zero-one programming formulation of the problem uses variables x_{ij} ($i, j = 1, \cdots, n$) defined by

$$x_{ij} = \begin{cases} 1 & \text{if } i \neq j, \text{ and task } T_i \text{ is scheduled before task } T_j, \\ 0 & \text{otherwise.} \end{cases}$$

The values of some x_{ij} are implied by the precedence constraints, while others need to be determined. Let $e_{ij} = 1$ when the precedence constraints specify that task T_i is a predecessor of task T_j and let $e_{ij} = 0$ otherwise. Now, since the completion of task T_j occurs at time $\Sigma_i p_i x_{ij} + p_j$, the problem can be written as

$$minimize \quad \sum_i \sum_j p_i x_{ij} w_j \tag{4.2.5}$$

$$subject\ to \quad x_{ij} \geq e_{ij} \qquad (i, j = 1, \cdots, n), \tag{4.2.6}$$

$$x_{ij} + x_{ji} = 1 \qquad (i, j = 1, \cdots, n; i \neq j), \tag{4.2.7}$$

$$x_{ij} + x_{jk} + x_{ki} \leq 2 \qquad (i,j,k = 1, \cdots, n; \ i \neq j; \ i \neq k; \ j \neq k), \qquad (4.2.8)$$

$$x_{ij} \in \{0,1\} \qquad (i,j = 1, \cdots, n), \qquad\qquad\qquad (4.2.9)$$

$$x_{ii} = 1 \qquad (i = 1, \cdots, n). \qquad\qquad\qquad (4.2.10)$$

The constraints (4.2.6) ensure that $x_{ij} = 1$ whenever the precedence constraints specify that task T_i is a predecessor of task T_j. The fact that any task T_i is to be scheduled either before or after any other task T_j is presented by (4.2.7). The matrix $X = [x_{ij}]$ may be regarded as the adjacency matrix of a complete graph G_X in which each edge has one of the two possible orientations, and where G is a sub-graph of G_X. As a matter of fact, each such graph G_X has the property that if it contains a cycle, then it contains a directed cycle with three edges. Thus the constraints (4.2.7) and (4.2.8) ensure that G_X contains no cycles. When all constraints are satisfied, G_X defines a complete ordering of the tasks in which case G_X is called the *order graph* of X.

Using (4.2.7) and (4.2.8), it is possible to derive more general cycle elimination constraints involving r edges. They are of the form

$$\sum_{h=1}^{r} x_{i_h i_{h+1}} \leq r - 1, \qquad\qquad\qquad (4.2.11)$$

where $i_1, \cdots, i_r$ correspond to r different tasks and where $i_{r+1} = i_1$. For example, adding the constraints $x_{hi} + x_{ij} + x_{jh} \leq 2$ and $x_{hj} + x_{jk} + x_{kh} \leq 2$, and using $x_{jh} + x_{hj} = 1$ yields $x_{hi} + x_{ij} + x_{jk} + x_{kh} \leq 3$.

The coefficient $p_i w_j$ of x_{ij} in (4.2.5) may be regarded as the cost of scheduling task T_i before T_j. It is convenient to define the cost matrix $C = [c_{ij}]$, where the cost of scheduling task T_i before task T_j is

$$c_{ij} = \begin{cases} p_i w_j & \text{if } e_{ji} = 0 \\ \infty & \text{if } e_{ji} = 1 \end{cases} \qquad (i,j = 1, \cdots, n, \ i \neq j). \qquad (4.2.12)$$

Whenever the precedence constraints specify that task T_j is a predecessor of task T_i, we have $c_{ij} = \infty$ which ensures that constraints (4.2.6) are satisfied without applying them explicitly. The problem can now be written as

$$minimize \ \sum_i \sum_j c_{ij} x_{ij}$$

subject to (4.2.7)-(4.2.10).

For special classes of precedence constraints optimization scheduling algorithms running in polynomial time are known. Horn [Hor72] and Adolphson and Hu [AH73] discussed the problem for tree-like precedence graphs. For series-parallel precedence graphs, Lawler [Law78] presented an $O(n \log n)$ time algorithm where an interchange relation similar to that presented in (4.2.1) is applied. Ichimori et al. [IIN81] considered classes of graphs for which Algorithm 4.2.10

has polynomial time complexity. They showed that if the precedence constraints $\prec$ are such that the ρ^*-minimal subsets for $(\mathcal{T}, \prec)$ are series-parallel, Algorithm 4.2.10 can be implemented to run in $O(n^5)$ time. In fact, Lawler [Law78] was able to prove the existence of exact algorithms for scheduling problems with far more general optimization criteria than $\Sigma w_j C_j$. Let again γ be a real-valued function on permutations. Note that a schedule for one processor is defined by a permutation of the given task set, hence, as the order of task execution is restricted by the given precedence constraints, only "feasible" permutations are allowed. A permutation π is called *feasible* if $T_i \prec T_k$ implies that task T_i precedes task T_k under π. The objective is to find a feasible permutation π^* such that

$$\gamma(\pi^*) = \min\{\gamma(\pi) \mid \pi \text{ feasible}\} .$$

Unfortunately, the task interchange relation introduced in (4.2.1) is not general enough to solve this problem. Instead considering pairs of tasks, we should rather deal with pairs of strings of tasks: A *string interchange relation* is a transitive and complete relation $\lessgtr$ on strings with the property that for any two disjoint strings δ and δ' of tasks, and any permutation of the form $\alpha\delta\delta'\beta$ we have

$$\delta \lessgtr \delta' \implies \gamma(\alpha\delta\delta'\beta) \leq \gamma(\alpha\delta'\delta\beta) .$$

Smith's *WSPT* rule can again be used to define a string interchange relation: for any string δ, define $\rho_\delta = \sum_{T_j \in \delta} p_j / \sum_{T_j \in \delta} w_j$, and $\delta \lessgtr \delta' \Leftrightarrow \rho_\delta \leq \rho_{\delta'}$. A reasoning similar to that following (4.2.2) proves that function $\Sigma w_j C_j$ admits this string interchange relation $\lessgtr$.

Clearly, a string interchange relation implies a task interchange relation; but it is not true in general that every function γ which admits a task interchange relation also has a string interchange relation. Lawler's result is the following.

Theorem 4.2.12 [Law78] *If γ admits a string interchange relation $\lessgtr$ and if the precedence constraints $\prec$ are series-parallel, then an optimal permutation π^* can be found by an algorithm which requires $O(n\log n)$ comparisons of strings with respect to $\lessgtr$.*   □

Recall from Section 2.3 that series-parallel precedences can be described by means of a decomposition tree (see for example Figure 2.3.4). Working from the bottom of the tree upward, we can compute a set of strings of tasks for each node of the tree from the sets of strings obtained for its children. The objective is to obtain a set of strings at the root such that concatenating these strings in order according to $\lessgtr$ yields an optimal feasible schedule. We will accomplish this objective if each set S of strings obtained satisfies two conditions.

(*i*) Any concatenation of the strings in a set S in order according to $\lessgtr$ does not contradict the order given by the precedence constraint $\prec$, and (*ii*) At any point in the computation, let $S_1, \cdots, S_k$ be the sets of strings computed for nodes such that sets have not yet been computed for their parents. Then some ordering of the strings in the set $S_1 \cup \cdots \cup S_k$ yields an optimal feasible sub-schedule.

If the strings computed at the root are concatenated in order according to $\lesssim$, then condition (*i*) ensures that the resulting schedule is feasible, and condition (*ii*) ensures that it is optimal.

There is another class of promising scheduling algorithms. These algorithms obtain optimal schedules by finding optimal sub-schedules for progressively larger *modules* of tasks until all tasks are scheduled. This idea can be, for example, applied to series-parallel graphs which can be built up recursively from modules as specified by the decomposition tree (see Section 2.3.2). Möhring and Radermacher [MR85] generalized the notion of a decomposition tree to arbitrary precedence graphs. For the class of all precedence graphs built up by substitution from *prime* (indecomposable) *modules* of size $\leq k$, k arbitrary, there is an optimization algorithm of complexity $O(n^{(k^2)})$ to minimize $\Sigma w_j C_j$. Sidney and Steiner [SS86] improved this algorithm to run in $O(n^{w+1})$ time, where w denotes the maximum width of a prime module.

The idea of decomposing posets into prime modules can also be applied to optimization criteria other than $\Sigma w_j C_j$, as for example for the exponential cost function criterion (see Section 4.4.2). Monma and Sidney [MS87] proved that if the objective function obeys certain interchange properties then the so-called *job module property* is satisfied. The job module property says that any optimal solution to a sub-problem defined by a task module is consistent with at least one optimal schedule for the entire problem.

Problems $1 \mid prec, r_j \mid \Sigma w_j C_j$ and $1 \mid prec, \tilde{d}_j \mid \Sigma w_j C_j$

Lenstra and Rinnooy Kan [LRK80] proved that the problems $1 \mid chains, r_j, p_j = 1 \mid \Sigma w_j C_j$ and $1 \mid chains, \tilde{d}_j, p_j = 1 \mid \Sigma w_j C_j$ of scheduling unit time tasks subject to chain-like precedence constraints and either arbitrary release dates or arbitrary deadlines so as to minimize $\Sigma w_j C_j$ are both NP-hard.

4.3 Minimizing Due Date Involving Criteria

In this section scheduling problems with optimization criteria involving due dates will be considered. These include: maximum lateness L_{max}, weighted number of tardy tasks $\Sigma w_j U_j$, mean weighted tardiness $\Sigma w_j D_j$, and a combination of earliness-tardiness criteria.

4.3.1 Maximum Lateness

Whereas the problem $1 \| L_{max}$ can easily be solved in polynomial time by Jackson's earliest due date algorithm, other cases turn out to be more complex. The problem $1 | r_j | L_{max}$ is strongly NP-hard [LRKB77]. For this, and for $1 | prec, r_j | L_{max}$ as well, solution methods based on branch and bound are known. If tasks are preemptable or have unit processing time, the problem is easy, even if the order of task execution is constrained by a precedence relation [Bla76, Sid78, Mon82].

Problem $1 \| L_{max}$

The *earliest due date algorithm* (*EDD* rule) of Jackson [Jac55] provides a simple and elegant solution to this problem. In this algorithm, tasks are scheduled in order of non-decreasing due dates. The optimality of this rule can be proved by a simple interchange argument. Let S be any schedule and S^* be an *EDD* schedule. If $S \ne S^*$ then there exist two tasks T_j and T_k with $d_k \le d_j$, such that T_j immediately precedes T_k in S, but T_k precedes T_j in S^*. Since $d_k \le d_j$, interchanging the positions of T_j and T_k in S cannot increase the value of L_{max}. A finite number of such changes transforms S into S^*, showing that S^* is optimal. The *EDD* rule minimizes maximum lateness and maximum tardiness as well.

Problem $1 | r_j | L_{max}$

The problem $1 | r_j | L_{max}$ is known to be NP-hard in the strong sense [LRKB77]. Many exact algorithms have been proposed for this problem, but they are all based on enumerative methods and their computation time grows exponentially with the size of the problem. Research on this problem has focused on reducing the computational time for scheduling large task sets. Achieving this goal will also improve the efficiency of algorithms used to solve the more difficult $Pm | r_j | L_{max}$ problem by using the optimal solutions to the $1 | r_j | L_{max}$ problem [LLRK76].

There is certain symmetry inherent in the problem which becomes apparent if the model is presented in an alternative way. In this *delivery time model*, there are three processors, P_1, P_2, and P_3, where P_1 and P_3 are assumed to be *non-bottleneck processors* of infinite capacity, and P_2 is a *bottleneck processor* of capacity 1 (i.e. only one task can be processed at a time). Each task T_j has to visit P_1, P_2, P_3 in that order and has to spend

- a *head* $T_j^{(1)}$ on P_1 during time interval $[0, r_j)$,
- a *body* $T_j^{(2)}$ on P_2 from time $s_j \ge r_j$ to $C_j = s_j + p_j$,
- a *tail* $T_j^{(3)}$ on P_3 from time C_j to $L_j' = C_j + q_j$,

where the processing time q_j of tail $T_j^{(3)}$ is assumed to be $K - d_j$ for some constant $K \geq \max_i \{d_i\}$. The objective is to minimize the maximum completion time $L'_{max} = \max_i \{L'_i\} = L_{max} + K$. Notice that this model is exactly the same as the *delivery time model* discussed in Section 4.1.2. Whereas the head part of a task simply realizes the release time, the body part corresponds to the actual task to be processed on the single processor, and the tail part represents the delivery time of the task.

We will refer to the delivery time model as (r, p, q) where r, p, q are vectors of dimension n specifying release times (heads), processing times (bodies), and tails, respectively, for the tasks. It is interesting to note that problem (r, p, q) can be reversed: the inverse problem is defined by (q, p, r), and an optimal schedule for (q, p, r) can be reversed to obtain an optimal schedule for (r, p, q), with the same value of L_{max}.

Of particular importance are the algorithms of Bratley et al. [BFR73], Baker and Su [BS74], and McMahon and Florian [MF75]. The algorithm of McMahon and Florian (in the following referred to as *MF* algorithm) follows a novel approach in the way it applies the branch and bound method to scheduling problems. It searches for an optimal schedule over a tree of all possible schedules. Unlike other branch and bound algorithms in which most nodes in the tree represent partial schedules, the *MF* tree defines a complete schedule on each node. The schedule is used to derive a lower bound (*LB*) and an upper bound (*UB*) on the optimal solution at that node. In addition, the value of the maximum lateness of all tasks (L_{max}) in the schedule is computed. The search strategy is of the *jumptracking* type and follows always the node with the current lowest *LB* (the current node). From that node, only schedules which can potentially reduce the value of L_{max} are generated. The current lowest upper bound (*LUB*) is continually updated, and a node is eliminated if its $LB \geq LUB$. The search stops when the current node passes an optimality test. The algorithm derives its efficiency from the procedures which perform the following functions:
(1) construct a complete schedule at each node, including the initial schedule;
(2) test each schedule for optimality and compute the lower bound if the current solution is not proven optimal;
(3) generate successor of a node.

The *MF* algorithm can be characterized as a forward scheduling procedure since it starts by placing a task in the first position and continues to place tasks in succeeding positions until it reaches the task in the last position. It turns out that the *MF* algorithm tends to be inefficient when the problem (r, p, q) has a particular structure, for example when the range of ready times is less than that of due dates. Recognizing this difficulty, Lenstra [Len77] reversed the problem to (q, p, r). Since the ready times (r_j) are exchanged with the values q_j, the ranges of ready times and due dates are exchanged, too. As a consequence, the performance of the *MF* algorithm was improved considerably.

Erschler et al. [EFMR83] introduced a new *dominance concept* which permits a restricted set of schedules (the "most feasible ones") to be established on the basis of the ordering of ready times and due dates only. In particular, this dominance property is independent of task processing times, which is especially attractive if the data are not reliable. Carlier [Car82] and Larson et al. [LDD85] improved the previous algorithms with approaches following the *MF* algorithm, where the principles of branching are quite different and fully exploiting the problems' symmetrical features.

Compared to the branch and bound algorithms known for the problem in question, heuristic algorithms such as special list scheduling algorithms can be extremely efficient and often provide solutions adequate for practical applications. They can also be used to provide upper bounds on the criterion values of optimal schedules. This practical and theoretical importance of the problem motivates the search for efficient approximation algorithms with guaranteed accuracies. Larson and Dessouky [LD78] considered eleven heuristic algorithms and compared them experimentally. Kise et al. [KIM79] discussed several heuristic strategies from a more theoretical point of view. Among them are simple heuristics such as Jackson's *EDD* rule, or an algorithm where tasks are scheduled in order of their ready times, or combinations of these two strategies. The main result of [KIM79] is that the relative deviation L_{max}/L^*_{max} of the approximate solutions is not larger than $2-2/p$ where p is the sum of processing times of the tasks. For an iterative version of Jackson's rule (*IJ*) Potts [Pot80b] was able to prove

$$L_{max}(IJ)/L^*_{max} \leq 3/2 .$$

Hall and Shmoys [HS88] proved that a modification of *IJ*, *MIJ*, where the roles of release times and delivery times are interchanged, guarantees

$$L_{max}(MIJ)/L^*_{max} \leq 4/3 .$$

In the same paper, the authors also presented two algorithms A_{1k} and A_{2k} that guarantee

$$L_{max}(A_{ik})/L^*_{max} \leq 1+1/k \text{ for } i = 1, 2 \text{ and natural } k .$$

A_{1k} runs in $O(nlogn+nk^{16k^2+8k})$ time, whereas A_{2k} runs in $O(2^{4k}(nk)^{4k+3})$ time.

The case of equal due dates is equivalent to $1|r_j|C_{max}$ which can be solved optimally by scheduling the tasks in order of non-decreasing release dates (see Section 4.1).

If all execution times p_j are equal (but due dates are different), a polynomial time method is not available, unless $p_j = 1$ for all tasks T_j. If all tasks have unit execution times ($1|r_j, p_j = 1|L_{max}$), an optimal schedule is generated in polynomial time by involving repeated application of Jackson's *EDD* rule.

Algorithm 4.3.1 *Modification of EDD rule for* $1 \mid r_j, p_j = 1 \mid L_{max}$ [LLRK76].
```
begin
t := 0;
while T ≠ ∅ do
  begin
```
$t := \max\{t, \min_{T_j \in T}\{r_j\}\};$
$T' := \{T_j \mid T_j \in T, r_j \leq t\};$
Choose $T_i \in \{T_j \mid T_j \in T' \text{ for which } d_j = \min\{d_k \mid T_k \in T'\}\};$
$T := T - \{T_i\};$
Schedule T_i at time t;
$t := t + 1;$
```
  end;
end;
```

The proof of this result is straightforward and depends on the fact that no task can become available during the processing of another one, so that it is never advantageous to postpone processing the selected task T_i (recall that all r_j's are assumed to be integer).

If $p_j = p$, where p is an arbitrary integer, Algorithm 4.3.1 is not exact if p does not divide all r_j. For example, if $n = p = 2$, $r_1 = 0$, $r_2 = 1$, $d_1 = 7$, $d_2 = 5$, postponing T_1 is clearly advantageous. Simons [Sim78] presented a more sophisticated approach to solve the problem $1 \mid r_j, p_j = p \mid L_{max}$, where p is an arbitrary integer.

Problem $1 \mid pmtn, r_j \mid L_{max}$

For the preemptive case, $1 \mid pmtn, r_j \mid L_{max}$, a modification of Jackson's rule due to Horn [Hor74] solves the problem optimally in polynomial time.

Algorithm 4.3.2 *for problem* $1 \mid pmtn, r_j \mid L_{max}$ [Hor74].
```
begin
repeat
```
$\rho_1 := \min_{T_j \in T}\{r_j\};$
```
  if all tasks are available at time ρ₁
  then ρ₂ := ∞
  else ρ₂ := min{rⱼ | rⱼ ≠ ρ₁};
```
$E := \{T_j \mid r_j = \rho_1\};$
Choose $T_k \in E$ such that $d_k = \min_{T_j \in E}\{d_j\};$
$l := \min\{p_k, \rho_2 - \rho_1\};$

Assign T_k to the interval $[\rho_1, \rho_1 + l)$;
```
if  p_k ≤ l
then  T := T − {T_k}
else  p_k := p_k − l;
for all  T_j ∈ E do  r_j := ρ_1 + l;
until  T = ∅;
end;
```

Problem $1 \mid prec, r_j \mid L_{max}$

We first emphasize that the considerations concerning symmetry of problems $1 \mid r_j \mid L_{max}$ can be generalized to the case of precedence constraints. If a problem is specified by a triple of vectors (r, p, q) and - in addition - a precedence relation $\prec$, this is clearly equivalent to the inverse problem defined by (q, p, r) and $\prec'$ with $T_i \prec' T_j$ if $T_j \prec T_i$. Again, an optimal schedule for a problem can be reversed to obtain an optimal schedule for the original problem, with the same criterion value.

Let us now examine the introduction of precedence constraints in the problem in detail. As a general principle, release times r_j and tails q_j may be replaced by

$$r_j = \max \{r_j, \max\{r_i + p_i \mid T_i \prec T_j\}\}$$

$$q_j = \min \{q_j, \min\{p_i + q_i \mid T_j \prec T_i\}\},$$

because in every feasible schedule $s_j \geq C_i \geq r_i + p_i$ for all T_j with $T_i \prec T_j$ and $L_i' \geq C_j + p_i + q_i$ for all T_j with $T_j \prec T_i$. Hence, if $T_i \prec T_j$, we may assume that $r_i + p_i \leq r_j$ and $q_i \geq q_j + p_j$.

It follows that the case in which all due dates are equal is again solved by ordering the tasks according to non-decreasing r_j. Such an ordering will respect all precedence constraints in view of the preceding argument. If we apply this method to the problem in which all r_j's are equal, i.e. for $1 \mid prec \mid L_{max}$, the resulting algorithm can be interpreted as a special case of Lawler's more general algorithm to minimize $\max_{j}\{G_j(C_j)\}$ for arbitrary non-decreasing cost functions G_j (cf. Section 4.5). A similar observation can be made with respect to the case $p_j = 1$ for all j, where Algorithm 4.3.1 will produce a schedule respecting the precedence constraints.

In the general case, however, the precedence constraints are not respected automatically. Consider for example five tasks with release times $r = [0, 2, 3, 0, 7]$, processing times $p = [2, 1, 2, 2, 2]$, and tales $q = [5, 2, 6, 3, 2]$, and the precedence constraint $T_4 \prec T_2$ (cf. [LRK73]); note that $r_4 + p_4 \leq r_2$ and $q_4 \geq p_2 + q_2$. If the constraint $T_4 \prec T_2$ is ignored, the unique optimal schedule is given by $(T_1,$

T_2, T_3, T_4, T_5) with value $L^*_{max} = 11$. Explicit inclusion of this constraint leads to $L^*_{max} = 12$.

The *MF* algorithm introduced by McMahon and Florian [MF75] can easily be adapted to deal with given precedence constraints. Since we may assume that $r_i < r_j$ and $q_i > q_j$ if $T_i \prec T_j$, they are respected by the *MF* algorithm, and obviously, the lower bound remains valid.

Problem 1 | pmtn, prec, r_j | L_{max}

This problem can be solved in $O(n^2)$ time by an application of the algorithm given in [Bla76], which combines the ideas of Lawler's approach to the solution of problem 1 | prec | L_{max} and these of Algorithm 4.3.2. We mention here that in fact the much larger class of problems 1 | pmtn, prec, r_j | G_{max}, where quite arbitrary cost functions are assigned to the tasks and maximum cost is to be minimized, can be optimally solved in time $O(n^2)$. This will be discussed in Section 4.5.

Minimizing Lateness Range

The usual type of scheduling problems considered in literature involves penalty functions which are non-decreasing in task completion times. Conway et al. [CMM67] refer to such functions as *regular performance criteria*. There are, however, many applications in which *non-regular criteria* are appropriate. One of them is the problem of minimizing the difference between maximum and minimum task lateness which is important in real life whenever it is desirable to give equal treatment to all customers (tasks). That is, the delays in filling the customer orders should be as nearly equal as possible for all customers. Another example are file organization problems the objective is to minimize the variance of retrieval times for records in a file.

In spite of the importance of non-regular performance measures, very little analytical work has been done in this area. Gupta and Sen [GS84] studied the problem 1 || $L_{max} - L_{min}$ where the tasks are pair-wise independent, ready at time zero, each having a due date d_j and processing time p_j. They used a heuristic rule in which tasks are ordered according to non-decreasing values of $d_j - p_j$ (*minimum slack time rule, MST*), and ties are broken according to earliest due dates. This heuristic allows to compute lower bounds for $L_{max} - L_{min}$ which are then used in a branch and bound algorithm to eliminate nodes from further consideration.

A more general objective function has been considered by Raiszadeh et al. [RDS87]. Their aim was to minimize the convex combination $Z = \lambda(L_{max} - L_{min}) + (1 - \lambda)L_{max}$, $0 \le \lambda \le 1$, of range of minimum and maximum lateness.

Let all the tasks be arranged in the earliest due date order (*EDD*) and indexed accordingly $(T_1, \cdots, T_n)$. Thus for any two tasks T_i, $T_j \in \mathcal{T}$, if $d_i < d_j$, we must have $i < j$. Ties are broken such that $d_i - p_i \le d_j - p_j$, i.e. in the minimum slack time (*MST*) order. If there is still a tie, it can be broken arbitrarily.

Let S be a schedule in which task T_i immediately precedes task T_j, and let S' be constructed from S by interchanging tasks T_i and T_j without changing the position of any other task in S. Then, due to [RDS87], we have the following result for the values Z of S and S'.

Lemma 4.3.3 (a) *If* $d_i - p_i \le d_j - p_j$, *then* $Z(S) \le Z(S')$.

(b) *If* $d_i - p_i > d_j - p_j$, *then* $Z(S) - Z(S') \le \lambda((d_i - p_i) - (d_j - p_j))$.   □

Lemma 4.3.3 can be used to find lower bounds for an optimal solution. This computation is illustrated in the following example.

Example 4.3.4 [RDS87]   Consider $n = 4$ tasks with processing times and due dates given by $p = [6, 9, 11, 10]$ and $d = [17, 18, 19, 20]$, respectively. For the *EDD* ordering $S = (T_1, T_2, T_3, T_4)$, the value of the optimization criterion is $Z(S) = 16 + 11\lambda$. Call this ordering "primary". A "secondary" ordering (this notation is due to Townsend [Tow78]) is obtained by repeatedly interchanging neighboring tasks T_i, T_j with $d_i - p_i > d_j - p_j$, until tasks are in *MST* order. From Lemma 4.3.3(b) we see that such an exchange operation will improve the criterion value of the schedule by at most $\lambda((d_i - p_i) - (d_j - p_j))$. For each interchange operation the maximum potential reduction (*MPR*) of the objective function is given in Table 4.3.1. Obviously, the value $Z(S)$ of the primary order can never be improved by more than 7λ, hence $Z(S) - 7\lambda = 16 + 4\lambda$ is a lower bound on the optimal solution.   □

Original Schedule	Interchange	Changed Schedule	MPR
(T_1, T_2, T_3, T_4)	T_1 and T_2	(T_2, T_1, T_3, T_4)	2λ
(T_2, T_1, T_3, T_4)	T_1 and T_3	(T_2, T_3, T_1, T_4)	3λ
(T_2, T_3, T_1, T_4)	T_1 and T_4	(T_2, T_3, T_4, T_1)	1λ
(T_2, T_3, T_4, T_1)	T_2 and T_3	(T_3, T_2, T_4, T_1)	1λ
		total	7λ

Table 4.3.1.

This bounding procedure is used in a branch and bound algorithm where a search tree is constructed according to the following scheme. A node at the rth level of the tree corresponds to a particular schedule with the task arrangement of the first r positions fixed. One of the remaining $n - r$ tasks is then selected for the

$(r+1)^{st}$ position. The lower bound for the node is then computed as discussed above. For this purpose the primary ordering will have the first $r+1$ positions fixed and the remaining $n-r-1$ positions in the MST order. Pairwise interchanges of tasks are executed among the last $n-r-1$ positions. At each step the branching is done from a node having the least lower bound.

A performance measure similar to the one considered above is the average deviation of task completion times. Under the restriction that tasks have a common due date d, a schedule which minimizes $\sum_{j=1}^{n} |C_j - d|$ has to be constructed.

This type of criterion has applications e.g. in industrial situations involving scheduling, where the completion of a task either before or after its due date is costly. It is widely recognized that completion after a due date is likely to incur costs in the loss of the order and of customer goodwill. On the other hand, completion before the due date may lead to higher inventory costs and, if goods are perishable, potential losses.

Raghavachari [Rag86] proved that optimal schedules are "V-shaped". Let T_k be a task with the smallest processing time among all the tasks to be scheduled. A schedule is *V-shaped* if all tasks placed before task T_k are in descending order of processing time and the tasks placed after T_k are in ascending order of processing time. For the special case of $d = \sum_{j=1}^{n} p_j$, an optimal schedule for $1 \mid\mid \sum |C_j - d|$ can be obtained in the following way.

Algorithm 4.3.5 *for problem* $1 \mid\mid \sum |C_j - d|$ [Kan81].

Method: The algorithm determines two schedules, $S^{\leq}$ and $S^{>}$. The tasks of $S^{\leq}$ are processed without idle times, starting at time $d - \sum_{T_j \in S^{\leq}} p_j$, the tasks of $S^{>}$ are processed without idle times, starting at time d.

```
begin
S≤ := ∅; S> := ∅;          -- initialization: empty schedules
while T ≠ ∅ do
   begin
   Choose Tₗ ∈ T such that pₗ = max {pⱼ | Tⱼ ∈ T };
                               j
   T := T - {Tₗ};  S≤ := S≤ ⊕ (Tₗ);
      -- Task Tₗ is inserted into the last position in sub-schedule S≤
   if T ≠ ∅ do
      begin
      Choose Tₗ ∈ T such that pₗ = max {pⱼ | Tⱼ ∈ T };
                                  j
      T := T - {Tₗ};  S> := (Tₗ) ⊕ S>;
         -- Task Tₗ is inserted before the first task of sub-schedule S≤
```

```
    end;
  end;
end;
```

Baghi et al. [BSC86] generalized this algorithm to the case

$$d \geq \begin{cases} p_1+p_3+\cdots+p_{n-1}+p_n & \text{if } n \text{ is even} \\ p_2+p_4+\cdots+p_{n-1}+p_n & \text{if } n \text{ is odd,} \end{cases}$$

where tasks are numbered in non-decreasing order of processing times, $p_1 \leq p_2 \leq \cdots \leq p_n$.

An interesting extension of the above criteria is a combination of the lateness and earliness. To be more precise, one of the possible extensions is minimizing the sum of earliness and absolute lateness for a set of tasks scheduled around a common restrictive due date. This problem, known also as the mean absolute deviation (MAD) one, is quite natural in just-in-time inventory systems or in scheduling a sequence of experiments that depends on a predetermined external event [GTW88]. In [KLS90] it has been proved that this problem is equivalent to the scheduling problem with the mean flow time criterion. Thus, all the algorithms valid for the latter problem can be also used for the MAD problem. On the other hand, however, if a due date is a restrictive one, the MAD problem starts to be NP-hard (in the ordinary sense) even for one processor [HKS91]. A pseudopolynomial time algorithm based on dynamic programming has been also given for this case [HKS91].

On the other hand, one may consider minimizing total squared deviation about a common unrestrictive due date. This problem is equivalent to the problem of minimizing the completion time variance and has been transformed to the maximum cut problem in a complete weighted graph [Kub95]. Its NP-hardness in the ordinary sense has been also proved [Kub93].

4.3.2   Number of Tardy Tasks

The problem of finding a schedule that minimizes the weighted number $\Sigma w_j U_j$ of tardy tasks is NP-hard [Kar72], whereas the unweighted case is simple. Given arbitrary ready times, i.e. in the case $1|r_j|\Sigma U_j$, the problem is strongly NP-hard, as was proved by Lenstra et al. [LRKB77]. If precedence constraints are introduced between tasks then the problem is NP-hard, even in the special case of equal processing times and chain-like precedence constraints. In [IK78], a heuristic algorithm for problem $1|tree, p_j = 1|\Sigma U_j$ is presented.

Problem $1||\Sigma U_j$

Several special cases do admit exact polynomial time algorithms. The most common special case occurs when all weights are equal. Moore [Moo68] published an optimization algorithm that solves the problem in polynomial time. This algorithm sorts the tasks according to *EDD* rule (tasks with earlier due dates first, also known as *Hodgson's algorithm*).

Algorithm 4.3.6 *Hodgson's algorithm for* $1 \parallel \Sigma U_j$ [Law82].

Input: Task set $\mathcal{T} = \{T_1, \cdots, T_n\}$.

Method: The algorithm operates in two steps: first, the subset $\mathcal{T}^{\leq}$ of tasks of $\mathcal{T}$ that can be processed on time is determined; then a schedule is constructed for the subsets $\mathcal{T}^{\leq}$, and $\mathcal{T}^{>} = \mathcal{T} - \mathcal{T}^{\leq}$.

begin
Sort tasks in *EDD* order;   -- w.l.o.g. assume that $d_1 \leq d_2 \leq \cdots \leq d_n$
$\mathcal{T}^{\leq} := \varnothing$;

$p := 0$;   -- p keeps track of the execution time of tasks of $\mathcal{T}^{\leq}$
for $j := 1$ **to** n **do**
  begin
  $\mathcal{T}^{\leq} := \mathcal{T}^{\leq} \cup \{T_j\}$;
  $p := p + p_j$;
  if $p > d_j$ -- i.e. task T_j doesn't meet its due date
  then
    begin
    Let T_k be a task in $\mathcal{T}^{\leq}$ with maximal processing time,
        i.e. with $p_k = \max\{p_i \mid T_i \in \mathcal{T}^{\leq}\}$;
    $p := p - p_k$;
    $\mathcal{T}^{\leq} := \mathcal{T}^{\leq} - \{T_k\}$;
    end;
  end;
Schedule the tasks in $\mathcal{T}^{\leq}$ according to *EDD* rule;
Schedule the remaining tasks $(\mathcal{T} - \mathcal{T}^{\leq})$ in an arbitrary order;
end;

Without proof we mention that this algorithm generates a schedule with the minimal number of tardy tasks. The algorithm can easily be implemented to run in $O(n\log n)$ time.

Example 4.3.7 Suppose there are eight tasks with processing times $p = [10, 6, 3, 1, 4, 8, 7, 6]$ and due dates $d = [35, 20, 11, 8, 6, 25, 28, 9]$. Set $\mathcal{T}^{\leq}$ will be $\{T_5, T_4, T_3, T_2, T_7, T_1\}$, and the schedule is $(T_5, T_4, T_3, T_2, T_7, T_1, T_6, T_8)$. Table 4.3.2 compares the due dates and completion times; note that the due dates of the last two tasks are violated.   □

	T_5	T_4	T_3	T_2	T_7	T_1	T_6	T_8
Due date d_j	6	8	11	20	28	35	25	9
Completion time C_j	4	5	8	14	21	31	39	45

Table 4.3.2 *Due dates and completion times in Example 4.3.7.*

Problem $1 \mid\mid \Sigma w_j U_j$

Karp [Kar72] included the decision version of minimizing the weighted sum of tardy tasks in his list of 21 NP-complete problems. Even if all the due dates d_j are equal, the problem is NP-hard; in fact, this problem is equivalent to the knapsack problem and thus is not strongly NP-hard. An optimal solution for $1 \mid\mid \Sigma w_j U_j$ can be specified by a partition of the task set $\mathcal{T}$ into two subsets $\mathcal{T}^{\leq}$ and $\mathcal{T}^{>}$ as defined above. Thus it suffices to find an optimal partition of the task set $\mathcal{T}$.

Sahni [Sah76] developed an exact pseudopolynomial time algorithm for $1 \mid\mid \Sigma w_j U_j$ with different due-dates which is based on dynamic programming and requires $O(n\Sigma w_j)$ time. Using digit truncation, depending from which digit on the weights are truncated, a series of approximation algorithms $A_1, \cdots, A_k$ (a so-called *approximation scheme*) with $O(n^3 k)$ running time can be derived such that

$$\Sigma (w_j \bar{U}_j(A_k))/\bar{U}_w^* \geq 1 - 1/k \,,$$

where $\bar{U}_j = 1 - U_j$. Note that $\Sigma w_j \bar{U}_j$ is the weighted sum of on-time tasks. It is possible to decide in polynomial time whether $\Sigma w_j U_j^* = 0$. Gens and Levner [GL78] developed an algorithm B_k with running time $O(n^3)$ such that

$$U_w^{B_k}/U_w^* \leq 1 + 1/k \,.$$

The same authors improved the implementation of algorithm B_k to run in $O(n^2 \log n + n^2 k)$ time [GL81].

When all processing times are equal, the problem $1 \mid p_j = 1 \mid \Sigma w_j U_j$ can easily be solved. For the more general case of $1 \mid\mid \Sigma w_j U_j$ where processing times and weights are *agreeable*, i.e. $p_i < p_j$ implies $w_i \geq w_j$, an exact $O(n\log n)$ time algorithm can be obtained by a simple modification of the Hodgson's algorithm [Law76]. We will present this algorithm below.

Suppose tasks are placed and indexed in *EDD* order. For $j \in \{1, \cdots, n\}$, $\mathcal{T}^{\leq}$ is said to be *j-optimal* if $\mathcal{T}^{\leq} \subseteq \{T_1, \cdots, T_j\}$ and the sum of weights of tasks in $\mathcal{T}^{\leq}$ is maximal with respect to all feasible sets having that property. Thus, an optimal solution is obtained from an *n*-optimal set $\mathcal{T}^{\leq}$ processed in non-decreasing due date order, and then executing the tasks of $T - T^{\leq}$ in any order. Lawler's algo-rithm is a variation of Algorithm 4.3.6 for the construction of j-optimal sets. The following algorithm uses a linear ordering $\lessdot$ induced on tasks by their *relative desirability*, for inclusion in an on-time set, i.e.:

$$T_i \lessdot T_j \text{ if and only if } \quad p_i > p_j, \text{ or}$$
$$p_i = p_j \text{ and } w_i < w_j, \text{ or}$$
$$p_i = p_j \text{ and } w_i = w_j \text{ and } i < j.$$

Algorithm 4.3.8 *for problem* $1 \,||\, \Sigma w_j U_j$ *with agreeable weights* [Law76].
```
begin
```
Sort tasks according to *EDD* rule; $-\!-$ w.l.o.g. assume that $d_1 \leq d_2 \leq \cdots \leq d_n$

$T^0 := \varnothing;$

```
for j := 0 to n − 1 do
    if  p(T^{j-1}) + p_j ≤ d_j
```
            $-\!-$ $p(T^{j-1})$ denotes the sum of processing times of tasks in T^{j-1}
```
    then  T^j := T^{j-1} ∪ {T_j}
    else
        begin
```
        Choose $T_l \in T^{j-1} \cup \{T_j\}$ minimal with respect to $\lessdot;$
        $T^j := (T^{j-1} \cup \{T_j\}) - \{T_l\};$
```
        end;
end;
```

It is easy to prove that for all j, T^j is a j-optimal set. Hence, T^n presents an exact solution in the sense that all tasks of T^n are completed on time, and the tasks of $T - T^n$ are tardy.

Another special case considered by Sidney [Sid73] assumes that the tasks of a given subset $T' \subseteq T$ must be completed on time. This problem can be formu-lated as $1 \,||\, \Sigma w_j U_j$ where the weights w_j are 0 or 1. Sidney presented two algo-rithms of polynomial time complexity which generalize the Hodgson's algorithm and solve the problem optimally.

Problem $1 \,|\, r_j \,|\, \Sigma w_j U_j$

This scheduling problem is known to be NP-hard in the strong sense [LRKB77]. If, however, all weights are 1 and there are certain dependencies between ready times and due dates, optimal schedules can be constructed in polynomial time. Kise et al. [KIM78] used a variation of Lawler's Algorithm 4.3.8, and Lawler [Law82] proved that the algorithm can be improved to run in $O(n\log n)$ time. For a given set of tasks, the release times and due dates are called *consistent*, if $r_i < r_j$ implies $d_i \le d_j$ for all tasks T_i, T_j. We start with ordering tasks according to both, non-decreasing ready times and non-decreasing due dates. Without loss of generality we may assume that the tasks are already indexed appropriately, i.e. $r_1 \le r_2 \le \cdots \le r_n$ and $d_1 \le d_2 \le \cdots \le d_n$. Any schedule S can again be described by a partition of task set $\mathcal{T}$ into on-time set $\mathcal{T}^{\le}$ and tardy set $\mathcal{T}^{>}$. Tasks of $\mathcal{T}^{\le}$ are processed in *EDD* order, so they are ordered according to their indices. Let $\mathcal{T}^{\le}$ $\{T_{k_1}, \cdots, T_{k_m}\}$, $k_1 < \cdots < k_m$. The completion time C_{k_i} of task T_{k_i} in this schedule is given by

$$C_{k_1} = r_{k_1} + p_{k_1}$$
$$C_{k_i} = \max\{C_{k_{i-1}}, r_{k_i}\} + p_{k_i} \qquad (i = 2, \cdots, m).$$

Then the last task of $\mathcal{T}^{\le}$ is completed at time $C(\mathcal{T}^{\le}) = C_{k_m}$.

The following algorithm generates optimal schedules for the subsets $\{T_1, \cdots, T_i\}$ in the order of $i = 1, \cdots, n$. Let an optimal schedule for $\{T_1, \cdots, T_i\}$ be specified by the subset $\mathcal{E}_i \subseteq \{T_1, \cdots, T_i\}$ of on-time tasks. Then, set $\mathcal{E}_n$ will yield an optimal schedule for $\mathcal{T}$.

Algorithm 4.3.9 *for computing* $\mathcal{E}_n$ [KIM78].
```
begin
Order tasks according to both, non-decreasing ready times and non-decreasing
    due dates;
E₀ := ∅;
for j := 1 to n do
  begin
  Eⱼ := Eⱼ₋₁ ∪ {Tⱼ};
    -- a sub-schedule S(Eⱼ) is obtained by sequencing the tasks of Eⱼ in EDD order
  if  C(Eⱼ) > dⱼ
    -- C(Eⱼ) denotes the completion time of the last task in S(Eⱼ)
  then
    begin
    Choose Tₖ ∈ Eⱼ such that the sub-schedule obtained for Eⱼ−{Tₖ} in EDD
        order is of minimal length;
    Eⱼ := Eⱼ−{Tₖ};
    end;
  end;
end;
```

We mention that, under the condition of consistent release times and due dates, Algorithm 4.3.9 determines an optimal schedule for problem $1 \mid r_j \mid \Sigma U_j$ in $O(n^2)$ time.

Example 4.3.10 Let 6 tasks already be ordered according to increasing ready times, $p = [4, 3, 3, 7, 7, 4]$, $r = [0, 0, 4, 4, 5, 8]$, $d = [4, 5, 7, 11, 14, 15]$. Algorithm 4.3.9 determines set $\mathcal{E}_n$ to be $\mathcal{E}_n = \{T_2, T_3, T_6\}$, and the corresponding optimal schedule is $(T_2, T_3, T_6, T_1, T_4, T_5)$ where the last three tasks are tardy. $\square$

If task preemptions are allowed, dynamic programming algorithms can be applied to solve $1 \mid pmtn, r_j \mid \Sigma U_j$ in $O(n^5)$ time, and $1 \mid pmtn, r_j \mid \Sigma w_j U_j$ in time $O(n^3 (\Sigma w_j)^2)$. We refer the interested reader to [Law82].

Problem $1 \mid prec \mid \Sigma w_j U_j$

Lenstra and Rinnooy Kan [LRK80] proved that the 3-PARTITION problem (see Section 4.1.1) is reducible to the decision version of the problem $1 \mid chains, p_j = 1 \mid \Sigma U_j$. Hence, scheduling unit time tasks on a single processor subject to chain-like precedence constraints so as to minimize the unweighted number of late tasks is NP-hard in the strong sense. For $1 \mid forest \mid \Sigma w_j U_j$, Ibarra and Kim [IK78] discussed an algorithm that finds for any positive integer k an approximate schedule S^k such that

$$U_w^k / U_w^* < 1 + \frac{1}{k+1} .$$

The approximate solution is found in $O(kn^{k+2})$ time. They give also examples showing that the algorithm is not applicable to tasks forming an arbitrary precedence graph.

4.3.3 Mean Tardiness

In this section we will consider problems concerned with the minimization of mean or mean weighted tardiness.

Problem $1 \mid\mid \Sigma w_j D_j$

McNaughton [McN59] has shown that preemption cannot reduce mean weighted tardiness for any given set of tasks. Thus, an optimal preemptive schedule has the same value of mean weighted tardiness as an optimal, non-preemptive schedule. It has been shown by Lawler [Law77] and by Lenstra et al. [LRKB77] that the problem of minimizing mean weighted tardiness is NP-hard in the strong sense. If all weights are equal, the problem is still NP-hard in the ordinary sense [DL90]. If unit processing times are assumed but weights are arbitrary, the problem can be formulated as a linear assignment problem, and hence it can be solved in $O(n^3)$ time [GLL+79]. If in addition all tasks have unit weights, simply sequencing tasks in non-decreasing order of their due dates minimizes the total tardiness, and hence this special problem can be solved in $O(n\log n)$ time.

In more detail we will consider another special problem of type $1\,||\,\Sigma w_j D_j$ where weights of tasks are *agreeable* (see Section 4.3.2) and processing times are integer. Lawler [Law77] presented a pseudopolynomial dynamic programming algorithm of the worst-case running time $O(n^4 p)$ or $O(n^5 p_{max})$, if $p = \Sigma p_j$, and $p_{max} = \max\{p_j\}$, respectively. The algorithm is pseudopolynomial because its time complexity is polynomial only with respect to an encoding in which p_j values are expressed in unary notation (see Section 2.2). We are going to present this algorithm.

Recall from Section 4.3.2 that weights of tasks of a set $\{T_1,\cdots,T_n\}$ are called *agreeable* iff $p_i < p_j$ implies $w_i \geq w_j$ for all $i, j \in \{1,\cdots,n\}$. The algorithm is based on the following theorem which claims an important property of an optimal schedule for $1\,||\,\Sigma w_j D_j$.

Theorem 4.3.11 [Law77] *Suppose the tasks are agreeably weighted and numbered in non-decreasing due date order, i.e. $d_1 \leq d_2 \leq \cdots \leq d_n$. Let task T_k be such that $p_k = \max\{p_j \,|\, j = 1,\cdots,n\}$. Then there is some index σ, $k \leq \sigma \leq n$, such that there exists an optimal schedule S in which T_k is preceded by all tasks T_j with $j \leq \sigma$ and $j \neq k$, and followed by all tasks T_j with $j > \sigma$.*   □

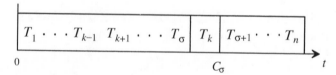

Figure 4.3.1   *An illustration of Theorem* 4.3.11.

Thus, if T_k is a task with largest processing time, then for some task T_σ, $k \leq \sigma \leq n$, there exists an optimal schedule where (see Figure 4.3.1)

(*i*) tasks T_1, $T_2,\cdots,T_{k-1}$, $T_{k+1},\cdots,T_\sigma$ form a partial schedule starting at time 0, which are followed by

(*ii*) task T_k, with completion time $C_\sigma = \sum_{j \leq \sigma} p_j$, followed by

(*iii*) tasks $T_{\sigma+1}$, $T_{\sigma+2},\cdots,T_n$, forming another partial schedule starting at time C_σ.

The overall schedule is optimal only if the partial schedules in (*i*) and (*iii*) are optimal, for starting times 0 and C_σ, respectively. This observation suggests a dynamic programming algorithm for the problem solution. For any given subset $\mathcal{T}'$ of tasks and starting time $t \geq 0$, there is a well-defined scheduling problem. An optimal schedule for problem $(\mathcal{T}, t)$ can be found recursively from the optimal schedules for problems of the form $(\mathcal{T}', t')$, where $\mathcal{T}'$ is a proper subset of $\mathcal{T}$, and $t' \geq t$.

Algorithm 4.3.12 *for problem* $1 \,||\, \Sigma w_j D_j$ [Law77].
Method: The algorithm calls the recursive procedure *sequence* with parameters t, denoting the start time of the sub-schedule to be determined, $\mathcal{T}'$ representing a subset of tasks numbered in non-decreasing due date order, and S' being an optimal schedule for the tasks in $\mathcal{T}'$.

Procedure *sequence*$(t, \mathcal{T}'$; var $S')$;

```
begin
if  T' = Ø  then  S' is the empty schedule
else
  begin
```
  Let $T_1, \cdots, T_m$ be the tasks of $\mathcal{T}'$, and $d_1 \leq d_2 \leq \cdots \leq d_m$;
  Choose T_k with maximum processing time among the tasks of $\mathcal{T}'$;
```
  for  σ := k to  m  do
    begin
```
    Let $\mathcal{T}^{\leq \sigma}$ be the subset $\{T_j \,|\, j \leq \sigma, j \neq k\}$ of $\mathcal{T}'$ tasks;
    Let $\mathcal{T}^{> \sigma}$ be the subset $\{T_j \,|\, j > \sigma\}$ of $\mathcal{T}'$ tasks;
    Call *sequence*$(t, \mathcal{T}^{\leq \sigma}, S^{\leq \sigma})$;
    $C_\sigma := t + \sum_{j \leq \sigma} p_j$;
    Call *sequence*$(C_\sigma, \mathcal{T}^{> \sigma}, S^{> \sigma})$;
         -- optimal sub-schedules for $\mathcal{T}^{\leq \sigma}$ and $\mathcal{T}^{> \sigma}$ are created
    $S_\sigma := S^{\leq \sigma} \oplus (T_k) \oplus S^{> \sigma}$;
         -- concatenation of sub-schedules and task T_k is constructed
    Compute value $\bar{D}_w^\sigma = \Sigma w_j D_j$ of sub-schedule S_σ;
```
    end;
```
  Choose S' with minimum value $\bar{D}_w^\sigma$ among the schedules S_σ, $k \leq \sigma \leq m$;
```
  end;
end;
```

```
begin      -- main algorithm
```
Order (and index) tasks of $\mathcal{T}$ in non-decreasing due date order;
$\mathcal{T} := (T_1, \cdots, T_n)$;

Call *sequence*$(0, \mathcal{T}, S)$;
　　　-- this call generates an optimal schedule S for $\mathcal{T}$, starting at time 0
end;

It is easy to establish an upper bound on the worst-case running time required to compute an optimal schedule for the complete set of n tasks. The subsets $\mathcal{T}'$ which enter into the recursion are of a very restricted type. Each subset consists of tasks whose subscripts are indexed consecutively, say from i to j, where possibly one of the indices, k, is missing, and where the processing times $p_i, \cdots, p_j$ of the tasks $T_i, \cdots, T_j$ are less than or equal to p_k. There are no more than $O(n^3)$ such subsets $\mathcal{T}'$, because there are no more than n values for each of the indices, i, j, k; moreover, several distinct choices of the indices may specify the same subset of tasks. There are surely no more than $p = \sum_{j=1}^{n} p_j \le np_{max}$ possible values of t. Hence there are no more than $O(n^3 p)$ or $O(n^4 p_{max})$ different calls of procedure *sequence* in Algorithm 4.3.12. Each call of *sequence* requires minimization over at most n alternatives, i.e. in addition $O(n)$ running time. Therefore the overall running time is bounded by $O(n^4 p)$ or $O(n^5 p_{max})$.

Example 4.3.13 [Law77] The following example illustrates performance of the algorithm. Let $\mathcal{T} = \{T_1, \cdots, T_8\}$, and processing times, due dates and weights be given by $p = [121, 79, 147, 83, 130, 102, 96, 88]$, $d = [260, 266, 269, 336, 337, 400, 683, 719]$ and $w = [3, 8, 1, 6, 3, 3, 5, 6]$, respectively. Notice that task weights are agreeable. Algorithm 4.3.12 calls procedure *sequence* with $\mathcal{T} = (T_1, \cdots, T_8)$, T_3 is the task with largest processing time, so in the **for**-loop procedure *sequence* will be called again for $\sigma = 3, \cdots, 8$. Table 4.3.3 shows the respective optimal schedules if task T_3 is placed in positions $\sigma = 3, \cdots, 8$.　　　□

Problem $1 \mid prec \mid \Sigma w_j D_j$

Lenstra and Rinnooy Kan [LRK78] studied the complexity of the mean tardiness problem when precedence constraints are introduced. They showed that $1 \mid prec, p_j = 1 \mid \Sigma D_j$ is NP-hard in the strong sense. For chain-like precedence constraints, they proved problem $1 \mid chains, p_j = 1 \mid \Sigma w_j D_j$ to be NP-hard.

σ	$sequence(C_\sigma, \mathcal{T}, S')$	optimal schedule	value $\overline{D}_w^\sigma$
3	$sequence(0, \{T_1,T_2\}, S^{\leq 3})$ $sequence(347, \{T_4,T_5,T_6,T_7,T_8\}, S^{>3})$	$(T_1,T_2,T_3,T_4,T_6,T_7,T_8,T_5)$	2565
4	$sequence(0, \{T_1,T_2,T_4\}, S^{\leq 4})$ $sequence(430, \{T_5,T_6,T_7,T_8\}, S^{>4})$	$(T_1,T_2,T_4,T_3,T_6,T_7,T_8,T_5)$	2084
5	$sequence(0, \{T_1,T_2,T_4,T_5\}, S^{\leq 5})$ $sequence(560, \{T_6,T_7,T_8\}, S^{>5})$	$(T_1,T_2,T_4,T_5,T_3,T_7,T_8,T_6)$	2007
6	$sequence(0, \{T_1,T_2,T_4,T_5,T_6\}, S^{\leq 6})$ $sequence(662, \{T_7,T_8\}, S^{>6})$	$(T_1,T_2,T_4,T_6,T_5,T_3,T_7,T_8)$	1928
7	$sequence(0, \{T_1,T_2,T_4,T_5,T_6,T_7\}, S^{\leq 7})$ $sequence(758, \{T_8\}, S^{>7})$	$(T_1,T_2,T_4,T_6,T_5,T_7,T_3,T_8)$	1785
8	$sequence(0, \{T_1,T_2,T_4,T_5,T_6,T_7,T_8\}, S^{\leq 8})$ $sequence(846, \varnothing, S^{>8})$	$(T_1,T_2,T_4,T_6,T_5,T_7,T_8,T_3)$	1111

Table 4.3.3 *Calls of procedure sequence in Example* 4.3.13.

4.3.4 Mean Earliness

It was pointed out by [DL90] that this problem is equivalent to the mean tardiness problem. To see this, we replace the given mean earliness problem by an equivalent mean tardiness scheduling problem.

Let $C = \sum_{j=1}^{n} p_j$. We construct an instance $\mathcal{T}' = \{T_1', \cdots, T_n'\}$ of the mean tardiness problem, where $p_j' = p_j$ for $j = 1, \cdots, n$, and where the due dates are defined by $d_j' = C - d_j + p_j$. Suppose S is an optimal schedule for $\mathcal{T}$. Define a schedule S' for $\mathcal{T}'$ as follows. If T_j is the kth task scheduled in S, then T_j' will be the $(n-k+1)$th task scheduled in S'. Clearly, we have $C_j' = C - C_j + p_j$, and hence

$$
\begin{aligned}
D_j' &= \max\{0, C_j' - d_j'\} \\
&= \max\{0, (C - C_j + p_j) - (C - d_j + p_j)\} \\
&= \max\{0, d_j - C_j\} = E_j.
\end{aligned}
$$

Thus, $\overline{E} = \overline{D}'$. Similarly, if S' is a schedule for $\mathcal{T}'$ such that $\overline{D}'$ is minimum we can construct a schedule S for $\mathcal{T}$ such that $\overline{E} = \overline{D}'$. Therefore, the minimum mean earliness of $\mathcal{T}$ is the same as the minimum mean tardiness for $\mathcal{T}'$. Hence, as we know that the mean tardiness problem on one processor is *NP*-hard, the mean earliness problem must also be *NP*-hard.

4.4 Minimizing Change-Over Cost

This section deals with the scheduling of tasks on a single processor where under certain circumstances a cost is inferred when the processor switches from one task to another. The reason for such "change-over" cost might be machine setup operations required if tasks of different types are processed in sequence.

First we present a more theoretical approach where a set of tasks subject to precedence constraints is given. In this section the purpose of the precedence relation $\prec$ is twofold: on one hand it defines the usual precedence constraints of the form $T_i \prec T_j$ where task T_j cannot be started before task T_i has been completed. On the other hand, if $T_i \prec T_j$, then we say that processing T_j immediately after T_i does not require any additional setup on the processor, so processing T_j does not incur any change-over cost. But if $T_i \nprec T_j$, i.e. T_j is not an immediate successor of T_i, then processing T_j immediately after T_i will require processor setup and hence will cause change-over cost.

The types of problems we are considering in Section 4.4.1 assume unit change-over cost for the setups. The problem then is to find schedules that minimize the number of setups.

In Section 4.4.2 we discuss a practically motivated model where jobs of different types are considered, and each job consists of a number of tasks. Processor setup is required, and consequently change-over cost is incurred, if the processor changes from one job type to another. Hence the tasks of each job type should be scheduled in sequences or *lots* of certain sizes on the processor. The objective is then to determine sizes of task lots, where each lot is processed non-preemptively, such that certain inventory and deadline conditions are observed, and change-over cost is minimized.

4.4.1 Setup Scheduling

Consider a finite partially ordered set $G = (\mathcal{T}, \prec^*)$, where $\prec^*$ is the reflexive, antisymmetric and transitive binary relation obtained from a given precedence relation $\prec$ as described in Section 2.3.2. Then, a *linear extension* of G is a linear order $(\mathcal{T}, \prec_L^*)$ that extends $(\mathcal{T}, \prec^*)$, i.e. for all $T', T'' \in \mathcal{T}$, $T' \prec^* T''$ implies $T' \prec_L^* T''$. For $\mathcal{T} = \{T_1, T_2, \cdots, T_n\}$, if the sequence $(T_{\alpha_1}, T_{\alpha_2}, \cdots, T_{\alpha_n})$ from left to right defines the linear order $\prec_L^*$, i.e. $T_{\alpha_1} \prec_L^* T_{\alpha_2} \prec_L^* \cdots \prec_L^* T_{\alpha_n}$, then $(T_{\alpha_1}, T_{\alpha_2}, \cdots, T_{\alpha_n})$ is obviously a schedule for $(\mathcal{T}, \prec)$.

Let $L = (T_{\alpha_1}, T_{\alpha_2}, \cdots, T_{\alpha_n})$ be a linear extension of $G = (\mathcal{T}, \prec^*)$ where $\prec^*$ is determined from precedence relation $\prec$. Two consecutive elements T_{α_i}, $T_{\alpha_{i+1}}$ of L are separated by a *jump* (or *setup*) if and only if $T_{\alpha_i} \not\prec T_{\alpha_{i+1}}$. The total number of jumps of L is denoted by $s(L, G)$. The *jump number* $s(G)$ of G is the minimum number of jumps in some linear extension, i.e.

$s(G) = \min\{s(L, G) \,|\, L$ is a linear extension of $G\}$.

A linear extension L of G with $s(L, G) = s(G)$ is called *jump-* (or *setup-*) *optimal*. The problem of finding a schedule with minimum number of setups is often called *jump number problem*.

If we assume that a jump causes change-over cost in the schedule, a jump-optimal schedule for $(\mathcal{T}, \prec)$ would obviously be one in which the total change-over cost is at minimum.

The notion of jump number has been introduced by Chein and Martin [CM72]. The problem of determining the setup number $s(G)$ and producing an optimal linear extension for any given ordered set G has been considered by numerous authors. While good algorithms have been found for certain restricted classes of ordered sets, it has been shown by W. R. Pulleyblank [Pul75] that finding the setup number even for partial orders of height one [2] is an NP-hard problem.

For a general poset $G = (\mathcal{T}, \prec^*)$, let $K_1, K_2, \cdots, K_r$ be any minimum family of disjoint chains (for definition of a chain we refer to Chapter 2.3.2) whose set union of tasks is $\mathcal{T}$. The concatenation $K_1 \oplus K_2 \oplus \cdots \oplus K_r$ of these chains obviously is not necessarily a linear extension of G. On the other hand, any linear extension L of a finite poset G can be expressed as a linear sum $K_1 \oplus K_2 \oplus \cdots \oplus K_r$ of chains, chosen so that in each chain neighboring tasks T_h and T_k are in relation $T_h \prec T_k$, and, for chains K_i, K_{i+1} $(i = 1, \cdots, r-1)$, the last task of K_i does not precede the first task of K_{i+1}. Notice that a linear extension represents a schedule for $(\mathcal{T}, \prec)$ in an obvious way. Setups occur exactly between two neighboring chains, i.e. between K_i and K_{i+1} for $i = 1, \cdots, r-1$.

The problem of scheduling precedence constrained tasks so that the number of setups is minimum is now formalized to the question of finding a linear extension that consists of a minimum number of chains.

One way of solving this problem heuristically is to determine so-called *greedy linear extensions*.

Algorithm 4.4.1 *Greedy linear extension of a partially ordered set* $(\mathcal{T}, \prec^*)$.
begin
$i := 0;$

---

[2] A partial order G is of height one if each directed path in G has at most two vertices.

```
while  T ≠ ∅  do
  begin  i := i+1;
```
  Let $T_i \in T$ be a task such that $T_k := \{ T \in T \mid T \prec^* T_i \}$ forms a maximal
     chain, i.e. there is no successor task T' of T_i for which $\{ T \in T \mid T \prec^* T' \}$ is
     a chain;
  Let K_i be the chain of tasks of T_i;
  $T := T - T_i$;
```
  end;
```
  $r := i$; $-- r$ is the number of chains obtained
  $L := K_1 \oplus K_2 \oplus \cdots \oplus K_r$;
```
end;
```

From the way the chains are constructed in this algorithm it is clear that $L = K_1 \oplus K_2 \oplus \cdots \oplus K_r$ is a linear extension of $G = (T, \prec^*)$, and hence it is a schedule for $(T, \prec)$. Greedy linear extensions can be characterized in the following way.

A linear extension L of G is *greedy* if and only if, for some r, L can be represented as $L = K_1 \oplus K_2 \oplus \cdots \oplus K_r$, where each K_i is a chain in G, the last task of K_i does not precede the first task of K_{i+1} (for $i = 1, \cdots, r-1$), and for each K_i and for any $T \in T$ which succeeds immediately the last task of K_i in G, there is a task $T' \in K_{i+1} \cup \cdots \cup K_r$ such that $T' \prec T$.

Example 4.4.2 To demonstrate how Algorithm 4.4.1 works, consider the precedence graph shown in Figure 4.4.1(a). The algorithm first chooses task T_3, thus getting the first chain $K_1 = (T_3)$. If the tasks chosen next are T_2 and T_1, then we get chains K_2 and K_3 shown encircled in Figure 4.4.1(a). The corresponding schedule is presented in Figure 4.4.1(b).   □

It can be shown that for any finite poset G there is a greedy linear extension L of G satisfying $s(G) = s(L, G)$. On the other hand, optimal linear extensions need not to be greedy. Also, greedy linear extensions may be far from optimum. So, for example, the setup number for the direct product of a two-element chain with an n-element chain is 1, yet there is a greedy linear extension with $n-1$ setups.

For some special classes of precedence graphs greedy linear extensions are known to be always optimal with respect to number of setups. Series-parallel graphs and N-free graphs are examples of such classes. For other examples and results we refer the interested reader to [ER85] and [RZ86].

Another important class of precedence graphs are interval orders (see Sections 2.1 and 2.3.2). Since interval orders model the sequential and overlapping structure of a set of intervals on the real line, they have many applications in several fields such as scheduling, VLSI routing in computer science, and in difference relations in measurement theory [Fis85, Gol80]. Faigle and Schrader

[FS85a and FS85b] presented a heuristic algorithm for the jump number problem for an interval order. But Ali and Deogun [AD90] were able to develop an optimization algorithm of time complexity $O(n^2)$ for n elements. They also presented a simple formula that allows to determine the minimal number of setups directly from the given interval order.

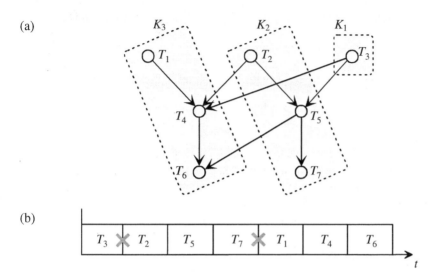

Figure 4.4.1 *An example for Algorithm 4.4.1*
(a) precedence graph and a chain decomposition,
(b) corresponding schedule; crosses (×) mark setups.

4.4.2   Lot Size Scheduling

The problem investigated in this subsection arises if tasks are scheduled in lots due to time and cost considerations. Let us consider for example the production of gearboxes of different types on a *transfer line*. The time required to manufacture one gearbox is assumed to be the same for all types. Changing from production of one gearbox type to another requires a change of machine (processor) installment to another state. As these change-overs are costly and time consuming the objective is to minimize the number of change-overs or the sum of their cost. The whole situation may be complicated by additional productional or environmental constraints. For example, there are varying demands of gearbox types over time. Storage capacity for *in-process inventory* of the produced items is limited. In-process inventory always increases if the production of a gearbox is finished; it is always decreased if produced items are delivered at given points in time where demand has to be fulfilled. A feasible schedule will assign gearbox productions to the processor in such a way that lots of gearboxes of the same type are manufactured without change-overs.

The problem can also be regarded as a special instance of the so-called *multi-product lot scheduling problem with capacity constraints*. For a detailed analysis of this problem and its various special modifications we refer e.g. to [BY82, FLRK80] and [Sch82]. All these models consider setup cost. Generally speaking, *setups* are events that may occur every time processing of a task or job is initiated again after a pause in processing. In many real processing systems such setups are connected with change-over costs.

Now, the lot size scheduling problem can be formulated as follows. Consider K deadlines and n different types of jobs. Set $\mathcal{J}_j$ includes all jobs of the jth type, $j = 1, \cdots, n$, and let $\mathcal{J} = \bigcup_{j=1}^{n} \mathcal{J}_j$ be the set of all jobs. Set $\mathcal{J}_j$ includes the jobs $J_j^1, \cdots, J_j^K$ with deadlines $\tilde{d}_{j1}, \cdots, \tilde{d}_{jK}$, respectively. Each job J_j^k itself consists of a number n_{jk} of unit processing time tasks. Whereas task preemption is not allowed, the processor may switch between jobs, even of different types. Only changing from a job of one type to that of another type is assumed to induce change-over cost. For each job type an upper bound $B_j \in \mathbb{N}_0$ on in-process inventory is given. Starting with some initial job inventory we want to find a feasible *lot size* schedule for the set $\mathcal{J}$ of jobs such that all deadlines are met, upper bounds on inventory are not exceeded, and the sum of all unit change-over cost is minimized.

For the above manufacturing example this model means that the transfer line is represented by the processor, and gearbox types relate to job types. Jobs J_j^k with deadlines $\tilde{d}_{jk}$ represent demands for gearbox types at different points in time. The number n_{jk} of tasks of each job J_j^k represents the number of items of gearbox type j required to be finished by time $\tilde{d}_{jk}$. Bound B_j relates to the limited storage capacity for in-process inventory of the different types of gearboxes. At each time $\tilde{d}_{jk}$ the in-process inventory of job type j is decreased by n_{jk}.

Let us assume that $H = \max_{jk} \{\tilde{d}_{jk}\}$ and that the processing capacity of the processor during the interval $[0, H]$ is decomposed in discrete *unit time intervals* (UTI) numbered by $h = 1, \cdots, H$. To ensure both, feasible production of all jobs and a feasible schedule without idle time we assume that $H = \sum_{j=1}^{n} n_j$ where $n_j = \sum_{k=1}^{K} n_{jk}$ represents the total number of tasks of $\mathcal{J}_j$. The lot size scheduling problem can now be formulated by the following mathematical programming problem. Let x_{jh} be a variable which represents the assignment of a job of type j to some UTI h such that $x_{jh} = 1$ if a job of this type is produced during interval h and $x_{jh} = 0$ otherwise. Let y_{jh} be a variable which represents unit change-over cost such that $y_{jh} = 1$ if jobs of different types are processed in UTI $h-1$ and UTI h, and

$y_{jh} = 0$ otherwise. Obviously, y_{jh} represents unit change-over cost. I_{jh} represents in-process inventory of job type j at the end of UTI h, and n_{jh} is the corresponding processing requirement (we set $n_{jh} = 0$ if there is no job with deadline $\tilde{d}_{jk} = h$). Let again B_j denote the upper bound on inventory of job type j.

$$\text{Minimize} \quad \sum_{j=1}^{n} \sum_{h=1}^{H} y_{jh} \tag{4.4.1}$$

$$\text{subject to} \quad I_{jh-1} + x_{jh} - I_{jh} = n_{jh} \qquad j = 1,\cdots,n;\ h = 1,\cdots,H, \tag{4.4.2}$$

$$\sum_{j=1}^{n} x_{jh} \leq 1 \qquad h = 1,\cdots,H, \tag{4.4.3}$$

$$0 \leq I_{jh} \leq B_j \qquad j = 1,\cdots,n;\ h = 1,\cdots,H, \tag{4.4.4}$$

$$x_{jh} \in \{0,1\} \qquad j = 1,\cdots,n;\ h = 1,\cdots,H, \tag{4.4.5}$$

$$y_{jh} = \begin{cases} 1 & \text{if } x_{jh} - x_{jh-1} > 0 \\ 0 & \text{otherwise} \end{cases} \qquad j = 1,\cdots,n;\ h = 1,\cdots,H. \tag{4.4.6}$$

The above constraints can be interpreted as follows. Equations (4.4.2) assure that the deadlines of all jobs are observed, (4.4.3) assure that at no time more than one job type is being processed, (4.4.4) restrict the in-process inventory to the given upper bounds. Equations (4.4.5) and (4.4.6) constrain all variables to binary numbers. The objective function (4.4.1) minimizes the total number of change-overs, respectively the sum of their unit cost. Note that for (4.4.1)-(4.4.6) a feasible solution only exists if the cumulative processing capacity up to each deadline is not less than the total number of tasks to be finished by this time.

The problem of minimizing the number of change-overs under the assumption that different jobs of different types have also different deadlines was first solved in [Gla68] by applying some enumerative method. There exist also dynamic programming algorithms for both, the problem with sequence-independent change-over cost [GL88, Mit72] and for the problem with sequence-dependent change-over cost [DE77]. For other enumerative methods see [MV90] and the references given therein. A closely related question to the problem discussed here has been investigated in [BD78], where each task has a fixed completion deadline and an integer processing time. The question studied is whether there exists a non-preemptive schedule that meets all deadlines and has minimum sum of change-over cost. For arbitrary integer processing times the problem is already NP-hard for unit change-over cost, three tasks per job type and two distinct deadlines, i.e. $K = 2$. Another similar problem was investigated in [HKR87] where the existence of unit change-over cost depends on some given order of tasks, i.e. tasks are indexed with $1, 2, \cdots$, and change-over cost occurs only if a task is followed by some other task with larger index. This problem is solvable in polynomial time.

Schmidt [Sch92] proved that the lot size scheduling problem formulated by
(4.4.1)-(4.4.6) is NP-hard for $n = 3$ job types. Now we show that it can be solved
in polynomial time if $n = 2$ job types have to be considered only. The algorithm
uses an idea which can be described by the rule "schedule all jobs such that no
unforced change-overs occur". This rule always generates an optimal schedule if
the earliest deadline has to be observed only by jobs of the same type. In case the
earliest deadline has to be observed by jobs of either type the rule by itself is not
necessarily optimal.

To find a feasible schedule with minimum sum of change-over cost we must
assign all jobs to a number $Z \leq H$ of non-overlapping production intervals such
that all deadlines are met, upper bounds on inventory are not exceeded, and the
number of all change-overs is minimum. Each production interval $z \in \{1, \cdots, Z\}$
represents the number of consecutive UTIs assigned only to jobs of the same
type, i.e. there exists only one setup for each z.

For simplicity reasons we now denote the two job types by q and r. Considering any production interval z, we may assume that a job of type q (r) is processed in UTIs h, $h+1$, h^*; if $h^* < H$ it has to be decided whether to continue
processing of jobs q (r) at h^*+1 or start a job of type r (q) in this UTI. Let

$$U_{rh^*} = \min\left\{ (i-h^*) - (\sum_{h=h^*+1}^{i} n_{rh} - I_{rh^*}) \mid i = h^*+1, \cdots, H \right\} \qquad (4.4.7)$$

be the remaining available processing capacity minus the processing capacity
required to meet all future deadlines of $\mathcal{J}_r$,

$$V_{qh^*} = \sum_{h=1}^{H} n_{qh} - \sum_{h=1}^{h^*} x_{qh} \qquad (4.4.8)$$

be the number of not yet processed tasks of $\mathcal{J}_q$, and

$$W_{qh^*} = B_q - I_{qh^*} \qquad (4.4.9)$$

be the remaining storage capacity available for job type q at the end of UTI h^*.
In-process inventory is calculated according to

$$I_{qh^*} = \sum_{h=1}^{h^*} (x_{qh} - n_{qh}). \qquad (4.4.10)$$

To generate a feasible schedule for job types q and r it is sufficient to change
the assignment from type q (r) to type r (q) at the beginning of UTI h^*+1, $1 \leq
h^* < H$, if $U_{rh^*} \cdot V_{qh^*} \cdot W_{qh^*} = 0$ for the corresponding job types in UTI h^*. Applying
this UVW-rule is equivalent to scheduling according to the above mentioned "no
unforced change-overs" strategy. The following algorithm makes appropriate use
of the UVW-rule.

Algorithm 4.4.3 *Lot size scheduling of two job types on a single processor*
[Sch92].
begin
$i := 1;\ \ x := r;\ \ y := q;$

```
while i < 3 do
  begin
  for h := 1 to H do
    begin
    Calculate U_xh, V_yh, W_yh according to (4.4.7)-(4.4.9);
    if U_xh·V_yh·W_yh = 0
    then
      begin
      Assign a job of type x;
      Calculate the number of change-overs;
      Exchange x and y;
      end
    else Assign a job of type y;
    end;
  i := i+1; x := q; y := r;
  end;
Choose the schedule with minimum number of change-overs;
end;
```

Using Algorithm 4.4.3 we generate for each job type $\mathcal{J}_j$ a number Z_j of production intervals $z_j = 1, \cdots, Z_j$ which are called q-intervals in case jobs of type q are processed, and r-intervals else, where $Z_q + Z_r = Z$. We first show that there is no schedule having less change-overs than the one generated by the UVW-rule, if the assignment of the first UTI ($h = 1$) and the length of the first production interval ($z = 1$; either a q- or an r-interval) are fixed. For $n = 2$ there are only two possibilities to assign a job type to $h = 1$. It can be shown by a simple exchange argument that there does not exist a schedule with less change-overs and the first production interval ($z = 1$) does not have UVW-length, if we fix the job type to be processed in the first UTI. Note that fixing the job type for $h = 1$ corresponds to an application of the UVW-rule considering an assignment of $h = 0$ to a job of types q or r. From this we conclude that if there is no such assignment of $h = 0$ then for finding the optimal schedule it is necessary to apply the UVW-rule twice and either assign job types q or r to $h = 1$. Let us first assume that $z = 1$ is fixed by length and job type assignment. We have the following lemmas [Sch92].

Lemma 4.4.4 *Changing the length of any production interval $z > 1$, as generated by the UVW*-rule, *cannot decrease the total number of change-overs.*   □

Lemma 4.4.5 *Having generated an UVW-schedule it might be possible to reduce the total number of production intervals by changing assignment and length of the first production interval $z = 1$.*   □

Using the results of Lemmas 4.4.4 and 4.4.5 we simply apply the UVW-rule twice, if necessary, starting with either job types. To get the optimal schedule we take that with less change-overs. This is exactly what Algorithm 4.4.3 does. As the resulting number of production intervals is minimal the schedule is optimal under the unit change-over cost criterion. For generating each schedule we have to calculate U, V, and W at most H times. The calculations of each V and W require constant time. Hence it follows that the time complexity of calculating all U is not more than $O(H)$ if appropriate data structures are used. The following example problem demonstrates the approach of Algorithm 4.4.3.

Example 4.4.6 $\mathcal{J} = \{\mathcal{J}_1, \mathcal{J}_2\}$, $\tilde{d}_{11} = 3$, $\tilde{d}_{12} = 7$, $\tilde{d}_{13} = 10$, $\tilde{d}_{21} = 3$, $\tilde{d}_{22} = 7$, $\tilde{d}_{23} = 10$, $B_1 = B_2 = 10$, $n_{11} = 1$, $n_{12} = 2$, $n_{13} = 1$, $n_{21} = 1$, $n_{22} = 1$, $n_{23} = 4$, and zero initial inventory. Table 4.4.1 shows the two schedules obtained when starting with either job type. Schedule S_2 has minimum number of change-overs and thus is optimal. □

h:	1	2	3	4	5	6	7	8	9	10
Schedule S_1:	$\mathcal{J}_1$	$\mathcal{J}_1$	$\mathcal{J}_2$	$\mathcal{J}_2$	$\mathcal{J}_2$	$\mathcal{J}_2$	$\mathcal{J}_1$	$\mathcal{J}_1$	$\mathcal{J}_2$	$\mathcal{J}_2$
Schedule S_2:	$\mathcal{J}_2$	$\mathcal{J}_2$	$\mathcal{J}_1$	$\mathcal{J}_1$	$\mathcal{J}_1$	$\mathcal{J}_1$	$\mathcal{J}_2$	$\mathcal{J}_2$	$\mathcal{J}_2$	$\mathcal{J}_2$

Table 4.4.1 *Two schedules for Example 4.4.6.*

4.5 Other Criteria

In this section we are concerned with single processor scheduling problems where each task T_j of the given task set $\mathcal{T} = \{T_1, \cdots, T_n\}$ is assigned a non-decreasing cost function G_j. Instead of a due date, function G_j specifies the cost $G_j(C_j)$ that is incurred by the completion of task T_j at time C_j. We will discuss two objective functions, maximum cost G_{max} and total cost $\Sigma G_j(C_j)$.

4.5.1 Maximum Cost

First we consider the problem of minimizing the maximum cost that is incurred by the completion of the tasks. We already know that the problem $1|r_j|G_{max}$ with $G_j(C_j) = L_j = C_j - d_j$ for given due dates d_j for the tasks, is NP-hard in the strong sense (cf. Section 4.3.1). On the other hand, if task preemptions are allowed, the problem becomes easy if the cost functions depend non-decreasingly on the task completion times. Also, the cases $1|prec|G_{max}$ and $1|pmtn, prec, r_j|G_{max}$ are solvable in polynomial time.

Problem $1 \mid pmtn, r_j \mid G_{max}$

Consider the case where task preemptions are allowed. Since cost functions are non-decreasing, it is never advantageous to leave the processor idle when unscheduled tasks are available. Hence, the time at which all tasks will be completed can be determined in advance by scheduling the tasks in order of non-decreasing release times r_j. This schedule naturally decomposes into blocks, where block $\mathcal{B} \subseteq \mathcal{T}$ is defined as the minimal set of tasks processed without idle time from time $r(\mathcal{B}) = \min\{r_j \mid T_j \in \mathcal{B}\}$ until $C(\mathcal{B}) = r(\mathcal{B}) + \sum_{T_j \in \mathcal{B}} p_j$ such that each task $T_k \notin \mathcal{B}$ is either completed not later than $r(\mathcal{B})$ (i.e. $C_k \leq r(\mathcal{B})$) or not released before $C(\mathcal{B})$ (i.e. $r_k \geq C(\mathcal{B})$).

It is easily seen that, when minimizing G_{max}, we can consider each block $\mathcal{B}$ separately. Let $G_{max}^*(\mathcal{B})$ be the value of G_{max} in an optimal schedule for the tasks in block $\mathcal{B}$. Then $G_{max}^*(\mathcal{B})$ satisfies the following inequalities:

$$G_{max}^*(\mathcal{B}) \geq \min_{T_j \in \mathcal{B}} \{G_j(C(\mathcal{B}))\},$$

and

$$G_{max}^*(\mathcal{B}) \geq G_{max}^*(\mathcal{B} - \{T_j\}) \text{ for all } T_j \in \mathcal{B}.$$

Let task $T_l \in \mathcal{B}$ be such that

$$G_l(C(\mathcal{B})) = \min_{T_j \in \mathcal{B}} \{G_j(C(\mathcal{B}))\}. \tag{4.5.1}$$

Consider a schedule for block $\mathcal{B}$ which is optimal subject to the condition that task T_l is processed only if no other task is available. This schedule consists of two complementary parts:

(i) An optimal schedule for the set $\mathcal{B} - \{T_l\}$ which decomposes into a number of sub-blocks $\mathcal{B}_1, \cdots, \mathcal{B}_b$,

(ii) A schedule for task T_l, where T_l is preemptively scheduled during the difference of time intervals given by $[r(\mathcal{B}), C(\mathcal{B})) - \bigcup_{j=1}^{b}[r(\mathcal{B}_j), C(\mathcal{B}_j))$.

For any such schedule we have

$$G_{max}(\mathcal{B}) = \max\{G_l(C(\mathcal{B})), G_{max}^*(\mathcal{B} - \{T_l\})\} \leq G_{max}(\mathcal{B}).$$

It hence follows that there is an optimal schedule in which task T_l is scheduled as described above.

The problem can now be solved in the following way. First, order the tasks according to non-decreasing r_j. Next, determine the initial block structure by scheduling the tasks in order of non-decreasing r_j. For each block $\mathcal{B}$, select task $T_l \in \mathcal{B}$ subject to (4.5.1). Determine the block structure for the set $\mathcal{B} - \{T_l\}$ by scheduling the tasks in this set in order of non-decreasing r_j, and construct the schedule for task T_l as described above. By repeated application of this procedure to each of the sub-blocks one obtains an optimal schedule. The algorithm is as follows.

Algorithm 4.5.1 *for problem* $1 \mid pmtn, r_j \mid G_{max}$ [BLL+83].
Method: The algorithm recursively uses procedure *oneblock* which is applied to blocks of tasks as described above.

Procedure *oneblock*$(\mathcal{B} \subseteq \mathcal{T})$;
begin
Select task $T_l \in \mathcal{B}$ such that $G_l(C(\mathcal{B})) = \min_{T_j \in \mathcal{B}} \{G_j(C(\mathcal{B}))\}$;

Determine sub-blocks $\mathcal{B}_1, \cdots, \mathcal{B}_b$ of the set $\mathcal{B} - \{T_l\}$;

Schedule task T_l in the intervals $[r(\mathcal{B}), C(\mathcal{B})) - \bigcup_{j=1}^{b} [r(\mathcal{B}_j), C(\mathcal{B}_j))$;

for $j := 1$ **to** b **do call** *oneblock*$(\mathcal{B}_j)$;
end;

begin   -- main algorithm
Order tasks so that $r_1 \le r_2 \le \cdots \le r_n$;
oneblock$(\mathcal{T})$;
end;

We just mention that the time complexity of Algorithm 4.5.1 can be proved to be $O(n^2)$. Another fact is that the algorithm generates at most $n-1$ preemptions. This is easily proved by induction: It is obviously true for $n = 1$. Suppose it is true for blocks of size smaller than $|\mathcal{B}|$. The schedule for block $\mathcal{B}$ contains at most $|\mathcal{B}_i| - 1$ preemptions for each sub-block $\mathcal{B}_i$, $i = 1, \cdots, b$, and at most b preemptions for the selected tasks T_l. Hence, and also considering the fact that $T_l \notin \bigcup_{i=1}^{b} \mathcal{B}_i$, we see that the total number of preemptions is no more than $\sum_{i=1}^{b} (|\mathcal{B}_i| - 1) + b = |\mathcal{B}| - 1$. This bound on the number of preemptions is best possible. It is achieved by the class of problem instances defined by $r_j = j$, $p_j = 2$, $G_j(t) = 0$ if $t \le 2n - j$, and $G_j(t) = 1$ otherwise $(j = 1, \cdots, n)$. The only way to incur zero cost is to schedule task T_j in the intervals $[j-1, j)$ and $[2n-j, 2n-j+1)$, $j = 1, \cdots, n$. This uniquely optimal schedule contains $n-1$ preemptions.

Note that the use of preemptions is essential in the algorithm. If no preemption is allowed, it is not possible to determine the block structure of an optimal schedule in advance.

Problem $1 \mid prec \mid G_{max}$

Suppose now that the order of task execution is restricted by given precedence constraints $\prec$ and tasks are processed without preemption. Problems of this type can be optimally solved by an algorithm presented by Lawler [Law73]. The basic idea of the algorithm is as follows: From among all tasks that are eligible to be scheduled last, i.e. those without successors under the precedence relation $\prec$, put that task last that will incur the smallest cost in that position. Then repeat this procedure on the set of $n-1$ remaining tasks, etc. This rule is justified as follows: Let $T = \{T_1, \cdots, T_n\}$ be the set of all tasks, and let $L \subseteq T$ be the subset of tasks without successors. For any $T' \in T$ let $G^*(T')$ be the maximum task completion cost in an optimal schedule for T'. If p denotes the completion time of the last task, i.e. $p = p_1 + p_2 + \cdots + p_n$, task $T_l \in L$ is chosen such that $G_l(p) = \min_{T_j \in L} \{G_j(p)\}$. Then the optimal value of a schedule subject to the condition that task T_l is processed last is given by $\max\{G^*(L-\{T_l\}), G_l(p)\}$. Since both, $G^*(L-\{T_l\}) \le G^*(L)$ and $G_l(p) \le G^*(L)$, the rule is proved.

The following algorithm finds a task that can be placed last in schedule S. Then, having this task removed from the problem, the algorithm determines a task that can be placed last among the remaining $n-1$ tasks and second-to-last in the complete schedule, and so on.

Algorithm 4.5.2 *for problem* $1 \mid prec \mid G_{max}$ [Law73].
```
begin
Let S be the empty schedule;
while T ≠ ∅ do
    begin
    p :=  Σ   pj;
         Tj ∈ T
    Let L ⊆ T be the subset of tasks with no successors;
    Choose task Tk ∈ L such that Gk(p) = min {Gj(p)};
                                           Tj ∈ L
    S := Tk ⊕ S;      -- task Tk is placed in front of the first element of schedule S
    T := T - {Tk};
    end;
end;
```

Notice that this algorithm requires $O(n^2)$ steps, where n is the number of tasks.

Example 4.5.3 Suppose there are five tasks $\{T_1, \cdots, T_5\}$ with processing times $p = [1, 2, 2, 2, 3]$ and precedence constraints as shown in Figure 4.5.1(a), and cost functions as indicated in Figure 4.5.1(b). The last task in a schedule for this problem will finish at time $p = 10$. Among the tasks having no successors the algorithm chooses T_3 to be placed last because $G_3(10)$ is minimum. Note that in the final schedule, T_3 will be started at time 8. Among the remaining tasks, $\{T_1, T_2, T_4, T_5\}$, T_4 and T_5 have no successors, so these two tasks are the candidates for being placed immediately before T_3. The algorithm chooses T_5 because at time 8 this task incurs lower cost to the schedule. Continuing this way Algorithm 4.5.2 will terminate with the schedule $(T_2, T_1, T_4, T_5, T_3)$. □

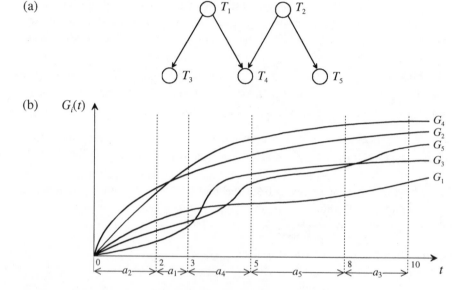

Figure 4.5.1 *An example problem for Algorithm 4.5.2*
(a) task set with precedence constraints,
(b) cost functions specifying penalties associated with task completion times.

Problem $1 \mid pmtn, prec, r_j \mid G_{max}$

In case $1 \mid pmtn, prec, r_j \mid G_{max}$, i.e. if preemptions are permitted, the problem is much easier. Baker et al. [BLL+83] presented an algorithm which is an extension of Algorithm 4.5.1. First, release dates are modified so that $r_j + p_j \leq r_k$ whenever T_j precedes T_k. This is being done by replacing r_k by $\max\{r_k, \max\{r_j + p_j \mid T_j \prec T_k\}\}$ for $k = 2, \cdots, n$. The block structures are obtained as in Algorithm 4.5.1. As the block structures are determined by scheduling tasks in order of non-decreasing values of r_j, this implies that we can ignore precedence constraints at that level. Then, for each block $\mathcal{B}$, the subset $\mathcal{L} \subseteq \mathcal{B}$ of tasks that have no successor in $\mathcal{B}$ is determined. The selection of task $T_l \in \mathcal{B}$ subject to equation (4.5.1) is replaced by the selection of task $T_l \in \mathcal{B}$ such that $G_l(C(\mathcal{B})) = \min_{T_j \in \mathcal{L}} \{G_j(C(\mathcal{B}))\}$.

This ensures that the selected task has no successors within block $\mathcal{B}$. We mention that this algorithm can still be implemented to run in $O(n^2)$ time.

Example 4.5.4 [BLL+83] To illustrate the last algorithm consider five tasks $\{T_1, \cdots, T_5\}$ whose processing times and release times are given by the vectors $p = [4, 2, 4, 2, 4]$ and $r = [0, 2, 0, 8, 14]$, respectively. The precedence constraints and cost functions are specified in Figure 4.5.2(a) and (b), respectively. From the precedence constraints we obtain the modified release dates $r' = [0, 2, 4, 8, 14]$. Taking modified release dates instead of r, Algorithm 4.5.1 determines two blocks, $\mathcal{B}_1 = \{T_1, T_2, T_3, T_4\}$ from time 0 to 12, and $\mathcal{B}_2 = \{T_5\}$ from 14 until 18 (Figure 4.5.2(c)). Block $\mathcal{B}_2$ consists of a single task and therefore represents an optimal part of the schedule. For block $\mathcal{B}_1$, we find the subset of tasks without successors $\mathcal{L}_1 = \{T_3, T_4\}$ and select task T_3 since $G_3(12) < G_4(12)$. By re-scheduling the tasks in $\mathcal{B}_1$ while processing task T_3 (only if no other task is available), we obtain two sub-blocks: $\mathcal{B}_{11} = \{T_1, T_2\}$ from time 0 to 6, and $\mathcal{B}_{12} = \{T_4\}$ from 8 until 10 (viz. Figure 4.5.2(c)). Block $\mathcal{B}_{12}$ needs no further attention. For block $\mathcal{B}_{11}$ we find $\mathcal{L}_{11} = \{T_1, T_2\}$ and select task T_1 since $G_1(6) < G_2(6)$. By rescheduling the tasks in $\mathcal{B}_{11}$ again we finally obtain an optimal schedule (Figure 4.5.2(c)). □

4.5.2 Total Cost

From [LRKB77] we know that the general problem $1 \| \Sigma G_j$ of scheduling tasks, such that the sum of values $G_j(C_j)$ is minimal, is NP-hard. If tasks have unit processing times, i.e. for $1|p_j = 1|\Sigma G_j$, the problem is equivalent to finding a permutation $(\alpha_1, \cdots, \alpha_n)$ of the task indices $1, \cdots, n$ that minimizes $\Sigma G_j(C_{\alpha_j})$. This is a weighted bipartite matching problem, which can be solved in $O(n^3)$ time [LLR+89]. For the case of arbitrary processing times, Rinnooy Kan et al. [RKLL75] presented a branch an bound algorithm. The computation of lower bounds on the costs of an optimal schedule follows an idea similar to that used in the $p_j = 1$ case.

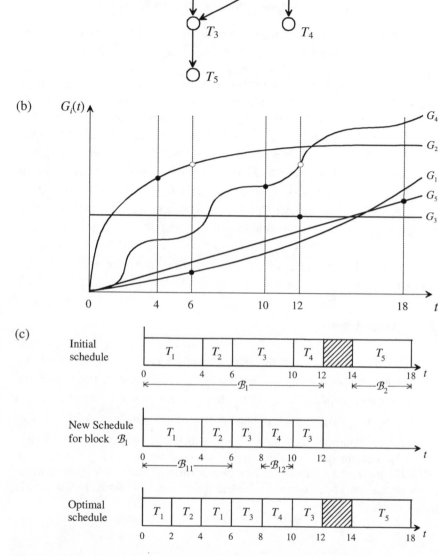

Figure 4.5.2 *An example problem $1 \mid pmtn, prec, r_j \mid G_{max}$*
  (a) *task set with precedence constraints,*
  (b) *cost functions specifying penalties associated with task completion times,*
  (c) *block schedules and an optimal preemptive schedule.*

Suppose that $p_1 \leq \cdots \leq p_n$, and define $t_k = p_1 + \cdots + p_k$ for $k = 1, \cdots, n$. Then $G_j(t_k)$ is a lower bound on the cost of scheduling T_j in position k, and an overall lower bound is obtained by solving the weighted bipartite matching problem with coefficients $G_j(t_k)$. In addition to lower bounds, elimination criteria are used to discard partial schedules in the search tree. These criteria are generally of the form: if the cost functions and processing times of T_i and T_j satisfy a certain relationship, then there is an optimal schedule in which T_i precedes T_j.

A number of results are available for special kinds of cost functions. If all cost functions depend linearly on the task completion times, Smith [Smi56] proved that an optimal schedule is obtained by scheduling the tasks in order of non-decreasing values of $G_j(p)/p_j$ where $p = \Sigma p_j$.

For the case that cost of each task T_j is a quadratic function of its completion time, i.e. $G_j(C_j) = c_j C_j^2$ for some constant c_j, branch and bound algorithms were developed by Townsend [Tow78] and by Bagga and Kalra [BK81]. Both make use of task interchange relations similar to those discussed in Section 4.2 (see equation (4.2.1)) to obtain sufficient conditions for the preference of a task T_i over another task T_j. For instance, following [BK81], if $c_i \geq c_j$ and $p_i \leq p_j$ for tasks T_i, T_j, then there will always be a schedule where T_i is performed prior to T_j, and whose total cost will not be greater than the cost of any other schedule where T_i is started later than T_j. Such a rule can be obviously used to reduce the number of created nodes in the tree of a branch and bound procedure.

A similar problem was discussed by Gupta and Sen [GS83] where each task has a given due date, and the objective is to minimize the sum of squares of lateness values, $\sum_{j=1}^{n} L_j^2$. If tasks can be arranged in a schedule such that every pair of adjacent tasks T_i, T_j (i.e. T_i is executed immediately before T_j) satisfies the conditions

$$p_i \leq p_j \quad \text{and} \quad \frac{d_i}{p_i} \leq \frac{d_j}{p_j},$$

then the schedule can be proved to be optimal. For general processing times and due dates, a branch and bound algorithm was presented in [GS83].

The problems of minimizing total cost are equivalent to maximization problems where each task is assigned a *profit* that urges tasks to finish as early as possible. The profit of a task is described by a non-increasing and concave function G_j on the finishing time of the task. Fisher and Krieger [FK84] discussed a class of heuristics for scheduling n tasks on a single processor to maximize the sum of profits $\Sigma G_j(C_j - p_j)$. The heuristic used in [FK84] is based on linear approximations of the functions G_j. Suppose several tasks have already been scheduled for processing in the interval $[0, t)$, and we must choose one of the remaining tasks to start at time t. Then the approximation of G_j is the linear function through the points $(t, C_j(t))$ and $(p, C_j(p))$ where $p = \sum_{j=1}^{n} p_j$. The task chosen maximizes $(C_j(t) - C_j(p))/t$. The main result presented in [FK84] is that the heuristic always obtains at least 2/3 of the optimal profit.

Finally we mention that there are few results available for the case that, in addition to the previous assumptions, precedence constraints restrict the order of task execution. For the total weighted exponential cost function criterion $\sum_{j=1}^{n} w_j \exp(-cC_j)$), where c is some given "discount rate", Monma and Sidney [MS87] were able to prove that the job module property (see end of Section 4.2) is satisfied. As a consequence, for certain classes of precedence constraints that are built up iteratively from prime modules, the problem $1 \,|\, prec \,|\, \Sigma\,(w_j \exp(-cC_j))$ can be solved in polynomial time. As an example, series-parallel precedence constraints are of that property. For more details we refer the reader to [MS87].

Dynamic programming algorithms for general precedence constraints and for the special case of series-parallel precedence graphs can be found in [BS78a, BS78b, and BS81], where each task is assigned an arbitrary cost function that is non-negative and non-decreasing in time.

References

AD90   H. H. Ali, J. S. Deogun, A polynomial algorithm to find the jump number of interval orders, Preprint, Univ. of Nebraska Lincoln, 1990.

AH73   D. Adolphson, T. C. Hu, Optimal linear ordering, *SIAM J. Appl. Math.* 25, 1973, 403-423.

BA87   U. Bagchi, R. H. Ahmadi, An improved lower bound for minimizing weighted completion times with deadlines, *Oper. Res.* 35, 1987, 311-313.

Ban80   S. P. Bansal, Single machine scheduling to minimize weighted sum of completion times with secondary criterion - a branch-and-bound approach, *European J. Oper. Res.* 5, 1980, 177-181.

BD78   J. Bruno, P. Downey, Complexity of task sequencing with deadlines, set-up times and changeover costs, *SIAM J. Comput.* 7, 1978, 393-404.

BFR71   P. Bratley, M. Florian, P. Robillard, Scheduling with earliest start and due date constraints, *Naval Res. Logist. Quart.* 18, 1971, 511-517.

BFR73   P. Bratley, M. Florian, P. Robillard, On sequencing with earliest starts and due dates with application to computing bounds for the (n/m/G/Fmax) problem, *Naval Res. Logist. Quart.* 20, 1973, 57-67.

BH89   V. Bouchitte, M. Habib, The calculation of invariants of ordered sets, in: I. Rival (ed.), *Algorithms and Order*, Kluwer, Dordrecht, 1989, 231-279.

BK81   P. C. Bagga, K. R. Kalra, Single machine scheduling problem with quadratic functions of completion time - a modified approach, *J. Inform. Optim. Sci.* 2, 1981, 103-108.

Bla76   J. Błażewicz, Scheduling dependent tasks with different arrival times to meet
          deadlines, in: E. Gelenbe, H. Beilner (eds.), *Modelling and Performance
          Evaluation of Computer Systems*, North Holland, Amsterdam, 1976, 57-65.

BLL+83   K. R. Baker, E. L. Lawler, J. K. Lenstra, A. H. G. Rinnooy Kan, Preemptive
          scheduling of a single machine to minimize maximum cost subject to release
          dates and precedence constraints, *Oper. Res.* 31, 1983, 381-386.

BM83   H. Buer, R. H. Möhring, A fast algorithm for the decomposition of graphs and
          posets, *Math. Oper. Res.* 8, 1983, 170-184.

BR82   L. Bianco, S. Ricciardelli, Scheduling of a single machine to minimize total
          weighted completion time subject to release dates, *Naval Res. Logist. Quar-
          terly.* 29, 1982, 151-167.

BS74   K. R. Baker, Z.-S. Su, Sequencing with due dates and early start times to
          minimize maximum tardiness, *Naval Res. Logist. Quart.* 21, 1974, 171-176.

BS78a   K. R. Baker, L. Schrage, Dynamic programming solution for sequencing prob-
          lems with precedence constraints, *Oper. Res.* 26, 1978, 444-449.

BS78b   K. R. Baker, L. Schrage, Finding an optimal sequence by dynamic program-
          ming: An extension to precedence related tasks, *Oper. Res.* 26, 1978, 111-120.

BS81   R. N. Burns, G. Steiner, Single machine scheduling with series-parallel prece-
          dence constraints, *Oper. Res.* 29, 1981, 1195-1207.

Bur76   R. N. Burns, Scheduling to minimize the weighted sum of completion times
          with secondary criteria, *Naval Res. Logist. Quart.* 23, 1976, 25-129.

BY82   G .R. Bitran, H. H. Yanasse, Computational complexity of the capacitated lot
          size problem, *Management Sci.* 28, 1982, 1174-1186.

Car82   J. Carlier, The one-machine sequencing problem, *European J. Oper. Res.* 11,
          1982, 42-47.

CM72   M. Chein, P. Martin, Sur le nombre de sauts d'une foret, *C. R. Acad. Sc. Paris*
          275, serie A, 1972, 159-161.

CMM67   R. W. Conway, W. L. Maxwell, L. W. Miller, *Theory of Scheduling*, Addison-
          Wesley, Reading, Mass., 1967.

Cof76   E. G. Coffman, Jr. (ed.), *Scheduling in Computer and Job Shop Systems*,
          J. Wiley, New York, 1976.

CS86   S. Chand, H. Schneeberger, A note on the single-machine scheduling problem
          with minimum weighted completion time and maximum allowable tardiness,
          Naval Res. Logist. Quart. 33, 1986, 551-557.

DD81   M. I. Dessouky, J. S. Deogun, Sequencing jobs with unequal ready times to
          minimize mean flow time, *SIAM J. Comput.* 10, 1981, 192-202.

DE77   W. C. Driscoll, H. Emmons, Scheduling production on one machine with
          changeover costs, *AIIE Trans.* 9, 1977, 388-395.

DL90   J. Du, J. Y.-T. Leung, Minimizing total tardiness on one machine is NP-hard,
          Math. Oper. Res. 15, 1990, 483-495.

EFMR83   J. Erschler, G. Fontan, C. Merce, F. Roubellat, A new dominance concept in
          scheduling n jobs on a single machine with ready times and due dates,
          Oper. Res. 31, 1983, 114-127.

Emm75   H. Emmons, One machine sequencing to minimize mean flow time with mini-
           mum number tardy, *Naval Res. Logist. Quart.* 22, 1975, 585-592.

ER85   M. H. El-Zahar, I. Rival, Greedy linear extensions to minimize jumps, *Discrete
           Appl. Math.* 11, 1985, 143-156.

Fis85   P. C. Fishburn, *Interval Orders and Interval Graphs*, J. Wiley, New York,
           1985.

FK84   M. L. Fisher, A. M. Krieger, Analysis of a linearization heuristic for single
           machine scheduling to maximize profit, *Math. Programming* 28, 1984, 218-
           225.

FLRK80   M. Florian, J. K. Lenstra, A. H. G. Rinnooy Kan, Deterministic production
           planning: algorithms and complexity, *Management Sci.* 26, 1980, 669-679.

FS85a   U. Faigle, R. Schrader, A setup heuristic for interval orders, *Oper. Res. Lett.* 4,
           1985, 185-188.

FS85b   U. Faigle, R. Schrader, Interval orders without odd crowns are defect optimal,
           Report 85382-OR, University of Bonn, 1985.

FTM71   M. Florian, P. Trepant, G. McMahon, An implicit enumeration algorithm for
           the machine sequencing problem, *Management Sci.* 17, 1971, B782-B792.

GJ76   M. R. Garey, D. S. Johnson, Scheduling tasks with non-uniform deadlines on
           two processors, *J. Assoc. Comput. Mach.* 23, 1976, 461-467.

GJ79   M. R. Garey, D. S. Johnson, *Computers and Intractability: A Guide to the The-
           ory of NP-Completeness*, W. H. Freeman, San Francisco, 1979.

GJST81   M. R. Garey, D. S. Johnson, B. B. Simons, R. E. Tarjan, Scheduling unit-time
           tasks with arbitrary release times and deadlines, *SIAM J. Comput.* 10, 1981,
           256-269.

GK87   S. K. Gupta, J. Kyparisis, Single machine scheduling research, *OMEGA Inter-
           nat. J. Management Sci.* 15, 1987, 207-227.

GL78   G. V. Gens, E. V. Levner, Approximation algorithm for some scheduling prob-
           lems, *Engrg. Cybernetics* 6, 1978, 38-46.

GL81   G. V. Gens, E. V. Levner, Fast approximation algorithm for job sequencing
           with deadlines, *Discrete Appl. Math.* 3, 1981, 313-318.

GL88   A. Gascon, R. C. Leachman, A dynamic programming solution to the dynamic,
           multi-item, single-machine scheduling problem, *Oper. Res.* 36, 1988, 50-56.

Gla68   C. R. Glassey, Minimum changeover scheduling of several products on one
           machine, *Oper. Res.* 16, 1968, 342-352.

GLL+79   R. L. Graham, E. L. Lawler, J. K. Lenstra, A. H. G. Rinnooy Kan, Optimiza-
           tion and approximation in deterministic sequencing and scheduling: a survey,
           Ann. Discrete Math. 5, 1979, 287-326.

Gol80   M. C. Golumbic, *Algorithmic Graph Theory and Perfect Graphs*, Academic
           Press, New York, 1980.

GS83   S. K. Gupta, T. Sen, Minimizing the range of lateness on a single machine,
           Engrg. Costs Production Economics 7, 1983, 187-194.

GS84   S. K. Gupta, T. Sen, Minimizing the range of lateness on a single machine, *J. Oper. Res. Soc.* 35, 1984, 853-857.

GTW88   M. R. Garey, R. E. Tarjan, G. T. Wilfong, One-processor scheduling with earliness and tardiness penalties, *Math. Oper. Res.* 13, 1988, 330-348.

HKR87   T. C. Hu, Y. S. Kuo, F. Ruskey, Some optimum algorithms for scheduling problems with changeover costs, *Oper. Res.* 35, 1987, 94-99.

HKS91   N. G. Hall, W. Kubiak, S. P. Sethi, Earliness-tardiness scheduling problems, II: Deviation of completion times about a restictive commen due date, *Oper. Res.* 39, 1991, 847-856.

Hor72   W. A. Horn, Single-machine job sequencing with tree-like precedence ordering and linear delay penalties, *SIAM J. Appl. Math.* 23, 1972, 189-202.

Hor74   W. A. Horn, Some simple scheduling algorithms, *Naval Res. Logist. Quart.* 21, 1974, 177-185.

HS88   L. A. Hall, D. B. Shmoys, Jackson's rule for one-machine scheduling: Making a good heuristic better, Department of Mathematics, Massachusetts Institute of Technology, Cambridge, 1988.

IIN81   T. Ichimori, H. Ishii, T. Nishida, Algorithm for one machine job sequencing with precedence constraints, *J. Oper. Res. Soc. Japan* 24, 1981, 159-169.

IK78   O. H. Ibarra, C. E. Kim, Approximation algorithms for certain scheduling problems, *Math. Oper. Res.* 3, 1978, 197-204.

Jac55   J. R. Jackson, Scheduling a production line to minimize maximum tardiness, Research Report 43, Management Sci. Res. Project, UCLA, 1955.

Kan81   J. J. Kanet, Minimizing the average deviation of job completion times about a common due date, *Naval Res. Logist. Quart.* 28, 1981, 643-651.

Kar72   R. M. Karp, Reducibility among combinatorial problems, in: R. E. Miller, J. W. Thatcher (eds.), *Complexity of Computer Computations*, Plenum Press, New York, 1972, 85-103.

KIM78   H. Kise, T. Ibaraki, H. Mine, A solvable case of a one-machine scheduling problem with ready and due times, *Oper. Res.* 26, 1978, 121-126.

KIM79   H. Kise, T. Ibaraki, H. Mine, Performance analysis of six approximation algorithms for the one-machine maximum lateness scheduling problem with ready times, *J. Oper. Res. Soc. Japan* 22, 1979, 205-224.

KK83   K. R. Kalra, K. Khurana, Single machine scheduling to minimize waiting cost with secondary criterion, *J. Math. Sci.* 16-18, 1981-1983, 9-15.

KLS90   W. Kubiak, S. Lou, S. Sethi, Equivalence of mean flow time problems and mean absolute deviation problems, *Oper. Res. Lett.* 9, 1990, 371-374.

Kub93   W. Kubiak, Completion time variance minimization on a single machine is difficult, *Oper. Res. Lett.* 14, 1993, 49-59.

Kub95   W. Kubiak, New results on the completion time varaince minimization, *Discrete Appl. Math.* 58, 1995, 157-168.

Law64   E. L. Lawler, On scheduling problems with deferral costs, *Management Sci.* 11, 1964, 280-288.

Law73   E. L. Lawler, Optimal sequencing of a single machine subject to precedence constraints, *Management Sci.* 19, 1973, 544-546.

Law76   E. L. Lawler, Sequencing to minimize the weighted number of tardy jobs, *RAIRO Rech. Opér.* 10, 1976, Suppl. 27-33.

Law77   E. L. Lawler, A 'pseudopolynomial' algorithm for sequencing jobs to minimize total tardiness, *Ann. Discrete Math.* 1, 1977, 331-342.

Law78   E. L. Lawler, Sequencing jobs to minimize total weighted completion time subject to precedence constraints, *Ann. Discrete Math.* 2, 1978, 75-90.

Law82   E. L. Lawler, Sequencing a single machine to minimize the number of late jobs, Preprint, Computer Science Division, University of California, Berkeley, 1982.

Law83   E. L. Lawler, Recent results in the theory of machine scheduling, in: A. Bachem, M. Grötschel, B. Korte (eds.), *Mathematical Programming: The State of the Art*, Springer, Berlin, 1983, 202-234.

LD78   R. E. Larson, M. I. Dessouky, Heuristic procedures for the single machine problem to minimize maximum lateness, *AIIE Trans.* 10, 1978, 176-183.

LDD85   R. E. Larson, M. I. Dessouky, R. E. Devor, A forward-backward procedure for the single machine problem to minimize maximum lateness, *IIE Trans.* 17, 1985, 252-260.

Len77   J. K. Lenstra, *Sequencing by Enumerative Methods*, Mathematical Centre Tract 69, Mathematisch Centrum, Amsterdam, 1977.

LLL+84   J. Labetoulle, E. L. Lawler, J. K. Lenstra, A. H. G. Rinnooy Kan, Preemptive scheduling of uniform machines subject to release dates, in: W. R. Pulleyblank (ed.), *Progress in Combinatorial Optimization*, Academic Press, New York, 1984, 245-261.

LLRK76   B. J. Lageweg, J. K. Lenstra, A. H. G. Rinnooy Kan, Minimizing maximum lateness on one machine: Computational experience and some applications, *Statist. Neerlandica* 30, 1976, 25-41.

LLRK82   E. L. Lawler, J. K. Lenstra, A. H. G. Rinnooy Kan, Recent development in deterministic sequencing and scheduling: a survey, in: M. A. H. Dempster, J. K. Lenstra, A. H. G Rinnooy Kan (eds.), *Deterministic and Stochastic Scheduling*, Reidel, Dordrecht. 1982, 35-73.

LLR+93   E. L. Lawler, J. K. Lenstra, A. H. G. Rinnooy Kan, D. B. Shmoys, Sequencing and scheduling: Algorithms and complexity, in: S. C. Graves, A. H. G. Rinnooy Kan, P. H. Zipkin (eds.), *Handbook in Operations Research and Management Science, Vol. 4: Logistics of Production and Inventory*, Elsevier, Amsterdam, 1993.

LM69   E. L. Lawler, J. M. Moore, A functional equation and its application to resource allocation and sequencing problems, *Management Sci.* 16, 1969, 77-84.

LRK73   J. K. Lenstra, A. H. G. Rinnooy Kan, Towards a better algorithm for the job-shop scheduling problem - I, Report BN 22, 1973, Mathematisch Centrum, Amsterdam.

LRK78   J. K. Lenstra, A. H. G. Rinnooy Kan, Complexity of scheduling under precedence constraints, *Oper. Res.* 26, 1978, 22-35.

LRK80   J. K. Lenstra, A. H. G. Rinnooy Kan, Complexity results for scheduling chains on a single machine, *European J. Oper. Res.* 4, 1980, 270-275.

LRKB77   J. K. Lenstra, A. H. G. Rinnooy Kan, P. Brucker, Complexity of machine scheduling problems, *Ann. Discrete Math.* 1, 1977, 343-362.

McN59   R. McNaughton, Scheduling with deadlines and loss functions, *Management Sci.* 6, 1959, 1-12.

MF75   G. B. McMahon, M. Florian, On scheduling with ready times and due dates to minimize maximum lateness, *Oper. Res.* 23, 1975, 475-482.

Mit72   S. Mitsumori, Optimal production scheduling of multicommodity in flow line, *IEEE Trans. Syst. Man Cybernet.* CMC-2, 1972, 486- 493.

Miy81   S. Miyazaki, One machine scheduling problem with dual criteria, *J. Oper. Res. Soc. Japan* 24, 1981, 37-51.

Moe89   R. H. Möhring, Computationally tractable classes of ordered sets, in: I. Rival (ed.), *Algorithms and Order*, Kluwer, Dordrecht, 1989, 105-193.

Mon82   C. L. Monma, Linear-time algorithms for scheduling on parallel processors, *Oper. Res.* 30, 1982, 116-124.

Moo68   J. M. Moore, An n job, one machine sequencing algorithm for minimizing the number of late jobs, *Management Sci.* 15, 1968, 102-109.

MR85   R. H. Möhring, F. J. Radermacher, Generalized results on the polynomiality of certain weighted sum scheduling problems, *Methods of Oper. Res.* 49, 1985, 405-417.

MS87   C. L. Monma, J. B. Sidney, Optimal sequencing via modular decomposition: Characterization of sequencing functions, *Math. Oper. Res.* 12, 1987, 22-31.

MS89   J. H. Muller, J. Spinrad, Incremental modular decomposition, *J. Assoc. Comput. Mach.* 36, 1989, 1-19.

MV90   T. L. Magnanti, R. Vachani, A strong cutting plane algorithm for production scheduling with changeover costs, *Oper. Res.* 38, 1990, 456-473.

Pos85   M. E. Posner, Minimizing weighted completion times with deadlines, *Oper. Res.* 33, 1985, 562-574.

Pot80a   C. N. Potts, An algorithm for the single machine sequencing problem with precedence constraints, *Math. Programming Study* 13, 1980, 78-87.

Pot80b   C. N. Potts, Analysis of a heuristic for one machine sequencing with release dates and delivery times, *Oper. Res.* 28, 1980, 1436-1441.

Pot85   C. N. Potts, A Lagrangian based branch and bound algorithm for a single machine sequencing with precedence constraints to minimize total weighted completion time, *Management Sci.* 31, 1985, 1300-1311.

Pul75   W. R. Pulleyblank, On minimizing setups in precedence constrained scheduling, Report 81105-OR, University of Bonn, 1975.

PW83   C. N. Potts, L. N. van Wassenhove, An algorithm for single machine sequencing with deadlines to minimize total weighted completion time, *European J. Oper. Res.* 12, 1983, 379-387.

Rag86   M. Raghavachari, A V-shape property of optimal schedule of jobs about a common due date, *European J. Oper. Res.* 23, 1986, 401-402.

RDS87   F. M. E. Raiszadeh, P. Dileepan, T. Sen, A single machine bicriterion scheduling problem and an optimizing branch-and-bound procedure, *J. Inform. Optim. Sci.* 8, 1987, 311-321.

RKLL75   A. H. G. Rinnooy Kan, B. J. Lageweg, J. K. Lenstra, Minimizing total costs in one-machine scheduling, *Oper. Res.* 23, 1975, 908-927.

RZ86   I. Rival, N. Zaguiga, Constructing greedy linear extensions by interchanging chains, *Order* 3, 1986, 107-121.

Sah76   S. Sahni, Algorithms for scheduling independent tasks, *J. Assoc. Comput. Mach.* 23, 1976, 116-127.

Sch71   L. E. Schrage, Obtaining optimal solutions to resource constrained network scheduling problems, *AIIE Systems Engineering Conference,* Phoenix, Arizona, 1971.

Sch82   L. E. Schrage, The multiproduct lot scheduling problem, in: M. A. H. Dempster, J. K. Lenstra, A. H. G Rinnooy Kan (eds.), *Deterministic and Stochastic Scheduling,* Reidel, Dordrecht, 1982.

Sch92   G. Schmidt, Minimizing changeover costs on a single machine, in: W. Bühler, F. Feichtinger, F.-J. Radermacher, P. Feichtinger (eds.), *DGOR Proceedings* 90, Vol. 1, Springer, 1992, 425-432.

Sid73   J. B. Sidney, An extension of Moore's due date algorithm, in: S. E. Elmaghraby (ed.), *Symposium on the Theory of Scheduling and Its Applications,* Springer, Berlin, 1973, 393-398.

Sid75   J. B. Sidney, Decomposition algorithms for single-machine sequencing with precedence relations and deferral costs, *Oper. Res.* 23, 1975, 283-298.

Sim78   B. Simons, A fast algorithm for single processor scheduling, *Proc. 19th Annual IEEE Symp. Foundations of Computer Science,* 1978, 50-53.

Smi56   W. E. Smith, Various optimizers for single-stage production, *Naval Res. Logist. Quart.* 3, 1956, 59-66.

SS86   J. B. Sidney, G. Steiner, Optimal sequencing by modular decomposition: polynomial algorithms, *Oper. Res.* 34, 1986, 606-612.

Tow78   W. Townsend, The single machine problem with quadratic penalty function of completion times: A branch and bound solution, *Management Sci.* 24, 1978, 530-534.

VB83   F. J. Villarreal, R. L. Bulfin, Scheduling a single machine to minimize the weighted number of tardy jobs, *AIIE Trans.* 15, 1983, 337-343.

5 Scheduling on Parallel Processors

This chapter is devoted to the analysis of scheduling problems in a parallel processor environment. As before the three main criteria to be analyzed are schedule length, mean flow time and lateness. Then, some more developed models of multiprocessor systems are described, imprecise computations and lot size scheduling. Corresponding results are presented in the four following sections.

5.1 Minimizing Schedule Length

In this section we will analyze the schedule length criterion. Complexity analysis will be complemented, wherever applicable, by a description of the most important approximation as well as enumerative algorithms. The presentation of the results will be divided into subcases depending on the type of processors used, the type of precedence constraints, and to a lesser extent task processing times and the possibility of task preemption.

5.1.1   Identical Processors

Problem $P \| C_{max}$

The first problem considered is $P \| C_{max}$ where a set of independent tasks is to be scheduled on identical processors in order to minimize schedule length. We start with complexity analysis of this problem which leads to the conclusion that the problem is not easy to solve, since even simple cases such as scheduling on two processors can be proved to be NP-hard [Kar72].

Theorem 5.1.1 *Problem $P2 \| C_{max}$ is NP-hard.*

Proof. As a known NP-complete problem we take PARTITION [Kar72] which is formulated as follows.

> *Instance:* Finite set $\mathcal{A}$ and a size $s(a_i) \in \mathbb{N}$ for each $a_i \in \mathcal{A}$.
> *Answer:* "Yes" if there exists a subset $\mathcal{A}' \subseteq \mathcal{A}$ such that
> $$\sum_{a_i \in \mathcal{A}'} s(a_i) = \sum_{a_i \in \mathcal{A} - \mathcal{A}'} s(a_i).$$
> Otherwise "No".

Given any instance of PARTITION defined by the positive integers $s(a_i)$, $a_i \in \mathcal{A}$, we define a corresponding instance of the decision counterpart of $P2 \| C_{max}$ by assuming $n = |\mathcal{A}|$, $p_j = s(a_j)$, $j = 1, 2, \cdots, n$, and a threshold value for the schedule length, $y = \frac{1}{2} \sum_{a_i \in \mathcal{A}} s(a_i)$. It is obvious that there exists a subset $\mathcal{A}'$ with the desired property for the instance of PARTITION if and only if, for the corresponding instance of $P2 \| C_{max}$, there exists a schedule with $C_{max} \leq y$ (cf. Figure 5.1.1). This proves the theorem. $\qquad\qquad\qquad\qquad\qquad\qquad\qquad\qquad\qquad$ □

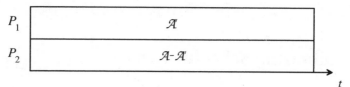

Figure 5.1.1 *A schedule for Theorem* 5.1.1.

Since there is no hope of finding an optimization polynomial time algorithm for $P \| C_{max}$, one may try to solve the problem along the lines presented in Section 3.2. First, one may try to find an approximation algorithm for the original problem and evaluate its worst case as well as its mean behavior. We will present such an analysis below.

One of the most often used general approximation strategies for solving scheduling problems is *list scheduling*, whereby a priority list of the tasks is given, and at each step the first available processor is selected to process the first available task on the list [Gra66] (cf. Section 3.2). The accuracy of a given list scheduling algorithm depends on the order in which tasks appear on the list. One of the simplest algorithms is the *LPT algorithm* in which the tasks are arranged in order of non-increasing p_j.

Algorithm 5.1.2 *LPT Algorithm for* $P \| C_{max}$.

begin
Order tasks on a list in non-increasing order of their processing times;
       -- i.e. $p_1 \geq \cdots \geq p_n$
for $i = 1$ **to** m **do** $s_i := 0$;
       -- processors P_i are assumed to be idle from time $s_i = 0$ on, $i = 1, \cdots, m$
$j := 1$;
repeat
       $s_k := \min\{s_i\}$;
       Assign task T_j to processor P_k at time s_k;
             -- the first non-assigned task from the list is scheduled on the first processor
             -- that becomes free

$s_k := s_k + p_j;\ j := j+1;$
until $j = n+1;$ -- all tasks have been scheduled
end;

It is easy to see that the time complexity of this algorithm is $O(n\log n)$ since its most complex activity is to sort the set of tasks. The worst case behavior of the *LPT* rule is analyzed in Theorem 5.1.3.

Theorem 5.1.3 [Gra69] *If the LPT algorithm is used to solve problem $P\|C_{max}$, then*

$$R_{LPT} = \frac{4}{3} - \frac{1}{3m}. \tag{5.1.1}$$

□

Space limitations prevent us from including here the proof of the upper bound in the above theorem. However, we will give an example showing that this bound can be achieved. Let $n = 2m+1$, $p = [2m-1, 2m-1, 2m-2, 2m-2, \cdots, m+1, m+1, m, m, m]$. For $m = 3$, Figure 5.1.2 shows two schedules, an optimal one and an *LPT* schedule.

We see that in the worst case an *LPT* schedule can be up to 33% longer than an optimal schedule. However, one is led to expect better performance from the *LPT* algorithm than is indicated by (5.1.1), especially when the number of tasks becomes large. In [CS76] another absolute performance ratio for the *LPT* rule was proved, taking into account the number k of tasks assigned to a processor whose last task terminates the schedule.

Theorem 5.1.4 *For the assumptions stated above, we have*

$$R_{LPT}(k) = 1 + \frac{1}{k} - \frac{1}{km}. \tag{5.1.2}$$

□

(a) 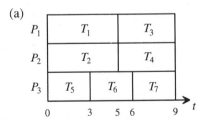   (b)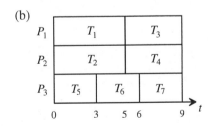

Figure 5.1.2   *Schedules for Theorem 5.1.3*
              (a) *an optimal schedule,*
              (b) *LPT schedule.*

This result shows that the worst-case performance bound for the *LPT* algorithm approaches one as fast as $1 + 1/k$.

On the other hand, it would be of interest to know how good the *LPT* algorithm is on the average. Such a result was obtained by [CFL84], where the relative error was found for two processors on the assumption that task processing times are independent samples from the uniform distribution on [0, 1].

Theorem 5.1.5 *Under the assumptions already stated, we have the following bounds for the mean value of schedule length for the LPT algorithm, $E(C_{max}^{LPT})$, for problem $P2 \| C_{max}$.*

$$\frac{n}{4} + \frac{1}{4(n+1)} \leq E(C_{max}^{LPT}) \leq \frac{n}{4} + \frac{e}{2(n+1)}, \tag{5.1.3}$$

where $e = 2.7\cdots$ is the base of the natural logarithm. □

Taking into account that $n/4$ is a lower bound on $E(C_{max}^*)$ we get

$$E(C_{max}^{LPT})/E(C_{max}^*) < 1 + O(1/n^2).$$

Therefore, as n increases, $E(C_{max}^{LPT})$ approaches the optimum no more slowly than $1 + O(1/n^2)$ approaches 1. The above bound can be generalized to cover also the case of m processors for which we have [CFL83]:

$$E(C_{max}^{LPT}) \leq \frac{n}{2m} + \left(\frac{m}{n}\right).$$

Moreover, it is also possible to prove [FRK86, FRK87] that $C_{max}^{LPT} - C_{max}^*$ almost surely converges to 0 as $n \to \infty$ if the task processing time distribution has a finite mean and a density function f satisfying $f(0) > 0$. It is also shown that if the distribution is uniform or exponential, the rate of convergence is $O(\log(\log n)/n)$. This result, obtained by a complicated analysis, can also be guessed from simulation studies. Such an experiment was reported by Kedia [Ked70] and we present the summary of the results in Table 5.1.1. The last column presents the ratio of schedule lengths obtained by the *LPT* algorithm and the optimal preemptive one. Task processing times are drawn from the uniform distribution of the given parameters.

To conclude the above analysis we may say that the *LPT* algorithm behaves quite well and may be useful in practice. However, if one wants to have better performance guarantees, other approximation algorithms should be used, as for example *MULTIFIT* introduced by Coffman et al. [CGJ78] or the algorithm proposed by Hochbaum and Shmoys [HS87]. A comprehensive treatment of approximation algorithms for this and related problems is given by Coffman et al. [CGJ84].

n, m		Intervals of task processing time distribution	C_{max}	$C_{max}^{LPT} / C_{max}^*$
6	3	1, 20	20	1.00
9	3	1, 20	32	1.00
15	3	1, 20	65	1.00
6	3	20, 50	59	1.05
9	3	20, 50	101	1.03
15	3	20, 50	166	1.00
8	4	1, 20	23	1.09
12	4	1, 20	30	1.00
20	4	1, 20	60	1.00
8	4	20, 50	74	1.04
12	4	20, 50	108	1.02
20	4	20, 50	185	1.01
10	5	1, 20	25	1.04
15	5	1, 20	38	1.03
20	5	1, 20	49	1.00
10	5	20, 50	65	1.06
15	5	20, 50	117	1.03
25	5	20, 50	198	1.01

Table 5.1.1   *Mean performance of the LPT algorithm.*

We now pass to the second way of analyzing problem $P || C_{max}$. Theorem 5.1.1 gave a negative answer to the question about the existence of an optimization polynomial time algorithm for solving $P2 || C_{max}$. However, we have not proved that our problem is NP-hard in the strong sense and we may try to find a pseudo-polynomial optimization algorithm. It appears that, based on a dynamic programming approach, such an algorithm can be constructed using ideas presented by Rothkopf [Rot66]. Below the algorithm is presented for $P || C_{max}$; it uses Boolean variables $x_j(t_1, t_2, \cdots, t_m)$, $j = 1, 2, \cdots, n$, $t_i = 0, 1, \cdots, C$, $i = 1, 2, \cdots, m$, where C denotes an upper bound on the optimal schedule length C_{max}^*. The meaning of these variables is the following

$$
x_j(t_1, t_2, \cdots, t_m) = \begin{cases} \textbf{true} & \text{if tasks } T_1, T_2, \cdots, T_j \text{ can be scheduled on} \\ & \text{processors } P_1, P_2, \cdots, P_m \text{ in such a way that } P_i \\ & \text{is busy in time interval } [0, t_i], i = 1, 2, \cdots, m, \\ \textbf{false} & \text{otherwise.} \end{cases}
$$

Now, we are able to present the algorithm.

Algorithm 5.1.6 *Dynamic programming for* $P \parallel C_{max}$ [Rot66].

```
begin
for all (t₁,t₂,···,tₘ) ∈ {0,1,···,C}ᵐ do
```
$(t_1, t_2, \cdots, t_m) \in \{0, 1, \cdots, C\}^m$ **do** $x_0(t_1, t_2, \cdots, t_m) :=$ **false;**
$x_0(0, 0, \cdots, 0) :=$ **true;**
　　-- initial values for Boolean variables are now assigned
for $j = 1$ **to** n **do**
　　for all $(t_1, t_2, \cdots, t_m) \in \{0, 1, \cdots, C\}^m$ **do**

$$x_j(t_1, t_2, \cdots, t_m) = \bigvee_{i=1}^{m} x_{j-1}(t_1, t_2, \cdots, t_{i-1}, t_i - p_j, t_{i+1}, \cdots, t_m); \qquad (5.1.4)$$

$$C_{max}^* := \min\{\max\{t_1, t_2, \cdots, t_m\} \mid x_n(t_1, t_2, \cdots, t_m) = true\}; \qquad (5.1.5)$$
　　-- optimal schedule length has been calculated
Starting from the value C_{max}^*, assign tasks $T_n, T_{n-1}, \cdots, T_1$ to appropriate
　　processors using formula (5.1.4) backwards;
end;

The above procedure solves problem $P \parallel C_{max}$ in $O(nC^m)$ time; thus for fixed m it is a pseudopolynomial time algorithm. As a consequence, for small values of m and C the algorithm can be used even in computer applications. To illustrate the use of the above algorithm let us consider the following example.

Example 5.1.7 Let $n = 3$, $m = 2$ and $p = [2, 1, 2]$. Assuming bound $C = 5$ we get the cube given in Figure 5.1.3(a) where particular values of variables $x_j(t_1, t_2, \cdots, t_m)$ are stored. In Figures 5.1.3(b) through 5.1.3(e) these values are shown, respectively, for $j = 0, 1, 2, 3$ (only true values are depicted). Following Figure 5.1.3(e) and equation (5.1.5), an optimal schedule is constructed as shown in Figure 5.1.3(f). □

The interested reader may find a survey of some other enumerative approaches for the problem in question in [LLR+89].

Problem $P \mid pmtn \mid C_{max}$

Now one may try the third way of analyzing the problem $P \parallel C_{max}$ (as suggested in Section 3.2), i.e. on may relax some constraints imposed on problem $P \parallel C_{max}$ and allow preemptions of tasks. It appears that problem $P \mid pmtn \mid C_{max}$ can be solved very efficiently. It is easy to see that the length of a preemptive schedule cannot be smaller than the maximum of two values: the maximum processing time of a task and the mean processing requirement on a processor [McN59], i.e.:

$$C_{max}^* = \max\{\max_j\{p_j\}, \frac{1}{m}\sum_{j=1}^{n} p_j\}. \qquad (5.1.6)$$

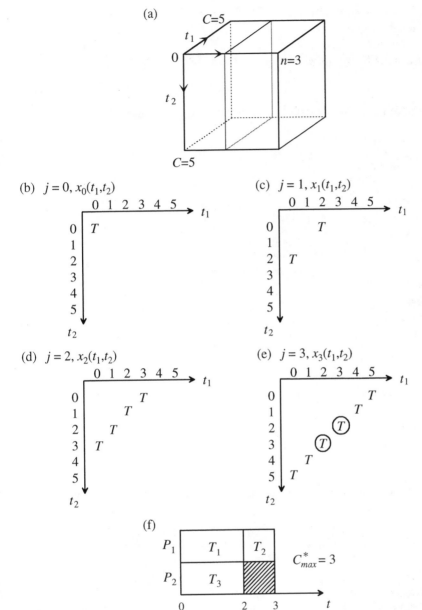

Figure 5.1.3 *An application of dynamic programming for Example 5.1.7*
(a) a cube of Boolean variables,
(b)-(e) values of $x_j(t_1,t_2)$ for $j = 0, 1, 2, 3$, respectively (here T stands for true),
(f) an optimal schedule.

The following algorithm given by McNaughton [McN59] constructs a schedule whose length is equal to C_{max}^* .

Algorithm 5.1.8 *McNaughton's rule for* $P \mid pmtn \mid C_{max}$ [McN59].

```
begin
```

$$C_{max}^* := \max\{ \sum_{j=1}^{n} p_j / m, \ \max_j \{p_j\} \}; \quad \text{-- minimum schedule length}$$

$t := 0; \ i := 1; \ j := 1;$

```
repeat
    if t+p_j ≤ C*max
    then
        begin
```
Assign task T_j to processor P_i , starting at time t;

$t := t+p_j; \ j := j+1;$

     -- task T_j can be fully assigned to processor P_i,

     -- assignment of the next task will continue at time $t + p_j$

```
        end
    else
        begin
```
Starting at time t, assign task T_j for $C_{max}^* - t$ units to processor P_i;

     -- task T_j is preempted at time C_{max}^*,

     -- processor P_i is now busy until C_{max}^*,

     -- assignment of T_j will continue on the next processor at time 0

$p_j := p_j - (C_{max}^* - t); \ t := 0; \ i := i+1;$

```
        end;
    until j = n+1;        -- all tasks have been scheduled
end;
```

Note that the above algorithm is an optimization procedure since it always finds a schedule whose length is equal to C_{max}^*. Its time complexity is $O(n)$.

We see that by allowing preemptions we made the problem easy to solve. However, there still remains the question of practical applicability of the solution obtained this way. One has to ask if this model of preemptive task scheduling can be justified, because it cannot be expected that preemptions are free of cost. Generally, two kinds of preemption costs have to be considered: time and finance. Time delays originating from preemptions are less crucial if the delay caused by a single preemption is small compared to the time the task continuously spends on the processor. Financial costs connected with preemptions, on the other hand, reduce the total benefit gained by preemptive task execution; but again, if the profit gained is large compared to the losses caused by the preemptions the schedule will be more useful and acceptable. These circumstances suggest the introduction of a scheduling model where task preemptions are only allowed after the tasks have been processed continuously for some given amount k of time. The value for k (*preemption granularity*) should be chosen large enough so that the time delay and cost overheads connected with preemptions are negligible. For given granularity k, upper bounds on the preemption overhead can easily be estimated since the number of preemptions for a task of processing time p is limited by $\lfloor p/k \rfloor$. In [EH93] the problem $P \mid pmtn \mid C_{max}$ with k-restricted preemptions is discussed: If the processing time p_j of a task T_j is less than or equal to k, then preemption is not allowed; otherwise preemption may take place after the task has been continuously processed for at least k units of time. For the remaining part of a preempted task the same condition is applied. Notice that for $k = 0$ this problem reduces to the "classical" preemptive scheduling problem. On the other hand, if for a given instance the granularity k is larger than the longest processing time among the given tasks, then no preemption is allowed and we end up with non-preemptive scheduling. Another variant is the *exact-k-preemptive* scheduling problem where task preemptions are only allowed at those moments when the task has been processed exactly an integer multiple of k time units. In [EH93] it is proved that, for $m = 2$ processors, both the k-preemptive and the exact-k-preemptive scheduling problems can be solved in time $O(n)$. For $m > 2$ processors both problems are NP-hard.

Problem $P \mid prec \mid C_{max}$

Let us now pass to the case of dependent tasks. At first tasks are assumed to be scheduled non-preemptively. It is obvious that there is no hope of finding a polynomial time optimization algorithm for scheduling tasks of arbitrary length since $P \mid\mid C_{max}$ is already NP-hard. However, one may try again list scheduling algorithms. Unfortunately, this strategy may result in an unexpected behavior of constructed schedules, since the schedule length for problem $P \mid prec \mid C_{max}$ (with arbitrary precedence constraints) may increase if:

− the number of processors increases,

− task processing times decrease,

− precedence constraints are weakened, or

− the priority list changes.

Figures 5.1.4 through 5.1.8 indicate the effects of changes of the above mentioned parameters. These *list scheduling anomalies* have been discovered by Graham [Gra66], who has also evaluated the maximum change in schedule length that may be induced by varying one or more problem parameters. We will quote this theorem since its proof is one of the shortest in that area and illustrates well the technique used in other proofs of that type. Let there be defined a task set $\mathcal{T}$ together with precedence constraints $\prec$. Let the processing times of the tasks be given by vector p, let $\mathcal{T}$ be scheduled on m processors using list L, and let the obtained value of schedule length be equal to C_{max}. On the other hand, let the above parameters be changed: a vector of processing times $p' \le p$ (for all the components), relaxed precedence constraints $\prec' \subseteq \prec$, priority list L' and the number of processors m'. Let the new value of schedule length be C'_{max}. Then the following theorem is valid.

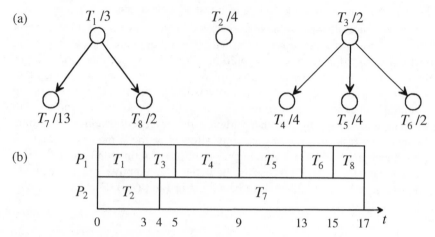

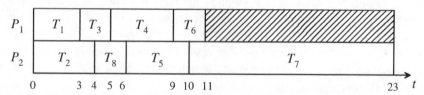

Figure 5.1.4 (a) *A task set, $m = 2$, $L = (T_1, T_2, T_3, T_4, T_5, T_6, T_7, T_8)$,*
(b) *an optimal schedule.*

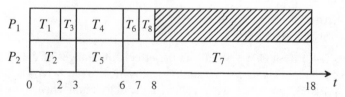

Figure 5.1.5 *Priority list changed: A new list $L' = (T_1, T_2, T_3, T_4, T_5, T_6, T_8, T_7)$.*

Figure 5.1.6 *Processing times decreased; $p'_j = p_j - 1, j = 1, 2, \cdots, n$.*

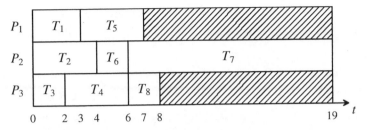

Figure 5.1.7 *Number of processors increased, m = 3.*

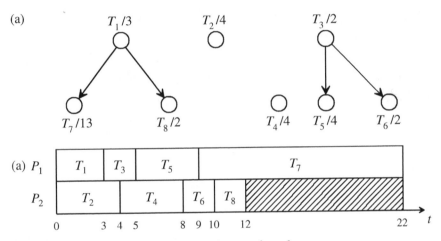

Figure 5.1.8 **(a)** *Precedence constraints weakened,*
(b) *a resulting list schedule.*

Theorem 5.1.9 [Gra66] *Under the above assumptions,*

$$\frac{C'_{max}}{C_{max}} \le 1 + \frac{m-1}{m'}. \tag{5.1.7}$$

Proof. Let us consider schedule S' obtained by processing task set $\mathcal{T}$ with primed parameters. Let the interval $[0, C'_{max})$ be divided into two subsets, $\mathcal{A}$ and $\mathcal{B}$, defined in the following way:

$$\mathcal{A} = \{t \in [0, C'_{max}) \mid \text{all processors are busy at time } t\}, \mathcal{B} = [0, C'_{max}) - \mathcal{A}.$$

Notice that both $\mathcal{A}$ and $\mathcal{B}$ are unions of disjoint half-open intervals. Let T_{j_1} denote a task completed in S' at time C'_{max}, i.e. $C_{j_1} = C'_{max}$. Two cases may occur:

1. The starting time s_{j_1} of T_{j_1} is an interior point of $\mathcal{B}$. Then by the definition of $\mathcal{B}$ there is some processor P_i which for some $\varepsilon > 0$ is idle during interval $[s_{j_1} - \varepsilon,\ s_{j_1})$. Such a situation may only occur if we have $T_{j_2} \prec' T_{j_1}$ and $C_{j_2} = s_{j_1}$ for some task T_{j_2}.

2. The starting time of T_{j_1} is not an interior point of $\mathcal{B}$. Let us also suppose that $s_{j_1} \neq 0$. Define $x_1 = \sup\{x \mid x < s_{j_1},\ \text{and}\ x \in \mathcal{B}\}$ or $x_1 = 0$ if set $\mathcal{B}$ is empty. By the construction of $\mathcal{A}$ and $\mathcal{B}$, we see that $x_1 \in \mathcal{A}$, and processor P_i is idle in time interval $[x_1 - \varepsilon, x_1)$ for some $\varepsilon > 0$. But again, such a situation may only occur if some task $T_{j_2} \prec' T_{j_1}$ is processed during this time interval.

It follows that either there exists a task $T_{j_2} \prec' T_{j_1}$ such that $y \in [C_{j_2}, s_{j_1})$ implies $y \in \mathcal{A}$ or we have: $x < s_{j_1}$ implies either $x \in \mathcal{A}$ or $x < 0$.

The above procedure can be inductively repeated, forming a chain T_{j_3}, $T_{j_4}, \cdots$, until we reach task T_{j_r} for which $x < s_{j_r}$ implies either $x \in \mathcal{A}$ or $x < 0$. Hence there must exist a chain of tasks

$$T_{j_r} \prec' T_{j_{r-1}} \prec' \cdots \prec' T_{j_2} \prec' T_{j_1} \tag{5.1.8}$$

such that at each moment $t \in \mathcal{B}$, some task T_{j_k} is being processed in S'. This implies that

$$\sum_{\phi' \in S'} p'_{\phi'} \le (m'-1) \sum_{k=1}^{r} p'_{j_k} \tag{5.1.9}$$

where the sum on the left-hand side is made over all idle-time tasks ϕ' in S'. But by (5.1.8) and the hypothesis $\prec' \subseteq \prec$ we have

$$T_{j_r} \prec T_{j_{r-1}} \prec \cdots \prec T_{j_2} \prec T_{j_1}. \tag{5.1.10}$$

Hence,

$$C_{max} \ge \sum_{k=1}^{r} p_{j_k} \ge \sum_{k=1}^{r} p'_{j_k}. \tag{5.1.11}$$

Furthermore, by (5.1.9) and (5.1.11) we have

$$C'_{max} = \frac{1}{m'}\left(\sum_{k=1}^{n} p'_k + \sum_{\phi' \in S'} p'_{\phi'}\right) \le \frac{1}{m'}(m\, C_{max} + (m'-1)\, C_{max}). \tag{5.1.12}$$

It follows that

$$\frac{C'_{max}}{C_{max}} \le 1 + \frac{m-1}{m'}$$

and the theorem is proved. $\square$

From the above theorem, the *absolute performance ratio* for an arbitrary list scheduling algorithm solving problem $P||C_{max}$ can be derived.

Corollary 5.1.10 [Gra66] *For an arbitrary list scheduling algorithm LS for $P||C_{max}$ we have*

$$R_{LS} = 2 - \frac{1}{m}. \tag{5.1.13}$$

Proof. The upper bound of (5.1.13) follows immediately from (5.1.7) by taking $m' = m$ and by considering the list leading to an optimal schedule. To show that this bound is achievable let us consider the following example: $n = (m-1)m+1$, $p = [1, 1, \cdots, 1, 1, m]$, $\prec$ is empty, $L = (T_n, T_1, T_2, \cdots, T_{n-1})$ and $L' = (T_1, T_2, \cdots, T_n)$. The corresponding schedules for $m = 4$ are shown in Figure 5.1.9. $\square$

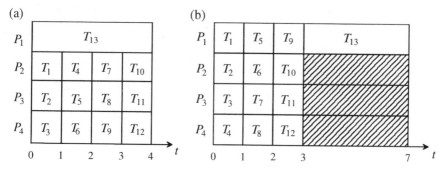

Figure 5.1.9 *Schedules for Corollary 5.1.10*
  (a) *an optimal schedule,*
  (b) *an approximate schedule.*

It follows from the above considerations that an arbitrary list scheduling algorithm can produce schedules almost twice as long as optimal ones. However, one can solve optimally problems with tasks of unit lengths.

Problem $P|prec, p_j = 1|C_{max}$

The first algorithm has been given for scheduling *forests*, consisting either of *in-trees* or of *out-trees* [Hu61]. We will first present Hu's algorithm for the case of an in-tree, i.e. for the problem $P|in$-$tree, p_j = 1|C_{max}$. The algorithm is based on the notion of a *task level* in an in-tree which is defined as the number of tasks on the path to the root of the graph. The algorithm by Hu, which is also called *level algorithm* or *critical path algorithm* is as follows.

Algorithm 5.1.11 *Hu's algorithm for $P \mid in\text{-}tree, p_j = 1 \mid C_{max}$* [Hu61].

begin
Calculate levels of the tasks;
$t := 0;$
repeat
    Construct list L_t consisting of all the tasks without predecessors at time t;
        -- all these tasks either have no predecessors
        -- or their predecessors have been assigned in time interval $[0, t{-}1]$
    Order L_t in non-increasing order of task levels;
    Assign m tasks (if any) to processors at time t from the beginning of list L_t;
    Remove the assigned tasks from the graph and from the list;
    $t := t+1;$
until all tasks have been scheduled;
end;

The algorithm can be implemented to run in $O(n)$ time. An example of its application is shown in Figure 5.1.10.

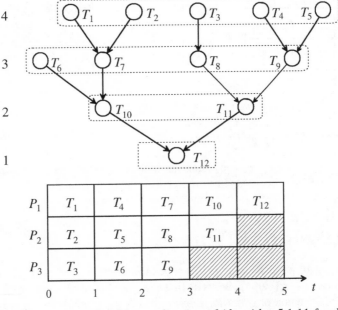

Figure 5.1.10 *An example of the application of Algorithm 5.1.11 for three processors.*

A *forest* consisting of in-trees can be scheduled by adding a dummy task that is an immediate successor of only the roots of in-trees, and then by applying Algorithm 5.1.11. A schedule for an *out-tree* can be constructed by changing the orientation of arcs, applying Algorithm 5.1.11 to the obtained in-tree and then reading the schedule backwards, i.e. from right to left.

It is interesting to note that the problem of scheduling *opposing forests* (that is, combinations of in-trees and out-trees) on an arbitrary number of processors is NP-hard [GJTY83]. However, if the number of processors is limited to 2, the problem is easily solvable even for arbitrary precedence graphs [CG72, FKN69, Gab82]. We present the algorithm given by Coffman and Graham [CG72] since it can be further extended to cover the preemptive case. The algorithm uses *labels* assigned to tasks, which take into account the levels of the tasks and the numbers of their immediate successors. The following algorithm assigns labels to the tasks, and then uses them to find the shortest schedule for problem $P2|prec$, $p_j = 1|C_{max}$.

Algorithm 5.1.12 *Algorithm by Coffman and Graham for $P2|prec, p_j = 1|C_{max}$* [CG72].

begin
Assign label 1 to any task T_0 for which $isucc(T_0) = \varnothing$;
        -- recall that $isucc(T)$ denotes the set of all immediate successors of T
$j := 1$;
repeat
    Construct set S of all unlabeled tasks whose successors are labeled;
    for all $T \in S$ **do**
        begin
        Construct list $L(T)$ consisting of labels of tasks belonging to $isucc(T)$;
        Order $L(T)$ in decreasing order of the labels;
        end;
    Order these lists in increasing lexicographic order $L(T_{[1]}) \prec \cdots \prec L(T_{[|S|]})$;
        -- see Section 2.1 for definition of $\prec$
    Assign label $j + 1$ to task $T_{[1]}$;
    $j := j + 1$;
until $j = n + 1$;   -- all tasks have been assigned labels
call Algorithm 5.1.11;
        -- here the above algorithm uses labels instead of levels when scheduling tasks
end;

A careful analysis shows that the above algorithm can be implemented to run in time which is almost linear in n and in the number of arcs in the precedence graph [Set76]; thus its time complexity is practically $O(n^2)$. An example of the application of Algorithm 5.1.12 is given in Figure 5.1.11.

It must be stressed that the question concerning the complexity of problem $Pm|prec, p_j = 1|C_{max}$ with a fixed number m of processors is still open despite the fact that many papers have been devoted to solving various subcases of precedence constraints. If tasks with unit processing times are considered, the following results are available. Problems $P3 \mid opposing\ forest, p_j = 1 \mid C_{max}$ and $Pk \mid opposing\ forest, p_j = 1 \mid C_{max}$ are solvable in time $O(n)$ [GJTY83] and $O(n^{2k-2} \log n)$ [DW85], respectively. On the other hand, if the number of available processors is variable, then this problem becomes NP-hard. Some results are also available for the subcases in which task processing times may take only two values. Problems $P2 \mid prec, p_j = 1\ or\ 2 \mid C_{max}$ and $P \mid prec, p_j = 1\ or\ k \mid C_{max}$ are NP-hard [DL88], while problems $P2 \mid tree, p_j = 1\ or\ 2 \mid C_{max}$ and $P2 \mid tree, p_j = 1\ or$ $3 \mid C_{max}$ are solvable in time $O(n \log n)$ [NLH81] and $O(n^2 \log n)$ [DL89], respectively. Arbitrary processing times result in strong NP-hardness even for the case of chains scheduled on two processors (problem $P2 \mid chains \mid C_{max}$) [DLY91].

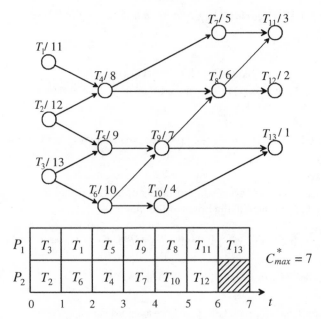

Figure 5.1.11   *An example of the application of Algorithm 5.1.12 (tasks are denoted by T_j/label).*

Furthermore, several papers deal with approximation algorithms for $P \mid prec, p_j = 1 \mid C_{max}$ and more general problems. We quote some of the most interesting results. The application of the level algorithm (Algorithm 5.1.11) to solve $P \mid prec, p_j = 1 \mid C_{max}$ has been analyzed by Chen and Liu [CL75] and Kunde [Kun76]. The following bound has been proved.

$$R_{level} = \begin{cases} \dfrac{4}{3} & \text{for } m = 2 \\[2mm] 2 - \dfrac{1}{m-1} & \text{for } m \geq 3 . \end{cases}$$

Algorithm 5.1.12 is slightly better, its bound is $R = 2 - \dfrac{2}{m} - \dfrac{m-3}{m \cdot C_{max}^{*}}$ for $m \geq 3$ [BT94]. In this context one should not forget the results presented in Theorems 5.1.9 and 5.1.10, where list scheduling anomalies have been analyzed.

Problem $P \,|\, pmtn, prec \,|\, C_{max}$

The analysis also showed that preemptions can be profitable from the viewpoint of two factors. First, they can make problems easier to solve, and second, they can shorten the schedule. Recently Coffman and Garey [CG91] proved that for problem $P2 \,|\, prec \,|\, C_{max}$ the least schedule length achievable by a non-preemptive schedule is no more than 4/3 the least schedule length achievable when preemptions are allowed. While the proof of this fact seems to be tedious, a very simple example showing that this bound is met can easily be given for a set of three independent tasks of equal length (cf. Figure 5.1.12).

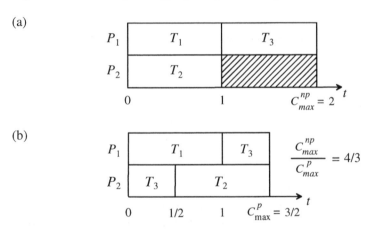

Figure 5.1.12 *An example of 4/3 conjecture*
   (a) *non-preemptive scheduling,*
   (b) *preemptive scheduling.*

In the general case of dependent tasks scheduled on processors in order to minimize schedule length, one can construct optimal preemptive schedules for tasks of arbitrary length and with other parameters the same as in Algorithm 5.1.11 or 5.1.12. The approach again uses the notion of the *level* of task T_j in a precedence graph, by which is now understood the sum of processing times (including p_j) of tasks along the longest path between T_j and a terminal task (a task with no successors). Let us note that the level of a task being executed is decreasing. We have the following algorithm [MC69, MC70] for the problems $P2\,|\,pmtn,$ $prec\,|\,C_{max}$ and $P\,|\,pmtn, forest\,|\,C_{max}$. The algorithm uses a notion of a *processor shared schedule*, in which a task receives some fraction β (≤ 1) of the processing capacity of a processor.

Algorithm 5.1.13 *Algorithm by Muntz and Coffman for* $P2\,|\,pmtn, prec\,|\,C_{max}$ *and* $P\,|\,pmtn, forest\,|\,C_{max}$ [MC69, MC70].

```
begin
for all T ∈ T do  Compute the level of task T;
t := 0; h := m;
repeat
    Construct set Z of tasks without predecessors at time t;
    while h > 0 and |Z| > 0 do
      begin
      Construct subset S of Z consisting of tasks at the highest level;
      if |S| > h
      then
        begin
        Assign β := h/|S| of a processing capacity to each of the tasks from S;
        h := 0;          -- a processor shared partial schedule is constructed
        end
      else
        begin
        Assign one processor to each of the tasks from S;
        h := h−|S|;   -- a "normal" partial schedule is constructed
        end;
      Z := Z−S;
      end;    -- the most "urgent" tasks have been assigned at time t
    Calculate time τ at which either one of the assigned tasks is finished or a
        point is reached at which continuing with the present partial assignment
        means that a task at a lower level will be executed at a faster rate β than a
        task at a higher level;
    Decrease levels of the assigned tasks by (τ−t)β;
    t := τ; h := m;
            -- a portion of each assigned task equal to (τ−t)β has been processed
until all tasks are finished;
call Algorithm 5.1.8 to re-schedule portions of the processor shared schedule
    to get a normal one;
end;
```

The above algorithm can be implemented to run in $O(n^2)$ time. An example of its application to an instance of problem $P2\,|\,pmtn, prec\,|\,C_{max}$ is shown in Figure 5.1.13.

At this point let us also consider another class of the precedence graphs for which the scheduling problem can be solved in polynomial time. To do this we have to present precedence constraints in the form of an activity network (task-on-arc precedence graph, viz. Section 3.1) whose nodes (events) are ordered in such a way that the occurrence of node i is not later than the occurrence of node j, if $i < j$.

(a)

(b)

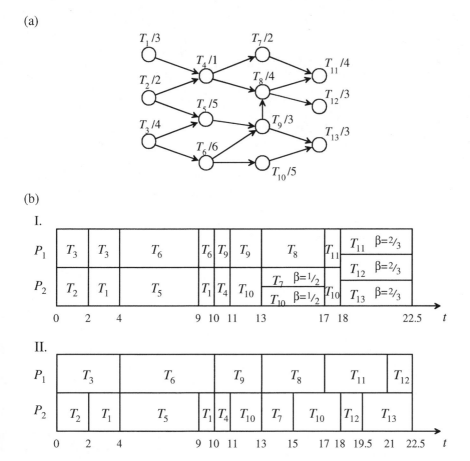

Figure 5.1.13 *An example of the application of Algorithm 5.1.13*
  (a) *a task set (nodes are denoted by* T_j/p_j*),*
  (b) I: *a processor-shared schedule,* II: *an optimal schedule.*

Now, let S_I denote the set of all the tasks which may be performed between the occurrence of event (node) I and $I+1$. Such sets will be called *main sets*. Let us consider *processor feasible sets*, i.e. those main sets and those subsets of the main sets whose cardinalities are not greater than m, and number these sets from 1 to some K. Now, let Q_j denote the set of indices of processor feasible sets in which task T_j may be performed, and let x_i denote the duration of the ith feasible set. Then, a linear programming problem can be formulated in the following way [WBCS77, BCSW76b] (another *LP* formulation for unrelated processors is presented in Section 5.1.2):

$$\text{Minimize} \qquad C_{max} = \sum_{i=1}^{K} x_i \qquad\qquad (5.1.14)$$

$$\text{subject to} \qquad \sum_{i \in Q_j} x_i = p_j, \quad j = 1, 2, \cdots, n, \qquad\qquad (5.1.15)$$

$$x_i \geq 0, \qquad i = 1, 2, \cdots, K.$$

It is clear that the solution of the *LP* problem depends on the order of nodes in the activity network; hence an optimal solution is found when this topological order is unique. Such a situation takes place for a *uniconnected activity network* (*uan*), i.e. one in which any two nodes are connected by a directed path in only one direction. An example of a uniconnected activity network together with the corresponding precedence graph is shown in Figure 5.1.14. On the other hand, the number of variables in the above *LP* problem depends polynomially on the input length, when the number of processors m is fixed. We may then use a non-simplex algorithm (e.g. from [Kha79] or [Kar84]) which solves any *LP* problem in time polynomial in the number of variables and constraints. Hence, we may conclude that the above procedure solves problem $Pm \mid pmtn, uan \mid C_{max}$ in polynomial time.

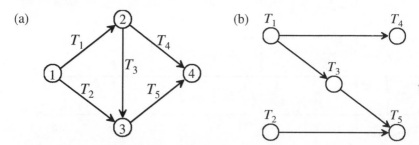

Figure 5.1.14   (a) *An example of a simple uniconnected activity network,*
                    (b) *The corresponding precedence graph.*
                    Main sets $S_1 = \{T_1, T_2\}$, $S_2 = \{T_2, T_3, T_4\}$, $S_3 = \{T_4, T_5\}$.

Recently another *LP* formulation has been proposed which enables one to solve problem $P \mid pmtn, uan \mid C_{max}$ in polynomial time, regardless of a number of processors [JMR+04].

As we already mentioned the uniconnected activity network has a task-on-node equivalent representation in a form of the interval order. Below, we present a sketch of the proof [BK02]. Let us start with the following theorem which will be given without a proof.

Theorem 5.1.14 *Let G be an activity network (task-on-arc graph). G is uniconnected if and only if G has a Hamiltonian path.*   □

Now, the following theorems may be proved [BK02].

Theorem 5.1.15 *If G is a uan, then G is a task-on-arc representation of an interval order.*

Proof. By Theorem 5.1.14, $G = (V, A)$ is composed of a Hamiltonian path $W = (v_1, \ldots, v_n)$ with possibly some additional arcs of the form (v_i, v_j) with $i < j$. The interval order we are looking for is defined by the following collection of intervals $(I_a)_{a \in A}$. For every arc $a = (v_i, v_j)$ of A, we put the interval $[i, j)$ into the collection.

We have now to show that $I_a = [i, j)$ is entirely to the left of $I_{a'} = [i', j')$ if and only if a has to precede a' in the task precedence constraints represented by G. This is easy to show, since:

$I_a = [i, j)$ is entirely to the left of $I_{a'} = [i', j')$

⇔ $j \le i'$

⇔ there is a path from v_j to $v_{i'}$ in G (along W)

⇔ a with head j has to precede a' with tail i'.   □

If dummy tasks are not allowed, an interval order does not necessarily have a task-on-arc representation. Indeed, if we consider the collection of intervals $\{[1,2), [1,3), [2,4), [3,4)\}$, its task-on-node representation is graph N in Figure 2.3.1. It implies that this partial order does not have a task-on-arc representation without dummy tasks. But the equivalence of task-on-node and task-on-arc representations can be obtained through the use of dummy tasks. Since we allow them also here, the following result can be proved.

Theorem 5.1.16 *Any interval order has a task-on-arc representation with a Hamiltonian path (and therefore corresponds to a uan).*

Proof. Consider any collection of intervals $(I_a)_{a \in A}$ with $I_a = [b_a, e_a)$. We define the following graph $G = (V, E)$. Set

$$V = \{ b_a \mid a \in A \} \cup \{ e_a \mid a \in A \}.$$

For any v in V, let $next(v)$ be the vertex $w > v$ such that there is no x in V with $v \prec x \prec w$ ($next(v)$ is not defined for the largest e_a). Set

$$A' = \{(v, next(v)) \mid v \in V \text{ and } next(v) \text{ defined}\}$$

and

$$E = A' \cup \{(b_a, e_a) \mid a \in A\}.$$

The arcs in A' represent dummy tasks. This graph G has indeed a Hamiltonian path, starting with the smallest b_a ($\min_{a \in A} e_a$), following the arcs in A' and ending at the largest e_a ($\max_{a \in A} e_a$). It remains to show that $I_a = [b_a, e_a)$ is entirely to the left of $I_{a'} = [b_{a'}, e_{a'})$ if and only if arc (b_a, e_a) has to precede arc $[b_{a'}, e_{a'})$ in the task precedence constraints represented by G. We do not have to deal with arcs in A' since they represent dummy tasks:

$I_a = [b_a, e_a)$ is entirely to the left of $I_{a'} = [b_{a'}, e_{a'})$

$\Leftrightarrow e_a \le b_{a'}$

$\Leftrightarrow$ there is a path from e_a to $b_{a'}$ in G (using the arcs in A')

$\Leftrightarrow (b_a, e_a)$ with head e_a has to precede $(b_{a'}, e_{a'})$ with tail $b_{a'}$.

$\square$

The following corollary is a direct consequence of Theorems 5.1.15 and 5.1.16

Corollary 5.1.17 *Let Q be a partial order. If dummy tasks are allowed, Q is an interval order if and only if Q can be represented as a uan.*

We may now conclude the above considerations with the following result:

$P \mid pmtn, interval\ order \mid C_{max}$ is solvable in polynomial time.

For general precedence graphs, however, we know from Ullman [Ull76] that the problem is NP-hard. In that case a heuristic algorithm such as Algorithm 5.1.13 my be chosen. The worst-case behavior of Algorithm 5.1.13 applied in the case of $P \mid pmtn, prec \mid C_{max}$ has been analyzed by Lam and Sethi [LS77]:

$$R_{Alg.5.1.13} = 2 - \frac{2}{m}, \qquad m \ge 2.$$

5.1.2 Uniform and Unrelated Processors

Problem $Q \mid p_j = 1 \mid C_{max}$

Let us start with an analysis of independent tasks and non-preemptive scheduling. Since the problem with arbitrary processing times is already NP-hard for identical processors, all we can hope to find is a polynomial time optimization algorithm for tasks with unit standard processing times only. Such an approach

has been given by Graham et al. [GLL+79] where a transportation network for-
mulation has been presented for problem $Q|p_j = 1|C_{max}$. We describe it briefly
below.

Let there be n sources j, $j = 1, 2, \cdots, n$, and mn sinks (i, k), $i = 1, 2, \cdots, m$ and
$k = 1, 2, \cdots, n$. Sources correspond to tasks and sinks to processors and positions
of tasks on them. Let $c_{ijk} = k/b_i$ be the cost of arc $(j, (i, k))$; this value corresponds
to the completion time of task T_j processed on P_i in the kth position. The arc flow
x_{ijk} has the following interpretation:

$$x_{ijk} = \begin{cases} 1 & \text{if } T_j \text{ is processed in the } k\text{th position on } P_i \\ 0 & \text{otherwise.} \end{cases}$$

The min-max transportation problem can be now formulated as follows:

$$\textit{Minimize} \qquad \max_{i, j, k} \{c_{ijk} x_{ijk}\} \tag{5.1.16}$$

$$\textit{subject to} \qquad \sum_{i=1}^{m} \sum_{k=1}^{n} x_{ijk} = 1 \quad \text{for all } j, \tag{5.1.17}$$

$$\sum_{j=1}^{n} x_{ijk} \leq 1 \qquad \text{for all } i, k, \tag{5.1.18}$$

$$x_{ijk} \geq 0 \qquad \text{for all } i, j, k. \tag{5.1.19}$$

This problem can be solved by a standard transportation procedure (cf. Section
2.3) which results in $O(n^3)$ time complexity, or by a procedure due to Sevastjanov [Sev91]. Below we sketch this last approach. It is clear that the minimum
schedule length is given as

$$C^*_{max} = \sup \{t \mid \sum_{i=1}^{m} \lfloor tb_i \rfloor < n\}. \tag{5.1.20}$$

On the other hand, a lower bound on the schedule length for the above problem is

$$C' = n / \sum_{i=1}^{m} b_i \leq C^*_{max}. \tag{5.1.21}$$

Bound C' can be achieved e.g. by a preemptive schedule. If we assign $k_i = \lfloor C'b_i \rfloor$
tasks to processor P_i, $i = 1, 2, \cdots, m$, respectively, then these tasks may be processed in time interval $[0, C']$. However, $l = n - \sum_{i=1}^{m} k_i$ tasks remain unassigned.
Clearly $l \leq m - 1$, since $C'b_i - \lfloor C'b_i \rfloor < 1$ for each i. The remaining l tasks are then
assigned one by one to those processors P_i for which $\min_i \{(k_i + 1) / b_i\}$ is attained
at a given stage, where, of course, k_i is increased by one after the assignment of a
task to a particular processor P_i. This procedure is repeated until all tasks are

assigned. We see that this approach results in an $O(m^2)$-algorithm for solving problem $Q \mid p_j = 1 \mid C_{max}$.

Example 5.1.18 To illustrate the above algorithm let us assume that $n = 9$ tasks are to be processed on $m = 3$ uniform processors whose processing speeds are given by the vector $\boldsymbol{b} = [3, 2, 1]$. We get $C' = 9/6 = 1.5$. The numbers of tasks assigned to processors at the first stage are, respectively, 4, 3, and 1. A corresponding schedule is given in Figure 5.1.15(a), where task T_9 has not yet been assigned. An optimal schedule is obtained if this task is assigned to processor P_1, cf. Figure 5.1.15(b).   □

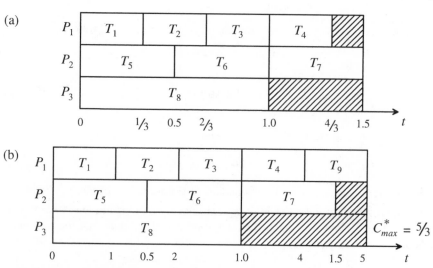

Figure 5.1.15   *Schedules for Example* 5.1.18
               (a) a *partial schedule,*
               (b) *an optimal schedule.*

Problem $Q \mid\mid C_{max}$

Since other problems of non-preemptive scheduling of independent tasks are NP-hard, one may be interested in applying certain heuristics. One heuristic algorithm which is a list scheduling algorithm, has been presented by Liu and Liu [LL74a]. Tasks are ordered on the list in non-increasing order of their processing times and processors are ordered in non-increasing order of their processing speeds. Now, whenever a machine becomes free it gets the first non-assigned task of the list; if there are two or more free processors, the fastest is chosen. The worst-case behavior of the algorithm has been evaluated for the case of an $m+1$ processor system, m of which have processing speed factor equal to 1 and the remaining processor has processing speed factor b. The bound is as follows.

$$R = \begin{cases} \dfrac{2(m+b)}{b+2} & \text{for } b \leq 2 \\[2mm] \dfrac{m+b}{2} & \text{for } b > 2. \end{cases}$$

It is clear that the algorithm does better if, in the first case ($b \leq 2$), m decreases faster than b, and if b and m decrease in case of $b > 2$. Other algorithms have been analyzed by Liu and Liu [LL74b, LL74c] and by Gonzalez et al. [GIS77].

Problem $Q\,|\,pmtn\,|\,C_{max}$

By allowing preemptions, i.e. for the problem $Q\,|\,pmtn\,|\,C_{max}$, one can find optimal schedules in polynomial time. We present an algorithm given by Horvath et al. [HLS77] despite the fact that there is a more efficient one by Gonzalez and Sahni [GS78]. We do this because the first algorithm covers also precedence constraints, and it generalizes the ideas presented in Algorithm 5.1.13. The algorithm is based on two concepts: the *task level*, defined as previously as processing requirement of the unexecuted portion of a task, but now expressed in terms of a standard processing time, and *processor sharing*, i.e. the possibility of assigning only a fraction β ($0 \leq \beta \leq \max\{b_i\}$) of processing capacity to some task. Let us assume that tasks are indexed in order of non-increasing p_j's and processors are in order of non-increasing values of b_i. It is quite clear that the minimum schedule length can be estimated by

$$C_{max}^* \geq C = \max\left\{ \max_{1 \leq k \leq m} \left\{\frac{X_k}{B_k}\right\}, \left\{\frac{X_n}{B_m}\right\} \right\} \tag{5.1.22}$$

where X_k is the sum of processing requirements (i.e. standard processing times p_j) of the first k tasks, and B_k is the collective processing capacity (i.e. the sum of processing speed factors b_i) of the first k processors. The algorithm presented below constructs a schedule of length equal to C for the problem $Q\,|\,pmtn\,|\,C_{max}$.

Algorithm 5.1.19 *Algorithm by Horvath, Lam and Sethi for* $Q\,|\,pmtn\,|\,C_{max}$ [HLS77].

```
begin
for all T ∈ T do  Compute level of task T;
t := 0; h := m;
repeat
  while h > 0 do
    begin
      Construct subset S of T consisting of tasks at the highest level;
        -- the most "urgent" tasks are chosen
```

```
if |S| > h
then
  begin
  Assign the tasks of set S to the h remaining processors to be processed
```
at the same rate $\beta = \sum\limits_{i=m-h+1}^{m} b_i/|S|$;
```
  h := 0;      -- tasks from set S share the h slowest processors
  end
else
  begin
  Assign tasks from set S to be processed at the same rate β on the fastest
    |S| processors;
  h := h - |S|;    -- tasks from set S share the fastest |S| processors
  end;
end;        -- the most urgent tasks have been assigned at time t
```
Calculate time moment τ at which either one of the assigned tasks is finished
or a point is reached at which continuing with the present partial assignment causes that a task at a lower level will be executed at a faster rate β
than a higher level task;
-- note, that the levels of the assigned tasks decrease during task execution
Decrease levels of the assigned tasks by $(\tau - t)\beta$;
$t := \tau$; $h := m$;
-- a portion of each assigned task equal to $(\tau - t)\beta$ has been processed
```
until
```
 all tasks are finished;
-- the schedule constructed so far consists of a sequence of intervals during each
-- of which certain tasks are assigned to the processors in a shared mode.
-- In the next loop task assignment in each of these intervals is determined
```
for each
```
 interval of the processor shared schedule
```
do
  begin
```
Let y be the length of the interval;
```
if
```
 g tasks share g processors
```
then
```
 Assign each task to each processor for y/g time units
```
else
  begin
```
    Let p be the processing requirement of each of the g tasks in the inter-
        val;
    Let b be the processing speed factor of the slowest processor;
```
if
```
 $p/b < y$
```
then call
```
 Algorithm 5.1.8
        -- tasks can be assigned as in McNaughton's rule,
        -- ignoring different processor speeds
```
else
  begin
```
    Divide the interval into g subintervals of equal lengths;

Assign the g tasks so that each task occurs in exactly h intervals, each
    time on a different processor;
      end;
    end;
  end;
    -- a normal preemptive schedule has now been constructed
end;

The time complexity of Algorithm 5.1.19 is $O(mn^2)$. An example of its application is shown in Figure 5.1.16.

(a)

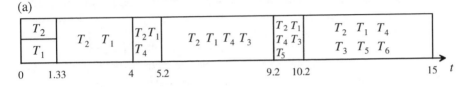

(b)

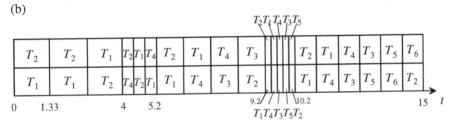

Figure 5.1.16 *An example of the application of Algorithm 5.1.19: $n = 6$, $m = 2$,*
        $p = [20, 24, 10, 12, 5, 4]$, $b = [4, 1]$
        (a) a processor shared schedule,
        (b) an optimal schedule.

Problem $Q \mid pmtn, prec \mid C_{max}$

When considering dependent tasks, only preemptive polynomial time optimization algorithms are known. Algorithm 5.1.19 also solves problem $Q2 \mid pmtn$,
$prec \mid C_{max}$, if the level of a task is understood as in Algorithm 5.1.13 where
standard processing times for all the tasks were assumed. When considering this
problem one should also take into account the possibility of solving it for uniconnected activity networks and interval orders via the slightly modified linear
programming approach (5.1.14)-(5.1.15). It is also possible to solve the problem
by using another *LP* formulation which is described for the case of $R \mid pmtn \mid$
C_{max}.

    It is also possible to solve problem $Q \mid pmtn, prec \mid C_{max}$ approximately by the
two machine aggregation approach, developed in the framework of flow shop

scheduling [RS83] (cf. Chapter 8). In this case the two fastest processors are used only, and the worst case bound is

$$
\frac{C_{max}}{C_{max}^*} \leq
\begin{cases}
\sum\limits_{i=1}^{m/2} \max\{b_{2i-1}/b_1, b_{2i}/b_2\} & \text{if } m \text{ is even,} \\
\sum\limits_{i=1}^{\lfloor m/2 \rfloor} \max\{b_{2i-1}/b_1, b_{2i}/b_2\} + b_m/b_1 & \text{if } m \text{ is odd.}
\end{cases}
$$

Problem $R \mid pmtn \mid C_{max}$

Let us pass now to the case of unrelated processors. This case is the most difficult. We will not speak about unit-length tasks, because unrelated processors with unit length tasks would reduce to the case of identical or uniform processors. Hence, no polynomial time optimization algorithms are known for problems other than preemptive ones. Also, very little is known about approximation algorithms for this case. Some results have been obtained by Ibarra and Kim [IK77], but the obtained bounds are not very encouraging. Thus, we will pass to the preemptive scheduling model.

Problem $R \mid pmtn \mid C_{max}$ can be solved by a *two-phase method*. The first phase consists in solving a linear programming problem formulated independently by Błażewicz et al. [BCSW76a, BCW77] and by Lawler and Labetoulle [LL78]. The second phase uses the solution of this *LP* problem and produces an optimal preemptive schedule.

Let $x_{ij} \in [0, 1]$ denote the part of T_j processed on P_i. The *LP* formulation is as follows:

$$
\textit{Minimize} \quad C_{max} \tag{5.1.23}
$$

$$
\textit{subject to} \quad C_{max} - \sum_{j=1}^{n} p_{ij} x_{ij} \geq 0, \quad i = 1, 2, \cdots, m \tag{5.1.24}
$$

$$
C_{max} - \sum_{i=1}^{m} p_{ij} x_{ij} \geq 0, \quad j = 1, 2, \cdots, n \tag{5.1.25}
$$

$$
\sum_{i=1}^{m} x_{ij} = 1, \quad j = 1, 2, \cdots, n. \tag{5.1.26}
$$

Solving the above problem, we get $C_{max} = C_{max}^*$ and optimal values x_{ij}^*. However, we do not know how to schedule the task parts, i.e. how to assign these parts to processors in time. A schedule may be constructed in the following way. Let $T = [t_{ij}^*]$ be the $m \times n$ matrix defined by $t_{ij}^* = p_{ij} x_{ij}^*$, $i = 1, 2, \cdots, m$, $j = 1$, $2, \cdots, n$. Notice that the elements of T reflect optimal values of processing times of particular tasks on the processors. The jth column of T corresponding to task T_j

will be called *critical* if $\sum\limits_{i=1}^{m} t^*_{ij} = C^*_{max}$. By Y we denote an $m \times m$ diagonal matrix

whose element y_{kk} is the total idle time on processor P_k, i.e. $y_{kk} = C^*_{max} - \sum\limits_{j=1}^{n} t^*_{kj}$.
Columns of Y correspond to dummy tasks. Let $V = [T,Y]$ be an $m \times (n+m)$ matrix. Now set $\mathcal{U}$ containing m positive elements of matrix V is defined as having exactly one element from each critical column and at most one element from other columns, and having exactly one element from each row. We see that $\mathcal{U}$ corresponds to a task set which may be processed in parallel in an optimal schedule. Thus, it may be used to construct a partial schedule of some length $\delta > 0$. An optimal schedule is then produced as the union of the partial schedules. This procedure is summarized in Algorithm 5.1.20 [LL78].

Algorithm 5.1.20 *Construction of an optimal schedule corresponding to LP solution for* $R \,|\, pmtn \,|\, C_{max}$.

```
begin
C := C*_max;
while C > 0 do
   begin
   Construct set U;
      -- thus a subset of tasks to be processed in a partial schedule has been chosen
   v_min := min  {v_ij};
           v_ij ∈ U

   v_max := max_{j ∈ {j' | v_ij' ∉ U for i = 1,...,m}} {Σ v_ij};
                                                       i
   if  C - v_min ≥ v_max
   then  δ := v_min
   else  δ := C - v_max;
      -- the length of the partial schedule is equal to δ
   C := C - δ;
   for each  v_ij ∈ U do  v_ij := v_ij - δ;
      -- matrix V is changed; notice that due to the way δ is defined,
      -- the elements of V can never become negative
   end;
end;
```

The proof of correctness of the algorithm can be found in [LL78].

Now we only need an algorithm that finds set $\mathcal{U}$ for a given matrix V. One of the possible algorithms is based on the network flow approach. In this case the network has m nodes corresponding to machines (rows of V) and $n+m$ nodes corresponding to tasks (columns of V), cf. Figure 5.1.17. A node i from the first group is connected by an arc to a node j of the second group if and only if $v_{ij} > 0$. Arc flows are constrained by b from below and by $c = 1$ from above, where the value of b is 1 for arcs joining the source with processor-nodes and critical task nodes with the sink, and $b = 0$ for the other arcs. Obviously, finding a feasible flow in this network is equivalent to finding set $\mathcal{U}$. The following example illustrates the second phase of the described method.

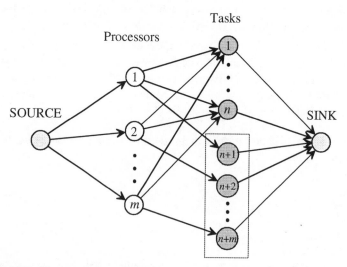

Figure 5.1.17 *Finding set $\mathcal{U}$ by the network flow approach.*

Example 5.1.21 Suppose that for a certain scheduling problem a linear programming solution of the two phase method has the form given in Figure 5.1.18(a). An optimal schedule is then constructed in the following way. First, matrix V is calculated.

$$
V = \begin{array}{c} \\ P_1 \\ P_2 \\ P_3 \\ \\ \end{array}
\begin{array}{c} T_1\,T_2\,T_3\,T_4\,T_5 \\ \left[\begin{array}{ccccc} \underline{3} & 2 & 1 & 4 & 0 \\ 2 & 2 & 0 & \underline{2} & 2 \\ 2 & 1 & \underline{4} & 0 & 1 \end{array}\right] \\ 7\ 5\ 5\ 6\ 3 \end{array}
\begin{array}{c} T_6\,T_7\,T_8 \\ \left[\begin{array}{ccc} 0 & 0 & 0 \\ 0 & 2 & 0 \\ 0 & 0 & 2 \end{array}\right] \\ 0\ 2\ 2 \end{array}
$$

Then elements constituting set $\mathcal{U}$ are chosen according to Algorithm 5.1.20, as depicted above. The value of a partial schedule length is $\delta = 2$. Next, the while-loop of Algorithm 5.1.20 is repeated yielding the following sequence of matrices V_i.

$$
V_1 = \begin{array}{c} \left[\begin{array}{ccccc} 1 & 2 & 1 & \underline{4} & 0 \\ 2 & \underline{2} & 0 & 0 & 2 \\ \underline{2} & 1 & 2 & 0 & 1 \end{array}\right] \end{array}
\begin{array}{c} \left[\begin{array}{ccc} 0 & 0 & 0 \\ 0 & 2 & 0 \\ 0 & 0 & 2 \end{array}\right] \end{array}
$$

$$V_2 = \begin{bmatrix} 1 & \underline{2} & 1 & 2 & 0 \\ \underline{2} & 0 & 0 & 0 & 2 \\ 0 & 1 & \underline{2} & 0 & 1 \end{bmatrix} \begin{bmatrix} 0 & 0 & 0 \\ 0 & 2 & 0 \\ 0 & 0 & 2 \end{bmatrix}$$

$$V_3 = \begin{bmatrix} 1 & 0 & 1 & \underline{2} & 0 \\ 0 & 0 & 0 & 0 & \underline{2} \\ 0 & \underline{1} & 0 & 0 & 1 \end{bmatrix} \begin{bmatrix} 0 & 0 & 0 \\ 0 & 2 & 0 \\ 0 & 0 & 2 \end{bmatrix}$$

$$V_4 = \begin{bmatrix} \underline{1} & 0 & 1 & 1 & 0 \\ 0 & 0 & 0 & 0 & \underline{1} \\ 0 & 0 & 0 & 0 & 1 \end{bmatrix} \begin{bmatrix} 0 & 0 & 0 \\ 0 & 2 & 0 \\ 0 & 0 & \underline{2} \end{bmatrix}$$

$$V_5 = \begin{bmatrix} 0 & 0 & \underline{1} & 1 & 0 \\ 0 & 0 & 0 & 0 & 0 \\ 0 & 0 & 0 & 0 & \underline{1} \end{bmatrix} \begin{bmatrix} 0 & 0 & 0 \\ 0 & \underline{2} & 0 \\ 0 & 0 & 1 \end{bmatrix}$$

$$V_6 = \begin{bmatrix} 0 & 0 & 0 & \underline{1} & 0 \\ 0 & 0 & 0 & 0 & 0 \\ 0 & 0 & 0 & 0 & 0 \end{bmatrix} \begin{bmatrix} 0 & 0 & 0 \\ 0 & \underline{1} & 0 \\ 0 & 0 & \underline{1} \end{bmatrix}.$$

A corresponding optimal schedule is presented in Figure 5.1.18(b).   □

(a)

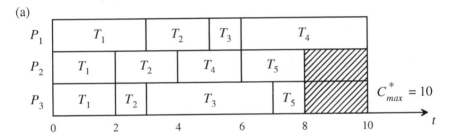

(b)

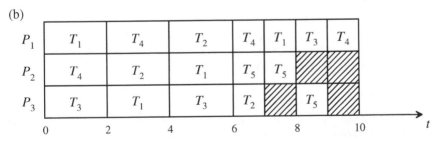

Figure 5.1.18 (a) *A linear programming solution for an instance of*
$R\,|\,pmtn\,|\,C_{max}$,
(b) *an optimal schedule.*

The overall complexity of the above approach is bounded from above by a poly-nomial in the input length. This is because the transformation to the *LP* problem is polynomial, and the *LP* problem may be solved in polynomial time using Kha-chiyan's algorithm [Kha79]; the loop in Algorithm 5.1.20 is repeated at most $O(mn)$ times and solving the network flow problem requires $O(z^3)$ time, where z is the number of network nodes [Kar74].

Problem *R | pmtn, prec | C_{max}*

If dependent tasks are considered, i.e. in the case $R | pmtn, prec | C_{max}$, linear pro-gramming problems similar to those discussed in (5.1.14)-(5.1.15) or (5.1.23)-(5.1.26) and based on the activity network presentation, can be formulated. For example, in the latter formulation one defines x_{ijk} as a part of task T_j processed on processor P_i in the main set S_k. Solving the *LP* problem for x_{ijk}, one then applies Algorithm 5.1.20 for each main set. If the activity network is uniconnected (a corresponding task-on-node graph represents an interval order), an optimal schedule is constructed in this way, otherwise only an approximate schedule is obtained.

We complete this chapter by remarking that introduction of ready times into the model considered so far is equivalent to the problem of minimizing maximum lateness. We will consider this type of problems in Section 5.3.

5.2   Minimizing Mean Flow Time

5.2.1   Identical Processors

Problem *P | | ΣC_j*

In the case of identical processors and equal ready times preemptions are not profitable from the viewpoint of the value of the mean flow time [McN59]. Thus, we can limit ourselves to considering non-preemptive schedules only.

When analyzing the nature of criterion ΣC_j, one might expect that, as in the case of one processor (cf. Section 4.2), by assigning tasks in non-decreasing or-der of their processing times the mean flow time will be minimized. In fact, a proper generalization of this simple rule leads to an optimization algorithm for $P | | \Sigma C_j$ (Conway et al. [CMM67]). It is as follows.

Algorithm 5.2.1 *SPT rule for problem $P | | \Sigma C_j$* [CMM67].
begin
Order tasks on list L in non-decreasing order of their processing times;

```
while L ≠ ∅ do
  begin
  Take the m first tasks from the list (if any) and assign these tasks arbitrarily to
    the m different processors;
  Remove the assigned tasks from list L;
  end;
Process tasks assigned to each processor in SPT order;
end;
```

The complexity of the algorithm is obviously $O(n\log n)$.

In this context let us also mention that introducing different ready times makes the problem strongly NP-hard even for the case of one processor (see Section 4.2 and [LRKB77]). Also, if we introduce different weights, then the 2-processor problem without release times, $P2\|\Sigma w_j C_j$, is already NP-hard [BCS74].

Problem $P\,|\,prec\,|\,\Sigma C_j$

Let us now pass to the case of dependent tasks. Here, $P\,|\,out\text{-}tree, p_j = 1\,|\,\Sigma C_j$ is solved by an adaptation of Algorithm 5.1.11 (Hu's algorithm) to the out-tree case [Ros–], and $P2\,|\,prec, p_j = 1\,|\,\Sigma C_j$ is strongly NP-hard [LRK78]. In the case of arbitrary processing times results by Du et al. [DLY91] indicate that even simplest precedence constraints result in computational hardness of the problem. That is problem $P2\,|\,chains\,|\,\Sigma C_j$ is already NP-hard in the strong sense. On the other hand, it was also proved in [DLY91] that preemptions cannot reduce the mean weighted flow time for a set of chains. Together with the last result this implies that problem $P2\,|\,chains, pmtn\,|\,\Sigma C_j$ is also NP-hard in the strong sense. Unfortunately, no approximation algorithms for these problems are evaluated from their worst-case behavior point of view.

5.2.2   Uniform and Unrelated Processors

The results of Section 5.2.1 also indicate that scheduling dependent tasks on uniform or unrelated processors is an NP-hard problem in general. No approximation algorithms have been investigated either. Thus, we will not consider this subject.

On the other hand, in the case of independent tasks, preemptions may be worthwhile, thus we have to treat non-preemptive and preemptive scheduling separately.

Problem $Q \| \Sigma C_j$

Let us start with uniform processors and non-preemptive schedules. In this case the flow time has to take into account processor speed; so the flow time of task $T_{i[k]}$ processed in the kth position on processor P_i is defined as $F_{i[k]} = \frac{1}{b_i} \sum_{j=1}^{k} p_{i[j]}$. Let us denote by n_i the number of tasks processed on processor P_i. Thus, $n = \sum_{i=1}^{m} n_i$. The mean flow time is then given by

$$\bar{F} = \frac{\sum_{i=1}^{m} \frac{1}{b_i} \sum_{k=1}^{n_i} (n_i - k + 1) p_{i[k]}}{n}. \tag{5.2.1}$$

It is easy to see that the numerator in the above formula is the sum of n terms each of which is the product of a processing time and one of the following coefficients:

$$\frac{1}{b_1} n_1, \frac{1}{b_1}(n_1 - 1), \cdots, \frac{1}{b_1}, \frac{1}{b_2} n_2, \frac{1}{b_2}(n_2 - 1), \cdots, \frac{1}{b_2}, \cdots, \frac{1}{b_m} n_m, \frac{1}{b_m}(n_m - 1), \cdots, \frac{1}{b_m}.$$

It is known from [CMM67] that such a sum is minimized by matching n smallest coefficients in non-decreasing order with processing times in non-increasing order. An $O(n \log n)$ implementation of this rule has been given by Horowitz and Sahni [HS76].

Problem $Q \| pmtn \| \Sigma C_j$

In the case of preemptive scheduling, it is possible to show that there exists an optimal schedule for $Q \| pmtn \| \Sigma C_j$ in which $C_j \leq C_k$ if $p_j < p_k$. On the basis of this observation, the following algorithm has been proposed by Gonzalez [Gon77].

Algorithm 5.2.2 *Algorithm by Gonzalez for $Q \| pmtn \| \Sigma C_j$ [Gon77].*
begin
Order processors in non-increasing order of their processing speed factors;
Order tasks in non-decreasing order of their standard processing times;
for $j = 1$ **to** n **do**
   begin
   Schedule task T_j to be completed as early as possible, preempting when
      necessary;
        -- tasks will create a staircase pattern "jumping" to a faster processor
        -- whenever a shorter task has been finished
   end;
end;

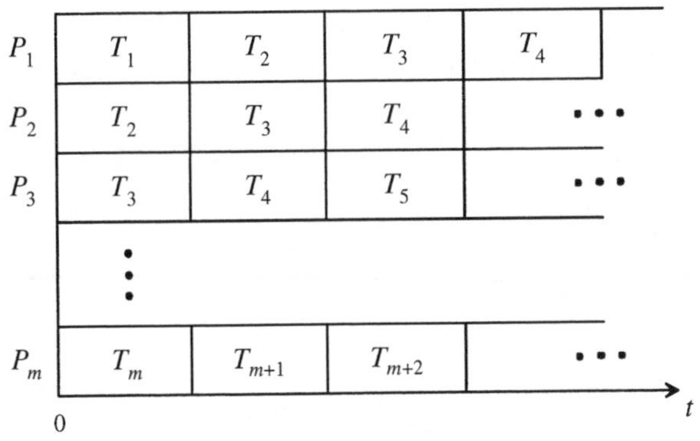

Figure 5.2.1 *An example of the application of Algorithm 5.2.2.*

The complexity of this algorithm is $O(n\log n + mn)$. An example of its application is given in Figure 5.2.1.

Problem $R\,||\,\Sigma C_j$

Let us now turn to the case of unrelated processors and consider problem $R\,||\,\Sigma C_j$. An approach to its solution is based on the observation that task $T_j \in \{T_1,\cdots,T_n\}$ processed on processor $P_i \in \{P_1,\cdots,P_m\}$ as the last task contributes its processing time p_{ij} to $\bar{F}$. The same task processed in the last but one position contributes $2p_{ij}$, and so on [BCS74]. This reasoning allows one to construct an $(mn)\times n$ matrix Q presenting contributions of particular tasks processed in different positions on different processors to the value of $\bar{F}$:

$$Q = \begin{bmatrix} [p_{ij}] \\ 2[p_{ij}] \\ \cdot \\ \cdot \\ \cdot \\ n[p_{ij}] \end{bmatrix}$$

The problem is now to choose n elements from matrix Q such that
- exactly one element is taken from each column,
- at most one element is taken from each row,
- the sum of selected elements is minimum.

We see that the above problem is a variant of the assignment problem (cf. [Law76]), which may be solved in a natural way via the transportation problem. The corresponding transportation network is shown in Figure 5.2.2. Careful analysis of the problem shows that it can be solved in $O(n^3)$ time [BCS74]. The following example illustrates this technique.

Example 5.2.3 Let us consider the following instance of problem $R||\Sigma C_j$: $n = 5$, $m = 3$, and matrix p of processing times

$$p = \begin{bmatrix} 3 & 2 & 4 & 3 & 1 \\ 4 & 3 & 1 & 2 & 1 \\ 2 & 4 & 5 & 3 & 4 \end{bmatrix}.$$

Using this data the matrix Q is constructed as follows:

$$Q = \begin{bmatrix} 3 & 2 & 4 & 3 & 1 \\ 4 & 3 & 1 & 2 & 1 \\ 2 & 4 & 5 & 3 & 4 \\ 6 & 4 & 8 & 6 & 2 \\ 8 & 6 & 2 & 4 & 2 \\ 4 & 8 & 10 & 6 & 8 \\ 9 & 6 & 12 & 9 & 3 \\ 12 & 9 & 3 & 6 & 3 \\ 6 & 12 & 15 & 9 & 12 \\ 12 & 8 & 16 & 12 & 4 \\ 16 & 12 & 4 & 8 & 4 \\ 8 & 16 & 20 & 12 & 16 \\ 15 & 10 & 20 & 15 & 5 \\ 20 & 15 & 5 & 10 & 5 \\ 10 & 20 & 25 & 15 & 20 \end{bmatrix}.$$

On the basis of this matrix a network as shown in Figure 5.2.2 is constructed.

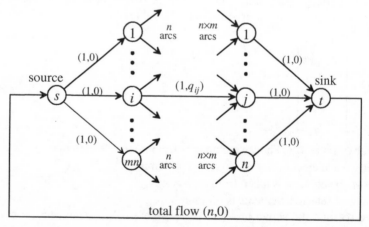

Figure 5.2.2 *The transportation network for problem $R||\Sigma C_j$: arcs are denoted by (c, y), where c is the capacity and y is the cost of unit flow.*

Solving the transportation problem results in the selection of the underlined elements of matrix Q. They correspond to the schedule shown in Figure 5.2.3.   □

A very surprising result has been recently obtained by Sitters. Problem $R \,|\, pmtn \,|\, \Sigma C_j$ has been proved to be strongly NP-hard [Sit05].

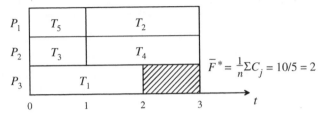

$$\overline{F}^* = \frac{1}{n} \Sigma C_j = 10/5 = 2$$

Figure 5.2.3   *An optimal schedule for Example 5.2.3.*

5.3   Minimizing Due Date Involving Criteria

5.3.1   Identical Processors

In Section 4.3 we have seen that single processor problems with due date optimization criteria involving due dates are NP-hard in most cases. In the following we will concentrate on minimization of L_{max} criterion. It seems to be quite natural that in this case the general rule should be to schedule tasks according to their earliest due dates (*EDD*-rule, cf. Section 4.3.1). However, this simple rule of Jackson [Jac55] produces optimal schedules under very restricted assumptions only. In other cases more sophisticated algorithms are necessary, or the problems are NP-hard.

Problem $P \,|\,|\, L_{max}$

Let us start with non-preemptive scheduling of independent tasks. Taking into account simple transformations between scheduling problems (cf. Section 3.4) and the relationship between the C_{max} and L_{max} criteria, we see that all the problems that are NP-hard under the C_{max} criterion remain NP-hard under the L_{max} criterion. Hence, for example, $P2 \,|\,|\, L_{max}$ is NP-hard. On the other hand, unit processing times of tasks make the problem easy, and $P \,|\, p_j = 1, r_j \,|\, L_{max}$ can be solved by an obvious application of the *EDD* rule [Bla77]. Moreover, problem $P \,|\, p_j = p, r_j \,|\, L_{max}$ can be solved in polynomial time by an extension of the single processor algorithm (see Section 4.3.1 and [GJST81]). Unfortunately very little is known about the worst-case behavior of approximation algorithms for the NP-hard problems in question.

Problem $P \mid pmtn, r_j \mid L_{max}$

The preemptive mode of processing makes the solution of the scheduling problem much easier. The fundamental approach in that area is testing feasibility of problem $P \mid pmtn, r_j, \tilde{d}_j \mid -$ via the network flow approach [Hor74]. Using this approach repetitively, one can then solve the original problem $P \mid pmtn, r_j \mid L_{max}$ by changing due dates (deadlines) according to a binary search procedure.

Let us now describe Horn's approach for testing feasibility of problem $P \mid pmtn, r_j, \tilde{d}_j \mid -$, i.e. deciding whether or not for a given set of ready times and deadlines there exists a schedule with no late task. Let the values of ready times and deadlines of an instance of $P \mid pmtn, r_j, \tilde{d}_j \mid -$ be ordered on a list in such a way that $e_0 < e_1 < \cdots < e_k$, $k < 2n$, where e_i stands for some r_j or $\tilde{d}_j$. We construct a network that has two sets of nodes, besides source and sink (cf. Figure 5.3.1). The first set corresponds to time intervals in a schedule, i.e. node w_i corresponds to interval $[e_{i-1}, e_i]$, $i = 1, 2, \cdots, k$. The second set corresponds to the task set. The capacity of an arc joining the source of the network to node w_i is equal to $m(e_i - e_{i-1})$ and thus corresponds to the total processing capacity of m processors in this interval. If task T_j could be processed in interval $[e_{i-1}, e_i]$ (because of its ready time and deadline) then w_i is joined to T_j by an arc of capacity $e_i - e_{i-1}$. Node T_j is joined to the sink of the network by an arc with capacity equal to p_j and with a lower bound on arc flow which is also equal to p_j. We see that finding a feasible flow pattern corresponds to constructing a feasible schedule and this test can be made in $O(n^3)$ time (cf. Section 2.3.3). A schedule is constructed on the basis of flow values on arcs between interval and task nodes. Let us consider the following example.

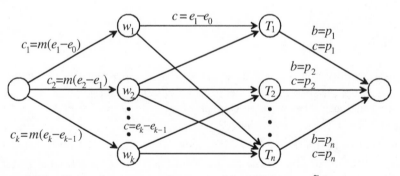

Figure 5.3.1   *Network corresponding to problem $P \mid pmtn, r_j, \tilde{d}_j \mid -$.*

Example 5.3.1 Let $n = 5$, $m = 2$, $p = [5, 2, 3, 3, 1]$, $r = [2, 0, 1, 0, 2]$, and $d = [8, 2, 4, 5, 8]$. The corresponding network is shown in Figure 5.3.2(a), and a feasible

flow pattern is depicted in Figure 5.3.2(b). On the basis of this flow the feasible
schedule shown in Figure 5.3.2(c) is constructed.   □

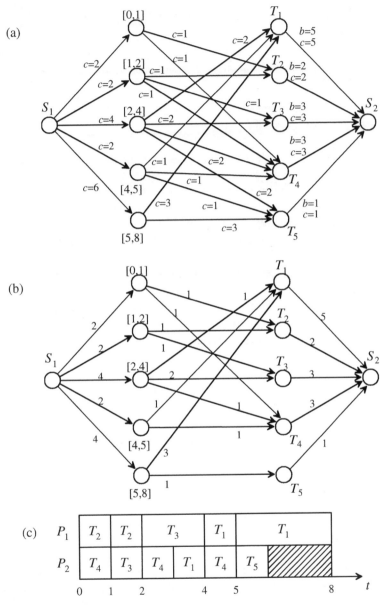

Figure 5.3.2 *Finding a feasible schedule via network flow approach (Example*
5.3.1)
(a) a corresponding network,

(**b**) *a feasible flow pattern,*
(**c**) *a schedule.*

In the next step a binary search can be conducted on the optimal value of L_{max}, with each trial value of L_{max} inducing deadlines which are checked for feasibility by means of the above network flow computation. This procedure can be implemented to solve problem $P \mid pmtn, r_j \mid L_{max}$ in $O(n^3 \min\{n^2, \log n + \log max\{p_j\}\})$ time [LLL+84].

Problem $P \mid prec, p_j = 1 \mid L_{max}$

Let us now pass to dependent tasks. A general approach in this case consists in assigning modified due dates to tasks, depending on the number and due dates of their successors. Of course, the way in which modified due dates are calculated depends on the parameters of the problem in question. If scheduling non-preemptable tasks on a multiple processor system only unit processing times can result in polynomial time scheduling algorithms. Let us start with in-tree precedence constraints and assume that if $T_i \prec T_j$ then $i > j$. The following algorithm minimizes L_{max} (isucc(j) denotes the immediate successor of T_j) [Bru76b].

Algorithm 5.3.2 *Algorithm by Brucker for $P \mid in\text{-}tree, p_j = 1 \mid L_{max}$* [Bru76b].
begin
$d_1^* := 1 - d_1;$ -- the due date of the root node is modified
for $k = 2$ **to** n **do**
   begin
   Calculate modified due date of T_k according to the formula
      $d_k^* := \max\ \{1 + d_{isucc(k)}^*,\ 1 - d_k\};$
   end;
Schedule tasks in non-increasing order of their modified due dates subject to
   precedence constraints;
end;

This algorithm can be implemented to run in $O(n\log n)$ time. An example of its application is given in Figure 5.3.3. Surprisingly *out-tree precedence constraints* result in the NP-hardness of the problem [BGJ77].

However, when we limit ourselves to two processors, a different way of computing modified due dates can be proposed which allows one to solve the problem in $O(n^2)$ time [GJ76]. In the algorithm below $g(k, d_i^*)$ is the number of successors of T_k having modified due dates not greater than d_i^*.

(a)

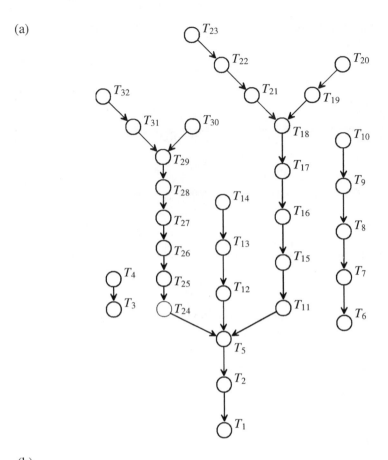

(b)

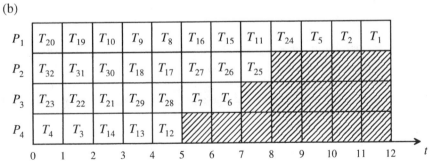

Figure 5.3.3 *An example of the application of Algorithm 5.3.2;*
*n = 32, m = 4, **d** = [16, 20, 4, 3, 15, 14, 17, 6, 6, 4, 10, 8, 9, 7, 10, 9, 10, 8,*
2, 3, 6, 5, 4, 11, 12, 9, 10, 8, 7, 5, 3, 5]
*(**a**) the task set,*
*(**b**) an optimal schedule.*

Algorithm 5.3.3 *Algorithm by Garey and Johnson for problem P2|prec,*
$p_j = 1 | L_{max}$ [GJ76].

```
begin
Z := T;
while Z ≠ ∅ do
  begin
```
Choose $T_k \in Z$ which is not yet assigned a modified due date and all of whose
    successors have been assigned modified due dates;
Calculate a modified due date of T_k as:

$$d_k^* := \min\{d_k, \min\{(d_i^* - \lceil \tfrac{1}{2} g(k, d_i^*) \rceil) \mid T_i \in \text{succ}(T_k)\}\};$$

$Z := Z - \{T_k\}$;
```
  end;
```
Schedule tasks in non-decreasing order of their modified due dates subject to
    precedence constraints;
```
end;
```

For $m > 2$ this algorithm may not lead to optimal schedules, as demonstrated in
the example in Figure 5.3.4. However, the algorithm can be generalized to cover
the case of different ready times too, but the running time is then $O(n^3)$ [GJ77]
and this is as much as we can get in non-preemptive scheduling.

Problem $P|pmtn, prec|L_{max}$

Preemptions allow one to solve problems with arbitrary processing times. In
[Law82b] algorithms have been presented that are preemptive counterparts of
Algorithms 5.3.2 and 5.3.3 and the one presented by Garey and Johnson [GJ77]
for non-preemptive scheduling and unit-length tasks. Hence problems $P|pmtn,$
$in\text{-}tree|L_{max}$, $P2|pmtn, prec|L_{max}$ and $P2|pmtn, prec, r_j|L_{max}$ are solvable in
polynomial time. Algorithms for these problems employ essentially the same
techniques for dealing with precedence constraints as the corresponding algorithms for unit-length tasks. However, the algorithms are more complex and will
not be presented here.

5.3.2 Uniform and Unrelated Processors

Problem $Q \| L_{max}$

From the considerations of Section 5.3.1 we see that non-preemptive scheduling to minimize L_{max} is in general a hard problem. Only for the problem $Q|p_j = 1|L_{max}$ a polynomial time optimization algorithm is known. This problem can be solved via a transportation problem formulation as in (5.1.16) - (5.1.19), where now $c_{ijk} = k/b_i - d_j$. Thus, from now on we will concentrate on preemptive scheduling.

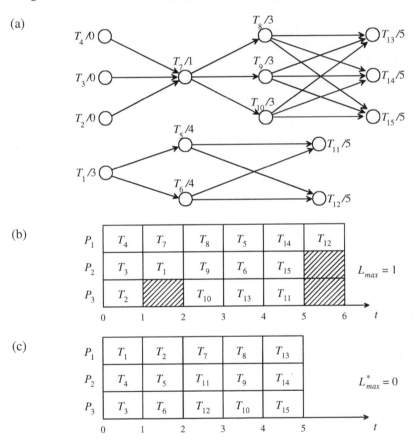

(a)

(b)

(c)

Figure 5.3.4 *Non-optimal schedules generated by Algorithm 5.3.3 for m=3, n=15, and all due dates $d_j = 5$*
(a) a task set (all tasks are denoted by T_j/d_j^*),
(b) a schedule constructed by Algorithm 5.3.3,
(c) an optimal schedule.

Problem $Q \mid pmtn \mid L_{max}$

One of the most interesting algorithms in that area has been presented for problem $Q \mid pmtn, r_j \mid L_{max}$ by Federgruen and Groenevelt [FG86]. It is a generalization of the network flow approach to the feasibility testing of problem $P \mid pmtn, r_j, \tilde{d}_j \mid$ – described above. The feasibility testing procedure for problem $Q \mid pmtn, r_j, \tilde{d}_j \mid$ – uses tripartite network formulation of the scheduling problem, where the first set of nodes corresponds to tasks, the second corresponds to processor-interval (period) combination and the third corresponds to interval nodes. The source is connected to each task node, the arc to the jth node having capacity p_j, $j = 1, 2, \cdots, n$. A task node is connected to all processor-interval nodes for all intervals during which the task is available. All arcs leading to a processor-interval node that corresponds to a processor of type r (processors of the same speed may be represented by one node only) and an interval of length τ, have capacity $(b_r - b_{r+1})\tau$, with the convention $b_{m+1} = 0$. Every node (w_i, r) corresponding to processor type r and interval w_i of length τ_i, $i = 1, 2, \cdots, k$, is connected to interval node w_i and has capacity $\sum_{j=1}^{r} m_j (b_r - b_{r+1})\tau_i$, where m_j denotes the number of processors of the jth type (cf. Figure 5.3.5). Finally, all interval nodes are connected to the sink with incapacitated arcs. Finding a feasible flow with value $\sum_{j=1}^{n} p_j$ in such a network corresponds to a construction of a feasible schedule for $Q \mid pmtn, r_j, \tilde{d}_j \mid$ –. This can be done in $O(mn^3)$ time.

Problem $Q \mid pmtn, prec \mid L_{max}$

In case of precedence constraints, $Q2 \mid pmtn, prec \mid L_{max}$ and $Q2 \mid pmtn, prec, r_j \mid L_{max}$ can be solved in $O(n^2)$ and $O(n^6)$ time, respectively, by the algorithms already mentioned [Law82b].

Problem $R \mid pmtn \mid L_{max}$

As far as unrelated processors are concerned, problem $R \mid pmtn \mid L_{max}$ can be solved by a linear programming formulation similar to (5.1.23) - (5.1.26) [LL78], where x_{ij}^k denotes the amount of T_j processed on P_i in time interval $[d_{k-1} + L_{max}, d_k + L_{max}]$, and where due dates are assumed to be ordered, $d_1 < d_2 < \cdots < d_n$. Thus, we have the following formulation:

$$Minimize \quad L_{max} \tag{5.3.1}$$

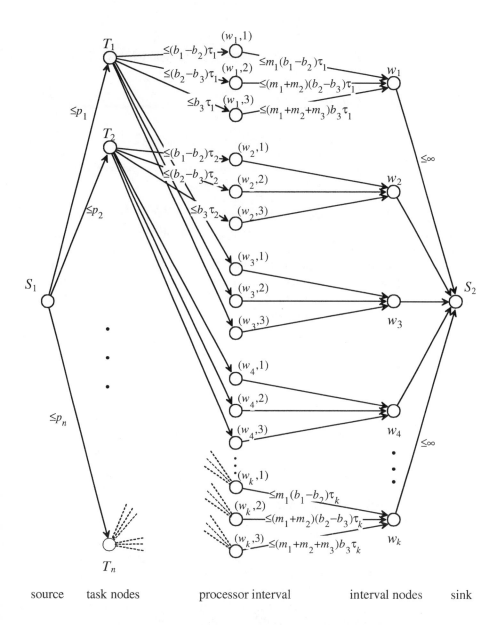

Figure 5.3.5 *A network corresponding to scheduling problem $Q \mid pmtn, r_j, \tilde{d}_j \mid -$ for three processor types.*

$$subject\ to\ \sum_{i=1}^{m} p_{ij}x_{ij}^{(1)} \le d_1 + L_{max}, \quad j = 1, 2, \cdots, n \tag{5.3.2}$$

$$\sum_{i=1}^{m} p_{ij}x_{ij}^{(k)} \le d_k - d_{k-1}, \quad j = k, k+1, \cdots, n;\ k = 2, 3, \cdots, n \tag{5.3.3}$$

$$\sum_{j=1}^{n} p_{ij}x_{ij}^{(1)} \le d_1 + L_{max}, \quad i = 1, 2, \cdots, m \tag{5.3.4}$$

$$\sum_{j=k}^{n} p_{ij}x_{ij}^{(k)} \le d_k - d_{k-1}, \quad i = 1, 2, \cdots, m;\ k = 2, 3, \cdots, n \tag{5.3.5}$$

$$\sum_{i=1}^{m} \sum_{k=1}^{j} x_{ij}^{(k)} = 1 \qquad j = 1, 2, \cdots, n. \tag{5.3.6}$$

Solving the *LP* problem we obtain n matrices $T^{(k)} = [t_{ij}^{(k)*}]$, $k = 1, \cdots, n$; then an optimal solution is constructed by an application of Algorithm 5.1.20 to each matrix separately.

In this context let us also mention that the case when precedence constraints form a uniconnected activity network (or interval order in a different presentation), can also be solved via the same modification of the *LP* problem as described for the C_{max} criterion [Slo81].

5.4 Other Models

In this section two more advanced models of scheduling on parallel processors will be considered. In Section 5.4.1 a scheduling model involving imprecise computations, resulting from tasks not finished before their deadlines, will be discussed and basic results will be presented. Section 5.4.2 analyzes a lot size scheduling problem with parallel processors.

5.4.1 Scheduling Imprecise Computations

In Section 5.3 scheduling problems have been considered with different criteria involving due-dates or deadlines. We already pointed out that the considered optimization criteria such as maximum lateness, and mean or mean weighted tardiness, are very useful for computer control systems, provided the optimal value of the criteria does not exceed zero (i.e. all the tasks are completed before their deadlines). Otherwise, the penalty is calculated with respect to the time at which the delayed tasks are completed. However, in these applications one would rather like to penalize the delayed portions of the tasks, no matter when they are completed. This is, for example, the case for a computer system that collects data from sensing devices. Exceeding a deadline causes the complete loss of uncollected data, and consequently reduces the precision of the measurement procedure. It follows that the *mean weighted information loss* criterion, introduced in [Bla84], is better suited for these situations as it takes into account the weighted loss of those parts of tasks which are unprocessed at their deadlines. Another example is to be found in agriculture [PW89; PW92], where different stretches of land are to be harvested by a single harvester. Any part of the crop not gathered by a given date (which differs according to the site) can no longer be used. Thus, in this case minimizing *total late work* (corresponding to the information loss above) corresponds to minimizing the quantity of wasted crop. Yet another example takes into account variations in processing times of dynamic algorithms and congestion on the communication network, that makes meeting all timing constraints at all times difficult. An approach to minimize this difficulty is to trade off the quality of the results produced by the tasks with the amounts of processing time required to produce the results. Such a tradeoff can be realized by using the *imprecise computation technique* [LLL87, LNL87, CLL90]. This technique prevents timing faults and achieves graceful degradation by making sure that an approximate result of an acceptable quality is available to the user whenever the exact results of the desired quality cannot be obtained in time. An example of a real-time application where one may prefer timely, approximate results, to late, exact results is image processing. It is often better to have frames of fuzzy images produced in time than perfect images produced too late. Another example is tracking. When a tracking system is overloaded and cannot compute all the accurate location data in time, one may, instead, choose to have their rough estimates that can be computed in less time.

In the following we will describe basic complexity results of the problem as discussed in [Bla84, BF87]. In order to unify different terminologies found in the literature we will use the name *execution time loss*.

For every task T_j in the schedule we denote by Y_j its *execution time loss*, defined as follows:

$$Y_j = \begin{cases} \text{amount of processing of } T_j \text{ exceeding } d_j \\ \qquad \text{if } T_j \text{ is not completed in time} \\ 0 \qquad \text{otherwise.} \end{cases}$$

The following criteria will be used to evaluate schedules:

Mean execution time loss: $\overline{Y} = \frac{1}{n} \sum_{j=1}^{n} Y_j$

or, more general,

Mean weighted execution time loss: $\overline{Y}_w = \sum_{j=1}^{n} w_j Y_j \Big/ \sum_{j=1}^{n} w_j$

According to the notation introduced in Section 3.4 we will denote these criteria by ΣY_j and $\Sigma w_j Y_j$, respectively. Let us start with non-preemptive scheduling.

Problem $P \mid r_j \mid \Sigma Y_j$

In [Bla84], the non-preemptive scheduling problem has been proved to be NP-hard in the strong sense, even for the one-processor case; thus we will concentrate on preemptive scheduling in the following.

Problem $P \mid pmtn, r_j \mid \Sigma w_j Y_j$

On the contrary to other total criteria involving due-dates (e.g. mean or mean weighted tardiness) the problem of scheduling preemptable tasks to minimize mean weighted execution time loss can be solved in polynomial time. Below we will present the approach of [BF87] transforming this problem to the one of finding a minimum cost flow in a certain network (cf. Section 5.3.1 where the algorithm by Horn for problem $P \mid pmtn, r_j, \tilde{d}_j \mid -$, has been described).

Let us order ready times r_j and due dates d_j in non-decreasing order and let there be k different values a_k (i.e., r_j or d_j), $k \leq 2n$. These values form $k-1$ time intervals $[a_1, a_2], ..., [a_{k-1}, a_k]$, the length of the ith interval being $t_i = a_{i+1} - a_i$. Now, the network $G = (V, E)$ is constructed as follows. Its set of vertices consists of source S_1 sink S_2, and two groups of vertices; the first corresponding to tasks $T_j, j = 1, 2, ..., n$, the second corresponding to time intervals $[a_i, a_{i+1}], i = 1, 2, ..., k-1$ (cf. Figure 5.4.1). The source S_1 is joined to the task vertex $T_j, j = 1, 2, ..., n$, by an arc of capacity equal to the processing time p_j. The task vertex $T_j, j = 1, 2, ..., n$, is joined by an arc to any interval node $[a_i, a_{i+1}]$ in which T_j can feasibly be processed, i.e. $r_j \leq a_i$ and $d_j \geq a_{i+1}$, and the capacity of the arc is equal to t_i. Moreover, vertex $T_j, j = 1, 2, ..., n$, is joined to the sink S_2 of the network by an arc of capacity equal to p_j. Each interval node $[a_i, a_{i+1}], i = 1, 2, ..., k-1$, is joined to the sink by an arc of capacity equal to mt_i, i.e., equal to the processing capacity of all m processors in that interval. Costs for the arcs directly joining task nodes T_j with sink S_2, are equal to corresponding weights w_j. All other arc costs are zero. The objective now is to find a flow pattern for which the value of the flow from S_1 to S_2 is equal to $\Sigma_{j=1}^{n} p_j$ and whose cost is minimum. It is clear that by solving the above network flow problem we also find an optimal solution to our scheduling problem. The total cost of the flow is exactly equal to the weighted sum of the processing times of those parts of tasks which are unprocessed at their due dates. An optimal schedule is then constructed step-by-step, separately in each interval, by taking the values of flows on arcs joining task nodes with interval nodes and using McNaughton's Algorithm 5.1.8 [McN59]. (Parts of tasks that exceed their respective due dates, if any, may be processed at the end of the schedule.)

Let us now calculate the complexity of the above approach. Clearly, it is predominated by the complexity of finding a minimum cost flow in the network having $O(n)$ nodes and $O(n^2)$ arcs. Using a result by Orlin [Orl88] who gave an $O(|\mathcal{E}|\cdot\log|\mathcal{V}|\cdot(|\mathcal{E}| + |\mathcal{V}|\cdot\log|\mathcal{V}|))$-time algorithm for the minimum cost maximum flow problem ($\mathcal{V}$ and $\mathcal{E}$ respectively denote the set of nodes and the set of edges), the overall complexity of the presented approach is $O(n^4\log n)$. In [BF87] an upper bound on the number of preemptions has been proved to be $(2n-1)(m-1)$.

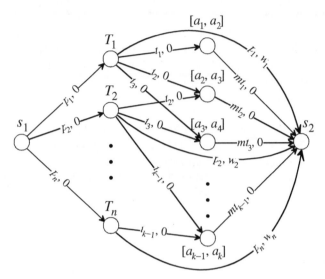

Figure 5.4.1 *A network corresponding to the scheduling problem.*

Problem $Q \mid pmtn, r_j \mid \Sigma w_j Y_j$

The above approach can be generalized to cover also the case of uniform processors [BF87]. The complexity of the algorithm in this case is $O(m^2 n^4 \log mn)$.

Dedicated Processors

Recently several papers have been devoted to analyze the imprecise model of computations in the context of dedicated processors [BPSW04], [BPSW05a], [BPSW05b]. The interested reader is referred to [LLS+91] and [Len04] for a survey of other models and algorithms in the area of imprecise computations.

5.4.2 Lot Size Scheduling

Consider the same problem as discussed in Section 4.4.2 but now instead of one processor there are m processors available for processing all tasks of all job types. Recall that the lot size scheduling problem can be solved in $O(H)$ time for one processor and two job types only. In the following we want to investigate the problem instance with two job types again but now allowing multiple identical processors. First we introduce some basic notation. Then the algorithm is presented without considering inventory restriction; later we show how to take these limitations into account.

Assume that m identical processors P_i, $i = 1, ..., m$ are available for processing the set of jobs $\mathcal{J}$ which consist of two types only; due to capacity restrictions we want to assume that the final schedule is tight. Considering a number $m > 1$ of processors we must determine to which unit time interval (UTI) on which processors a job has to be assigned. Because of continuous production requirements we might also assume an assignment of UTI $h = 0$ to some job type; this can be interpreted as an assignment of some job type to the last UTI of the preceding schedule.

The idea of the algorithm is to assign task after task of the two job types, now denoted by q and r, to empty UTI such that all deadlines are met and no other assignment can reduce change-over cost. In order to do this we have to classify UTIs appropriately. Based on this classification we will present the algorithm. With respect to each deadline d_k we define a "sequence of empty UTI" (SEU) as a processing interval $[h^*, h^*+u-1]$ on some processor consisting of u consecutive and empty UTI. UTI h^*-1 is assigned to some job; UTI h^*+u is either also assigned to some job or it is the first UTI after the occurrence of the deadline. Each SEU can be described by a 3-tuple (i, h^*, u) where i is the number of the processor on which the SEU exists, h^* the first empty UTI and u the number of the UTI in this SEU.

We differentiate between "classes" of SEU by considering the job types assigned to neighboring UTI h^*-1 and h^*+u of each SEU. In case h^*+u has no assignment we denote this by "E"; all other assignments of UTI are denoted by the number of the corresponding job type. Now a "class" is denoted by a pair $[x, y]$ where $x, y \in \{q, r, E\}$. This leads to nine possible classes of SEU from which only classes $[q, r]$, $[q, E]$, $[r, q]$, and $[r, E]$, have to be considered.

Figure 5.4.2 illustrates these definitions using an example with an assignment for UTI $h = 0$. For $d_1 = 6$ we have a SEU $(2,6,1)$ of class $[1, E]$; for $d_2 = 11$ we have $(1, 9, 3)$ of class $[1, E]$, $(2, 6, 2)$ of class $[1, 2]$, $(2, 10, 2)$ of class $[2, E]$.

For each d_k we have to schedule $n_{qk} \geq 0$ and $n_{rk} \geq 0$ tasks. We schedule the corresponding jobs according to non-decreasing deadlines with positive time orientation starting with $k = 1$ up to $k = K$ by applying the following algorithm.

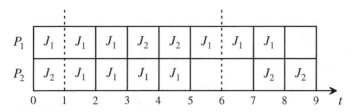

Figure 5.4.2 *Example schedule showing different SEU.*

Algorithm 5.4.1 *Lot size scheduling of two job types on identical processors (LIM)* [PS96].

```
begin
for k := 1 to K do
    while tasks required at d̃_k are not finished do
        begin
        if class [j,E] is not empty
        then Assign job type j to UTI h* of a SEU (i,h*,u) of class [j,E] with
                minimum u
        else
            if classes [q,r] or [r,q] are not empty
            then Assign job type q(r) to UTI h* of a SEU (i,h*,u) of class
                    [q,r] ([r,q]) or if this class is empty to UTI h*+u−1 of a
                    SEU (i,h*,u) of class [r,q] ([q,r])
            else Assign job type q(r) to UTI h*+u−1 of a SEU (i,h*,u) of
                    class [r,E] ([q,E]) with maximum u;
        Use new task assignment to calculate SEU of classes [r,E], [r,q], [q,r],
            and [q,E];
        end;
end;
```

In case the "**while**"-loop cannot be carried out no feasible schedule for the problem under consideration exists. It is necessary to update the classes after each iteration because after a task assignment the number u of consecutive and empty UTI of the concerned SEU decreases by one and thus the SEU might even disappear. Furthermore an assignment of UTI h^* or h^*+u-1 might force the SEU to change the class.

Let us demonstrate the approach by the following example. Let $m = 3$, $\mathcal{J} = \{J_1, J_2\}$, $\tilde{d}_1 = 4$, $\tilde{d}_2 = 8$, $\tilde{d}_3 = 11$, $n_{11} = 3$, $n_{12} = 7$, $n_{13} = 5$, $n_{21} = 5$, $n_{22} = 6$, $n_{23} = 7$ and zero initial inventory. Let us assume that there is a pre-assignment for $h = 0$ such that J_1 is processed by P_1 and J_2 is processed by P_2 and P_3. In Figure 5.4.3 the optimal schedule generated by Algorithm 5.4.1 is given.

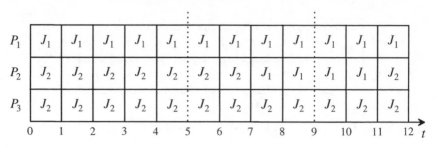

Figure 5.4.3 *Optimal schedule for the example problem.*

It can be shown that Algorithm 5.4.1 generates an optimal schedule if one exists. Feasibility of the algorithm is guaranteed by scheduling the job types according to earliest deadlines using only free UTI of the interval $[0, d_k]$. To prove optimality of the algorithm one has to show that the selection of the UTI for assigning the task under consideration is best possible. These facts have been proved in the following lemmas [PS96] which are formulated and proved for job type q, but they also hold in case of job type r.

Lemma 5.4.2 *There exists an optimal solution that can be built such that job type q is assigned to UTI h^* on processor P_i in case the selected SEU belongs to classes $[q, E]$ or $[q, r]$. If the SEU belongs to class $[r, E]$ or $[r, q]$ then q is assigned to UTI $h^* + u - 1$ on processor P_i.*   □

Lemma 5.4.3 *Algorithm 5.4.1 generates schedules with a minimum number of change-overs for two types of jobs.*   □

The complexity of Algorithm 5.4.1 is $O(Hm)$.

Let us now investigate how we can consider inventory restrictions for both job types, i.e. for each job type an upper bound B_j on in-process inventory is given. If there are only two job types, limited in-process storage capacity can be translated to updated demands of unit time tasks referring to given deadlines d_k. If processing of some job type has to be stopped because of storage limitations, processing of the other job has to be started as $Hm = \sum_{j=1,...,n} n_j$. This can be achieved by increasing the demand of the other job type, appropriately.

Assume that a demand and inventory feasible and tight schedule exists for the problem instance. Let N_{jk} be the updated demand after some preprocessing step now used as input for the algorithm. To define this input more precisely let us first consider how many unit time tasks of some job type, e.g. q, have to be processed up to some deadline d_k:

– at most the number of tasks of job type q which does not exceed storage limit, i.e. $L_q = B_q - \sum_{i=1,...,k-1} (N_{qi} - n_{qi})$;

– at least the number of required tasks of job type q, i.e.

$$D_q = n_{qk} - \sum_{i=1,...,k-1} (N_{qi} - n_{qi});$$

– at least the remaining processing capacity reduced by the number of tasks of job type r which can be processed feasibly. From this we get $R_q = c_k - \sum_{i=1,...,k-1} (N_q^i + n_{qi}) - (B_r - \sum_{i=1,...,k-1} (N_{ri} + n_{ri}))$, where $c_k = md_k$ is the total processing capacity in the intervals $[0, d_k]$ on m processors.

The same considerations hold respectively for the other job type r.

With the following lemmas we show how the demand has to be updated such that not only feasibility (Lemma 5.4.4) but also optimality (Lemma 5.4.6) concerning change-overs is retained. We start with showing that L_j can be omitted if we calculate N_{jk}.

Lemma 5.4.4 *In case that a feasible and tight schedule exists, $L_j = B_j - \sum_{i=1,...,k-1} (N_{ji} - n_{ji})$ can be neglected.* □

From the result of Lemma 5.4.4 we can define N_{jk} more precisely by

$$N_{qk} := \max\{ n_{qk} - \sum_{i=1,...,k-1} (N_{qi} - n_{qi}),$$
$$c_k - \sum_{i=1,...,k-1} (N_{qi} + N_{ri}) - (B_r - \sum_{i=1,...,k-1} (N_{ri} - n_{ri}))\} \qquad (5.4.1)$$

$$N_{rk} := \max\{ n_{rk} - \sum_{i=1,...,k-1} (N_{ri} - n_{ri}),$$
$$c_k - \sum_{i=1,...,k-1} (N_{ri} + N_q^i) - (B_q - \sum_{i=1,...,k-1} (N_{qi} - n_{qi}))\} \qquad (5.4.2)$$

One may show [PS96] that after updating all demands of unit time jobs of type q according to (5.4.1) the new problem instance is equivalent to the original one. We omit the case of job type r and (5.4.2), which directly follows in an analogous way. Notice that the demand will only be updated, if inventory restrictions limit assignment possibilities up to a certain deadline d_k. Only in this case the kth interval will be completely filled with jobs. If no inventory restrictions have to be considered equations (5.4.1) and (5.4.2) result in the original demand pattern.

Lemma 5.4.5 *After adapting N_{qk} according to (5.4.1) the feasibility of the solution according to the inventory constraints on r is guaranteed.* □

Lemma 5.4.6 *If*

(i) $\quad n_{qk} - \sum_{i=1,...,k-1} (N_{qi} - n_{qi}) \geq$
$$c_k - \sum_{i=1,...,k-1} (N_{qi} + N_{ri}) - (B_r - \sum_{i=1,...,k-1} (N_{ri} - n_{ri}))$$

or

(ii) $\quad n_{qk} - \sum_{i=1,...,k-1} (N_{qi} - n_{qi}) <$
$$c_k - \sum_{i=1,...,k-1} (N_{qi} + N_{ri}) - (B_r - \sum_{i=1,...,k-1} (N_{ri} - n_{ri}))$$

for some deadline d_k then a demand feasible and optimal schedule can be constructed.   □

The presented algorithm also solves the corresponding problem instance with arbitrary positive change-over cost because for two job types only, minimizing the number of change-overs is equivalent to minimizing the sum of their positive change-over cost. In order to solve the practical gear-box manufacturing problem where more than two job types have to be considered a heuristic has been implemented which uses the ideas of the presented approach. The corresponding scheduling rule is considered to be that no unforced change-overs should occur. The resulting algorithm is part of a scheduling system, which incorporates a graphical representation scheme using Gantt-charts and further devices to give the manufacturing staff an effective tool for decision support. For more results on the implementation of scheduling systems on the shop floor we refer to Chapter 16.

References

AH73   D. Adolphson, T. C. Hu, Optimal linear ordering, *SIAM J. Appl. Math.* 25, 1973, 403-423.

AHU74   A. V. Aho, J. E. Hopcroft, J. D. Ullman, *The Design and Analysis of Computer Algorithms*, Addison-Wesley, Reading, Mass., 1974.

Ash72   S. Ashour, *Sequencing Theory*, Springer, Berlin, 1972.

Bak74   K. Baker, *Introduction to Sequencing and Scheduling*, J. Wiley, New York, 1974.

BCS74   J. Bruno, E. G. Coffman, Jr., R. Sethi, Scheduling independent tasks to reduce mean finishing time, *Comm. ACM* 17, 1974, 382-387.

BCSW76a   J. Błażewicz, W. Cellary, R. Słowiński, J. Węglarz, Deterministyczne problemy szeregowania zadań na równoległych procesorach, Cz. I. Zbiory zadań nie zależnych, *Podstawy Sterowania* 6, 1976, 155-178.

BCSW76b   J. Błażewicz, W. Cellary, R. Słowiński, J. Węglarz, Deterministyczne problemy szeregowania zadań na równoległych procesorach, Cz. II. Zbiory zadań zależnych, *Podstawy Sterowania* 6, 1976, 297-320.

BCW77   J. Błażewicz, W. Cellary, J. Węglarz, A strategy for scheduling splittable tasks to reduce schedule length, *Acta Cybernet.* 3, 1977, 99-106.

BF87   J. Błażewicz, G. Finke, Minimizing mean weighted execution time loss on identical and uniform processors, *Inform. Process. Lett.* 24, 1987, 259-263.

BGJ77   P. Brucker, M. R. Garey, D. S. Johnson, Scheduling equal-length tasks under treelike precedence constraints to minimize maximum lateness, *Math. Oper. Res.* 2, 1977, 275-284.

BK00   J. Błażewicz, D. Kobler, On the ties between different graph representation for scheduling problems, Report, Poznan University of Technology, Poznan, 2000.

BK02   J. Błażewicz, D.Kobler, Review of properties of different precedence graphs for scheduling problems, *European J. of Oper. Res.* 142 (2002) 435-443.

Bla77   J. Błażewicz, Simple algorithms for multiprocessor scheduling to meet deadlines, *Inform. Process. Lett.* 6, 1977, 162-164.

Bla84   J. Błażewicz, Scheduling preemptible tasks on parallel processors with information loss, *Technique et Science Informatiques* 3, 1984, 415-420.

Bru76a   J. Bruno, Scheduling algorithms for minimizing the mean weighted flow-time, in: E. G. Coffman, Jr. (ed.), *Computer and Job-Shop Scheduling Theory*, J. Wiley, New York, 1976.

BPSW04   J. Błażewicz, E. Pesch, M. Sterna, F. Werner, Flow shop scheduling with late work criterion – choosing the best solution strategy, *Lecture Notes in Computer Sci.* 3285, 2004, 68-75.

BPSW05a   J. Błażewicz, E. Pesch, M. Sterna, F. Werner, The two-machine flow-shop problem with weighted late work criterion and common due-date, *European J. of Oper. Res.* 165, 2005, 408-415.

BPSW05a   J. Błażewicz, E. Pesch, M. Sterna, F. Werner, A comparison of solution procedures for two-machine flow-shop scheduling with late work criterion, *Computers and Industrial Eng.* 49, 2005, 611-624.

Bru76b   P. J. Brucker, Sequencing unit-time jobs with treelike precedence on m processors to minimize maximum lateness, *Proc. IX International Symposium on Mathematical Programming*, Budapest, 1976.

BT94   B. Braschi, D. Trystram, A new insight into the Coffman-Graham algorithm, *SIAM J. Comput.* 23, 1994, 662-669.

CD73   E. G. Coffman, Jr., P. J. Denning, *Operating Systems Theory*, Prentice-Hall, Englewood Cliffs, N. J., 1973.

CFL83   E. G. Coffman, Jr., G. N. Frederickson, G. S. Lueker, Probabilistic analysis of the LPT processor scheduling heuristic, unpublished paper, 1983.

CFL84   E. G. Coffman, Jr., G. N. Frederickson, G. S. Lueker, A note on expected makespans for largest-first sequences of independent task on two processors, *Math. Oper. Res.* 9, 1984, 260-266.

CG72   E. G. Coffman, Jr., R. L. Graham, Optimal scheduling for two-processor systems, *Acta Inform.* 1, 1972, 200-213.

CG91   E. G. Coffman, Jr., M. R. Garey, Proof of the 4/3 conjecture for preemptive versus nonpreemptive two-processor scheduling, Report Bell Laboratories, Murray Hill, 1991.

CGJ78   E. G. Coffman, Jr., M. R. Garey, D. S. Johnson, An application of bin-packing to multiprocessor scheduling, *SIAM J. Comput.* 7, 1978, 1-17.

CGJ84   E. G. Coffman, Jr., M. R. Garey, D. S. Johnson, Approximation algorithms for bin packing - an updated survey, in: G. Ausiello, M. Lucertini, P. Serafini (eds.), *Algorithm Design for Computer System Design*, Springer, Vienna, 1984, 49-106.

CL75   N.-F. Chen, C. L. Liu, On a class of scheduling algorithms for multiprocessor computing systems, in: T.-Y. Feng (ed.), *Parallel Processing, Lecture Notes in Computer Sci.* 24, Springer, Berlin, 1975, 1-16.

CLL90   J. Y. Chung, J. W. S. Liu, K. J. Lin, Scheduling periodic jobs that allow imprecise results, *IEEE Trans. Comput.* 19, 1990, 1156-1173.

CMM67   R. W. Conway, W. L. Maxwell, L. W. Miller, *Theory of Scheduling*, Addison-Wesley, Reading, Mass., 1967.

Cof73   E. G. Coffman, Jr., A survey of mathematical results in flow-time scheduling for computer systems, *GI - 3. Jahrestagung, Hamburg*, Springer, Berlin, 1973, 25-46.

Cof76   E. G. Coffman, Jr. (ed.), *Scheduling in Computer and Job Shop Systems*, J. Wiley, New York, 1976.

CS76   E. G. Coffman, Jr., R. Sethi, A generalized bound on LPT sequencing, *RAIRO-Informatique* 10, 1976, 17-25.

DL88   J. Du, J. Y.-T. Leung, Scheduling tree-structured tasks with restricted execution times, *Inform. Process. Lett.* 28, 1988, 183-188.

DL89   J. Du, J. Y.-T. Leung, Scheduling tree-structured tasks on two processors to minimize schedule length, *SIAM J. Discrete Math.* 2, 1989, 176-196.

DLY91   J. Du, J. Y.-T. Leung, G. H. Young, Scheduling chain structured tasks to minimize makespan and mean flow time, *Inform. and Comput.* 92, 1991, 219-236.

DW85   D. Dolev, M. K. Warmuth, Scheduling flat graphs, *SIAM J. Comput.* 14, 1985, 638-657.

EH93   K. H. Ecker, R. Hirschberg, Task scheduling with restricted preemptions. *Proc. PARLE93 - Parallel Architectures and Languages*, Munich, 1993.

FB73   E. B. Fernandez, B. Bussel, Bounds on the number of processors and time for multiprocessor optimal schedules, *IEEE Trans. Comput.* 22, 1973, 745-751.

FG86   A. Federgruen, H. Groenevelt, Preemptive scheduling of uniform processors by ordinary network flow techniques, *Management Sci.* 32, 1986, 341-349.

FKN69   M. Fujii, T. Kasami, K. Ninomiya, Optimal sequencing of two equivalent processors, *SIAM J. Appl. Math.* 17, 1969, 784-789, Err: *SIAM J. Appl. Math.* 20, 1971, 141.

Fre82   S. French, *Sequencing and Scheduling: An Introduction to the Mathematics of the Job-Shop*, Horwood, Chichester, 1982.

FRK86   J. B. G. Frenk, A. H. G. Rinnooy Kan, The rate of convergence to optimality of the LPT rule, *Discrete Appl. Math.* 14, 1986, 187-197.

FRK87   J. B. G. Frenk, A. H. G. Rinnooy Kan, The asymptotic optimality of the LPT rule, *Math. Oper. Res.* 12, 1987, 241-254.

Gab82   H. N. Gabow, An almost linear algorithm for two-processor scheduling, *J. Assoc. Comput. Mach.* 29, 1982, 766-780.

Gar -   M. R. Garey, Unpublished result.

Gar73   M. R. Garey, Optimal task sequencing with precedence constraints, *Discrete Math.* 4, 1973, 37-56.

GG73   M. R. Garey, R. L. Graham, Bounds on scheduling with limited resources, *Operating System Review,* 1973, 104-111.

GG75   M. R. Garey, R. L. Graham, Bounds for multiprocessor scheduling with re- source constraints, *SIAM J. Comput.* 4, 1975, 187-200.

GIS77   T. Gonzalez, O. H. Ibarra, S. Sahni, Bounds for LPT schedules on uniform processors, *SIAM J. Comput.* 6, 1977, 155-166.

GJ76   M. R. Garey, D. S. Johnson, Scheduling tasks with nonuniform deadlines on two processors, *J. Assoc. Comput. Mach.* 23, 1976, 461-467.

GJ77   M. R. Garey, D. S. Johnson, Two-processor scheduling with start-times and deadlines, *SIAM J. Comput.* 6, 1977, 416-426.

GJ79   M. R. Garey, D. S. Johnson, *Computers and Intractability: A Guide to the Theory of NP-Completeness*, W. H. Freeman, San Francisco, 1979.

GJST81   M. R. Garey, D. S. Johnson, B. B. Simons, R. E. Tarjan, Scheduling unit time tasks with arbitrary release times and deadlines, *SIAM J. Comput.* 10, 1981, 256-269.

GJTY83   M. R. Garey, D. S. Johnson, R. E. Tarjan, M. Yannakakis, Scheduling oppos- ing forests, *SIAM J. Algebraic Discrete Meth.* 4, 1983, 72-93.

GLL+79   R. L. Graham, E. L. Lawler, J. K. Lenstra, A. H. G. Rinnooy Kan, Optimiza- tion and approximation in deterministic sequencing and scheduling theory: a survey, *Ann. Discrete Math.* 5, 1979, 287-326.

Gon77   T. Gonzalez, Optimal mean finish time preemptive schedules, Technical Re- port 220, Computer Science Department, Pennsylvania State Univ., 1977.

Gra66   R. L. Graham, Bounds for certain multiprocessing anomalies, *Bell System Tech. J.* 45, 1966, 1563-1581.

Gra69   R. L. Graham, Bounds on multiprocessing timing anomalies, *SIAM J. Appl. Math.* 17, 1969, 263-269.

Gra76   R. L. Graham, Bounds on performance of scheduling algorithms, Chapter 5 in: E. G. Coffman, Jr. (ed.), *Scheduling in Computer and Job Shop Systems*, J. Wiley, New York, 1976.

GS78   T. Gonzalez, S. Sahni, Preemptive scheduling of uniform processor systems, *J. Assoc. Comput. Mach.* 25, 1978, 92-101.

HLS77   E. G. Horvath, S. Lam, R. Sethi, A level algorithm for preemptive scheduling, *J. Assoc. Comput. Mach.* 24, 1977, 32-43.

Hor73   W. A. Horn, Minimizing average flow time with parallel processors, *Oper. Res.* 21, 1973, 846-847.

Hor74   W. A. Horn, Some simple scheduling algorithms, *Naval Res. Logist. Quart.* 21, 1974, 177-185.

HS76   E. Horowitz, S. Sahni, Exact and approximate algorithms for scheduling non- identical processors, *J. Assoc. Comput. Mach.* 23, 1976, 317-327.

HS87   D. S. Hochbaum, D. B. Shmoys, Using dual approximation algorithms for scheduling problems: theoretical and practical results, *J. Assoc. Comput. Mach.* 34, 1987, 144-162.

Hu61   T. C. Hu, Parallel sequencing and assembly line problems, *Oper. Res.* 9, 1961, 841-848.

IK77   O. H. Ibarra, C. E. Kim, Heuristic algorithms for scheduling independent tasks on nonidentical processors, *J. Assoc. Comput. Mach.* 24, 1977, 280-289.

Jac55   J. R. Jackson, Scheduling a production line to minimize maximum tardiness, Res. Report 43, Management Research Project, University of California, Los Angeles, 1955.

JMR+04   J.Jozefowska, M.Mika, R.Rozycki, G.Waligora, J.Weglarz, An almost optimal heurisitc for preemptive Cmax scheduling of dependent tasks on parallel identical machines, Annals of Operations Research 129, 205-216, 2004.

Joh83   D. S. Johnson, The NP-completeness column: an ongoing guide, *J. Algorithms* 4, 1983, 189-203.

Kar72   R. M. Karp, Reducibility among combinatorial problems, in: R. E. Miller, J. W. Thatcher (eds.), *Complexity of Computer Computations*, Plenum Press, New York, 1972, 85-104.

Kar74   A. W. Karzanov, Determining the maximal flow in a network by the method of preflows, *Soviet Math. Dokl.* 15, 1974, 434-437.

Kar84   N. Karmarkar, A new polynomial-time algorithm for linear programming, *Combinatorica* 4, 1984, 373-395.

KE75   O. Kariv, S. Even. An $O(n^{2.5})$ algorithm for maximum matching in general graphs, *16th Annual Symposium on Foundations of Computer Science IEEE*, 1975, 100-112.

Ked70   S. K. Kedia, A job scheduling problem with parallel processors, Unpublished Report, Dept. of Ind. Eng., University of Michigan, Ann Arbor, 1970.

Kha79   L. G. Khachiyan, A polynomial algorithm for linear programming (in Russian), *Dokl. Akad. Nauk SSSR*, 244, 1979, 1093-1096.

KK82   N. Karmarkar, R. M. Karp, The differing method of set partitioning, Report UCB/CSD 82/113, Computer Science Division, University of California, Berkeley, 1982.

Kun76   M. Kunde, Beste Schranke beim LP-Scheduling, Bericht 7603, Institut für Informatik und Praktische Mathematik, Universität Kiel, 1976.

Law73   E. L. Lawler, Optimal sequencing of a single processor subject to precedence constraints, *Management Sci.* 19, 1973, 544-546.

Law76   E. L. Lawler, *Combinatorial optimization: Networks and Matroids*, Holt, Rinehart and Winston, New York, 1976.

Law82a   E. L. Lawler, Recent results in the theory of processor scheduling, in: A. Bachem, M. Grötschel, B. Korte (eds.) *Mathematical Programming: The State of Art*, Springer, Berlin, 1982, 202-234.

Law82b   E. L. Lawler, Preemptive scheduling in precedence-constrained jobs on parallel
          processors, in: M. A. H. Dempster, J. K. Lenstra, A. H. G. Rinnooy Kan (eds.),
          Deterministic and Stochastic Scheduling, Reidel, Dordrecht, 1982, 101-123.

Lee91   C.-Y. Lee, Parallel processor scheduling with nonsimultaneous processor
          available time, *Discrete Appl. Math.* 30, 1991, 53-61.

Len77   J. K. Lenstra, *Sequencing by Enumerative Methods*, Mathematical Centre Tract
          69, Mathematisch Centrum, Amsterdam, 1977.

Len04   J. Y-T. Leung (ed.), *Handbook of Scheduling. Algorithms, Models and Per-
          formance Analysis*, Chapman & Hall/CRC, Boca Raton, 2004.

LL74a   J. W. S. Liu, C. L. Liu, Performance analysis of heterogeneous multiprocessor
          computing systems, in E. Gelenbe, R. Mahl (eds.), *Computer Architecture and
          Networks*, North Holland, Amsterdam, 1974, 331-343.

LL74b   J. W. S. Liu, C. L. Liu, Bounds on scheduling algorithms for heterogeneous
          computing systems, Technical Report UIUCDCS-R-74-632, Dept. of Com-
          puter Science, University of Illinois at Urbana-Champaign, 1974.

LL78   E. L. Lawler, J. Labetoulle, Preemptive scheduling of unrelated parallel proc-
          essors by linear programming, *J. Assoc. Comput. Mach.* 25, 1978, 612-619.

LLL87   J. W. S. Liu, K. J. Lin, C. L. Liu, A position paper for the IEEE Workshop on
          real-time operating systems, Cambridge, Mass, 1987.

LLL+84   J. Labetoulle, E. L. Lawler, J. K. Lenstra, A. H. G. Rinnooy Kan, Preemptive
          scheduling of uniform processors subject to release dates, in: W. R. Pulley-
          blank (ed.), *Progress in Combinatorial Optimization*, Academic Press, New
          York, 1984, 245-261.

LLRK82   E. L. Lawler, J. K. Lenstra, A. H. G. Rinnooy Kan, Recent developments in
          deterministic sequencing and scheduling: a survey, in M. A. H. Dempster, J. K.
          Lenstra, A. H. G. Rinnooy Kan (eds.), *Deterministic and Stochastic Schedul-
          ing*, Reidel, Dordrecht, 1982, 35-73.

LLR+93   E. L. Lawler, J. K. Lenstra, A. H. G. Rinnooy Kan, D. B. Shmoys, Sequencing
          and scheduling: Algorithms and complexity, in: S. C. Graves, A. H. G. Rin-
          nooy Kan, P. H. Zipkin (eds.), *Handbook in Operations Research and Man-
          agement Science, Vol. 4: Logistics of Production and Inventory*, Elsevier, Am-
          sterdam, 1993.

LLS+91   J. W. S. Liu, K. J. Lin, W. K. Shih, A. C. Yu, J. Y. Chung, W. Zhao, Algo-
          rithms for scheduling imprecise computations, in: A. M. Van Tilborg, G. M.
          Koob (eds.) *Foundations of Real-Time Computing: Scheduling and Resource
          Management,* Kluwer, Boston, 1991.

LNL87   K. J. Lin, S. Natarajan, J. W. S. Liu, Imprecise results: utilizing partial compu-
          tations in real-time systems, *Proc. of the IEEE 8th Real-Time Systems Sympo-
          sium*, San Jose, California, 1987.

LRK78   J. K. Lenstra, A. H. G. Rinnooy Kan, Complexity of scheduling under prece-
          dence constraints, *Oper. Res.* 26, 1978, 22-35.

LRK84   J. K. Lenstra, A. H. G. Rinnooy Kan, Scheduling theory since 1981: an anno-
          tated bibliography, in: M. O'h Eigearthaigh, J. K. Lenstra, A. H. G. Rinnooy

Kan (eds.), *Combinatorial Optimization: Annotated Bibliographies,* J. Wiley, Chichester, 1984.

LRKB77   J. K. Lenstra, A. H. G. Rinnooy Kan, P. Brucker, Complexity of processor scheduling problems, *Ann. Discrete Math.* 1, 1977, 343-362.

LS77   S. Lam, R. Sethi, Worst case analysis of two scheduling algorithms, *SIAM J. Comput.* 6, 1977, 518-536.

MC69   R. Muntz, E. G. Coffman, Jr., Optimal preemptive scheduling on two-processor systems, *IEEE Trans. Comput.* 18, 1969, 1014-1029.

MC70   R. Muntz, E. G. Coffman, Jr., Preemptive scheduling of real time tasks on multiprocessor systems, *J. Assoc. Comput. Mach.* 17, 1970, 324-338.

McN59   R. McNaughton, Scheduling with deadlines and loss functions, *Management Sci.* 6, 1959, 1-12.

NLH81   K. Nakajima, J. Y-T. Leung, S. L. Hakimi, Optimal two processor scheduling of tree precedence constrained tasks with two execution times, *Performance Evaluation* 1, 1981, 320-330.

Orl88   J. B. Orlin, A faster strongly polynomial minimum cost flow algorithm, *Proc.20th ACM Symposium on the Theory of Computing,* 1988, 377-387.

PS96   M. Pattloch, G. Schmidt, Lotsize scheduling of two job types on identical processors, *Discrete Appl. Math.,* 1996, 409-419.

PW89   C. N. Potts, L. N. Van Wassenhove, Single processor scheduling to minimize total late work, Report 8938/A, Econometric Institute,Erasmus University, Rotterdam, 1989.

PW92   C. N. Potts, L. N. Van Wassenhove, Approximation algorithms for scheduling a single processor to minimize total late work, *Oper. Res. Lett.* 11,1992, 261-266.

Rin78   A. H. G. Rinnooy Kan, *Processor Scheduling Problems: Classification, Complexity and Computations,* Nijhoff, The Hague, 1978.

RG69   C. V. Ramamoorthy, M. J. Gonzalez, A survey of techniques for recognizing parallel processable streams in computer programs, AFIPS Conference Proceedings, Fall Joint Computer Conference, 1969, 1-15.

Ros–   P. Rosenfeld, unpublished result.

Rot66   M. H. Rothkopf, Scheduling independent tasks on parallel processors, *Management Sci.* 12, 1966, 347-447.

RS83   H. Röck, G. Schmidt, Processor aggregation heuristics in shop scheduling, *Methods Oper. Res.* 45, 1983, 303-314.

Sah79   S. Sahni, Preemptive scheduling with due dates, *Oper. Res.* 5, 1979, 925-934.

SC80   S. Sahni, Y. Cho, Scheduling independent tasks with due times on a uniform processor system, *J. Assoc. Comput. Mach.* 27, 1980, 550-563.

Sch84   G. Schmidt, Scheduling on semi-identical processors, *ZOR* A28, 1984, 153-162.

Sit05   R. Sitters, Complexity of preemptive minsum scheduling on unrelated parallel machines, *J. Algorithms* 57, 2005, 37-48.

Sch88   G. Schmidt, Scheduling independent tasks with deadlines on semi-identical processors, *J. Oper. Res. Soc.* 39, 1988, 271-277.

Set76   R. Sethi, Algorithms for minimal-length schedules, Chapter 2 in: E. G. Coffman, Jr. (ed.), *Scheduling in Computer and Job Shop Systems*, J. Wiley, New York, 1976.

Set77   R. Sethi, On the complexity of mean flow time scheduling, *Math. Oper. Res.* 2, 1977, 320-330.

Sev91   S. V. Sevastjanov, Private communication, 1991.

Sit05   R.Sitters, Complexity of preemptive minsum scheduling on unrelated parallel machines, *J. Algorithms* , 2005, 37-48.

Slo78   R. Słowiński, Scheduling preemptible tasks on unrelated processors with additional resources to minimise schedule length, in G. Bracci, R. C. Lockemann (eds.), *Lecture Notes in Computer Sci.* 65, Springer, Berlin, 1978, 536-547.

SW77   R. Słowiński, J. Węglarz, Minimalno-czasowy model sieciowy z różnymi sposobami wykonywania czynnosci, *Przeglad Statystyczny* 24, 1977, 409-416.

Ull76   J. D. Ullman, Complexity of sequencing problems, Chapter 4 in: E. G. Coffman, Jr. (ed.), *Scheduling in Computer and Job Shop Systems*, J. Wiley, New York, 1976.

WBCS77   J. Węglarz, J. Błażewicz, W. Cellary, R. Słowiński, An automatic revised simplex method for constrained resource network scheduling, *ACM Trans. Math. Software* 3, 1977, 295-300.

Wer84   D. de Werra, Preemptive scheduling linear programming and network flows, *SIAM J. Algebra Discrete Math.* 5, 1984, 11-20.

6 Communication Delays and Multiprocessor Tasks

6.1 Introductory Remarks

One of the assumptions imposed in Chapter 3 was that each task is processed on at most one processor at a time. However, in recent years, with the rapid development of manufacturing as well as microprocessor and especially multi-microprocessor systems, the above assumption has ceased to be justified in some important applications. There are, for example, self-testing multi-microprocessor systems in which one processor is used to test others, or diagnostic systems in which testing signals stimulate the tested elements and their corresponding outputs are simultaneously analyzed [Avi78, DD81]. When formulating scheduling problems in such systems, one must take into account the fact that some tasks have to be processed on more than one processor at a time. On the other hand, communication issues must be also taken into account in systems where tasks (e. g. program modules) are assigned to different processors and exchange information between each other.

Nowadays, parallel and distributed systems are distinguished. In parallel systems, processors work cooperatively on parts of the same "big" job. A set of processors is tightly coupled to establish a large multiprocessor system. Due to rather short communication links between processors, communication times are small as compared to that in distributed systems, where several independent computers are connected via a local or wide area network.

In general, one may distinguish two approaches to handle processor assignment problems arising in the above context [1] [BEPT00, Dro96a, Vel93]. The first approach, the so-called *load balancing and mapping*, assumes a program to be partitioned into tasks forming an undirected graph [Bok81]. Adjacent tasks communicate with each other, thus, their assignment to different processors causes certain communication delay. The problem is to allocate tasks to processors in such a way that the total interprocessor communication is minimized, while processor loads are balanced. This approach is based on graph theoretic methods and is not considered in the following.

---

[1] We will neither be concerned here with the design of proper partition of programs into module tasks, nor with programming languages parallelism.

The other approach (discussed in this Chapter) assumes that, as in the classical scheduling theory, a program is partitioned into tasks forming a directed graph. Here, nodes of the graph represent tasks and directed arcs show a one way communication between predecessor tasks and successor tasks. Basically, there exist three models describing the communication issues in scheduling.

In the first model [BDW84, BDW86, Llo81] each task may require more than one processor at a time. During an execution of these *multiprocessor* tasks communication among processors working on the same task is implicitly hidden in a "black box" denoting an assignment of this task to a subset of processors during some time interval. The second model assumes that *uniprocessor* tasks, each assigned to one of the processors, need to communicate. If two such tasks are assigned to different processors communication is explicitly performed via links of the processor network [Ray87a, Ray87b, LVV96], and connected with this some communication delay occurs. The last model is a combination of the first two models and involves the so-called divisible tasks [CR88, SRL95]. A *divisible* task (*load*) is a task that can be partitioned into smaller parts that are distributed among the processors. Communication and computation phases are interleaved during the execution of a task. Such a model is particularly suited in cases where large data files are involved, such as in image processing, experimental data processing or Kalman filtering.

In the following three sections basic results concerning the above three models will be presented. Before doing this processor, communication and task systems respectively, will be described in a greater detail.

As far as processor systems are concerned, they may be divided (as before) into two classes: parallel and dedicated. Usually each processor has its *local memory*. *Parallel processors* are functionally identical, but may differ from each other by their speeds. On the other hand, *dedicated processors* are usually specialized to perform specific computations (functions). Several models of processing task sets on dedicated processors are distinguished; flow shop, open shop and job shop being the most representative. As defined in Chapter 3, in these cases the set of tasks is partitioned into subsets called *jobs*. However, in the context of the considered multiprocessor systems, dedication of processors may also denote a preallocation feature of certain tasks to functionally identical processors. As will be discussed later a task can be preallocated to a single processor or to several processors at a time.

The property of a processor network with respect to communication performance depends greatly on its specific topological structure. Examples of standard network topologies are: *linear arrays (processor chains)*, *rings*, *meshes*, *trees*, *hypercubes* (Figure 6.1.1). Other, more recent network structures are the *de Bruijn* [Bru46] or the *Benes multiprocessor networks* [Ben64]. Besides these, a variety of so-called *multistage interconnection networks* [SH89] had been introduced; examples are *perfect shuffle networks* [Sto71], *banyans* [Gok76], and *delta networks* [KS86]. Finally, many more kinds of network structures like combinations of the previous network types such as the *cube connected cycles networks* [PV81], or network hierarchies (e.g. *master-slave networks*) can be found in the literature.

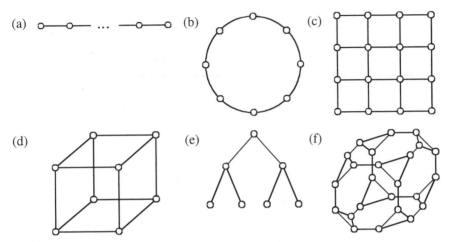

Figure 6.1.1 *Basic interconnection networks:*
    (a) *linear array,*
    (b) *ring,*
    (c) *mesh,*
    (d) *hypercube,*
    (e) *tree*
    (f) *cube connected cycles network.*

An important feature of a communication network is the ability (or lack) of a single processing element to compute tasks and to communicate at the same time. If this is possible we will say that each processing element has a *communication coprocessor*. Another important feature of the communication system is the ability (or lack) of each processing element to communicate with other processing elements via several communication links (or several ports) at the same time. We will not discuss here message routing strategies in communication networks. However, we refer the interested reader to [BT89] where such strategies like *store-and-forward*, *wormhole routing*, *circuit switching*, and *virtual-cut-through* are presented and their impact on communication delays is discussed.

We will now discuss differences between the model of a task system as presented in Chapter 3 and the one proposed for handling communication issues in multiprocessor systems. The main difference is caused by the fact that tasks (or a task) processed by different processors must exchange information and such an exchange introduces communication delays. These delays may be handled either implicitly or explicitly. In the first case, the communication times are already included in the task processing times. Usually, a task requires then more than one processor at a time, thus, we may call it a *multiprocessor task*. Multiprocessor tasks may specify their processor requirements either in terms of number of simultaneously required processors, or in terms of an explicit specification of a processor subset (or processor subsets) which is or are required for processing. In the first case we will speak about *parallel processor requirement*, whereas in the second we will speak about *dedicated processor requirement*.

An interesting question is the specification of task processing times. In case of parallel processors one may define several functions describing the dependency of this time on the size of the processor system required by a task. In the following we will assume that the task processing time is inversely proportional to the processor set size, i.e. $p_j^k = p_j^1/k$, where k is the size of a required processor set. We refer the interested reader to [Dro96a] where more complicated models are analyzed. In this context let us mention the processing systems where task processing times are arbitrary functions of a number of processors allocated. Two tasks are distinguished: *malleable* and *moldable task models* [Len04]. In the first case, a number of processors allocated to a task may change during the execution of this task. In the second case, this number is constant during the task execution.

In case of dedicated processors each task processing time must be explicitly associated with the processor set required, i.e. with each processor set $\mathcal{D}$ which can be assigned to task T_j processing time $p_j^{\mathcal{D}}$ is associated. As in the classical scheduling theory a preemptable task is completed if all its parts processed on different sets of processors sum up to 1 if normalized to fractions of a unit.

As far as an explicit handling of communication delays is concerned one may distinguish two subcases. In the first case each *uniprocessor task* requires only one processor at a time and after its completion sends a message (results of the computations) to all its successors in the precedence graph. We assume here that each task represents some amount of computational activity. If two tasks, T_i and T_j are in precedence relation, i.e. $T_i \prec T_j$, then T_j will partly use information produced by task T_i. Thus, only if both tasks are processed on different processors, transmission of the required data will be due, and a *communication delay* will occur.

In the deterministic scheduling paradigm, we assume that communication delays are predictable. If required we may assume that the delays depend on various parameters like the amount of data, or on the distance between source and target processor.

The transfer of data between tasks T_i and T_j can be represented by a data set associated with the pair (T_i, T_j). Each transmission activity causes a certain delay that is assumed to be fixed. If we assume that T_i and T_j are processed on P_k and P_l, respectively, let $c(T_i, P_k; T_j, P_l)$ denote the delay due to transmitting the required data of T_i from P_k to P_l. That means we assume that after this time elapsed the data are available at processor P_k. This delay takes into account the involved tasks and the particular processors the tasks are processed on. However, it is normally assumed to be independent of the actual communication workload in the network.

The third case is concerned with the allocation of *divisible tasks* to different processors. In fact, this model combines multiprocessor task scheduling with explicit communication delays.

The $\alpha \mid \beta \mid \gamma$ - notation of scheduling problems introduced in Section 3.4 will now be enhanced to capture the different features of multiprocessor systems discussed above. Such changes were first introduced in [Vel93, Dro96a].

In the first field (α) describing processor environment two parameters α_3 and α_4 are added. Parameter $\alpha_3 \in \{\varnothing, conn, linear\ array, ring, tree, mesh, hypercube\}$ describes the architecture of a communication network. The network types mentioned here are only examples; the notation is open for other network types.

$\alpha_3 = \varnothing$: denotes a communication network in which communication delays
  - either are negligible, or
  - they are included in processing times of multiprocessor tasks, or
  - for a given network they are included in communication times between two dependent tasks assigned to different processors.

$\alpha_3 = conn$: denotes an arbitrary network.

$\alpha_3 = linear\ array, ring, tree, mesh, hypercube$: denotes respectively a linear array, ring, tree, mesh, or hypercube.

Parameter $\alpha_4 \in \{\varnothing, no\text{-}overlap\}$ describes the ability of a processor to communicate and to process a task in parallel.

$\alpha_4 = \varnothing$: parallel processing and communications are possible,

$\alpha_4 = no\text{-}overlap$: no overlapping of processing and communication.

In the second field (β) describing task characteristics, parameter β_1 takes more values than described in Section 3.4, and three new parameters β_9, β_{10} and β_{11} are added. Parameter $\beta_1 \in \{\varnothing, pmtn, div\}$ indicates the possibility of task preemptability or divisibility.

$\beta_1 = \varnothing$: no preemption is allowed,

$\beta_1 = pmtn$: preemptions are allowed,

$\beta_1 = div$: tasks are divisible (by definition preemptions are also possible).

Parameter $\beta_9 \in \{spdp\text{-}lin, spdp\text{-}any, size_j, cube_j, mesh_j, fix_j, set_j, \varnothing\}$ describes the type of a multiprocessor task. The first five-symbols are concerned with parallel processors, the next two with dedicated ones.

$\beta_9 = spdp\text{-}lin$: denotes multiprocessor tasks which processing times are inversely proportional to the number of processors assigned,

$\beta_9 = spdp\text{-}any$: processing times arbitrarily depend on the number of processors granted (malleable and moldable are subproblems of this problem),

$\beta_9 = size_j$: means that each task requires a fixed number of processors at a time,

$\beta_9 = cube_j$: each task requires a sub-hypercube of a hypercube processor network,

$\beta_9 = mesh_j$: each task requires a submesh for its processing,

$\beta_9 = fix_j$: each task can be processed on exactly one subgraph of the multi-processor system,

$\beta_9 = set_j$: each task has its own collection of subgraphs of the multiprocessor system on which it can be processed,

$\beta_9 = \varnothing$: each task can be processed on any single processor.

Parameter $\beta_{10} \in \{com,\ c_{jk}.\ c_{j*},\ c_{*k},\ c,\ c=1,\ \varnothing\}$ concerns the communication delays that occur due to data dependencies.

$\beta_{10} = com$: communication delays are arbitrary functions of data sets to be transmitted,

$\beta_{10} = c_{jk}$: whenever $T_j < T_k$, and T_j and T_k are assigned to different processors, a communication delay of a given duration c_{jk} occurs,

$\beta_{10} = c_{j*}$: the communication delays depend on the broadcasting task only,

$\beta_{10} = c_{*k}$: the communication delays depend on the receiving task only,

$\beta_{10} = c$: the communication delays are equal,

$\beta_{10} = c = 1$: each communication delay takes unit time,

$\beta_{10} = \varnothing$: no communication delays occur.

Parameter $\beta_{11} \in \{dup, \varnothing\}$.

$\beta_{11} = dup$: task duplication is allowed.

$\beta_{11} = \varnothing$: task duplication is not allowed.

The third field γ describing criteria is not changed.

Some additional changes describing more deeply architectural constraints of multiprocessor systems have been introduced in [Dro96a] and we refer the readers to this position.

In the following three sections scheduling multiprocessor tasks, scheduling uniprocessor tasks with communication delays and scheduling divisible tasks (i.e. multiprocessor tasks with communication delays) respectively, are discussed.

6.2 Scheduling Multiprocessor Tasks

In this Section we will discuss separately parallel and dedicated processors.

6.2.1 Parallel Processors

We start a discussion with an analysis of the simplest problems in which each task requires a fixed number of parallel processors during execution, i.e. processor requirement denoted by $size_j$. Following Błażewicz et al. [BDW84, BDW86] we will set up the subject more precisely. Tasks are to be processed on a set of identical processors. The set of tasks is divided into k subsets $\mathcal{T}^1 = \{T_1^1, T_2^1, \cdots, T_{n_1}^1\}$, $\mathcal{T}^2 = \{T_1^2, T_2^2, \cdots, T_{n_2}^2\}, \cdots, \mathcal{T}^k = \{T_1^k, T_2^k, \cdots, T_{n_k}^k\}$ where $n = n_1 + n_2 + \cdots + n_k$. Each task T_i^1, $i = 1, \cdots, n_i$, requires exactly one of the processors for its processing and its processing time is equal to p_i^1. Similarly, each task T_i^l, where $1 < l \le k$, requires l arbitrary processors simultaneously for its processing during a period of time whose length is equal to p_i^l. We will call tasks from $\mathcal{T}^l$ *width-l tasks* or $\mathcal{T}^l$ *-tasks*. For the time being tasks are assumed to be *independent*, i.e. there are no precedence constraints among them. A schedule will be called feasible if, besides the usual conditions, each $\mathcal{T}^l$-task is processed by l processors at a time, $l = 1, \cdots, k$. Minimizing the schedule length is taken as optimality criterion.

Problem $P \mid p_j = 1, size_j \mid C_{max}$

Let us start with non-preemptive scheduling. The general problem is NP-hard (cf. Section 5.1), and starts to be strongly NP-hard for $m = 5$ processors [DL89]. Thus, we may concentrate on unit-length tasks. Let us start with the problem of scheduling tasks which belong to two sets only: $\mathcal{T}^1$ and $\mathcal{T}^k$, for arbitrary k, i.e. problem $P \mid p_j = 1, size_j \in \{1, k\} \mid C_{max}$. This problem can be solved optimally by the following algorithm.

Algorithm 6.2.1 *Scheduling unit tasks from sets $\mathcal{T}^1$ and $\mathcal{T}^k$ to minimize C_{max}* [BDW86].

begin
Calculate the length of an optimal schedule according to the formula

$$C_{max}^* = \max\left\{ \left\lceil \frac{n_1 + kn_k}{m} \right\rceil, \left\lceil n_k / \left\lfloor \frac{m}{k} \right\rfloor \right\rceil \right\};$$ (6.2.1)

Schedule $\mathcal{T}^k$-tasks in time interval $[0, C_{max}^*]$ using first-fit algorithm;
        -- see Section 12.1 for the description of the first-fit algorithm
Assign $\mathcal{T}^1$-tasks to the remaining free processors;
end;

It should be clear that (6.2.1) gives a lower bound on the schedule length of an optimal schedule and this bound is always met by a schedule constructed by Algorithm 6.2.1.

If tasks belong to sets $\mathcal{T}^1, \mathcal{T}^2, \cdots, \mathcal{T}^k$, where k is a fixed integer, the problem can be solved by an approach similar to that for the problem of non-preemptive scheduling of unit processing time tasks under fixed resource constraints [BE83]. We will describe that approach in Section 12.1.

Problem $P \mid pmtn, size_j \mid C_{max}$

Now, we will pass to preemptive scheduling. First, let us consider the problem of scheduling tasks from sets $\mathcal{T}^1$ and $\mathcal{T}^k$ in order to minimize schedule length, i.e. problem $P \mid pmtn, size_j \in \{1, k\} \mid C_{max}$. In [BDW84, BDW86] it has been proved that among minimum-length schedules for the problem there always exists a feasible *normalized schedule*, i.e. one in which first all $\mathcal{T}^k$-tasks are assigned in time interval $[0, C_{max}^*]$ using McNaughton's rule (Algorithm 5.1.8), and then all $\mathcal{T}^1$-tasks are assigned, using the same rule, in the remaining part of the schedule (cf. Figure 6.2.1).

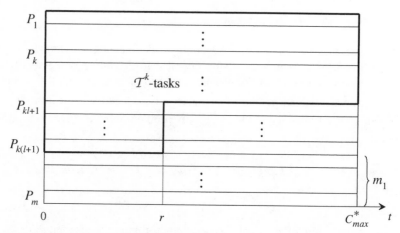

Figure 6.2.1 *An example normalized schedule.*

Following the above result, we will concentrate on finding an optimal schedule among normalized ones. A lower bound on the schedule length C_{max} can be obtained as follows. Define

$$X = \sum_{i=1}^{n_1} p_i^1, \quad Y = \sum_{i=1}^{n_k} p_i^k, \quad Z = X + kY,$$

$$p_{max}^1 = \max_{T_i^1 \in \mathcal{T}^1} \{p_i^1\}, \quad p_{max}^k = \max_{T_i^k \in \mathcal{T}^k} \{p_i^k\}.$$

Then,

$$C_{max} \geq C = \max\{Z/m, Y/\lfloor m/k \rfloor, p_{max}^1, p_{max}^k\}. \tag{6.2.2}$$

It is clear that no feasible schedule can be shorter than the maximum of the above values, i.e. mean processing requirement on one processor, mean processing requirement of $\mathcal{T}^k$-tasks on k processors, the maximum processing time among $\mathcal{T}^1$-tasks, and the maximum processing time among $\mathcal{T}^k$-tasks. If $mC > Z$, then in any schedule there will be an idle time of minimum length $IT = mC - Z$. On the basis of bound (6.2.2) and the reasoning preceding it one can try to construct a preemptive schedule of minimum length equal to C. However, this will not always be possible, and one has to lengthen the schedule. Below we present the reasoning that allows finding the optimal schedule length. Let $l = \lfloor Y/C \rfloor$. It is quite clear that the optimal schedule length C_{max}^* must obey the inequality

$$C \leq C_{max}^* \leq Y/l.$$

We know that there exists an optimal normalized schedule where tasks are arranged in such a way that kl processors are devoted entirely to $\mathcal{T}^k$-tasks, k processors are devoted to $\mathcal{T}^k$-tasks in time interval $[0, r]$, and $\mathcal{T}^1$-tasks are scheduled in the remaining time (cf. Figure 6.2.1). Let m_1 be the number of processors that can process $\mathcal{T}^1$-tasks during time interval $[0, r]$, i.e. $m_1 = m - (l+1)k$. In a normalized schedule which completes all tasks by some time B, where $C \leq B \leq Y/l$, we will have $r = Y - Bl$. Thus, the optimum value C_{max}^* will be the smallest value of B ($B \geq C$) such that the $\mathcal{T}^1$-tasks can be scheduled on m_1 processors available during the interval $[0, B]$ and on $m_1 + k$ processors available in the interval $[r, B]$. Below we give necessary and sufficient conditions for the unit width tasks to be scheduled. To do this, let us assume that these tasks are ordered in such a way that $p_1^1 \geq p_2^1 \geq \cdots \geq p_{n_1}^1$. For a given pair B, r with $r = Y - Bl$, let $p_1^1, p_2^1, \cdots, p_j^1$ be the only processing times greater than $B - r$. Consider now two cases.

Case 1: $j \leq m_1 + k$. Then $\mathcal{T}^1$-tasks can be scheduled if and only if

$$\sum_{i=1}^{j} [p_i^1 - (B-r)] \le m_1 r. \tag{6.2.3}$$

To prove that this condition is indeed necessary and sufficient, let us first observe that if (6.2.3) is violated the T^1-tasks cannot be scheduled. Suppose now that (6.2.3) holds. Then one should schedule the excesses (exceeding $B-r$) of "long" tasks $T_1^1, T_2^1, \cdots, T_j^1$, and (if (6.2.3) holds without equality) some other tasks on m_1 processors in time interval $[0, r]$ using McNaughton's rule. After this operation the interval is completely filled with unit width tasks on m_1 processors.

Case 2: $j > m_1 + k$. In that case T^1-tasks can be scheduled if and only if

$$\sum_{i=1}^{m_1+k} [p_i^1 - (B-r)] \le m_1 r. \tag{6.2.4}$$

Other long tasks will have enough space on the left hand side of the schedule because condition (6.2.2) is obeyed.

Next we describe how the optimum value of schedule length (C_{max}^*) can be found. Let $W_j = \sum_{i=1}^{j} p_i^1$. Inequality (6.2.3) may then be rewritten as

$$W_j - j(B-r) \le m_1(Y-Bl).$$

Solving it for B we get

$$B \ge \frac{(j-m_1)Y + W_j}{(j-m_1)l + j}.$$

Define

$$H_j = \frac{(j-m_1)Y + W_j}{(j-m_1)l + j}.$$

Thus, we may write

$$C_{max}^* = \max\{C, H_{m_1+1}, H_{m_2+1}, \cdots, H_{n_1}\}.$$

Let us observe that we do not need to consider values $H_1, H_2, \cdots, H_{m_1}$ since the m_1 longest T^1-tasks will certainly fit into the schedule of length C (cf. (6.2.2)). Finding the above maximum can clearly be done in $O(n_1 \log n_1)$ time by sorting the unit width tasks by p_i^1. But one can do better by taking into account the following facts.

1. $H_i \le C$ for $i \ge m_1 + k$.

2. H_i has no local maximum for $i = m_1 + 1, \cdots, m_1 + k - 1$.

Thus, to find a maximum over $H_{m_1+1}, \cdots, H_{m_1+k-1}$ and C we only need to apply a linear time median finding algorithm [AHU74] and a binary search. This will result in an $O(n_1)$ algorithm that calculates C_{max}^*. (Finding the medians takes $O(n_1)$ the first time, $O(n_1/2)$ the second time, $O(n_1/4)$ the third time, etc. Thus the total time to find the medians is $O(n_1)$.)

Now we are in the position to present an optimization algorithm for scheduling width-1 and width-k tasks.

Algorithm 6.2.2 *Scheduling preemptable tasks from sets* $\mathcal{T}^1$ *and* $\mathcal{T}^k$ *to minimize* C_{max} [BDW86].

begin
Calculate the minimum schedule length C_{max}^*;
Schedule $\mathcal{T}^k$-tasks in the interval $[0, C_{max}^*]$ using McNaughton's rule (Algorithm
    5.1.8);
$l := \lfloor Y/C_{max}^* \rfloor$; $m_1 := m-(l+1)k$; $r := Y-C_{max}^* l$;
Calculate the number j of long $\mathcal{T}^1$-tasks that exceed $C_{max}^* - r$;
if $j \le m_1 + k$ **then**
    begin
    Schedule the excesses of the long tasks and possibly some other parts of tasks
        on m_1 processors using McNaughton's rule to fill interval $[0, r]$ com-
        pletely;
    Schedule the remaining processing requirement in interval $[r, C_{max}^*]$ on
        $m_1 + k$ processors using McNaughton's rule;
    end
else
    begin
    Schedule part $\left((m_1+k)(C_{max}^*-r)/\sum_{i=1}^{j} p_i^1\right) p_h^1$ of each long task (plus possibly
        parts of smaller tasks T_z^1 with processing times p_z^1, $r < p_z^1 \le C_{max} - r$) in in-
        terval $[r, C_{max}^*]$ on $m_1 + k$ processors using McNaughton's rule;
    -- if among smaller tasks not exceeding $(C_{max}^* - r)$ there are some tasks longer than r,
    -- then this excess must be taken into account in the denominator of the above rate
    Schedule the rest of the task set in interval $[0, r]$ on m_1 processors using
        McNaughton's algorithm;
    end;
end;

The optimality of the above algorithm follows from the preceding discussion. Its time complexity is $O(n_1 + n_k)$, thus we get $O(n)$.

Considering the general case of preemptively scheduling tasks from sets $\mathcal{T}^1$, $\mathcal{T}^2, \cdots, \mathcal{T}^k$, i.e. the problem $Pm \mid pmtn, size_j \in \{1, \cdots, k\} \mid C_{max}$, we can use the very useful linear programming approach presented in equations (5.1.14)-(5.1.15) to solve this problem in polynomial time.

Problem $P \mid prec, size_j \mid C_{max}$

Let us now consider the case of non-empty precedence constraints. Arbitrary processing times result in the strong NP-hardness of the problem, even for chains and two processors [DL89]. In case of unit processing times the last problem can be solved for arbitrary precedence constraints using basically the same approach as in the Coffman-Graham algorithm (Algorithm 5.1.12) [Llo81]. On the other hand, three processors result in a computational hardness of the problem even for chains, i.e. problem $P3 \mid chain, p_j = 1, size_j \mid C_{max}$ is strongly NP-hard [BL96]. However, if task requirements of processors are either uniform or monotone decreasing (or increasing) in each chain then the problem can be solved in $O(n\log n)$ time even for an arbitrary number m of processors ($m < 2size_j$ for the case of monotone chains) [BL96, BL02].

Problem $Q \mid pmtn, size_j \mid C_{max}$

In [BDSW94] a scheduling problem has been considered for a multiprocessor built up of uniform k-tuples of identical parallel processors. The processing time of task T_i is the ratio p_i/b_i, where b_i is the speed of the slowest processor that executes T_i. It is shown that this problem is solvable in $O(nm + n\log n)$ time if the sizes are such that $size_j \in \{1, k\}, j = 1, 2, \cdots, n$. For a fixed number of processors, a linear programming formulation is proposed for solving this problem in polynomial time for sizes belonging to $\{1, 2, \cdots, k\}$.

Problem $Pm \mid pmtn, r_j, size_j \mid L_{max}$

Minimization of other criteria has not been considered yet, except for maximum lateness. In this context problem $Pm \mid pmtn, r_j, size_j \mid L_{max}$ has been formulated as a modified linear programming problem (5.1.14) - (5.1.15). Thus, it can be solved in polynomial time for fixed m [BDWW96].

Let us consider now a variant of the above problem in which each task requires a fixed *number* of processors being a power of 2, thus requiring a cube of a certain dimension. Because of the complexity of the problem we will only consider the preemptive case.

Problem $P \mid pmtn, cube_j \mid C_{max}$

In [CL88b] an $O(n^2)$ algorithm is proposed for building the schedule (if any exists) for tasks with a common deadline C. This algorithm builds so-called stairlike schedules. A schedule is said to be *stairlike* if there is a function $f(i)$ ($i = 1, \cdots, m$) such that processor P_i is busy before time moment $f(i)$ and idle after, and f is non-increasing. Function f is called a *profile* of the schedule. Tasks are scheduled in the order of non-increasing number of required processors. A task is scheduled in time interval $[C-p_j, C]$, utilizing the first of the subcubes of the task's size on each "stair" of the stairlike partial schedule. Using a binary search, the C^*_{max} is calculated in time $O(n^2(\log n + \log(\max\{p_j\})))$. The number of preemptions is at most $n(n-1)/2$.

In [Hoe89], a feasibility-testing algorithm of the complexity $O(n \log n)$ is given. To calculate C^*_{max} with this algorithm $O(n \log n (\log n + \log(\max\{p_j\})))$ time is needed. This algorithm uses a different method for scheduling; it builds so-called *pseudo-stairlike* schedules. In this case $f(i) < f(j) < C$, for $i, j = 1, \cdots, m$, implies that $i > j$. Each task is feasibly scheduled on at most two subcubes. Thus the number of generated preemptions is at most $n-1$.

A similar result is presented in [AZ90], but the number of preemptions is reduced to $n-2$ because the last job is scheduled on one subcube, without preemption. This work was the basis for the paper [SR91] in which a new feasibility testing algorithm is proposed, with running time $O(mn)$. The key idea is to schedule tasks in the order of non-increasing execution times, in sets of tasks of the same size (number of required processors). Based on the claim that there exists some task in each optimal schedule utilizing all the remaining processing time on one of the processors in the schedule profile, an $O(n^2m^2)$ algorithm is proposed to calculate C^*_{max}.

Problem $P \mid pmtn, cube_j \mid L_{max}$

Again minimization of L_{max} can be solved via linear programming formulation [BDWW96].

Let us consider now the most complicated case of parallel processor requirements specified by numbers of processors required, where task processing times depend on numbers of processors assigned.

Problem $P \mid spdp-any \mid C_{max}$

A *dynamic programming* approach leads to the observation that $P2 \mid spdp\text{-}any \mid C_{max}$ and $P3 \mid spdp\text{-}any \mid C_{max}$ are solvable in pseudopolynomial time [DL89]. Arbitrary schedules for instances of these problems can be transformed into so called *canonical schedules*. A canonical schedule on two processors is one that first processes the tasks using both processors. It is completely determined by three numbers: the total execution times of the single-processor tasks on processor P_1 and P_2, respectively, and the total execution time of the biprocessor tasks. For the case of three processors, similar observations are made. These characterizations are the basis for the development of the pseudopolynomial algorithms. The problem $P4 \mid spdp\text{-}any \mid C_{max}$ remains open; no pseudopolynomial algorithm is given.

Surprisingly preemptions do not result in polynomial time algorithms [DL89].

Problem $P \mid pmtn, spdp\text{-}any \mid C_{max}$

Problem $P \mid pmtn, spdp\text{-}any \mid C_{max}$ is proved to be strongly NP-hard by a reduction from 3-Partition [DL89]. With restriction to two processors, $P2 \mid pmtn, spdp\text{-}any \mid C_{max}$ is still NP-hard, as is shown by a reduction from PARTITION. Using Algorithm 6.2.2 [BDW86], Du and Leung [DL89] show that for any fixed number of processors $Pm \mid pmtn, spdp\text{-}any \mid C_{max}$ is also solvable in pseudopolynomial time. The basic idea of the algorithm is as follows. For each schedule S of $Pm \mid pmtn, size_j \mid C_{max}$, there is a corresponding instance of $Pm \mid pmtn, spdp\text{-}any \mid C_{max}$ with sizes belonging to $\{1, \cdots, k\}$, in which task T_i is an l-processor task if it uses l processors with respect to S. An optimal schedule for the latter problem can be found in polynomial time by Algorithm 6.2.2. What remains to be done is to generate optimal schedules for instances of $Pm \mid pmtn, size_j \mid C_{max}$ that correspond to schedules of $Pm \mid pmtn, spdp\text{-}any \mid C_{max}$, and choose the shortest among all. It is shown by a dynamic programming approach that the number of schedules generated can be bounded from above by a pseudopolynomial function of the size of $Pm \mid pmtn, spdp\text{-}any \mid C_{max}$.

If in the above problem one assumes a linear model of dependency of task processing times on a number of processors assigned, the problem starts to be solvable in polynomial time. That is problem $P \mid pmtn, spdp\text{-}lin \mid C_{max}$ is solvable in $O(n)$ time [DK99] and $P \mid pmtn, r_j, spdp\text{-}lin \mid C_{max}$ is solvable in $O(n^2)$ time [Dro96b].

On the other hand, the special case of *malleable tasks* received recently quite considerable attention. It was proved that in the case of convex speed functions (relating processing speed to the number of processors allocated), an optimal schedule can be constructed by a sequential performance of tasks (being assigned all available processors) [BKM+04]. The case of concave functions for all the tasks is solvable in polynomial time for a fixed number of processors [BKM+06].

6.2.2 Dedicated Processors

In this section we will consider dedicated processor case. Following the remarks of Section 6.1 we will denote here by $\mathcal{T}^{\mathcal{D}}$ the set of tasks each of which requires set $\mathcal{D}$ of processors simultaneously. Task $T_i \in \mathcal{T}^{\mathcal{D}}$ has processing time $p_i^{\mathcal{D}}$. For the sake of simplicity we define by $p^{\mathcal{D}} = \sum_{T_i \in \mathcal{T}^{\mathcal{D}}} p_i^{\mathcal{D}}$ the total time required by all tasks which use set of processors $\mathcal{D}$. Thus, e.g. $p^{1,2,3}$ is the total processing time [2] of tasks each of which requires processors P_1, P_2, and P_3 simultaneously. We will start with task requirements concerning only one subset of processors for each task, i.e. fix_j requirements.

Problem $P \mid p_j = 1, fix_j \mid C_{max}$

The problem with unit processing times can be proved to be strongly NP-hard for an arbitrary number of processors [KK85]. Moreover, in [HVV94] it has been proved that even the problem of deciding whether an instance of the problem has a schedule of length at most 3 is strongly NP-hard. As a result there is no polynomial time algorithm with worst case performance ratio smaller than 4/3 for $P \mid p_j = 1, fix_j \mid C_{max}$, unless $P = NP$. On the other hand, if the number of processors is fixed, then again an approach for non-preemptive scheduling of unit length tasks under fixed resource constraints [BE93] (cf. Section 12.1) can be used to solve the problem in polynomial time.

Problem $P \mid fix_j \mid C_{max}$

It is trivial to observe that the problem of non-preemptive scheduling tasks on two processors under fixed processor requirements is solvable in polynomial time. On the other hand, if the number of processors is increased to three, the problem starts to be strongly NP-hard [BDOS92]. Despite the fact that the general problem is hard we will show below that there exist polynomially solvable cases of three processor scheduling [BDOS92]. Let us denote by R^i the total time processor P_i processes tasks. For instance, $R^1 = p^1 + p^{1,2} + p^{1,3}$.

Moreover, let us denote by RR the total time during which two processors must be used simultaneously, i.e. $RR = p^{1,2} + p^{1,3} + p^{2,3}$. We obviously have the following:

Lemma 6.2.3 [BDOS92] $C_{max} \geq \max\{\max_i\{R^i\}, RR\}$ for problem $P3 \mid fix_j \mid C_{max}$. $\square$

---

[2] For simplicity reasons we write $p^{1,2,3}$ instead of $p^{\{1,2,3\}}$.

Now we consider the case for which $p^1 \leq p^{2,3}$. The result given below also covers cases $p^2 \leq p^{1,3}$ and $p^3 \leq p^{1,2}$, if we take into account renumbering of processors.

Theorem 6.2.4. [BDOS92] *If $p^1 \leq p^{2,3}$, then $P3 \,|\, fix_j \,|\, C_{max}$ can be solved in polynomial time. The minimum makespan is then*

$$C_{max} = \max\{\max_i\{R^i\}, RR\}.$$

Proof. The proof is constructive in nature and we consider four different subcases.

Case a: $p^2 \leq p^{1,3}$ and $p^3 \leq p^{1,2}$. In Figure 6.2.2 a schedule is shown which can always be obtained in polynomial time for this case. The schedule is such that $C_{max} = RR$, and thus, by Lemma 6.2.3, it is optimal.

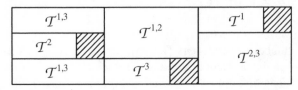

Figure 6.2.2 *Case a of Theorem 6.2.4.*

Case b: $p^2 \leq p^{1,3}$ and $p^3 > p^{1,2}$. Observe that in this case $R^3 = \max_i\{R^i\}$. Hence, a schedule which can be found in polynomial time is shown in Figure 6.2.3. The length of the schedule is $C_{max} = R^3 = \max_i\{R^i\}$, and thus this schedule is optimal (cf. Lemma 6.2.3).

Case c: $p^2 \geq p^{1,3}$ and $p^3 \geq p^{1,2}$. Observe that $R^1 \leq R^2$ and $R^1 \leq R^3$. Two subcases have to be considered here.

   Case c': $R^2 \leq R^3$. The schedule which can be obtained in this case in polynomial time is shown in Figure 6.2.4(a). Its length is $C_{max} = R^3 = \max_i\{R^i\}$.

   Case c'': $R^2 > R^3$. The schedule which can be obtained in this case in polynomial time is shown in Figure 6.2.4(b). Its length is $C_{max} = R^2 = \max_i\{R^i\}$.

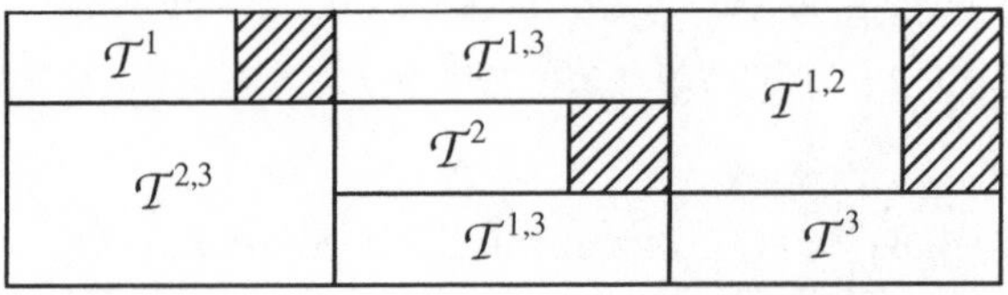

Figure 6.2.3 *Case b of Theorem 6.2.4.*

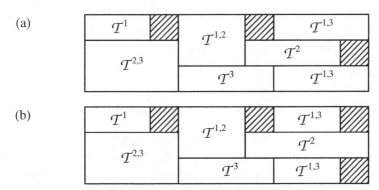

Figure 6.2.4 *Case c of Theorem* 6.2.4.

Case d: $p^2 \ge p^{1,3}$ and $p^3 \le p^{1,2}$. Note that the optimal schedule would be the same as in Case *b* if we renumbered the processors. □

It follows that the hard problem instances are those for which $p^1 > p^{2,3}$, $p^2 > p^{1,3}$ and $p^3 > p^{1,2}$. Let us call these cases the *Hard-C* subcases. However, also among the problem instances which satisfy the *Hard-C* property, some particular cases can be found which are solvable in polynomial time.

Theorem 6.2.5 [BDOS92] *If Hard-C holds and*

$$R^1 \ge p^2 + p^3 + p^{2,3} \text{ or } p^1 \ge p^2 + p^{2,3} ,$$

then problem $P3\,|\,fix_j\,|\,C_{max}$ *can be solved in polynomial time, and* $C_{max} = \max_i\{R^i\}$.

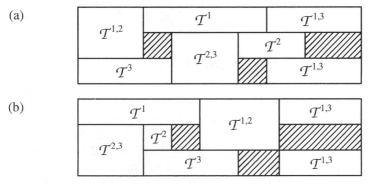

Figure 6.2.5 *Two cases for Theorem* 6.2.5.

Proof. Observe that if $R^1 \geq p^2 + p^3 + p^{2,3}$ then $R^1 \geq R^2$ and $R^1 \geq R^3$. The schedule which can be immediately obtained in this case is shown in Figure 6.2.5(a). As $C_{max} = R^1$, the schedule is optimal by Lemma 6.2.3.

If $p^1 \geq p^2 + p^{2,3}$, the optimal schedule is as shown in Figure 6.2.5(b). In this case $C_{max} = \max\{R^1, R^3\}$. $\square$

Observe that the optimal schedules found for the polynomial cases in Theorems 6.2.4 and 6.2.5 are all *normal schedules*, i.e. those in which all tasks requiring the same set of processors are scheduled consecutively. Let us denote by C_{max}^S the schedule length of the best normal schedule for $P3\,|fix_j|\,C_{max}$ and by C_{max}^* the value of the minimum schedule length for the same instance of the problem. Then [BDOS92]

$$\frac{C_{max}^S}{C_{max}^*} < \frac{4}{3} \, .$$

Since the best normal schedule can be found in polynomial time [BDOS92], we have defined a polynomial time approximation algorithms with the worst case behavior not worse than 4/3. Recently this bound has been improved. In [OST93] and in [Goe95] new approximation algorithms have been proposed with bounds equal to 5/4 and 7/6, respectively.

An interesting approach to the solution of the above problem is concerned with the graph theoretic approach. The computational complexity of the problem $P\,|fix_j|\,C_{max}$ where $|fix_j| = 2$ is analyzed in [CGJP85]. The problem is modeled by the use of the so-called *file transfer graph*. In such a graph, nodes correspond to processors, and edges correspond to tasks. A weight equal to the execution time is assigned to each edge. A range of computational-complexity results have been established. For example, the problem $P\,|p_j = 1,\, fix_j|\,C_{max}$ is easy when the file transfer graph is one of the following types of a graph: bipartite, tree, one cycle graph, star, caterpillar, cycle, path; but, in general, the problem is NP-hard. It is proved that the makespan C_{max}^{LS} of the schedule obtained with any list-scheduling algorithm satisfies $C_{max}^{LS} \leq 2C_{max}^* + \max\{0,\ \max\{p_j\}(1 - 2/d)\}$, where d is the maximum degree of any vertex in the graph.

In [Kub87], the problem is modeled by means of weighted edge coloring. The graph model is extended in such a way that a task requiring one processor is represented as a loop which starts and ends in the same node. The problem $P\,|fix_j|\,C_{max}$, where $|fix_j| \in \{1,2\}$ is NP-hard for the graph, which is either a star with a loop at each non-central vertex or a caterpillar with only one loop. In [BOS91] the problem $P\,|fix_j|\,C_{max}$ is also analyzed with the use of the graph approach. This time, however, the graph reflecting the instance of the problem, called a *constraint graph*, has nodes corresponding to tasks, and edges join two tasks which cannot be executed in parallel. An execution time is associated with each node. It is shown that the problem $P \mid p_j = 1, fix_j \mid C_{max}$ is NP-hard but can be solved polynomially for $m = 4$ ($P4 \mid p_j = 1, fix_j \mid C_{max}$). Next, when the constraint graph is a comparability graph, the transitive orientation of such a graph gives an optimal solution. In this case, C_{max} is equal to the weight of the maximum weight clique. The transitive orientation (if any exists) can be found in time $O(dn^2)$. For the general case, when the constraint graph is not a comparability graph, a branch and bound algorithm is proposed. In this case, the problem consists in adding (artificial) edges such that the new graph is a comparability graph. The search is directed by the weight of the heaviest clique in the new graph. Results of computational experiments are reported.

Another constrained version of the problem is considered in [HVV94]. It is assumed that in problem $P3 \mid fix_j \mid C_{max}$ all biprocessor tasks that require the same processors are scheduled consecutively (the so-called *block constraint*). Under this assumption this problem is solvable in pseudopolynomial time.

Problem $P \mid pmtn, fix_j \mid C_{max}$

Let us consider now preemptive case. In general the problem $P \mid pmtn, fix_j \mid C_{max}$ is strongly NP-hard [Kub90]. For simpler cases of the problem, however, linear time algorithms have been proposed [BBDO94]. An optimal schedule for the problem $P2 \mid pmtn, fix_j \mid C_{max}$ does not differ from a schedule for the non-preemptive case. For three processors ($P3 \mid pmtn, fix_j \mid C_{max}$), an optimal schedule has the following form: biprocessor tasks of the same type are scheduled consecutively (without gaps and preemptions) and uniprocessor tasks are scheduled in the free-processing capacity, in parallel with appropriate biprocessor tasks. The excess of the processing time of uniprocessor tasks is scheduled at the end of the schedule. The complexity of the algorithm is $O(n)$. In a similar way, optimal schedules can be found for $P4 \mid pmtn, fix_j \mid C_{max}$. When m is limited, the problem $Pm \mid pmtn, fix_j \mid C_{max}$ can be solved in polynomial time using feasible sets and a linear programming approach [BBDO94].

Problem $P \mid prec, fix_j \mid C_{max}$

If we consider precedence constrained task sets, the problem immediately starts to be computationally hard, since even problem $P2 \mid chain, p_j = 1, fix_j \mid C_{max}$ is strongly NP-hard [HVV94] (cf.[BLRK83] where a similar proof has been used for the resource constrained scheduling).

Problem $P \mid fix_j \mid \Sigma C_j$

Let us consider now minimization of mean flow time. Problem $Pk \mid p_j = 1, fix_j \mid \Sigma C_j$ is still open. The general version of the problem has been considered in [HVV94]. The main result is establishing NP-hardness in the ordinary sense for $P2 \mid fix_j \mid \Sigma C_j$. The question whether this problem is solvable in pseudopolynomial time or NP-hard in the strong sense still has to be resolved. The weighted version, however, is shown to be NP-hard in the strong sense. The problem with unit processing times is NP-hard in the strong sense if the number of processors is a part of the problem instance, but the complexity is still open in case of a fixed number of processors. As could be expected, the introduction of precedence constraints does not simplify the computational complexity. It is shown that even the problem with two processors, unit processing times, and chain-type precedence constraints, is NP-hard in the strong sense.

Problem $P \mid pmtn, fix_j \mid L_{max}$

Since the non-preemptive scheduling problem with the L_{max} criterion is already strongly NP-hard for two processors [HVV94], more attention has been paid to preemptive case. In [BBDO97a] linear time algorithms have been proposed for problem $P2 \mid pmtn, fix_j \mid L_{max}$ and for some special subcases of three and four processor scheduling. More general cases can be solved by linear programming approach [BBDO97b] even for different ready times for tasks.

Problem $P \mid set_j \mid C_{max}$

Let us consider now the case of set_j processor requirements. In [CL88b] this problem is restricted to single-processor tasks of unit length. In the paper matching technique is used to construct optimal solutions in $O(n^2 m^2)$ time. In [CC89] the same problem is solved in $O(\min\{\sqrt{n}, m\} nm \log n)$ time by use of network flow approach. More general cases have been considered in [BBDO95].

Problem $P2 \mid set_j \mid C_{max}$ is NP-hard, but can be solved in pseudopolynomial time by a dynamic programming procedure. In this case, the schedule length depends on three numbers: total execution time of tasks requiring P_1 (p^1), tasks requiring P_2 (p^2), and tasks requiring P_1 and P_2 (p_2^1). In an optimal schedule, T_2^1 tasks are executed first, then uniprocessor tasks are processed in any order without gaps. A similar procedure is proposed for a restricted version of $P3 \mid set_j \mid C_{max}$ in which one type of dual-processor task is absent (e.g., T_3^1). On the other hand, for the problem $P \mid set_j \mid C_{max}$, the shortest processing time (SPT) heuristic is proposed. Thus, tasks are scheduled in any order, in their shortest-time processing configuration. A tight performance bound for this algorithm is m.

Some other cases, mostly restricted to $|set_j| = 1$, are analyzed in [Bru95].

Problem $Pm \mid pmtn, set_j \mid C_{max}$

In case of preemptions again linear programming approach can be used to solve the problem in polynomial time for fixed m [BBDO95].

Problem $Pm \mid pmtn, set_j \mid L_{max}$

Following the results for fix_j processor requirements, most of the other cases for set_j requirements are computationally hard. One of the few exceptions is problem $Pm \mid pmtn, set_j \mid L_{max}$ which can be solved by linear programming formulation [BBDO97b].

6.2.3 Refinement Scheduling

Usually deterministic scheduling problems are formulated on a single level of abstraction. In this case all information about the processor system and the tasks is known in advance and can thus be taken into account during the scheduling process. We also know that generating a schedule that is optimal under a given optimization criterion, is usually very time consuming in case of large task systems, due to the inherent computational complexity of the problem.

On the other hand, in many applications it turns out that detailed information about the task system is not available at the beginning of the scheduling process. One example is the construction of complex real-time systems that is only possible if the dependability aspects are taken into account from the very beginning; adding non-functional constraints at a later stage does not work. Another example are large projects that run over long periods; in order to settle the contract, reasonable estimates must be made in an early stage when probably only the coarse structure of the project and a rough estimate of the necessary resources are known. A similar situation occurs in many manufacturing situations; the delivery time must be estimated as part of the order although the detailed shop floor and machine planning is not yet known.

A rather coarse grained knowledge of the task system in the beginning is refined during later planning stages. This leads to a stepwise refinement technique where intermediate results during the scheduling process allows to recognize trends in

− processing times of global steps
− total execution time

- resource requirements
- feasibility of the task system.

In more detail, we first generate a schedule for the coarse grained (global) tasks. For these we assume that estimates of processing times and resource requirements are known. Next we go into details and analyze the structure of the global tasks (refinement of tasks). Each global task is considered as a task system by itself consisting of a set of sub-tasks, each having certain (may be estimated) processing time and resource requirement. For each such sub-task a schedule is generated which then replaces the corresponding global task. Proceeding this way from larger to smaller tasks we are able to correct the task completion times from earlier planning stages by more accurate values, and we get more reliable information about the feasibility of the task system.

Algorithm 6.2.6 *Refinement scheduling* [EH94].

begin
Define the task set in terms of its structure (precedence relations), its estimated
    resource consumption (execution times) and its deadlines;
        -- Note that the deadlines are initially defined at the highest level (level 0)
        -- as part of the external requirements.
        -- During the refinement process it might be convenient to refine these
        -- deadlines too; i.e. to assign deadlines to lower level tasks.
        -- Depending on the type of problem, it might, however, be sufficient to prove
        -- that an implementation at a particular level of refinement obeys the deadlines
        -- at some higher level.
Schedule the given task set according to some discipline, e.g. earliest deadline
    first;
repeat
    Refine the task set, again in terms of its structure, its resource consumption
        and possibly also its deadlines;
    Try to schedule the refinement of each task within the frame that is defined by
        the higher level schedule;
        -- Essentially this boils down to introducing a virtual deadline that is the
        -- finishing time of the higher level task under consideration.
        -- Note also that the refined schedule might use more resources (processors)
        -- than the initial one.
        -- If no feasible schedule can be found, backtracking must take place.
        -- In our case this means, that the designer has to find another task structure,
        -- e.g. by redefining the functionality of the tasks or by exploiting additional
        -- parallelism [VWHS95] (introduction of additional processors,
        -- cloning of resources and replacement of synchronous by asynchronous
        -- procedure calls).
    Optimize the resulting refined schedule by shifting tasks;
        -- This step aims at restructuring (compacting) the resulting schedule
        -- in such a way that the number of resources (processors) is as small as possible.

until the finest task level is reached;
end;

The algorithm essentially defines a first schedule from the initial given task set and refines it recursively. At each refinement step a preliminary schedule is developed that is refined (detailed) in the next step. This way, we will finally end up with a schedule at the lowest level. The tasks are now the elementary actions that have to be taken in order to realize all the initially given global tasks.

6.3 Scheduling Uniprocessor Tasks with Communication Delays

The following simple example serves as an introduction to the problems we deal with in the present section. Let there be given three tasks with precedences as shown in Figure 6.3.1(a). The computational results of task T_1 are needed by both successor tasks, T_2 and T_3. We assume unit processing times. For task execution there are two identical processors, connected by a communication link. To transmit the results of computation T_1 along the link takes 1.5 units of time. The schedule in Figure 6.3.1(b) shows a schedule where communication delays are not considered. The schedule (c) is obtained from (b) by introducing a communication delay between T_1 and T_3. Schedule (d) demonstrates that there are situations where a second processor does not help to gain a shorter schedule. The fourth schedule, (e), demonstrates another possibility: if task T_1 is processed on both processors, an even shorter schedule is obtained. The latter case is usually referred to as *task duplication*.

The problems considered in this area are often simplified by assuming that communication delays are the same for all tasks (so-called *uniform delay scheduling*). Other approaches distinguish between *coarse grain* and *fine grain* parallelism: In contrast to the latter, a high computation-communication ratio can be expected in coarse grain parallelism. As pointed out before, *task duplication* often leads to shorter schedules; this is in particular the case if the communication times are large compared to the processing times.

In Section 6.3.1, we discuss briefly recent results concerning algorithms and complexity of task scheduling with communication delays, but without task duplication. The corresponding problems with task duplication are considered in Section 6.3.2. Finally, in Section 6.3.3, we discuss the influence of particular network structures on schedules.

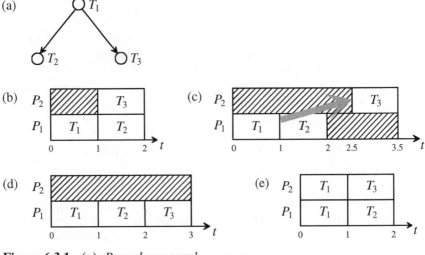

Figure 6.3.1 **(a)** *Precedence graph*
(b) *Schedule without consideration of communication delays*
(c) *Schedule considering communication from T_1 to T_3*
(d) *Optimal schedule without task duplication*
(e) *Optimal schedule with task duplication.*

6.3.1 Scheduling without Task Duplication

Problem $P \mid prec, c = 1, p_j = 1 \mid C_{max}$

This problem was first discussed by Rayward-Smith in [Ray87a] who established its strong NP-hardness. The question if $P \mid prec, c = 1, p_j = 1 \mid C_{max}$ is NP-hard for fixed $m \geq 2$ is still open. If the width of the precedence graph, i.e. the largest number of incomparable tasks in $(\mathcal{T}, \prec)$, is bounded, then the problem can be solved in polynomial time [Moh89, Vel93]. Picouleau [Pic91] proved that the problem of deciding whether an instance has a schedule of makespan at most 3 can be decided in polynomial time. From Hoogeveen et al. [HLV94] we know, however, that the same problem for schedules with C_{max} at most equal to 4 is NP-complete, even for bipartite [3] precedence relations. As a consequence of this result we see that there is no polynomial-time algorithm with performance bound <

---

[3] A precedence relation is *bipartite* if the longest path in the corresponding precedence graph has length 2.

5/4, unless $P = NP$. Otherwise, for instances with $C_{max} = 4$ the polynomial-time algorithm would construct a schedule of length < 5 which would be optimal.

There are also results available for cases where the number m of processors is unrestricted, i.e. if $n \leq m$. In such a case deciding whether $C_{max} < l$ can be done in polynomial time for $l \leq 5$, but is NP-complete for $l \geq 6$ [Vel93].

Rayward-Smith [Ray87a] discussed the performance of demand schedules (called *greedy*). The makespan C_{max}^G of a demand schedule as compared to that of an optimal schedule can be proved to be $\dfrac{C_{max}^G}{C_{max}^*} \leq 3 - \dfrac{2}{m}$.

In case of tree-like precedences, Lenstra et al. [LVV96] proved the problem $P \mid tree, c = 1, p_j = 1 \mid C_{max}$ to be NP-hard. However, for a fixed number of processors, i.e. for the problem $Pm \mid tree, c = 1, p_j = 1 \mid C_{max}$, an optimal algorithm of polynomial time complexity exists [VRK92]. In case of an unbounded number of processors ($m \geq n$), the problem can be solved in $O(n)$ time [Chr89a].

In [BBGT96], the problem $Q2 \mid in\text{-}tree, c = 1, p_j = 1 \mid C_{max}$ with two uniform processors of speeds 2 and 1, respectively, was considered. Thus, the execution of a task takes two units of time on the slower processor (P_1) and one unit of time on the faster processor (P_2). The in-tree is assumed to be complete. If two tasks being in relation $\prec$ are processed on different processors, the communication delay is one unit of time. An $O(h)$ time algorithm is presented for this problem for trees of height h.

For the case of *interval order* precedences, the problem $P \mid interval\ order, c = 1, p_j = 1 \mid C_{max}$ can be solved in polynomial time [Pic92].

Problem $P \mid prec, c, p_j = 1 \mid C_{max}$

Assuming first an unbounded number of processors, we know from Jakoby et al. [JR92] that the problem is NP-hard, in contrast to Chretienne's linear-time solution [CHR89a] for the same type of problem with $c = 1$.

If the number of processors is finite the problem of course remains hard. In situations where the communication delays are large it can be useful to get information about how far the makespan is influenced by the largest communication delay. From [BGK96] it is known that the problem $P \mid prec, c, p_j = 1 \mid C_{max}$ with $C_{max} \leq c+2$ can be solved in polynomial time, whereas problem $P \mid prec, c, p_j = 1 \mid C_{max}$ with the question "$C_{max} = \leq c+3$" is NP-complete, even if $prec = bipartite$ graph. Furthermore, there is a lower bound on the performance of approximation algorithms. There exists no polynomial-time algorithm with performance bound smaller than $1 + 1/(c+3)$ for $P \mid prec, c, p_j = 1 \mid C_{max}$, unless $P = NP$. This result holds even in the special case that the precedence relation is of bipartite type.

For the same problem but where the precedence relation is a complete k-ary tree, Jakoby and Reischuk [JR92] presented an $O(n^2 \log n)$-time algorithm. If completeness of the tree is not guaranteed, the problem is NP-hard even for in-trees where each task has in-degree at most 2.

Problem $P \mid prec, c_{jk} \mid C_{max}$

The problem is shown to be strongly NP-hard for binary tree precedences by a transformation from Exact-3-Cover [JR92]. Only for the very restricted problem $P \mid tree, c_{jk} \mid C_{max}$ where trees are of depth 1, and under the assumption of infinite number of processors, Picouleau [Pic92] was able to present an $O(n \log n)$ time algorithm. For general tree-like precedences and unit time tasks, the problem is NP-hard even if the number of processors is unlimited.

Several other results for the general communication delay case make assumptions on the relationship between the sizes of processing times and communication times: The *granularity* g of an instance can be defined as $g := \min_i \{p_i\} / \max_{i,j} \{c_{ij}\}$. An instance is said to be *coarse grained* if $g \geq 1$. Another also useful definition is that of the *grain* of a task: The *grain* g_j of task T_j is defined by $g_j = \min_{i \in ipred(T_j)} \{p_i\} / \max_{i \in ipred(T_j)} \{c_{ij}\}$. Based on this notion, Chretienne and Picouleau [CP91] use a less restrictive definition of instance granularity: An instance is of *coarse-grained type* if $g_j \geq 1$ for all $j = 1, \cdots, n$. We will distinguish between these two definitions by writing "$g \geq 1$" in the first and "$g_j \geq 1$" in the second case.

The problem $P \mid prec, c_{jk}, g \geq 1 \mid C_{max}$ with an unlimited number of processors was independently studied by Gerasoulis and Yang [GY92] and Picouleau [Pic91]. Both presented approximation algorithms of performance $1 + 1/g$. For tree-like precedences, this problem can be solved in $O(n)$ time [Chr89a, AHC90]. For $P \mid prec, c, g \geq 1 \mid C_{max}$ with $c \leq 1$ and again an unlimited number of processors we know from [Pic91] and [Vel93] that it is NP-complete to decide whether an instance has a schedule of length $C_{max} \leq 5 + 3c$ or $C_{max} \leq 6c$, respectively.

From [CP91] we know that problem $P \mid prec, c_{jk}, g_j \geq 1 \mid C_{max}$ with an unlimited number of processors and bipartite or series-parallel precedence constraints is solvable in polynomial time.

Problem $P \mid pmtn, c \mid C_{max}$

Rayward-Smith [Ray87b] studied problem $P \mid pmtn, c \mid C_{max}$ with unlimited number of processors, and where preemptions are allowed at integer points. It turns out that the communication delays cause an increase of C_{max}^* by at most $c - 1$ units of time. Thus problem $P \mid pmtn, c = 1 \mid C_{max}$ can be solved optimally by

McNaughton's rule (see Section 5.1). Surprisingly, the problem is strongly NP-hard for any fixed $c \geq 2$.

6.3.2 Scheduling with Task Duplication

In this section the problem of scheduling single processor tasks with communication delays and task duplications will be considered. Most results available here are obtained under the unrealistic assumption that the number of processors is unlimited. This assumption allows to process copies of a task as often as required in order to avoid communication, so that the maximum makespan is minimized. A more realistic approach, however, would be one where the number of processors is $m < n$. The then required careful decision between the two options of task duplication vs. acceptation of communication delay makes the problem much more difficult.

Problem $P \mid prec, c, dup \mid C_{max}$

The NP-completeness proof of Hoogeveen et al. [HLV94] for deciding whether $P \mid prec, c = 1, p_j = 1 \mid C_{max}$ has a schedule of the length is ≤ 4 implies that answering the same question for problem $P \mid prec, c = 1, p_j = 1, dup \mid C_{max}$ is NP-complete, too.

Papadimitriou and Yannakakis showed that the problem $P \mid prec, c, p_j = 1, dup \mid C_{max}$, with an unlimited number of processors is NP-hard [PY90]. Even more, the problem of deciding whether an instance of $P \mid prec, c, p_j = 1, dup \mid C_{max}$ has a solution with makespan $C_{max} \leq c + 3$ is NP-complete [BGK96]. Following [JKS89], $P \mid prec, c, dup \mid C_{max}$ can be solved via a dynamic programming approach in time $O(n^{c+1})$.

Problem $P \mid prec, c_{jk}, dup \mid C_{max}$

An approximation algorithm proposed in [PY90] for problem $P \mid prec, c_{jk}, dup \mid C_{max}$, under the assumption that the number of processors is unlimited, brings out quite interesting ideas on the way to design heuristics that take communication times into account. The algorithm proposed has time complexity $O(n^2(e + n\log n))$ where e denotes the number of precedence constraint task pairs. An interesting fact is that this method can also be used to solve coarse-grained problems ($g_j \geq 1$ for $j = 1, \cdots, n$) optimally in $O(n^2)$ time [CC91].

For out-tree precedences, scheduling tasks on an unlimited number of processors can be done in polynomial time. In the case of in-tree precedence constraints, the problem can be transformed into an equivalent problem with out-tree precedence constraints and duplications not allowed. From [Chr94] this problem is known to be NP-hard.

6.3.3 Scheduling in Processor Networks

Picouleau [Pic92] studied a variant of problem $P \mid tree, c_{jk} \mid C_{max}$ where the number of processors is unlimited, the precedence relation can be represented as a tree of depth 1, and a distance function is specified. For a pair of processors, P_h, P_i, their distance d_{hi} is defined by $d_{hi} = |h - i|$. The communication time $c(T_j, P_h, T_k, P_i)$ is assumed as $c_{jk}d_{hi}$, provided that $T_j \prec T_k$. The problem is shown to be NP-hard by a transformation from PARTITION.

The distance function defined by Picouleau can be motivated as being appropriate for linear array networks. For general network structures a distance between two processors may be defined as the length of a shortest path that connects a pair of processors. Such a model has been considered in [EH96]. El-Rewini and Lewis [ERL90] also considered a distance function, and in addition took contention into account. By *contention* we understand the event that two or more data transmissions simultaneously have to pass a single communication channel whose limited capacity enforces serialized transmission.

Hwang et al. [HCAL89] studied approximation list algorithms for scheduling problems where the communication times depend both on the involved tasks and on the processors which execute the tasks. The underlying communication system model allows covering several types of systems such as fully connected systems, hypercubes, or local area networks. The communication is assumed to be contention free. The authors examined a simple strategy called *extended list scheduling*, *ELS*, which is a straightforward extension of list scheduling. The ELS method adopts a two-phase strategy. First tasks are allocated to processors by applying list scheduling as if the underlying system were free of communication overhead. Second, the necessary communication is added to the schedule obtained in the first phase. Denoting by C_{ELS} the makespan of a schedule derived by applying the *ELS* method, and by C_{nocomm}^* the makespan of an optimal schedule for the same instance where communication delays are not considered, Hwang et al. proved the following bound

$$C_{ELS} \le \left(2 - \frac{1}{n}\right) C_{nocomm}^* + \tau_{max} \sum_{\substack{T \in T \\ T' \in \text{isucc}(T)}} \eta(T, T').$$

Here, $\tau_{max} = \max\{\tau(P, P')\}$ is the maximum time to transmit a packet of unit length from one processor the another, $\eta(T, T')$ is the length of a packet of data to be sent from T to T'. It is also shown that this bound cannot be improved in general.

Since the performance of *ELS* is unsatisfactory, an improved strategy called the *earliest task first, (ETF)* was proposed in [HCAL89]. This algorithm uses a greedy strategy where the earliest ready task is scheduled first. The strategy is improved by the ability to postpone a scheduling decision to the next decision moment if a task completion between two decision points may make a more urgent task schedulable. The performance ratio obtained from a detailed analysis is

$$C_{ETF} \leq (2 - \frac{1}{n})\, C^*_{nocomm} + c_{max}$$

where c_{max} is the maximum communication requirement along all chains of $\mathcal{T}$, that is

$$c_{max} = \max\left\{ \tau_{max} \sum_{k=1}^{l-1} \eta(T_{c_i}, T_{c_{i+1}}) \mid (T_{c_1}, ..., T_{c_l}) \text{ is a chain in } \mathcal{T} \right\}.$$

Very little is known about the design of efficient approximation algorithms for the scheduling of task on a real multiprocessor topology. Of particular practical interest are also problems with constraints on communication channel capacities or unequal processor distances.

Ecker and Hodam [EH96] considered a similar model where communication packets are of various lengths and transmission times per unit length message depend on the given network topology. If T_j and T_k are processed on processors P_h and P_i, respectively, the time for transmitting the required data of T_j to T_k is $c(T_j, P_h, T_k, P_i) = \kappa(T_j, T_k) d(P_h, P_i)$, where $d(P_h, P_i)$, the *distance* between processors P_h and P_i, is the length of a shortest path from P_h to P_i. As further simplification it is assumed that for $T_j \prec T_k$ the length of transmitted data $\kappa(T_j, T_k)$ depends only on task T_j. Moreover, in many applications it makes sense to say that the amount of data produced by a task increases linearly with the processing time of the task. This assumptions lead to the simplification $\kappa(T_j, T_k) = \pi_L p_j$ where $\pi_L \in \mathbb{R}_{\geq 0}$ is a *packet length coefficient* that measures the amount of data produced in unit time by T_j, and p_j is the processing time of this task. Six heuristic algorithms were empirically compared. These include: hill-climbing, threshold accepting, great deluge algorithm, record-to-record travel, simulated annealing, and tabu search. The seed solution used in these heuristics is obtained by generating a critical path schedule. The paper [EH96] compares the performance of the heuristic strategies for different network topologies.

To obtain a model that takes the network structure more accurately into account, Ecker and Hirschberg [EH93] introduced a formal description of networks. It was shown that hypergraphs are a useful tool to model the structural and behavioral properties of networks that are needed for optimal organization of communication in networks. The approach is very general in the sense that it can be applied to different kinds of networks under various assumptions about the transmission capabilities of links. Essentially the communication hypergraph informs about maximal sets of simultaneous communication in the network. In [EH93] this approach was used to develop scheduling strategies for "pure" communication problems, i.e. problems where each task merely has to transmit data of a certain amount from one processor to another. Several approximation algorithms for this problem have been defined, and their performances have been compared against each other. Notice that such problems occur in different areas. In synchronized multiprocessor systems, for example, processes are usually organized in alternating phases of *computation* and *communication* [BT89]. Here the hypergraph approach can be applied to optimize the communication phase.

6.4   Scheduling Divisible Tasks

In this section we will consider the problem of scheduling divisible tasks. The study of divisible load theory started from the consideration of intelligent sensor networks by Cheng and Robertazzi [CR88]. An intelligent sensor is a single processor based unit which can make measurements, compute and communicate with other intelligent sensors. Later this intelligent sensor network application was replaced by the application of load sharing in a multiprocessor environment. The main problem in this research is to determine the optimal fraction of the workload to assign to each processor. That is, when a network receives a burst of data to process, one must decide what portion of the entire workload should be kept by the distributing processor and what portion of the entire workload should be distributed to each processor in order to minimize the total processing time.

In [CR88], recursive expressions for calculating the optimal load allocation for linear daisy chains of processors were presented. This is based on the assumption that for an optimal allocation of load, all processors must stop processing at the same time. Intuitively, this is because otherwise some processors would be idle while others were still busy. Analogous solutions were developed for tree networks [CR90] and bus networks [BR91]. In [Rob93], the concept of an equivalent processor that behaves identically to a collection of processors in the context of a linear daisy chain of processors and a proof that, for such a network structure, the optimal solution involves all processors stopping at the same time were introduced. An analytic proof for bus networks that for a minimal solution time all processors must finish computing at the same time was shown in [SR96, BGM92].

In [SR94], a more sophisticated load sharing strategy was proposed for bus networks that exploit the special structure of divisible load theory to yield a smaller solution time when a series of tasks are submitted to the network. In [SR95], a deterministic analysis is provided for the case when the processor speed and the channel speed are time-varying due to the background tasks submitted to a distributed system. A stochastic analysis which makes use of Markovian queuing theory was also introduced for the case when the arrival and departure times of the background tasks are not known. The equivalence of first distributing load either to left or to the right form a point in the interior of a linear daisy chain is demonstrated in [GM94]. Optimal sequences of load distribution in tree networks are described in [BGM94, KJL95]. In [BD95], a deterministic approach to find an optimal distribution of the load on a hypercube of processors was proposed. Simple formulae were found to determine the distribution of the task's load and the equivalent speed of the whole network of processors for two-dimensional mesh architecture in [BD96]. A uniform methodology was presented in [BD97] to achieve minimum completion time for a wide range of interconnection architectures, assuming that the communication time is equal to some start-up value plus some amount proportional to the value of transferred data. An example of optimal load allocation in a real time system appeared in [Had94]. Finally, in [BDGT99] a new broadcasting scheme to distribute parts of the task to processors in a minimum time [PS96] was analyzed in the context of divisible task processing.

We will illustrate the technique used in the analysis of divisible task processing by an example of a hypercube communication network [BD95]. The goal of the analysis is to find bounds for the performance of the above network expressed as:

— processing time of a task (processed by the whole network),

— processing speed of the whole network,

— speedup,

— processor utilization.

Problem $P, cube \mid div, cube_j \mid C_{max}$

A computer system to be considered consists of a set of identical processing elements (PE's): processors with local memories connected by a network of communication links. The architecture of the interconnection network is assumed to be a hypercube (see Figure 6.4.1), i.e. each processor is a node of a multidimensional cube. For the hypercube of dimension d there are 2^d processors in the system. Each of the processors has direct links to d neighbors. The label of a processor is a binary number from the interval $[0, 2^d - 1]$. Note, that each of the processor's neighbors has a label differing on exactly one position.

At time 0 a task arrives at processor P_0. Some part α_0 of the total load is processed by processor P_0, the rest of the load $(1-\alpha_0)$ is transmitted in equal parts to its d neighbors for processing. Immediate neighbors of processor P_0 take some part α_1 of the total load and retransmit the rest to the still idle neighbors. This process is continued until the last idle processor in the hypercube is reached. We assume that the processing time of the task on a standard processor is p, while on processor with a different speed it is wp where w is proportional to the reciprocal of the processor's speed. On the other hand, the transmission time of the whole task's data is t for a standard data link, while for a link with different capabilities it is zt where z is the reciprocal of the link bandwidth. We assume two things about the processing element: it must receive its entire load before transmitting the proper part to the neighbors and it is capable of simultaneous transmitting and computing.

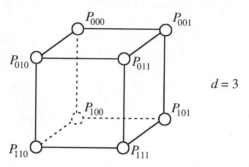

Figure 6.4.1 *A hypercube architecture.*

When processor P_0 receives a burst of data to process (cf. Figure 6.4.2), it takes α_0 of it for local processing; $(1-\alpha_0)$ of the load is transmitted to d neighbors. Since processor P_0 has no 1 in its address, its neighbors have exactly one 1 in their addresses. The part $(1-\alpha_0)$ transmitted from processor P_0 is fairly divided among all d neighbors. Then, each of the processors with only one 1 in the address takes α_1 of the whole load for local processing from the part it receives from processor P_0. The rest is transmitted, in equal shares, to its $d-1$ idle neighbors. Processors with one 1 in the address have $d-1$ idle neighbors with exactly two 1's in the address. Note, that processors with one 1 in the address can be reached from the originator of the load via only one link, while processors with two 1's can be accessed via two links. The process of data dissemination is repeated until the last processing element with address $11\cdots1$ is reached via d links. Let us call by *layer i* a set of all processors reached in the same number i of hops, starting from layer 0 consisting of the originator only (processor P_0). The last layer d consists of a single processor. Note, that data transmission does not cause contention in use of any communication link because each link is used only once and each communication path is one link long. As we already mentioned the computations must finish on all exploited processors at the same time. The following lemma describes useful topological properties of a hypercube.

Lemma 6.4.1 [BD95]. *In each layer i of a d-dimensional hypercube there are $\binom{d}{i}$ processing elements each of which can be accessed through i communication links and is capable of transmitting to $d-i$ still idle processors.* □

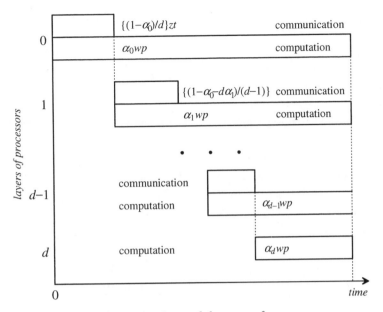

Figure 6.4.2 *Process of computation and data transfer*

Let us consider now layer i and denote by Vol_i the amount of data received by a processor in this layer, and by $\hat{\alpha}_i$ the part of the received load that is intercepted for local processing by this processor. We see that

$$\alpha_i = \hat{\alpha}_i \, Vol_i \, .$$

The processor in layer i works the same amount of time as it takes to transmit data to $d-i$ processors in layer $i+1$ and to computer in layer $i+1$ (cf. Figure 6.4.2). What is more, processors in layer $i+1$ receive data to process from $i+1$ links. Thus,

$$\hat{\alpha}_i Vol_i wp = \frac{(1-\hat{\alpha}_i)Vol_i((i+1)w_{i+1}p + zt)}{d-i} \qquad \text{for } i = 0, \cdots, d-1, \qquad (6.4.1)$$

where w_{i+1} is equivalent to a reciprocal of the speed for all processors in layers $i+1, \cdots, d$ which receive some part of the load from the considered processor in layer i. Hence , $\hat{\alpha}_i$ is equal to

$$\hat{\alpha}_i = \frac{1}{1 + \dfrac{(d-1)wp}{(i+1)w_{i+1}p + zt}} \qquad \text{for } i = 0, \cdots, d-1 , \qquad (6.4.2)$$

The value of w_i can be calculated according to the expression:

$$w_i = \frac{\hat{\alpha}_i Vol_i wp}{Vol_i p} = \hat{\alpha}_i w$$

For $i = 0$ equation (6.4.1) has the following form

$$\hat{\alpha}_0 wp = \frac{(1-\hat{\alpha}_0)(w_1p + zt)}{d},$$

hence,

$$Vol_0 = \alpha_0 = \hat{\alpha}_0 = \frac{1}{1 + \dfrac{dwp}{w_1p + zt}}. \qquad (6.4.3)$$

Equations (6.4.2) and (6.4.3) form a set of expressions which can be solved for $\hat{\alpha}_i$, recursively starting from $i = d-1$ (for which we know that $\hat{\alpha}_d = 1$, $w_d = w$) until $i = 1$. Then, the portions α_i of the whole load can be calculated. Thus, the originator processes locally α_0 of the whole load and sends to each of its d neighbors a share of load equal to $(1-\alpha_0)/d$. Each of processors in layer i ($i = 1, \cdots, d$) receives through i links a share of load equal to $\dfrac{(1-\hat{\alpha}_{i-1})Vol_{i-1}}{d-i+1}$, thus to-tally $\dfrac{i(1-\hat{\alpha}_{i-1})Vol_{i-1}}{d-i+1}$, from which $\alpha_i = \dfrac{i(1-\hat{\alpha}_{i-1})Vol_{i-1}\hat{\alpha}_i}{d-i+1}$ is intercepted for local processing.

Now, one can calculate an equivalent reciprocal of the speed for the whole hypercube of processors:

$$w^{eq} = \frac{\alpha_0 wp}{p} = \alpha_0 w.$$

Then, the *speedup S*, measured as a ratio of the sequential computation time, i.e. on the sole originator, to the working time of the originator embedded in the hypercube, and the average *processor utilization* of (U) can be found:

$$S = \frac{wp}{\alpha_0 wp} = \frac{1}{\alpha_0} = 1 + \frac{dwp}{w_1p + zt},$$

$$U = \frac{S}{2^d} = \frac{1}{2^d \alpha_0},$$

where w_1 is calculated according to the above recursive procedure.

From the above formulae one can derive several qualitative conclusions. The speedup depends on the dimension d of the hypercube but depends also on the w_1, which would have to decrease at least linearly in order to preserve linear speedup. The average utilization of the processors has 2^d in the denominator; then again, to preserve linear speedup and utilization close to 1, α_0 must decrease very fast.

Using the above formulae one can analyze a performance of the hypercube depending on such parameters as dimension of the hypercube (d), reciprocal of the communication speed (z), reciprocal of the processing speed (w) and size of the computing task (p) [BD95].

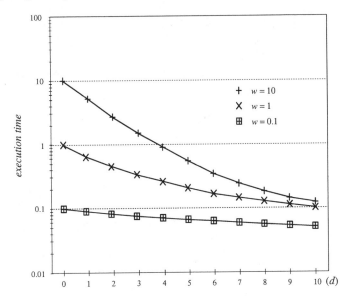

Figure 6.4.3 *Execution time vs. processor speed and dimension*

As the first performance parameter we will analyze an execution time of the task and its dependence on w, z, p, and d. The execution time of the task decreases with the dimension of the hypercube. However, for faster processors ($w = 0.1$) this reduction is relatively smaller than for slow processors ($w = 10$) where execution time can be reduced by two orders of magnitude (cf. Figure 6.4.3). Conclusion is that gain from parallel processing on slow processors is higher than on fast processors. In Figure 6.4.4 we see that fast communication network ($z = 0.1$) is more reasonable than the slow one ($z = 10$). In this picture a curve for $z = 0$ is also included which is a case of the ideal network (without transportation delays). Finally, Figure 6.4.5 demonstrates relative processing time as a function of p and d. The relative execution time is equal to the quotient of the actual processing time and processing time on the processor of the same speed. We see that the gain from parallel processing is bigger for long tasks ($p = 10$).

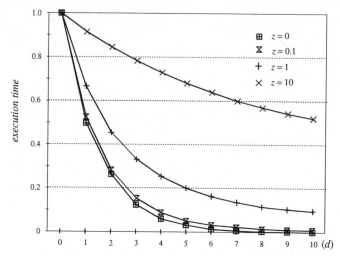

Figure 6.4.4 *Execution time vs. communication speed and dimension.*

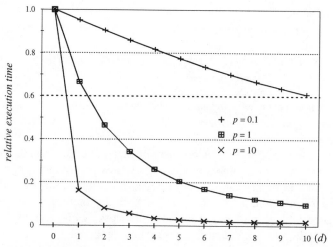

Figure 6.4.5 *Execution time vs. size of the computing task p and dimension.*

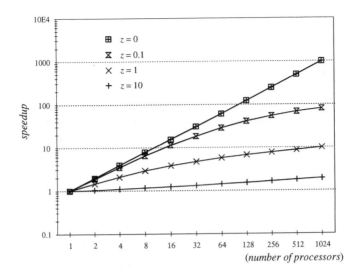

Figure 6.4.6   *Speedup for different z vs. number of processors.*

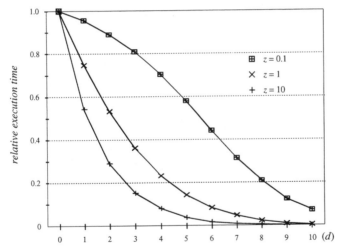

Figure 6.4.7   *Utilization of processors for different z vs. dimension.*

Next we analyze the impact of network parameters on the speedup (cf. Figure 6.4.6) and the utilization of processors (cf. Figure 6.4.7). In both figures curves are presented for different values of z, including $z = 0$. The network with $z = 0$ represents an ideal network. As can be seen, the linear speedup can be achieved when the communication medium is perfect. In more realistic cases ($z > 0$) the speedup curve levels off very fast, especially for slow networks ($z = 10$). The utilization of processors decreases with the size of the network. Only for the perfect communication network, utilization equal to 1 can be achieved. On the other hand, when we compare this result with Figure 6.4.4, the tendency to preserve linear speedup by significantly improving of the network speed, may not be justified in practice.

To sum up one may say that the faster are the processors and the communication links, the more efficient is the hypercube. On the other hand, the gain from parallel processing is more significant on slower (cheaper) processors than on fast processors. Finally, as we demonstrated the linear speedup in the hypercube can be obtained only in a perfect network with no communication delays at all.

References

AHC90   F. D. Anger, J. Hwang, Y. Chow, Scheduling with sufficiently loosely coupled processors, *J. Parallel Distributed Comput.* 9, 87-92, 1990.

AHU74   A. V. Aho, J. E. Hopcroft, J. D. Ullman, *The Design and Analysis of Computer Algorithms*, Addison-Wesley, Reading, Mass., 1974.

Avi78   A. Avizienis, Fault tolerance: the survival attribute of digital systems, *Proc. IEEE* 66, 1978, 1109-1125.

AZ90   M. Ahuja, Y. Zhu, An O(logn) feasibility algorithm for preemptive scheduling of n independent jobs on a hypercube, *Inform. Process. Lett.* 35, 1990, 7-11.

BBDO94   L. Bianco, J. Błażewicz, M. Drozdowski, P. Dell'Olmo, Scheduling preemptive multiprocessor tasks on dedicated processors, *Performance Evaluation* 20, 1994, 361-371.

BBDO95   L. Bianco, J. Błażewicz, M. Drozdowski, P. Dell'Olmo, Scheduling multiprocessor tasks on a dynamic configuration of dedicated processors, *Ann. Oper. Res.* 58, 1995, 493-517

BBDO97a   L. Bianco, J. Błażewicz, M. Drozdowski, P. Dell'Olmo, Linear algorithms for preemptive scheduling of multiprocessor tasks subject to minimal lateness, *Discrete Appl. Math.* 72, 1997, 25-46.

BBDO97b   L. Bianco, J. Błażewicz, M. Drozdowski, P. Dell'Olmo, Preemptive multiprocessor task scheduling with release times and time windows, *Ann. Oper. Res.*70, 1997, 43-55.

BBGT96   J. Błażewicz, P. Bouvry, F. Guinand, D. Trystram, Scheduling complete intrees on two uniform processors with communication delays, *Inform. Process. Lett.* 58, 1996, 255-263.

BD95   J. Błażewicz, M. Drozdowski, Scheduling jobs on hypercubes, *Parallel Computing* 21, 1995, 1946-1956.

BD96   J. Błażewicz, M. Drozdowski, The performance limits of a two dimensional network of load-sharing processors, *Foundations Comput. Dec. Sci* 21, 1996, 3-15.

BD97   J. Błażewicz, M. Drozdowski, Distributed processing of divisible jobs with communication startup costs, *Discrete Appl. Math.* 76, 1997, 21-41.

BDGT99   J. Błażewicz, M. Drozdowski, F. Guinand, D. Trystram, Scheduling under architectural constraints, *Discrete Appl. Math.* 94, 1999, 35-50.

BDOS92   J. Błażewicz, M. Drozdowski, P. Dell'Olmo, M. G. Speranza, Scheduling multiprocessor tasks on three dedicated processors, *Inform. Process. Lett.* 41, 1992, 275-280. Corrigendum: *Inform. Process. Lett.* 49, 1994, 269-270.

BDSW90   J. Błażewicz, M. Drozdowski, G. Schmidt, D. de Werra, Scheduling independent two processor tasks on a uniform duo-processor system, *Discrete Appl. Math.* 28, 1990, 11-20.

BDSW94   J. Błażewicz, M. Drozdowski, G. Schmidt, D.de Werra, Scheduling independent multiprocessor tasks on a uniform *k*-processor system, *Parallel Computing* 20, 1994, 15-28.

BDW84   J. Błażewicz, M. Drabowski, J. Weglarz, Scheduling independent 2-processor tasks to minimize schedule length, *Inform. Process. Lett.* 18, 1984, 267-273.

BDW86   J. Błażewicz, M. Drabowski, J. Weglarz, Scheduling multiprocessor tasks to minimize schedule length, *IEEE Trans. Comput.* C-35, 1986, 389-393.

BDWW96   J. Błażewicz, M. Drozdowski, D. de Werra, J. Weglarz, Deadline scheduling of multiprocessor tasks, *Discrete Appl. Math.* 65, 1996, 81-96.

BE83   J. Błażewicz, K. Ecker, A linear time algorithm for restricted bin packing and scheduling problems, *Oper. Res. Lett.* 2, 1983, 80-83.

BEPT00   J. Błażewicz, K. Ecker, B. Plateau, D. Trystram, *Handbook on Parallel and Distributed Processing*, Springer, Berlin, New York, 2000.

Ben64   V. E. Benes, Permutation groups, complexes, and rearrangeable connecting networks, *Bell Syst. Tech. J.* 43, 1964, 1619-1640.

BGK96   E. Bampis, A. Giannokos, J.-C. König, On the complexity of scheduling with large communication delays, *European J. Oper. Res.* 94, 1996, 252-260.

BGM92   V. Bharadwaj, D. Ghose, V. Mani, A study of optimality conditions for load distribution in tree networks with communication delays, Tech.Report 423/GI/02-92, Dept. of Aerospace Engineering, Indian Institute of Science, Bangalore, 1992.

BGM94   V. Bharadwaj, D. Ghose, V. Mani, Optimal sequencing and arrangement in distributed single-level tree networks with communication delays, *IEEE Trans. Parallel Distrib. Syst.* 5, 1994, 968-976.

BKM+04   J. Błażewicz, M. Kovalyov, M. Machowiak, D. Trystram, J. Weglarz, Scheduling malleable tasks on parallel processors o minimize the makespan, *Annals of Oper. Res.* 129, 2004, 65-80.

BKM+06   J. Błażewicz, M. Kovalyov, M. Machowiak, D. Trystram, J. Weglarz, On preemptable scheduling malleable tasks, *IEEE T. Comput.* 2006, to appear.

BL96   J. Błażewicz, Z. Liu, Scheduling multiprocessor tasks with chain constraints, *European J. Oper. Res.* 94, 1996, 231-241.

BLRK83   J. Błażewicz, J. K. Lenstra, A. H. G. Rinnoy Kan, Scheduling subject to resource constraints : classification and complexity, *Discrete Appl. Math.* 5, 1983, 11-24

Bok81   S. H. Bokhari, On the mapping problem, *IEEE Trans. Comput.* C-30, 1981, 207-214.

BOS91   L. Bianco, P. Dell'Olmo, M. G. Speranza, Nonpreemptive scheduling of independent tasks with dedicated resources, Report, IASI, Roma, 1991

BR91   S. Bataineh, T. G. Robertazzi, Bus oriented load sharing for a network of sensor driven processors, *IEEE Trans. Syst. Man. Cybernet.* 21, 1991, 1202-1205.

Bru46   N. G. de Bruijn, A combinatorial problem, *Koninklijke Netherlands: Academie van Wetenschappen, Proc.* 49, 1946, 758-764.

Bru95   P. Brucker, *Scheduling Algorithms,* Springer, Berlin, 1995.

BT89   D. Bertsekas, J. Tsitsiklis, *Parallel and Distributed Computation*, Prentice Hall, Englewood Cliffs, N.J., 1989.

CC89   Y. L. Chen, Y. H. Chin, Scheduling unit-time jobs on processors with different capabilities, *Comput. Oper. Res.* 16, 1989, 409-417.

CC91   I. Y. Colin, P. Chretienne, C.P.M. scheduling with small communication delays and task duplication, *Oper. Res.* 39, 1991, 680-684.

CGJP85   E. G. Coffman, Jr., M. R. Garey, D. S. Johnson, A. S. La Paugh, Scheduling file transfers, *SIAM J. Comput.* 14, 1985, 744-780.

Chr89a   P. Chretienne, A polynomial algorithm to optimally schedule tasks over a virtual distributed system under tree-like precedence constraints, *European J. Oper. Res.* 43, 1989, 225-230.

Chr89b   P. Chretienne, Task scheduling over distributed memory machines, in: *Proceedings of the International Workshop on Parallel and Distributed Algorithms*, North-Holland, Amsterdam, 1989.

Chr94   P. Chretienne, Tree scheduling with communication delays, *Discrete Appl. Math.* 49, 1994, 129-141.

CL88a   G. I. Chen, T. H. Lai, Preemptive scheduling of independent jobs on a hypercube, *Inform. Process. Lett.* 28, 1988, 201-206.

CL88b   R. S. Chang, R. C. T. Lee, On a scheduling problem where a job can be executed only by a limited number of processors, *Comput. Oper. Res.* 15, 1988, 471-478

CP91   P. Chretienne, C. Picouleau, The basic scheduling problem with interprocessor communication delays, MASI Report 91/6, Institut Blaise Pascal, Paris, 1991.

CR88   Y. C. Cheng, T. G. Robertazzi, Distributed computation with communication delays, *IEEE Trans. Aerospace Electr. Syst.* 24, 1988, 511-516

CR90   Y. C. Cheng, T. G. Robertazzi, Distributed computation with communication delays, *IEEE Trans. Aerospace Electr. Syst.* 26, 1990, 511-516.

DD81   M. Dal Cin, E. Dilger, On the diagnostability of self-testing multimicroprocessor systems, *Microprocessing and Microprogramming* 7, 1981, 177-184.

DK99   M. Drozdowski, W. Kubiak, Scheduling parallel tasks with sequential heads and tails, *Annals of Oper. Res.* 90, 1999, 211-246.

DL89 J. Du, J. Y-T. Leung, Complexity of scheduling parallel tasks systems, *SIAM J. Discrete Math.* 2, 1989, 473-487.

Dro96a M. Drozdowski, *Selected Problems of Scheduling Tasks in Multiprocessor Computer Systems,* Poznań University of Technology Press, Poznań, 1996.

Dro96b M. Drozdowski, Real-time scheduling of linear speedup parallel tasks, *Inform. Process. Lett.* 57, 1996, 35-40.

EH93 K. H. Ecker, R. Hirschberg, Scheduling communication demands in networks, *Proc. of the Workschop on Parallel and Distributed Real-Time Systems,* Newport Beach, 1993.

EH94 K. H. Ecker, D. Hammer, Integrated scheduling for CIM systems. *Proc. TIMS XXXII,* Anchorage, 1994.

EH96 K. H. Ecker, H. Hodam, Heuristic algorithms for the task scheduling under consideration of communication delays, Report, T. U. Clausthal,1996.

ERL90 H. El-Rewini, T. G. Lewis, Scheduling parallel program tasks onto arbitrary target machines, *J. Parallel Distributed Comput.* 9, 1990, 138-153.

GM94 D. Ghose, V. Mani, Distributed computation in a linear network : Closed-form solutions and computational techniques, *IEEE Trans. Aerospace Electr. Syst.* 30, 1994, 471-483.

Goe95 M. Goemans, An approximation algorithm for scheduling on three dedicated processors, *Discrete Appl. Math.* 61, 1995, 49-60.

Gok76 R. L. Goke, *Banyan networks for partitioning multiprocessor systems*, Ph.D. Thesis, Univ. Florida, Gainesville, 1976.

GT93 F. Guinand, D. Trystram, Optimal scheduling of UECT trees on two processors, Report, Universite Joseph Fourier, Grenoble, 1993.

GY92 A. Gerasoulis, T. Yang, On the granularity and clustering of directed acyclic task graphs, Report TR-153, Dept. Comput. Sci., Rutgers University, 1992.

Had94 E. Haddad, Communication protocol for optimal redistribution of divisible load in distributed real-time systems, *Proc. of the ISMM Internat.Conf. on Intelligent Information Management Systems,* Washington, 1994, 39-42.

HCAL89 J.-J. Hwang, Y.-C. Chow, F. D. Anger, C.-Y. Lee, Scheduling precedence graphs in systems with interprocessor communication times, *SIAM J. Comput.* 18, 1989, 244-257.

HLV94 J. A. Hoogeveen, J. K. Lenstra, B. Veltman, Three, four, five, six or the complexity of scheduling with communication delays, *Oper. Res. Lett.* 16, 1994, 129-136.

Hoe89 C. P. M. van Hoesel, Preemptive scheduling on a hypercube, Report 8963/A, Econometric Institute, Erasmus University, Rotterdam, 1989.

HVV94 J. A. Hoogeven, S. L. van de Velde, B. Veltman, Complexity of scheduling multiprocessor tasks with prespecified processor allocation, *Discrete Appl. Math.* 55, 1994, 259-272.

JM84 D. S. Johnson, C. L. Monma, A scheduling problem with simultaneous machine requirement, *Proc., TIMS XXVI,* Copenhagen, 1984.

JR92   A. Jakoby, R. Reischuk, The complexity of scheduling problems with commu-
            nication delays for trees, *Proc. Scandinavian Workshop on Algorithmic Theory*
            3, 1992, 165-177.

KJL95   H. J. Kim, G. I. Jee, J. G. Lee, Optimal load distribution for tree network proc-
            essors, 1995, submitted for publication.

KK85   H. Krawczyk, M. Kubale, An approximation algorithm for diagnostic test
            scheduling in multicomputer systems, *IEEE Trans. Comput.* C-34, 1985, 869-
            872.

KL88   G. A. P. Kindervater, J. K. Lenstra, Parallel computing in combinatorial opti-
            mization, *Ann. Oper. Res.* 14, 1988, 245-289.

KS86   C. P. Kruskal, M. Snir, A unified theory of interconnection network structure,
            Preprint, 1986.

Kub87   M. Kubale, The complexity of scheduling independent two-processor tasks on
            dedicated processors, *Inform. Proc. Lett.* 24, 1987, 141-147.

Kub90   M. Kubale, Preemptive scheduling of two-processor tasks on dedicated proces-
            sors (in Polish), *Zeszyty Naukowe Politechniki Ślaskiej, Automatyka* 100, 1990,
            145-153.

Len04   J. Y-T. Leung (ed.), *Handbook of Scheduling. Algorithms, Models and Per-
            formance Analysis*, Chapman & Hall/CRC, Boca Raton, 2004.

Llo81   E. L. Lloyd, Concurrent task systems, *Oper. Res.* 29, 189-201.

LS77   S. Lam, R. Sethi, Worst case analysis of two scheduling algorithms, *SIAM J.
            Comput.* 6, 1977, 518-536.

LVV96   J. K. Lenstra, M. Veldhorst, B. Veltman, The complexity of scheduling trees
            with communication delays, *Journal of Algorithms* 20, 1996, 157-173.

Moh89   R. H. Möhring, Computationally tractable classes of ordered sets, I. Rival (ed.),
            Algorithms and Order, 105-193, Kluwer, Dordrecht, 1989.

OST93   P. Dell'Olmo, M. G. Speranza, Z. S. Tuza, Easy and hard cases of a scheduling
            problem on 3 dedicated processor, Report, IASI, Rome, 1995.

Pic91   C. Picouleau, Two new NP-complete scheduling problems with communica-
            tion delays and unlimited number of processors. Report RP91/24, MASI, Insti-
            tut Blaise Pascal, Universite Paris VI, 1991.

Pic92   C. Picouleau, *Etude de problèmes d'optimization dans les systèmes distribués*,
            Ph.D. Thesis, Université Paris VI, 1992.

PS96   J. G. Peters, M. Syska, Circuit-switched broadcasting in torus networks, *IEEE
            Trans. Parallel Distrib. Syst.,* 1996, to appear.

PV81   F. P. Preparata, J. Vuillemin, The cube-connected cycles: A versatile network
            for parallel computation, *Commun. ACM*, May 1981, 300-309.

PY90   C. H. Papadimitriou, M. Yannakakis, Towards an architecture-independent
            analysis of parallel algorithms, *SIAM J. Comput.* 19, 1990, 322-328.

Rob93   T. G. Robertazzi, Processor equivalence for a linear daisy chain of load sharing
            processors, *IEEE Trans. Aerospace Electr. Syst.* 29, 1993, 1216-1221.

Ray87a   V. J. Rayward-Smith, UET scheduling with unit interprocessor communication delays, *Discrete Appl. Math.* 18, 1987, 55-71.

Ray87b   V. J. Rayward-Smith, The complexity of preemptive scheduling given inter-processor communication delays, *Inform. Process. Lett.* 25, 1987, 123-125.

SH89   T. H. Szymansky, V. C. Hamacher, On the universality of multipath multistage interconnection networks, *J. Parallel Distributed Comput.* 7, 1989, 541-569.

SR91   X. Shen, E. M. Reingold, Scheduling on a hypercube, *Inform. Process. Lett.* 40, 1991, 323-328.

SR94   J. Sohn, T. G. Robertazzi, A multi-job load sharing strategy for divisible jobs on bus networks, Tech. Report 697, SUNY at Stony Brook College of Eng. and Appl. Sci., 1994

SR95   S. Sohn, T. G. Robertazzi, An optimal load sharing strategy for divisible jobs with time-varying processor speed and channel speed, *Proc. of the ISCA International Conf. on Parallel and Distributed Computing Systems,* Orlando, 1995, 27-32.

SR96   J. Sohn, T. G. Robertazzi, Optimal divisible job load sharing for bus networks, *IEEE Trans. Aerospace Electr. Syst.* 32, 1996.

SRL95   J. Sohn, T. G. Robertazzi, S. Luryi, Optimizing computing costs using divisible load analysis, Report CEAS 719, University at Stony Brook, 1995.

Sto71   H. S. Stone, Parallel processing with the perfect shuffle, *IEEE Trans. Comput.* C-20, 1971, 153-161.

Vel93   B. Veltman, *Multiprocessor Scheduling with Communication Delays*, Ph.D. Thesis, CWI-Amsterdam, 1993.

VRK92   T. A. Varvarigou, V. P. Roychowdhury, T. Kailath, unpublished manuscript, 1992.

VWHS95   J. P. C. Verhoosel, L. R. Welch, E. Liut, D. K. Hammer, A. D. Stoyenko, A model for scheduling of object-based, distributed real-time systems, *J. Real-Time Systems* 8, 1995, 5-34.

7 Scheduling in Hard Real-Time Systems

In Chapters 4 and 5 we analyzed scheduling problems in which the task performance is subject to temporal restrictions such as release times or deadlines. The present chapter deals with a similar problem, but where the tasks are to be processed repeatedly, and each execution is restricted by release times and deadlines. The release times are regularly distributed over time with equal distances called the task period. Such tasks are called periodic. The deadline is usually assumed to coincide with the release time of the next period. In many applications such as real-time systems we find problems where sets of periodic tasks are to be processed on a single processor or on a distributed or parallel processor system.

The area of real-time systems is a field of applications that differs considerably from the type of applications we have seen so far. A real-time system can roughly be described as a computing system designed for controlling some technical facility. Before we deal with the scheduling problem in such systems, we start with a short introduction to real-time systems, what we understand by them, present some applications and discuss characteristic properties and general functional requirements of such systems.

Section 7.1 introduces the main characteristics of real-time systems, presents some application examples, and discusses the functional requirements for real-time systems. A coarse idea of structuring a real-time system and about the nature of the tasks is given in Section 7.2. Section 7.3 deals with the scheduling of periodic tasks on a single processor. The rate monotonic and earliest deadline first strategies are analyzed from both, properties and performance points of view. The generalization to multiprocessor systems is discussed in Section 7.4. In Section 7.5 we review shortly runtime problems caused by blockings due to the use of non-preemptable (non-withdrawable) resources. We are not able to present the details of synchronization protocols, but just touch the subject, and discuss how blocking delays can be handled. Finally, in Section 7.6 we introduce several variants of the periodic task model that are considered in literature in order to gain higher flexibility as compared to the simple periodic task model.

7.1 Introduction

This chapter gives an overview on the peculiarities of real-time systems. An area like real-time systems is in fact the merging point of many disciplines, ranging form hardware and operating systems, to requirement analysis and design methodologies. Its many facets cannot be presented in the required detail, but our aim is to give an overview including the definition and general characterization.

7.1.1 What is a Real-Time System?

The role played by real-time systems today is increasing with a surprising speed, since our everyday life becomes more and more dependent on them. Real-time systems are used to control industrial, medical, scientific, consumer, environmental and other processes. They operate in close connection with technical systems called "*environment*" or "*external system*" (viz. Figure 7.1.1), such as production facilities, power plants, and as well in embedded system applications. The purpose of the real-time system (also called "*internal system*", in contrast to the external system) is to enforce some specific behavior of the environment. This is maintained by proper reactions within strict time limits in case the status of the environment deviates from the required specification. Depending on the nature of the applications, the internal system has many processes that are executed repeatedly. Examples are the periodic measurement by instruments that inform about the status of the controlled system, the evaluation of the measured data, the computation of new data derived from the latter for e.g. monitoring purposes and long term storage, and the generation of signals for controlling the environment by adjusting the actuators appropriately, to ensure the required or expected functioning of the environment.

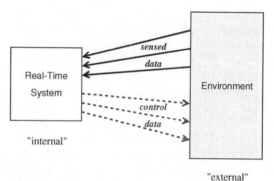

Figure 7.1.1   *Communication between a controlled environment and a real-time system.*

There are several attempts to define real-time systems. In our opinion, the following definition given by [KKZ88] comprises the characteristics of a real-time system most closely:

> "*A real-time system is defined as an interaction system that maintains an ongoing relationship with an asynchronous environment, i.e. an environment that progresses irrespectively of the real-time system in an uncooperative manner. The real-time system is fully responsi-*

ble for the proper synchronization of its operation with respect to its environment."

Closely related, or often even synonymously used, is the notion of an *embedded system*. The computer is interfaced directly to the physical equipment of the environment. There is no exact definition available, but real-time systems are mostly to be considered as being embedded in larger environments. From the way this term is used we may deduct that embedded systems can be understood as working autonomously and pertaining the complete internal system, including the sensors, actuators and alarms. Similar to embedded systems are *mechatronic systems*, which can be understood as mechanical systems like machines, enhanced by closely attached control devices. In such systems, the functionality of a machine is directly dependent on the real-time control system. The machine is useless if the computer system does not work properly. Typical examples of embedded and mechatronic system applications are industrial robots, NC-machines and car engine control, besides many others (viz. [WS98]).

One of the main characteristic of a real-time system is a strong interaction with its environment. Technical systems demand from their control systems significant computation and control processing, and a guarantee of predictable, reliable and timely operations. The problems the designer of a real-time system has to face are manifold:

− Technical processes as presented by an external system are usually very complex.

− The corresponding real-time systems consist of a large number of mutually dependent components.

− The notion of "system" is difficult to handle. Essential dependencies between parts of the system have to be realized.

− Both, the external system and the real-time system have to be partitioned into smaller components ("partial systems", "partial processes").

− Knowledge about partial systems is usually incomplete.

7.1.2 Examples of Real-Time Systems

To guide the reader towards a better understanding of the subject we give some insight to specific real-time applications. There are fundamental differences between non-real-time applications and real-time applications. Looking, for example, at the differences between a compiler and a chemical process control program we see that a program compilation can be fast or slow, depending on the speed of the machine, the used language, and the size of the program. The user more or less tolerates compilation time. A chemical process control program, in contrast, controls an external process. Therefore the program steps must be synchronized with the external events of the process. A careless design can result in the danger of damages in the environment, as can easily be imaged when thinking on heat control in a chemical reactor.

– An environment such as a chemical factory consists in general of many components that need to be controlled. A simple example regarding just one component of a larger system is the control of an even flow of liquid in a pipe by a valve. The computation to calculate the new valve angle may be quite complex. The computer interacts with the facility using sensors and actuators. Depending on the criticality of the application, there are hard or soft real-time requirements.

– *Mechatronics* is an interdisciplinary cooperation of mechanical engineering, electronic engineering, and software engineering. Classical mechanical systems, enhanced by electronic components allow the creation of new products of greater functionality and adaptability. Typically, electronic system monitors the status of the mechanical system via sensors and computes actuator signals for enforcing certain optimal performance. Examples of mechatronic systems are automated household appliances such as CD players, microwaves, and dishwashers. Traffic control, complex car control systems, transportation systems and computer aided manufacturing systems belong as well to the wide spectrum of real-time applications.

– *Manufacturing*: The entire manufacturing process from product design to fabrication is controlled by a large real-time system, usually distributed over many computing resources. Here the soft real-time character is emphasized.

– *Communication*, *command* and *control*: Wide range of disparate applications exhibit similar characteristics. They include airline seat reservation, air traffic control, remote bank accounting and many others. Devices and instruments for gathering information that is required for decision making are often distributed over a wide geographical area. Mostly, the real-time requirements are of soft character, with due dates and cost functions depending on the particular application

To summarize, we see that the real-time requirements in these applications differ considerably. The correctness of a real-time system depends not only on the *logical result* of the computation but also on the *time* at which the results are produced. The timing requirements can range from msec to hours, days, $\cdots$, even within the same application. Another important point regards the urgency of control actions. In hard real-time control, the deadlines have to be observed, and the responses must occur within specified deadlines. In soft real-time control, in contrast, actions may by delayed up to a certain extent (due dates), or the response times, though hard, may be occasionally missed.

7.1.3 Characteristics of Real-Time Systems

In real-time systems time is a central issue. The operating system is responsible for the timeliness of operations. Therefore, not only the time behavior of programs has to be well understood, but also scheduling theory turns out to be a key discipline. In this connection, issues such as real-time mode of computer operations, management of different types of processes, time-, event-, and priority-based interrupts, synchronization primitives, concept of tasks as a parallel program thread in a multiprogramming environment, task scheduling operations, and timing of task executions are fundamental. On the other hand, processor utilization is less relevant, because the cost of any processor involved in a failure of an external process is negligible as compared to the cost of damages caused by the failed process. This is true even in comparatively inexpensive environments

What is instead required is dependable and predictable fulfillment of the requirements. If a computer cannot guarantee reaction times, it may be unable to cope appropriately with exceptional and emergency situations arising in the environment and may thus violate the latter's safety requirements. The dynamics of the physical system under control impose timing constraints that must be met and therefore dictate the temporal behavior to be achieved. Therefore, the assurance of the (functional and temporal) correctness of a real-time system, mainly for those embedded in safety-critical systems, must be possible.

There are environments with hard and with soft timing constraints. They are distinguished by the consequences of violating the timeliness requirement: soft real-time environments are characterized by costs rising with increasing lateness of results. In hard real-time environments such lateness is not permitted under any circumstances, because late computer reactions are either useless or dangerous. In other words, the cost for missing deadlines in hard real-time environments are - from the application's point of view - considered as infinitely high. Hard time constraints have to be determined precisely, and typically result from the physical laws governing the technical processes being controlled.

One can already see that general questions from different knowledge areas arise in connection with the realization of a real-time system. The construction of real-time systems includes the requirement analysis and specification, formal methods and models, refinement, language design, compilers, operating systems and scheduling, hardware aspects, etc.. Solving the real-time system implementation problem thus needs a synergetic exertion of experts in various knowledge areas. Our objective here is to concentrate on the scheduling aspect in real-time systems.

7.1.4 Functional Requirements for Real-Time Systems

Technical systems demand from their control systems significant computation and control processing, and the guarantee of *predictable*, *reliable* and *timely operation*. Design problems to meet these requirements are manifold:

Timeliness: There are two general kinds of requirements: In *relative* timing constraints, actions may have to be performed within a given interval of time relative to the occurrence of an external or internal event. *Absolute* timing constraints specify the system behavior for globally given points in time.

Predictability: The controlling real-time system must nevertheless handle each external event predictably within the associated time constraints. The reactions to be carried out by the computer must therefore be precisely planned in order to get a fully predictable behavior.

Dependability refers to the general requirement of trustworthiness. The system has to produce the right control signals at the right time (correctness). Robustness refers to the requirement that the system remains in a predictable state, even if the environment does not correspond to the specification, e.g., if inputs are not within a given range. Another condition is permanent readiness, meaning that the real-time system does not terminate (e.g. as a result of a failure), and tolerance against software or hardware faults.

7.2   Basic Notions

The design of real-time systems is an iterative process in which solution concepts and their verification is modified until a hopefully satisfying final solution is obtained. Unfortunately, there is no closed design theory available. First we give a coarse idea about the structure of a real-time system and about the nature of the tasks. Then we introduce a formal task model and discuss general scheduling issues.

7.2.1   Structure of a Real-Time System

A coarse view of the activities that should be performed by the real-time system is presented in Figure 7.2.1. A 5-stage structure distinguishes

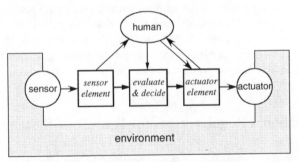

Figure 7.2.1   *Five stage structure of a real-time system.*

– *sensors*, which are devices informing about the status of the environment,
– *sensor elements*, which consist of the software modules that perform some pre-processing on the measured data,
– an *evaluation and decision stage*, where the validity of pre-processed data is checked and put into relation with other data such as historical data, and requirements for changes in the environment are computed. In larger systems a database or knowledge base will be needed for performing the required functions,
– *actuator elements*, which convert the required changes into control data for the actuators, and

– *actuators*, which are the final devices to change the behavior of the environment.

A real-time system has in general several threads of control. Several types of threads (also called end-to-end paths [Ger95]) are distinguished. A *periodic* path is executed repeatedly, once in a period of fixed length, and the completion of one execution must not exceed a given deadline. A typical example is to read sensor data and update the current state of internal variables and outputs. An *asynchronous* or *aperiodic* path responds to internal or external events, but precise request times for the execution are not known in advance. Usually, the minimum amount of time between two consecutive requests and a deadline, by which an execution must be completed, are available.

We adopt here a simplified model in which each path is represented by a task whose runtime conditions are specified by a number of attributes such as a period or the requirement of a repeated execution guided by a given maximal time lag, with or without deadlines (hard) or due dates (soft).

In the next section we present a more formal view of the task system and its particular properties.

7.2.2 The Task Model

Suppose we are given a set of tasks, $\mathcal{T} = \{ T_1, \cdots, T_n \}$. There are no precedences between the tasks, and each task is characterized by a set of parameters that are assumed to be integers. The notion of *real-time scheduling* refers to the condition that given deadlines is observed. As discussed in Chapter 3, every task T_j characterized by the following data:

– *Processing time* (p_j) is usually be assumed to be the worst case execution time.

– *Ready times* or *release times* (r_j) and *deadlines* ($\tilde{d}_j$) may be specified for each task.

– A *periodic task* T_j is characterized by a sequence of equally distant release times defining its period π_j. In each interval one instance of the task is started. A periodic task can thus be characterized as a potentially infinite sequence of instances, each with a release time and a deadline. The ready times are usually the left interval boundaries. The simplest condition for the deadlines is to assume the right interval boundaries.

– *Aperiodic tasks:* The execution of a task may be triggered by an aperiodic event such as a hardware interrupt, or by another task in case of a particular computation result. Interrupts can take place at random intervals and be devastating if the system does not anticipate and correctly handle them.

– A *sporadic task* is an aperiodic task that is repeatedly executed. Instead of a period, it is characterized by a minimum time t_j between releases. The inverse of

t_j is referred as the *maximum arrival rate*. This allows us to handle sporadic tasks as periodic with a period being the inverse of the maximum arrival rate.

− *Task offset* or *phase*: The first period of task T_j starts with an offset $offset_j$ (usually $offset_j = 0$). The ith instance of T_j has then the release time $offset_j + (i-1)\pi_j$. If the deadline is at the end of the period it is given by $offset_j + i\pi_j$.

− The importance of task T_j is expressed by a numerical weight (*priority*) w_j. In case of task contention, higher critical tasks must be processed prior to less critical tasks. We assume here that tasks are immediately preempted at the release of a higher priority task.

− Tasks may require *additional renewable resources* $R(T_j)$ besides processors (cf. Chapter 12). Because of their possible impact on the runtime behavior, we are here only interested in resources that are exclusively used and non-preemptable, where a task will keep hold on an assigned resource until its completion, even if the task is preempted. Examples of such resources are communication channels, buffers, storage devices, etc..

7.2.3 Schedules [1]

Feasibility

As before a schedule must obey all the conditions specified in Section 3.1. There are two ways to specify schedules. An *explicit* specification contains the complete and detailed description of the schedule with all the timing parameters and processors which the tasks are assigned to for execution. For the non-preemptive tasks, it suffices to specify the start time and the processor, for example by a sequence of pairs (start times, a processor). Alternatively, one may specify a list of pairs (task, start time) for each processor. In the preemptive case we need to define the start times and duration of all processing intervals. *Explicit* schedules are used in *pre-runtime* or *off-line* scheduling. An *implicit* description is based on a scheduling rule, according to which the tasks are sequenced. This can be done by means of priorities where tasks of higher priority are given preference. When a schedule is constructed implicitly we talk about *on-line* scheduling.

For solving the scheduling problem we may have to make assumptions about the number and type of processors. If processors have local memory there would be no need to provide all processors with all the code segments. As a consequence, a processor will only be able to process those tasks whose code is stored locally. The problems we are confronted with, are manifold:

− How many processors are required, and how should the code be distributed.

− Find a feasible and safe schedule, in which also communication delays between pairs of dependent tasks are taken into account.

---

[1] An interesting historical overview on real-time scheduling can be found in [SAA04].

– Which strategy should the runtime system follow in order to ensure correct behavior.

General strategies for scheduling periodic tasks

Based on the priorities a simple *priority-driven scheduling rule* can be defined by preempting an executed task immediately if an instance of a higher priority task is requested. Conversely, a lower priority task is never able to preempt a higher priority task. Of course, preempting a task and continuing it later causes delays due to the context switches, but for simplicity reasons we neglect the task switching times.

For a given task set with priorities, a schedule can easily be determined. A priority assignment is called feasible if the schedule resulting from that assignment is feasible. Priorities can be

– *fixed* (or *static*): each task has a user- or system-defined priority that remains constant for the lifetime of the task. Let $T = \{T_1, \cdots, T_n\}$ be a task set with respective priorities $w_1, \cdots, w_n$. W.l.o.g. we assume that the priority assignment is injective, i.e., any two tasks have different priorities. The (uniquely defined) *priority list* is the sequence of tasks ordered decreasingly with the priorities.

– *dynamic*: depending on execution parameters such as upcoming deadlines or other runtime conditions, the priorities vary at runtime.

We distinguish two versions of priority driven task executions, pre-runtime (off-line) and runtime (on-line) scheduling.

Off-line versus on-line scheduling

We give a short discussion of the pros and cons of the off-line and on-line scheduling paradigm.

In off-line scheduling the schedule for the periodic processes is computed during and explicitly specified the system design. The time efficiency of the scheduling algorithm is not a critical concern. In most cases off-line scheduling would provide a better chance to satisfy all the timing and resource constraints. Then, at runtime, the periodic processes are executed according to the previously computed schedule. As a further advantage of the pre-runtime approach, it is relatively easy to take into account additional constraints, such as arbitrary release times, deadlines, and precedences.

In on-line scheduling decisions about which task is to be executed next are made at runtime. Compared to the off-line scheduling paradigm, the scheduler has more work to perform at runtime, in particular if processes contain critical sections. Another drawback is that an on-line algorithm may fail to provide a feasible solution though it could be solved with the pre-runtime approach. Additional application constraints are likely to conflict with the priorities that are assigned at runtime to the processes. As an advantage, on-line scheduling generally allows more flexible reaction on unforeseen or exceptional situations than off-line schedules.

7.3 Single Processor Scheduling

In this section we concentrate on problems of scheduling periodic tasks on a single processor. The tasks are assumed to be preemptable, and additional resources besides processors are not considered. Hence the question we are dealing with is: Given a set of n periodic tasks $T = \{T_1, \cdots, T_n\}$ with respective processing times (worst case execution times) $p_1, \cdots, p_n$ and request periods $\pi_1, \cdots, \pi_n$, is it possible to process the tasks preemptively on a single processor? To answer this question it is important to realize that each task T_i utilizes the fraction $u_i := p_i/\pi_i$ of time the processor uses for execution. The total processor *utilization*

$$W := \sum_{i=1}^{n} p_i/\pi_i$$

represents hence the fraction of time needed for executing the whole set of tasks. Obviously, if $W > 1$ the processor is over-utilized and no feasible schedule will exist. Hence $W \leq 1$ is a necessary condition for schedulability.

The schedule construction is guided by the special way how the priorities are defined. Well-known examples of priority rules are the *rate monotonic* (*RM*) priority assignment defined by $w_j := 1/\pi_j$. This is a fixed priority rule, easy to implement, and easy to manage. The *earliest deadline first* (*EDF*) priority assignment is dynamic. In this section we discuss the properties of both rules.

As an introduction we first consider the special and particularly simple case of harmonic task sets.

Harmonic task sets

Let $T = \{T_1, \cdots, T_n\}$ be a set of periodic tasks, indexed in order of increasing periods: $\pi_1 \leq \pi_2 \leq \cdots \leq \pi_n$. Then T is called harmonic if, for each $i \in \{2, \cdots, n\}$, π_i is an integer multiple of π_j for all $j < i$. The following Theorem shows that a set of harmonic task can easily be scheduled, as long as the total utilization does not exceed 1.

Theorem 7.3.1 *Any harmonic task set with total utility $W = \sum p_i/\pi_i \leq 1$ can be feasibly scheduled by the rate monotonic priority rule.*

Proof. By an inductive argument we show that a feasible schedule can directly be constructed, and the arrangement of the task instances follows exactly the given priority rule. All task offsets are set to 0.

Starting with T_1, its instances are initiated at the instants $0, \pi_1, 2\pi_1, \cdots$. Since T_1 has highest priority, the instances are not preempted.

The tasks T_i, $i = 2, 3, \cdots, n$, are scheduled in that order of increasing periods. Suppose the tasks $T_2, \cdots, T_{i-1}$ have already been successfully scheduled.

For T_i, the instances are scheduled at earliest possible instants after the times $0, \pi_i, 2\pi_i, \cdots$. They are filled preemptively in the gaps between the instances of the tasks $T_1, \cdots, T_{i-1}$. For schedulability, consider the first period of T_i: The interval $[0, \pi_i]$ contains π_1/π_i instances of T_1, π_2/π_i instances of $T_2, \cdots$, and π_{i-1}/π_i instances of T_{i-1}. Notice that these are all integers due to the harmonic assumption. So the processor time consumed by the higher priority tasks during this interval is $(\pi_i/\pi_1)p_1 + (\pi_i/\pi_2) + \cdots + (\pi_i/\pi_{i-1})p_{i-1}$, and the processor utilization during $[0, \pi_i]$ is $p_1/\pi_1 + p_1/\pi_2 + \cdots + p_{i-1}/\pi_{i-1} < W$. Obviously, because of $W \leq 1$, there is enough room left for the first instance of T_i. It is easy to see that the same situation appears in the following periods of T_i. It follows by induction that, since $\sum p_i/\pi_i \leq 1$, all tasks can be feasibly scheduled.   □

7.3.1   Static Priority Schedules

We now turn to sets of general periodic tasks. Our aim is to find an optimal priority rule in the sense that if some priority assignment is feasible then the priority rule is feasible as well. The following example shows that the way how priorities are chosen may have consequences regarding the schedulability of a given task set.

Example 7.3.2 Given tasks T_1 and T_2 with respective periods and processing times $\pi_1 = 10$, worst case execution time $p_1 = 5$, and $\pi_2 = 20$, $p_2 = 6$. The tasks are schedulable if we choose $pr_1 < pr_2$. If $pr_1 > pr_2$ then no schedule exists (see Figure 7.3.1).   □

We first introduce some basic definitions. The start time of the first period, t, is called the task offset. The offset of task $T_j \in \mathcal{T}$ is denoted by *offset*$_j$. It is assumed that $0 \leq$ *offset*$_j \leq \pi_j$ for $j = 1, \cdots, n$. The *response time* of a request for a task T_j is the time span between the request and the completion time of the response to that request. A *critical instant* for T_j is an instant at which a request for T_j has maximum response time. The *critical zone* is the time interval between a critical instant and the completion of the response.

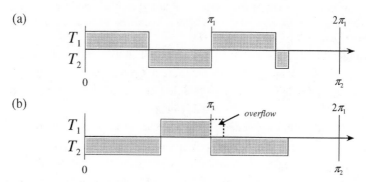

Figure 7.3.1 *Influence of priorities:* **(a)** $pr_1 < pr_2$;   **(b)** $pr_2 < pr_1$.

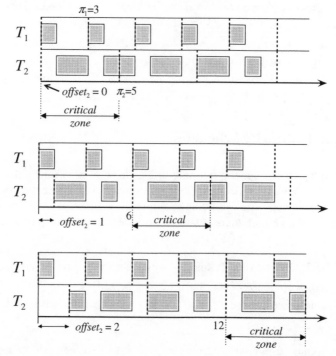

Figure 7.3.2 *Dependency of the critical instant on the offset of* T_2.

The positions of the critical zones may depend on the task offsets. Consider as an example two tasks, T_1 and T_2 with $\pi_1 = 3$, $p_1 = 1$, $\pi_2 = 5$, $p_1 = 3$, and assume $pr_1 > pr_2$, and $offset_1 = 0$. The higher priority task T_1 is scheduled at time instants $0, 3, 6$, etc. There are three different offsets for T_2 : $0, 1, 2$ ($offset_2 = 3$ is obviously equivalent to $offset_2 = 0$). Figure 7.3.2 shows three schedules for the different offsets of T_2. We see that the maximum response time (i.e., the length of a critical zone) of T_2 is 5. In the case of $offset_2 = offset_1 = 0$ the time 0 is a critical instant.

Lemma 7.3.3 *Given tasks T_1 and T_2 with respective periods π_1 and π_2. Let $pr_1 > pr_2$. Then the maximum response time of T_2 is gained if $offset_1 = offset_2$.*

Proof. Assume first $\pi_1 < \pi_2$. Consider a request of T_2 between t_1 and $t_1 + \pi_2$ (see Figure 7.3.3). In this interval task T_1 will occur at times t_2 ($\geq t_1$), $t_2 + \pi_1, \cdots, t_2 + k\pi_1 \leq t_1 + \pi_2$. Unless T_2 is completed before time $t_2 + \pi_1$, T_2 will experience certain delays caused by preemptions of the higher priority task T_1. We see that making t_2 smaller will not decrease the completion time of T_2. Hence the delay of T_2 will be largest if $t_2 = t_1$.

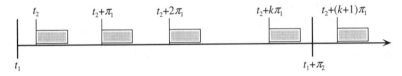

Figure 7.3.3 *Requests of T_2 during one period of T_1.*

If $\pi_1 > \pi_2$ (and still $pr_1 > pr_2$) and a feasible schedule exists then the maximum response time of T_2 is $p_1 + p_2$ which is gained if $offset_1 = offset_2$. □

Theorem 7.3.4 (Critical Instant Theorem, Liu et al. [LL73]) *A critical instant of any task occurs whenever the task is requested simultaneously with requests of all higher priority tasks.*

Proof. Let the set $\mathcal{T} = \{T_1, \cdots, T_m\}$ of tasks be indexed in order of decreasing priority: $pr_1 \geq pr_2 \geq \cdots \geq pr_m$. The theorem follows if, for all $j = 2, \cdots, m$, the previous lemma is repeatedly applied for $i = 1, \cdots, j-1$. □

Corollary 7.3.5 *Consider a schedule for $T_1, \cdots, T_m$.*
(i) *If the requests for all tasks at their critical instants are fulfilled before their respective deadlines, then the schedule is feasible.*
(ii) *Assume $offset_j = 0$ for $j = 1, \cdots, n$, and let π be the maximum period. Then the schedule is feasible, iff all task instances between 0 and π can be completed before their respective deadlines.* □

The next lemma establishes the preparation for the proof that the rate monotonic rule is optimal, in the sense that if a task set that can be scheduled by any priority assignment can also be scheduled by the rate monotonic assignment.

Lemma 7.3.6 *Let $\mathcal{T} = \{T_1, T_2\}$ and $\pi_1 < \pi_2$. If there exists a feasible schedule with $pr_2 > pr_1$, then there exists also a feasible schedule with $pr_1 > pr_2$.*

Proof. From Theorem 7.3.4 we know the existence of a schedule with $pr_1 > pr_2$ implies that a critical instant for T_2 occurs if it is requested simultaneously with T_1. In other words, both tasks have the same offset. In this case we see that, while T_2 is executed once, at least $\lfloor \pi_2/\pi_1 \rfloor$ instances of T_1 will be executed. Hence a necessary condition for schedulability with $pr_1 > pr_2$ is

$$\lfloor \pi_2/\pi_1 \rfloor \cdot p_1 + p_2 \leq \pi_2. \tag{7.3.1}$$

We have to show that (7.3.1) is true if T_1 and T_2 can be feasibly scheduled with $pr_2 > pr_1$. Obviously we have $p_1 + p_2 \leq \pi_1$. Condition (7.3.1) follows from this condition because

$$\lfloor \pi_2/\pi_1 \rfloor p_1 + \lfloor \pi_2/\pi_1 \rfloor p_2 \leq \lfloor \pi_2/\pi_1 \rfloor \pi_1 \leq \pi_2 \text{ and } p_2 \leq \lfloor \pi_2/\pi_1 \rfloor p_2.$$

Hence we can conclude: If the schedule with $pr_2 > pr_1$ is feasible, then the schedule with $pr_1 > pr_2$ is as well.   □

Interpretation of this result: If $\pi_1 < \pi_2$, and if p_1 and p_2 are such that the schedule is feasible with $pr_2 > pr_1$, it is also feasible with $pr_1 > pr_2$. The opposite is not true in general, as we have already seen from the Example 7.3.2. This is generalized by the following theorem:

Theorem 7.3.7 *If a feasible priority assignment exists for some task set, the rate monotonic priority assignment is feasible for that task set.*

Proof. We use an adjacent pair-wise interchange property: Suppose the tasks are indexed in order of increasing periods. Choose any priority assignment that defines a feasible schedule. If the priorities are not rate monotonic there will be a pair of "adjacent" tasks (T_i, T_j) in the priority list such that $\pi_i > \pi_j$. It is easy to see from Lemma 7.3.6 that interchanging the priorities of T_i and T_j does not violate feasibility of the schedule. The theorem follows because the rate monotonic order can be obtained from any other order by a sequence of pair-wise interchanges.   □

From this theorem we conclude that the rate monotonic priority assignment can be considered as the best among all priority lists. Despite this fact, it can easily be seen that *RM* scheduling is not necessarily feasible, though the total utilization is smaller than 1. For example, the two tasks T_1: $\pi_1 = 12$, $p_1 = 6$, and T_2: $\pi_2 = 18$, $p_2 = 7$ have total utilization $W = 8/9$, but it cannot be feasibly scheduled by *RM*.

The question one my want to have answered is: What is the least upper bound for W such that the rate monotonic rule can safely be blindly applied. Based on the concept of critical instant both, Serlin [Ser67] and Liu and Layland [LL73] proved a sufficient utilization-based condition for feasibility of the *RM* policy.

Consider a schedule for a given feasible priority list. If the processor utilization is sufficiently small it will be possible that feasibility is still kept even if processing times are increased or periods are decreased. For particular problem settings we may be able to reach 100 % processor utilization by such changes, but in general it has to be expected that - sooner or later, depending on the priority list - we end up with some smaller utilization value.

In this sense we call the task set *extreme* (with respect to the priority list) (in [LL73] extreme task set is called *fully utilizing the processor*) if increasing any processing time or decreasing any period makes the priority list infeasible.

The question is to which extent the processor utilization can be increased by such parameter changes, before the schedule becomes infeasible. Hence for a given static priority-based scheduling rule, one would like to know the *least upper bound* of the utilization factor which is defined as the minimum of the utilization factors over all extreme task sets. Such bound allows formulating a simple sufficient schedulability criterion: As long as a task set has smaller utilization than the least upper bound, feasibility is guaranteed. In fact, in view of the optimality of the *RM* rule we are interested in such a bound for *RM*.

Theorem 7.3.8 [LL73] *For a set of m tasks with RM priority order, the least upper bound for the processor utilization is* $W^{(n)} = n(2^{1/n} - 1)$.

Proof. We present the proof for $n = 2$ tasks T_1, T_2. Assuming $\pi_1 < \pi_2$, RM schedules T_1 with higher priority than T_2. Hence, when starting the schedule at time $0, T_1$ will be processed non-preemptively at instants $0, \pi_1, 2\pi_1, 3\pi_1, \ldots$

It follows from Theorem 7.3.4 that the time 0 is a critical instant of T_2. The idea is to increase p_2 until $\{T_1, T_2\}$ is extreme, and then estimate the infimum value for the utilization factor. Two cases are discussed separately:

(i) During the first period of T_2 there are $\lceil \pi_2/\pi_1 \rceil$ requests for T_1. All requests of T_1 in the critical time zone of T_2 are completed before the next request of T_2: Then we must have $p_1 \leq \pi_2 - \pi_1 \lfloor \pi_2/\pi_1 \rfloor$, and consequently the largest possible value of p_2 is $p_2 = \pi_2 - p_1 \lceil \pi_2/\pi_1 \rceil$. The corresponding utilization factor calculates to

$$W^{(2)} = 1 + p_1 \cdot [(1/\pi_1) - (1/\pi_2) \cdot \lceil \pi_2/\pi_1 \rceil].$$

We see that $W^{(2)}$ is monotonically decreasing in p_1

(ii) The execution of the $\lceil \pi_2/\pi_1 \rceil^{th}$ request of T_1 overlaps with the next request of T_2: Then we must have $p_1 \geq \pi_2 - \pi_1 \lfloor \pi_2/\pi_1 \rfloor$, and consequently the largest possible value of p_2 is $p_2 = -p_1 \lceil \pi_2/\pi_1 \rceil + \pi_1 \lfloor \pi_2/\pi_1 \rfloor$. The corresponding utilization factor calculates to

$$W^{(2)} = (\pi_1/\pi_2) \cdot \lfloor \pi_2/\pi_1 \rfloor + p_1 \cdot [(1/\pi_1) - (1/\pi_2) \cdot \lfloor \pi_2/\pi_1 \rfloor].$$

Now $W^{(2)}$ is monotonically increasing in p_1.

The minimum of $W^{(2)}$ in these two cases obviously is reached for $p_1 = \pi_2 - \pi_1 \cdot \lfloor \pi_2/\pi_1 \rfloor$ which gives us the expression

$$W^{(2)} = 1 - (\pi_1/\pi_2) \cdot [\lceil \pi_2/\pi_1 \rceil - (\pi_1/\pi_2)] \cdot [(\pi_1/\pi_2) - \lfloor \pi_2/\pi_1 \rfloor].$$

Using abbreviations $I := \lfloor \pi_2/\pi_1 \rfloor$ and $f := (\pi_2/\pi_1) - \lfloor \pi_2/\pi_1 \rfloor$ for the integer and fractional part of π_1/π_2 we get

$$W^{(2)} = 1 - f(1 - f)/(I + f).$$

Since $W^{(2)}$ is monotonic increasing with I, the minimum of $W^{(2)}$ is obtained for $I = 1$. $W^{(2)}$ is minimized for $f = 2^{1/2} - 1$, which gives us the result $W^{(2)} = 2 \cdot (2^{1/2} - 1) \approx 0.83$. This completes the proof for $n = 2$ tasks.   □

The proof of the general case is omitted because it is based on the essentially same idea, but requires more technical effort: showing that for determining the least upper bound it suffices to restrict to $\pi_i/\pi_j < 2$ for all i and j (this corresponds to the above condition $I = 1$), and stepwise maximizing the completion times p_2, $\cdots, p_n$ until $p_i = \pi_{i+1} - \pi_i$ for $i = 2, \cdots, n$, which is again the minimization condition for $W^{(n)}$, and finally minimizing a multi-dimensional equation for $W^{(n)}$.

It should be mentioned that R. Devillers and J. Goossens [DG00] found out that the proof of the $n > 2$ case is incomplete, but the bound is correct.

The above results lead us to the following remarks.

(i)   For harmonic tasks $\{T_1, T_2\}$, since the period of T_2 is an integer multiple of the period of T_2, we get $W^{(2)} = 1$ (see also Theorem 7.3.1).

(ii)   Notice that $W^{(n)} < W^{(n-1)}$, and in the limit $n \to \infty$, $W^{(n)}$ tends to $\ln(2) \approx 0.693$. Moreover, except for the trivial case $n = 1$, the bound $W^{(n)}$ is never reached, since it is irrational, while from our assumptions W is always rational.

(iii) Task sets with utilization smaller than $W^{(n)}$ can always be scheduled via the RM rule. In this sense, the RM strategy can be considered as robust. This encourages one to use a thumb rule in practice: If a task set utilizes the processor not more than 70 %, the RM strategy can be even used as an on-line scheduling strategy.

It should be emphasized that the upper bound $W^{(n)} = n(2^{1/n} - 1)$ is a sufficient but not necessary condition for schedulability by the RM rule. On the other hand, in special cases with larger utilization, the RM rule may still allow to construct feasible schedules.

In practice, it is often possible to replace a given task set by a harmonic one, where periods are slightly reduced, or task splitting is applied [SG90]. In task splitting a task with period π and processing time p is replaced by $k \geq 2$ new tasks, each with period π/k and processing time p/k. For example, if there are two tasks with periods 11 and 15, the first period can be reduced to 10 (step (i)), and replacing this task by two tasks, each with period 5, we end up with harmonic tasks. As a drawback, the number of preemptions can be expected to be larger than in a schedule for the original task set.

We realize that the *RM* algorithm is often able to produce a feasible schedule though the total utilization is higher than $W^{(n)}$. For example, the tasks T_1, T_2 with the respective periods and processing times $\pi_1 = 5$, $\pi_2 = 8$; $p_1 = 2$, $p_2 = 4$ have the utilization factor $W \approx 0.9 > W^{(3)} \approx 0.779$. The *RM schedule* is shown in Figure 7.3.4.

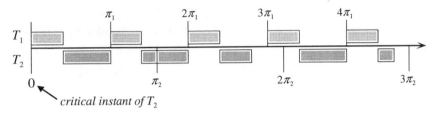

Figure 7.3.4 *Initial part of a feasible preemptive schedule.*

For practical reasons one is interested in a simple criterion that decides upon applicability of *RM*. A necessary and sufficient characterization of the *RM* was given by Lehoczky et al. [LSD89]. The basis of the idea of their analysis is the following:

In a given task set $\{T_1, \cdots, T_n\}$ with $\pi_1 \le \pi_2 \le \cdots \le \pi_n$, T_j $(1 < j \le n)$ can only be preempted by tasks of higher priority, that is by the tasks $T_1, \cdots, T_{j-1}$. Therefore, for determining schedulability of T_j, only the task set $\{T_1, \cdots, T_j\}$ needs to be considered.

Another point regards the task offsets: Though it is shown in [LL73] that the critical instant 0 is sufficient for calculating bound $W^{(n)}$, critical instants > 0 may also have to be considered for deriving a necessity condition.

We start with introducing some useful notion. To determine if a task T_j can meet its deadline under worst case offsets, the processor demand made by the task set is considered as a function of time (t). The cumulative workload on the processor caused by the tasks $T_1, \cdots, T_j$ over the interval $[0, t]$ if 0 is a critical instant is denoted by

$$W_j(t) := \sum_{i=1}^{j} p_i \left\lceil \frac{t}{\pi_j} \right\rceil.$$

Furthermore, denote by $\widetilde{W}_j(t) := W_j(t)/t$ the average workload of the first j tasks per time unit in $[0, t]$, and let $\widetilde{W}_j := \min\{\widetilde{W}_j(t) \mid 0 < t \le \pi_j\}$, and $\widetilde{W} := \max\{\widetilde{W}_j(t) \mid 1 \le j \le n\}$. An exact characterization for schedulability of task T_j by the *RM* algorithm is:

Theorem 7.3.9 [LSD89] *Let $T_1, \cdots, T_n$ be periodic tasks and $\pi_1 \leq \pi_2 \leq \cdots \leq \pi_n$.*

(i) T_j can be scheduled for all offsets $offset_j \in [0, \pi_j)$ by the RM algorithm if and only if $\widetilde{W}_j \leq 1$.

(ii) The entire task set can be scheduled by the RM algorithm if and only if $\widetilde{W} \leq 1$.

Proof. (i) Assume $offset_i = 0$ for $i = 1, \cdots, n$. T_j completes its first computation at time $t \in [0, \pi_j]$ if and only if all the requests from all higher priority tasks are completed at time t. The total processing request in $[0, t]$ is given by $W_j(t)$, and hence T_j is completed at time t if and only if $W_j(t) = t$, or equivalently, $\widetilde{W}_j(t) = 1$. Since furthermore $W_j(s) > s$ for $s \in [0, t)$, it follows that a necessary and sufficient condition for T_j to meet its deadline is the existence of a time $t \in [0, \pi_j]$ such that $\widetilde{W}_j = 1$.

Using Theorem 7.3.4, we conclude that, under general offsets, a necessary and sufficient condition for T_j to meet its deadline is $\widetilde{W}_j \leq 1$.

(ii) follows directly from (i).　　　　　　　　　　　　　　　　　　□

For practical application of this theorem, let us analyze the properties of $\widetilde{W}_j(t)$ in greater detail:

$$\widetilde{W}_j(t) = \sum_{i=1}^{j} \frac{p_i}{t} \left\lceil \frac{t}{\pi_j} \right\rceil$$

is a piecewise monotonically decreasing function that is strictly decreasing except at a finite set of values, called *RM scheduling points*, and denoted by S_j.

When t is a multiple of one of the periods π_i, $\widetilde{W}_j(t)$ has a local minimum and jumps to a higher value to the right (see Figure 7.3.5). Hence $S_j = \{ k \cdot \pi_i \mid i = 1, \cdots, j,$ and $k = 1, \cdots, \lfloor \pi_j / \pi_i \rfloor \}$. Consequently, for determining the minimum of $\widetilde{W}_j$ one needs to check the points of the finite set S_j.

This observation leads to the following

Corollary 7.3.10 (Theorem 2 in [LSD89]). *Given a set of periodic tasks as in the above Theorem.*

(i) T_j can be scheduled for arbitrary offsets by the RM algorithm if and only if

$$\widetilde{W}_j := \min \left\{ \sum_{i=1}^{j} \frac{p_i}{t} \left\lceil \frac{t}{\pi_j} \right\rceil \mid t \in S_j \right\}$$

(ii) $T_1, \cdots, T_n$ can be scheduled by the RM algorithm for arbitrary offsets if and only if

$$\widetilde{W} := \max\{\,\widetilde{W}_j \mid 1 \le j \le n\,\} \le 1. \qquad\qquad\qquad\qquad \square$$

Example 7.3.11 T_1: $p_1 = 2$, $\pi_1 = 5$; T_2: $p_2 = 4$, $\pi_2 = 14$; T_3: $p_3 = 9$, $\pi_3 = 33$. The total utilization is $0.96 > W^{(3)} = 0.779$. The scheduling points are $S_1 = \{5\}$, $S_2 = \{5, 10, 14\}$, $S_3 = \{5, 10, 14, 15, 20, 25, 28, 30, 33\}$. For example, the minimum of $\widetilde{W}_2$ in the interval $[0, 14]$ is at $t = 14$ (see Figure 7.3.5). $\qquad \square$

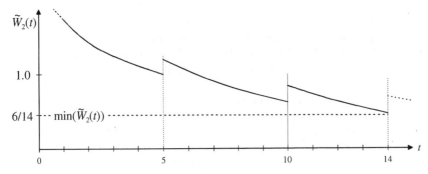

Figure 7.3.5 *Graph of function $\widetilde{W}_2$ in the interval $[0,14]$.*

The following algorithm is based on Corollary 7.3.10:

Algorithm 7.3.12 *Check_Schedulability.*
Input: m periodic tasks $T_1, \cdots, T_n$ with respective integer periods $\pi_1, \cdots, \pi_n$ and integer processing times $p_1, \cdots, p_n$.
Output: "*schedulable*" or "*not schedulable*"

```
begin
sort the tasks increasingly with the periods;    -- let π₁ ≤ π₂ ≤ ... ≤ πₙ
failed := false;
for j := 1 to n do
  for i := 1 to j do
    begin
```
$$\widetilde{W} := \text{infinity} ; \qquad -- \text{ set } \widetilde{W} \text{ to a value} \ge \sum_{i=1}^{n} \frac{2p_i}{\pi_j}$$
```
    for k := 1 to ⌊πⱼ/πᵢ⌋ do
      begin
      t := k·πᵢ ;
```
$$\text{if } \sum_{i=1}^{j} \frac{p_i}{t}\left\lceil \frac{t}{\pi_j} \right\rceil < \widetilde{W} \text{ then } \widetilde{W} := \sum_{i=1}^{j} \frac{p_i}{t}\left\lceil \frac{t}{\pi_j} \right\rceil ;$$
```
      end
    end;
```

if $\widetilde{W} < 1$ **then** print "*schedulable*" else print "*not schedulable*" ;
end.

The correctness of Algorithm *Check_Schelulability* follows directly from the corollary. The time complexity is $O(n^2 \cdot \pi_n / \pi_1)$.

7.3.2   Dynamic Priority Scheduling

The second priority rule to be presented is earliest deadline First (*EDF*) : at each point of time, the next process to run is the one with the closest deadline. In contrast to *RM*, the *EDF* rule is dynamic because each time a new instance is released it has to be decided which of the current unfulfilled instances has closest completion request. In the following Lemma, an *overflow* denotes an instant at which an instance misses its deadline.

Lemma 7.3.13 (Theorem 6 in [LL73]). *When the deadline driven scheduling algorithm is used to schedule a set of tasks on a processor, and the tasks have all offset = 0, there is no processor idle time prior to an overflow.*

Proof. Given an *EDF* schedule and assume that an overflow occurs at time t_3. Suppose there is an idle interval before t_3; let $[t_1, t_2]$ be the last such interval.

Modify the schedule: For each task T_j whose first instance after the idle interval is requested at a time $t > t_2$, move its and the following requests forward such that the first of these is a time t_2. By this move, processor load is not decreased, and hence

 - the overflow will stay at t_3 or be earlier,
 - the time span between t_2 and t_3 will stay idle-free,

Hence in the modified schedule, each task instance is requested at the same time t_2, there is an overflow at some time t_3', and there is no idle interval between t_2 and t_3'. This however is a contradiction to the assumption that all tasks have equal offsets and there is an idle interval before the overflow.   □

Case of general offsets: With the same argument as in the previous proof we conclude that if an idle interval exists before an overflow, then the same task set will have an overflow if all offsets are set to 0. Therefore we restrict w.l.o.g. our considerations to task sets with offsets = 0.

Theorem 1.3.14 [LL73]. *The tasks* $\{T_1, \cdots, T_n\}$ *can be scheduled preemptively by EDF if and only if* $W = \sum p_i / \pi_i \le 1$.

Proof. $W \le 1$ is necessary because otherwise a feasible schedule cannot exist because of processor overload.

For sufficiency, assume there is a task set with $W \le 1$, where *EDF* is not feasible. Let $\pi_C := \mathrm{lcm}\{\pi_1, \pi_2, \cdots, \pi_n\}$ be the cycle length of the schedule [2]. Then there is an overflow between the time 0 and π_C. Assuming offsets 0, then π_C is the first time where the request times coincide again. Then an overflow must occur at some time t_0 between 0 and π_C.

For a detailed analysis, let *a-instances* denote the subset of instances with request time t_0, and *b-instances* all instances with a deadline beyond t_0.

Case 1: None of b-instances are started before t_0. Then, since there is no idle period in $[0, t_0]$, the processing load is $\sum t_0/\pi_i > t_0$. Since furthermore $x \ge \lfloor x \rfloor$, we get $W = \sum p_i/\pi_i > 1$, which contradicts $W \le 1$. For an illustration see the example below.

Case 2: Some of the b-instances are already processed before t_0, though their deadline is beyond t_0. At the overflow time t_0, exactly one instance misses its deadline, which must be one of the a-instances because the deadline coincides with the next request time. This means that the overflowing a-instance must have been processed for some time before t_0. This is only possible if it has highest priority. In fact, all the a-instances have the same priority between some time $t' < t_0$ and t_0. Consequently, the b-instances processed before t_0 must have been processed before t' (see Figure 7.3.6). Even more, it is not possible that other a-instances released before t' are processed after t' because their deadline (t_0) is before those of the b-instances. Hence we can summarize: The interval $[t', t_0]$ contains only task instances that are initiated and have deadlines in this interval. Therefore, the total processing demand in $[t', t_0]$ is $\sum \lfloor (t_0 - t')/\pi_i \rfloor$, and, since there is an overflow, $\sum \lfloor (t_0 - t')/\pi_i \rfloor > t_0 - t'$. This implies again that $W = \sum p_i/\pi_i > 1$, which contradicts $W \le 1$. $\square$

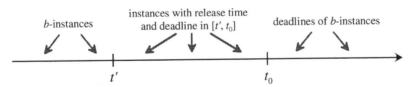

Figure 7.3.6 *Proof of Theorem 7.3.14: location of b-instances.*

---

[2] *lcm* is the least common multiple.

An example for case 1 in the proof consider the tasks T_1 with $\pi_1 = 8$, $p_1 = 4$ and T_2 with $\pi_2 = 5$, $p_2 = 3$. The second instance of T_1 is an a-instance which overflows at time 16; the third instance of T_2 is a b-instance. Figure 7.3.7 shows the schedule.

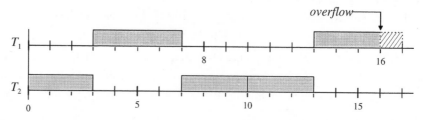

Figure 7.3.7 *Proof of Theorem 7.3.14: An example to case 1.*

Dertouzos [Der74] showed that EDF is optimal among all preemptive scheduling algorithms: If, for a given set of periodic tasks, a feasible schedule exists then EDF is also feasible.

7.4   Scheduling Periodic Tasks on Parallel Processors

The real-time system is structured as collection of interconnected processors, for executing a given set of periodic tasks. Besides processing times, delays caused by communication should be taken into account, but for simplicity reasons we assume here that *communication delays are small* compared to process run times and can be neglected. After the set of tasks is properly distributed among the processors, scheduling strategies as discussed in Section 7.3 can be applied *separately* to each processor. There are two principal kinds of strategies:

– *static binding*, where each task is assigned to one specific processor, and

– *dynamic binding*, where tasks compete greedily for the use of the processors.

Dhall and Liu [DL78] showed that the global application of *RM* scheduling on m processors cannot guarantee schedulability. In the following example, on-line allocation (dynamic binding) of *RM* performs poorly.

Example 7.4.1 Given tasks $T_1, \cdots, T_n$ to be processed on $m = n-1$ processors, with processing times and periods $p_j = 2\varepsilon < 1$, $\pi_j = 1$ ($j = 1, \cdots, n-1$), and $p_n = 1$, $\pi_n = 1+\varepsilon$. The rate-monotonic strategy with *dynamic binding* assigns first tasks $T_1, \cdots, T_{n-1}$ to processors $P_1, \cdots, P_{n-1}$. T_n cannot be scheduled and misses its deadline. The rate-monotonic strategy with *static binding* (off-line) could bound

T_n to one of the processors. Since ε is sufficiently small, the rest can distributed among the other processors. □

Distributing the tasks *off-line* (static binding) can be based on variations of the well known bin packing strategy. Given periodic tasks $T_1, \cdots, T_n$ with utilizations $u_i = p_i/\pi_i \in (0, 1]$. The problem is to partition the tasks into subsets such that the sum of utilizations is not beyond 1, while the number of subsets is minimized. The bin packing problem is known to be NP-hard [GJ79] (cf. Section 12.1). One of the simplest strategies is First Fit (*FF*), in which the tasks $T_1, T_2, \cdots$ are assigned to the fist processor as long as the sum of utilizations is ≤ 1. Then the filling continues on the second processor, etc..

The *Rate Monotonic First Fit* algorithm (*RMFF*) [DL78] is based on the first fit allocation strategy and applies then *RM* on each processor separately. It is known from Dhall Liu [DL78] that a safe use of *RMFF* may require between 2.4 and 2.67 times as many processors as an optimal partition.

Oh and Baker [OB98] proved that *RMFF* can schedule any task set on m processors as long as the total utilization is not beyond $m(2^{1/2} - 1)$. This result was improved by Lopez et al. [LDG01]: If the total utilization is bounded by $(m+1)(2^{1/(m+1)} - 1)$ then the task set is schedulable.

A more general result was obtained by Andersson et al. [ABJ01] who showed that for any fixed priority multiprocessor scheduling algorithm, schedulability is guaranteed if the total utilization is not higher than $(m+1)/2$. This holds for both, static and dynamic binding.

Global *RM* scheduling seems to work well with small task utilizations. Let $\lambda \in [0, 1]$ be an upper bound on the individual utilizations, then smaller values of λ would allow for a larger total system utilization. For example, if $\lambda = m/(3m-2)$, a total system utilization of at least $m^2/(3m-1)$ can be guaranteed [ABJ01]. Baruah and Goossens [BG03] proved that if $\lambda = 1/3$, a system utilization $\geq m/3$ can be gained, and Baker [Bak03] showed that for any $\lambda \leq 1$, a system utilization of $(m/2)(1-\lambda) + 1$ can be guaranteed. On the other hand, if there are also tasks with arbitrary task utilizations, an algorithm called *RM-US(Θ)* gives highest priority to tasks with utilization $\geq \Theta$, and schedules the remaining tasks with *RM*. Then a system utilization of at least $(m+1)/3$ for $\Theta = 1/3$ can be guaranteed [Bak03].

7.5 Resources

As pointed out in Section 7.2, tasks may need resources of limited availability that can only be exclusively and non-preemptably accessed, thus leading to mutual exclusions of tasks. A task holding a resource may block another task that tries to access the same resource.

We first consider the situation for a fixed priority scheme such as rate monotonic: Suppose a high priority task T_h needs access to a resource that is locked by another task T_l with lower priority. Since the resource is non-preemptable and can hence not be withdrawn from T_l, T_h has to wait regardless of its high priority. This situation is called *priority inversion*. The problem is that T_h may be delayed for an undefined amount of time if T_l is preempted by other tasks not requiring the resource.

There are special run-time protocols that organize the task execution in case of priority inversion. The inheritance protocol and other synchronization protocols for both, the single and the multiprocessor case, were introduced by, among others, Sha et al. [SLR87] and Rajkumar et al. [RSLR94]. The basic inheritance protocol, for example, gives the lower priority task temporarily (i.e., until it releases the resource) the (higher) priority of the blocked task.

In a feasibility analysis one needs to know how long a task can be delayed by tasks of lower priority. Depending on the synchronization protocol, upper bounds for blocking times can be derived and taken into account. The worst case execution times are simply enlarged by these blocking times.

For case of dynamic priorities such as earliest deadlines there are variations of the previously mentioned run-time protocols, as for example the dynamic priority inheritance protocol. For details we refer to the book of Stankovic et al. [SRSB98].

7.6   Variations of the Periodic Task Model

The introduced task model was generalized in many ways. A generalization is discussed by Sorensen and Hamacher [Sor74, SH75] and similarly by Teixeira [Tei78], in which the maximum response times need not be confined by the right end of the period. Ramamrithram and Stankovic [RS84, RSC85] consider a distributed hard real-time model with one CPU per node and periodic and sporadic processes. The periodic processes are assigned to CPUs initially and guaranteed to meet their maximum response times. The sporadic tasks arrive randomly with deadlines and unrestricted arrival rate. Accepted sporadic processes are locally scheduled according to the preemptive earliest-deadline-first rule.

Another alternative is the model discussed by Chen et al. [CA94, CA95] which assumes periodic tasks together with the additional, as they call it, "relative timing constraints" of a *low and high jitter* (viz. Figure 7.6.1) for the distance between two consecutive task instances.

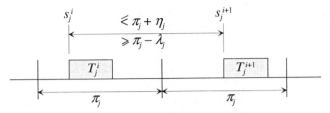

Figure 7.6.1 *Task execution with low jitter λ_j and high jitter η_j.*

Halang & Stoyenko [HS91] present a *"frame superimposition"* model for periodic processes with known processing time characteristics, release times and deadlines. In this model, one of the processes is chosen to start with its frame at some time t_0. The frames of the other processes are then positioned in various ways along the time line, relative to time t_0. Their algorithm shifts the frames exhaustively and checks feasibility for every possible combination of frames.

Other generalizations regard processing times. Choi and Agrawala [CA97a] assume that each task has a given lower and upper bound for the processing time. Mok et al. [MC96a, MC96b] consider a model for real-time tasks, called *multi-frame model* where the tasks are instantiated periodically, but with different execution times in each interval.

The model discussed in the Ph.D. thesis of Choi [Cho97] and in [CA97a] assumes a cyclic execution of a set of tasks with precedences, relative inter-task constraints in form of min/max conditions between start and finish times of any two tasks. Furthermore, upper and lower bounds for task execution times are assumed.

A similar model is *end-to-end scheduling*, as considered e.g. by Gerber [Ger95] and Gerber et al. [GHS95 and GPS95], which deals with the scheduling of sets of tasks with precedences, deadlines; various inter-task constraints, and communication delays. Another model modification is discussed in [CAS97] where repeating processes are considered, and the time between the processes is newly determined at each iteration step by a so-called dynamic temporal controller.

References

ABJ01   B. Andersson, S. Baruah, J. Johnsson, static-priority scheduling on multiprocessors, in: *Proceedings of the 22nd IEEE Real-Time Systems Symposium*, London, UK, 2001, 193-202.

Bak03   T. P. Baker, Multiprocessor EDF and deadline monotonic schedulability analysis, in: *Proceedings of the 24th IEEE Real-Time Systems Symposium*, 2003, 120-129.

BG03   S. Baruah, J. Goossens, Rate monotonic scheduling on uniform multiprocesors, *IEEE Trans. Comput.* 52, 2003, 966-970.

CA94   S.-T. Cheng, A. K. Agrawala, Scheduling periodic tasks with relative timing constraints, U. of Maryland, CS-TR-3392, 1994.

CA95   S.-T. Cheng, A. K. Agrawala, Allocation and scheduling of real-time periodic tasks with relative timinig constraints. U. of Maryland, CS-TR-3402, 1995.

CA97a   S. Choi, A. K. Agrawala, Dynamic dispatching of cyclic real-time tasks with relative constraints, U. of Maryland, CS-TR-3370, 1997.

CAS97   S. Choi, A. K. Agrawala, L. Shi, Designing dynamic temporal controls for critical systems, U. of Maryland, CS-TR-3804, 1997.

Cho97   S. Choi, *Dynamic Time-Based Scheduling for Hard Real-Time Systems*. Ph.D. Thesis, University of Maryland, 1997.

Der74   M. L. Dertouzos, Control Robotics: The Procedural Control of Physical Processors *Proc. IFIP Congress,* 1974, 807-813.

DG00   R. Devillers and J. Goossens, Liu and Layland's Schedulability Test Revisited, *Inform. Process. Lett.* 73, 2000, 157-161.

DL78   S. K. Dhall and C. L. Liu, On a real-time scheduling problem, *Oper. Res.* 26, 1978, 127-140.

Ger95   R. Gerber, Guaranteeing end-to-end timing processes, *Proc. IEEE Real-Time System*, 1995, 192-203.

GHS95   R. Gerber, S. Hong, M. Saksena, Guaranteeing real-time requirements with resource-based calibration of periodic processes', *IEEE Trans. Software Engineering*, 21, 1995, 579-592.

GJ79   M. R. Garey, D. S. Johnson, *Computers and Intractability: A Guide to the Theory of NP-Completeness*. W. H. Freeman, San Francisco, 1979.

GPS95   R. Gerber, W. Pugh, M. Saksena, Parametric dispatching of hard real-time tasks, *IEEE Trans. Comput.* 44, 1995, 471-479.

HS91   W. A. Halang, A. D. Stoyenko, *Constructing Predictable Real-Time Systems*. Kluwer Academic Publishers, Boston, 1991.

KKZ88   R. Koymans, R. Kuiper, E. Zijlstra, Paradigms for real-time systems, in: M. Joseph (ed.), *Formal Techniques in Real-Time and Fault-Tolerant Systems*, Lecture Notes in Computer Science 331, Springer, 1988.

LDG01   J. M. Lopez, J. L. Diaz, D. F. Garcia, Minimum and maximum utilization bounds for multiprocessor RM scheduling, in: *Proceedings of the Euromicro Conference on Real-Time Systems*, Delft, Netherlands, 2001, 67-75.

LL73   C. L. Liu, J. W, Layland, Scheduling algorithms for multiprogramming in a hard-real-time environment, *J. ACM* 20, 1973, 46-61.

LSD89   J. Lehoczky, L. Sha, and Y. Ding. The Rate Monotonic Scheduling Algorithm: Exact Characterization and Average Case Behavior, in: *Proc. of the IEEE Real-Time Systems Symposium*, 1989, 166-171.

MC96a   A. K. Mok, D. Chen, A multiframe model for real-time tasks. University of Texas at Austin, 1996.

MC96b   A. K. Mok, D. Chen, A general model for real-time tasks. University of Texas at Austin, 1996.

OB98   D. I. Oh, T. P. Baker, Utilization for N-processor rate monotone scheduling with stable processor assignment. *Real-Time Systems* 15, 1989, 183-193.

RS84   K. Ramamrithram, J. Stankovic, Dynamic task scheduling in distributed hard real-time systems, in *Proc. 4th IEEE International Conference on Distributed Computing Systems*, 1984, 96-107.

RSC85   K. Ramamrithram, J. Stankovic, S. Cheng, Evaluation of a flexible task scheduling algorithm for distributed hard real-time systems, *IEEE Trans. Comput.* 34, 1985, 1130-1143.

RSLR94   R. Rajkumar, L. Sha, J. P. Lehoczki, K. Ramamritham, An optimal priority inheritance policy for synchronization ion real-time systems, in: S. H. Son (ed.), *Advances in Real-Time Systems*, Englewood Cliffs, NJ, Prentice-Hall, 1994, 249-271.

SAA04   L. Sha, T, Abdelzaher, K.E. Årzén, A. Cervin, Th, Baker, A. Burns, G. Butazzo, M. Caccamo, J. Lehoczky, A. Mok, Real time scheduling theory: A historical perspective, *Real-Time Systems* 28, 2004, 101-155.

SG90   L. Sha, J. Goodenough, Real-time scheduling theory and Ada, *IEEE Computer* 23, 1990, 53-62.

SH75   P. Sorensen, V. Hamacher, A real-time system design methodology, *INFOR* 13, 1975, 1-18.

SLR87   L. Sha, J. P. Lehoczki, R. Rajkumar, Task scheduling in distributed real-time systems, in: *Proceedings of the IEEE Industrial Electronics Conference*, 1987.

Sor74   P. Sorensen, *A Methodology for Real-Time System Development*. PhD Thesis, University of Toronto, 1974.

SRSB98   J. A. Stankovic, K. Ramamritham, M. Spuri, G. Buttazzo, *Deadline Scheduling for Real-Time Systems*, Kluwer, Boston-Dortrecht-London, 1998.

Tei78   T. Teixeira, Static priority interrupt scheduling, in *Proc. 7th Texas Conference on Computing Systems*, 15, 1978, 5.13-5.18.

WS98   J. Wikander, B. Svensson (guest eds.), Special Issue on Real-Time Systems in Mechatronic Applications. *Real-Time Systems* 14, 1998, 217-218

8 Flow Shop Scheduling

Consider scheduling tasks on dedicated processors or machines. We assume that tasks belong to a set of n jobs, each of which is characterized by the same machine sequence. For convenience, let us assume that any two consecutive tasks of the same job are to be processed on different machines. The type of factory layout in the general case - handled in Chapter 10 - is the job shop; the particular case where each job is processed on a set of machines in the same order is the flow shop. The most commonly used performance measure will be makespan minimization.

8.1 Introduction

8.1.1 The Flow Shop Scheduling Problem

A flow shop consists of a set of different machines (processors) that perform tasks of jobs. All jobs have the same processing order through the machines, i.e. a job is composed of an ordered list of tasks where the ith task of each job is determined by the same machine required and the processing time on it. Assume that the order of processing a set of jobs $\mathcal{J}$ on m different machines is described by the machine sequence $P_1, \cdots, P_m$. Thus job $J_j \in \mathcal{J}$ is composed of the tasks $T_{1j}, \cdots, T_{mj}$ with processing times p_{ij} for all machines P_i, $i = 1, \cdots, m$. There are several constraints on jobs and machines: (*i*) There are no precedence constraints among tasks of different jobs; (*ii*) each machine can handle only one job at a time; (*iii*) each job can be performed only on one machine at a time. While the machine sequence of all jobs is the same, the problem is to find the job sequences on the machines which minimize the makespan, i.e. the maximum of the completion times of all tasks. It is well known that - in case of practical like situations - the problem is NP-hard [GJS76].

Most of the literature on flow shop scheduling is limited to a particular case of flow shop - the *permutation flow shops*, in which each machine processes the jobs in the same order. Thus, in a permutation flow shop once the job sequence on the first machine is fixed it will be kept on all remaining machines. The resulting schedule will be called *permutation schedule*.

By a simple interchange argument we can easily see that there exists an optimal flow shop schedule with the same job order on the first two machines P_1 and P_2 as well as the same job order on the last two machines P_{m-1} and P_m. Consider an optimal flow shop schedule. Among all job pairs with different processing orders on the first two machines, let J_i and J_k be two jobs such that the number of tasks scheduled between T_{1i} and T_{1k} is minimum. Suppose T_{1i} is processed before T_{1k} (while T_{2i} is processed after T_{2k}). Obviously, T_{1k} immediately follows T_{1i} and no other job is scheduled on machine P_1 in between. Hence, interchanging T_{1i} and T_{1k} has no effect on any of the remaining tasks' start times. Repetitious application of this interchange argument yields the same job order on the first two machine (and analogously for the last two machines). Consequently, any flow shop scheduling problem consisting of at most three machines has an optimal schedule which is a permutation schedule. This result cannot be extended any further as can be shown by a 2-job 4-machine example with $p_{11} = p_{41} = p_{22} = p_{32} = 4$ and $p_{21} = p_{31} = p_{12} = p_{42} = 1$. Both permutation schedules have a makespan of 14 while job orders (J_2, J_1) on P_1 and P_2 and (J_1, J_2) on P_3 and P_4 lead to a schedule with a makespan of 12. Although it is common practice to focus attention on permutation schedules, Potts et al. [PSW91] showed that this assumption can be costly in terms of the deviation of the maximum completion times, i.e. the makespans, of the optimal permutation schedule and the optimal flow shop schedule. They showed that there are instances for which the objective value of the optimal permutation schedule is much worse (in a factor more than $1/2\sqrt{m}$) than that of the optimal flow shop schedule.

Any job shop model (see Chapter 10) can be used to model the flow shop scheduling problem. We present a model basically proposed by Wagner [Wag59, Sta88] in order to describe the permutation flow shop. The following decision variables are used (for $i, j = 1, \cdots, n; k = 1, \cdots, m$):

$$z_{ij} = \begin{cases} 1 & \text{if job } J_i \text{ is assigned to the } j^{th} \text{ position in the permutation,} \\ 0 & \text{otherwise;} \end{cases}$$

$x_{jk} = $ idle time (waiting time) on machine P_k before the start of the job in position j in the permutation of jobs;

$y_{jk} = $ idle time (waiting time) of the job in the jth position in the permutation, after finishing processing on machine P_k, while waiting for machine P_{k+1} to become free ;

$C_{max} = $ makespan or maximum flow time of any job in the job set.

Hence we get the model:

Minimize C_{max}

subject to $\displaystyle\sum_{j=1}^{n} z_{ij} = 1,$ $\qquad\qquad i = 1, \cdots, n$ $\qquad\qquad$ (8.1.1)

$\displaystyle\sum_{i=1}^{n} z_{ij} = 1,$ $\qquad\qquad j = 1, \cdots, n$ $\qquad\qquad$ (8.1.2)

$$\sum_{i=1}^{n} p_{ri} z_{ij+1} + y_{j+1\,r} + x_{j+1\,r} = y_{jr} + \sum_{i=1}^{n} p_{r+1\,i} z_{ij} + x_{j+1\,r+1}, \quad (8.1.3)$$

$$j = 1, \cdots, n-1; \, r = 1, \cdots, m-1$$

$$\sum_{j=1}^{n} \sum_{i=1}^{n} p_{mi} z_{ij} + \sum_{j=1}^{n} x_{jm} = C_{max}, \qquad\qquad (8.1.4)$$

$$\sum_{r=1}^{k-1} \sum_{i=1}^{n} p_{ri} z_{i1} = x_{1k}, \qquad\qquad k = 2, \cdots, m \qquad\qquad (8.1.5)$$

$$y_{1k} = 0, \qquad\qquad k = 1, \cdots, m-1 \qquad\qquad (8.1.6)$$

Equations (8.1.1) and (8.1.2) assign jobs and permutation positions to each other. Equations (8.1.3) provide Gantt chart accounting between all adjacent pairs of machines in the m-machine flow shop. Equation (8.1.4) determines the makespan. Equations (8.1.5) account for the machine idle time of the second and the following machines while they are waiting for the arrival of the first job. Equations (8.1.6) ensure that the first job in the permutation would always pass immediately to each successive machine.

8.1.2 Complexity

The minimum makespan problem of flow shop scheduling is a classical combinatorial optimization problem that has received considerable attention in the literature. Only a few particular cases are efficiently solvable, cf. [MRK83]:

(*i*) The two machine flow shop case is easy [Joh54]. In the same way the case of three machines is polynomially solvable under very restrictive requirements on the processing times of the intermediate machine [Bak74].

(*ii*) The two machine flow shop scheduling algorithm of Johnson can be applied to a case with three machines if the intermediate machine is no bottleneck, i.e. it can process any number of jobs at the same time, cf. [CMM67]. An easy consequence is that the two machine variant with time lags is solvable in polynomial time. That means for each job J_i there is a minimum time interval l_i between the completion of job J_i on the first machine P_1 and its starting time on the second machine P_2. The time lags can be viewed as processing times on an intermediate machine without limited capacity. Application of Johnson's algorithm to the problem with two machines P_1 and P_2, and processing times $p_{1i}+l_i$ and $p_{2i}+l_i$ on P_1 and P_2, respectively, yields an optimal schedule, cf. [Joh58, Mit58, MRK83].

(*iii*) Scheduling two jobs by the graphical method as described in [Bru88] and first introduced by Akers [Ake56]. (Actually this method also applies in the more general case of a job shop, cf. Chapter 10.)

(*iv*) Johnson's algorithm solves the preemptive two machine flow shop $F2 \mid pmtn \mid C_{max}$.

(*v*) If the definition of precedence constraints $J_i \prec J_j$ specifies that job J_i must complete its processing on each machine before job J_j may start processing on that machine then the two machine flow shop problem with tree or series-parallel precedence constraints and makespan minimization is solvable in polynomial time, cf. [Mon79, Sid79, MS79].

Slight modifications, even in the case of two machines, turn out to be difficult, see [TSS94]. For instance, $F3 \mid\mid C_{max}$ [GJS76], $F2 \mid r_j \mid C_{max}$ [LRKB77], $F2 \mid\mid L_{max}$ [LRKB77], $F2 \mid\mid \Sigma\, C_j$ [GJS76], $F2 \mid pmtn,\ r_j \mid C_{max}$ [CS81], $F2 \mid pmtn \mid L_{max}$ [TSS94], $F3 \mid pmtn \mid C_{max}$ [GS78], $F3 \mid pmtn \mid \Sigma\, C_j$ [LLR+93], $F2 \mid prec \mid C_{max}$ [Mon80], and $F2 \mid pmtn \mid \Sigma\, C_j$ [DL93] are strongly NP-hard.

8.2 Exact Methods

In this section we will be concerned with a couple of polynomially solvable cases of flow shop scheduling and continue to the most successful branch and bound algorithms. A survey on earlier approaches in order to schedule flow shops exactly can be found in [Bak75, KK88]. Dudek et al. [DPS92] review flow shop sequencing research since 1954.

8.2.1 The Algorithms of Johnson and Akers

An early idea of Johnson [Joh54] turned out to influence the development of solution procedures substantially. Johnson's algorithm solves the $F2 \mid\mid C_{max}$ to optimality constructing an optimal permutation schedule through the following approach:

Algorithm 8.2.1 *Johnson's algorithm for* $F2 \mid\mid C_{max}$ [Joh54].

`begin`

Let S_1 contain all jobs $J_i \in \mathcal{J}$ with $p_{1i} \le p_{2i}$ in a sequence of non-decreasing order of their processing times p_{1i};

Let S_2 contain the remaining jobs of $\mathcal{J}$ (not in S_1) in a sequence of non-increasing order of their processing times p_{2i};

Schedule all jobs on both machines in order of the concatenation sequence
(S_1, S_2);

`end;`

As Johnson's algorithm is a sorting procedure its time complexity is $O(n\log n)$. The algorithm is based on the following sufficient optimality condition.

Theorem 8.2.2 [Joh54] *Consider a permutation of n jobs where job J_i precedes job J_j if* $\min\{\,p_{1i}, p_{2j}\,\} \leq \min\{\,p_{2i}, p_{1j}\,\}$ *for* $1 \leq i, j \leq n$. *Then the induced permutation schedule is optimal for* $F2 \parallel C_{max}$.

Proof. Let π be a permutation defining a schedule of the flow shop problem with n jobs. We may assume $\pi = (1, 2, \cdots, n)$. Then, there is an $s \in \{1, 2, \cdots, n\}$ such that the makespan $C_{max}(\pi)$ of the schedule equals

$$\sum_{i=1}^{s} p_{1i} + \sum_{i=s}^{n} p_{2i} = \sum_{i=1}^{s} p_{1i} - \sum_{i=1}^{s-1} p_{2i} + \sum_{i=1}^{n} p_{2i}\,.$$

Hence minimization of the makespan $\min_{\pi}\{C_{max}(\pi)\}$ is equivalent to

$$\min_{\pi}\{\max_{1\leq s\leq n} \Delta_s(\pi)\} \text{ where } \Delta_s(\pi) = \sum_{i=1}^{s} p_{1i} - \sum_{i=1}^{s-1} p_{2i}\,.$$

Let π' be another permutation different from π in exactly two positions j and $j+1$, i.e. the jobs' order defined by π' is $J_1, J_2, \cdots, J_{j-1}, J_{j+1}, J_j, J_{j+2}, J_{j+3}, \cdots, J_n$. As $\Delta_s(\pi) = \Delta_s(\pi')$ for $s = 1, \cdots, j - 1, j+2, \cdots, n$, we get, that $\max_{1\leq s\leq n} \Delta_s(\pi) \leq \max_{1\leq s\leq n} \Delta_s(\pi')$ holds if $\max\{\Delta_j(\pi), \Delta_{j+1}(\pi)\} \leq \max\{\Delta_j(\pi'), \Delta_{j+1}(\pi')\}$. The latter is equivalent to

$$\max\{\,p_{1j}, p_{1j}-p_{2j}+p_{1j+1}\,\} \leq \max\{\,p_{1j+1}, p_{1j+1}-p_{2j+1}+p_{1j}\,\}$$

which is equivalent to

$$p_{1j} \leq p_{1j+1} \text{ and } p_{1j}-p_{2j}+p_{1j+1} \leq p_{1j+1}$$

or

$$p_{1j} \leq p_{1j+1}-p_{2j+1}+p_{1j} \text{ and } p_{1j}-p_{2j}+p_{1j+1} \leq p_{1j+1}-p_{2j+1}+p_{1j}\,.$$

Thus, if $p_{1j} \leq \min\{p_{1j+1}, p_{2j}\}$ or $p_{2j+1} \leq \min\{p_{1j+1}, p_{2j}\}$, or equivalently, if $\min\{p_{1j}, p_{2j+1}\} \leq \min\{p_{1j+1}, p_{2j}\}$ then permutation π defines a schedule at least as good as π'.

Among all permutations defining an optimal schedule, assume π is a permutation satisfying J_i precedes J_j if $\min\{p_{1i}, p_{2j}\} \leq \min\{p_{2i}, p_{1j}\}$, for any two jobs J_i and J_j where one is an immediate successor of the other in the schedule. It remains to verify transitivity, i.e. if $\min\{p_{1i}, p_{2j}\} \leq \min\{p_{2i}, p_{1j}\}$ implies J_i precedes J_j and $\min\{p_{1j}, p_{2k}\} \leq \min\{p_{2j}, p_{1k}\}$ implies J_j precedes J_k then $\min\{p_{1i}, p_{2k}\} \leq \min\{p_{2i}, p_{1k}\}$ implies J_i precedes J_k in π. There are 16 different cases to distinguish according to the relative values of the four processing time pairs p_{1i},

p_{2j}; p_{2i}, p_{1j}; p_{1j}, p_{2k} and p_{2j}, p_{1k}. Twelve of the cases are easy to verify. The remaining four cases,

(1) $p_{1i} \geq p_{2j} \leq p_{1k}$ and $p_{2i} \leq p_{1j} \leq p_{2k}$;

(2) $p_{1i} \geq p_{2j} \leq p_{1k}$ and $p_{2i} \geq p_{1j} \leq p_{2k}$;

(3) $p_{1i} \geq p_{2j} \geq p_{1k}$ and $p_{2i} \leq p_{1j} \leq p_{2k}$; and

(4) $p_{1i} \geq p_{2j} \geq p_{1k}$ and $p_{2i} \geq p_{1j} \leq p_{2k}$

imply that J_i may precede J_j or J_j may precede J_i. Hence, there is an optimal schedule satisfying the condition of the theorem for any pair of jobs. Finally, observe that this schedule is uniquely defined in case of strict inequalities $\min\{p_{1i}, p_{2i+1}\} < \min\{p_{2i}, p_{1i+1}\}$ for all pairs i, $i+1$ in π. If $\min\{p_{1i}, p_{2i+1}\} = \min\{p_{2i}, p_{1i+1}\}$ for a pair i, $i+1$ in π then an interchange of J_i and J_{i+1} will not increase the makespan. This proves that the theorem describes a sufficient optimality condition.   □

Johnson's algorithm can be used as a heuristic when $m > 2$. Then the set of machines is divided into two subsets each of which defines a pseudo-machine having a processing time equal to the processing time on the real machines assigned to that subset. Johnson's algorithm can be applied to this n-job 2-pseudo-machine problem to obtain a permutation schedule. The quality of the outcome heavily depends on the splitting of the set of jobs into two subsets. If $m = 3$ an optimal schedule can be found from the two groups $\{P_1, P_2\}$ and $\{P_2, P_3\}$ if $\max_i p_{2i} \leq \min_i p_{1i}$ or $\max_i p_{2i} \leq \min_i p_{3i}$. Thus, for the pseudo machines $\{P_1, P_2\}$ and $\{P_2, P_3\}$ the processing times are defined as $p_{\{P_1,P_2\},i} = p_{1i} + p_{2i}$ and $p_{\{P_2,P_3\},i} = p_{2i} + p_{3i}$.

The problem of scheduling only two jobs on an arbitrary number of machines can be solved in polynomial time using the graphical method proposed by [Bru88] and first introduced by Akers [Ake56].

Assume to process two jobs J_1 and J_2 (not necessarily in the same order) in an m-machine flow shop. The problem can be formulated as a shortest path problem in the plane with rectangular objects as obstacles. The processing times of the tasks of J_1 (J_2) on the machines are represented as intervals on the x-axis (y-axis) which are arranged in order (next to each other) in which the corresponding tasks are to be processed. An interval I_{i1} (I_{i2}) on the x-axis (y-axis) is associated to a machine P_i on which the job J_1 (J_2) is supposed to be processed. Let x_F (y_F) denote the sum of the processing times of job J_1 (J_2) on all machines. Let $F = (x_F, y_F)$ be that point in the plane with coordinates x_F and y_F. Any rectangular $I_{i1} \times I_{i2}$ defines an obstacle in the plane. A feasible schedule corresponds to a path from the origin $O = (0,0)$ to F avoiding passing through any obstacle. Such a path consists of a couple of segments parallel to one of the axis or diagonal in the plane. A segment parallel to the x-axis (y-axis) can be interpreted in such a way that only job J_1 (J_2) is processed on a particular machine while J_2 (J_1) is waiting for that machine, because parallel segments are only required if the path from O to F touches the border of an obstacle. An obstacle defined by some machine P_i and forcing the path from O to F to continue in parallel to one of the axis implies an avoidance of a conflict among both jobs. Hence, an obstacle means to sequence both jobs with respect to P_i. Minimization of the makespan corresponds to finding a shortest path from O to F avoiding all obstacles. The problem can be reduced to the problem of finding an unrestricted shortest path in an appropriate network $G = (\mathcal{V}, \mathcal{E})$. The set of vertices consists of O, F and all north-west and south-east corners of all rectangles. Each vertex v (except F) has at most two outgoing edges. These edges are obtained as follows: We are going from the point in the plane corresponding to vertex v diagonally until we hit the border of an obstacle or the boundary of the rectangle defined by O and F. In the latter case F is a neighbor of v. The length d_{vF} of the edge connecting v and F equals the length of the projection of the diagonal part of the v and F connecting path plus the length of the parallel to one of the axis part of this path. In other words, if v is defined in the plane by the coordinates (x_v, y_v) then $d_{vF} = \max\{ x_F - x_v, y_F - y_v \}$. If we hit the border of an obstacle, we introduce two arcs connecting the north-west corner (say vertex u defined by coordinates (x_u, y_u)) and the south-east corner (say vertex w defined by coordinates (x_w, y_w)) to v. The length of the edge connecting v to u is $d_{vu} = \max\{ x_u - x_v, y_u - y_v \}$. Correspondingly the length of the edge connecting v and w is $d_{vw} = \max\{ x_w - x_v, y_w - y_v \}$. Thus an application of a shortest path algorithm yields the minimum makespan. In our special case the complexity of the algorithm reduces to $O(m \log m)$, cf. [Bru88].

8.2.2 Dominance and Branching Rules

One of the early branch and bound procedures used to find an optimal permutation schedule is described by Ignall and Schrage [IS65] and, independently by Lomnicki [Lom65]. Associated with each node of the search tree is a partial permutation π defining a partial permutation schedule S_π on a set of jobs. Let $\mathcal{J}_\pi$ be the set of jobs from the schedule S_π. A lower bound is calculated for any completion τ of the partial permutation π to a complete permutation $(\pi\tau)$. The lower bound is obtained by considering the work remaining on each machine. The number of branches departing from a search tree node (with a minimum lower bound) equals the number of jobs not in S_π, i.e. for each job J_i with $i \notin \pi$ a branch is considered extending the partial permutation π by one additional position to a new partial permutation (πi). Moreover extensions of the algorithm use some dominance rules under which certain completions of partial permutations π can be eliminated because there exists a schedule at least as good as π among the completions of another partial permutation π'. Let $C_k(\pi)$ denote the completion time of the last job in S_π on machine P_k, i.e. $C_k(\pi)$ is the earliest time at which some job not in $\mathcal{J}_\pi$ could begin processing on machine P_k. Then π' dominates π if for any completion τ of π there exists a completion τ' of π' such that $C_m(\pi'\tau') \le C_m(\pi\tau)$. An immediate consequence is the following transitive *dominance criterion*.

Theorem 8.2.3 [IS65] *If $\mathcal{J}_\pi = \mathcal{J}_{\pi'}$ and $C_k(\pi') \le C_k(\pi)$ for $k = 1, 2, \cdots, m$, then π' dominates π.*   □

There are other dominance criteria reported in [McM69] and [Szw71, Szw73, Szw78] violating transitivity. In general these dominance criteria consider sets $\mathcal{J}_\pi \ne \mathcal{J}$. We can formulate

Theorem 8.2.4 *If $C_{k-1}(\pi j i) - C_{k-1}(\pi i) \le C_k(\pi j i) - C_k(\pi i) \le p_{kj}$ for $k = 2, \cdots, m$, then $(\pi j i)$ dominates (πi).*   □

8.2.3 Lower Bounds

Next we consider different types of lower bounds that apply in order to estimate the quality of all possible completions τ of partial permutations π to a complete permutation $(\pi\tau)$.

The amount of processing time yet required on the first machine is $\sum_{j \in \tau} p_{1j}$. Suppose that a particular job J_j will be the last one in the permutation schedule. Then after completion of job J_j on P_1 an interval of at least $\sum_{k=2}^{m} p_{kj}$ must elapse before the whole schedule can be completed. In the most favorable situation the last job will be the one which minimizes the latter sum. Hence a lower bound on the makespan is

$$LB_1 = C_1(\pi) + \sum_{i \in \tau} p_{1i} + \min_{j \in \tau} \{ \sum_{k=2}^{m} p_{kj} \}\, .$$

Similarly we obtain lower bounds (with respect to the remaining machines)

$$LB_p = C_p(\pi) + \sum_{i \in \tau} p_{pi} + \min_{j \in \tau} \{ \sum_{k=p+1}^{m} p_{kj} \}\, , \text{ for } p = 2, \cdots, m - 1.$$

And on the last machine we get

$$LB_m = C_m(\pi) + \sum_{i \in \tau} p_{mi}\, .$$

The lower bound proposed by Ignall and Schrage is the maximum of these m bounds.

To illustrate the procedure let us consider a 4–job, 3-machine instance from [Bak74]. The processing times p_{ij} can be found in Table 8.2.1.

p_{ij}	J_1	J_2	J_3	J_4
P_1	3	11	7	10
P_2	4	1	9	12
P_3	10	5	13	2

Table 8.2.1 *Processing times of a 4-job, 3-machine instance.*

Initially the permutation π is empty and four branches are generated from the initial search tree node. Each branch defines the next (first) position 1, 2, 3, or 4 in π. The partial permutations π, the values $C_p(\pi)$, and the lower bounds LB_p, for $p = 1, 2, 3$, and the maximum LB of the lower bounds obtained throughout the search are given in Table 8.2.2.

π	$C_1(\pi)$	$C_2(\pi)$	$C_3(\pi)$	LB_1	LB_2	LB_3	LB
1	3	7	17	37	31	37	37
2	11	12	17	45	39	42	45
3	7	16	29	37	35	46	46
4	10	22	24	37	41	52	52
1, 2	14	15	22	45	38	37	45
1, 3	10	19	32	37	34	39	39
1, 4	13	25	27	37	40	45	45

Table 8.2.2 *Search tree nodes of the Ignall / Schrage [IS65] branch and bound.*

Two additional branches are generated from that node associated with permutation $(1, 4)$. These branches immediately lead to feasible solutions $(1, 3, 2, 4)$ and $(1, 3, 4, 2)$ with makespans equal to 45 and 39, respectively. Hence, $(1, 3, 4, 2)$ is a permutation defining an optimal schedule.

The calculation of lower bound can be strengthened in a number of ways. On each machine P_k, except the first one, there may occur some idle time of P_k between the completion of job J_i and the start of its immediate successor J_j. The idle time arises if J_j is not ready "in time" on the previous machine P_{k-1}, in other words $C_{k-1}(\pi j) > C_k(\pi)$. Thus we can improve the aforementioned bounds if we replace the earliest start time on P_r of the next job not in $\mathcal{I}_\pi$ by

$$\bar{C}_1(\pi) = C_1(\pi) \text{ and } \bar{C}_r(\pi) = \max_{k=1,2,\cdots,r} \{C_k(\pi) + \min_{j\in\pi}\{\sum_{q=k}^{r-1} p_{qj}\}\}, \text{ for } r = 2,\cdots,m.$$

Besides the above machine based bound another job based bound can be calculated as follows: Consider a partial permutation π and let τ be an extension to a complete schedule $S_{\pi\tau}$. For any job J_j with $j\in\tau$ we can calculate a lower bound

on the makespan of $S_{\pi\tau}$ as $C_1(\pi) + \sum_{k=1}^{m} p_{kj} + \sum_{J_i\in\mathcal{I}_1} p_{1i} + \sum_{J_i\in\mathcal{I}_2} p_{mi}$ where $\mathcal{I}_1$ ($\mathcal{I}_2$) are the

sets of jobs processed before (after) J_j in schedule $S_{\pi\tau}$, respectively. Since $\sum_{J_i\in\mathcal{I}_1} p_{1i}$

$+ \sum_{J_i\in\mathcal{I}_2} p_{mi} \geq \sum_{\substack{i\in\tau \\ i\neq j}} \min\{p_{1i}, p_{mi}\}$ we get the following lower bounds:

$$LB_{J_j} = \max\{\max_{\substack{j\in\tau \\ 1\leq r\leq s\leq m}} \{C_r(\pi) + \sum_{q=r}^{s} p_{qj} + \sum_{\substack{i\in\tau \\ i\neq j}} \min\{p_{ri}, p_{si}\}\}\}$$

Let us consider the computation of lower bounds within a more general framework which can be found in [LLRK78]. The makespan of an optimal solution of any sub-problem consisting of all jobs and a subset of the set of machines defines a lower bound on the makespan of the complete problem. In general these bounds are costly to compute (the problem is NP-hard if the number of machines is at least 3) except in the case of two machines where we can use Johnson's algorithm. Therefore let us restrict ourselves to the case of any two machines P_u and P_v. That means only P_u and P_v are of limited capacity and can process only one job at a time. P_u and P_v are said to be bottleneck machines, while the remaining machines $P_1,\cdots,P_{u-1}$, $P_{u+1},\cdots,P_{v-1}$, $P_{v+1},\cdots,P_m$, the non-bottleneck machines, are available with unlimited capacity. In particular, a non-bottleneck machine may process jobs simultaneously. Since the three (at most) sequences of non-bottleneck machines $P_{1u} = P_1,\cdots,P_{u-1}$; $P_{uv} = P_{u+1},\cdots,P_{v-1}$, and $P_{vm} = P_{v+1}$, $\cdots,P_m$ can be treated as one machine each (because we can process the jobs on the non-bottleneck machines without interruption), it follows that (in our lower bound computation) each partial permutation π defines a partial schedule for a problem with at most five machines P_{1u}, P_u, P_{uv}, P_v, P_{vm}, in that order. Of course, the jobs' processing times on P_{1u}, P_{uv}, and P_{vm} have still to be defined. We define for any job J_i the processing times

$$p_{1u\,i} = \max_{r=1,2,\cdots,u-1}\{C_r(\pi) + \sum_{k=r+1}^{u-1} p_{ki}\}; \quad p_{uv\,i} = \sum_{k=u+1}^{v-1} p_{ki}; \quad p_{vm\,i} = \sum_{k=v+1}^{m} p_{ki};$$

processing times on bottleneck machines are unchanged. Thus, the processing times $p_{1u\,i}, p_{uv\,i}$, and $p_{vm\,i}$ are the earliest possible start time of processing of job J_i on machine P_u, the minimum time lag between completion time of J_i on P_u and start time of J_i on P_v, and a minimum remaining flow time of J_i after completion on machine P_v, respectively. If $u = v$ we have a problem of at most three machines with only one bottleneck machine. Note, we can drop any of the machines P_{1u}, P_{uv}, or P_{vm} from the (at most) five machine modified flow shop problem through the introduction of a lower bound r_{1u}, r_{uv} on the start time of the successor machine, or a lower bound r_{vm} on the finish time of the whole schedule, respectively. In that case $r_{1u} = \min_{i \in \pi}\{p_{1u\,i}\}$, $r_{uv} = \min_{i \in \pi}\{p_{uv\,i}\}$; $r_{vm} = \min_{i \in \pi}\{p_{vm\,i}\}$. If $u = 1$, $v = u + 1$, or $v = m$ we have $r_{1u} = C_1(\pi)$, $r_{uv} = 0$, or $r_{vm} = 0$, respectively. The makespan $LB_\pi(\alpha, \beta, \gamma, \delta, \varepsilon)$ of an optimal solution for each of the resulting problems defines a lower bound on the makespan of any completion τ to a permutation schedule $(\pi\tau)$. Hereby α equals P_{1u} or r_{1u} reflecting the cases whether the start times on P_u are depending on the completion on a preceding machine P_{1u} or an approximation of them, respectively. In analogy we get $\gamma \in \{P_{uv}, r_{uv}\}$ and $\varepsilon \in \{P_{vm}, r_{vm}\}$. Parameters β and δ correspond to P_u and P_v, respectively.

Let us consider the bounds in detail (neglecting symmetric cases):

(1) $LB_\pi(r_{1u}, P_u, r_{um}) = r_{1u} + \sum_{\substack{i=1 \\ i \in \pi}}^{n} p_{ui} + r_{um}$.

(2) The computation of $LB_\pi(r_{1u}, P_u, P_{um})$ amounts to minimization of the maximum completion time on machine P_{um}. The completion time of J_i on machine P_{um} equals the sum of the completion time of J_i on machine P_u and the processing time $p_{um\,i}$. Hence, minimizing maximum completion time on machine P_{um} corresponds to minimizing maximum lateness on machine P_u if the due date of job J_i is defined to be $-p_{um\,i}$. This problem can be solved optimally using the earliest due date rule, i.e. ordering the jobs according to non-decreasing due dates. In our case this amounts to ordering the jobs according to non-increasing processing times $p_{um\,i}$. Adding the value r_{1u} to the value of an optimal solution of this one-machine problem with due dates yields the lower bound $LB_\pi(r_{1u}, P_u, P_{um})$.

(3) The bound $LB_\pi(P_{1u}, P_u, r_{um})$ leads to the solution of a one-machine problem with release date $p_{1u\,i}$ for each job J_i. Ordering the jobs according to non-decreasing processing time $p_{1u\,i}$ yields an optimal solution. Once again, adding r_{um} to the value of this optimal solution gives the lower bound $LB_\pi(P_{1u}, P_u, r_{um})$.

(4) The computation of $LB_\pi(P_{1u}, P_u, P_{um})$ corresponds to minimizing maximum lateness on P_u with respect to due dates $-p_{um\,i}$ and release dates $p_{1u\,i}$. The problem is NP-hard, cf. [LRKB77]. Anyway, the problem can be solved quickly if the number of jobs is reasonable, see the one-machine lower bound on the job shop scheduling problem described in Chapter 10.

(5) Computation of $LB_\pi(r_{1u}, P_u, r_{uv}, P_v, r_{um})$ leads to the solution of a flow shop scheduling problem on two machines P_u and P_v. The order of the jobs obtained from Johnson's algorithms will not be affected if P_v is unavailable until $C_v(\pi)$. Adding r_{1u} and r_{um} to the makespan of an optimal solution of this two machine flow shop scheduling problem yields the desired bound.

(6) Computation of $LB_\pi(r_{1u}, P_u, P_{uv}, P_v, r_{um})$ leads to the solution of a 3-machine flow shop problem with a non-bottleneck machine between P_u and P_v. The same procedure as described under (5) yields the desired bound. The only difference being that Johnson's algorithm is used in order to solve a 2-machine flow shop with processing times $p_{ui} + p_{uv\,i}$ and $p_{vi} + p_{uv\,i}$ for all $i \notin \pi$.

Computation of the remaining lower bounds require to solve NP-hard problems, cf. [LRKB77] and [LLRK78].

$LB_\pi(r_{1u}, P_u, P_{uv}, P_v, r_{um})$ and $LB_\pi(P_{1u}, P_u, P_{um})$ turned out to be the strongest lower bounds. Let us consider an example taken from [LLRK78]: Let $n = m = 3$; let $p_{11} = p_{12} = 1$, $p_{13} = 3$, $p_{21} = p_{22} = p_{23} = 3$, $p_{31} = 3$, $p_{32} = 1$, $p_{33} = 2$. We have $LB_\pi(P_{1u}, P_u, P_{um}) = 12$ and $LB_\pi(r_{1u}, P_u, P_{uv}, P_v, r_{um}) = 11$. If $p_{21} = p_{22} = p_{23} = 1$ and all other processing times are kept then $LB_\pi(P_{1u}, P_u, P_{um}) = 8$ and $LB_\pi(r_{1u}, P_u, P_{uv}, P_v, r_{um}) = 9$.

In order to determine the minimum effort to calculate each bound we refer the reader to [LLRK78].

8.3 Approximation Algorithms

8.3.1 Priority Rule and Local Search Based Heuristics

Noteworthy flow shop heuristics for the makespan criterion are those of Campbell et al. [CDS70] and Dannenbring [Dan77]. Both used Johnson's algorithm, the former to solve a series of two machine approximations to obtain a complete schedule. The second method locally improved this solution by switching adjacent jobs in the sequence. Dannenbring constructed an artificial two machine flow shop problem with processing times $\sum_{j=1}^{m}(m-j+1)p_{ji}$ on the first artificial machine and processing times $\sum_{j=1}^{m} jp_{ji}$ on the second artificial machine for each job J_i, $i = 1, \cdots, n$. The weights of the processing times are based on Palmer's [Pal65] 'slope index' in order to specify a job priority. Job priorities are chosen so that jobs with processing times that tend to increase from machine to machine will receive higher priority while jobs with processing times that tend to decrease from machine to machine will receive lower priority. Hence the slope index, i.e.

the priority to choose for the next job J_i is $s_i = \sum_{j=1}^{m} (m - 2j + 1)p_{ji}$ for $i = 1, \cdots, n$.

Then a permutation schedule is constructed using the job ordering with respect to decreasing s_i. Hundal and Rajgopal [HR88] extended Palmer's heuristic by computing two other sets of slope indices which account for machine $(m + 1)/2$ when m is odd. Two more schedules are produced and the best one is selected. The two sets of slope indices are $s_i = \sum_{j=1}^{m} (m - 2j + 2)p_{ji}$ and $s_i = \sum_{j=1}^{m} (m - 2j)p_{ji}$ for $i = 1, \cdots, n$.

Campbell et al. [CDS70] essentially generate a set of $m - 1$ two machine problems by splitting the m machines into two groups. Then Johnson's two machine algorithm is applied to find the $m - 1$ schedules, followed by selecting the best one. The processing times for the reduced problems are defined as p_{1ki} $= \sum_{j=1}^{k} p_{ji}$ and $p_{2ki} = \sum_{j=m-k+1}^{m} p_{ji}$ for $i = 1, \cdots, n$, where p_{1ki} (p_{2ki}) represents the processing time for job J_i on the artificial first (second) machine in the kth problem, $k = 1, \cdots, m - 1$.

Gupta [Gup71] recognizes that Johnson's algorithm is in fact a sorting algorithm which assigns an index to each job and sorts the jobs in ascending order by these indices. He generalized the index function to handle also cases of more than three machines. The index of job J_i is defined as

$$s_i = \lambda / \min_{1 \le j \le m-1} \{p_{ji} + p_{j+1\,i}\} \text{ for } i = 1, \cdots, n$$

where

$$\lambda = \begin{cases} 1 & \text{if } p_{ji} \le p_{1i}, \\ -1 & \text{otherwise.} \end{cases}$$

The idea of [HC91] is the heuristical minimization of gaps between successive jobs. They compute the differences $d_{kij} = p_{k+1\,i} - p_{kj}$ for $i, j = 1, \cdots, n; k = 1, \cdots, m - 1$ and $i \ne j$. If job J_i precedes job J_j in the schedule, then the positive value d_{kij} implies that job J_j needs to wait on machine P_{k+1} at least d_{kij} units of time until job J_i finishes. A negative value of d_{kij} implies that there exist d_{kij} units of idle time between job J_i and job J_j on machine P_{k+1}. Ho and Chang define a certain factor to discount the negative values. This factor assigns higher values to the first machines and lower values to last ones in order to reduce accumulated positive gaps effectively. The discount factor is defined as follows:

$$\delta_{kij} = \begin{cases} \dfrac{0.9\,(m - k - 1)}{m - 2} + 0.1 & \text{if } d_{kij} < 0, \\ 1 & \text{otherwise} \end{cases} \qquad \begin{array}{l} (\text{for } i, j = 1, \cdots, n; \\ \text{and } k = 1, \cdots, m - 1). \end{array}$$

Combining the d_{kij} and the discount factor, Ho and Chang define the overall revised gap:

$$d_{Rij} = \sum_{k=1}^{m-1} d_{kij}\delta_{kij}, \text{ for } i,j = 1,\cdots,n.$$

Let $J_{[i]}$ be the job in the ith position of a permutation schedule defined by permutation π. Then the heuristic works as follows:

Algorithm 8.3.1 *Gap minimization heuristic* [HC91].
begin
Let S be a feasible solution (schedule);
Construct values d_{Rij} for $i,j = 1,\cdots,n$;
$a := 1; \; b := n$;
repeat
   $S' := S$;
   Let $d_{R[a][u]} = \max\limits_{a<j<b}\{d_{R[a][j]}\}$;
   Let $d_{R[v][b]} = \min\limits_{a<j<b}\{d_{R[j][b]}\}$;
   if $d_{R[a][u]} < 0$ **and** $d_{R[v][b]} > 0$ **and** $|d_{R[a][u]}| \le |d_{R[v][b]}|$
   then
     begin
     $a = a+1$;
     Swap the jobs in the positions a and u of S;
     end;
   if $d_{R[a][u]} < 0$ **and** $d_{R[v][b]} > 0$ **and** $|d_{R[a][u]}| > |d_{R[v][b]}|$
   then
     begin
     $b = b-1$;
     Swap the jobs in the positions b and v of S;
     end;
   if $|d_{R[a][u]}| > |d_{R[v][b]}|$
   then
     begin
     $a = a+1$;
     Swap the jobs in the positions a and u of S;
     end;
   if the makespan of S increased **then** $S = S'$;
   until $b = a+2$
end;

Simulation results show that the heuristic [HC91] improves the best heuristic (among the previous ones) in three performance measures, namely makespan, mean flow time and mean utilization of machines.

An initial solution can be obtained using the following fast insertion method proposed in [NEH83].

Algorithm 8.3.2 *Fast insertion* [NEH83].

```
begin
    Order the n jobs by decreasing sums of processing times on the machines;
    Use Aker's graphical method to minimize the makespan of the first two jobs
        on all machines;
        -- The schedule defines a partial permutation schedule for the whole problem.
    for i = 3 to n do
        Insert the i^th job of the sequence into each of the i possible positions in the
            partial permutation and keep the best one defining an increased partial
            permutation schedule;
end;
```

Widmer and Hertz [WH89] and Taillard [Tai90] solved the permutation flow shop scheduling problem using tabu search. Neighbors are defined mainly as in the traveling salesman problem by one of the following three neighborhoods:

(1) Exchange two adjacent jobs.

(2) Exchange the jobs placed at the ith position and at the kth position.

(3) Remove the job placed at the ith position and put it at the kth position.

   Werner [Wer90] provides an improvement algorithm, called path search, and shows some similarities to tabu search and simulated annealing. The tabu search described in [NS96] resembles very much the authors' tabu search for job shop scheduling. Therefore we refer the reader to the presentation in the job shop chapter. There are other implementations based on the neighborhood search, for instance, the simulated annealing algorithm [OP89] or the genetic algorithm [Ree95] or the parallel genetic algorithm [SB92].

8.3.2 Worst-Case Analysis

As mentioned earlier the polynomially solvable flow shop cases with only two machines are frequently used to generate approximate schedules for those problems having a larger number of machines.

   It is easy to see that for any active schedule (a schedule is active if no job can start its processing earlier without delaying any other job) the following relation holds between the makespan $C_{max}(S)$ of an active schedule and the makespan C_{max}^* of an optimal schedule:

$$C_{max}(S) / C_{max}^* \leq \max_{\substack{1 \leq i \leq m \\ 1 \leq j \leq n}} \{p_{ij}\} / \min_{\substack{1 \leq i \leq m \\ 1 \leq j \leq n}} \{p_{ij}\}$$

Gonzales and Sahni [GS78] showed that $C_{max}(S)/C_{max}^* \leq m$ which is tight. They also gave a heuristic H_1 based on $\lfloor m/2 \rfloor$ applications of Johnson's algorithm with $C_{max}(S)/C_{max}^* \leq \lceil m/2 \rceil$ where S is the schedule produced by H_1.

Other worst-case performance results can be found in [NS93].

In [Bar81] an approximation algorithm has been proposed whose absolute error does not depend on n and is proved to be

$$C_{max}(S) - C_{max}^* = 0.5\,(m-1)\,(3m-1)\,\max_{\substack{1\leq i\leq m \\ 1\leq j\leq n}}\{p_{ij}\}\;.$$

where S is the produced schedule.

Potts [Pot85] analyzed a couple of approximation algorithms for $F2\,|\,r_j\,|$ C_{max}. The best one, called RJ', based on a repeated application of a modification of Johnson's algorithm has an absolute performance ratio of $C_{max}(S)/C_{max}^* \leq 5/3$ where S is the schedule obtained through RJ'.

In the following we concentrate on the basic ideas of *machine aggregation heuristics* using pairs of machines as introduced by Gonzalez and Sahni [GS78] and Röck and Schmidt [RS83]. These concepts can be applied to a variety of other NP-hard problems with polynomially solvable two-machine cases (cf. Sections 5.1 and 12.1). They lead to worst case performance ratios of $\lceil m/2 \rceil$, and the derivation of most of the results may be based on the following more general lemma which can also be applied in cases of open shop problems modeled by unrelated parallel machines.

Lemma 8.3.3 [RS83] *Let S be a non-preemptive schedule of a set $\mathcal{T}$ of n tasks on $m \geq 3$ unrelated machines P_i, $i = 1, \cdots, m$. Consider the complete graph $(\mathcal{P},$ $\mathcal{E})$ of all pairs of machines, where $\mathcal{E} = \{\{P_i, P_j\} \mid i, j = 1, \cdots, m,$ and $i \neq j\}$. Let $\mathcal{M}$ be a maximum matching for $(\mathcal{P}, \mathcal{E})$. Then there exists a schedule S' where*

(1) each task is processed on the same machine as in S, and S' has at most n preemptions,

(2) all ready times, precedence and resource constraints under which S was feasible remain satisfied,

(3) no pair $\{P_i, P_j\}$ of machines is active in parallel at any time unless $\{P_i, P_j\} \in \mathcal{M}$, and

(4) the finish time of each task increases by a factor of at most $\lceil m/2 \rceil$.

Proof. In case of odd m add an idle dummy machine P_{m+1} and match it with the remaining unmatched machine, so that an even number of machines can be assumed. Decompose S into sub-schedules $S(q, f)$, $q \in \mathcal{M}$, $f \in \{f_q^1, f_q^2, \cdots, f_q^{K_q}\}$ where $f_q^1 < f_q^2 < \cdots < f_q^{K_q}$ is the sequence of distinct finishing times of the tasks which are processed on the machine pair q. Without loss of generality we assume that $K_q \geq 1$. Let $f_q^0 = 0$ be the start time of the schedule and let $S(q, f_q^k)$ denote the sub-schedule of the machine pair q during interval $[f_q^{k-1}, f_q^k]$. The schedule S' which is obtained by arranging all these sub-schedules of S one after the other in the order of non-decreasing endpoints f, has the desired properties because (1): each task can preempt at most one other task, and this is the only source of pre-emption. (2) and (3): each sub-schedule $S(q, f)$ is feasible in itself, and its position in S' is according to non-decreasing endpoints of f. (4): the finish time C_j' of task $T_j \in \mathcal{T}$ in S' is located at the endpoint of the corresponding sub-schedule $S(q(j), C_j)$ where $q(j)$ is the machine pair on which T_j was processed in S, and C_j is the completion time of T_j in S. Due to the non-decreasing endpoint order of the sub-schedules it follows that $C_j' \leq \lceil m/2 \rceil C_j$. $\qquad\square$

For certain special problem structures Lemma 8.3.3 can be specialized so that preemption is kept out. Then, the aggregation approach can be applied to problems $F \mid\mid C_{max}$ and $O \mid\mid C_{max}$, and to some of their variants which remain solvable in case of $m = 2$ machines. We assume that for flow shops the machines are numbered that reflects the order each job is assigned to the machines.

We present two aggregation heuristics that are based on special conditions restricting the use of machines.

Condition C1: No pair $\{P_i, P_j\}$ of machines is allowed to be active in parallel at any time unless $\{P_i, P_j\} \in \mathcal{M}_1 = \{\{P_{2l-1}, P_{2l}\} \mid l = 1, 2, \cdots, \lfloor m/2 \rfloor\}$.

Condition C2: Let $(\mathcal{P}, \mathcal{E})$ be a bipartite graph where $\mathcal{E} = \{\{P_a, P_b\} \mid a \in \{1, 2, \cdots, \lceil m/2 \rceil\}, b \in \{\lceil m/2 \rceil + 1, \cdots, m\}$, and let $\mathcal{M}_2$ be a maximal matching for $(\mathcal{P}, \mathcal{E})$. Then no pair $\{P_i, P_j\}$ of machines is allowed to be active in parallel at any time unless $\{P_i, P_j\} \in \mathcal{M}_2$.

The following Algorithms 8.3.4 and 8.3.5 are based on conditions C1 and C2, respectively.

Algorithm 8.3.4 *Aggregation heuristic H_1 for $F \mid\mid C_{max}$ [GS78].*

```
begin
for each  pair q_i = {P_{2i-1}, P_{2i}} ∈ M_1
    begin
    Find an optimal sub-schedule S*_i for the two machines P_{2i-1} and P_{2i};
    if  m is odd
    then
        S*_{⌈m/2⌉} := an arbitrary schedule of the tasks on the remaining unmatched ma-
            chine P_m;
```

$$S := S_1^* \oplus S_2^* \oplus \cdots \oplus S_{\lceil m/2 \rceil}^*;$$
   end;
end;

As already mentioned, for $F \,|\,|\, C_{max}$ this heuristic was shown in [GS78] to have the worst case performance ratio of $C_{max}(H_1)/C_{max}^* \leq \lceil m/2 \rceil$. The given argument can be extended to $F \,|\, pmtn \,|\, C_{max}$ and $O \,|\,|\, C_{max}$, and also to some resource constrained models. Tightness examples which reach $\lceil m/2 \rceil$ can also be constructed, but heuristic H_1 is not applicable if permutation flow shop schedules are required.

In order to be able to handle this restriction consider the following Algorithm 8.3.5 which is based on condition $C2$. Assume for the moment that all machines with index less than or equal $\lceil m/2 \rceil$ are represented as a virtual machine P_1', and those with an index larger than $\lceil m/2 \rceil$ as a virtual machine P_1'. We again consider the given scheduling problem as a two machine problem.

Algorithm 8.3.5 *Aggregation heuristic H_2 for $F \,|\,|\, C_{max}$ and its permutation variant* [RS83].

begin
Solve the flow shop problem for two machines P_1', P_2' where each job J_j has
     processing time $a_j = \sum\limits_{i=1}^{\lceil m/2 \rceil} p_{ij}$ on P_1' and processing time $b_j = \sum\limits_{i=\lceil m/2 \rceil+1}^{m} p_{ij}$ on P_2',
     respectively;
Let S be the two-machine schedule thus obtained;
Schedule the jobs on the given m machines according to the two machine
    schedule S;
end;

The worst case performance ratio of Algorithm 8.3.5 can be derived with the following Lemma 8.3.6.

Lemma 8.3.6 *For each problem $F \,|\,|\, C_{max}$ (permutation flow shops included) and $O \,|\,|\, C_{max}$, the application of H_2 guarantees $C_{max}(H_2)/C_{max}^* \leq \lceil m/2 \rceil$.*

Proof. Let S be an optimal schedule of length C_{max}^* for an instance of the problem under consideration. As $\mathcal{M}_2$ from condition $C2$ is less restrictive than $\mathcal{M}_1$, it follows from Lemma 8.3.3 that there exists a preemptive schedule S' which remains feasible under $C2$, and whose length is $C_{max}'/C_{max}^* \leq \lceil m/2 \rceil$. By construction of $\mathcal{M}_2$, S' can be interpreted as a preemptive schedule of the job set on the two virtual machines P_1', P_2', where P_1' does all processing which is required on the machines $P_1, \cdots, P_{\lceil m/2 \rceil}$, and P_2' does all processing which is required on the machines $P_{\lceil m/2 \rceil+1}, \cdots, P_m$. Since on two machines preemptions are not advanta-

geous the schedule S generated by algorithm H_2 has length $C_{max}(H_2) \leq C'_{max} \leq$
$\lceil m/2 \rceil C^*_{max}$.   □

H_2 can be implemented to run in $O(n(m + \log n))$ for $F \| C_{max}$ and also for its
permutation variant using Algorithm 8.2.1. It is easy to adapt Lemma 8.3.3 to a
given preemptive schedule S so that the $\lceil m/2 \rceil$ bound for H_2 extends to $F \mid pmtn \mid$
C_{max} as well.

The following example shows that the $\lceil m/2 \rceil$ bound of H_2 is tight for $F \|$
C_{max}. Take m jobs J_j, $j = 1, \cdots, m$, with processing times $p_{ij} = p > 0$ for $i = j$,
whereas $p_{ij} = \varepsilon \to 0$ for $i \neq j$. H_2 uses the processing times $a_j = p + (\lceil m/2 \rceil - 1)\varepsilon$,
$b_j = \lfloor m/2 \rfloor \varepsilon$ for $j \leq \lceil m/2 \rceil$, and $a_j = \lceil m/2 \rceil \varepsilon$, $b_j = p + (\lceil m/2 \rceil - 1)\varepsilon$ for $j > \lceil m/2 \rceil$.
Consider job sets $\mathcal{J}^k$ which consist of k copies of each of these m jobs. For an
optimal flow shop schedule for $\mathcal{J}^k$ we get $C^*_{max} = kp + (m-1)(k+1)\varepsilon$. The optimal two machine flow shop schedule for $\mathcal{J}^k$ produced by H_2 may start with all k
copies of $J_{\lceil m/2 \rceil + 1}$, $J_{\lceil m/2 \rceil + 2}, \cdots$, J_m and then continue with all k copies J_1, $J_2, \cdots$,
$J_{\lceil m/2 \rceil}$. On m machines this results in a length of $C_{max}(H_2) =$
$(m - 1 + k\lfloor m/2 \rfloor)\varepsilon + \lceil m/2 \rceil pk$. It follows that $C_{max}(H_2) / C^*_{max}$ approaches $\lceil m/2 \rceil$ as
$\varepsilon \to 0$.

8.3.3 No Wait in Process

An interesting sub-case of flow shop scheduling is that with *no-wait constraints*
where no intermediate storage is considered and a job once finished on one machine must immediately be started on the next one.

The two-machine case, i.e. problem $F2 \mid no\text{-}wait \mid C_{max}$, may be formulated as
a special case of scheduling jobs on one machine whose *state* is described by a
single real valued variable x (the so-called *one state-variable machine problem*)
[GG64, RR72]. Job J_i requires a starting state $x = A_i$ and leaves with $x = B_i$.
There is a cost for changing the machine state x in order to enable the next job to
start. The cost c_{ij} of J_j following J_i is given by

$$
c_{ij} = \begin{cases} \displaystyle\int_{B_i}^{A_j} f(x)dx & \text{if } A_j \geq B_i, \\[2em] \displaystyle\int_{A_j}^{B_i} f(x)dx & \text{if } B_i > A_j, \end{cases}
$$

where $f(x)$ and $g(x)$ are integrable functions satisfying $f(x) + g(x) \geq 0$. The objective is to find a minimal cost sequence for the n jobs. Let us observe that problem

$F2 \mid no\text{-}wait \mid C_{max}$ may be modeled in the above way if $A_j = p_{1j}$, $B_j = p_{2j}$, $f(x) = 1$ and $g(x) = 0$. Cost c_{ij} then corresponds to the idle time on the second machine when J_j follows J_i, and hence a minimal cost sequence for the one state-variable machine problem also minimizes the completion time of the schedule for problem $F2 \mid no\text{-}wait \mid C_{max}$. On the other hand, the first problem corresponds also to a special case of the traveling salesman problem which can be solved in $O(n^2)$ time [GG64]. Unfortunately, more complicated assumptions concerning the structure of the flow shop problem result in its NP-hardness. So, for example, $Fm \mid no\text{-}wait \mid C_{max}$ is unary NP-hard for fixed $m \geq 3$ [Röc84].

As far as approximation algorithms are concerned H_1 is not applicable here, but H_2 turns out to work [RS83].

Lemma 8.3.7   *For $F \mid no\text{-}wait \mid C_{max}$, the application of H_2 guarantees* $C_{max}(H_2) / C_{max}^* \leq \lceil m/2 \rceil$.

Proof. It is easy to see that solving the two machine instance by H_2 is equivalent to solving the given instance of the m machine problem under the additional condition C2. It remains to show that for each no-wait schedule S of length C_{max} there is a corresponding schedule S' which is feasible under C2 and has length $C_{max}' \leq \lceil m/2 \rceil C_{max}$. Let $J_1, J_2, \cdots, J_n$ be the sequence in which the jobs are processed in S and let s_{ij} be the start time of job J_j, $j = 1, \cdots, n$, on machine J_i, $i = 1, \cdots, m$. As a consequence of the no-wait requirement, the successor J_{j+1} of J_j cannot start to be processed on machine P_{i-1} before J_j starts to be processed on machine P_i. Thus for $q = \lceil m/2 \rceil$ we have $s_j^{q+1} \leq s_{j+1}^{q} \leq \ldots \leq s_{j+q}^{1}$ and for the finish time C_j of job J_j we get $C_j \leq s_{j+1}^{m} \leq s_{j+2}^{m-1} \leq \ldots \leq s_{j+m-q}^{q+1} \leq s_{j+q}^{q+1}$. This shows that if we would remove the jobs between J_j and J_{j+q} from S, then S would satisfy C2 in the interval $[s_j^{q+1}, s_{j+q}^{q+1}]$. Hence, for each $k = 1, \ldots, q$ the sub-schedule S_k of S which covers only the jobs of the subsequence $J_k, J_{k+q}, J_{k+2q}, \cdots, J_{k+\lfloor (n-k)/q \rfloor q}$ of $J_1, \cdots$, J_n satisfies C2. Arrange these sub-schedules in sequence. None is longer than C_{max}, and each job appears in one of them. The resulting schedule S' is feasible and has length $C_{max}' \leq qC_{max}$.   □

Using the algorithm of Gilmore and Gomory [GG64], H_2 runs in $O(n(m+\log n))$ time. The tightness example given above applies to the no-wait flow shop as well, since the optimal schedule is in fact a no-wait schedule. Moreover, on two machines it is optimal to have any alternating sequence of jobs J_b, J_a, J_b, J_a with $a \in \{1, \cdots, \lceil m/2 \rceil\}$ and $b \in \{\lceil m/2 \rceil + 1, \cdots, m\}$, and in case of odd m this may be followed by all copies of J_1. When ε tends to zero, the length of such a schedule on m machines approaches $\lceil m/2 \rceil kp$, thus $C_{max}(H_2) / C_{max}^*$ approaches $\lceil m/2 \rceil$.

An interesting fact about the lengths of no-wait and normal flow shop schedules, respectively, has been proved by Lenstra. It appears that the no-wait constraint may lengthen the optimal flow shop schedule considerably, since $C^*_{max}(no\text{-}wait)/C^*_{max} < m$ for $m \geq 2$.

8.4 Scheduling Flexible Flow Shops

8.4.1 Problem Formulation

The hybrid or flexible flowshop problem is a generalization of the flowshop in such a way that every job can be processed by one among several machines on each machine stage. In recent years a number of effective exact methods have been developed. A major reason for this progress is the development of new job and machine based lower bounds as well as the rapidly increasing importance of constraint programming.

We consider the problem of scheduling n parts or jobs J_j, $j = 1, 2, \cdots, n$, through a manufacturing system that will be called a *flexible flow shop* (FFS), to minimize the schedule length. An FFS consists of $m \geq 2$ machine stages or centers with stage l having $k_l \geq 1$ identical parallel machines $P_{l1}, P_{l2}, \cdots, P_{lk_l}$ (see Figure 8.4.1). For job J_j vector $[p_{1j}, p_{2j}, \cdots, p_{mj}]^T$ of processing times is known, where $p_{lj} \geq 0$ for all l, j. Task T_{lj} of job J_j may be processed on any of the k_l machines. This is the generalization of the standard flow shop scheduling problem, whereas all the remaining assumptions remain unchanged.

The jobs have to visit the stages in the same order starting from stage 1 through stage m. A machine can process at most one job at a time and a job can be processed by at most one machine at a time. Preemption of processing is not allowed. The scheduling problem consists of assigning jobs to machines at each stage and sequencing the jobs assigned to the same machine so that some optimality criterion C is minimized.

Note that the processing time p_{lj} does not depend on the machine assigned to job J_j at stage l. This notation is applied when stage l consists of identical parallel machines. The completion time (which is a decision variable) of job J_j at stage l will be denoted by $C_j^{(l)}$.

A *partial schedule* S assigns some jobs to machines and fixes the processing order of another subset of jobs. S can be modeled by a directed graph $G = (V, A)$, where V consists of $nm+2$ nodes, i.e., one node (j, l) for each job J_j at each stage l and two additional nodes, 0 and *. A contains the arcs (directed edges) $(0,(j, l))$ and $((j, l),*)$ for all nodes (j, l). Moreover, the arcs $((j, l), (j, l-1))$ belong to A for all j and $1 \leq l \leq m-1$. Finally, whenever S fixes that job J_i precedes job J_j

at some stage l then arc $((i,l),(j,l))$ belongs to A. The *length of an arc* $((i,l),x) \in A$ is p_{li}, where x is a node of G. The length of any arc $(0,(i,l)) \in A$ is null. A *path* ρ in G is a sequence of nodes $(\rho_1, ..., \rho_e)$ such that ρ contains no node twice and $(\rho_u, \rho_{u+1}) \in A$ for all $1 \le u \le e-1$. The *length of a path* ρ is the sum of the lengths of the arcs (ρ_u, ρ_{u+1}), $1 \le u \le e-1$, along the path. Let $h(x,y)$ represent the length of the longest path between nodes x and y. If no path exists between x and y in G, then $h(x,y) = \infty$. Finally, the *release date* or *head* $r_j^{(l)}$ of job J_j at stage l is $h(0,(j,l))$, while its *delivery time* or *tail* $q_j^{(l)}$ is $h((j,l),*) - p_{lj}$, see Blazewicz et al. [BDP96]. Figure 8.4.1 is an example with m stages and k_l machines at each stage l.

Machines may remain idle and in-process inventory is allowed. This is important, since a restricted version of the problem was studied already by Salvador [Sal73] who presented a branch and bound algorithm for FFS with no-wait schedules and $p_{lj} > 0$ for all l, j. He identified the problem in the polymerization process where there are several effectively identical and thus interchangeable plants each of which can be considered as a flow shop. Of course, all situations where a parallel machine(s) is (are) added at least one stage of a flow shop to solve a bottleneck problem or to increase the production capacity lead to the FFS scheduling. Another interesting application of the problem was described by Brah and Hunsucker [BH91] and concerns the running of a program on a computer where the three steps of compiling, linking and running are performed in a fixed sequence and we have several processors (software) at each step. Other real life examples exist in the electronics manufacturing.

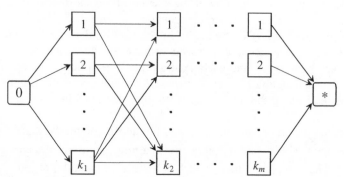

Figure 8.4.1   *Schematic representation of a flexible flow shop.*

Heuristics for the general FFS scheduling problem (in the sense stated above) were developed by Wittrock [Wit85, Wit88], and by Kochbar and Morris [KM87]. The first paper deals with a periodic algorithm where a small set of jobs is scheduled and the schedule is repeated many times, whereas the second one presents a non-periodic algorithm. The basic approach in both cases is to decompose the problem into three sub-problems: machine allocation, sequencing and timing. The first sub-problem is to determine which jobs will visit which machine at each stage. The second sub-problem sequences jobs on each machine, and the third one consists of finding the times at which the jobs should enter the system. The heuristic algorithm developed by Kochbar and Morris considers setup times, finite buffers, blocking and starvation, machine down time, and current and subsequent state of the system. The heuristics tend to minimize the effect of setup times and blocking.

The standard $\alpha \,|\, \beta \,|\, \gamma$ notation for classifying scheduling problems by Graham et al. [GLL+79] has been extended by Vignier et al. [VBP99] to take the new machine environment into consideration. Here we will consider only models with identical parallel machines at the stages and the objective is to minimize the makespan, denoted by $Fm \,|\, k_1, k_2, \cdots, k_m \,|\, C_{max}$, and the mean flow time, $Fm \,|\, k_1,$ $k_2, \cdots, k_m \,|\, \sum C_i$, respectively. In fact, we are not aware of efficient exact solution procedures for the general m-stage problem with other processing environments. By "general m-stage problem" we mean that m is not restricted to a small constant.

The general m-stage multiprocessor flowshop scheduling problem is strongly NP-hard for all traditional optimality criteria, since the special cases $F3 \,|\,| C_{max}$ and $F2 \,|\,| \sum C_i$ having only one machine at each stage are NP-hard in the strong sense, as shown in Garey et al. [GJS76]. Moreover, the makespan minimization problem is already NP-hard in the strong sense when $m = 2$ and $\max\{k_1, k_2\} > 1$ as shown by Gupta [Gup88]. Note that Hoogeveen et al. [HLV96] have proven that $F2 \,|\, 2, 1 \,|\, C_{max}$ is at least NP-hard in the ordinary sense, while its preemptive version, $F2 \,|\, 2, 1, pmtn \,|\, C_{max}$, has been shown NP-hard in the strong sense.

In the following sections we will present some heuristics for simple subproblems of our problem for which the worst and average case performance is known.

Then we provide a comprehensive and uniform overview on exact solution methods for flexible flowshops with branching, bounding and propagation of constraints under two different objective functions: minimizing the makespan of a schedule and the mean flow time. This part is based on Kis and Pesch [KP05].

We do not discuss the large body of work on the two-stage special case. The review by Vignier et al. [VBP99] offers an exhaustive overview on two-stage problems.

We present a mixed integer-linear program modeling the constraints of both the minimum makespan and the minimum mean flow time problems, respectively. Then we consider the minimum makespan problem, followed by a discussion on approaches for minimizing the mean flow time. The latter two sections have a common structure: first various lower bounds are presented and compared if possible, then branching schemes and their merits are discussed.

8.4.2 Heuristics and Their Performance

The results presented in this section were obtained by Sriskandarajah and Sethi [SS89]. In the sequel the FFS scheduling problem with m machine stages and k_i machines at stage i will be denoted by $Fm \mid k_1, k_2, \cdots, k_m \mid C_{max}$.

Let us start with the problem $F2 \mid k_1 = 1, k_2 = k \geq 2 \mid C_{max}$, and let us assume that the buffer between the machine stages has unlimited capacity. First, consider the list scheduling algorithm in which a list of the job indices $1, 2, \cdots, n$ is given. Jobs enter the first machine stage (i.e. machine P_1) in the order defined by the list, form the queue between the stages and are processed in center 2, whenever a machine at this stage becomes available. C_{max} denotes the schedule length of the set of jobs when the list scheduling algorithm is applied, and C_{max}^* is the minimum possible schedule length of this set of jobs. Then the following theorem holds.

Theorem 8.4.1 [SS89] *For the list scheduling algorithm applied to the* problem $Fm \mid k_{m-1} = 1, k_m = k \geq 2 \mid C_{max}$ *we have*

$$C_{max}/C_{max}^* \leq m + 1 - \frac{1}{k},$$

and this is the best possible bound.   □

The proof of this theorem is based on Grahams result [Gra66] for algorithms applied to the problem $Pm \mid\mid C_{max}$ (or $F1 \mid k_1 = k \geq 2 \mid C_{max}$). The bound is, as we remember from Section 5.1, $C_{max}/C_{max}^* \leq 2 - 1/k$.

Consider now Johnson's algorithm which, as we remember, is optimal for problem $F2 \mid\mid C_{max}$. The following can be proved.

Theorem 8.4.2 [SS89] *For Johnson's algorithm applied to problems* $F2 \mid k_1 = 1,$ $k_2 = k = 2 \mid C_{max}$ *and* $F2 \mid k_1 = 1, k_2 = k \geq 3 \mid C_{max}$ *with* $C_{max} \leq \sum_{j=1}^{n} p_{1j} + \max_{j}\{p_{2j}\}$ *the following holds:*

$$C_{max}/C_{max}^* \leq 2,$$

and this is the best possible bound.   □

Theorem 8.4.3 [SS89] *For Johnson's algorithm applied to the problem* $F2 \mid k_1 = 1, k_2 = k \geq 3 \mid C_{max}$ *with* $C_{max} > \sum_{j=1}^{n} p_{1j} + \max_{j}\{p_{2j}\}$ *we have*

$$C_{max}/C_{max}^* \leq 1 + (2 - \frac{1}{k})(1 - \frac{1}{k}).$$ □

Notice that the bounds obtained in Theorems 8.4.2 and 8.4.3 are better than those of Theorem 8.4.1.

Let us now pass to the problem $F2\,|\,k_1 = k_2 = k \geq 2\,|\,C_{max}$. The basic algorithm is the following.

Algorithm 8.4.4 *Heuristic H_a for $F2\,|\,k_1 = k_2 = k \geq 2\,|\,C_{max}$ [SS89].*

begin
Partition the set of machines into k pairs $\{P_{11}, P_{21}\}, \{P_{12}, P_{22}\}, \cdots, \{P_{1k}, P_{2k}\}$, treating each pair as an artificial machine P'_i, $i = 1, 2, \cdots, k$, respectively;
for each job $J_j \in \mathcal{J}$ **do** $p'_j := p_{1j} + p_{2j}$;
call List scheduling algorithm;
    -- this problem is equivalent to the NP-hard problem $Pk\,||\,C_{max}$ (see Section 5.1),
    -- where a set of jobs with processing times p'_j is scheduled non-preemptively on a set
    -- of k artificial machines; list scheduling algorithm solves this problem heuristically
for $i = 1$ **to** k **do call** Algorithm 8.2.1;
    -- this loop solves optimally each of the k flow shop problems
    -- with unlimited buffers, i.e. for each artificial machine P'_i the processing times p'_i
    -- assigned to it are distributed among the two respective machines P_{1i} and P_{2i}
end;

Let us note, that in the last **for** loop one could also use the Gilmore-Gomory algorithm, thus solving the k flow shop problems with the no-wait condition. The results obtained from hereon hold also for the FFS with no-wait restriction, i.e. for the case of no buffer between the machine stages. On the basis of the Graham's reasoning, in [SS89] the same bound as in Theorem 8.4.1 has been proved for H_a, and this bound remains unchanged even if a heuristic list scheduling algorithm is used in the last **for** loop. Since an arbitrary list scheduling algorithm has the major influence on the worst case behavior of Algorithm H_a, in [SS89] another Algorithm, H_b, was proposed in which the *LPT* algorithm is used. We know from Section 5.1 that in the worst case *LPT* is better than an arbitrary list scheduling algorithm for $Pm\,||\,C_{max}$. Thus, one can expect that for H_b a better bound exists than for H_a.

The exact bound R_{H_b} for H_b is not yet known, but Srishkandarajah and Sethi proved the following inequality,

$$\frac{7}{3} - \frac{2}{3k} \leq R_{H_b} \leq 3 - \frac{1}{k}.$$

The same authors proved that if *LPT* in H_b is replaced by a better heuristic or even by an exact algorithm, the bound would still be $R_{H_b} \geq 2$. The bound 2 has also been obtained in [Lan87] for a heuristic which schedules jobs in non-increasing order of p_{2j} in FFS with $m = 2$ and an unlimited buffer between the stages.

Computational experiments performed in [SS89] show that the average performance of the algorithms presented above is much better than their worst case behavior. However, further efforts are needed to construct heuristics with better bounds (i.e. less than 2).

8.4.3 A Model

A mixed integer programming formulation for $Fm \mid k_1, k_2, \cdots, k_m \mid C_{max}$ is given by Guinet et al. [GSKD96]. The decision variables specify the order of jobs on the machines and the completion times of the jobs at each stage:

x_{ijkl} = 1, if job J_j is processed directly after job J_i on machine P_k in stage l,
    0 otherwise,

x_{0ikl} = 1, if job J_i is the first job on machine P_k at stage l,
    0 otherwise,

x_{i0kl} = 1, if job J_i is the last job on machine P_k at stage l,
    0 otherwise,

$C_j^{(l)}$ = completion time of job J_j at stage l,

C_{max} = completion time of all jobs.

The mixed integer programming formulation in [GSKD96] is as follows:

$$\text{minimize} \ \ C_{max} \tag{8.4.1}$$

subject to

$$\sum_{i=0, i \neq j}^{n} \sum_{k=1}^{k_l} x_{ijkl} = 1 \qquad \forall \ j = 1, \cdots, n, \ l = 1, \cdots, m \tag{8.4.2}$$

$$\sum_{j=0}^{n} x_{ijkl} \leq 1 \qquad \begin{array}{l} \forall \ h = 0, \cdots, n, \ k = 1, \cdots, k_l, \\ l = 1, \cdots, L \end{array} \tag{8.4.3}$$

$$\sum_{i=0, i \neq h}^{n} x_{ihkl} - \sum_{j=0, j \neq h}^{n} x_{hjkl} = 0 \qquad \begin{array}{l} \forall \ h = 1, \cdots, n, \ k = 1, \cdots, k_l, \\ l = 1, \cdots, L \end{array} \tag{8.4.4}$$

$$C_i^{(l)} + \sum_{k=1}^{k_l} x_{ijkl} \cdot p_{lj} + \left(\left(\sum_{k=1}^{k_l} x_{ijkl} \right) - 1 \right) B \leq C_j^{(l)} \qquad \begin{array}{l} \forall \ i = 1, \cdots, n, \\ j = 1, \cdots, n, \\ l = 1, \cdots, m \end{array} \tag{8.4.5}$$

$$C_j^{(l-1)} + p_{lj} \leq C_j^{(l)} \qquad \forall \ j = 1, \cdots, n, l = 2, \cdots, m \tag{8.4.6}$$

$$C_j^{(l)} \leq C_{max} \qquad\qquad \forall\ j = 1,\cdots,n, l = 1,\cdots,m \qquad (8.4.7)$$

$$x_{ijkl} \in \{0,1\} \qquad\qquad \begin{array}{l} \forall\ i = 0,\cdots,n, j = 0,\cdots,n, \\ k = 1,\cdots,\ k_l, l = 1,\cdots,m \end{array} \qquad (8.4.8)$$

$$C_j^{(l)} \geq 0 \qquad\qquad \forall\ j = 1,\cdots,n, l = 1,\cdots,m \qquad (8.4.9)$$

In this program B is a very big constant, i.e., greater than the sum of all job processing times.

The makespan minimization aspect of the problem is expressed by (8.4.1). Constraints (8.4.2), (8.4.3) and (8.4.4) ensure that each job is processed precisely once at each stage. In particular, (8.4.2) guarantees that at each stage l for each job J_j there is a unique machine such that either J_j is processed first or after another job on that machine. The inequalities (8.4.3) imply that at each stage there is a machine on which a job has a successor or is processed last. Finally, at each stage for each job there is one and only one machine satisfying both of the previous two conditions by (8.4.4). Constraints (8.4.5) and (8.4.6) take care of the completion times of the jobs. Inequalities (8.4.5) ensure that the completion times $C_i^{(l)}$ and $C_j^{(l)}$ of jobs J_i and J_j scheduled consecutively on the same machine respect this order. On the other hand, inequalities (8.4.6) imply that jobs go through the stages in the right order, i.e. from stage 1 through stage m. The constraint that the makespan is not smaller than the completion time of any job is expressed by (8.4.7). The last two constraints specify the domains of the decision variables.

To minimize the mean flow time instead of the makespan it is enough to replace the objective function (8.4.1) with the following one:

$$\min \sum_{i=1}^{n} C_i^{(m)} \qquad\qquad (8.4.10)$$

Moreover, the variable C_{max} and all constraints involving it can be dropped.

8.4.4 The Makespan Minimization Problem

First we discuss various techniques for obtaining lower bounds, then we present branching schemes and also implementations and computational results.

Lower Bounds

Although we are concerned with the general m-stage problem, it is worth to recapitulate lower bounds for the two-stage special case, since several ideas stem from studying the latter problems. We will highlight the key ideas and cite papers that appear to propose them. If not mentioned otherwise, we assume that at each stage there are identical parallel machines.

I) *Reduction to classical flowshop.* Consider the classical flowshop scheduling problem obtained by dividing the job processing times at each stage by the number of machines. That is, define the new processing time of job J_j at stage l as $\tilde{p}_{lj}$ $= p_{lj}/k_l$, $l = 1,\cdots,m$, $j = 1,\cdots,n$. The n new jobs with processing times $\tilde{p}_{lj}$ at the different stages constitute a classical flowshop scheduling problem. When $L = 2$, this flowshop scheduling problem can be solved to optimality by Johnson's rule [Joh54]. The optimum makespan C_{LB} of the latter problem is a lower bound on the optimum makespan of the original problem, as it is observed in Lee and Vairaktarakis [LV94]. To see this, let S^* be an optimal schedule for the two-stage multiprocessor problem and suppose the jobs are indexed in non-decreasing order of their completion time at stage 1, i.e., $i < j$ iff $C_i^{(1)} \le C_j^{(1)}$. Consider the first i jobs at stage 1 and the last $n-i+1$ jobs at stage 2. Since the last $n-i+1$ jobs cannot start earlier at the second stage than the completion of the first i jobs at the first stage, the completion time $C_{max}(S^*)$ satisfies

$$C_{max}(S^*) \ge \frac{1}{k_1} \sum_{j=1}^{i} p_{1j} + \frac{1}{k_2} \sum_{j=1}^{n} p_{2j}, \quad 1 \le i \le n.$$

Now consider the two-stage flowshop scheduling problem with n jobs having the above processing times. When sequencing the jobs at both stages in increasing order of their indices we obtain a feasible schedule and a longest path in that schedule with length $\bar{C}_{max}$. On this longest past there is a job h such that

$$\bar{C}_{max} = \sum_{j=1}^{h} \frac{p_{1j}}{k_1} + \sum_{j=h}^{n} \frac{p_{2j}}{k_2}.$$

We immediately see that $\bar{C}_{max} \le C_{max}(S^*)$. Moreover, as $C_{LB} \le \bar{C}_{max}$, the statement follows.

II) *Aggregation.* This is a very rich class of lower bounds based on computing the total amount of work on some stages or machines. Again, we begin with the case $m = 2$ and the following two lower bounds, $LB(1)$ and $LB(2)$, are enhancements of those suggested by Sriskandarajah and Sethi [SS89], generalizations of the bounds proposed by Gupta and Tunc [GT91] and Gupta [Gup88] and are reported in their present form by Guinet et al. [GSKD96].

$$LB(l) = \min_{i=1,\ldots,n} p_{3-l\,i} + \max\left\{ \left(\sum_{i=1}^{n} C_i^{(m)}\right)/k_l, \max_{i=1,\ldots,n} p_{mi} \right\}, \quad l = 1,2. \quad (8.4.11)$$

This bound is based on aggregating the work at stage l. Consider e.g., $LB(1)$. The processing of all jobs at the first stage cannot complete sooner than the max in (8.4.11). In addition to that, the last job, say job J_j, finished at this stage must be completed at the second stage too. The minimum amount of time spent by job j at the second stage is expressed by the min in (8.4.11). Hence, $LB(1)$ is a lower bound on the makespan. By reversing the time the same argument shows that $LB(2)$ is a lower bound on the makespan as well.

Lee and Vairaktarakis introduced a different set of lower bounds for the $m = 2$ case in [LV94]. Suppose the jobs are indexed in non-decreasing order of stage 1 processing times, i.e., $p_{11} \le \ldots \le p_{1n}$. Let $P_{lq} = \sum_{j=1}^{q} p_{lj}$ denote the summation of the q shortest job processing times at stage l. If $k_1 \ge k_2$ then

$$LB_1 = \frac{P_{1k_2} + P_{2n}}{k_2} \tag{8.4.12}$$

is a lower bound on C_{max}. Namely, because of the flowshop constraints, on each machine at stage 2 there will be some idle time before processing may start, i.e., there will be a machine with idle time at least p_{11}, a machine with idle time p_{12}, $\ldots$, and a machine with idle time p_{1k_2}. Consequently, the makespan is no less than the average idle time plus the average workload at stage 2.

However, if $k_1 < k_2$ the above lower bound can be improved. Certainly, on each machine at stage 2 there will be idle time before processing starts and these idle times are at least $p_{11}, \cdots, p_{1k_2}$, respectively. Moreover, on $k_2 - k_1$ of these machines processing cannot start until at least two jobs are completed at stage 1. Hence, an additional idle time of at least p_{11} units is unavoidable on $k_2 - k_1$ machines at stage 2. Consequently, the following

$$LB_2 = \frac{P_{1k_2} + (k_2 - k_1)P_{11} + P_{2n}}{k_2} \tag{8.4.13}$$

is a lower bound on the makespan. By exploiting the symmetry of the two-stage multiprocessor flowshop problem we obtain another two bounds by interchanging the roles of stage 1 and stage 2. The new bounds will be

$$LB_3 = \frac{P_{2k_1} + (k_1 - k_2)P_{21} + P_{1n}}{k_1} \quad \text{if } k_1 \ge k_2, \tag{8.4.14}$$

$$LB_4 = \frac{P_{2k_1} + P_{1n}}{k_1} \quad \text{if } k_1 < k_2. \tag{8.4.15}$$

These lower bounds can be combined to obtain the following lower bound:

$$LB = \begin{cases} \max\{LB_1, LB_3, C_{LB}\} & \text{if } k_1 \ge k_2 \\ \max\{LB_2, LB_4, C_{LB}\} & \text{if } k_1 < k_2 \end{cases}$$

where C_{LB} is the lower bound of Lee and Vairaktarakis obtained by reduction to classical flowshop.

Brah and Hunsucker proposed two bounds for the general m-stage problem, one based on machines and another based on jobs [BH91]. Suppose all jobs are sequenced on stages 1 through $l-1$ and a subset $\mathcal{A}$ of jobs is already scheduled at stage l. Before describing the two bounds, we introduce additional notation.

$\mathcal{J}$ = set of all jobs,

$\mathcal{A}$ = set of jobs already scheduled at stage l,

$S^{(l)}(\mathcal{A})$ = partial schedule of jobs in $\mathcal{A}$ at stage l,

$C[S^{(l)}(\mathcal{A})]_k$ = completion time of the partial sequence on machine k.

Notice that in order to compute $C[S^{(l)}(\mathcal{A})]_k$ we have to fix the schedule of the upstream stages.

Having fixed the schedule of all jobs on the first $l-1$ stages and that of the jobs in $\mathcal{A}$ at stage l, the average completion time of all jobs at stage l, $ACT[S^{(l)}(\mathcal{A})]$, can be computed as follows:

$$ACT[S^{(l)}(\mathcal{A})] = \frac{\sum_{k=1}^{k_l} C[S^{(l)}(\mathcal{A})]_k}{k_l} + \frac{\sum_{j \in \mathcal{J} - \mathcal{A}} p_{lj}}{k_l} \tag{8.4.16}$$

It is worth mentioning that in any complete schedule of all jobs at stage l that contains the partial schedule $S^{(l)}(\mathcal{A})$, there will be a job completing not sooner than $ACT[S^{(l)}(\mathcal{A})]$.

The maximum completion time of jobs in $\mathcal{A}$ at stage l, $MCT[S^{(l)}(\mathcal{A})]$, is given by

$$MCT[S^{(l)}(\mathcal{A})] = \max_{1 \leq k \leq k_l} C[S^{(l)}(\mathcal{A})]_k . \tag{8.4.17}$$

The machine based lower bound, LBM, is given by

$$LBM[S^{(l)}(\mathcal{A})] = \begin{cases} ACT[S^{(l)}(\mathcal{A})] + \min_{i \in \mathcal{J} - \mathcal{A}} \left\{ \sum_{l'=l+1}^{m} p_{l'i} \right\} \\ \quad \text{if } ACT[S^{(l)}(\mathcal{A})] \geq MCT[S^{(l)}(\mathcal{A})] , \\ MCT[S^{(l)}(\mathcal{A})] + \min_{i \in \mathcal{A}} \left\{ \sum_{l'=l+1}^{m} p_{l'i} \right\} \\ \quad \text{otherwise} \end{cases} \tag{8.4.18}$$

The rationale behind separating the two cases stems from the following observation. If $ACT[S^{(l)}(\mathcal{A})] \geq MCT[S^{(l)}(\mathcal{A})]$ then the last job finished at stage l will be a job in $\mathcal{J} - \mathcal{A}$. If $ACT[S^{(l)}(\mathcal{A})] < MCT[S^{(l)}(\mathcal{A})]$ then the last job scheduled at stage l may come from $\mathcal{A}$ or from $\mathcal{J} - \mathcal{A}$.

The job based lower bound, LBJ, is defined by

$$LBJ[S^{(l)}(\mathcal{A})] = \min_{1 \le k \le k_l} \{ C[S^{(l)}(\mathcal{A})]_k \} + \max_{i \in \mathcal{J}-\mathcal{A}} \Big\{ \sum_{l'=l}^{m} p_{l'i} \Big\}. \tag{8.4.19}$$

Finally, the composite lower bound, LBC, is given by

$$LBC[S^{(l)}(\mathcal{A})] = \max\{ LBM[S^{(l)}(\mathcal{A})], LBJ[S^{(l)}(\mathcal{A})] \}. \tag{8.4.20}$$

The LBM bound (8.4.18) is improved in Portmann et al. [PVDD98]. Namely, if $ACT[S^{(l)}(\mathcal{A})] = MCT[S^{(l)}(\mathcal{A})]$ and $\mathcal{J}-\mathcal{A} \ne \varnothing$ then it may happen that

$$\min_{i \in \mathcal{A}} \Big\{ \sum_{l'=l+1}^{m} p_{l'i} \Big\} > \min_{i \in \mathcal{J}-\mathcal{A}} \Big\{ \sum_{l'=l+1}^{m} p_{l'i} \Big\} \tag{8.4.21}$$

holds, for the processing times of the jobs in $\mathcal{A}$ and in $\mathcal{J}-\mathcal{A}$ are unrelated. In this case LBM can be improved by the difference of the left and right hand sides of (8.4.21). That is, if $\mathcal{J}-\mathcal{A} \ne \varnothing$, the improved lower bound becomes

$$LBM[S^{(l)}(\mathcal{A})] = \begin{cases} ACT[S^{(l)}(\mathcal{A})] + \min_{i \in \mathcal{J}-\mathcal{A}} \Big\{ \sum_{l'=l+1}^{m} p_{l'i} \Big\} \\ \quad \text{if } ACT[S^{(l)}(\mathcal{A})] > MCT[S^{(l)}(\mathcal{A})], \\[4pt] MCT[S^{(l)}(\mathcal{A})] + \min_{i \in \mathcal{A}} \Big\{ \sum_{l'=l+1}^{m} p_{l'i} \Big\} \\ \quad \text{if } ACT[S^{(l)}(\mathcal{A})] < MCT[S^{(l)}(\mathcal{A})], \\[4pt] ACT[S^{(l)}(\mathcal{A})] + \\ \quad \max\Big\{ \min_{i \in \mathcal{J}-\mathcal{A}} \Big\{ \sum_{l'=l+1}^{m} p_{l'i} \Big\}, \min_{i \in \mathcal{A}} \Big\{ \sum_{l'=l+1}^{m} p_{l'i} \Big\} \Big\} \\ \quad \text{if } ACT[S^{(l)}(\mathcal{A})] = MCT[S^{(l)}(\mathcal{A})]. \end{cases} \tag{8.4.22}$$

III) *Bounds with heads and tails.* The set of bounds in this category share the property that they can be computed for any stage l and it is not assumed that all jobs are completely scheduled on all upstream stages. This is in contrast with bounds (8.4.18), (8.4.19) and (8.4.22) that heavily rely on this assumption. Lower bounds based on heads and tails can easily be updated whenever a scheduling decision has been made either through branching in a branch and bound procedure or through propagation of constraints. While the basic idea of the bounds (8.4.18), (8.4.19) and (8.4.22) is calculation of average processing times or average machine in process times, the main idea of the subsequent bounds is the calculation and subsequent reduction of the domains of start times of the jobs at each stage, i.e. the interval limited by the earliest and latest possible start and completion times of the jobs, see [VHHL05].

To simplify notation, we fix stage l. Assume that a partial schedule S already exists (maybe S is empty). We define a set $\mathcal{B}$ of $\bar{n}$ tasks with processing times $\tilde{p}_j = p_{lj}$ for each $T_j \in \mathcal{B}$, noting that j refers also to job J_j of the multiprocessor flowshop scheduling problem. In addition to that, a ready time $r_j = r_j^{(l)}$ and a delivery time $q_j = q_j^{(l)}$ are defined for each task T_j, where $r_j^{(l)}$ and $q_j^{(l)}$ are determined with respect to the graph representation of the partial solution S.

The problem of scheduling n tasks on $\bar{m} = k_l$ identical parallel machines subject to release dates and delivery times to minimize the makespan, $P\bar{m} \mid r_j, q_j \mid C_{max}$, is a relaxation of the multiprocessor flowshop scheduling problem, as it is pointed out by Carlier and Pinson [CP98]. Since this problem is NP-hard in the strong sense, as the one machine special case $1 \mid r_j, q_j \mid C_{max}$ already is [GJ77], various lower bounds are proposed in Carlier [Car87]. All of these lower bounds are lower bounds for the multiprocessor flowshop problem as well [CN00].

The most basic lower bound for the $P\bar{m} \mid r_j, q_j \mid C_{max}$ problem is

$$LB_1 = \max_{T_i \in \mathcal{B}} \{r_i + p_i + q_i\} \qquad (8.4.23)$$

Now consider a subset $\mathcal{B}'$ of $\mathcal{B}$ and define the quantity

$$G(J) = \min_{T_j \in \mathcal{B}'} \{r_j\} + \frac{1}{\bar{m}} \Big(\sum_{T_j \in \mathcal{B}'} p_j \Big) + \min_{T_j \in \mathcal{B}'} \{q_j\}$$

Clearly, $G(\mathcal{B}')$ is a lower bound for the $P\bar{m} \mid r_j, q_j \mid C_{max}$ problem with respect to any $\mathcal{B}' \subseteq \mathcal{B}$. Consequently, taking the maximum over all subsets $\mathcal{B}'$ of $\mathcal{B}$ we obtain another lower bound:

$$LB_2 = \max_{T_j \in \mathcal{B}'} \{G(\mathcal{B}')\}, \qquad (8.4.24)$$

which can be computed in $O(\bar{n} \cdot \log \bar{n})$ time generating Jackson's preemptive schedule for the one-machine scheduling problem with heads $\bar{m}r_j$, processing times p_j and tails $\bar{m}q_j$, see Carlier [Car87]. The optimal value of a preemptive solution of the one machine problem is $\bar{m}LB_2$.

The next lower bound tries to take into account the heads and tails of different operations in a more efficient way. Namely, let $\mathcal{B}'$ be a subset of $\mathcal{B}$ with $|\mathcal{B}'| \geq \bar{m}$. Denote $r_{i_1}, \cdots, r_{i_{\bar{m}}}$ and $q_{j_1}, \cdots, q_{j_{\bar{m}}}$ the $\bar{m}$ smallest release times and delivery times, respectively, of jobs in $\mathcal{B}'$. Define the quantity $G'(\mathcal{B}')$ by

$$G'(\mathcal{B}') = \frac{1}{\bar{m}} \Big(\sum_{u=1}^{\bar{m}} r_{i_u} + \sum_{T_j \in \mathcal{B}'} p_j + \sum_{u=1}^{\bar{m}} q_{i_u} \Big) \qquad (8.4.25)$$

It is shown in [Car87] that LB_3, as defined by (8.4.26) below, is a lower bound for the $Pm \mid r_j, q_j \mid C_{max}$ problem.

$$LB_3 = \max_{\mathcal{B}' \subseteq \mathcal{B}, |\mathcal{B}'| \geq \bar{m}} \{G'(\mathcal{B}')\} \qquad (8.4.26)$$

In order to show this, we may assume that each machine is used from jobs of $\mathcal{B}'$.

A (every) machine is idle from time 0 to time r_{i_1}, a second machine is idle from time 0 to time r_{i_2} and the machine $\overline{m}$ is idle from 0 to $r_{i_{\overline{m}}}$. Similarly, one machine is idle after processing for q_{j_1} time units, a second machine is idle for q_{j_2} time units, etc. Adding processing and idle times for all machines it is obvious that (8.4.25) is a lower bound for any subset $\mathcal{B}'$ of $\mathcal{B}$.

Bound (8.4.26) can straightforwardly be computed in $O(\overline{n}^3)$ time [Van94], [Per95]. However, it is shown in [CP98] that the stronger lower bound

$$LB_4 = \max\{LB_1, LB_3\} \tag{8.4.27}$$

can be computed in $O(\overline{n} \cdot \log \overline{n} + \overline{n}m \cdot \log \overline{m})$ time using Jackson's Pseudo Preemptive schedule. In such a schedule, an operation may be processed on more than one machine at a time. Moreover, it can be shown that the distance between the non-preemptive optimal makespan and LB_4 is at most $2p_{max}$ [Car87].

For the sake of completeness we mention that when schedule S is empty then $r_i^{(l)} = \sum_{l'=1}^{l-1} p_{l'i}$ and symmetrically $q_i^{(l)} = \sum_{l'=l+1}^{L} p_{l'i}$ hold for each job J_i. For this special case Santos et al. [SHD95] has proven that $G'(\mathcal{B}')$ (cf. equation (8.4.25)) is a lower bound when $\mathcal{B}'$ consists of all jobs. Computational results show that, on average, the lower bound is within 8% of the optimum.

An even stronger lower bound can be obtained by solving the preemptive version of the $P\overline{m} \mid r_j, q_j \mid C_{max}$ problem using a network flow model. Fix a makespan C and define deadlines $d_j = C - q_j$ for each job J_j. Job J_j must be processed in the interval $[r_j, d_j]$ in order to complete all jobs by time C. There are at most $h \le 2n$ different r_j and d_j values and let $v_1, \ldots, v_h$ represent these values arranged increasingly, i.e., $v_1 < v_2 < \ldots < v_h$. Let $I_t = [v_t, v_{t+1})$, $t = 1, \cdots, h-1$, represent $h-1$ intervals with lengths $l_1, \cdots, l_{h-1}$. If $I_t \subseteq [r_j, d_j]$ then a part $\min\{l_t, p_j\}$ of job J_j can be processed in interval I_t. Hence we form a capacitated network with $\overline{n} + (h-1) + 2$ nodes, having one source node s, one sink node r, $\overline{n}$ nodes for representing the jobs and $h-1$ nodes for representing the intervals. Source s is connected to each job node j with an arc of capacity p_j. Each job node j is connected to each interval I_t with $I_t \subseteq [r_j, d_j]$ using an arc of capacity $\min\{l_t, p_j\}$, and finally each interval I_t is connected to the sink by an arc of capacity $\overline{m}l_t$. In this network there is a flow of value $\sum_j p_j$ if and only if the preemptive $P\overline{m} \mid r_j, q_j, pmtn \mid C_{max}$ problem has a solution with makespan C. Using dichotomic search, the smallest C admitting a compatible flow of value $\sum_j p_j$ can be found in polynomial time. It is shown in Hoogeveen et al. [HHLV95] that the difference between the preemptive makespan and LB_4 is not more than $\overline{m}/(\overline{m}-1)p_{max}$. Nonetheless, this gap is claimed to vanish in practice [CP98]. The drawback of this method is the relatively high computation time for finding the maximum flow.

We close this section by a rather tricky lower bound of Carlier and Néron.

Let $R_1, \cdots, R_{\bar{m}}$ denote the $\bar{m}$ smallest increasingly ordered machine availability times at stage l, noting that they depend on the partial schedule S. Let

$$G'_{machine}(\mathcal{B}') = \frac{1}{m}(\max(R_1, r_{i_1}) + \ldots + \max(R_{\bar{m}}, r_{i_{\bar{m}}}) + \sum_{J_j \in \mathcal{B}'} p_j + q_{i_1} + \ldots + q_{i_m}).$$

(8.4.28)

Now, if $R_m + q_{j_m} < UB$, then $G'_{machine}(\mathcal{B}')$ is a lower bound on UB.

Let us briefly sketch the main ideas of the proof which can be found in [CN00]. Let S be a schedule with a makespan of at most UB. If there exists a machine P_h different from $P_{\bar{m}}$ without any job from $\mathcal{B}'$ to process then the jobs from $\mathcal{B}'$ scheduled on machine $P_{\bar{m}}$ can be scheduled on machine P_h. From $R_l \leq R_{\bar{m}}$ we know this will not increase the makespan of S. If no job from $\mathcal{B}'$ is processed on machine $P_{\bar{m}}$ consider the difference $\delta = UB - R_{\bar{m}} - q_{i_{\bar{m}}} > 0$. A part δ of job $J_{i_{\bar{m}}}$ (with release date $r_{i_{\bar{m}}}$ and tail $q_{i_{\bar{m}}}$) which is scheduled on some machine can be scheduled in the interval $[UB - q_{i_{\bar{m}}} - \delta, UB - q_{i_{\bar{m}}}]$ on machine $P_{\bar{m}}$ without increasing the makespan. Thus, we can conclude, if there is a schedule with a makespan of at most UB then there is also a (preemptive) schedule with a makespan of at most UB in which all machines have to process at least a part of a job from $\mathcal{B}'$.

What remains is to sum up idle times and processing times of all machines with respect to the (preemptive) schedule. The first machine is idle from 0 to R_1 but also from 0 to r_{i_1}. Therefore it is idle from 0 to $\max\{R_1, r_{i_1}\}$. There is also a machine idle from time $UB - q_{i_1}$ until time UB. Similar conclusions for the remaining machines yield the desired result.

Branch-and-Bound Methods

We have introduced several lower bounds in the previous section. Below we discuss branching schemes and search strategies.

The first branch-and-bound procedure for the $Fm \mid k_1, \cdots, k_m \mid C_{max}$ problem is proposed in Brah and Hunsucker [BH91].

This procedure is a modification of the method developed by Bratley et al. [BFR75] for scheduling on parallel machines. At each stage l two decisions must be made: the assignment of the jobs to a machine P_{li}, and the scheduling of jobs on every machine at stage l. The enumeration is accomplished by generating a tree with two types of nodes: node (j) denotes that job J_j is scheduled on the current machine, whereas node $\boxed{j}$ denotes that J_j is scheduled on a new machine, which now becomes the current machine. The number of $\boxed{}$ nodes on each branch is equal to the number of parallel machines used by that branch, and thus must be less than or equal to k_l at stage l. The number of possible branches at each stage l was established by Brah in [Bra88] as

$$N(n, k_l) = \binom{n-1}{k_l-1} \frac{n!}{k_l!}.$$

Consequently, the total number of possible end nodes is equal to

$$S(n, m, \{k_l\}_{l=1}^{m}) = \prod_{l=1}^{m} \binom{n-1}{k_l-1} \frac{n!}{k_l!}.$$

For the construction of a tree for the problem, some definitions and rules at each stage l are useful. Let the level 0_l represent the root node at stage l, and 1_l, $2_l, \cdots, z_l$ represent different levels of the stage, with z_l being the terminal level of this stage. Of course, the total number of levels is nm. The necessary rules for the procedure generating the branching tree are the following.

Rule 1 Level 0_i contains only the dummy root node of stage l, $l = 1, 2, \cdots, m$ (each l is starting of a new stage).

Rule 2 Level 1_l contains the nodes $\boxed{1}, \boxed{2}, \cdots, \boxed{x}$, where $x = n - k_l + 1$ (any number larger than x would violate Rules 5 and 7).

Rule 3 A path from level 0_l to level j_l, $i = 1, 2, \cdots, m$, $j = 1, 2, \cdots, n$, may be extended to the level $(j+1)_l$ by any of the nodes $\boxed{1}, \boxed{2}, \cdots, \boxed{n}, \textcircled{1}, \textcircled{2}, \cdots, \textcircled{n}$ provided the rules 4 to 7 are observed (all unscheduled jobs at stage l are candidates for $\boxed{}$ and $\bigcirc$ nodes as long as they do not violate Rules 4 to 7).

Rule 4 If $\boxed{a}$ or $\textcircled{a}$ has previously appeared as a node at level j_l, then a may not be used to extend the path at that level (this assures that no job is scheduled twice at one stage).

Rule 5 $\boxed{a}$ may not be used to extend a path at level j_l, which already contains some node $\boxed{r}$ with $r > a$ (this is to avoid duplicate generation of sequences in the tree).

Rule 6 No path may be extended in such a way that it contains more than $k_l \boxed{}$ nodes at each stage l (this guarantees that no more than k_l machines are used at stage l).

Rule 7 No path may terminate in such a way that it contains less than $k_l \boxed{}$ nodes at each stage l unless the number of jobs is less than k_l (there is no advantage in keeping a machine idle if the processing cost is the same for all of the machines).

A sample tree representation of a problem with 4 jobs and 2 parallel machines is given in Figure 8.4.2. All of the end nodes can serve as a starting point for the next stage 0_{l+1} ($l < m$). All of the nodes at a subsequent stage may not be candidates due to their higher value of lower bounds, and thus not all of the nodes need to be explored. It may also be observed that all of the jobs at stage l will not be readily available at the next stage, and thus inserted idle time will increase their lower bounds and possibly remove them from further considerations. This will help to reduce the span of the tree. The number of search nodes could be further reduced, if the interest is in the subclass of active schedules called *non-delay* schedules. These are schedules in which no machine is kept idle when it could start processing some task.

The use of these schedules does not necessarily provide an optimal schedule, but the decrease in the number of the nodes searched gives a strong empirical motivation to do that, especially for large problems [Fre82].

Finally we describe the idea of the branch and bound algorithm for the problem. It uses a variation of the depth-first least lower bound search strategy, and is as follows.

LEVEL

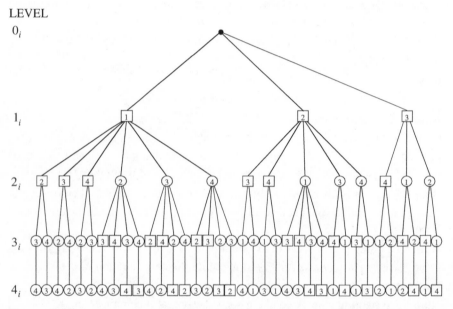

Figure 8.4.2   *Tree representation of four jobs on two parallel machines.*

Step 1   Generate $n - k_1 + 1$ ☐ nodes at stage 1 and compute their lower bounds. Encode the information about the nodes and add them to the list of unprocessed nodes. Initialize counters (number of iterations, time) defining end of computation.

Step 2   Remove a node from the list of unprocessed nodes with the priority given to the deepest node in the tree with the least lower bound. Break ties arbitrarily.

Step 3 Procure all information about the retrieved node. If this is one of the end nodes of the tree go to Step 5, while if this is the last node in the list of unprocessed nodes then go to Step 6.

Step 4 Generate branches from the retrieved node and compute their lower bounds. Discard the nodes with lower bounds larger than the current upper bound. Add the remaining nodes to the list of unprocessed nodes and go to Step 2.

Step 5 Save the current complete schedule, as the best solution. If this is the last branch of the tree, or if the limit on the number of iterations or computation time has reached, then pass to the next step, otherwise go to Step 2.

Step 6 Print the results and stop.

As we see, the algorithm consists of three major parts: the branching tree generation, the lower bound computing, and the list processing part. The first two parts are based on the concepts described earlier with some modifications utilizing specific features of the problem. For the list processing part, the information is first coded for each branching node. If the lower bound is better than the best available C_{max} value of a complete solution (i.e. the current upper bound), provided it is available at the moment, the node is stored in the list of unprocessed nodes. The information stored for each branching node is the following:

$$KODE = NPR \times 1\,000\,000 + NPS \times 10\,000 + LSN \times 100 + JOB$$

$$LBND = NS \times 10\,000\,000 + NSCH \times 100\,000 + LB$$

where *NPR* is the machine number in use, *NPS* is the sequence number of this machine, *LSN* is number of the last ☐ nodes, *JOB* is the index of the job, *NS* is the index of the stage, *NSCH* is the number in the processing sequence, and *LB* is the lower bound of the node.

The stage and the level numbers are coded in the opposite manner to their position in the tree (the deepest node has the least value). Thus, the deepest node is stored on top of the list and can be retrieved first. If two or more nodes are at the same stage and level, the one with the least lower bound is retrieved first and processed. Once a node is retrieved, the corresponding information is decoded and compared with the last processed node data. If the node has gone down a step in the tree, the necessary information, like sequence position and completion time of the job on the retrieved node, is established and recorded. However, if the retrieved node is at a higher or the same level as the previous node, the working sequence and completion time matrix of the nodes lower than the present level and up to the level of the last node are re-initialized. The lower bound is then compared with the best known one, assuming it is available, and is either eliminated or branched on except when this is the last node in the tree. The qualifying nodes are stored in the list of unprocessed nodes according to the priority rule described in Step 2 of the algorithm. However, in case this is the last node in the tree, and it satisfies the lower bound comparison test, the working sequence position and job completion time matrix along with the completion time of the schedule is saved as the best known solution.

Of course, the algorithm described above is only a basic framework for further improvements and generalizations. For example, in order to improve the computation speed for large problems some elimination criteria, like the ones developed in [Bra88] can be used together with the lower bounds. The lower bound in Step 1 and Step 4 of the algorithm are computed according to (8.4.20), using (8.4.18) and (8.4.19). The algorithm could also be applied for schedule performance measures other than the schedule length, if corresponding lower bounds would be elaborated. Moreover, the idea of the algorithm can be used in a heuristic way, e.g. by setting up a counter of the number of nodes to be fully explored or by defining a percentage improvement index on each new feasible solution.

The algorithm of Portmann et al. [PVDD98] extends that of Brah and Hunsucker in several ways. First, it uses the improved machine based lower bound (8.4.22) instead of (8.4.18) when computing (8.4.20). Moreover, it computes an upper bound before starting to schedule the jobs at a new stage l. The upper bound is computed by a genetic algorithm (GA) that determines a schedule of all jobs at stages l through m. The schedule of the jobs at the first $l-1$ stages is fixed and is given by the path from the root of the branching tree to the root node of stage l. For details of GA we refer the reader to [RC92].

The results of a detailed computational study show that the method of Portmann et al. is able to solve problems to optimality with up to five stages and ten or fifteen jobs. However, it seems that the method is very sensitive to the pattern of the number of parallel machines at the stages. Another conclusion is that the algorithm proves the optimality of solutions, within a given time limit, more frequently when GA is used.

A Method Based on Constraint Propagation

The method of Carlier and Néron [CN00] is significantly different from that of Brah and Hunsucker and of Portmann et al. The novelty of the approach consists in working on all m parallel machine problems at the same time. Namely, instead of solving the parallel machine problem completely at a stage, like in the branch-and-bound algorithm of Brah and Hunsucker, the method selects a stage and the next job to be processed at that stage. Having scheduled the selected job, heads and tails are adjusted and the method proceeds with selecting a new stage.

First we discuss how to select the job to be scheduled next at some stage with respect to a fixed upper bound UB. To simplify notation fix a stage l and consider the $\bar{m}$ machine problem with $\bar{m} = k_l$. We identify the processing of job J_i at stage l with task T_i. The processing time of task T_i is $\tilde{p}_i = p_{li}$ and its starting time (to be determined) will be t_i. Let $\mathcal{B}$ denote the set of all $\bar{m}$ tasks. A central notion is that of selection. A *selection A* for an $\bar{m}$-machine problem is an ordered list of tasks $\{T_{i_1}, T_{i_2}, \cdots, T_{i_{h-1}}, T_{i_h}\}$ such that: if T_i precedes T_j in A then $t_i < t_j$, or $t_i = t_j$ and $i < j$. A selection is complete if $\mathcal{B}$ is totally ordered. To complete the definition note that in a selection more than one task can be processed at the same time, but the total number of tasks processed simultaneously cannot exceed m.

Carlier [Car84] has proposed a simple list scheduling algorithm, the *Strict* algorithm, to schedule tasks with respect to a selection at their earliest possible date, the result is called *strict schedule*. It is shown that strict schedules dominate all other schedules. Consequently, it is enough to work with strict schedules.

Let us fix an upper bound *UB* for the *m*-machine problem. A task $T_i \in \mathcal{B}$ is an *input* (*output*) of the $\bar{m}$-machine problem if and only if there exists a schedule $S = \{ t_j \mid T_j \in \mathcal{B} \}$ with makespan at most *UB* and verifying $t_j \geq t_i$ (respectively $t_j + \tilde{p}_j \leq t_i + \tilde{p}_i$) for all $T_j \in \mathcal{B} - \{T_i\}$. Inputs and outputs will be selected by computing lower bounds after fixing a task T_i to be scheduled before or after all other unscheduled tasks. However, lower bounds may not detect that no schedule of the remaining tasks with makespan at most *UB* exists.

For solving the makespan minimization problem, Carlier and Néron solve the decision version of the problem and apply a dichotomic search to find the smallest UB for which a solution exists.

The decision problem is solved by branch-and-bound in which branching consists of fixing a task as input (or output) of a stage. More concretely, the branch-and-bound method proceeds as follows:

Step 1. Determine the most critical (machine) center, which is the set of parallel machines on some stage that will most likely create a bottleneck when scheduling all jobs (see below). Decide if the selection is built according to inputs or outputs. If selection based on outputs is chosen then reverse the problem.

Step 2. If *bestsolution* $\leq$ *UB* then answer YES and stop. Otherwise, if all nodes are explored then answer NO and stop. Otherwise proceed with Step 3.

Step 3. Choose the node *N* in the branch-and-bound tree to be explored. If the current center, i.e. the parallel machine problem under consideration in node *N*, is completely selected then proceed with Step 4, otherwise proceed with Step 6.

Step 4. If all centers are completely selected in *N* and *solution* $\leq$ *UB* then answer YES and stop. Otherwise, if there exists a center in *N* not completely selected then choose the most critical center among the not completely selected ones as the current center of *N* and proceed with Step 5. In all other cases proceed with Step 6.

Step 5. Determine a solution for *N*. If the makespan of the solution found is not greater than *UB* then answer YES and stop.

Step 6. Compute lower bounds with respect to the current center of *N*. If *lowerbounds* > *UB* then discard node *N* and go to Step 2. Otherwise proceed with Step 7.

Step 7. Apply local enumerations to N and proceed with Step 8.

Step 8. Determine the list of feasible inputs for the current center of N. For each feasible input i create a new node by adding i to the partial selection of the current center of N and adjust heads and tails. Go to Step 2.

Below we provide some details of this algorithm:

- *The most critical center*: a lower bound is computed for each $\bar{m} = k_l$-machine problem. The $\bar{m}$-machine problem with the largest lower bound defines the most critical center which will be selected first.
- *The current center*: the center where the selection is built and it is always the most critical center.
- The search tree is visited in a depth-first manner such that, among the children of a node, the child with the smallest release date of its input is chosen for exploration.
- *Solutions* are generated during the exploration of the tree using the Strict list scheduling algorithm. The (ordered) list of operations for each center is determined by either a complete selection, if available, or by sorting the operations in decreasing tail order (steps 4 and 5).
- Lower bounds are computed using eq. (8.4.25) and also eq. (8.4.28).
- Local enumerations at Step 7 refer to two things. On the one hand, unscheduled operations are selected in all possible ways while respecting UB in order to improve their heads and tails. On the other hand, a restricted multiprocessor flowshop problem is solved during the construction of the selection of the most critical center.
- The selection of inputs at Step 8 consists in finding jobs that can be scheduled next (before all other unscheduled jobs) without augmenting a lower bound beyond UB.

The adjustments of heads and tails start from the current center and are propagated through the other centers. The efficiency of the head and tail based lower bounds heavily depends on this propagation phase. Moreover it influences the number of feasible inputs and therefore the size of the branching tree.

Assume task T_e has been detected as a possible input in the current machine center. There might be a partial selection within this center which is not yet complete. Let $\tilde{\mathcal{B}}$ be the set of unselected tasks. If T_e is an input all other tasks cannot start before the release time r_e of e, i.e. $r_i := \max\{r_i, r_e\}$ for all tasks T_i of $\tilde{\mathcal{B}}$. For all machines P_k of the current center the machine availability time can be updated to $R_k := \max\{R_k, r_e\}$. The adjustment of the machine availability times again might cause an increase of the release dates of all tasks from $\tilde{\mathcal{B}}$, i.e. $r_i := \max\{r_i, R_1\}$, because a task cannot start before a machine becomes available.

In the domain of possible start times for task e on any machine the latest possible start time is limited by

$$\max\left\{ UB - p_e - q_e \, , \, UB - \frac{1}{m} \cdot \left(\sum_{T_i \in \widetilde{\mathcal{B}}} p_i + p_e + q_{j_1} + \ldots + q_{j_{\overline{m}}} \right) \right\}$$

which implies the updating of the tail

$$q_e := \max\left\{ q_e \, , \, \frac{1}{m} \cdot \left(\sum_{T_i \in \widetilde{\mathcal{B}}} p_i + p_e + q_{j_1} + \ldots + q_{j_{\overline{m}}} \right) - p_e \right\}.$$

The complexity of this adjustment is at most $O(\overline{m}n)$.

Consider a partial selection and set $\widetilde{\mathcal{B}}$ of unselected tasks at a node in the branch-and-bound tree. Let $G'(\widetilde{\mathcal{B}})$ be the lower bound (8.4.25) and UB the current upper bound. Let δ be the smallest non-negative integer such that

$$\frac{1}{m} \cdot \left(\sum_{u=1}^{\overline{m}} r_{i_u} + \delta + \sum_{T_j \in \widetilde{\mathcal{B}}} p_j + \sum_{u=1}^{\overline{m}} q_{i_u} \right) > UB \quad \text{and} \quad r_{i_{\overline{m}+1}} - r_{i_1} \geq \delta.$$

Thus, we can conclude that $t_{i_1} < r_{i_1} + \delta$ otherwise the aforementioned new value $G'(\widetilde{\mathcal{B}})$, where the release date of T_{i_1} has been increased by δ, will be strictly greater than UB. As $t_{i_1} + p_{i_1} + q_{i_1} \leq C_{i_1} + q_{i_1} \leq UB$ we can set $q_{i_1} := \max\{q_{i_1}, UB - (r_{i_1} + \delta + p_{i_1}) + 1\}$. If the new q_{i_1} is greater than the previous one, the same deduction can be applied to q_{i_2}. Similarly, adjustments can be derived for the release dates leading to the following updates $r_{i_1} := \max\{r_{i_1}, UB - (q_{i_1} + \delta' + p_{i_1}) + 1)$.

The modification of the release dates of the tasks of the current center are propagated to the subsequent machine centers and the new tails are propagated to the previous machine centers.

For more details see [CN00].

As far as the benefits of this method are concerned, most of the problems reported hard by Vignier [Vig97] are very easy to solve by constraint propagation. Those that are not solved immediately, are hard for the new method as well. The method seems to perform well on problem instances in which there is a "bottleneck" center having one machine only.

8.4.5 The Mean Flow Time Problem

We are aware of only very few results on solving multiprocessor flowshop with respect to the mean flow time objective. The general problem has been studied by Azizoglu et al. [ACK01] and a special case where an optimal permutation schedule is sought has been studied by Rajendran and Chaudhuri [RC92]. We commence with a lower bound for the optimal permutation schedule problem and continue with that for the general case. Then a branch-and-bound method for each of the two problems will be presented.

Lower Bounds

Permutation flowshops

Before presenting the lower bound proposed by Rajendran and Chaudhuri for the permutation flowshop problem we introduce additional notation.

σ = a permutation of jobs (indices of the jobs) that defines an available partial schedule,

n' = the number of scheduled jobs in σ,

$\mathcal{U}$ = the set of unscheduled jobs,

$R_k^{(l)}(\sigma)$ = the release time of machine P_k at stage l w.r.t. σ,

$C(\sigma j, l)$ = the completion time of an unscheduled job $J_j \in \mathcal{U}$ at stage l when appended to σ,

$F(\sigma)$ = the total flow-time of jobs in σ,

$LBC_j^{(l)}(\sigma)$ = the lower bound on the completion time of job J_j at stage l,

$LB(\sigma)$ = the lower bound on the total flow-time of all schedules beginning with partial schedule σ.

In a *permutation schedule* the completion times of the jobs at the stages are determined w.r.t. a permutation σ of jobs. The completion times of the jobs in σ are determined iteratively by using the processing times and machine assignments. Namely, assuming that $\sigma = \sigma' j$ and that job J_j is assigned to machine $P_{k(j,l)}$ at stage l, the completion time of job J_j at the first stage is

$$C^{(1)}(\sigma' j) = R_{k(j,1)}^{(1)} + p_{1j}.$$

The completion time at each stage $l = 2, \cdots, m$ (in this order) is determined by

$$C^{(l)}(\sigma' j) = \max\{ C^{(l-1)}(\sigma' j), R_{k(j,l)}^{(l)} \} + p_{lj}.$$

Finally, the release times of the machines at the stages $l = 2, \cdots, m$ are given by

$$R_k^{(l)}(\sigma' j) = \begin{cases} C^{(l)}(\sigma' j) & \text{if } k = k(j,l) \\ R_k^{(l)}(\sigma) & \text{otherwise} \end{cases}.$$

Now we turn to the lower bound. To this end we need other expressions that are defined next. The earliest time when an unscheduled job in $\mathcal{U}$ becomes available at stage l can be computed as follows:

$$s^{(l)} = \max \left\{ \begin{array}{l} \min\{\,R_k^{(1)}(\sigma) \mid 1 \le k \le k_1\,\} + \min_{J_j \in \,\mathcal{U}} \left\{\sum_{q=1}^{l-1} p_{qj}\right\}, \\[2ex] \min\{\,R_k^{(2)}(\sigma) \mid 1 \le k \le k_2\,\} + \min_{J_j \in \,\mathcal{U}} \left\{\sum_{q=1}^{l-1} p_{qj}\right\}, \\[1ex] \cdots \\[1ex] \min\{\,R_k^{(l-1)}(\sigma) \mid 1 \le k \le k_{l-1}\,\} + \min_{J_j \in \,\mathcal{U}} \left\{p_{l-1\,j}\right\} \end{array} \right\} \qquad (8.4.30)$$

Therefore, the earliest starting time of an unscheduled job is given by

$$\max\left\{\,\min\{\,R_k^{(l)}(\sigma) \mid 1 \le k \le k_l\,\},\, s^{(l)}\,\right\}$$

Let $R_r^{(l)}$ denote $\min\{\,R_k^{(l)}(\sigma) \mid 1 \le k \le k_l\,\}$. With this notation the lower bound on the completion time of job J_{j_1}, where $J_{j_1} \in \mathcal{U}$, at stage l is given by

$$LBC_{j_1}^{(l)}(\sigma) = \max\left\{R_r^{(l)}, s^{(l)}\right\} + p_{l j_1} + \sum_{q=l+1}^{m} p_{q j_1}. \qquad (8.4.31)$$

We place tentatively J_{j_1} on machine P_r and update the machine's release time as

$$R_r^{(l)} = \max\left\{R_r^{(l)}, s^{(l)}\right\} + p_{l j_1}. \qquad (8.4.32)$$

Now, updating $R_r^{(l)}$ is correct only if J_{j_1} is an unscheduled job with smallest processing time. Let $j_1, j_2, \cdots, j_{n-n'}$ be a permutation of the indices of all unscheduled jobs in $\mathcal{U}$ satisfying $p_{j_1}^{(l)} \le p_{j_2}^{(l)} \le \ldots \le p_{j_{n-n'}}^{(l)}$, we compute $LBC_{j_t}^{(l)}(\sigma)$ for $t = 1, \ldots, n-n'$, in this order. Then we obtain the lower bound

$$LB^{(l)}(\sigma) = F(\sigma) + \sum_{J_j \in \,\mathcal{U}} LBC_j^{(l)}(\sigma),$$

at stage l on the total flow time of all permutation schedules beginning with partial schedule σ.

Finally, a lower bound on the total flow time of any schedule beginning with σ is obtained by computing $LB^{(l)}(\sigma)$ for all stages $1 \le l \le m$ and taking the maximum:

$$LB(\sigma) = \max_{1 \le l \le m} LB^{(l)}(\sigma). \qquad (8.4.33)$$

The general case

When schedules are not restricted to permutation schedules Azizoglu et al. [ACK01] propose two other lower bounds obtained by solving two different relaxations of the following parallel machine problem. Let $\Pi^{(l)}$ be the total flow time problem on k_l identical parallel machines and n tasks with processing times

$p_{l1}, \ldots, p_{ln}$ and ready times $r_1^{(l-1)}, \ldots, r_n^{(l-1)}$. If $F^{(l)}$ is the optimal flow time of problem $\Pi^{(l)}$ then $F^{(l)}$ is a lower bound on the optimal solution to the L-stage flowshop problem. However, $\Pi^{(l)}$ is NP-hard as is its single machine special case. Hence, we compute lower bounds on $F^{(l)}$.

The first lower bound, LB_1, is the optimum of a relaxation of $\Pi^{(l)}$ when all job ready times are set to $\min_j \{r_j^{(l-1)}\}$. This problem can be solved to optimality in polynomial time by the SPT rule, cf. Blazewicz et al. [BEPSW01].

In the second lower bound, LB_2, job ready times are kept, but instead of solving a parallel machine problem, a single machine problem is considered. More precisely, define n new tasks with processing times p_{lj}/k_l and ready times $r_j^{(l-1)}$, $j = 1, \ldots, n$. Total flow time minimization on a single machine with ready times is NP-hard, Lenstra et al. [LRKB77], however its preemptive version can be solved with the shortest remaining processing time (*SRPT*) rule, Schrage [Sch68]. The preemptive optimum is a lower bound on the non-preemptive single machine problem, therefore on $F^{(l)}$.

In the next section we describe algorithms using the bounds presented in this section.

Branch-and-Bound Procedures

Permutation flowshops

Rajendran and Chaudhuri propose a very simple algorithm for solving the permutation flowshop problem. Let k denote the minimum number of parallel machines over all stages, that is, $k = \min_l \{k_l\}$. The algorithm starts by generating $\binom{n}{k}$ nodes, one for each subset of k jobs out of the set of n jobs. In each of these nodes the k jobs are placed on k distinct machines in every stage, and the partial schedule σ is defined accordingly. Then, the lower bound $LB(\sigma)$ (eq. 8.4.33) is computed for each node and the node with the smallest lower bound is selected for exploration. Exploring a node consists in generating $n-n'$ new nodes, one for each of the $n-n'$ unscheduled jobs. When generating a new node using an unscheduled job J_j, then the operations of job J_j are joined to σ starting with the operation at stage 1 and finishing with the operation at stage m. An operation is always placed on a machine having the smallest release time. After computing a lower bound for each child generated, the procedure proceeds by choosing the next node to branch from. The algorithm stops when a node with $n-1$ scheduled jobs is chosen for exploration. Notice that in this case the lower bound matches the flow time of the schedule obtained by scheduling the only unscheduled job. Consequently, when the algorithm stops the node chosen augmented with the unscheduled job constitutes an optimal solution to the permutation flowshop problem.

As far as the power of the above method is concerned, instances up to 10 jobs, 15 stages and up to 4 machines at each stage are solved while exploring only small search trees of less than 10.000 nodes in a short computation time (less than a minute) on a mainframe computer.

The general case

For the general case, Azizoglu et al. propose a new branching scheme which is different from that of Brah and Hunsucker (described in Section 8.4.4) developed for the makespan minimization problem. In each stage, there are n nodes at the first level of the tree, each node representing the assignment of a particular job to the earliest available machine. A node at the n'^{th} level of the tree corresponds to a partial sequence with n' jobs scheduled. Each node at level n' branches to $(n-n')$ nodes each is representing the assignment of an unscheduled job to the earliest available machine.

The number of possible branches is thus $n!$ at each stage. Therefore the total number of leaves at the mth stage is $(n!)^m$.

In fact, the branching scheme of Azizoglu et al. generates only a subset of nodes generated by that of Brah and Hunsucker. The following example of Azizoglu et al. illustrates the difference between the two branching schemes. Suppose there are four jobs satisfying $p_{11} \leq p_{12} + p_{13}$. At the first stage the branching scheme of Brah and Hunsucker would consider to assign job J_1 to the first machine and jobs J_2, J_3 and J_4 to the second machine in this order. In contrast, the new branching scheme under the assumption on job processing times at the first stage would not process job J_4 on the second machine after jobs J_2 and J_3, for processing job J_4 on the first machine would dominate the former partial schedule.

Another dominance relation between schedules comes from the following observation. If $\max\{R_r^{(l)}, r_i^{(l-1)}\} + p_{li} \leq r_j^{(l-1)}$, where J_i and J_j are distinct jobs not yet scheduled at stage l and $R_r^{(l)}$ is the earliest time point when a machine becomes available at stage l, then processing job J_i next dominates any schedule in which job J_j is processed next. The branch-and-bound tree generated contains only non-dominated nodes. A lower bound is computed for each node not eliminated using either LB_1 or LB_2 (defined in the previous section).

Computational results show that the new branching scheme with LB_1 outperforms the algorithm using the new branching scheme and LB_2 and also the algorithms using the branching scheme of Brah and Hunsucker with either lower bound. The largest problem instances on which the methods were tested consisted of 15 jobs, 2 stages and at most 5 parallel machines at a stage, and 12 jobs, 5 stages and 4 machines at a stage. Moreover, a general observation is that the larger the number of machines at the first stage, the more difficult the problem becomes. The results are in contrast with the permutation schedule case where instances with considerably more stages can easily be solved.

For several other interesting conclusions about the properties of the proposed algorithm and also that of the lower bounds we refer the interested reader to [ACK01].

References

ACK01   M. Azizoglu, E. Cakmak, S. Kondakci, A flexible flowshop problem with total flow time minimization, *European J. Oper. Res.* 132, 2001, 528-538.

Ake56   S. B. Akers, A graphical approach to production scheduling problems, *Oper. Res.* 4, 1956, 244-245.

Bak74   K. R. Baker, *Introduction to Sequencing and Scheduling*, J. Wiley, New York, 1974.

Bak75   K. R. Baker, A comparative study of flow shop algorithms, *Oper. Res.* 23, 1975, 62-73.

Bar81   I. Barany, A vector-sum theorem and its application to improving flow shop guarantees, *Math. Oper. Res.* 6, 1981, 445-452.

BDP96   J. Blazewicz, W. Domschke, E. Pesch, The job shop scheduling problem: Conventional and new solution techniques, *European J. Oper. Res.* 93, 1996, 1-33.

BFR75   P. Bratley, M. Florian, P. Robillard, Scheduling with earliest start and due date constraints on multiple machines, *Naval Res. Logist. Quart.* 22, 1975, 165-173.

BH91   S. A. Brah, J. L. Hunsucker, Branch and bound algorithm for the flow shop with multiple processors, *European J. Oper. Res.* 51, 1991, 88-99.

Bru88   P. Brucker, An efficient algorithm for the job-shop problem with two jobs, *Computing* 40, 1988, 353-359.

Car82   J. Carlier, The one machine sequencing problem, *European J. Oper. Res.* 11, 1982, 42-47.

Car84   J. Carlier, *Problèmes d'ordonnancement à contraintes de ressources: algorithmes et complexité*, Thèse d'État, MASI, 1984.

Car87   J. Carlier, Scheduling jobs with release dates and tails on identical machines to minimize makespan, *European J. Oper. Res.* 29, 1987, 298-306.

CDS70   H. G. Campbell, R. A. Dudek, M. L. Smith, A heuristic algorithm for the n job, m machine sequencing problem, *Management Sci.* 16B, 1970, 630-637.

CMM67   R. W. Conway, W. L. Maxwell, L. W. Miller, *Theory of Scheduling*, Addison Wesley, Reading, Mass., 1967.

CP98   J. Carlier, E. Pinson, Jackson's pseudo preemptive schedule for the $Pm|r_i, q_i|C_{max}$ problem, *Ann. Oper. Res.* 83, 1998, 41-58.

CN00   J. Carlier, E. Néron, An exact method for solving the multi-processor flow-shop, *RAIRO Rech. Oper.* 34, 2000, 1-25.

CS81   Y. Cho, S. Sahni, Preemptive scheduling of independent jobs with release and due times on open, flow and job shops, *Oper. Res.* 29, 1981, 511-522.

Dan77   D. G. Dannenbring, An evaluation of flow shop sequencing heuristics, *Management Sci.* 23, 1977, 1174-1182.

DL93   J. Du, J. Y.-T- Leung, Minimizing mean flow time in two-machine open shops and flow shops, *J. Algorithms* 14, 1993, 24-44.

DPS92   R. A. Dudek, S. S. Panwalkar, M. L. Smith, The lessons of flowshop scheduling research, *Oper. Res.* 40, 1992, 7-13.

GG64   P. C. Gilmore, R. E. Gomory, Sequencing a one-state variable machine: A solvable case of the traveling salesman problem, *Oper. Res.* 12, 1964, 655-679.

GJ77   M. Garey, D. S. Johnson, Two-processor scheduling with start times and deadlines, *SIAM J. Comput.* 6, 1977, 416-426.

GJS76   M. R. Garey, D. S. Johnson, R. Sethi, The complexity of flowshop and jobshop scheduling, *Math. Oper. Res.* 1, 1976, 117-129.

GLL+79   R. L. Graham, E. L. Lawler, J. K. Lenstra, A. H. G. Rinnooy Kan, Optimization and approximation in deterministic sequencing and scheduling theory: a survey, *Ann. Discrete Math.* 5, 1979, 287-326.

GS78   T. Gonzalez, S. Sahni, Flowshop and jobshop schedules: Complexity and approximation, *Oper. Res.* 26, 1978, 36-52.

GSKD96   A. Guinet, M. M. Solomon, P. K. Kedia, A. Dussauchoy, A computational study of heuristics for two-stage flexible flowshops, *Int. J. Prod. Res.* 34, 1996, 1399-1415.

GT91   J. N. D. Gupta, E. Tunc, Schedules for the two-stage hybrid flowshop with parallel machines at the second stage, *Int. J. Prod. Res.* 29, 1991, 1489-1502.

Gup71   J. N. D. Gupta, A functional heuristic algorithm for the flow-shop scheduling problem, *Oper. Res. Quart.* 22, 1971, 39-47.

Gup88   J. N. D. Gupta, Two-stage hybrid flowshop scheduling problem, *J. Oper. Res. Soc.* 39, 1988, 359-364.

HC91   J. C. Ho, Y.-L- Chang, A new heuristic for the *n*-job, M-machine flow-shop problem, *European J. Oper. Res.* 52, 1991, 194-202.

HHLV95   H. Hoogeveen, C. Hurkens, J. K. Lenstra and A. Vandevelde, Lower bounds for the multiprocessor flow shop, In: *Proceedings of the 2nd Workshop on Models and Algorithms for Planning and Scheduling*, Wernigerode, May 22-26, 1995.

HLV96   J. A. Hoogeveen, J. K. Lenstra, B. Veltman, Preemptive scheduling in a two-stage multiprocessor flow shop is NP-hard, *European J. Oper. Res.* 89, 1996, 172-175.

HR88   T. S. Hundal, J. Rajgopal, An extension of Palmer's heuristic for the flow-shop scheduling problem, *Internat. J. Prod. Res.* 26, 1988, 1119-1124.

IS65   E. Ignall, L. E. Schrage, Application of the branch-and-bound technique to some flow-shop scheduling problems, *Oper. Res.* 13, 1965, 400-412.

Joh54   S. M Johnson, Optimal two- and three-stage production schedules with setup times included, *Naval Res. Logist. Quart.* 1, 1954, 61-68.

Joh58   S. M. Johnson, Discussion: Sequencing *n* jobs on two machines with arbitrary time lags, *Management Sci.* 5, 1958, 299-303.

KK88   T. Kawaguchi, S. Kyan, Deterministic scheduling in computer systems: A survey, *J. Oper. Res. Soc. Japan* 31, 1988, 190-217.

KM87   S. Kochbar, R. J. T. Morris, Heuristic methods for flexible flow line scheduling, *J. Manuf. Systems* 6, 1987, 299-314.

KP05   T. Kis and E. Pesch, A review of exact solution methods for the non-preemptive multiprocessor flowshop problem, *European J. Oper. Res.* 164, 2005, 592-608.

Lan87   M. A. Langston, Improved LPT scheduling identical processor systems, *RAIRO Technique et Sci. Inform.* 1, 1982, 69-75.

LLRK78   B. J. Lageweg, J. K. Lenstra, A. H. G. Rinnooy Kan, A general bounding scheme for the permutation flow-shop problem, *Oper. Res.* 26, 1978, 53-67.

LLR+93   E. L. Lawler, J. K. Lenstra, A. H. G. Rinnooy Kan, D. B. Shmoys, Sequencing and scheduling: algorithms and complexity, in: S. C. Graves, A. H. G. Rinnooy Kan, P. H. Zipkin, (eds.), *Handbooks in Operations Research and Management Science*, Vol. 4: Logistics of Production and Inventory, Elsevier, Amsterdam, 1993.

Lom65   Z. A. Lomnicki, A branch-and-bound algorithm for the exact solution of the three-machine scheduling problem, *Oper. Res. Quart.* 16, 1965, 89-100.

LRKB77   J. K. Lenstra, R. H. G. Rinnooy Kan, P. Brucker, Complexity of machine scheduling problems, *Ann. Discrete Math.* 1, 1977, 343-362.

LV94   C. Y. Lee and G. L. Vairaktarakis, Minimizing makespan in hybrid flowshop, *Oper. Res. Letters* 16, 1994, 149-158.

McM69   G. B. McMahon, Optimal production schedules for flow shops, *Canadian Oper. Res. Soc. J.* 7, 1969, 141-151.

Mit58   L. G. Mitten, Sequencing *n* jobs on two machines with arbitrary time lags, *Management Sci.* 5, 1958, 293-298.

Mon79   C. L. Monma, The two-machine maximum flow time problem with series-parallel precedence relations: An algorithm and extensions, *Oper. Res.* 27, 1979, 792-798.

Mon80   C. L. Monma, Sequencing to minimize the maximum job cost, *Oper. Res.* 28, 1980, 942-951.

MRK83   C. L. Monma, A. H. G. Rinnooy Kan, A concise survey of efficiently solvable special cases of the permutation flow-shop problem, *RAIRO Rech. Opér.* 17, 1983, 105-119.

MS79   C. L. Monma, J. B. Sidney, Sequencing with series-parallel precedence, *Math. Oper. Res.* 4, 1979, 215-224.

NEH83   M. Nawaz, E. E. Enscore, I. Ham, A heuristic algorithm for the *m*-machine, *n*-job flow-shop sequencing problem, *Omega* 11, 1983, 91-95.

NS93   E. Nowicki, C. Smutnicki, New results in the worst-case analysis for flow-shop
            scheduling, *Discrete Appl. Math.* 46, 1993, 21-41.

NS96   E. Nowicki, C. Smutnicki, A fast tabu search algorithm for the permutation
            flow-shop problem, *European J. Oper. Res.* 91, 1966, 160-175

OP89   I. H. Osman, C. N. Potts, Simulated annealing for permutation flow-shop
            scheduling, *Omega* 17, 1989, 551-557.

OS90   F. A. Ogbu, D. K. Smith, The application of the simulated annealing algorithm
            to the solution of the $n/m/C_{max}$ flowshop problem, *Comput. Oper. Res.* 17,
            1990, 243-253.

Pal65   D. S. Palmer, Sequencing jobs through a multi-stage process in the minimum
            total time - a quick method of obtaining a near optimum, *Oper. Res. Quart.* 16,
            1965, 101-107.

Per95   M. Perregaard, *Branch and bound methods for the multiprocessor jobshop
            and flowshop scheduling problems*, Department of Computer Science, Univer-
            sity of Copenhagen, Master's Thesis, 1995.

Pot85   C. N. Potts, Analysis of heuristics for two-machine flow-shop sequencing sub-
            ject to release dates, *Math. Oper. Res.* 10, 1985, 576-584.

PSW91   C. N. Potts, D. B. Shmoys, D. P. Williamson, Permutation vs. non-permutation
            flow shop schedules, *Oper. Res. Lett.* 10, 1991, 281-284.

PVDD98   M.-C. Portmann, A. Vignier, D. Dardilhac and D. Dezalay, Branch and bound
            crossed with GA to solve hybrid flowshops, *European J. Oper. Res.* 107, 1998,
            389-400.

RC92   C. Rajendran and D. Chaudhuri, A multi-stage parallel-processor flowshop
            problem with minimum flowtime, *European J. Oper. Res.* 57, 1992, 111-122.

Ree95   C. Reeves, A genetic algorithm for flowshop sequencing, *Comput. Oper. Res.*
            22, 1995, 5-13.

Röc84   H. Röck, The three-machine no-wait flow shop problem is NP-complete, *J.
            Assoc. Comput. Mach.* 31, 1984, 336-345.

RR72   S. S. Reddi, C. V. Ramamoorthy, On the flow shop sequencing problem with
            no wait in process, *Oper. Res. Quart.* 23, 1972, 323-331.

RS83   H. Röck, G. Schmidt, Machine aggregation heuristics in shop scheduling,
            Methods Oper. Res. 45, 1983, 303-314.

Sal73   M. S. Salvador, A solution of a special class of flowshop scheduling problems,
            *Proceedings of the Symposium on the theory of Scheduling and its Applica-
            tions*, Springer, Berlin, 1975, 83-91.

SB92   S. Stöppler, C. Bierwirth, The application of a parallel genetic algorithm to the
            $n/m/P/\ C_{max}$ flowshop problem, in: T. Gulledge, A. Jones, (eds.), *New Direc-
            tions for Operations Research in Manufacturing*, Springer, Berlin, 1992, 161-
            175.

Sch68   L. Schrage, A proof of the optimality of shortest remaining processing time
            discipline, *Operations Research* 16, 1968, 687-690.

SHD95   D. L. Santos, J. L. Hunsucker and D. E. Deal, Global lower bounds for flow
             shops with multiple processors, *European J. Oper. Res.* 80, 1995, 112-120.

Sid79   J. B. Sidney, The two-machine maximum flow time problem with series-
             parallel precedence relations, *Oper. Res.* 27, 1979, 782-791.

SS89   C. Sriskandarajah and S. Sethi, Scheduling algorithms for flexible flow-shops:
             worst and average performance, *European J. Oper. Res.* 43, 1989, 143-160.

Sta88   E. F. Stafford, On the development of a mixed-integer linear programming
             model for the flowshop sequencing problem, *J. Oper. Res. Soc.* 39, 1988,
             1163-1174.

Szw71   W. Szwarc, Elimination methods in the $m \times n$ sequencing problem. *Naval Res.
             Logist. Quart.* 18, 1971, 295-305.

Szw73   W. Szwarc, Optimal elimination methods in the $m \times n$ sequencing problem,
             Oper. Res. 21, 1973, 1250-1259.

Szw78   W. Szwarc, Dominance conditions for the three-machine flow-shop problem,
             Oper. Res. 26, 1978, 203-206.

Tai90   E. Taillard, Some efficient heuristic methods for the flow shop sequencing
             problem, *European J. Oper. Res.* 47, 1990, 65-74.

TSS94   V. S. Tanaev, Y. N. Sotskov, V. A. Strusevich, *Scheduling Theory. Multi-
             Stage Systems*, Kluwer, Dordrecht, 1994.

Van94   A. Vandevelde, *Minimizing the makespan in a multiprocessor flow shop*,
             Eindhoven University of Technology - Master's Thesis, 1994.

Vig97   A. Vignier, *Contribution à la résolution des problèmes d'ordonnancement de
             type monogamme, multimachines*, PhD Thesis, 1997.

VBP99   A. Vignier, J.-C. Billaut and C. Proust, Les problèmes d'ordonnancement de
             type flow shop hybride: état de l'art, *RAIRO Rech. Oper.* 33, 1999, 117-182.

VHHL05   A. Vandevelde, H. Hoogeveen, C. Hurkens, J. K. Lenstra, Lower bounds for
             the head-body-tail problem on parallel machines: a computational study of the
             multiprocessor flow shop, *INFORMS J. Comput.* 17, 2005, 305-320.

Wag59   H. M. Wagner, An integer linear-programming model for machine scheduling,
             Naval Res. Logist. Quart. 6, 1959, 131-140.

Wer90   F. Werner, On the combinatorial structure of the permutation flow shop prob-
             lem, *ZOR* 35, 1990, 273-289.

WH89   M. Widmer, A. Hertz, A new heuristic method for the flow shop sequencing
             problem, *European J. Oper. Res.* 41, 1989, 186-193.

Wit85   R. J. Wittrock, Scheduling algorithms for flexible flow lines, *IBM J. Res. De-
             velop.* 29, 1985, 401-412.

Wit88   R. J. Wittrock, An adaptable scheduling algorithms for flexible flow lines,
             Oper. Res. 33, 1988, 445-453.

9 Open Shop Scheduling

The formulation of an open shop scheduling problem is the same as for the flow shop problem except that the order of processing tasks comprising one job may be arbitrary.

Thus, the open shop scheduling problem (OSP) can be described as follows: a finite set of tasks has to be processed on a given set of machines. Each task has a specific processing time during which it may not be interrupted, i.e. preemption is not allowed. Tasks are grouped to jobs (sets of tasks), so that each task belongs to exactly one job. Furthermore, each task requires exactly one machine for processing. The objective of the OSP is to schedule all tasks, i.e. determine their start times, so as to minimize the maximum completion time (makespan) given the additional constraints that (a) tasks which belong to the same job and (b) tasks which use the same machine cannot be processed simultaneously.

9.1 Complexity Results

Problem $O2 \mid\mid C_{max}$

Let us consider non-preemptive scheduling first. Problem $O2 \mid\mid C_{max}$ can be solved in $O(n)$ time [GS76]. We give here a simplified description of the algorithm presented in [LLRK81]. For convenience let us denote $a_j = p_{1j}$, $b_j = p_{2j}$, $\mathcal{A} = \{J_j \mid a_j \geq b_j\}$, $\mathcal{B} = \{J_j \mid a_j < b_j\}$, $K_1 = \Sigma\, a_j$ and $K_2 = \Sigma\, b_j$.

Algorithm 9.1.1 *Gonzalez-Sahni algorithm for $O2 \mid\mid C_{max}$ [GS76].*

begin
Choose any two jobs J_k and J_l for which $a_k \geq \max\limits_{J_j \in \mathcal{A}} \{b_j\}$ and $b_l \geq \max\limits_{J_j \in \mathcal{B}} \{a_j\}$;

Set $\mathcal{A}' := \mathcal{A} - \{J_k\}$;

Set $\mathcal{B}' := \mathcal{B} - \{J_l\}$;

Construct separate schedules for $\mathcal{B}' \cup \{J_l\}$ and $\mathcal{A}' \cup \{J_k\}$ using patterns
    shown in Figure 9.1.1;   -- other tasks from $\mathcal{A}'$ and $\mathcal{B}'$ are scheduled arbitrarily
Join both schedules in the way shown in Figure 9.1.2;
Move tasks from $\mathcal{B}' \cup \{J_l\}$ processed on P_2 to the right;
     -- it has been assumed that $K_1 - a_l \geq K_2 - b_k$; the opposite case is symmetric

Change the order of processing on P_2 in such a way that T_{2k} is processed first on
this machine;
end;

The above problem becomes NP-hard as the number of machines increases to 3.
As far as heuristics are concerned we refer to the machine aggregation algorithms
introduced in Section 8.3.2 which use Algorithm 9.1.1 in the case of open shop.

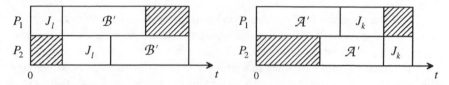

Figure 9.1.1 *A schedule for Algorithm* 9.1.1

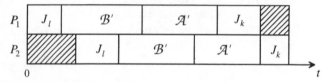

Figure 9.1.2 *A schedule for Algorithm 9.1.1.*

Problem $O \,|\, pmtn \,|\, C_{max}$

Again preemptions result in a polynomial time algorithm. That is, problem $O \,|\,$
$pmtn \,|\, C_{max}$ can be optimally solved by taking

$$C_{max}^* = \max \{ \max_j \{ \sum_{i=1}^{m} p_{ij} \}, \max_i \{ \sum_{j=1}^{n} p_{ij} \} \}$$

and then by applying Algorithm 5.1.16 [GS76].

Problems $O2 \,||\, \Sigma C_j$ and $O2 \,||\, L_{max}$

Let us mention here that problems $O2 \,||\, \Sigma C_j$ and $O2 \,||\, L_{max}$ are *NP*-hard, as
proved in [AC82] and [LLRK81], respectively, and problem $O \,|\, pmtn, r_j \,|\, L_{max}$ is
solvable via the linear programming approach [CS81].

As far as heuristics are concerned, arbitrary list scheduling and the *SPT* algo-
rithm have been evaluated for $O \,||\, \Sigma C_j$ [AC82]. Their asymptotic performance
ratios are $R_L^\infty = n$ and $R_{SPT}^\infty = m$, respectively. Since the number of tasks is usually
much larger than the number of machines, the bounds indicate the advantage of
SPT schedules over arbitrary ones.

A survey of results in open shop scheduling may be found in [DPP01, KSZ91].

9.2 A Branch and Bound Algorithm for Open Shop Scheduling

Only few exact solution methods are available for the open shop scheduling problem. We describe a branch-and-bound algorithm of Dorndorf et al. [DPP01] for solving this problem which performs better than other existing algorithms. The key to the efficiency of the algorithm lies in the following approach: instead of analyzing and improving the search strategies for finding solutions, the authors focus on constraint propagation based methods for reducing the search space. Extensive computational experiments on several sets of well-known benchmark problem instances are reported. For the first time, many problem instances are solved to optimality in a short amount of computation time.

9.2.1 The Disjunctive Model of the OSP

Studies have shown that within the class of intractable problems the OSP belongs to the especially hard ones [BHJW97, GJP00]. As an example, the famous job shop scheduling problem (JSP) which is a close relative of the OSP is easily solvable by now for problem instances with up to 100 tasks, see e.g. [AC91, CP94, CL95, MS96], while there still remain unsolved instances of the OSP with less than 50 tasks.

In this chapter, we describe a branch-and-bound algorithm for solving the OSP. Instead of analyzing and improving the search strategies, we especially focus on constraint propagation based methods for reducing the search space. As a positive side-effect, the constraint propagation algorithm implicitly calculates strong lower bounds so that an explicit computation is not necessary. Extensive computational experiments on several sets of well-known benchmark problem instances show that this algorithm outperforms other exact solution methods for the OSP. With this algorithm, for the first time, many problem instances were solved to optimality within a very short amount of computation time.

The remainder of this chapter is organized as follows. Next we describe the well-known disjunctive model of the OSP that is due to Roy and Sussmann [RS64] and its extension by Błażewicz et al. [BPS00] and give a short review on solution methods for the OSP. Section 9.2.2 introduces the basic concepts of constraint propagation and presents several consistency tests which are used for the reduction of the search space. These consistency tests are embedded in a branch-and-bound algorithm that uses a branching scheme that is due to Brucker et al. [BHJW97]. A short description of this algorithm is given in section 9.2.3.

Extensive computational results of the branch-and-bound algorithms that use different consistency tests are then presented in the last section.

The disjunctive model that has been introduced by Roy and Sussmann [RS64] for the job shop scheduling problem can be easily adapted to the OSP. Let $\mathcal{T} = \{T_1, \cdots, T_n\}$ be the set of tasks to be scheduled. The processing time of task $T_i \in \mathcal{T}$ is denoted with p_i. Choosing sufficiently small time units, we can always assume that the processing times are positive integer values. Each task is associated a start time variable st_i with domain set $\mathbb{N}_0$. Since we want to minimize the makespan, i.e. the maximum completion time of all tasks, the objective function is $C_{max}(st_1, \cdots, st_n) = \max_{T_i \in \mathcal{T}}\{st_i + p_i\}$.

Let $job(i)$ denote the job associated to a task T_i. Further, let $mach(i)$ be the machine required by task T_i. Obviously, two tasks T_i and T_j cannot be processed simultaneously at any time, if $job(i) = job(j)$ or $mach(i) = mach(j)$. These two tasks (as a pair) will belong to set D of forbidden pairs. However, if T_i and T_j cannot be processed in parallel then *either* T_i must finish before T_j can start *or* T_j must be completed before T_i is started. Thus, given

$$D = \{\{i,j\} \mid T_i, T_j \in \mathcal{T}, i \neq j, job(i) = job(j) \vee mach(i) = mach(j)\},$$

the OSP can be written as the following model with disjunctive constraints

$$\begin{aligned}
&\min\{C_{max}(st_1, \cdots, st_n)\}, & st_i \in \mathbb{N}_0 \quad T_i \in \mathcal{T}, \\
&(st_i + p_i \leq st_j) \vee (st_j + p_j \leq st_i) & \{i, j\} \in D.
\end{aligned} \tag{9.2.1}$$

A *schedule* is an assignment $S = (st_1, \cdots, st_n) \in \mathbb{N}_0 \times \cdots \times \mathbb{N}_0$ of all start time variables. For the sake of simplicity, we will use the same notation for variables and their assignments. Schedule S is *feasible* if it satisfies all constraints given by (9.2.1). Reformulating the OSP, the goal is to find a feasible schedule with minimal objective function value $C_{max}(S)$.

The significance of the disjunctive scheduling model for the development of efficient solution methods is revealed if we consider its graph theoretical interpretation. The *disjunctive graph* associated to an OSP instance is a weighted graph $G = (\mathcal{T}, D, W)$ with the node set $\mathcal{T}$, arc set D and the weight set $W = \{w_{ij} = p_i \mid \{i,j\} \in D\}$. D is also called the set of disjunctive arcs. Since D is symmetric, we will represent disjunctive arcs as doubly directed arcs. From now on, we will further use the suggestive notation $i \leftrightarrow j$ for pairs (i,j), (j,i) of disjunctive arcs, and $i \rightarrow j$ to specify one of the arc orientations.

A disjunctive graph is transformed into a directed graph by choosing one arc orientation of each disjunctive arc pair $i \leftrightarrow j \in D$. We obtain a complete (partial) selection if (at most) one arc orientation is chosen from each disjunctive arc pair. The selection is acyclic if after the removal of all remaining undirected disjunctions the resulting directed graph is acyclic.

There exists a simple and well-known many-to-one relationship between feasible schedules and complete, acyclic selections which allows us to restate the OSP as a graph theoretical problem: find a complete and acyclic selection, so that the longest path in the associated directed graph has a minimum length. Thus, it is sufficient to search through the space of all selections which is of cardinality $2^{|D|}$ instead of the space of all schedules which is of cardinality $|N_0|^n$.

Most solution methods for the OSP are based on this fundamental observation. However, due to the exceptionally intractable nature of the OSP, mainly heuristic solution methods have been proposed. Simple list scheduling heuristics based on priority dispatching rules have been examined by Guéret and Prins [GP98a]. Matching algorithms are discussed by Bräsel et al. [BTW93] and Guéret and Prins [GP98a, GP98b]. The shifting bottleneck procedure, originally designed for the JSP, has been adapted by Ramudhin and Marier [RM96] to the OSP. Another important class of heuristics are the insertion algorithms which have been introduced by Werner and Winkler [WW95] for the JSP and generalized by Bräsel et al. [BTW93] for the OSP. Local search approaches (tabu search) and genetic algorithms have been examined by Taillard [Tai93], Liaw [Lia98] and Prins [Pri00]. Colak and Agarwal [CA05] developed a neural network based meta-heuristic approach that allows integration of domain specific knowledge. Learning strategies imply improved neighbour solutions. Blum and Sampels [BS04, Blu05] applied ant colony optimization to shop scheduling. Some of these heuristics, especially the genetic algorithm of Prins and the ant colony optimization of Blum and Sampels, show a very good performance, and for specific classes of OSP instances they often are able to find optimal solutions. However, in general, the solutions found for arbitrary OSP instances are of course of a suboptimal nature.

Only few exact solution methods are available for the OSP. A branch-and-bound algorithm which applies a block-oriented branching scheme and some basic constraint propagation methods for reducing the search tree has been proposed by Brucker et al. [BHJW97]. Guéret et al. [GJP00] improved this algorithm by using an intelligent backtracking technique which replaces the simple depth-first search used by the former. They further applied some additional search tree reduction methods in their branch-and-bound algorithm based on *forbidden intervals* (see Chapter 4.1), i.e. time intervals in which no task can start or end in an optimal solution [GP98b]. All these exact solution methods are capable of solving smaller OSP instances for which they naturally show a better performance than the heuristic methods. However, even for simple, but larger OSP instances for which the heuristic methods easily find an optimal solution, the performance of the exact solution methods is rather poor, since the search space reduction methods applied are not sufficient to handle the combinatorial explosion. In the next section, we will therefore examine additional concepts for reducing the search space which have been described in Dorndorf et al. [DPP01]. It will turn out that these constraint propagation based methods are very efficient and allow solving a large number of simple, hard and very hard OSP benchmark instances which up to now have not been solved.

9.2.2   Constraint Propagation and the OSP

Constraint propagation is an elementary method of search space reduction which has become more and more important in the last decades. The basic idea of constraint propagation is to evaluate implicit constraints through the repeated analysis of the variables, domains and constraints that describe a specific problem instance. This analysis makes it possible to detect and remove *inconsistent* variable assignments that cannot participate in any solution by a merely partial problem analysis. A whole theory is devoted to the definition of different concepts of consistency which, roughly speaking, define the maximal search space reduction that is possible regarding some specific criteria and may serve as a theoretical background for propagation techniques. An exhaustive study of the theory of constraint propagation can be found in [Tsa93]. Dorndorf et al. [DPP99, DPP00] examine constraint propagation techniques for disjunctive and cumulative scheduling problems; for the details we refer to Chapter 13.

Removing *all* inconsistent assignments is in general not possible due to an exponentially increasing computational complexity, so we usually have to content ourselves with approximations. The main issue is to describe simple rules which allow efficient search space reductions, but at the same time can be implemented efficiently. These rules are called *consistency tests*. In the disjunctive scheduling community, some of them are also known as *immediate selection* or *edge-finding* rules.

Consistency tests are generally described through a condition and a search space reduction rule. Whenever the condition is satisfied, the reduction rule is executed. In order to describe the basic concepts of constraint propagation more precisely, we will focus on *domain consistency tests* for the time being. Similar results, however, apply for other types of consistency tests.

A domain consistency test is a consistency test which deduces domain reductions. Let Δ_i be the *current domain* of the start time variable st_i. If UB is an upper bound of the optimal makespan, then we can initially set $\Delta_i := [0, UB - p_i]$. This is necessary, since most consistency tests can only deduce domain reductions if the current domains are finite. The upper bound UB can be found by applying a simple heuristic method or by choosing the trivial value $\sum_{T_i \in \mathcal{T}} p_i$. Given a current domain for each start time variable, a domain consistency test maps a set $\Delta = \{ \Delta_i \mid T_i \in \mathcal{T} \}$ of current domains into a set $\Delta' = \{ \Delta_i' \mid T_i \in \mathcal{T} \}$ of hopefully, but not necessarily reduced current domains. Of course, a domain consistency test only removes values, for which provably no feasible schedule S exists that could be developed from Δ.

In order to obtain the maximal domain reduction possible, it is not sufficient to apply each of these tests only once. The reason for this is that after the reduction of several domains, additional domain adjustments could possibly be derived using some of the tests which previously have failed in deducing any reductions.

Thus, all consistency tests have to be applied in an iterative fashion rather than only once until no more updates are possible. This is equivalent to the computation of a fixed point. Notice that this fixed point does not have to be unique and in general depends upon the order of the application of the consistency tests. Thus, for some application orders the domain reductions obtained may be stronger than for others. Fortunately, it is possible to show that for consistency tests which satisfy a quite natural monotony property, the fixed point computed is always unique [DPP00, Chapter 13]. Since the consistency tests studied are all monotonous in this sense, the application order is irrelevant regarding the extent of the domain reduction. Regarding the complexity of the fixed point computation, however, the application order does play a very crucial role. Notice that the revision of a single domain already forces all consistency tests to be reapplied in the next iteration even though only a small number of constraints and variables are possibly affected by this reduction. Thus, choosing an intelligent order can decrease the computation time to a large extent. However, we will not deal with this issue more closely, but choose a quite naive propagation order.

In the next subsections, we will describe the set of consistency tests used in the algorithm. In addition to domain consistency tests, the disjunctive scheduling model and its graph theoretical interpretation allow the definition of consistency tests which operate on the set of complete selections. These consistency tests reduce the set of complete selections by detecting sequences of tasks which must occur in every optimal solution. Since this is done by selecting disjunctive arc orientations, the latter approach has been often labeled *immediate selection* (see e.g. [CP89, BJK94]) or *edge-finding* (see e.g. [AC91]). We will use the term *sequence consistency test* as opposed to domain consistency tests and as used in [DPP99,DPP00]. Domain and sequence consistency tests are two different concepts which complement each other. Often, a situation occurs in which either only reductions of the current domains or only arc orientations are deducible. The best results, in fact, are obtained by applying both types of consistency tests, as fixing disjunctive arcs may initiate additional domain reductions and vice versa, cf. Chapter 13.

Input/Output Consistency Tests

Quite important for the development of efficient consistency tests for the OSP is the concept of *disjunctive cliques* or *cliques* for short. We will say that $O_c \subseteq \mathcal{T}$ is a clique if any pair of tasks in O_c cannot be processed in parallel, i.e. if all tasks in O_c either belong to the same job or require the same machine. A clique O_c is said to be maximal, if no true superset of O_c is a clique. Therefore, there exist $|\mathcal{J}|$ maximal job cliques, where $\mathcal{J}$ denotes the set of jobs, and $|\mathcal{P}|$ maximal machine cliques, where $\mathcal{P}$ denotes the set of machines (processors).

For the rest of this section, we will assume that $O_c \subseteq \mathcal{T}$ is a maximal clique and that *all* subsets A (tasks T_i) are subsets (elements) of this clique. Without loss of generality we will number the indices of the elements of O_c by $1, 2, \cdots, |O_c|$. Let further $est_i := \min \Delta_i$ and $lst_i := \max \Delta_i$ denote the earliest and latest start time of task T_i, and let $ect_i := est_i + p_i$ and $lct_i := lst_i + p_i$ denote the earliest and latest completion time of task T_i. Finally, for a subset $A \subset O_c$ of tasks, let $EST_{min}(A)$ $:= \min_{T_i \in A} est_i$, $LCT_{max}(A) := \max_{T_i \in A} lct_i$, and $p(A) := \sum_{T_i \in A} p_i$.

Given a clique of tasks $A \subset O_c$ and an additional task $T_i \in O_c \setminus A$, Carlier and Pinson [CP89] were the first to derive conditions which imply that T_i has to be processed *before* or *after* all tasks $T_j \in A$. In the first case, they called T_i the *input* of A, in the second case, the *output* of A, and so Dorndorf et al. [DPP00] have chosen the name *input/output conditions*.

Theorem 9.2.1 *(Input/Output Sequence Consistency Tests). Let $A \subset O_c$ and $T_i \in O_c \setminus A$. If the input condition*

$$LCT_{max}(A \cup \{T_i\}) - EST_{min}(A) < p(A \cup \{T_i\}) \qquad (9.2.2)$$

is satisfied then task T_i has to be processed before all tasks in A, for short, $T_i \to A$. Likewise, if the output condition

$$LCT_{max}(A) - EST_{min}(A \cup \{T_i\}) < p(A \cup \{T_i\}) \qquad (9.2.3)$$

is satisfied then task T_i has to be processed after all tasks in A, $A \to T_i$. □

Domain consistency tests that are based on the input/output conditions can now be simply derived. We will only examine the adjustment of the earliest start times, as the adjustment of the latest start times can be handled analogously. Obviously, if task T_i is the output of a clique A then T_i can only start if all tasks in A have finished. Thus, the earliest start time of T_i is at least the maximum completion time of all tasks in A being scheduled without preemption. Unfortunately, however, the computation of this makespan requires the solution of an NP-hard single-machine scheduling problem. Therefore, if the current domains are to be updated efficiently, we have to content ourselves with approximations of this bound. The following theorem is due to Carlier and Pinson [CP89, CP90].

Theorem 9.2.2 *(Output Domain Consistency Tests, part 1). If the output condition is satisfied for $A \subset O_c$ and $T_i \in O_c \setminus A$ then the earliest start time of T_i can be adjusted to $est_i := \max\{est_i, C_{max}^{pr}(A)\}$, where $C_{max}^{pr}(A)$ is the maximum completion time of all tasks in A being scheduled with preemption allowed.* □

Notice that the computation of $C_{max}^{pr}(A)$ has time complexity $O(|A| \log |A|)$ [Jac56].

It has already been mentioned that applying both sequence and domain consistency tests together can lead to better search space reductions. Quite evidently, any domain reductions deduced by Theorem 9.2.2 can lead to additional arc orientations deduced by Theorem 9.2.1. We will now discuss the case in which the inverse is also true. Imagine a situation in which $A \to T_i$ can be deduced for a subset of tasks, but in which the output condition does not hold for the couple (A, T_i). Such a situation can actually occur as can be seen in the following example.

In Figure 9.2.1, an example with three tasks is shown. The earliest start time of T_i is $est_i = 4$, while its latest completion time is $lct_i = 9$. The earliest start and latest completion times of T_j and T_k are $est_j = est_k = 0$ and $lct_j = lct_k = 9$, respectively. The processing times of T_i, T_j and T_k are $p_i = p_j = p_k = 3$. Notice that we can both deduce $T_j \to T_i$ and $T_k \to T_i$ using the input conditions for the couple $(\{T_i\}, T_j)$ and $(\{T_i\}, T_k)$, since e.g. $LCT_{max}(\{T_i, T_j\}) - est_i = 5 < 6 = p_i + p_j$. Thus, we know that $\{T_j, T_k\} \to T_i$. However, the output condition is not satisfied for the couple $(\{T_j, T_k\}, T_i)$ because $LCT_{max}(\{T_j, T_k\}) - EST_{min}(T_i, T_j, T_k) = 9 = p_i + p_j + p_k$.

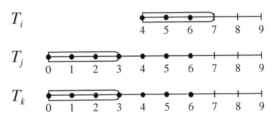

$$
\begin{array}{l}
T_i \\
T_j \\
T_k
\end{array}
$$

Figure 9.2.1 *An example with three tasks.*

This example motivates the following theorem as an extension of Theorem 9.2.2.

Theorem 9.2.3 (*Input/Output Domain Consistency Tests, part 2*). *Let $A \subset O_c$ and $T_i \in O_c \backslash A$. If $A \to T_i$ then the earliest start time of T_i can be adjusted to*

$$est_i := \max\{est_i, C^{pr}_{max}(A)\} . \qquad \square$$

Here, the reader should recall once more that the subset A mentioned in the last theorem does not have to coincide with the subset for which the input or the output condition is satisfied.

An important question to answer now is whether there exist efficient algorithms that implement the input/output consistency tests. An efficient implementation is obviously not possible if all pairs (A, T_i) of subsets $A \subset O_c$ and tasks $T_i \in O_c \backslash A$ are to be tested separately. Fortunately, it is not necessary to do so as has been first shown by Carlier and Pinson [CP90] who have developed an $O(|O_c|^2)$

algorithm for applying the input/output consistency tests described in Theorem 9.2.1 and Theorem 9.2.2 (It is common practice to only report the time complexity for applying the consistency tests *once* for all couples (A, T_i). In general, however, the number of iterations necessary for computing the fixed point of current domains has to be considered as well. This accounts for an additional factor c which depends upon the size of the current domains, but is omitted here.) Several years later, $O(|O_c| \log |O_c|)$ algorithms have been proposed by Carlier and Pinson [CP94] and Brucker et al. [BJK94] which until now have the best asymptotic performance, but are less efficient for smaller problem instances and require quite complex data structures.

Nuijten [Nui94], Caseau and Laburthe [CL95] and Martin and Shmoys [MS96] have chosen a solely domain oriented approach and have derived different algorithms for implementing the input/output consistency tests that are based on Theorem 9.2.2. Nuijten developed an algorithm with time complexity $O(|O_c|^2)$ which can be generalized to scheduling problems with discrete resource capacity. Caseau and Laburthe presented an $O(|O_c|^3)$ algorithm which works in an incremental fashion, so that $O(|O_c|^3)$ is a worst case, since not all consistency tests are applied within an iteration of the fixed point computation. The algorithm proposed by Martin and Shmoys [MS96] also has a time complexity of $O(|O_c|^2)$.

Dorndorf et al. [DPP01] have implemented the input/output tests described in Theorems 9.2.1 and 9.2.2. They could have used the $O(|O_c|^2)$ algorithm of Carlier and Pinson for implementing Theorem 9.2.1 and then adjusted the current domains according to Theorem 9.2.3. However, the algorithm of Carlier and Pinson already requires the adjustment of some of the domains and, in fact, is a combination of the consistency tests described in Theorems 9.2.1 and 9.2.2. Thus, many of these domain adjustments would be recomputed if Dorndorf et al. afterwards applied the consistency tests described in Theorem 9.2.3. They have therefore developed two algorithms which work in a purely sequential fashion, one of which has a time complexity of $O(|O_c|^3)$, while the other has a time-complexity of $O(|O_c|^2)$. These algorithms are based on the definition of *task sets* as introduced by Caseau and Laburthe [CL95]. A detailed description of the algorithms is given in [Pha00].

Given the arc orientations derived, the domain adjustments of Theorem 9.2.3 can then be applied with effort $O(|O_c|^2 \log |O_c|)$ using Jackson's famous algorithm [Jac56].

Note that the approach by Dorndorf et al. of first deducing the arc orientations and then applying the domain adjustments implies a higher time complexity than for algorithms based on the purely domain oriented approach. However, stronger domain reductions may be achieved, as demonstrated by the previous example.

Input/Output Negation Consistency Tests

In the last subsection, a condition has been described which implies that a task has to be processed before (after) another set of tasks. In this subsection, the inverse situation, that a task *cannot* be processed first (last), is studied. The following theorem is due to Carlier and Pinson [CP89, CP90]. For reasons near at hand, Dorndorf et al. have chosen the name input/output negation for the conditions described in this theorem.

Theorem 9.2.4 *(Input/Output Negation Sequence Consistency Tests). Let $A \subset O_c$ and $T_i \in O_c \setminus A$. If the input negation condition*

$$LCT_{max}(A) - est_i < p(A \cup \{T_i\}) \tag{9.2.4}$$

is satisfied then task T_i cannot be processed before all tasks in A. Likewise, if the output negation condition

$$LCT_{max} - EST_{min}(A) < p(A \cup \{T_i\}) \tag{9.2.5}$$

is satisfied then task T_i cannot be processed after all other tasks in A.   □

This theorem allows a reduction of the current domains which, in general, is weaker than the one that has been described in Theorem 9.2.3. However, since the input/output negation conditions are more often satisfied than the input/output conditions, they will turn out to be quite important for solving the OSP efficiently.

Let us study the input negation condition and the adjustments of earliest start times. If (9.2.4) is satisfied for $A \subset O_c$ and $T_i \in O_c \setminus A$, there must be a task in A which starts and finishes before T_i, although we generally do not know which one. This proves the following theorem [CP89, CP90].

Theorem 9.2.5 *(Input/Output Negation Domain Consistency Tests). If the input negation condition is satisfied for $A \subset O_c$ and $T_i \in O_c \setminus A$ then the earliest start time of task T_i can be adjusted to $est_i := \max\{est_i, \min_{T_u \in A} ect_u\}$.*   □

Input/output negation consistency tests have been applied by Nuijten [Nui94], Baptiste and Le Pape [BL95] and Caseau and Laburthe [CL95] for the JSP. All these algorithms only test some, but not all interesting couples (A, T_i). An algorithm which deduces all domain reductions with a time complexity of $O(|O_c|^2)$ has only recently been developed by Baptiste and Le Pape [BL96]. Dorndorf et al. [DPP01] have developed another algorithm which also performs all possible domain adjustments in $O(|O_c|^2)$. This algorithm uses some main ideas of Baptiste and Le Pape, but can be better combined with the algorithms that Dorndorf et al. have developed for the other consistency tests, since some computations can be reused, see again [Pha00] for the details.

Shaving

A closer look at the consistency tests presented so far reveals that they all share the following common and simple idea: a hypothesis (e.g. task T_i starts at time st_i) can be refuted, if there exists no schedule so that this hypothesis is satisfied. Consistency tests only differ in the kind of hypotheses that are made and the proof for showing that no schedule can exist under these hypotheses. The input negation consistency test, for instance, verifies for a given clique A of tasks whether there exists a schedule in which some task T_i is started within the time interval $[est_i, \min_{T_u \in A} ect_u - 1]$. This verification is carried out through a simple test which compares the length of the time interval $[est_i, LCT_{max}(A)]$ with the sum of processing times $p(A \cup \{T_i\})$. Replacing this simple test with other and possibly more sophisticated tests leads to different and probably more powerful consistency tests.

A general approach in which all hypotheses are of the kind: "task T_i starts at its earliest start time" or "task T_i starts at its latest start time" has been proposed by Martin and Shmoys under the name *shaving* [MS96]. In *exact one-machine shave* the verification is carried out by solving an instance of a one-machine scheduling problem in which $st_i := est_i$ or, alternatively, $st_i := lst_i$. *One-machine shave* relaxes the non-preemption requirement and tests whether a possibly preemptive schedule exists under the aforementioned hypothesis. Carlier and Pinson [CP94] and Martin and Shmoys [MS96] both proposed the computation of fixed points as a method for proving that a feasible schedule cannot exist under a certain hypothesis. More precisely, the hypothesis is falsified if a current domain becomes empty during the fixed point computation.

Dorndorf et al, [DPP01] apply shaving by testing the hypotheses $st_i > t \in \Delta_i$ and $st_i < t \in \Delta_i$. Test values for t are chosen during a combination of bisection and incremental search. Apparently, the application of shaving techniques can be very costly. However, the search space reduction obtained by shaving by far offsets these costs.

9.2.3   The Algorithm and Its Performance

In general, the single application of constraint propagation is not sufficient for solving the OSP. Although for certain problem instances the search space reduction may be of a considerable size, a branch-and-bound search is usually still necessary for finding an optimal solution. In this section, we give a short description of the block branching scheme which has been described by Brucker et al. [BJS94] for the JSP and, for instance, by [BHJW97] for the OSP and which we have used as well in our branch-and-bound algorithm. A deeper insight into the nature of the block branching scheme is given by Phan Huy [Pha00], who discusses a generalization for shop scheduling problems with arbitrary disjunctive constraints.

The block branching scheme requires the computation of a heuristic solution (complete and acyclic selection) in each node of the branching tree that is compatible with the arc orientations already chosen. Dorndorf et al. [DPP01] chose as a heuristic solution method the priority rule based dispatching heuristic that has been described by Brucker et al. [BHJW97] in their algorithm B&B1. Given a complete and acyclic selection, a critical path B, i.e. a path of maximal length is chosen within the associated directed graph. This critical path is then decomposed into so-called maximal blocks, the definition of which is given in the following. A subpath $B' = u_1 \rightarrow \cdots \rightarrow u_l$ of B of length $l \geq 2$ is a *block* iff, for all $i \neq j$, we have $u_i \leftrightarrow u_j \in D$, i.e. iff two pairwise different tasks in B' are always in disjunction, since they belong to the same job or require the same machine. A block B' is said to be *maximal*, iff extending B' by even only one node (task) already violates the block condition. Obviously, given a critical path, there always exists a unique decomposition into maximal blocks. Given this block decomposition, the block branching scheme as described by Brucker et al. [BJS94] is based on the following observation:

Let S be a complete and acyclic selection and B a critical path in the corresponding directed graph. If S is not optimal, i.e. there exists a selection S' with a smaller makespan, then there is a maximal block in B so that in S' a task within this block is processed before the first or after the last task of this block.

Thus, child nodes are created by moving tasks of a block to the beginning or end of the block. Consequently, $2 \cdot (l-1)$ child nodes are generated for each block of length $l > 2$, while for blocks of length 2 obviously only one child node is generated. Improving this branching scheme, Brucker et al. [BJS94] described how to fix additional arcs depending on the search nodes that have been already visited prior to the generation of the actual search node. Further they described the particular role played by the first and the last block of the maximal block decomposition, since the number of tasks to be moved to the beginning or end of these blocks can be reduced. The search strategy of their branching algorithm has been organized in a depth-first manner. For further details on the block branching scheme we refer the reader to the work of Brucker et al. [BJS94, BHJW97]. Dorndorf et al. [DPP01] have used this branching scheme except for some minor modifications regarding the branching order, i.e. the sequence in which the child nodes are generated, see [Pha00] for the technical details.

Upon finding an improved solution in a node (initial solution in the root node) of the branching tree, the makespan of this solution is, of course, used as upper bound *UB*. The lower bounds used within the branch-and-bound algorithm are the preemptive one-machine (one-job) lower bounds which are computed using Jackson's algorithm [Jac56]. Notice, however, that stronger bounds are calculated in an implicit manner by the application of constraint propagation: whenever an inconsistency is detected, for instance, if a current domain becomes empty, we know that no solution can be generated from the actual search tree node with a makespan of UB and, therefore, UB is indeed a lower bound for this search tree node.

Dorndorf et al. [DPP01] have implemented the branch-and-bound algorithm together with the constraint propagation techniques in C on Pentium II (333 MHz) in MSDOS environment. They have tested the algorithm on a large set of benchmark problems that have been generated by Taillard [Tai93] (Tai-n-*) and Brucker et al. [BHJW97] (Hur-n-*). All test instances are quadratic of size n jobs and n machines, with n ranging from 6 to 20. We will see below that, on the one hand, most of the quite large instances of Taillard are easily solved by Dorndorf et al.'s algorithm. They have solved all the 10×10 instances, something which none of the current exact algorithms is capable of, and even do so with an average run time of less than a minute. Further, they have solved most of the 15×15 instances in several minutes and most of the 20×20 instances in less than an hour. Among these instances, three instances (Tai-15-5, Tai-15-9, Tai-20-6) have not been solved prior the start of their experiments. On the other hand, the rather small instance Hur-7-1 of size 7×7 still remains open, although they have been able to improve the current best lower bound from 1000 to 1021.

Brucker et al. [BHJW97] have proposed an explanation for this phenomenon which is based on the *work load* of a problem instance. The work load of an OSP instance is defined as follows: given a set of jobs $\mathcal{J}$ and a set of machines $\mathcal{P}$. Let O^{ι} be the maximal clique of tasks belonging to job J_{ι} and O_{μ} be the maximal clique of tasks requiring machine P_{μ}. Let, further, $LB := \max\{ \max\{ p(O^{\iota}) \mid J_{\iota} \in \mathcal{J} \}, \max\{ \{ p(O_{\mu}) \mid P_{\mu} \in \mathcal{P} \} \}$ define the trivial lower bound which is the maximum of the job and machine bounds (the sum of processing times of tasks belonging to a job or machine clique). The *average work load WL* is then defined as

$$WL = \frac{\sum_{J_{\iota} \in \mathcal{J}} p(O^{\iota}) + \sum_{P_{\mu} \in \mathcal{P}} p(O_{\mu})}{(|\mathcal{J}| + |\mathcal{P}|) \cdot LB}$$

If the work load of an OSP instance is close to 1 then all job and machine bounds are not much smaller than LB, so that finding a solution with a makespan close to LB is not very probable. On the contrary, an OSP instance with a low work load tends to have an optimal solution with a makespan equal to the lower bound LB. These problem instances are less hard to solve, since an optimal solution can be more easily verified.

Considering this intuitive interpretation, it is possible to use the work load to guide the choice of an appropriate solution strategy. The alternatives that are given are a *top-down* and a *bottom-up* strategy. Both strategies use the branch-and-bound algorithm and only differ in the way of choosing the initial upper bound(s). The top-down strategy starts with a *real* upper bound which, in [DPP01], is determined by the heuristic solution method and tries to improve (decrease) this upper bound by applying the branch-and-bound algorithm. The bottom-up approach uses a lower bound as a hypothetical upper bound and, whenever the branch-and-bound algorithm does not find a solution which is consistent with this upper bound, increases it by one time unit. This process is repeated until a solution is found.

Notice, that the top-down approach only applies the branch-and-bound algorithm once, but that constraint propagation is less effective since the current domains are less tight due to the high initial upper bound. Hence, searching the whole search tree may require a higher computation time. The bottom-up approach, on the contrary, reinitializes the branch-and-bound algorithm several times, but allows more constraint propagation since the current domains are smaller. Therefore, the search trees that are created are smaller. Altogether, the top-down approach seems to be more suited, if the optimal makespan is far from the lower bound LB, since the multiple application of the branch-and-bound algorithm within a bottom-up approach would offset its propagation advantages. Also, according to this logic, the bottom-up approach is to be preferred if the optimal makespan is near to the lower bound LB. Thus, it is straightforward to choose the top-down approach whenever the work load of an OSP instance is high and the bottom-up approach whenever the work load is low.

At first, however, we will only evaluate the top-down approach since this allows to analyze better the impact of the different consistency tests. It seems justified to say that instances of the OSP, especially those with a high work load, are generally more difficult to solve than instances of the JSP with the same number of tasks, jobs and machines. To one part, this is due to the larger solution space: not only machine sequences, but also job sequences have to be determined. Thus, Dorndorf et al. [DPP01] have often encountered a situation in which the search process was trapped in an unfavorable region of the search space from which it could not escape within a reasonable amount of time. Another reason for the intractability of the OSP, however, is the lack of strong lower bounds. In fact, if no search is carried out, the lower bound LB is already the best bound one is able to find. Thus, constraint propagation plays a more important role in reducing the search space.

In the beginning, the experiments for the two different classes of consistency tests (input/output and input/output negation consistency tests), have been carried out for a set of smaller instances, namely, the instances Tai-7-* and Hur-6-*. The results are shown in Table 9.2.1. CP1 applies the input/output tests as described in Theorem 9.2.1 and Theorem 9.2.3, while CP2 applies both the input/output tests and the input/output negation tests. Since [DPP01] have applied a top-down strategy for CP1 and CP2, they report for each problem instance the initial upper bound found by the heuristic solution method (UB_{init}) in addition to the optimal makespan (UB_{best}). They further report the number of search tree nodes generated by each of the algorithms and the total run time. All of the instances have naturally been solved to optimality.

problem	UB_{best}	UB_{init}	CP1		CP2	
			nodes	time	nodes	time
Hur-6-1	1056	1528	55634	149.7 s	36876	133.0 s
Hur-6-2	1045	1377	3291	7.3 s	1711	5.2 s
Hur-6-3	1063	1536	9737	23.9 s	5401	18.0 s
Hur-6-4	1005	1481	8553	20.3 s	4356	14.4 s
Hur-6-5	1021	1647	2983	6.4 s	1562	4.6 s
Hur-6-6	1012	1276	8406	19.7 s	4263	13.8 s
Hur-6-7	1000	1454	4557	11.5 s	3205	10.7 s
Hur-6-8	1000	1636	169	0.4 s	132	0.4 s
Hur-6-9	1000	1524	525	1.2 s	326	1.0 s
Tai-7-1	435	609	147	0.4 s	130	0.4 s
Tai-7-2	443	614	309	1.1 s	225	0.9 s
Tai-7-3	468	632	8789	36.6 s	5661	30.9 s
Tai-7-4	463	664	1892	7.5 s	1040	5.3 s
Tai-7-5	416	551	521	2.0 s	409	2.0 s
Tai-7-6	451	581	28347	124.5 s	16464	95.8 s
Tai-7-7	422	693	61609	254.5 s	30101	167.7 s
Tai-7-8	424	637	1467	5.9 s	961	5.0 s
Tai-7-9	458	551	237	0.8 s	194	0.8 s
Tai-7-10	398	576	25837	107.2 s	9427	53.2 s

Table 9.2.1   *Results for some smaller instances (top-down).*

Obviously, CP2 generates less search tree nodes and has a lower total run time than CP1, although more constraint propagation is applied in each of the single nodes. On average, CP2 generates approximately 40 % less search tree nodes than CP1 and has a run time which is lower by about 25 %. Note, that a different observation has been made for the JSP, see [Pha00]: although the number of search tree nodes decreases as well, the total run time increases (for smaller instances) due to the additional propagation effort. Thus, the additional application of the input/output negation tests is more efficient for the OSP than for the JSP. This can be explained as follows: the input/output tests, if applied on their own, deduce only few arc orientations for the OSP in the beginning of the branch-and-bound process, because at that time most of the current domains are just too large and coincide with the trivial interval $[0, UB - p_i]$. Only at a certain depth of the search tree, more arc orientations are deduced, however, the portion of the search tree that can be pruned by that time is rather small. Consequently, the additional application of the input/output negation tests improves the efficiency of the input/output tests since the former are a relaxation of the latter and so are capable of deducing domain reductions at an earlier stage of the branching process.

Next Dorndorf et al. [DPP01] tested the better algorithm CP2 on the larger OSP instances Tai-10-* and the harder instances Hur-7-*. They further tested a shaving variant of CP2, i.e. in each of the search tree nodes they applied the shaving procedure described in Section 9.2.2 and used the input/output and input/output negation tests for detecting inconsistencies. The results for CP2 are shown in Table 9.2.2 and those for the branch-and-bound algorithm with shaving

CPS2 in Table 9.2.3. In addition to the usual information listed further above, they report for each problem instance the best upper bound found (UB_{found}) within a time limit of 5 hours. Upper bounds shown in parentheses are either non optimal or optimal, but could not be verified. As an example, 1048 is the best upper bound known for the instance Hur-7-1 and 1052 is the best bound found by CP2 within 5 hours of computation time.

Problem	UB_{best}	UB_{init}	CP2		
			UB_{found}	nodes	time
Hur-7-1	(1048)	1487	(1052)	2677448	18000.0 s
Hur-7-2	1055	1839	(1055)	2916573	18000.0 s
Hur-7-3	1056	1839	(1056)	2993406	18000.0 s
Hur-7-4	1013	1418	1013	960092	5796.4 s
Hur-7-5	1000	1188	1000	775960	4420.6 s
Hur-7-6	1011	1545	(1011)	2897640	18000.0 s
Hur-7-7	1000	1419	1000	1628	8.8 s
Hur-7-8	1005	1510	1005	208340	1197.6 s
Hur-7-9	1003	1435	1003	1807635	10797.5 s
Tai-10-1	637	949	637	418594	5455.7 s
Tai-10-2	588	751	588	123104	2219.9 s
Tai-10-3	598	854	(607)	1025042	18000.0 s
Tai-10-4	577	856	577	64244	1175.8 s
Tai-10-5	640	1057	640	8173	126.0 s
Tai-10-6	538	770	(555)	1012887	18000.0 s
Tai-10-7	616	904	(827)	2136045	18000.0 s
Tai-10-8	595	853	595	164977	2255.7 s
Tai-10-9	595	880	595	6036	98.1 s
Tai-10-10	596	894	(639)	1162480	18000.0 s

Table 9.2.2 *Results for some larger instances (top-down).*

Regarding the instances of Taillard, CP2 solves 6 of them. The run times for all the instances that have been solved have a high standard deviation and vary from 2 minutes to 2 hours. This is because the optimal makespan may be hard to find, but once found it is easily verified and in all cases coincides with the trivial lower bound *LB*. Regarding the instances of Brucker et al., CP2 solves 5 instances, but none of the very hard instances Hur-7-1, Hur-7-2 and Hur-7-3. It finds, however, the optimal makespans of Hur-7-2 and Hur-7-3 without proof of optimality.

The results for CPS2 are much better. To the best of our knowledge, it is the first exact algorithm which solves all 10×10 OSP instances of Taillard. It even does so with an average run time of slightly above 10 minutes starting with a rather high upper bound. Further, CPS2 solves nearly all instances of Brucker et al. except the instance Hur7-1 which is still unsolved. The quality of CPS2 relies on the fact that the extensive application of constraint propagation results in a drastic reduction of the search tree. Quite impressively, the number of search tree nodes generated by CPS2 on average only amounts to 0.1% of the number of nodes generated by CP2 . Therefore, the probability of getting lost in unfavourable regions of the search tree is significantly cut down. This underlines the importance and effectiveness of enhanced constraint propagation techniques for solving the OSP.

Problem	UB_{best}	UB_{init}	CPS2		
			UB_{found}	nodes	time
Hur-7-1	(1048)	1487	(1058)	4575	18000.0
Hur-7-2	1055	1839	1055	3364	9421.8
Hur-7-3	1056	1839	1056	3860	9273.5
Hur-7-4	1013	1418	1013	1123	2781.9
Hur-7-5	1000	1188	1000	742	1563.0
Hur-7-6	1011	1545	1011	5195	15625.1
Hur-7-7	1000	1419	1000	88	48.8
Hur-7-8	1005	1510	1005	209	318.8
Hur-7-9	1003	1435	1003	788	2184.9
Tai-10-1	637	949	637	612	1398.6
Tai-10-2	588	751	588	396	981.7
Tai-10-3	598	854	598	520	2664.3
Tai-10-4	577	856	577	496	847.1
Tai-10-5	640	1057	640	392	724.5
Tai-10-6	538	770	538	415	1101.5
Tai-10-7	616	904	616	565	982.8
Tai-10-8	595	853	595	461	837.3
Tai-10-9	595	880	595	222	655.1
Tai-10-10	596	894	596	562	993.8

Table 9.2.3 *Results for some larger instances using shaving (top-down).*

Up to now, we have applied a top-down solution approach which starts with an initial upper bound and tries to improve, i.e decrease this upper bound. As an alternative, we will now consider a bottom-up approach which starts with a lower bound as hypothetical upper bound and increases this bound by one time unit until a solution is found. The trivial job and machine based lower bound LB is chosen as an initial lower bound. For the computation of more sophisticated lower bounds which involves some search, we refer the reader to the work of Guéret and Prins [GP99].

The results for this approach are shown in Table 9.2.4. There are only the results for the best algorithm, namely CPS2. UB_{best} denotes the best lower bound found within a maximum run time of 5 hours. If LB_{best} is not written in parentheses then it also has been verified to be an upper bound. Again, all 10×10 instances of Taillard are solved, however, this time with an average run time of less than a minute. The results for the instances of Brucker are less impressive if compared with the top-down approach. Studying the work load WL of each instance, we can observe that the bottom-up approach is more efficient for instances with a lower work load, while the top-down approach shows better results for those with a higher work load. This perfectly fits with the intuitive remarks made at the beginning of this section: instances with a lower work load tend to have an optimal makespan close or equal to LB, so that only a few lower bounds have to be tested in a bottom-up approach. On the contrary, instances with a higher work load tend to have an optimal makespan which is far from the initial lower bound. For these instances, the top-down approach is more efficient. Dorndorf et al. propose that the bottom-up approach is the favourite choice for problem instances with a work load of less than 0.9, while the top-down approach is to be preferred for instances with a work load greater than 0.95. For problem instances with a work load between 0.9 and 0.95, the situation is less clear, see e.g. the problem instance Hur-7-5 with a work load of 0.944 (bottom-up performs better) and Hur-7-9 with a work load of 0.925 (top-down performs better).

Problem	UB_{best}	LB	WL	CPS2		
				LB_{best}	nodes	time
Hur-7-1	(1048)	1000	1.000	(1021)	3974	18000.0
Hur-7-2	1055	1000	1.000	(1045)	5988	18000.0
Hur-7-3	1056	1000	1.000	(1042)	7057	18000.0
Hur-7-4	1013	1000	0.958	1013	5692	15178.1
Hur-7-5	1000	1000	0.944	1000	146	314.7
Hur-7-6	1011	1000	0.951	(1006)	5797	18000.0
Hur-7-7	1000	1000	0.879	1000	10	5.0
Hur-7-8	1005	1000	0.931	1005	194	625.5
Hur-7-9	1003	1000	0.925	1003	1376	4073.0
Tai-10-1	637	637	0.861	637	12	30.2
Tai-10-2	588	588	0.834	588	22	70.6
Tai-10-3	598	598	0.850	598	23	185.5
Tai-10-4	577	577	0.828	577	21	29.7
Tai-10-5	640	640	0.834	640	17	32.0
Tai-10-6	538	538	0.857	538	17	32.7
Tai-10-7	616	616	0.838	616	18	30.9
Tai-10-8	595	595	0.823	595	17	44.1
Tai-10-9	595	595	0.846	595	14	39.8
Tai-10-10	596	596	0.834	596	14	29.1

Table 9.2.4   *Results for some larger instances using shaving (bottom-up).*

Dorndorf et al. have also tested the bottom-up variant of CPS2 on the remaining 15×15 and 20×20 instances of Taillard for which no other results of exact solution approaches have been reported in literature. These results are shown in Table 9.2.5. Again, the bottom-up approach shows very good results. All 15×15 instances except one and 7 of the 20×20 instances have been solved, among others the instances Tai-15-5, Tai-15-9 and Tai-20-6 that have not been solved before. Except for the unsolved instances, the run time is always less than 12 minutes for the 15×15 instances and about an hour for the 20×20 instances.

Let us finally compare the results of the algorithms from [DPP01] with those of some other branch-and-bound algorithms for the OSP. B&B1 of Brucker et al.

[BHJW97] is a typical representative of their 6 slightly different algorithms and the algorithm B&Bi of Guéret et al. [GJP00], where 'i' stands for intelligent backtracking. These algorithms are compared with Dorndorf et al.'s combined top-down/bottom-up approach which works as follows: whenever the work load of an instance is at most 0.9, the bottom-up version of CPS2 is applied; for instances with a work load greater than 0.9, on the contrary, the top-down version of CPS2 is used.

Problem	UB_{best}	LB	WL	CPS2		
				LB_{best}	nodes	time
Tai-15-1	937	937	0.800	937	42	481.4 s
Tai-15-2	918	918	0.834	(918)	193	18000.0 s
Tai-15-3	871	871	0.824	871	44	611.6 s
Tai-15-4	934	934	0.794	934	45	570.1 s
Tai-15-5	946	946	0.842	946	34	556.3 s
Tai-15-6	933	933	0.795	933	51	574.5 s
Tai-15-7	891	891	0.828	891	52	724.6 s
Tai-15-8	893	893	0.813	893	46	614.0 s
Tai-15-9	899	899	0.830	899	36	646.9 s
Tai-15-10	902	902	0.824	902	34	720.1 s
Tai-20-1	1155	1155	0.820	1155	59	3519.8 s
Tai-20-2	1241	1241	0.838	(1241)	69	18000.0 s
Tai-20-3	1257	1257	0.803	1257	77	4126.3 s
Tai-20-4	1248	1248	0.825	(1248)	92	18000.0 s
Tai-20-5	1256	1256	0.809	1256	56	3247.3 s
Tai-20-6	1204	1204	0.810	1204	65	3393.0 s
Tai-20-7	1294	1294	0.807	1294	48	2954.8 s
Tai-20-8	(1171)	1169	0.854	(1169)	69	18000.0 s
Tai-20-9	1289	1289	0.800	1289	69	3593.8 s
Tai-20-10	1241	1241	0.817	1241	65	4936.2 s

Table 9.2.5   *Results for even larger instances using shaving (bottom-up).*

The results have been summarized in Table 9.2.6. A dash indicates that the corresponding data have not been available. Brucker et al. chose a time limit of 50 hours on a Sun 4/20 workstation, whereas Guéret et al. stopped the search after 250000 backtracks which according to their time measurements corresponds to approximately 3 hours on a Pentium PC with a clock pulse of 133 MHz. As the results have been established on different platforms, they have to be interpreted with care. However, especially regarding the Taillard instances, it seems fair to say that the algorithm of Dorndorf et al. has a much better performance. While neither B&B1 and B&Bi solve more than 3 of the 10 × 10 instances of Taillard to optimality, they solve all 10 instances in an average run time of less than a minute. Notice further that even the version of CPS2 which works in a purely top-down fashion as well solves all ten instances and that CP2 which does not use shaving all the same solves 6 instances to optimality. Since the branching schemes employed by all these exact algorithms are basically the same (except for the branching order in the algorithm of Dorndorf et al. and the intelligent backtracking component in the algorithm of Guéret et al.), we can conclude that the application of strong constraint propagation techniques sheds a new light on the solvability of the OSP and allows to cope with instances of the OSP that formerly seemed intractable. Computational experiments on some famous test sets of benchmark problem instances taken from literature demonstrate the efficiency of this approach. For the first time, many problem instances are solved in a short amount of computation time.

Problem	UB_{best}	B&B1[a]		B&B1[b]		CPS2[c]	
		nodes	time	nodes	time	nodes	time
Hur-7-1	(1048)	-	>50 h	-	-	4575	>5 h
Hur-7-2	1055	-	35451.5 s	-	-	3364	9421.8 s
Hur-7-3	1056	-	176711.1 s	-	-	3860	9273.5 s
Hur-7-4	1013	-	77729.2 s	-	-	1123	2781.9 s
Hur-7-5	1000	-	6401.6 s	-	-	742	1563.0 s
Hur-7-6	1011	-	277271.1 s	-	-	5195	15625.1 s
Tai-10-1	637	-	>50 h	>250000	>3 h	12	30.2 s
Tai-10-2	588	44332	10671.5 s	>250000	>3 h	22	70.6 s
Tai-10-3	598	-	>50 h	>250000	>3 h	23	185.5 s
Tai-10-4	577	163671	40149.4 s	26777	-	21	29.7 s
Tai-10-5	640	-	>50 h	>250000	>3 h	17	32.0 s
Tai-10-6	538	-	>50 h	>250000	>3 h	17	32.7 s
Tai-10-7	616	-	>50 h	4843	-	18	30.9 s
Tai-10-8	595	-	>50 h	>250000	>3 h	17	44.1 s
Tai-10-9	595	97981	24957.0 s	245100	-	14	39.8 s
Tai-10-10	596	-	>50 h	>250000	>3 h	14	29.1 s

[a] Run time on a Sun 4/20 Workstation
[b,c] Run time on a Pentium II/133

Table 9.2.6 *A comparison of computational results.*

References

AC82   J. O. Achugbue, F. Y. Chin, Scheduling the open shop to minimize mean flow time, *SIAM J. Comput.* 11, 1982, 709-720.

AC91   D. Applegate and W. Cook, A computational study of the job shop scheduling problem, *ORSA J. Comput.* 3, 1991, 149-156.

BHJW97 P. Brucker, J. Hurink, B. Jurisch and B. Wöstmann, A branch and bound algorithm for the open shop problem, *Discrete Appl. Math.* 76, 1997, 43-59.

BJK94   P. Brucker, B. Jurisch and A. Krämer, The job shop problem and immediate selection, *Annals Oper. Res.* 50, 1994, 73-114.

BJS94   P. Brucker, B. Jurisch, and B. Sievers, A fast branch and bound algorithm for the job shop scheduling problem, *Discrete Appl. Math.* 49, 1994, 107-127.

BL95   P. Baptiste and C. Le Pape, A theoretical and experimental comparison of constraint propagation techniques for disjunctive scheduling, In: *Proceedings of the 14th International Joint Conference on Artificial Intelligence*, Montreal, 136-140, 1995.

BL96   P. Baptiste and C. Le Pape, Edge-finding constraint propagation algorithms for disjunctive and cumulative scheduling, In: *Proceedings of the 15th Workshop of the U. K. Planning Special Interest Group*, Liverpool, 1996.

Blu05   C. Blum, Beam-ACO – Hybridizing ant colony optimization with beam search: An application to open shop scheduling, *Comput. Oper. Res.* 32, 2005, 1565-1591.

BPS00   J. Błażewicz, E. Pesch, M. Sterna, The disjunctive graph machine representation of the job shop scheduling problem, *European J. Oper. Res.* 127, 2000, 317-331.

BS04   C. Blum, M. Sampels, An ant colony optimization algorithm for shop scheduling problems, *J. Math. Modelling and Algorithms* 3, 2004, 285-308.

BTW93   H. Bräsel, T. Tautenhahn and F. Werner, Constructive heuristic algorithms for the open shop problem, *Computing* 51, 1993, 95-110.

CA05   S. Colak, A. Agarwal, Non-greedy heuristics and augmented neural networks for the open-shop scheduling problem, *Naval Res. Logist.* 52, 2005, 631-644.

CL95   Y. Caseau and F. Laburthe, Disjunctive scheduling with task intervals, Technical Report 95-25, Laboratoire d'Informatique de l'Ecole Normale Superieure, Paris, 1995.

CP89   J. Carlier and E. Pinson, An algorithm for solving the job shop problem, *Management Sci.* 35, 1989, 164-176.

CP90   J. Carlier and E. Pinson, A practical use of Jackson's preemptive schedule for solving the job shop problem, *Annals Oper. Res.* 26, 1990, 269-287, 1990.

CP94   J. Carlier and E. Pinson, Adjustments of heads and tails for the job shop problem, *European J. Oper. Res.* 78, 1994, 146-161.

CS81   Y. Cho, S. Sahni, Preemptive scheduling of independent jobs with release and due times on open, flow and job shops, *Oper. Res.* 29, 1981, 511-522.

DPP99   U. Dorndorf, T. Phan-Huy and E. Pesch, A survey of interval capacity consistency tests for time and resource constrained scheduling, In: J. Weglarz, editor, *Project Scheduling - Recent Models, Algorithms and Applications*, 213-238, Kluwer Academic Publishers, Boston, 1999.

DPP00   U. Dorndorf, E. Pesch and T. Phan-Huy, Constraint propagation techniques for disjunctive scheduling problems, *Artificial Intelligence* 122, 2000, 189-240.

DPP01   U. Dorndorf, E. Pesch and T. Phan-Huy, Solving the open shop scheduling problem, *Journal of Scheduling* 4, 2001, 157-174.

GJP00   C. Gueret, N. Jussien, and C. Prins, Using intelligent backtracking to improve branch and bound methods: An application to open shop problems, *European J. Oper. Res.* 127, 2000, 344-354.

GP98a   C. Gueret and C. Prins. Classical and new heuristics for the open shop prob-
         lem: A computational evaluation, *European J. Oper. Res.* 107, 1998, 306-314,
         1998.

GP98b   C. Gueret and C. Prins. Forbidden intervals for open-shop problems, Research
         Report 98/10/AUTO, Ecole de Mines de Nantes, Nantes, 1998.

GP99   C. Gueret and C. Prins, A new lower bound for the open-shop problem, *Annals
         Oper. Res.* 92, 1999, 165-183.

GS76   T. Gonzalez, S. Sahni, Open shop scheduling to minimize finish time, *J. Assoc.
         Comput. Mach.* 23, 1976, 665-679.

Jac56   J. Jackson, An extension on Johnson's results on job lot scheduling, *Naval Res.
         Logist. Quart.* 3, 1956, 201-203.

KSZ91   W. Kubiak, C. Srishkandarajah, K. Zaras, A note on the complexity of open
         shop scheduling problems, *INFOR* 29, 1991, 284-294.

Lia98   C. F. Liaw, An iterative improvement approach for nonpreemptive open shop
         scheduling problem, *European J. Oper. Res.* 111, 1998, 509-517.

LLRK81   E. L. Lawler, J. K. Lenstra, A. H. G. Rinnooy Kan, Minimizing maximum
         lateness in a two-machine open shop, *Math. Oper. Res.* 6, 1981, 153-158; Erra-
         tum, *Math. Oper. Res.* 7, 1982, 635.

MS96   P. Martin and D. B. Shmoys, A new approach to computing optimal schedules
         for the job shop scheduling problem, In: *Proceedings of the 5th International
         IPCO Conference*, 1996.

Nui94   W. P. M. Nuijten, *Time and Resource Constrained Scheduling: A Constraint
         Satisfaction Approach*, PhD thesis, Eindhoven University of Technology,
         1994.

Pha00   T. Phan-Huy, *Constraint Propagation in Flexible Man*ufacturing, Springer,
         Heidelberg, 2000.

Pri00   C. Prins, Competitive genetic algorithms for the open-shop scheduling problem
         Math. Methods of Oper. Res. 52, 2000, 389-411.

RM96   A. Ramudhin and P. Marier, The generalized shifting bottleneck procedure,
         European J. Oper. Res. 93, 1996, 34-48.

RS64   B. Roy and B. Sussmann, Les problemes d'ordonnancement avec contraintes
         disjonctives. Note D. S. 9, SEMA, Paris, 1964.

Tai93   E. Taillard, Benchmarks for basic scheduling problems, *European J. Oper.
         Res.*, 64, 1993, 278-285.

Tsa93   E. Tsang, *Foundations of Constraint Satisfaction*. Academic Press, Essex,
         1993.

WW95   F. Werner and A. Winkler, Insertion techniques for the heuristic solution of the
         job shop problem, *Discrete Appl. Math.* 58, 1995, 191-211.

10 Scheduling in Job Shops

In this chapter we continue scheduling of tasks on dedicated processors or machines. We assume that tasks belong to a set of jobs, each of which is characterized by its own machine sequence. We will assume that any two consecutive tasks of the same job are to be processed on different machines. The type of factory layout is the job shop. It provides the most flexible form of manufacturing, however, frequently accepting unsatisfactory machine utilization and a large amount of work-in-process. Hence, makespan minimization is one of the objectives in order to schedule job shops effectively, see e.g. [Pin95].

10.1 Introduction

10.1.1 The Problem

A job shop (cf. Section 3.1) consists of a set of different machines (like lathes, milling machines, drills etc.) that perform tasks of jobs. Each job has a specified processing order through the machines, i.e. a job is composed of an ordered list of tasks each of which is determined by the machine required and the processing time on it. There are several constraints on jobs and machines: (*i*) There are no precedence constraints among tasks of different jobs; (*ii*) tasks cannot be interrupted (non-preemption) and each machine can handle only one job at a time; (*iii*) each job can be performed only on one machine at a time. While the machine sequence of the jobs is fixed, the problem is to find the job sequences on the machines which minimize the makespan, i.e. the maximum of the completion times of all tasks. It is well known that the problem is NP-hard [LRK79], and belongs to the most intractable problems considered, cf. [LLR+93].

10.1.2 Modeling

There are different problem formulations, those in [Bow59, Wag59], and the mixed integer formulation [Man60] are the first ones published; see also [BDW91, BHS91, [MS96]. We have adopted the one presented in [ABZ88].

Let $\mathcal{T} = \{ T_0, T_1, \cdots, T_n \}$ denote the set of tasks where T_0 and T_n are considered as dummy tasks "start" (the first task of all jobs) and "end" (the last task of all jobs), respectively, both of zero processing time. Let $\mathcal{P}$ denote the set of m machines and $\mathcal{A}$ be the set of ordered pairs (T_i, T_j) of tasks constrained by the precedence relations $T_i < T_j$ for each job. For each machine P_k, set $\mathcal{E}_k$ describes the set of all pairs of tasks to be performed on this machine, i.e. tasks which cannot overlap (cf. (ii)). For each task T_i, its processing time p_i is fixed, and the earliest possible starting time of T_i is t_i, a variable that has to be determined during the optimization. Hence, the job shop scheduling problem can be modeled as:

$$Minimize \quad t_n$$

$$subject\ to \quad t_j - t_i \geq p_i \qquad\qquad \forall\, (T_i, T_j) \in \mathcal{A}, \qquad\qquad (10.1.1)$$

$$t_j - t_i \geq p_i \ or \ t_i - t_j \geq p_j \quad \forall\, \{T_i, T_j\} \in \mathcal{E}_k, \forall\, P_k \in \mathcal{P}, \quad (10.1.2)$$

$$t_i \geq 0 \qquad\qquad\qquad \forall\, T_i \in \mathcal{T}. \qquad\qquad (10.1.3)$$

Restrictions (10.1.1) ensure that the processing sequence of tasks in each job corresponds to the predetermined order. Constraints (10.1.2) demand that there is only one job on each machine at a time, and (10.1.3) assures completion of all jobs. Any feasible solution to the constraints (10.1.1), (10.1.2), and (10.1.3) is called a schedule.

An illuminating problem representation is the *disjunctive graph* model due to [RS64]. It has mostly replaced the solution representation (within algorithms) by Gantt charts as described in [Gan19, Cla22, Por68]. The latter, however, is useful in user interfaces to graphically represent a solution to a problem.

In the edge-weighted graph there is a vertex for each task, additionally there exist two dummy vertices 0 and n, representing the start and end of a schedule, respectively. For every two consecutive tasks of the same job there is a directed arc; the start vertex 0 corresponds to the first task T_0 of every job and the vertex n corresponds to the last task T_n of every job. For each pair of tasks $\{T_i, T_j\} \in \mathcal{E}_k$ that require the same machine there are two arcs (i, j) and (j, i) with opposite directions. The tasks T_i and T_j are said to define a *disjunctive arc pair* or a *disjunctive edge*. Thus, single arcs between tasks represent the precedence constraints on the tasks of the same job and a pair of opposite directed arcs between two tasks represents the fact that each machine can handle at most one task at the same time. Each arc (i, j) is labeled by a weight p_i corresponding to the processing time of task T_j. All arcs from vertex 0 have label 0.

Figure 10.1.1 illustrates the disjunctive graph for a problem instance with 3 machines P_1, P_2, P_3 and 3 jobs J_1, J_2, J_3 with together 8 tasks/operations. The machine sequences of jobs J_1, J_2, and J_3 (see the rows of Figure 10.1.1(a)) are $P_1 \rightarrow P_2 \rightarrow P_3$, $P_3 \rightarrow P_2$ and $P_2 \rightarrow P_1 \rightarrow P_3$, respectively. The processing times are presented in Table 10.1.1.

P_1	3	–	3
P_2	2	4	6
P_3	3	3	2
	J_1	J_2	J_3

Table 10.1.1 *Processing times of a 3 job 3 machine instance.*

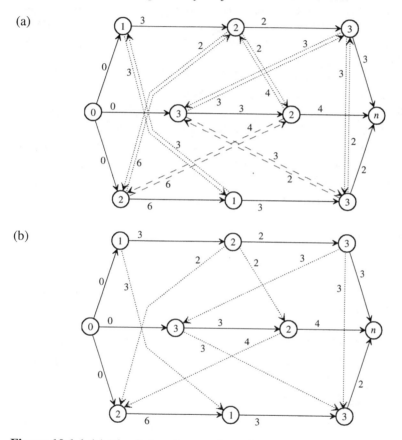

Figure 10.1.1 (a) *The disjunctive graph, and*
(b) *a feasible schedule for the problem instance of Table 10.1.1.*

The job shop scheduling problem requires to find an order of the tasks on each machine, i.e. to select one arc among all opposite directed arc pairs such that the resulting graph G is acyclic (i.e. there are no precedence conflicts between tasks) and the length of the maximum weight path between the start and end vertex is minimal. Obviously, the length of a maximum weight or longest path in G connecting vertices 0 and i equals the earliest possible starting time t_i of task T_i; the makespan of the schedule is equal to the length of the *critical path*, i.e. the weight of a longest path from start vertex 0 to end vertex n. Any arc (i, j) on a critical path is said to be *critical*; if T_i and T_j are tasks from different jobs then (i, j) is called a *disjunctive critical arc*, otherwise it is a *conjunctive critical arc*. We agree on the convention, that, if vertex i is on a critical path, then task T_i is said to be a critical task or on a critical path. For convenience we sometimes identify a feasible job shop schedule and its disjunctive graph representation.

In order to improve a current schedule, we have to modify the machine order of jobs (i.e. the sequence of tasks) on longest paths. Therefore a neighborhood structure can be defined by (*i*) reversing a disjunctive critical arc, i.e. selecting the opposite arc, or (*ii*) reversing a disjunctive critical arc such that this arc is incident to an arc of the arc set $\mathcal{A}$, cf. [MSS88, ALLU94, LAL92, VAL96].

For the problem instance of Table 10.1.1 and Figure 10.1.1(a) let us consider the schedule defined by the job processing sequence $J_1 \rightarrow J_3$ on machine P_1, and $J_1 \rightarrow J_2 \rightarrow J_3$ on machine P_2 and P_3. Hence all tasks are lying on a longest path of length 26, see Figure 10.1.1(b). Reversing the processing order of jobs J_2 and J_3 on machine P_2 yields a reduced makespan of 16 for the new schedule. Extensions of the disjunctive graph representation including additional job shop constraints are discussed in [BPS99, BPS00, WR90].

10.1.3 Complexity

The minimum makespan problem of job shop scheduling is a classical combinatorial optimization problem that has received considerable attention in the literature. It belongs to the most intractable problems considered. Only a few particular cases are efficiently solvable:

- Scheduling two jobs by the graphical method as described in [Bru88] and first introduced by Akers [Ake56] (see Section 7.2). In general this idea can be used to compute good lower bounds sometimes superior to the one-machine bounds [Car82, BJ93].
- The two machine flow shop case, i.e. the machine sequences of all jobs are the same [Joh54, GS78], see Section 7.2.
- The two machine job shop problem where each job consists of at most two tasks [Jac56].
- The two machine job shop case with unit processing times [HA82, KSS94].
- The two machine job shop case with a fixed number of jobs (and, of course, repetitive processing of jobs on the machines, [Bru94]).

Slight modifications turn out to be difficult. The two machine job shop problem where each job consists of at most three tasks, the three machine job shop problem where each job consists of at most two tasks, the three machine job shop problem with three jobs are NP-hard, see [LRKB77, GS78, SS95]. The job shop problems with two and three machines and task processing times equal to 1 or 2, and equal to 1, respectively, are NP-hard even in the case of preemption, see [LRK79].

10.1.4 The History

The history of the job shop scheduling problem, starting more than 40 years ago, is also the history of two well known benchmark problems consisting of 10 jobs and 10 machines as well as of 20 jobs and 5 machines and introduced by Fisher and Thompson [FT63]. The data of these instances is presented in Table 10.1.2. While the 20 job 5 machine instance turned out to be a challenge for ten years the particular instance of a 10 job 10 machine problem opposed its solution for 25 years leading to a competition among researchers for the most powerful solution procedure. Since then branch and bound procedures have received substantial attention from numerous researchers. Early work was presented [BW65], followed by [Gre68], whose method was based on Manne's integer programming formulation. Further papers included [Bal69, CD70, FTM71, AH73], and [Fis73] who obtained lower bounds by the use of Lagrange multipliers.

For long time the algorithm in [MF75] was the best exact solution method. Instead of using worse bounds of [CD70] they combined the bounds for the one machine scheduling problem with task arrival time and the objective function to minimize maximum lateness with the enumeration of active schedules (see [GT60]) among which are also optimal ones. An alternative approach whereby at each stage one disjunctive arc of some crucial pair is selected leads to a computationally inferior method, [LLRK77].

Considerable effort has been invested in the empirical testing of various priority rules, see [Ger66], and the survey papers [DH70, PI77, Hau89].

During the 80's substantial algorithmic improvements were achieved and accurately reflected by the stepwise optimum approach for the notorious 10-job 10-machine problem. [FLL+83] applied computationally costly surrogate duality relaxations, weighting and aggregating into a single constraint, either machine capacity constraints or job-task precedence constraints. A first attempt to obtain bounds by polyhedral techniques was made in [Bal85]. The neighborhood structure used in some recent local search algorithms is also mainly employed as branching structure in the exact method of [BM85]. They rearrange tasks on a longest path if the tasks use the same machine. [LLR+93] report that, with respect to the famous 10×10 problem "Lageweg (1984) found a schedule of 930, without proving optimality; he also computed a number of multi-machine lower bounds, ranging from a three-machine bound of 874 to a six-machine bound of 907". So he was the first who found an optimal solution. Optimality of a schedule of length 930 was first proved by Carlier and Pinson [CP89]. Their algorithm is based on bounds obtained for the one machine problems with precedence constraints, task arrival times and allowed preemptions. This problem is polynomially solvable. Additionally, they used several simple but effective inference rules on task subsets.

(a)

J_1	1, 29	2, 78	3, 9	4, 36	5, 49	6, 11	7, 62	8, 56	9, 44	10, 21
J_2	1, 43	3, 90	5, 75	10, 11	4, 69	2, 28	7, 46	6, 46	8, 72	9, 30
J_3	2, 91	1, 85	4, 39	3, 74	9, 90	6, 10	8, 12	7, 89	10, 45	5, 33
J_4	2, 81	3, 95	1, 71	5, 99	7, 9	9, 52	8, 85	4, 98	10, 22	6, 43
J_5	3, 14	1, 6	2, 22	6, 61	4, 26	5, 69	9, 21	8, 49	10, 72	7, 53
J_6	3, 84	2, 2	6, 52	4, 95	9, 48	10, 72	1, 47	7, 65	5, 6	8, 25
J_7	2, 46	1, 37	4, 61	3, 13	7, 32	6, 21	10, 32	9, 89	8, 30	5,55
J_8	3, 31	1, 86	2, 46	6, 74	5, 32	7, 88,	9, 19	10, 48	8, 36	4, 79
J_9	1, 76	2, 69	4, 76	6, 51	3, 85	10, 11	7, 40	8, 89	5, 26	9, 74
J_{10}	2, 85	1, 13	3, 61	7, 7	9, 64	10, 76	6, 47	4, 52	5, 90	8, 45

(b)

J_1	1, 29	2, 9	3, 49	4, 62	5, 44
J_2	1, 43	2, 75	4, 69	3, 46	5, 72
J_3	2, 91	1, 39	3, 90	5, 12	4, 45
J_4	2, 81	1, 71	5, 9	3, 85	4, 22
J_5	3, 14	2, 22	1, 26	4, 21	5, 72
J_6	3, 84	2, 52	5, 48	1, 47	4, 6
J_7	2, 46	1, 61	3, 32	4, 32	5, 30
J_8	3, 31	2, 46	1, 32	4, 19	5, 36
J_9	1, 76	4, 76	3, 85	2, 40	5, 26
J_{10}	2, 85	3, 61	1, 64	4, 47	5, 90
J_{11}	2, 78	4, 36	1, 11	5, 56	3, 21
J_{12}	3, 90	1, 11	2, 28	4, 46	5, 30
J_{13}	1, 85	3, 74	2, 10	4, 89	5, 33
J_{14}	3, 95	1, 99	2, 52	4, 98	5, 43
J_{15}	1, 6	2, 61	5, 69	3, 49	4, 53
J_{16}	2, 2	1, 95	4, 72	5, 65	3, 25
J_{17}	1, 37	3, 13	2, 21	4, 89	5, 55
J_{18}	1, 86	2, 74	5, 88	3, 48	4, 79
J_{19}	2, 69	3, 51	1, 11	4, 89	5, 74
J_{20}	1, 13	2, 7	3, 76	4, 52	5, 45

Table 10.1.2 (a) *The 10 job 10 machine instance* [FT63].
        (b) *The 20 job 5 machine instance* [FT63].
           Row j contains the order of the tasks of job J_j;
           each entry (i, p) contains the index of machine P_i
           and the processing time p_{ij} on it.

Some algorithms developed in the 90's are still the job shop champions among the exact methods. Besides the branch and bound implementations of Applegate and Cook [AC91], Martin and Shmoys [MS96], and Perregaard and Clausen [PC95], there are the branch and bound algorithms [CL95, BLN95, CP90, CP94, BJS92, BJS94, BJK94]. The power of their methods basically results from some inference rules which describe simple cuts, and a branching scheme such that tasks which belong to a block (a sequence of tasks on a machine) on the longest path are moved to the block ends, hence improving an idea described in [GNZ86].

Throughout the chapter, experimental results are reported mainly for the 10×10 problem. Techniques that are giving good results on the 10×10 problem need not necessarily perform well on other instances of the job shop scheduling problem, even for instances of the same size (10×10) like [LA19, LA20, ORB2, ORB3, ORB4] (a description of these instances may be found e.g. in [AC91]). Applegate and Cook's algorithm is very efficient on the 10×10 problem, but much less on other instances, in particular [LA19, ORB2 and ORB3]. On the contrary, a lot of the successful approaches mentioned use important ideas from Applegate and Cook's paper. An interesting benchmark is [LA21]. Vaessens [Vae95] solved it with a modified version of Applegate and Cook's algorithm (~40,000,000 nodes). Then it was solved by Baptiste et al. [BPN95] using ~4,000,000 nodes in ~48 hours, by Caseau and Laburthe [CL95] using ~2,000,000 nodes in ~24 hours, and by Martin and Shmoys [MS96] in about one hour.

Tailored approximation methods viewed as an opportunistic (greedy-type) problem solving process can yield optimal or near-optimal solutions even for problem instances up to now considered as difficult, cf. [ABZ88, OS88, Sad91, BLV95, DL93, BV98]. Hereby opportunistic problem solving or opportunistic reasoning characterizes a problem solving process where local decisions on which tasks, jobs, or machines should be considered next, are concentrated on the most promising aspects of the problem, e.g. job contention on a particular machine. Hence sub-problems often defining bottlenecks are extracted and separately solved and serve as a basis from which the search process can expand. Breaking down the whole problem into smaller pieces takes place until, eventually, sufficiently small sub-problems are created for which effective exact or heuristic procedures are available. However the way in which a problem is decomposed affects the quality of the solution reached. Not only the type of decomposition such as machine/resource [ABZ88], job/order [DPP02], or event based [Sad91] has a dramatic influence onto the outcome but also the number of sub-problems and the order of their consideration. In fact, an opportunistic view suggests that the initial decomposition be reviewed in the course of problem solving to see if changes are necessary. The shifting bottleneck heuristic from [ABZ88] and its improving modifications from [BLV95] and [DL93] are typical representatives of opportunistic reasoning. It is resource based as there are sequences of one machine schedules successively solved and their solutions introduced into the overall schedule. In the 90's local search based scheduling became very popular; see the surveys [GPS92, AGP97, VAL96]. These algorithms are all based on a certain neighborhood structure and some rules defining how to obtain a new solution from existing ones. The first efforts to implement powerful general problem solvers such as simulated annealing [LAL92, MSS88, Kol99, EAZ07], tabu search [DT93], parallel tabu search [Tai94], and genetic algorithms [ALLU94, NY91, YN92, SWV92a], finally culminated in the excellent tabu search implementation of Nowicki and Smutnicki [NS96, NS05] and Balas and Vazacopoulos [BV98]. Among the genetic based methods only a few, e.g. [YN92, DP93a, DP93b, Mat96], could solve the notorious 10 job 10 machine problem optimally. Most of the current local search approaches rely on naive search neighborhoods which fail to exploit problem specific knowledge. Applications of local and probabilistic search methods to sequencing problems are based on neighborhoods defined in the solution space of the problem. The method in [SWV92a] is based on problem perturbation neighborhoods, i.e. the original data is genetically perturbed and a neighbor is defined as a solution which is obtained when a base heuristic is applied to the perturbed problem. The obtained solution sequence for the perturbed problem is mapped to the original data, i.e. the non-perturbed tasks are scheduled in the same way and the makespan of the solution to the original problem data defines the quality of the perturbed problem.

The local search heuristics like simulated annealing, tabu search, and genetic algorithms are modestly robust under different problem structures and require only a reasonable amount of implementation work with relatively little insight into the combinatorial structure of the problem. Problem specific characteristics are mainly introduced via some improvement procedures, the kind of representation of solutions as well as their modifications based on some neighborhood structure.

In recent years job shop problems with additional specifics motivated from practice have been investigated by few authors, e.g. job shop problems with transport robots [BK06].

In the next section we will go into detail and present ideas of some exact algorithms and heuristic approaches, see [BDP96], [JM99], and [Bru04].

10.2 Exact Methods

In this section we will be concerned with branch and bound algorithms, exploring specific knowledge about the critical path of the job shop scheduling problem.

10.2.1 Branch and Bound

The principle of branch and bound is the enumeration of all feasible solutions of a combinatorial optimization problem, say a minimization problem, such that properties or attributes not shared by any optimal solution are detected as early as possible. An attribute (or branch of the enumeration tree) defines a subset of the set of all feasible solutions of the original problem where each element of the subset satisfies this attribute. In general, attributes are chosen such that the union of all attribute-defined subsets equals the set of all feasible solutions of the problem and any two of these subsets do not intersect. For each subset the objective value of its best solution is estimated by a lower bound (bounding). An optimal solution of a relaxation of the original problem such that this optimal solution also satisfies the subset defining attribute, serves as a lower bound. In case the lower bound exceeds the value of the best (smallest) known upper bound (a heuristic solution of the original problem) the attribute-defined subset can be dropped from further consideration. Otherwise, search is continued departing from the most promising subset which is divided into smaller subsets through the definition of additional attributes. Hence, at any search stage a subset of solutions is defined by a set of attributes all of which are satisfied by these solutions.

We shall see that the attributes of a branch and bound process exactly correspond to attributes forbidding moves in tabu search. Branching from one solution subset to a new smaller one can be associated with a tabu search move.

10.2.2 Lower Bounds

One of the main drawbacks of all branch and bound methods is the lack of strong lower bounds in order to cut branches of the enumeration tree as early as possible. Several types of lower bounds are applied in the literature, for instance, bounds based on Lagrangian relaxation, see [Vel91], bounds based on the optimal solution of a sub-problem consisting of only two or three jobs and all machines, see [Ake56, Bru88, BJ93]. However the most prominent bounding procedure has been described in [Car82, Pot80b]. Consider any task T_i in the job shop respectively its associated vertex i in the disjunctive graph that may include already a partial selection of arcs from disjunctive arc pairs. Then there is a longest path from the artificial vertex 0 to i of length r_i as well as a longest path of length q_i connecting the end of vertex i to the last one, the dummy vertex n. Task T_i cannot start to be processed earlier than its arrival time r_i (also called release time or *head*) and its processing has to be finished at the latest until its due date $t_n - q_i$ in order to cause no schedule delay. The time q_i is said to be the *tail* of task T_i. There exist m one-machine lower bounds for the optimal makespan of the job shop scheduling problem where each bound is obtained from the exact solution of a one-machine scheduling problem with release times, due dates, and minimization of the makespan. Although this problem is NP-complete Carlier's algorithm quickly solves the one machine problems optimally for all problem sizes in the job shop under consideration. In [BLV95] there is an even better branch and bound procedure that can yield improved lower bounds. The method additionally takes minimum delays between pairs of tasks into account. That means, if there is a directed path connecting vertices i and j in the disjunctive graph, then

$$t_j - t_i \geq L(i, j) \tag{10.2.1}$$

where $L(i, j)$ is the i and j connecting path's length.

While the one machine scheduling problem with heads r_i and tails q_i, for all tasks T_i, can be solved in $O(n\log n)$ time if $r_i = r_j$, for all T_i, T_j, (use the longest tail rule, i.e. schedule the jobs in order of decreasing tails) or if $q_i = q_j$, for all T_i, T_j, (use the shortest head rule, i.e. schedule the jobs in order of increasing heads) this is not true any longer if time lags $L(i, j)$ are imposed, see [BLV95].

The branch and bound algorithms [Car82] and [BLV95] extensively make use of the fact that the shortest makespan of a one machine schedule cannot fall below

$$LB1(C) := \min\{r_i \mid T_i \in C\} + \sum_{T_i \in C} p_i + \min\{q_i \mid T_i \in C\} \tag{10.2.2}$$

for any subset C of all tasks which have to be scheduled on a particular machine, where p_i is the processing time of task T_i. This lower bound can be calculated in $O(n \cdot \log n)$ time by solving the preemptive one machine problem without time lags. It is well known, that the strongest bound $LB1(C)$ equals the minimum makespan of the preemptive version of Algorithm 4.1.2. Let us consider the idea of branching. Consider a schedule produced by the longest tail rule, i.e. among the released jobs schedule that one with longest tail. Let $C := \{T_0, T_{i_1}, T_{i_2}, \cdots, T_{i_z}, T_n\}$ be a sequence of tasks constituting a critical path. Further, let T_c be the last task encountered in C such that $q_c < q_{i_z}$, i.e. all tasks in C between T_c and T_n have tails at least q_{i_z}. Let C' denote the set of these tasks excluding T_c and T_n. Then branching basically is based on the following observation: If $r_i \geq \max\{t_{i_1}, t_c\}$ for all $T_i \in C'$, and if the part $C(c, i_z)$ of the critical path C connecting c to i_z contains no precedence relation then the longest tail schedule is optimal in case $c = 0$. Otherwise, if $c > 0$, in any schedule better than the current one, task T_c either precedes or succeeds all tasks in C', cf. [BLV95].

There are a lot of additional inference rules - several are summarized in the sequel - in order to cut the enumeration tree during a preprocessing or the search phase, see [Car82, BLV95, CP89, CP90, CP94, BJS92, BJ94, CL95, AC91, DPP00, DPP02, BB01].

10.2.3 Branching

Consider once more the one machine scheduling problem consisting of the set $\mathcal{N}$ of tasks, release times r_i and tails q_i for all $T_i \in \mathcal{N}$. Let C_{max} be the maximum completion time of a feasible job shop schedule, i.e. C_{max} is an upper bound for the makespan of an optimal one machine schedule. Let $\mathcal{E}_C$, $\mathcal{S}_C$ and C be subsets of $\mathcal{N}$ such that $\mathcal{E}_C, \mathcal{S}_C \subseteq C$, and any task $T_j \in C$ also belongs to $\mathcal{E}_C(\mathcal{S}_C)$ if there is an optimal single machine schedule such that T_j is first (respectively last) among all tasks in C. Then the following conditions hold for any task T_k of $\mathcal{N}$:

If $r_k + \sum\limits_{T_i \in C} p_i + \min\{q_i \mid T_i \in \mathcal{S}_C, T_i \neq T_k\} > C_{max}$ then $T_k \notin \mathcal{E}_C$, $\qquad$ (10.2.3)

If $\min\{r_i \mid T_i \in \mathcal{E}_C, T_i \neq T_k\} + \sum\limits_{T_i \in C} p_i + q_k > C_{max}$ then $T_k \notin \mathcal{S}_C$, $\qquad$ (10.2.4)

If $T_k \notin \mathcal{E}_C$ and $LB1(C - \{T_k\}) + p_k > C_{max}$ then $T_k \in \mathcal{S}_C$, $\qquad$ (10.2.5)

If $T_k \notin \mathcal{S}_C$ and $LB1(C - \{T_k\}) + p_k > C_{max}$ then $T_k \in \mathcal{E}_C$. $\qquad$ (10.2.6)

The preceding results tell us that, if C contains only two tasks T_i and T_k such that $r_k + p_k + p_i + q_i > C_{max}$ then task T_i is processed before T_k, i.e. from the disjunctive arc pair connecting i and k arc (i, k) is selected. Moreover (10.2.5) and (10.2.6) can be used in order to adapt heads and tails within the branch and bound process, see [Car82, CP89]. If (10.2.5) or (10.2.6) holds then one can fix all arcs (i, k) or (k, i), respectively, for all tasks $T_i \in C$, $T_i \neq T_k$. Application of (10.2.3) to (10.2.6) guarantees an immediate selection of certain arcs from disjunctive arc pairs before branching into a sub-tree. There are problem instances such that conditions (10.2.3) to (10.2.6) cut the enumeration tree substantially.

The branching structure of [CP89, CP90, CP94] is based on the disjunctive arc pairs which define exactly two sub-trees. Let T_i and T_j be such a pair of tasks which have to be scheduled on a critical machine, i.e. a machine with longest initial lower bound (= the preemptive one machine solution). Then, roughly speaking, according to [BR65], both sequences of the two tasks are checked with respect to their regrets if they increase the best lower bound LB. Let

$$d_{ij} := \max \{0, r_i + p_i + p_j + q_j - LB\} ,$$
$$d_{ji} := \max \{0, r_j + p_j + p_i + q_i - LB\} , \qquad (10.2.7)$$
$$a_{ij} := \min \{d_{ij}, d_{ji}\}, \text{ and } b_{ij} := |d_{ij} - d_{ji}| .$$

Among all possible candidates of disjunctive arc pairs with respect to the critical machine that one is chosen that maximizes b_{ij} and, in case of a tie, the pair is chosen with the maximum a_{ij}. Carlier and Pinson were the first to prove that an optimal solution of the 10×10 benchmark has a makespan of 930. In order to reach this goal a lot of work had to be done. In 1971 Florian et al. [FTM71] proposed a branch and bound algorithm where at a certain time the set of available tasks is considered, i.e. all tasks without any unscheduled predecessor are possible candidates for branching. At each node of the enumeration tree the number of branches generated corresponds to the number of available tasks competing for a particular machine. Branching continues from that node with smallest lower bound regarding the node associated to the partial schedule. As lower bounds Florian et al. used a one machine lower bound without considering tails, i.e. the optimal sequencing of the tasks on this particular machine is in increasing order of the earliest possible start times. They could find a solution of 1041 for the 10×10 benchmark, thus, a slight improvement compared to Balas' best solution of 1177 obtained two years earlier. His work was based on the disjunctive graph concept where he considered two successor nodes in the enumeration tree instead of as many nodes as there are conflicting tasks. McMahon and Florian [MF75] laid the foundation for Carlier's one machine paper. Contrary to earlier approaches where the nodes of the enumeration tree corresponded to incomplete (partial) schedules, they associated a complete solution with each search tree node. An initial solution and upper bound is obtained by Schrage's algorithms, i.e. among all available (released) tasks choose always that one with longest tail (earliest due date). Their objective is to minimize maximum lateness on one machine where each task is described by its release time, the processing time, and its due date. At any search node an MF-critical task T_j (with respect to the node associated schedule) is defined to be a task that realizes the value of the maximum lateness in the given schedule. Hence, an improvement is only possible if T_j is scheduled earlier. The idea of branching is to consider those tasks having greater due dates than the MF-critical task (i.e. having smaller tails than the MF-critical task) and to schedule these tasks after the MF-critical one. They continuously apply Schrage's algorithm in order to obtain a feasible schedule while the heads are adapted appropriately. McMahon and Florian also used their branching structure in order to solve the minimum makespan job shop scheduling problem and reached a value of 972 for the 10×10 problem. Moreover, they were the first to solve the Fisher and Thompson 5×20 benchmark to optimality, i.e. a makespan of 1165.

In [LLRK77] the one machine lower bound is introduced, hence extending the previously used lower bounds. They generated all active schedules branching over the conflict set in Giffler and Thompson's algorithm (see Section 3) or branching over the disjunctive arcs. A priority rule at each node of the search tree delivers an upper bound. There is no report on the notorious 10 jobs 10 machines problem.

Barker and McMahon [BM85] associated with each node in their enumeration tree a sub-problem whose solutions are a subset of the solution set of the original problem, a complete schedule, a BM-critical block in the schedule which is used to determine the descendant sub-problems, and a lower bound on the value of the solutions of the sub-problem. The lower bound is a single machine lower bound as computed in [MF75]. Each node of the search tree is associated with a different sub-problem. Hence at each node in the search tree there is a complete schedule containing a BM-critical task (a BM-critical task is the earliest scheduled task T_i where $t_i + q_i$ is at least the value of the best known solution) and an associated BM-critical block (a continuous sequence of tasks on a single machine ending with a BM-critical task). The BM-critical task must be scheduled earlier if this sub--problem is to yield an improved solution. Thus, a set of sub-problems is explored, in each of which a different member of the BM-critical block is made to precede all other members or to be the last of the tasks in the block to be scheduled. Earliest start times and tails are adapted accordingly. While Barker and McMahon reached a value of 960 for the 10×10 problem they were not able to solve the 5×20 problem to optimality. Only a value of 1303 is obtained.

Branching in the algorithm [BJS92, BJS94] is also restricted to moves of tasks which belong to a critical path of a solution obtained by a heuristic based on dispatching rules. For a *block* $\mathcal{B}$, i.e. successively processed tasks on the same machine, that belongs to a critical path, new sub-trees are generated if a task is moved to the very beginning or the very end of this block. In any case the critical path is modified and additional disjunctive arcs are selected according to formulae (10.2.2)-(10.2.6) proposed in [CP89]. Brucker et al. [BJS92, BJS94] calculated different lower bounds: one machine relaxations and two jobs relaxation, cf. [BJ93]. Moreover, if task T_i is moved before the block $\mathcal{B}$, all disjunctive arcs $\{(i, j) \mid T_j \in \mathcal{B} \text{ and } T_j \neq T_i\}$ are fixed. Hence,

$$r_i + p_i + \max \{ \max_{T_j \in \mathcal{B}, T_j \neq T_i} (p_j + q_j), \sum_{T_j \in \mathcal{B}, T_j \neq T_i} p_j + \min_{T_j \in \mathcal{B}, T_j \neq T_i} q_j \}$$

is a simple lower bound for the search tree node. Similarly, the value

$$\max \{ \max_{T_j \in \mathcal{B}, T_j \neq T_i} (r_j + p_j), \sum_{T_j \in \mathcal{B}, T_j \neq T_i} p_j + \min_{T_j \in \mathcal{B}, T_j \neq T_i} r_j \} + p_i + q_i$$

is a lower bound for the search tree node if task T_i is moved to the very end position of block $\mathcal{B}$. In order to keep the generated sub-problems non-intersecting it is necessary to fix some additional arcs. Promising sub-problems are heuristically detected. The branch and bound [BJS92, BJS94] improves and accelerates the branch and bound algorithm [CP89] substantially and easily reaches an optimal schedule for the 10×10 problem. However, to find an optimal solution for the 5×20 problem within a reasonable amount of time was impossible.

If we add a value δ at the left hand side of the head and tail update rules (10.2.3) and (10.2.4) then they can be considered as equations. Depending on the choice of C_{max} integer δ can also be positive. This results in the assignment of time windows $[r_k, r_k + \delta]$ of possible start times of tasks T_k to task sets supposed to be scheduled on the same machine. The branching idea of Martin and Shmoys

[MS96] uses the tightness of these windows as a branching criterion. For tight or almost tight windows, where the window size equals (almost) the sum of the processing times of the task set C, branching depends on which task in C is processed first. When a task is chosen to be first the size of its window is reduced. The size of the windows of the other tasks in C are updated in order to reflect the fact that they cannot start until the chosen task is completed. Martin and Shmoys needed about 9 minutes for finding an optimal schedule for the 10×10 problem. Comparable propagation ideas (after branching on disjunctive arcs) based on time window assignments to tasks are considered in [CL95]. They found an optimal schedule to the 10×10 problem within less than 3 minutes. In both papers, the updating of windows on tasks of one machine causes further updates on all other machines. This iterated one machine window reduction algorithm generated lower bounds superior to the one machine lower bound.

Perregaard and Clausen [PC95] obtained excellent results through a parallel branch and bound algorithm on a 16-processor system based on Intel i860 processors each with 16 MB internal memory. There is a peak performance of about 500 MIPS. As a lower bound Jackson's preemptive schedule is used. A branching strategy is the one from [CP89] where a new disjunctive arc pair describes the branches originating from a node, this is done in analogy to the rules (10.2.3)-(10.2.7). Another branching strategy considered is the one described in [BJS94], i.e. moving tasks to block ends. Perregaard and Clausen easily found optimal solutions to the 10×10 and 5×20 problems, both in time much less than a minute (of course including the optimality proof). For some other even more difficult problems they could prove optimality or obtained results unknown up to now.

10.2.4 Valid Inequalities

Among the most efficient algorithms for solving the job shop scheduling problem exactly is the branch and bound approach by Applegate and Cook [AC91]. In order to obtain good lower bounds they developed cutting plane procedures for both the disjunctive and the mixed integer problem formulation, see [Man60]. In the latter case the disjunctive constraints can be modeled by introducing a binary variable y_{ij}^k for any task pair T_i, T_j supposed to be processed on the same machine P_k. The interpretation is, y_{ij}^k equals 1 if T_i is scheduled before T_j on machine P_k, and 0 if T_j is scheduled before T_i. Let Ω be some large constant. Then the following inequalities hold for all tasks T_i and T_j on machine P_k:

$$t_i \geq t_j + p_j - \Omega y_{ij}^k \tag{10.2.8}$$

$$t_j \geq t_i + p_i - \Omega(1 - y_{ij}^k) \tag{10.2.9}$$

Starting from the LP-relaxation, valid inequalities involving variables y_{ij}^k as well as inequalities developed for the disjunctive programming formulation are generated. Consider a feasible solution and set C of tasks processed on the same machine P_k. Let $T_i \in C$ be a task processed on P_k and assume all members of the subset C_i of C are scheduled before T_i on P_k. Then the start time t_i of task $T_i \notin C_i$ satisfies

$$t_i \geq \min \{r_j \mid T_j \in C_i\} + \sum_{T_j \in C_i} p_j \geq \min \{r_j \mid T_j \in C\} + \sum_{T_j \in C_i} p_j . \qquad (10.2.10)$$

In order to become independent of the considered schedule we multiply this inequality by p_i and take the sum over all members of C. Hence, we get the *basic cuts* from [AC91]:

$$\sum_{T_i \in C} t_i p_i \geq (\sum_{T_i \in C} p_i) \min\{r_j \mid T_j \in C\} + \frac{1}{2} \sum_{\substack{T_i \in C \\ T_i \neq T_j}} \sum_{\substack{T_j \in C \\ T_j \neq T_i}} p_j p_i . \qquad (10.2.11)$$

With the addition of the variables y_{ij}^k we can easily reformulate (10.2.10) and obtain the *half cuts*

$$t_i \geq \min \{r_j \mid T_j \in C\} + \sum_{\substack{T_j \in C \\ T_j \neq T_i}} y_{ji}^k p_j \qquad (10.2.12)$$

because $y_{ji}^k = 1$ if and only if $T_j \in C_i$.

Let T_i and T_j be two tasks supposed to be scheduled on the same machine. Let α and β be any two nonnegative parameters. Assume T_i is supposed to be processed before T_j, then $\alpha t_i + \beta t_j \geq \alpha r_i + \beta (r_i + p_i)$ because $t_i \geq r_i$ and task T_j cannot start before T_i is finished. Similarly, under the assumption that T_j is scheduled before T_i, we get $\alpha t_i + \beta t_j \geq (r_j + p_j) + \beta r_j$. Thus, both inequalities hold, for instance, if the right hand sides of these inequalities are equal. Both inequalities are satisfied if $\alpha = p_i + r_i - r_j$ and $\beta = p_j + r_j - r_i$. Hence, the *two-job cuts* [Bal85]

$$(p_i + r_i - r_j) t_i + (p_j + r_j - r_i) t_j \geq p_i p_j + r_i p_j + r_j p_i \qquad (10.2.13)$$

sharpen (10.2.11) if $r_i + p_i > r_j$ and $r_j + p_j > r_i$.

The lower bounds in the branch and bound algorithm [AC91] are based on these inequalities, the corresponding reverse ones, i.e. considering the jobs in reverse order, and a couple of additional cuts. The cutting plane based lower bounds are superior to the one machine lower bounds, however, at the cost of additional computation time. For instance, for the 10×10 problem Applegate and Cook were able to improve the one machine bound of 808 obtained in less than one second up to 824 (in 300 seconds) or 827 in 7500 seconds. The branch and bound tree is established continuing the search from a tree node where the preemptive one machine lower bound is minimum. The branching scheme results from the disjunctive model, i.e. each disjunctive edge defines two sub-problems according to the corresponding disjunctive arcs. However, among all possible disjunctive edges the one is chosen, connecting tasks T_i and T_j, which maximizes the minimum $\{LB(i \rightarrow j), LB(j \rightarrow i)\}$ where $LB(i \rightarrow j)$ and $LB(j \rightarrow i)$ are the two preemptive one machine lower bounds for the generated sub-problems where the disjunctive arcs (i, j) or (j, i), respectively, are selected. Furthermore, in a more sophisticated branch and bound method branching is realized on the basis of the values a_{ij} and b_{ij} of (10.2.7). Then, based on the work [CP89], inequalities (10.2.3) and (10.2.4) are applied to all task subsets on each particular machine in order to eliminate disjunctive edges, simplifying the problem. In order to get a high quality feasible solution Applegate and Cook modified the shifting bottle-neck procedure [ABZ88]. After scheduling all but s machines, for the remaining s machines the bottleneck criterion (see below) is replaced by complete enumeration. Not necessarily the machine with largest makespan is included into the partial schedule but each of the remaining s machines is considered to be the one introduced next into the partial schedule. The value s has to be small in order to keep the computation time low.

In order to obtain better feasible solutions they proceed as follows. Given a complete schedule the processing order of the jobs on a small number s of machines is kept fixed. The processing order on the remaining machines is skipped. The resulting partial schedule can be quickly completed to a new schedule using the aforementioned branch and bound procedure. If the new schedule is shorter than the original, then this process is repeated with the restriction that the set of machines whose schedules are kept fixed is modified. The number s of machines to fix follows the need to have enough structure to rapidly fill in the rest of the schedule and leave a sufficient amount of freedom for improving the processing orders (see [AC91], for additional information on how to choose s). Applegate and Cook easily found an optimal solution to the 10×10 job shop in less than 2 minutes (including the proof of optimality).

10.3 Approximation Algorithms

10.3.1 Priority Rules

Priority rules are probably the most frequently applied heuristics for solving (job shop) scheduling problems in practice because of their ease of implementation and their low time complexity. The algorithm of Giffler and Thompson [GT60] can be considered as a common basis of all priority rule based heuristics. Let $Q(t)$ be the set of all unscheduled tasks at time t. Let r_i and C_i denote the earliest possible start and the earliest possible completion time, respectively, of task T_i.

The algorithm of Giffler and Thompson assigns available tasks to machines, i.e. tasks which can start being processed. Conflicts, i.e. tasks competing for the same machine, are solved randomly. A brief outline of the algorithm is given in Algorithm 10.3.1.

Algorithm 10.3.1 *The algorithm of Giffler and Thompson* [GT60].
```
begin
```
$t := 0; Q(t) := \{T_1, \cdots, T_{n-1}\};$
```
repeat
```
Among all unscheduled tasks in $Q(t)$ let T_{j*} be the one with smallest completion time, i.e. $C_{j*} = \min \{C_j \mid T_j \in Q(t), C_j = \max\{t, r_j\} + p_j\}$. Let P_{k*} denote
the machine T_{j*} has to be processed on;

Randomly choose a task T_i from the conflict set $\{ T_j \in Q(t) \mid T_j$ has to be processed on machine P_{k*} and $r_j < C_{j*}\}$;

$Q(t) := Q(t) - \{T_i\}$;

Modify C_j for all tasks $T_j \in Q(t)$ supposed to be processed one machine P_{k*};

Set t to the next possible task to machine assignment, i.e. $C_{i*} = \min\{C_i \mid T_i$ is
in process on some machine P_k and there is at least one task in $Q(t)$ that requires $P_k\}$;

$r_{j*} := \min\{r_j \mid T_j \in Q(t)\};$
$t := \max\{ C_{i*}, r_{j*}\};$
```
until Q(t) is empty
end;
```

The Giffler-Thompson algorithm can generate all *active schedules* (a schedule is said to be active, if no task can start its processing without delaying any other task) among which are also optimal schedules. As the conflict set consists only of tasks, i.e. jobs, competing for the same machine, the random choice of a task or job from the conflict set may be considered as the simplest version of a priority rule where the priority assigned to each task or job in the conflict set corresponds to a certain probability. Many other priority rules can be considered, e.g. the total processing time of all tasks succeeding T_i in a given job. A couple of rules are collected in Table 10.3.1; for an extended summary and discussion see [PI77, BPH82, Hau89]. The first column of Table 10.3.1 contains an abbreviation and name of the rule while the last column describes which task or job in the conflict set gets highest priority.

	rule	*description*
1.	STT-rule (shortest task time)	A task with a shortest processing time on the considered machine.
2.	LTT-rule (longest task time)	A task with a longest processing time on the machine considered.
3.	LRPT-rule (longest remaining processing time)	A task with a longest remaining job processing time.
4.	SRPT-rule (shortest remaining processing time)	A task with a shortest remaining job processing time.
5.	LTRPT-rule (longest task remaining processing time)	A task with a highest sum of tail and task processing time.
6.	Random	A task for the considered machine is randomly chosen.
7.	FCFS-rule (first come first served)	The first task in the queue of jobs waiting for the same machine.
8.	SPT-rule (shortest processing time)	A job with a smallest total processing time.
9.	LPT-rule (longest processing time)	A job with a longest total processing time.
10.	LTS-rule (longest task successor)	A task with a longest subsequent task processing time.
11.	SNRT-rule (smallest number of remaining tasks)	A task with a smallest number of subsequent tasks in the job.
12.	LNRT-rule (largest number of remaining tasks)	A task with a largest number of subsequent tasks in the job.

Table 10.3.1 *Priority rules.*

10.3.2 The Shifting Bottleneck Heuristic

The shifting bottleneck heuristic ([ABZ88], [BLV95 and DMU97, PM00]) is one of the most powerful procedures among heuristics for the job shop scheduling problem. The idea is to solve for each machine a one machine scheduling problem to optimality under the assumption that a lot of arc directions in the optimal one machine schedules coincide with an optimal job shop schedule. Consider all tasks of a job shop scheduling instance that have to be scheduled on machine P_k. In the (disjunctive) graph including a partial selection among opposite directed arcs (corresponding to a partial schedule) there exists a longest path of length r_i

from dummy vertex 0 to each vertex i corresponding to T_i scheduled on machine P_k. Processing of task T_i cannot start before r_i. There is also a longest path of length q_i from i to the dummy node n. Obviously, when T_i is finished it will take at least q_i time units to finish the whole schedule. Although the one machine scheduling problem with heads and tails is NP-hard, there is the powerful branch and bound method (see Section 10.2.2) proposed by Potts [Pot80b] and Carlier [Car82, Car87] which dynamically changes heads and tails in order to improve the tasks sequence.

The shifting bottleneck heuristic consists of two subroutines. The first one (SB_1) repeatedly solves one machine scheduling problems while the second one (SB_2) builds a partial enumeration tree where each path from the root to a leaf is similar to an application of SB_1. As its name suggests, the shifting bottleneck heuristic always schedules bottleneck machines first. As a measure of the bottleneck quality of machine P_k, the value of an optimal solution of a one machine scheduling problem on machine P_k is used. The one machine scheduling problems considered are those which arise from the disjunctive graph model when certain machines are already sequenced. The task orders on sequenced machines are fully determined. Hence sequencing an additional machine probably results in a change of heads and tails of those tasks of which the machine order is still open. For all machines not sequenced, the maximum makespan of the corresponding optimal one machine schedules, where the arc directions of the already sequenced machines are fixed, determines the bottleneck machine. In order to minimize the makespan of the job shop scheduling problem, the bottleneck machine should be sequenced first. A brief statement of the shifting bottleneck procedure is given in Algorithm 10.3.2.

Algorithm 10.3.2 *Shifting bottleneck (SB_1) heuristic.*
begin
Let $\mathcal{P}$ be the set of all machines and let $\mathcal{P}' := \varnothing$ be the - initially empty - set of
    all sequenced machines;
repeat
   for $P_k \in \mathcal{P} - \mathcal{P}'$ **do**
      begin
      Compute head and tail for each task T_i that has to be scheduled on
         machine P_k;
      Solve the one machine scheduling problem to optimality for machine P_k;
      Let $C(k)$ be the resulting makespan for this machine;
      end;
   Let P_{k*} be the bottleneck machine, i. e. $C(k^*) \geq C(k)$ for all $P_k \in \mathcal{P} - \mathcal{P}'$;
   $\mathcal{P}' := \mathcal{P}' \cup \{ P_{k*} \}$;
   for $P_k \in \mathcal{P}'$ in the order of its inclusion **do**   -- local re-optimization

```
    begin
    Delete all arcs between tasks on P_k while all arc directions between tasks
        on machines from P' - {P_k} are fixed;
    Compute heads and tails of all tasks on machine P_k and solve the one
        machine scheduling problem and reintroduce the obtained task orders on
        P_k;
    end;
  until  P = P'
end;
```

The one machine scheduling problems, although they are NP-hard (contrary to the preemptive case, cf. [BLL+83]), can quickly be solved using the algorithm [Car82]. Unfortunately, adjusting heads and tails does not take into account a possible already fixed processing order of tasks connecting two tasks T_i and T_j on the same machine, whereby this particular machine is still unscheduled. So, we get one machine scheduling problems with heads, tails, and time lags (minimum delay between two tasks), problems which cannot be handled with Carlier's algorithm. In order to overcome these difficulties an improved SB_1 version is suggested by Dauzere-Peres and Lasserre [DL93] using approximate one machine solutions. Balas et al. [BLV95] solved the one machine problems exactly. So, there is a SB_1-heuristic superior to the SB_1-heuristic proposed in [ABZ88]. On average, the SB_1-heuristic results from [BLV95] are slightly worse than those obtained by the SB_2-heuristic from [BLV95].

During the local re-optimization part of the SB_1-heuristic, the task sequence is re-determined for each machine, keeping the sequences of all other already scheduled machines untouched. As the one machine problems use only partial knowledge of the whole problem, it is not surprising, that optimal solutions will not be found easily. This is even more the case because Carlier's algorithm considers the one machine problem as consisting of independent tasks while some dependence between tasks of a machine might exist in the underlying job shop scheduling problem. Moreover, a monotonic decrease of the makespan is not guaranteed in the re-optimization step of Adams et al. [ABZ88]. Dauzere-Peres and Lasserre [DL93] were the first to improve the robustness of SB_1 and to ensure a monotonic decrease of the makespan in the re-optimization phase and eliminate sensitivity to the number of local re-optimization cycles. Contrary to Carlier's algorithm, they update the task release time s each time they select a new task by Schrage's procedure. They obtained a solution of 950 for the 10×10 problem using this modified version of the SB_1-heuristic.

The quality of the schedules obtained by the SB_1-heuristic heavily depends on the sequence in which the one machine problems are solved and thus on the order these machines are included in the set P'. Sequence changes may yield substantial improvements. This is the idea behind the second version of the shifting bottleneck procedure, i.e. the SB_2-heuristic, as well as behind the second genetic algorithm approach by Dorndorf and Pesch [DP95]. The SB_2-heuristic applies a slightly modified SB_1-heuristic to the nodes of a partial enumeration tree. A node corresponds to a set $\mathcal{P}'$ of machines that have been sequenced in a particular way. The root of the search tree corresponds to $\mathcal{P}' = \varnothing$. A branch corresponds to the inclusion of machine P_k into $\mathcal{P}'$, thus the branch leads to a node representing an extended set $\mathcal{P}' \cup \{P_k\}$. At each node of the search tree a single step of the SB_1-heuristic is applied, i.e. machine P_k is included followed by a local re-optimization. Each node in the search tree corresponds to a particular sequence of inclusions of the machines into set $\mathcal{P}'$. Thus, the bottleneck criterion no longer determines the inclusion into $\mathcal{P}'$. Obviously a complete enumeration of the search tree is not acceptable. Therefore a breadth-first search up to depth l is followed by a depth-first search. In the former case, for a search node corresponding to set $\mathcal{P}'$ all possible branches are considered which result from inclusion of machine $P_k \notin \mathcal{P}'$. Hence the successor nodes of node $\mathcal{P}'$ correspond to machine sets $\mathcal{P}' \cup \{P_k\}$ for all $P_k \in \mathcal{P} - \mathcal{P}'$. Beyond the depth l an extended bottleneck criterion is applied, i.e. instead of $|\mathcal{P} - \mathcal{P}'|$ successor nodes there are several successor nodes generated corresponding to the inclusion of the bottleneck machine as well as several other machines P_k to $\mathcal{P}'$.

10.3.3 Opportunistic Scheduling

For long time priority rules were the only possible way to tackle job shops of at least 100 tasks [CGTT63]. Recently, generally applicable approximation procedures such as tabu search, simulated annealing or genetic algorithm learning strategies became very attractive and successive solution strategies. Their general idea is to modify current solutions in a certain sense, where the modifications are defined by a neighborhood operator, such that new feasible solutions are generated, the so called neighbors, which hopefully have an improved or at most limited deterioration of their objective function value. In order to reach this goal problem specific knowledge, incorporated by problem specific heuristics, has to be introduced into the local search process of the general problem solvers (see Section 2.5.2).

Knowledge based scheduling systems have been built by various people using various techniques. Some of them are rule based systems others are based on frame representations. Some of them use heuristic rules only to construct a schedule, others conduct a constraint directed state space search. ISIS [Fox87, FS84] is a constraint directed reasoning system for the scheduling of factory job shops. The main feature is that it formalizes various scheduling influences in the form of constraints on the system's knowledge base and uses these constraints to guide the search in order to generate heuristically the schedule. In each scheduling cycle it first selects an order of tasks to be scheduled according to priority rules and then proceeds through a level of analysis of existing schedules, a level of constraint directed search and a level of detailed assignment of resources and time intervals for each task in order. A large amount of work done by ISIS actually involves the extraction and organization of constraints that are created specifically for the problem under consideration. Scheduling relies only on order based problem decomposition. The system OPIS [OS88, SFO86] which is a direct descendant of ISIS attempts to make some progress by concentrating more on bottlenecks and scheduling under the perspective of resource based decomposition, cf. [ABZ88, CPP92, BLV95, DL93]. The term "opportunistic reasoning" has been used to characterize a problem-solving process whereby activity is consistently directed toward those actions that appear most promising in terms of the current problem-solving state. The strategy is to identify the most "solvable" aspects of the problem (e.g. those aspects with the least number of choices or where powerful heuristics are known) and develop candidate solutions to these sub-problems. However the way in which a problem is decomposed affects the quality of the solution reached. No sub-problem contains all the information of the original problem. Sub-problems should be as independent as possible in terms of effects of decisions on other sub-problems. OPIS is an opportunistic scheduling system using a genetic opportunistic scheduling procedure. For instance, it constantly redirects the scheduling effort towards those machines that are likely to be the most difficult to schedule (so-called bottleneck machines). Decomposing the job shop into single machine scheduling problems bottleneck machines might get a higher priority for being scheduled first. Hence, dynamically revised decision making based on heuristic rules focuses on the most critical decision points and the most promising decisions at these points, cf. [Sad91]. The average complexity of the procedures is kept on a very low level by interleaving the search with application of consistency enforcing techniques and a set of look-ahead techniques that help to decide which task to schedule next (i.e. so-called variable-ordering and value-ordering techniques). Clearly, start times of tasks competing for highly contended machines are more likely to become unavailable than those of other tasks. A critical variable is one that is expected to cause backtracking, i.e. one which remaining possible values are expected to conflict with the remaining possible values of other variables. A good value is one that is expected to participate in many solutions. Contention between unscheduled tasks for a machine over some time interval is determined by the number of unscheduled tasks competing for that machine/time interval and the reliance of each one of these tasks on the availability of this machine/time interval. Typically, tasks with few possible starting times left will heavily rely on the availability of any one of these remaining starting times in competition, whereas tasks with many remaining starting times will rely much less on any one of these times. Each starting time is assigned a subjective probability to be assigned to a particular task. The task with the highest contribution to the demand for the most contended machine/time interval is considered the most likely to violate a constraint, cf. [CL95, PT96].

Very recent solution approaches use ant colony optimization [MFP06, BS04] and artificial immune systems [CAK+06] as successful and competitive algorithms.

10.3.4 Local Search

An important issue is the extent to which problem specific knowledge must be used in the construction of learning algorithms (in other words the power and quality of inferencing rules) capable to provide significant performance improvements. Very general methods having a wide range of applicability in general are weak with respect to their performance. Problem specific methods achieve a highly efficient learning but with little use in other problem domains. Local search strategies are falling somewhat in between these two extremes, where genetic algorithms or neural networks tend to belong to the former category while tabu search or simulated annealing etc. are counted as instances of the second category. Anyway, these methods can be viewed as tools for searching a space of legal alternatives in order to find a best solution within reasonable time limitations. When sufficient knowledge about the search space is available a priori, one can often exploit that knowledge (inference) in order to introduce problem specific search strategies capable to find rapidly solutions of higher quality. Whiteout such a priori knowledge, or in cases where close to optimum solutions are indispensable, information about the problem has to be accumulated dynamically during the search process. Likewise obtained long-term as well as short-term memorized knowledge constitutes one of the basic parts in order to control the search process and in order to avoid getting stuck in a locally optimal solution. In random search finding an acceptable solution within a reasonable amount of time is impossible because any kind of random search is not using any knowledge generated during the search process in order to improve its performance. Any global information assessed during the search will not be exploited.

Local search algorithms (see section 2.5) provide general problem solving strategies incorporating and exploiting problem-specific knowledge capable even to explore search spaces containing an exponentially growing number of local optima with respect to the problem defining parameters.

Tabu Search and Simulated Annealing Based Job Shop Scheduling

In the 90's local search based scheduling of job shops became very popular; for a survey see [VAL96, Vae95, AGP97, WW95]. These algorithms are all based on a certain neighborhood structure. A simple neighborhood structure (N_1) has been used in the simulated annealing procedure of [LAL92]:

N_1: *Transition from a current solution to a new one is generated by replacing in the disjunctive graph representation of the current solution a disjunctive arc (i, j) on a critical path by its opposite arc (j, i).*

In other words, N_1 means reversing the order in which two tasks T_i and T_j (or jobs) are processed on a machine where these two tasks belong to a longest path. This parallels the early branching structures of exact methods. It is possible to construct a finite sequence of transitions leading from a locally optimal solution to the global optimum, i.e. the neighborhood is connected. This is a necessary and sufficient condition for asymptotic convergence of simulated annealing. On average (on VAX 785 over 5 runs on each instance) it took about 16 hours to solve the 10×10 benchmark to optimality. A value of 937 was reached within almost 100 minutes. The 5×20 benchmark problem was solved to optimality within almost 18 hours.

Lourenço [Lou93, Lou95] introduces a combination of small step moves based on the neighborhood N_1 and large step moves in order to reach new search areas. The small steps are responsible for search intensification in a relatively narrow area. Therefore a simple hill-climbing as well as simulated annealing are used, both with respect to neighborhood N_1. The large step moves modify the current schedule and drive the search to a new region. Simultaneously a modest optimization is performed to obtain a schedule reasonably close to a local optimum by local search such as hill-climbing or simulated annealing. The large steps considered are the following: Randomly select two machines and remove all disjunctive arcs connecting tasks on these two machines in the current schedule. Then solve the two one machine problems - using Carlier's algorithm or allowing preemption and considering time lags - and return the obtained arcs according to their one machine solutions into the whole schedule. Starting solutions are generated through some randomized dispatching rules, one for an instance, in the same way as in [Bie95] (see below).

More powerful neighborhood definitions are necessary. A neighborhood N_2 defined in [MSS88], has been also applied in the local search improvement steps of the genetic algorithms in [ALLU94]:

N_2: *Consider a feasible solution and a critical arc (i, j) defining the processing order of tasks T_i and T_j on the same machine, say machine P_k. Define $T_{ipred(i)}$ and $T_{isucc(i)}$ to be the immediate predecessor and immediate successor of T_i, respectively, on machine P_k. Restrict the choice of arc (i, j) to those vertices for which at least one of the arcs $(ipred(i), i)$ or $(j, isucc(j))$ is not on a longest path, i.e. i or j are block end vertices (cf. the branching structure in BJS94). Reverse (i, j) and, additionally also reverse $(ipred(h), h)$ and $(l, isucc(l))$ - provided they exist - where T_h directly precedes T_i in the job, and T_l is the immediate successor of T_j in the job. The latter arcs are reversed only if a reduction of the makespan can be achieved.*

Thus, a neighbor of a solution with respect to N_2 may be found by reversing more than one arc. Within a time bound of 99 seconds the results of two simulated annealing algorithms based on the two different neighborhood structures N_1 and N_2 were 969 and 977, respectively, for the 10×10 problem as well as 1216 and 1245, respectively, for the 5×20 problem, see [ALLU94].

Dell'Amico and Trubian [DT93] considered the problem as being symmetric and scheduled tasks bi-directionally, i.e. from the beginning and from the end, in order to obtain a priority rule based feasible solution. The resulting two parts finally are put together in order to constitute a complete solution. The neighborhood structure (N_3) employed in their tabu search extends the connected neighborhood structure N_1:

N_3: *Let (i, j) be a disjunctive critical arc. Consider all permutations of the three vertices $\{ipred(i), i, j\}$ and $\{i, j, isucc(j)\}$ in which (i, j) is reversed.*

Again, it is possible to construct a finite sequence of moves with respect to N_3 which leads from any feasible solution to an optimal one. In a restricted version N_3' of N_3 arc (i, j) is chosen such that either T_i or T_j is the end vertex of a block. In other words, arc (i, j) is not considered as candidate when both $(ipred(i), i)$ and $(j, isucc(j))$ are on a longest path in the current solution. N_3' is not any longer a connected neighborhood. Another branching scheme is considered to define a neighborhood structure N_4:

N_4: *For all tasks T_i in a block move T_i to the very beginning or to the very end of this block.*

Once more, N_4 is connected, i.e. for each feasible solution it is possible to construct a finite sequence of moves, with respect to N_4, leading to a globally optimal solution. For a while the tabu search [DT93] was the most powerful method to solve job shops. They were able to find an optimal solution to the 5×20 problem within 2.5 minutes and a solution of 935 to the 10×10 problem in about the same amount of time.

N_1 and N_4 are also the two neighborhood structures used in the tabu search of [SBL95]. In 40 benchmark problems they always obtained better solutions or reduced running times compared to the shifting bottleneck procedure. For instance, they generated an optimal solution to the 10×10 problem within 157 seconds.

In the parallel tabu search Taillard [Tai94] used the N_1 neighborhood. Every 15 iterations the length of the tabu list is randomly changed between 8 and 14. He obtained high quality solutions even for very large problem instances up to 100 jobs and 20 machines.

Barnes and Chambers [BC95] also used N_1 in their tabu search algorithm. They fixed the tabu list length and whenever no feasible move is available the list entries are deleted. Start solutions are obtained through dispatching rules.

Nowadays, the most efficient tabu search implementations are described in [NS96, NS05] and [BV98]. The size of the neighborhood N_1 depends on the number of critical paths in a schedule and the number of tasks on each critical path. It can be pretty large. Nowicki and Smutnicki [NS96] consider a smaller neighborhood (N_5) restricting N_1 (or N_4) to reversals on the border of a block. Moreover, they restrict to a single critical path arbitrarily selected

N_5: *A move is defined by the interchange of two successive tasks T_i and T_j, where either T_i or T_j is the first or last task in a block that belongs to a critical path. In the first block only the last two tasks and symmetrically in the last block of the critical path only the first two tasks are swapped.*

The set of moves is not empty only if the number of blocks is more than one and if at least one block consists of more than one task. In other words, if the set of moves is empty then the schedule is optimal. If we consider neighborhoods N_1 and N_5 in more detail, then we can deduce: A schedule obtained from reversing any disjunctive arc which is not critical cannot reduce the makespan; a move that belongs to N_1 but not to N_5 cannot reduce the makespan. Let us go into more detail of [NS96], see also [JRM00].

The neighborhood search strategy includes an aspiration criterion and reads as follows:

Algorithm 10.3.3 *Neighborhood search strategy of Nowicki-Smutnicki,* [NS96].

begin

Let x be a current schedule (feasible solution) with makespan C_{max}^x;

$\mathcal{N}(x)$ denotes the set of all neighbors of x;

C_{max} is the makespan of the currently best solution;

T is a tabu list;

Let $\mathcal{A}$ be the set $\{x' \in \mathcal{N}(x) \mid \text{Move}(x \to x') \in T$ and $C_{max}^{x'} < C_{max}\}$;

    -- i.e. all schedules in $\mathcal{A}$ satisfy the aspiration criterion
    -- to improve the currently best makespan.

if $\{\mathcal{N}(x) \mid \text{Move}(x \to x')$ is not tabu$\} \cup \mathcal{A}$ is not empty
then

Select y such that

$C_{max}^y = \min\{C_{max}^{x'} \mid x' \in \mathcal{N}(x)$ or if $\text{Move}(x \to x')$ is tabu then $x' \in \mathcal{A}\}$

else

  repeat

Drop the "oldest" entry in T and append a copy of the last element in T

  until there is a non-tabu move $\text{Move}(x \to x')$;

Let $\text{Move}(x \to x')$ be defined by arc (i, j) in the disjunctive graph of x, then append arc (j, i) to T

end;

The design of a classical tabu search algorithm is straightforward. A stopping criterion is when the optimal schedule is detected or the number of iterations without any improvement exceeds a certain limit. The initial solution can be generated using an insertion technique, e.g. as described in [NEH83]. Nowicki and Smutnicki note that the essential disadvantage of this approach consists of loosing information about previous runs. Therefore they suggest to build up a list of the l best solutions and their associated tabu lists during the search. Whenever the classical tabu search has finished go back to the most recent entry, i.e. the best schedule x from this list of at most l solutions, and restart the classical tabu search. Whenever a new best solution is encountered the list of best solutions is updated. This extended tabu search "with backtracking" continues until the list of best solutions is empty. Nowicki and Smutnicki obtained very good results; for instance, they could solve the notorious 10×10 problem within 30 seconds to optimality, even on a small personal computer. They solved the 5×20 problem within 3 seconds to optimality.

The idea of Balas and Vazacopoulos [BV98] of the guided local search procedure is based on reversing more than one disjunctive arc at a time. This leads to a considerably larger neighborhood than in the previous cases. Moreover, neighbors are defined by interchanging a set of arcs of varying size, hence the search is of variable depth and supports search diversification in the solution space. The employed neighborhood structure (N_6) is an extension of all previously encountered neighborhood structures. Consider any feasible schedule x and any two tasks T_i and T_j to be performed on the same machine, such that i and j are on the same critical path, say $CP(0, n)$, but not necessarily adjacent. Assume T_i is processed before T_j. Besides $T_{ipred(i)}$, $T_{ipred(j)}$ and $T_{isucc(i)}$, $T_{isucc(j)}$, the immediate machine predecessors and machine successors of T_i and T_j in x, let $T_{a(i)}$, $T_{a(j)}$ and $T_{b(i)}$ and $T_{b(j)}$ denote the job predecessors and job successors of tasks T_i and T_j, respectively. Moreover, let $r(i) := r_i + p_i$ and $q(i) := p_i + q_i$ be the length of a longest path (including the processing time p_i of T_i) connecting 0 and i, or i and n. An interchange on T_i and T_j either is a move of T_i right after T_j (forward interchange) or a move of T_j right before T_i (backward interchange). We have seen that schedule x cannot be improved by an interchange on T_i and T_j if both tasks are adjacent and none of the vertices corresponding to them is the first or the last one of a block in $CP(0, n)$. In other words, in order to achieve an improvement either $a(i)$ or $b(j)$ must be contained in $CP(0, n)$. This statement can easily be generalized to the case where T_i is not an immediate predecessor of T_j. Thus for an interchange on T_i and T_j to reduce the makespan, it is necessary that the critical path $CP(0, n)$ containing i and j also contains at least one of the vertices $a(i)$ or $b(j)$. Hence, the number of "attractive" interchanges reduces drastically and the question remains, under which conditions an interchange on T_i and T_j is guaranteed not to create a cycle in the graph. It is easy to derive that a forward interchange on T_i and T_j yields a new schedule x' (obtained from x) if there is no directed path from $b(i)$ to j in x. Similarly, a backward interchange on T_i and T_j will not create a cycle if there is no directed path from i to $a(j)$ in x.

Now, the neighborhood structure N_6 can be introduced.

N_6: A neighbor x' of a schedule x is obtained by an interchange of two tasks T_i and T_j in one block of a critical path. Either task T_j is the last one in the block and there is no directed path in x connecting the job successor of T_i to T_j, or, task T_i is the first one in the block and there is no directed path in x connecting T_i to the job predecessor of T_j.

Whereas the neighborhood N_1 involves the reversal of a single arc (i, j) on a critical path the more general move defined by N_6 involves the reversal of potentially a large number of arcs.

Assume that an interchange on a task pair T_i, T_j results in a makespan increase of the new schedule x' compared to the old one x. Then it is obvious that every critical path in x' contains arc (j, i). The authors make use of this fact in order to further reduce the neighborhood size. Consider a forward interchange resulting in a makespan increase: Since (j, i) is a member of any critical path in x' the arc $(i, b(i))$ is as well (because T_i became the last task in its block). We have to distinguish two cases. Either the length of a longest path from $b(i)$ to n in x, say $q(b(i))$, exceeds the length of a longest path from $b(j)$ to n in x, say $q(b(j))$ or $q(b(i)) \le q(b(j))$. In the former case $q(b(i))$ is responsible for the makespan increase. In the latter case $isucc(i)$ is the first task in its block in x'. Hence, the length $r(j)$ of a longest path in x connecting 0 to j is smaller than the length $r'(i)$ of a longest path in x' connecting 0 to i. Thus, the number of interchange candidates can be reduced defining some guideposts. In a forward interchange a right guidepost h is reached if $q(b(h)) < q(b(i))$; a left guidepost h is reached if $r(j) < r'(h)$ holds. Equivalently, in a backward interchange that worsens the makespan of schedule x a left guidepost h is reached if $r(ipred(h)) < r(ipred(j))$; a right guidepost h is reached if $q(i) < q'(h)$ holds.

After an interchange that increased the makespan, if a left guidepost is reached the list of candidates for an interchange is restricted to those task pairs on a critical path in x' between 0 and j. If a right guidepost is reached candidates for an interchange are chosen from the segment on a critical path in x' between j and n. If both guideposts are reached the set of candidates is not restricted. Thus, in summary, if the makespan increases after an interchange, available guideposts restrict the neighborhood.

The guided local search procedure by Balas and Vazacopoulos [BV98] uses the neighborhood structure N_6 including the restrictions aforementioned. The procedure builds up an incomplete enumeration (called neighborhood) tree. Each node of the tree corresponds to a schedule, an edge of the tree joins two schedules x and x' where descendant x' is obtained through an interchange on two tasks T_i and T_j lying on a critical path in x. The arc (j, i) is fixed in all schedules corresponding to the nodes of the sub-tree rooted at x'. The number of direct descendants of x', i.e. the number of possible moves, is the entire neighborhood if x' is a shorter schedule than x. It is the restricted (with respect to the guideposts) neighborhood if the makespan of x' is worse than the one of x. The children of a node corresponding to schedule x are ranked by their evaluations. The number of children is limited by a decreasing function of the depth in the neighborhood tree. After an interchange on T_i, T_j leading from x to schedule x' the arc (j,i) remains fixed in all schedules of the sub-tree rooted in x'. Additionally, the arc (j,i) is also fixed in all schedules corresponding to brothers of x' (i.e. children of x) having a makespan worse than x' (in the sequence of the ranked list). Finally, besides arc fixing and limits on the number of children a third factor is applied to keep the size of the tree small. The depth of the tree is limited by a logarithmic function of the number of tasks on the tree's level. Altogether, the size of the neighborhood tree is bounded by a linear function of the number of tasks.

The number of neighborhood trees generated is governed by some rules. The root of a new neighborhood tree corresponds to the best schedule available if it is generated in the current tree. Otherwise, if the current tree is not a step into a better local optimum the root of the new tree is randomly chosen among the nodes of the current tree.

In order to combine local search procedures operating on different neighborhoods (which makes it more likely to escape local optima and explore regions not available by any single neighborhood structure) Balas and Vazacopoulos combined their guided local search with the shifting bottleneck procedure. Remember, every time a new machine has been sequenced the shifting bottleneck procedure re-optimizes the sequence of each previously processed machine, by again solving a one machine problem with the sequence on the other machines held fixed. The idea of Balas and Vazacopoulos is to replace the re-optimization cycle of the shifting bottleneck procedure with the neighborhood trees of the guided local search procedure. Whenever there are l fixed machine sequences defining a partial schedule the shifting bottleneck guided local search (SB-GLS) generates $2l|\mathcal{J}|$ neighborhood trees instead of starting a re-optimization cycle. The root of the first tree is defined by the partial schedule of the l already sequenced machines. The roots of the other trees are obtained as described above. The best schedule obtained from this incorporated guided local search is then used as a starting point for continuation of the shifting bottleneck procedure. A couple of modifications of SB-GLS ideas are applied which basically differ from SB-GLS in the number of sequenced machines (hence the root of the first neighborhood tree) in the shifting bottleneck part, cf. [BV98].

SB-GLS and its modifications is currently the most powerful heuristic to solve job shop scheduling problems. It outperforms many others in solution quality and computation time. Needless to say that all versions of GLS and SB-GLS easily could solve the 10×10 problem to optimality in time between 12 seconds up to a couple of minutes (see [BV98] for the results of an extensive computational work).

Excellent results are also presented in [ZLRG06]. The authors describe a combination of tabu search and simulated annealing and use above mentioned neighborhoods in their local search.

[HL06] combine an ant colony approach with the tabu search approach of Nowicki and Smutnicki. The ant colony idea is based on the shifting bottleneck idea, i.e., the ants are generating feasible one-machine schedules.

Genetic Based Job Shop Scheduling

As described in Section 2.5, a genetic algorithm aims at producing near-optimal solutions by letting a population of random solutions undergo a sequence of transformations governed by a selection scheme biased towards high-quality solutions. The effect of the transformations is that implicitly good properties are identified and combined into a new population which hopefully has the property that the best solution and the average value of the solutions are better than in previous populations. The process is then repeated until some stopping criteria are met.

A solution of a combinatorial optimization problem may be considered as a sequence of local decisions. A local decision for the job shop scheduling problem might be the choice of a task to be scheduled next. In an enumeration tree of all possible decision sequences a solution of the problem is represented as a path corresponding to the different decisions from the root of the tree to some leaf. Genetics can guide a search process in order to learn to find the most promising decisions, see Algorithm 2.5.4.

In case of an interpretation of an individual solution as a sequence of decision rules as described first in [DP95], an individual of a population is considered to be a subset of feasible schedules from the set of all feasible schedules.

Each individual of the *priority rule based genetic algorithm* (*P-GA*) is a string of $n-1$ entries $(f_1, f_2, \cdots, f_{n-1})$ where $n-1$ is the number of tasks in the underlying problem instance. An entry f_i represents one rule of the set of priority rules described in Table 10.3.1. The entry in the ith position says that a conflict in the ith iteration of the Giffler-Thompson algorithm should be resolved using priority rule f_i. More precisely, a task from the conflict set has to be selected by rule f_i; ties are broken by a random choice. Within a genetic framework a best sequence of priority rules has to be determined. An analogous encoding scheme has been used in [DTV95]. An individual is divided into sub-strings of preference lists. A sub-string defines preferences for task's selection for a particular machine.

The crossover operator is straightforward. Obviously, the simple crossover applies, where the sub-strings of two cut strings are exchanged, and which always yields feasible offspring. Heuristic information already occurs in the encoding scheme and a particular improvement step - contrary to genetic local search approaches, cf. [ALLU94] or [UAB+91] - is dropped. The mutation operator applied with a very small probability simply switches a string position to another one, i.e. the priority rule of a randomly chosen string entry is replaced by a new rule randomly chosen among the remaining ones. The approach in [DP95] to search a best sequence of decision rules for selecting tasks is just in line with the ideas described in [FT63] on probabilistic learning of sequences consisting of two priority rules, and [CGTT63], or [GH85] on learning how to find promising linear combinations of basic priorities. Fisher and Thompson [FT63] were amongst the first to suggest an adaptive approach by using a combination of rules in a sequencing system. They proposed using two separate sequencing criteria and, when a decision was taken, a random choice of a rule was made. Initially, there was an equal probability of selecting each rule but as the system progressed these probabilities were adjusted according to a predefined learning procedure. The following rules: *STT*, *LTT*, *LRPT*, *FCFS*, least remaining job slack per task, least remaining machine slack were considered in [CGTT63]. Their idea was to create a rule (as a linear combination of the above mentioned priority rules) capable of decisions which cannot be specified by any of the rules in isolation. Furthermore, the projection of the combined rule should yield each individual rule (see also [GH85]). In their experiments they restricted consideration to *STT* and *LRT*.

Besides using the genetic algorithm as a meta-strategy to optimally control the use of priority rules, another genetic algorithm described in [DP95] controls the selection of nodes in the enumeration tree of the shifting bottleneck heuristic (*shifting bottleneck based genetic algorithm, SB-GA*). Remember that the $SB2$-heuristic is only a repeated application of a part of the SB_1-heuristic where the sequence in which the one machine problems are solved is predetermined. Up to some depth l, a complete enumeration tree is generated and a partial tree for the remaining search levels. The SB_2-heuristic tries to determine the best single machine sequence for the SB_1-heuristic within a reasonable amount of time. This can also be achieved by a genetic strategy, even in a more effective way.

The length of a string representation of an individual in the population equals the number of machines in the problem which is equal to the depth of the enumeration tree in the SB_2-heuristic. Hence, an individual is encoded over the alphabet from 1 to the number of machines and a partial string from the first to the k th entry just describes the sequence in which the single machines are considered in the SB_1-heuristic. As a crossover operator one can use any traveling salesman crossover; Dorndorf and Pesch [DP95] chose the cycle crossover as described in [Gol89]. The best solutions found for the 10×10 and 5×20 problem, were 960 (*P−GA*)/938 (*SB−GA*) and 1249 (*P−GA*)/1178 (*SB−GA*), respectively. The running times are about 15 (*P−GA*)/2 (*SB−GA*) and 25 (*P−GA*)/1.5 (*SB−GA*) minutes.

Another genetic local search approach based on representation of the selected disjunctive arcs is described in [ALLU94] or in [NY91]. Their ideas are stimulated by the encouraging results obtained for the traveling salesman problem (cf. [UAB+91]). Aarts et al. [ALLU94] devise a multi-start local search embedded into a genetic framework; hence the name genetic local search. Each solution in the population is replaced by a locally optimal one with respect to moves based on the neighborhoods N_1 and N_2. The crossover idea is to implant a subset of arcs from one solution to another. The parent solutions are randomly chosen. The algorithm terminates when either all solutions in the population have equal fitness, or the best makespan in the population has not changed for 10 generations. Within a time bound of 99 or 88 seconds for the 10×10 or 5×20 problem the results of the genetic local search algorithms are worse than those from simulated annealing. However the results are better than a multi-start local search on randomly generated initial solutions.

The basic contribution of [NY91] is the representation of individuals in the population. An individual representing a schedule is described by a 0-1 matrix consisting of a column for each machine and a row for each pair of different jobs. Hence, the number of rows is limited to $\frac{1}{2}|\mathcal{J}|(|\mathcal{J}|-1)$ ordered job pairs. Entry 1 in row (i, j) and column k indicates that job i is supposed to be processed before job J_j on machine P_k. Otherwise, the entry is 0. The simple crossover (a random individual cut and tail exchange) and the simple mutation operators (flip an 0-1 entry) are applied. A harmonization algorithm turns a possibly inconsistent result through cycle elimination into a feasible schedule. Even for a population size of 1000 and 150 generations the 10×10 problem and the 5×20 problem could not be solved better than 965 and 1215, respectively.

A different approach has been followed in [SWV92]. The authors map the original data of the underlying problem instance to slightly disturbed and genetically controlled data representing new problem instances. The latter are solved heuristically and the solutions, i.e. the tasks' processing orders, are considered to be solutions of the original problem. Thus, the proposed neighborhood definition is based on the fact that a heuristic algorithm is a mapping of a problem to a solution; hence a heuristic algorithm problem pair is an encoding of a solution. A subset of solutions may be generated by the application of a single heuristic algorithm to perturbed versions of the original problem. That is, neighboring solutions are generated by applying the base heuristic to the perturbed problem, which is obtained through adding uniformly distributed random numbers to the job shop data. Then the solution is evaluated using the original problem data. The simple crossover applies. Their results are 976 for the 10×10 and 1186 for the 5×20 problem.

Yamada and Nakano [YN92] were first to use the Giffler-Thompson algorithm as crossover operator. The random selection of a next task is replaced by a choice of the task with respect to one of the parent schedules. That is, in order to resolve a conflict (i.e. choice of a next task from a set of tasks competing for the same machine) randomly choose one parent schedule. Select that task from the set of tasks in conflict which is also the first one processed from the conflict set of the parent schedule. A huge population size of 2000 individuals led them find an optimal schedule for the 10×10 problem. Their result on the 5×20 problem was not better than 1184.

The representation in [Bie95] is motivated by the idea to employ the traveling salesman crossover operators also in a job shop framework. He represented an individual as a string of length equal to the number of tasks in the job shop. An entry in this string is a job identification. The number of tasks of a job is the number of not necessarily consecutive string entries with the same job identification. For instance, if there are three jobs J_a, J_b, J_c having 3, 4, 3 tasks, respectively, then a randomly generated string (b,a,b,b,c,a,c,c,b,a) says, that string entries 1, 3, 4, and 9 correspond to the 1st, 2nd, 3rd, and 4th task of job J_b. Further, if the first task of job J_c and the last task of job J_b happen to need the same machine then J_c will come first. Now a TSP-crossover (cf. [KP94]) can be used to implant a substring of one parent schedule to another one. Within run times of about 9 to 10 minutes he reached a makespan of 936 and 1181 for the 10×10 and 5×20 problem. The population's size is 100.

Constraint Propagation, Decomposition and Edge-Guessing

The job shop scheduling problem is a typical representative of a binary constraint satisfaction problem (CSP), i.e., generally speaking, there is a set of variables each of which has its own domain of values. Find an assignment of values to variables such that a set of constraints on variable pairs is satisfied, see Chapter 13 [DP88, Mes89, MJPL92]. Assume that there is an upper bound on the makespan of an optimal schedule of the underlying job shop scheduling problem. Then computing heads and tails assigns to each task an interval of possible start times. Considering variable domains as possible task start times where the variables define the tasks in a schedule then the disjunctive graph illustrates the job shop scheduling constraint satisfaction problem, hence it corresponds to the constraint graph, [Mon74]. A set of k variables is said to be *k-consistent* if it is $k-1$-consistent and for each subset of $k-1$ variables holds: if a set of $k-1$ values each of which belongs to another of the $k-1$ variable domains violates none of the constraints on the considered $k-1$ variables, then there is a value in the domain of the remaining variable such that the set of all k values satisfies the set of constraints on the k variables. Let us assume that 0-consistency is always satisfied by definition. A set of variables is k-consistent if each subset of k variables is k-consistent. A 2-consistent set of variables is also said to be arc-consistent in order to emphasize the relation with the edges in the constraint graph (cf. [HDT92]). Consider a pair T_i, T_j of tasks. If for any two start times t_i and t_j of tasks T_i and T_j, respectively, and any third task T_k there exists a start time t_k of task T_k such that t_i, t_j, t_k satisfy constraints (10.1.1) to (10.1.3) then tasks T_i and T_j are said to be *path consistent*. Hence, consistency checks, or roughly speaking propagation of constraints will make implicitly defined constraints more visible and will prune the search tree in a branch and bound algorithm. The job shop scheduling problem is said to be path consistent if all task pairs are path consistent (cf. [Mac77, MH86, HL88]). Obviously, n-consistency, where n is the number of tasks, immediately implies that a feasible schedule can be generated easily, however, achieving n-consistency is in general not practicable. Moreover, worse upper bounds on the makespan of an optimal schedule will hardly reduce variable domains, i.e. only a few arc directions are fixed during the constraint propagation process. The better the bounds the more arc directions can be fixed. A detailed description of different levels of consistency for disjunctive scheduling problems can be found in Chapter 13.

Pesch [Pes94] introduced other genetic approaches. In the first one, *the one machine constraint propagation based genetic algorithm* (*1MCP-GA*), each entry of an individual is an upper bound on the makespan of the corresponding one-machine problem. In the second approach, the *two job constraint propagation based genetic algorithm* (*2JCP-GA*), each entry of an individual of the *2J-GA* is replaced by an upper bound on the makespan of a sub-problem consisting of a job pair. Whenever a new population is generated a local decision rule in the sense of constraint propagation in order to achieve arc- and path-consistency is applied simultaneously to each sub-problem (corresponding to an entry of an individual) with respect to its upper bound which is $\alpha\%$ above the optimal makespan of the sub-problem. The number of newly fixed arc directions divided by the number of arcs which were included into a cycle during the constraint propagation process on the sub-problems, defines the fitness of an individual. An individual of the population corresponds to a partial schedule. However each population is transformed to a population of feasible solutions, where each individual of a population is assessed in order to judge its contribution to a schedule. Therefore Giffler-Thompson's algorithm is applied with respect to the partial schedule representing individual. Ties are broken with respect to the complete schedules that are attached to the parents (partial schedules) of the considered offspring (partial schedule). Hence, the next task is chosen as in one of the parents corresponding complete schedules with the same probability. In the first population of complete schedules ties were broken randomly. For both problems, the 10×10 and 5×20, the optimal solution was reached rather quickly using the *1MCP-GA*. It was impossible to reach an optimum using *2JCP-GA*. Only values of 937 and 1175 could be found.

Dorndorf et al. [DPP02] took these ideas of combining constraint propagation with a problem decomposition approach in order to simplify the solution of the job shop scheduling problem a step further. Based on the observation that constraint propagation is more effective for 'small' problem instances the algorithm consists of deducing task sequences that are likely to occur in an optimal solution of the job shop scheduling problem.

The algorithm for which the name edge-guessing procedure has been chosen - since with respect to the job shop scheduling problem the deduction of machine sequences is mainly equivalent to orienting edges in a disjunctive graph - can be applied in a preprocessing step, reducing the solution space, thus speeding up the overall solution process. In spite of the heuristic nature of edge-guessing, it still leads to near-optimal solutions. If combined with a heuristic algorithm, they demonstrate that given the same amount of computation time, the additional application of edge-guessing leads to better solutions. This has been tested on a set of well-known job shop benchmark problem instances. Let us go into detail.

The solution of the job shop scheduling problem could be considerably simplified if some 'important' disjunctive edges were oriented the right way in advance. Pesch and Tetzlaff [PT96], for instance, showed that regarding the famous 10×10 instance of the job shop scheduling problem, one single `difficult' edge orientation exists which for the most part contributes to its intractability. Orienting this edge in the right direction, the optimal solution is found and verified in a fraction of the original time. Unfortunately, however, finding important edge orientations which occur in an optimal solution must, by definition, be a difficult task. Indeed, if an edge orientation can be easily found then orienting this edge cannot simplify the solution of the problem a lot, because it can be easily found. This dilemma can only be resolved if accuracy is sacrificed for efficiency, i.e. if we accept that some edge orientations derived may be wrong. The simplest heuristic method is a random selection of some edge orientations, however, the 'deduced' edge orientations will seldomly be oriented in the right direction.

A more sophisticated method for deducing edge orientations has been developed by Dorndorf et al. [DPP02] and Phan-Huy [PhH00], see also [PT96, Pes94]. This method reduces the number of wrong edge orientations through the combination of problem decomposition and constraint propagation techniques. In the sequel, we will present some new developments of the edge-guessing procedure for the job shop scheduling problem.

The next subsections motivate and describe the basic idea of edge-guessing followed by a short description of the original procedure applied in [PhH00] which decomposes problem instances in a parallel fashion. This parallel strategy is less suited in case of stronger constraint propagation techniques. As a consequence, this parallel approach very often leads to large cycle structures in the corresponding disjunctive graph of a given problem instance. A major improvement is obtained by the application of a sequential strategy which avoids the generation of cycles and, at the same time, leads to better edge orientations. Several versions of this sequential approach are presented in the last subsection.

Edge-Guessing - The Basic Idea: The combination of problem decomposition with constraint propagation is motivated by two observations: constraint propagation deduces more edge orientations if (a) the problem instance is small, i.e. contains a small number of tasks to be scheduled and (b) the initial upper bound is tight and, thus, leads to smaller current domains. The basic idea of edge-guessing is therefore to decompose a job shop scheduling problem instance into smaller sub-problem instances, then to choose some appropriate upper bounds *UB* for these sub-problem instances for which constraint propagation is then applied.

The definition of a sub-problem instance is quite straightforward. Let I denote an instance of the job shop scheduling problem and A be a subset of tasks. This allows us to derive a sub-problem $I(A)$ with heads and tails which is obtained by first removing all tasks $T_i \notin A$ and all constraints involving some task $T_i \notin A$ from I. Heads and tails are then added to each remaining task which basically is a consideration of the earliest start time est_i and latest completion time lct_i of each task T_i. This is necessary since otherwise no or only a few deductions will be achieved through constraint propagation. The terms heads and tails are more commonly used, since they allow a symmetric interpretation. The head r_i of T_i coincides with its earliest start time and can be interpreted as a lower bound of the total processing time of tasks that must finish before T_i can start. Likewise, the tail $q_i := UB - lct_i$ can be interpreted as a lower bound of the total processing time of tasks that must start after T_i has finished. Given a task $T_i \in A$, its heads and tails are considered by inserting a predecessor with processing time r_i and a successor with processing time q_i.

Constraint propagation is then applied to this sub-problem instance using some upper bound UB that still has to be specified. If the edge orientations deduced are inserted in the original problem instance I, i.e. if in the disjunctive graph edge orientations of some disjunctive edge pairs are chosen, we obtain a partial selection which hopefully simplifies the solution of I. We obtain a *complete (partial) selection* if (at most) one edge orientation is chosen from each disjunctive edge pair. The selection is acyclic, if after the removal of all remaining undirected pairs of disjunctions the resulting directed graph is acyclic.

The essential feature of this procedure is a suitable choice of the upper bounds. We will now describe this in more detail.

Let I be a job shop scheduling problem instance and I' a sub-problem instance with heads and tails. Let $C_{max}(I)$ and $C_{max}(I')$ denote the respective optimal makespan of I and I'. Quite evidently, for each optimal selection S of I, there exists a partial selection $S' \subseteq S$ which is a complete selection of I' with a makespan that will be denoted with $C_{max}(S')$. Applying constraint propagation to the sub-problem instance I' given the upper bound $C_{max}(S')$ now has two consequences. The first consequence is due to what has been said further above: since I' is 'smaller' than I and $C_{max}(S') \leq C_{max}(I)$, it is likely that more edges can be fixed than if constraint propagation is directly applied to I. The second consequence is due to the particular choice of the upper bound $C_{max}(S')$. Trivially, the edge orientations deduced define a *partially optimal selection* of I, i.e. must be contained in an optimal selection of I, namely the original selection S itself which has been assumed to be optimal. Thus, by decomposing I into many sub-problem instances, we could fix a high number of edges and approximate the optimal selection S quite well.

Unfortunately, this line of reasoning has a major flaw, since in general $C_{max}(S')$ is not known in advance. Thus, the problem is to find a good approximation of $C_{max}(S')$. To start with, a possible choice is the optimal makespan $C_{max}(I')$ of I'. However, since S' does not have to be an optimal solution of I', $C_{max}(I')$ may be much smaller than $C_{max}(S')$, and constraint propagation may deduce wrong edge orientations. We therefore choose a proportional increase of $C_{max}(I')$ by $\alpha\%$, that is, we apply constraint propagation to the sub-problem instance I' using the hypothetical upper bound

$$UB(I',\alpha) := C_{max}(I')\cdot(1 + \alpha/100) .$$

Since, in general, the computation of $C_{max}(I')$ is NP-hard, we will only choose subproblem instances, for which the computation can be efficiently carried out (e.g. single machine instances [1]).

Choosing the parameter α, the following trade-off between efficiency and accuracy of constraint propagation has to be considered: the greater α, the lower the error probability, but the less edges are fixed; the smaller α, the more edges are fixed, but the higher the probability that among these edges some are wrongly oriented.

In general, there is no means to efficiently test whether a selection (as a whole) is partially optimal, let alone to detect the edges that have been oriented in the wrong direction. The only exception is when the insertion of the oriented edges for a set of sub-problems results in a cycle. In this case, at least one upper bound chosen and one edge orientation is wrong. After removing the cycle, however, wrong edge orientations still may exist. Likewise, if no cycle has been created, this does not imply that the selection found is partially optimal. We cannot conclude that the bounds chosen and the edges fixed are correct, but can only deduce the trivial fact that constraint propagation has created no cycle. This reasserts the heuristic nature of edge-guessing.

In the original version of edge-guessing, a brute force approach has been applied which removes all edge orientations on all cycles. This simple approach, however, also removes a high number of possibly correct edge orientations. An improved edge-guessing procedure avoids the generation of cycles. To better understand this procedure, we start with a description of the original edge-guessing procedure.

A Parallel Strategy: The edge-guessing procedure presented by Pesch and Tetzlaff [PT96] and Phan Huy [PhH00] decomposes a job shop scheduling problem instance I into sub-problem instances $I(A_1),\cdots,I(A_d)$, where $A_1,\cdots, A_d \subseteq \mathcal{T}$ are some subsets of tasks. Constraint propagation is separately applied to each of these instances with the upper bound $UB(I(A_{d'}),\alpha_{d'})$, $d' = 1,\cdots, d$. A static choice

---

[1] Notice that here the consideration of heads and tails is crucial, since otherwise no edge orientations can be derived at all.

of α, i.e. all $\alpha_{d'}$ are set to a constant value α [PT96], and a choice which is guided by a genetic algorithm, a modification of what has been described above, have been studied. The edge orientations deduced define partial and acyclic selections $S_1, \cdots, S_d$ of the corresponding sub-problem instances. Setting $S := S_1 \cup \cdots \cup S_d$, however, we might not obtain an acyclic selection of I. In this case, all disjunctive edge orientations in S that belong to a cycle are removed.

We finally obtain a partial and acyclic selection which hopefully simplifies the solution of I. The complete parallel edge-guessing procedure is shown in Algorithm 10.3.4.

Algorithm 10.3.4 *Parallel edge-guessing* [DPP02].
begin
$A_1, \cdots, A_d \subseteq T$ are subsets of tasks;
$\alpha_1, \cdots, \alpha_d$ are non-negative real numbers;
for $d' := 1$ **to** d **do**
  begin
  $UB_{d'} := UB(I(A_{d'}), \alpha_{d'})$;
  constraint_propagation$(I(A_{d'}))$;
  $S_{d'} := \{$new edge orientations$\}$;
  end;
$S := S_1 \cup \cdots \cup S_d$;
$E := \{$edge orientations that are contained in a cycle$\}$;
return $(S \setminus E)$;
end;

A Sequential Strategy: The main problem in applying a parallel approach is that we do not know which of the sub-problem instances has been responsible for the creation of cycles. Thus, we run the risk of removing too many edge orientations. The solution is to adopt a sequential approach which allows us to much better control the generation and removal of cycles: only some edge orientations, but not all that are contained in a cycle have to be removed. This will be described in more detail in the following.

Let $A_1, \cdots, A_d$ be subsets of tasks. Start with $I(A_1)$ and apply constraint propagation to this instance with an upper bound $UB(I(A_1), \alpha_1)$. Observe that we obtain a partial selection S_1 which must be acyclic due to the choice of the upper bound. Therefore continue with the next sub-problem instance $I(A_2)$ to which constraint propagation is applied using the upper bound $UB(I(A_2), \alpha_2)$, and obtain a partial selection S_2. If the selection $S := S_1 \cup S_2$ induces a cycle then some of the newly derived edge orientations in S_2 have created this cycle. Thus, all edges in S_2 are removed and constraint propagation is reapplied with a higher upper bound. Dorndorf et al. opted for increasing α_2 by one percentage point, since this showed the best results. This procedure is repeated until no more cycles are generated. Then proceed with the remaining sub-problem instances $I(A_3), \cdots, I(A_d)$ in the same manner. The complete procedure is shown in Algorithm 10.3.5.

Algorithm 10.3.5 *Sequential edge-guessing 1* [DPP02].

```
begin
A₁,···, A_d ⊆ T are subsets of tasks;
α₁,···, α_d are non-negative real numbers;
S := ∅;
for d' := 1 to d do
   begin
   S_new := S;
   repeat
      cycle := false;
      UB_d' := UB(I(A_d'), α_d');
      constraint_propagation(I(A_d'));
      S_new := S_new ∪ {new edge orientations};
      if S_new induces a cycle then
         begin
         S_new := S;
         α_d' := α_d' + 1;
         cycle := true;
         end;
   until not cycle;
   S := S_new;
   end;
return (S);
end;
```

Some remarks have to be made. First, this algorithm terminates, because in the worst case, choosing a sufficiently high upper bound in the d'^{th} iteration will not deduce any edge orientations. Therefore, it must end up in a situation without cycles, as in the beginning of each iteration none existed. Second, even if the same subsets are taken, the sub-problem instances $I(A_{d'})$, $d' = 1, \cdots, d$, usually differ from the ones defined in the last section in spite of the similar notation. This is due to the fact that the edge orientations derived by $I(A_1), \cdots, I(A_{d'-1})$ are considered in the d'^{th} iteration which leads to stronger heads and tails and, by this, to the deduction of a greater number of edge orientations.

The last observation leads to an improvement of the sequential edge-guessing procedure. If any edge orientations are deduced in the d'^{th} iteration, this consequently has as well an effect on the heads and tails of the sub-problem instances $I(A_1), \cdots, I(A_{d'-1})$. Therefore, constraint propagation is re-applied to these modified instances using the upper bounds $UB(I(A_1),\alpha_1), \cdots, UB(I(A_{d'-1}),\alpha_{d'-1})$, that have been determined in the previous iterations, and re-continue with applying constraint propagation to $I(A_{d'})$ whenever its heads and tails have changed. This process is repeated until a fixed point is reached or a cycle is created. In the latter case, all edges are removed that have been deduced in the current iteration and restart the process with a greater upper bound for the d'^{th} sub-problem instance, while the upper bounds for all problem instances with a lower index are left unchanged. This procedure is shown in Algorithm 10.3.6.

Algorithm 10.3.6 *Sequential edge-guessing 2* [DPP02].
```
begin
```
$A_1, \cdots, A_d \subseteq \mathcal{T}$ are subsets of tasks;

$\alpha_1, \cdots, \alpha_d$ are non-negative real numbers;

$S := \varnothing$;

```
for d' := 1 to d do
    begin
```
    $S_{new} := S$;
```
    repeat
```
      fixed_point := *true*;

      cycle := *false*;
```
        for d" := d' downto 1 do
            begin
```
        $UB_{d''} := UB(I(A_{d''}),\alpha_{d''})$;

        constraint_propagation$(I(A_{d''}))$;
```
            if some heads and tails have changed then fixed_point := false;
```
        $S_{new} := S_{new} \cup \{\text{new edge orientations}\}$;
```
            if S_new induces a new cycle then
                begin
```
          $S_{new} := S$;

          $\alpha_{d'} := \alpha_{d'} + 1$;
```
                cycle := true;
                break;
                end;
            end;
    until fixed_point and not(cycle);
```
    $S := S_{new}$;
```
    end;
return (S);
end;
```

All that is left is to specify the order $A_1, \cdots, A_d$ in which the sub-problem instances are traversed. Instead of choosing a static order as indicated in the Algorithms 10.3.5 and 10.3.6, Dorndorf et al. actually implemented a dynamic rule which chooses the sub-problem instance with the maximal optimal makespan. This is justified by better results.

The introduced general method which combines constraint propagation with a problem decomposition approach can be applied in a preprocessing step before the actual solution of a problem, reducing the solution space and thus speeding up the overall solution process.

Since the solution of the job shop scheduling problem is mainly equivalent to orienting edges in a disjunctive graph, [DPP02] have named the preprocessing step the edge-guessing procedure. Several strategies in which sub-problem instances are examined in parallel or in a sequential manner are proposed. While the parallel approach analyzes sub-problem instances separately, so that constraint propagation only deduces information within each sub-problem, the sequential approach propagates information throughout the whole problem graph. Thus, more processing sequences (edge orientations) are deduced than with a parallel approach. The stronger consistency tests cause a further increase in the number of edge orientations that have been derived. Additionally, they have not only been able do derive more but also better edge orientations.

This has been verified by combining edge-guessing with a truncated branch-and-bound algorithm. Especially for larger and harder problem instances, the hybrid algorithm performs better than the pure truncated branch-and-bound algorithm, since it finds better solutions within a smaller or comparable amount of computation time.

However, for even larger and harder instances, truncated branch-and-bound may not be the best choice as a solution method. Therefore, in advanced research studies Dorndorf et al. have combined edge-guessing with local search algorithms which up to now provide the best solutions for the job shop scheduling problem. More precisely, they have combined popular tabu search algorithms with edge-guessing by incorporating the derived edges in a tabu list. Again, they have been able to produce better results for the combined algorithm than for the isolated tabu search algorithm. These encouraging results emphasize the potential of edge-guessing.

The main interest of constraint programming is the enormous flexibility that results from the fact that each constraint propagates independently from the existence or non-existence of other constraints. It appears that, within each constraint, considered separately, any type of technique (in particular OR algorithms) can be used. It appears that the propagation process can be organized to guarantee that propagation steps will occur in an order consistent with Ford's flow algorithm (hence with the same time complexity) [CL95]. Aggoun and Beldiceanu [AB93] present a construct called the "cumulative" constraint, incorporated in the CHIP constraint programming language. Using the cumulative constraint, Aggoun and Beldiceanu find the optimal solution of the 10×10 problem [MT63] in about 30 minutes (but cannot prove its optimality). Nuijten [NA96, Nui94] presents a variant of the algorithm by Carlier and Pinson [CP90] to update time-bounds of activities. It appears that this variant can easily be incorporated in a constraint satisfaction framework. Baptiste and Le Pape [BP95] explore various techniques based on "edge-finding" and "energetic reasoning" with the aim of integrating such techniques in Ilog Schedule, an industrial software tool for con-straint-based scheduling. In all of these cases, the flexibility inherent to constraint programming is maintained, but more efficient techniques can be archived using the wealth of the OR algorithmic work [BPS98, BPS00, DPP00].

Ejection Chains

Variable depth procedures (see Section 2.5) have had an important role in heuris-tic procedures for optimization problems.

An application with respect to neighborhood structure N_1 describes a move to a neighboring solution in which the processing order of tasks T_i and T_j is changed. The considered neighborhood structure is connected, i.e. for any two solutions (including the optimal one) x and y there is a sequence of moves, with respect to N_1, connecting x to y. The *gain* $g(i, j)$ affected by such a move from x to y can be estimated based on considerations about the minimal length of the critical path of the resulting disjunctive graph $G(y)$. Finding the exact gain of a move would generally involve a longest path calculation. The gain of a move can be negative, thus leading to a deterioration of the objective function. In [DP94] a local search procedure is presented based on a compound neighborhood struc-ture, each component consists of the neighborhood defined above. It is a variable depth search or ejection chain consisting of a simple neighborhood structure at each depth which is composed to complex and powerful moves. The basic idea is similar to the one used in tabu search, the main difference being that the list of forbidden (tabu) moves grows dynamically during a variable depth search itera-tion and is reset at the beginning of the next iteration. The algorithm is outlined in the following where $\gamma(x)$ is the objective function value (makespan).

Algorithm 10.3.7 *Ejection chain job shop scheduling* [DP94].

begin
Start with an initial solution x^* and the corresponding acyclic graph $G(x^*)$;
$x := x^*$;
repeat
    $TL := \varnothing$;   -- *TL* is the tabu list
    $d := 0$;   -- *d* is the current search depth
    while there are non-tabu critical arcs in $G(x(d))$ **do**
      begin
      $d := d + 1$;

Find the best move, i.e. the disjunctive critical arc (i^*, j^*) for which

$$g(i^*, j^*) = \max\{ g(i, j) \mid (i, j) \text{ is a disjunctive critical arc}$$
$$\text{which is not in } TL\};$$
$$\text{-- note that } g(i^*, j^*) \text{ can be negative}$$

Make this move, i.e. replace arc (i^*, j^*), thus obtaining the solution $x(d)$
  and its acyclic graph $G(x(d))$ at the search depth d;
$TL := TL \cup \{(j^*, i^*)\};$
end;
Let d^* denote the search depth at which the best solution $x(d^*)$ with
  $\gamma(x(d^*)) = \min \{\gamma(x(d)) \mid 0 < d \leq n\,(k-1)/2\}$ has been found;
if $d^* > 0$ **then begin** $x^* := x^*(d^*); x := x^*$ **end**;
until $d^* = 0$;
end;

Starting with an initially best solution $x^*(0)$, the procedure looks ahead for a certain number of moves and then sets the new currently best solution $x^*(d)$ for the
next iteration to the best solution found in the look-ahead phase at depth d^*.
These steps are repeated as long as an improvement is possible. The maximal
look-ahead depth is reached if all critical disjunctive arcs in the current solution
are set tabu. The step leading from a solution x in iteration k to a new solution in
the next iteration consists of a varying number d^* of moves in the neighborhood,
hence the name variable depth search where a complex compound move results
from a sequence of compressed simpler moves. The algorithm can escape local
optima because moves with negative gain are possible. A continuously increasing
growing tabu list avoids cycling of the search procedure. As an extension of the
algorithm, the whole **repeat** $\cdots$ **until** part could easily be embedded in yet
another control loop (not shown here) leading to a multi-level (parallel) search
algorithm.

A genetic algorithm with variable depth search has been implemented in
[DP93], i.e. each individual of a population is made locally optimal with respect
to the ejection chain based embedded neighborhood described in Algorithm
2.5.3. The algorithm has run five times on each problem instance, and all instances have been solved to optimality within a CPU time of ten minutes for a
single run. The algorithm has always solved the notoriously difficult 10×10 instance.

10.4 Conclusions

Although the 10×10 problem is not any longer a challenge it provides a way to
briefly get an impression of how powerful a certain method can be. For detailed
comparisons of solution procedure - if this is possible at all under different ma-

chine environments - this is obviously not enough and there are many other benchmark problems some of them with unknown optimal solution, see [Tai93]. It is apparent from the discussion that local search methods are the most powerful tool to schedule job shops. However, a stand alone local search cannot be competitive to those methods incorporating problem specific knowledge either by problem decomposition, special purpose heuristics, constraints and propagation of variables, domain modification, neighborhood structures (e.g. neighborhoods where each neighbor of a feasible schedule is locally optimal, cf. [BHW96, BHW97]), etc. or any composition of these tools.

The analogy of branching structures in exact methods and neighborhood structures reveals parallelism that is largely unexplored.

References

AB93   A. Aggoun, N. Beldiceanu, Extending CHIP in order to solve complex scheduling and placement problems, *Math. Comput. Model.* 17, 1993, 57-73.

ABZ88   J. Adams, E. Balas, D. Zawack, The shifting bottleneck procedure for job shop scheduling, *Management Sci.* 34, 1988, 391-401.

AC91   D. Applegate, W. Cook, A computational study of the job-shop scheduling problem, *ORSA J. Comput.* 3, 1991, 149-156.

AGP97   E. J. Anderson, C. A. Glass, C. N. Potts, Local search in combinatorial optimization: applications in machine scheduling, in: E. Aarts, J. K. Lenstra (eds.), *Local Search in Combinatorial Optimization*, Wiley, New York, 1997.

AH73   S. Ashour, S. R. Hiremath, A branch-and-bound approach to the job-shop scheduling problem, *Internat. J. Prod. Res.* 11, 1973, 47-58.

Ake56   S. B. Akers, A graphical approach to production scheduling problems, *Oper. Res.* 4, 1956, 244-245.

ALLU94   E. H. L. Aarts, P. J. P. van Laarhoven, J. K. Lenstra, N. L. J. Ulder, A computational study of local search shop scheduling, *ORSA J. Comput.* 6, 1994, 118-125.

Bal69   E. Balas, Machine sequencing via disjunctive graphs: An implicit enumeration algorithm, *Oper. Res.* 17, 1969, 941-957.

Bal85   E. Balas, On the facial structure of scheduling polyhedra, *Math. Programming Study* 24, 1985, 179-218.

BB01   W. Brinkköter, P. Brucker, Solving open benchmark problems for the job shop problem, *Journal of Scheduling* 4, 2001, 53-64.

BC95   J. W. Barnes, J. B. Chambers, Solving the job shop scheduling problem using tabu search, *IIE Transactions* 27, 1995, 257-263.

BDP96   J. Błażewicz, W. Domschke, E. Pesch, The job shop scheduling problem: Conventional and new solution techniques, *European J. Oper. Res.* 93, 1996, 1-33.

BDW91   J. Błażewicz, M. Dror, J. Weglarz, Mathematical programming formulations for machine scheduling: A survey. *European J. Oper. Res.* 51, 1991, 283-300 .

BHS91   S. Brah, J. Hunsucker, J. Shah, Mathematical modeling of scheduling problems, *J. Inform. Opt. Sci.* 12, 1991, 113-137.

BHW96   P. Brucker, J. Hurink, F. Werner, Improving local search heuristics for some scheduling problems, *Discrete Appl. Math.* 65, 1996, 97-122.

BHW97   P. Brucker, J. Hurink, F. Werner, Improving local search heuristics for some scheduling problems: Part II, *Discrete Appl. Math* 72, 1997, 47-69.

Bie95   C. Bierwirth, A generalized permutation approach to job shop scheduling with genetic algorithms, *OR Spektrum* 17, 1995, 87-92.

BJ93   P. Brucker, B. Jurisch, A new lower bound for the job-shop scheduling problem, *European J. Oper. Res.* 64, 1993, 156-167.

BJK94   P. Brucker, B. Jurisch, A. Krämer, The job-shop problem and immediate selection, *Annals of OR* 50, 1994, 73-114.

BJS92   P. Brucker, B. Jurisch, B. Sievers, Job-shop (C codes), *European J. Oper. Res.* 57, 1992, 132-133.

BJS94   P. Brucker, B. Jurisch, B. Sievers, A branch and bound algorithm for the job-shop scheduling problem, *Discrete Appl. Math.* 49, 1994, 107-127.

BK06   P. Brucker, S. Knust, *Complex Scheduling*, Springer, Berlin 2006.

BLL+83   K. R. Baker, E. L. Lawler, J. K. Lenstra, A. H. G. Rinnooy Kan, Preemptive scheduling of a single machine to minimize maximum cost subject to release dates and precedence constraints, *Oper. Res.* 31, 1983, 381-386.

BLV95   E. Balas, J. K. Lenstra, A. Vazacopoulos, One machine scheduling with delayed precedence constraints, *Management Sci.* 41, 1995, 94-109.

BM85   J. R. Barker, G. B. McMahon, Scheduling the general job-shop, *Management Sci.* 31, 1985, 594-598.

Bow59   E. H. Bowman, The scheduling sequencing problem, *Oper. Res.* 7, 1959, 621-624.

BP95   P. Baptiste, C. Le Pape, A theoretical and experimental comparison of constraint propagation techniques for disjunctive scheduling, *Proc. of the 14th Internat. Joint Conf. on Artificial Intelligence* (IJCAI), Montreal, 1995.

BPH82   J. H. Blackstone, D. T Phillips, G. L. Hogg, A state of the art survey of dispatching rules for manufacturing job shop tasks, *Internat. J. Prod. Res.* 20, 1982, 27-45.

BPN95a   P. Baptiste, C. Le Pape, W. Nuijten, Constraint-based optimization and approximation for job-shop scheduling, *Proc. of the AAAI-SIGMAN Workshop on Intelligent Manufacturing Systems*, IJCAI, Montreal, 1995.

BPN95b   P. Baptiste, C. Le Pape, W. Nuijten, Incorporating efficient operations research algorithms in constraint-based scheduling, *Proc. of the 1st. Joint Workshop on Artificial Intelligence and Operations Research*, Timberline Lodge, Oregon, 1995.

BPS98   J. Błażewicz, E. Pesch, M. Sterna, A branch and bound algorithm for the job shop scheduling problem, in: A. Drexl, A. Kimms (eds.) *Beyond Manufacturing Resource Planning (MRPII)*, Springer, 1998, 219-254.

BPS99   J. Błażewicz, E. Pesch, M. Sterna, A note on disjunctive graph representation, *Bulletin of the Polish Academy of Sciences* 47, 1999, 103-114.

BPS00   J. Błażewicz, E. Pesch, M. Sterna, The disjunctive graph machine representation of the job shop problem, *European J. Oper. Res.* 127, 2000, 317-331.

BR65   P. Bertier, B. Roy, Trois examples numeriques d'application de la procedure SEP, Note de travail No. 32 de la Direction Scientifique de la SEMA, 1965.

Bru88   P. Brucker, An efficient algorithm for the job-shop problem with two jobs, *Computing* 40, 1988, 353-359.

Bru94   P. Brucker, A polynomial algorithm for the two machine job-shop scheduling problem with a fixed number of jobs, *OR Spektrum* 16, 1994, 5-7.

Bru04   P. Brucker, *Scheduling Algorithms*, Springer, 4. edition, Berlin 2004.

BS04   C. Blum, M. Sampels, An ant colony optimization algorithm for shop scheduling problems, *Journal of Mathematical Modelling and Algorithms* 3, 2004, 285-308.

BV98   E. Balas, A. Vazacopoulos, Guided local search with shifting bottleneck for job shop scheduling, *Management Sci.* 44, 1998, 262-275.

BW65   G. H. Brooks, C. R. White, An algorithm for finding optimal or near-optimal solutions to the production scheduling problem, *J. Industrial Eng.* 16, 1965, 34-40.

CAK+06 M. Chandrasekaran, P. Asokan, S. Kumanan, T. Balamurugan, S. Nickolas, Solving job shop scheduling problems using artificial immune system, *International Journal of Advanced Manufacturing Technology* 31, 2006, 580-593.

Car82   J. Carlier, The one machine sequencing problem, *European J. Oper. Res.* 11, 1982, 42-47.

Car87   J. Carlier, Scheduling jobs with release dates and tails on identical machines to minimize the makespan, *European J. Oper. Res.* 29, 1987, 298-306.

CD70   J. M. Charlton, C. C. Death, A generalized machine scheduling algorithm, *Oper. Res. Quart.* 21, 1970, 127-134.

CGTT63 W. B. Crowston, F. Glover, G. L. Thompson, J. D. Trawick, Probabilistic and parametric learning combinations of local job shop scheduling rules, ONR Research Memorandum No. 117, GSIA, Carnegie-Mellon University, Pittsburg, 1963.

CL95   Y. Caseau, F. Laburthe, Disjunctive scheduling with task intervals, Working paper, Ecole Normale Supérieure, Paris, 1995.

Cla22   W. Clark, *The Gantt Chart: A Working Tool of Management*, The Ronald Press (3rd ed.), Pittman, New York, 1922.

CP89   J. Carlier, E. Pinson, An algorithm for solving the job-shop problem, *Management Sci.* 35, 1989, 164-176.

CP90   J. Carlier, E. Pinson, A practical use of Jackson's preemptive schedule for solving the job shop problem, *Ann. Oper. Res.* 26, 1990, 269-287.

CP94   J. Carlier, E. Pinson, Adjustments of heads and tails for the job-shop problem, *European J. Oper. Res.* 78, 1994, 146-161.

CPN96   Y. Caseau, C. Le Pape, W. P. M. Nuijten, private communication, 1996.

CPP92   C. Chu, M. C. Portmann, J. M. Proth, A splitting-up approach to simplify job-shop scheduling problems, *Internat. J. Prod. Res.* 30, 1992, 859-870.

DH70   J. E. Day, P. M. Hottenstein, Review of sequencing research, *Naval Res. Logistics Quart.* 17, 1970, 11-39.

DL93   S. Dauzere-Peres, J. -B. Lasserre, A modified shifting bottleneck procedure for job-shop scheduling, *Internat. J. Prod. Res.* 31, 1993, 923-932.

DMU97   E. Demirkol, S. Mehta, R. Uzsloy, A computational study of the shifting bottleneck procedure for job shop scheduling problems, *Journal of Heuristics* 3, 1997, 111-137.

DP88   R. Dechter, J. Pearl, Network-based heuristics for constraint satisfaction problems, *Artificial Intelligence* 34, 1988, 1-38.

DP93   U. Dorndorf, E. Pesch, Combining genetic and local search for solving the job shop scheduling problem, *Proc. Symposium on Appl. Mathematical Programming and Modeling - APMOD93*, Budapest, 1993, 142-149.

DP94   U. Dorndorf, E. Pesch, Variable depth search and embedded schedule neighborhoods for job shop scheduling, *Proc. 4th Internat. Workshop on Project Management and Scheduling*, 1994, 232-235.

DP95   U. Dorndorf, E. Pesch, Evolution based learning in a job shop scheduling environment, *Comput. Oper. Res.* 22, 1995, 25-40.

DPP99   U. Dorndorf, T. Phan Huy, E. Pesch, A survey of interval capacity consistency tests for time- and resource-constrained scheduling, in: J. Węglarz (ed.) *Project Scheduling - Recent Models, Algorithms and Applications*, Kluwer Academic Publ., 1999, 213-238.

DPP00   U. Dorndorf, E. Pesch, T. Phan Huy, Constraint propagation techniques for disjunctive scheduling problems, *Artificial Intelligence* 122, 2000, 189-240.

DPP02   U. Dorndorf, E. Pesch, T. Phan-Huy, Constraint propagation and problem decomposition: A preprocessing procedure for the job shop problem, *Ann. Oper. Res.* 115, 2002, 125-145.

DT93   M. Dell'Amico, M. Trubian, Applying tabu-search to the job shop scheduling problem, *Ann. Oper. Res.* 41, 1993, 231-252.

DTV95   F. Della Croce, R. Tadei, G. Volta, A genetic algorithm for the job shop problem. *Comput. Oper. Res.* 22, 1995, 15-24.

EAZ07   A. El-Bouri, N. Azizi, S. Zolfaghri, A comparative study of a new heuristic based on adaptive memory programming and simulated annealing: The case of job shop scheduling, *European J. Oper. Res.* 177, 2007, 1894-1910.

Fis73   M. L. Fisher, Optimal solution of scheduling problems using Lagrange multipliers: Part I, *Oper. Res.* 21, 1973, 1114-1127.

FLL+83   M. L. Fisher, B. J. Lageweg, J. K. Lenstra, A. H. G. Rinnooy Kan, Surrogate duality relaxation for job shop scheduling, *Discrete Appl. Math.* 5, 1983, 65-75.

Fox87   M. S. Fox, *Constraint-Directed Search: A Case Study of Job Shop Scheduling*, Pitman, London, 1987.

FS84   M. S. Fox, S. F. Smith, ISIS - a knowledge based system for factory scheduling, *Expert Systems* 1, 1984, 25-49.

FT63   H. Fisher, G. L. Thompson, Probabilistic learning combinations of local job-shop scheduling rules, in: J. F. Muth, G. L. Thompson (eds.), *Industrial Scheduling*, Prentice Hall, Englewood Cliffs, N.J., 1963.

FTM71   M. Florian, P. Trépant, G. McMahon, An implicit enumeration algorithm for the machine sequencing problem, *Management Sci.* 17, 1971, B782-B792.

Gan19   H. L. Gantt, Efficiency and democracy, *Trans. Amer. Soc. Mech. Engin.* 40, 1919, 799-808.

Ger66   W. S. Gere, Heuristics in job-shop scheduling, *Management Sci.* 13, 1966, 167-190.

GH85   P. J. O. Grady, C. Harrison, A general search sequencing rule for job shop sequencing, *Internat. J. Prod. Res.* 23, 1985, 951-973.

GNZ85   J. Grabowski, E. Nowicki, S. S. Zdrzalka, A block approach for single machine scheduling with release dates and due dates, *European J. Oper. Res.* 26, 1985, 278-285.

Gol89   D. E. Goldberg, *Genetic Algorithms in Search, Optimization and Machine Learning*, Addison-Wesley, Reading, Mass., 1989.

GPS92   C. A. Glass, C. N. Potts, P. Shade, Genetic algorithms and neighborhood-neighborhood search for scheduling unrelated parallel machines, Working paper No. OR47, University of Southampton.

Gre68   H. H. Greenberg, A branch and bound solution to the general scheduling problem, *Oper. Res.* 16, 1968, 353-361.

GS78   T. Gonzalez, S. Sahni, Flowshop and jobshop schedules: Complexity and approximation, *Oper. Res.* 20, 1978, 36-52.

GT60   B. Giffler, G. L. Thompson, Algorithms for solving production scheduling problems, *Oper. Res.* 8, 1960, 487-503.

HA82   N. Hefetz, I. Adiri, An efficient optimal algorithm for the two-machines unit-time job-shop schedule length problem, *Math. Oper. Res.* 7, 1982, 354-360.

Hau89   R. Haupt, A survey of priority-rule based scheduling, *OR Spektrum* 11, 1989, 3-16.

HDT92   P. van Hentenryck, Y. Deville, C. - M. Teng, A generic arc-consistency algorithm and its specializations, *Artificial Intelligence* 57, 1992, 291-321.

HL88   C. C. Han, C. H. Lee, Comments on Mohr and Hendersons path consistency algorithm, *Artificial Intelligence* 36, 1988, 125-130.

HL06   K.-L. Huang, C.-J. Liao, Ant colony optimization combined with taboo search
            for the job shop scheduling problem, Working paper, 2006, National Taiwan
            University of Science and Technology, Taipei.

Jac56   J. R. Jackson, An extension of Johnson's results on job lot scheduling, *Naval
            Res. Logist. Quart.* 3, 1956, 201-203.

JM99   A.S. Jain, S. Meeran, Deterministic job shop scheduling: past, present and
            future, *European J. Oper. Res.* 113, 1999, 390-434.

Joh54   S. M. Johnson, Optimal two- and three-stage production schedules with setup
            times included, *Naval Res. Logist. Quart.* 1, 1954, 61-68.

JRM00   A.S. Jain, B. Rangaswamy, S. Meeran, New and "stronger" job-shop neigh-
            bourhoods: A focus on the method of Nowicki and Smtnicki (1996), *Journal of
            Heuristics* 6, 2000, 457-480.

KP94   A. Kolen, E. Pesch, Genetic local search in combinatorial optimization, *Dis-
            crete Appl. Math.* 48, 1994, 273-284.

Kol99   M. Kolonko, Some new results on simulated annealing applied to the job shop
            scheduling problem, *European J. Oper. Res.* 113, 1999, 123-136.

KSS94   W. Kubiak, S. Sethi, C. Srishkandarajah, An efficient algorithm for a job shop
            problem, *Math. Industrial Syst.* 1, 1995, 203-216.

LAL92   P. J. M. van Laarhoven, E. H. L. Aarts, J. K. Lenstra, Job shop scheduling by
            simulated annealing, *Oper. Res.* 40, 1992, 113-125 .

LLRK77   B. Lageweg, J. K. Lenstra, A. H. G. Rinnooy Kan, Job-shop scheduling by
            implicit enumeration, *Management Sci.* 24, 1977, 441-450.

LLR+93   E. L. Lawler, J. K. Lenstra, A. H. G. Rinnooy Kan, D. B. Shmoys, Sequencing
            and scheduling: algorithms and complexity, in: S. C. Graves, A. H. G. Rinnooy
            Kan, P. H. Zipkin (eds.), *Handbooks in Oper. Res. and Management Sci.*,
            Vol. 4: Logistics of Production and Inventory, Elsevier, Amsterdam, 1993.

Lou93   H. R. Lourenço, A computational study of the job-shop and flow shop schedul-
            ing problems, Ph.D. thesis, Cornell University, 1993.

Lou95   H. R. Lourenço, Job-shop scheduling: Computational study of local search and
            large-step optimization methods, *European J. Oper. Res.* 83, 1995, 347-364.

LRK79   J. K. Lenstra, A. H. G. Rinnooy Kan, Computational complexity of discrete
            optimization problems, *Ann. Discrete Math.* 4, 1979, 121-140.

LRKB77   J. K. Lenstra, R. H. G. Rinnooy Kan, P. Brucker, Complexity of machine
            scheduling problems, *Ann. Discrete Math.* 4, 1977, 121-140.

Mac77   A. K. Mackworth, Consistency in networks of relations, *Artificial Intelligence*
            8, 1977, 99-118.

Man60   A. S. Manne, On the job shop scheduling problem, *Oper. Res.* 8, 1960, 219-
            223.

Mat96   D. C. Mattfeld, *Evolutionary Search and the Job Shop*, Physica, Heidelberg,
            1996.

Mes89   P. Meseguer, Constraint satisfaction problems: An overview, *AICOM* 2, 1989, 3-17.

MF75   G. B. McMahon, M. Florian, On scheduling with ready times and due dates to minimize maximum lateness, *Oper. Res.* 23, 1975, 475-482.

MFP06   J. Montgomery, C. Fayad, S. Petrovic, Solution representation for job shop scheduling problems in ant colony optimization, *Lecture Notes in Computer Science* 4150, 2006, 484-491.

MH86   R. Mohr, T. C. Henderson, Arc and path consistency revisited, *Artificial Intelligence* 28, 1986, 225-233.

MJPL92   S. Minton, M. D. Johnston, A. B. Philips, P. Laird, Minimizing conflicts: A heuristic repair method for constraint satisfaction and scheduling problems, *Artificial Intelligence* 58, 1992, 161-205.

Mon74   U. Montanari, Networks of constraints: fundamental properties and applications to picture processing, *Inform. Sci.* 7, 1974, 95-132.

MS96   P. Martin, D. Shmoys, A new approach to computing optimal schedules for the job shop scheduling problem, *Proceedings of the 5th International IPCO Conference*, 1996.

MSS88   H. Matsuo, C. J. Suh, R. S. Sullivan, A controlled search simulated annealing method for the general job shop scheduling problem, working paper 03-04-88, University of Texas Austin, 1988.

MT63   J. F. Muth, G. L. Thompson (eds.), *Industrial Scheduling*, Prentice Hall, Englewood Cliffs, N.J., 1963.

NA96   W. P. M. Nuijten, E. H. L. Aarts, A computational study of constraint satisfaction for multiple capacitated job shop scheduling, *European J. Oper. Res.* 90, 1996, 269-284.

NEH83   M. Nawaz, E. E. Enscore, I. Ham, A heuristic algorithm for the m-machine, n-job flow-shop sequencing problem, *Omega* 11, 1983, 91-95.

NS96   E. Nowicki, C. Smutnicki, A fast taboo search algorithm for the job shop problem, *Management Sci.* 42, 1996, 797-813.

NS05   E. Nowicki, C. Smutnicki, An advanced tabu search algorithm for the job shop problem, *Journal of Scheduling* 8, 2005, 145-159.

Nui94   W. P. M. Nuijten, *Time and Resource Constrained Scheduling*, Ponsen & Looijen, Wageningen, 1994.

NY91   R. Nakano, T. Yamada, Conventional genetic algorithm for job shop problems, in: R. K. Belew, L. B. Booker (eds.), *Proc. 4th. Internat. Conf. on Genetic Algorithms*, Morgan Kaufmann, 1991, 474-479.

OS88   P. S. Ow, S. F. Smith, Viewing scheduling as an opportunistic problem-solving process, *Ann. Oper. Res.* 12, 1988, 85-108.

PC95   M. Perregaard, J. Clausen, Parallel branch-and-bound methods for the job-shop scheduling problem, Working paper, University of Copenhagen, 1995.

Pes94   E. Pesch, *Learning in Automated Manufacturing*, Physica, Heidelberg, 1994.

PhH00   T. Phan-Huy, *Constraint Propagation in Flexible Manufacturing*, Springer, Berlin, 2000.

PI77   S. S. Panwalkar, W. Iskander, A survey of scheduling rules, *Oper. Res.* 25, 1977, 45-61.

Pin95   P. Pinedo, *Scheduling Theory, Algorithms and Systems*, Prentice Hall, Englewood Cliffs, N.J., 1995.

PM00   F. Pezzella, E. Merelli, Tabu search method guided by shifting bottleneck for the job shop scheduling problem, *European J. Oper. Res.* 120, 2000, 297-310.

Por68   D. B. Porter, The Gantt chart as applied to production scheduling and control, *Naval Res. Logist. Quart.* 15, 1968, 311-317.

Pot80   C. N. Potts, Analysis of a heuristic for one machine sequencing with release dates and delivery times, *Oper. Res.* 28, 1980, 1436-1441.

PT96   E. Pesch, U. Tetzlaff, Constraint propagation based scheduling of job shops, *Journal on Computing* 8, 1996, 144-157.

RS64   B. Roy, B. Sussmann, Les problémes d´ordonnancement avec contraintes disjonctives, SEMA, Note D. S. No. 9., Paris, 1964.

Sad91   N. Sadeh, Look-ahead techniques for micro-opportunistic job shop scheduling, Ph.D. thesis, Carnegie Mellon University, Pittsburgh, 1991.

SBL95   D. Sun, R. Batta, L. Lin, Effective job shop scheduling through active chain manipulation, *Comput. Oper. Res.* 22, 1995, 159-172.

SFO86   S. F. Smith, M. S. Fox, P. S. Ow, Constructing and maintaining detailed production plans: investigations into the development of knowledge-based factory scheduling systems, *AI Magazine*, 1986, 46-61.

SS95   Y. N. Sotskov, N. V. Shaklevich, NP-hardness of shop scheduling problems with three jobs, *Discrete Appl. Math.* 59, 1995, 237-266.

SWV92   R. H. Storer, S. D. Wu, R. Vaccari, New search spaces for sequencing problems with application to job shop scheduling, *Management Sci.* 38, 1992, 1495-1509.

Tai94   E. Taillard, Parallel tabu search technique for the job shop scheduling problem, *ORSA J. Comput.* 6, 1994, 108-117.

UAB+91   N. L. J. Ulder, E. H. L. Aarts, H. -J. Bandelt, P. J. P. van Laarhoven, E. Pesch, Genetic local search algorithms for the traveling salesman problem, *Lecture Notes in Computer Sci.* 496, 1991, 109-116.

Vae95   R. J. P. Vaessens, Generalized job shop scheduling: complexity and local search, Ph.D. thesis, University of Technology Eindhoven, 1995.

VAL96   R. J. P. Vaessens, E. H. L. Aarts, J. K. Lenstra, Job shop scheduling by local search, *Journal on Computing* 8, 1996, 302-317.

Vel91   S. van de Velde, Machine scheduling and lagrangian relaxation, Ph.D. thesis, CWI Amsterdam, 1991.

Wag59   H. P. Wagner, An integer linear programming model for machine scheduling, *Naval Res. Logist. Quart.* 6, 1959, 131-140.

WR90   K. P. White, R. V. Rogers, Job-shop scheduling: Limits of the binary disjunc-
            tive formulation, *Internat. J. Prod. Res.* 28, 1990, 2187-2200.

WW95   F. Werner, A. Winkler: Insertion techniques for the heuristic solution of the job
            shop problem, *Discrete Appl. Math.* 50, 1995, 191-211.

YN92   T. Yamada, R. Nakano, A genetic algorithm applicable to large-scale job-shop
            problems, in: R. Männer, B. Manderick (eds.), *Parallel Problem Solving from
            Nature* 2, Elsevier, 1992, 281-290.

ZLRG06   C.Y. Zhang, P.G. Li, Y.Q. Rao, Z.L. Guan, A very fast TS/SA algorithm for
            the job shop scheduling problem, Working paper, 2006, Huazhong University,
            Wuhan.

11 Scheduling with Limited Processor Availability[1]

In scheduling theory the basic model assumes that all machines are continuously available for processing throughout the planning horizon. This assumption might be justified in some cases but it does not apply if certain maintenance requirements, breakdowns or other constraints that cause the machines not to be available for processing have to be considered. In this chapter we discuss results related to deterministic scheduling problems where machines are not continuously available for processing.

Examples of such constraints can be found in many areas. Limited availabilities of machines may result from pre-schedules which exist mainly because most of the real world resources planning problems are dynamic. A natural approach to cope with a dynamic environment is to trigger a new planning horizon when the changes in the data justify it. However, due to many necessities, as process preparation for instance, it is mandatory to take results of earlier plans as fixed which obviously limits availability of resources for any subsequent plan. Consider e.g. ERP (Enterprise Resource Planning) production planning systems when a rolling horizon approach is used for customer order assignment on a tactical level. Here consecutive time periods overlap where planning decisions taken in earlier periods constrain those for later periods. Because of this arrangement orders related to earlier periods are also assigned to time intervals of later periods causing the resources not to be available during these intervals for orders arriving after the planning decisions have been taken. The same kind of problem may be repeated on the operational level of production scheduling. Here processing of some jobs is fixed in terms of starting and finishing times and machine assignment. When new jobs are released to the shop floor there are already jobs assigned to time intervals and machines while the new ones have to be processed within the remaining free processing intervals.

Another application of limited machine availability comes from operating systems for mono- and multi-processors, where subprograms with higher priority will interfere with the current program executed. A similar problem arises in multi-user computer systems where the load changes during the usage. In big massively parallel systems it is convenient to change the partition of the processors among different types of users according to their requirements for the machine. Fluctuations related to the processing capacity can be modeled by intervals of different processor availability. Numerous other examples exist where the investigation of limited machine availability is of great importance and the prac-

---

[1] This paper is based on O. Braun, J, Breit, G. Schmidt, Deterministic Machine Scheduling with Limited Machine Availability, Discussion paper B0403, Saarland University, 2004.

tical need to deal with this type of problem has been proven by a growing demand for commercial software packages. Thus, recently the analysis of these problems has attracted many researchers.

In the following we will investigate scheduling problems with limited machine availability in greater detail. The research was started by G. Schmidt [Sch84]. The review focuses on deterministic models with information about the availability constraints. Earlier surveys of this research area can be found in [SS98, Sch00, Lee04]. For stochastic scheduling problems with limited machine availability and prior distributions of the problem parameters see [GGN00, LS95b, LS97]. We will survey results for one machine, parallel machine and shop scheduling problems in terms of intractability and polynomial time algorithms. In some places also results from enumerative optimization algorithms and heuristics are analyzed. Doing this we will distinguish between non-preemptive and preemptive scheduling. We will finish with some conclusions and some suggestions for future research.

11.1 Problem Definition

A machine system with limited availability is a set of machines (processors) which does not operate continuously; each machine is ready for processing only in certain time intervals of availability. Let $\mathcal{P} = \{P_i \mid i = 1, \cdots, m\}$ be the set of machines with machine P_i only available for processing within S_i given time intervals $[B_i^s, F_i^s)$, $s = 1, \cdots, S_i$ and $B_i^{s+1} > F_i^s$ for all $s = 1, \cdots, S_{i-1}$. B_i^s denotes the start time and F_i^s the finish time of sth interval of availability of machine P_i.

We want to find a feasible schedule if one exists, such that all tasks can be processed within the given intervals of machine availability optimizing some performance criterion. Such measures considered here are completion time and due date related and most of them refer to the maximum completion time, the sum of completion times, and the maximum lateness.

The term preemption is used as defined before. Often the notion of resumability is used instead of preemption. Under a resumable scenario a task may be interrupted when a machine becomes unavailable and resumed as the machine becomes available again without any penalty. Under the non-resumable scenario task preemption is generally forbidden. The most general scenario is semi-resumability. Let x_j denote the part of task T_j processed before an interruption and let $\delta \in [0,1]$ be a given parameter. Under the semi-resumable scenario δx_j time units of task T_j have to be re-processed after the non-availability interval. The total processing time for task T_j is given by $x_j + \delta x_j + (p_j - x_j) = \delta x_j + p_j$.

In the following we base the discussion on the three field $\alpha \mid \beta \mid \gamma$ classification introduced in Chapter 3. We add some entry denoting machine availability and we omit entries which are not relevant for the problems investigated here.

The first field $\alpha = \alpha_1\alpha_2\alpha_3$ describes the machine (processor) environment. In [Sch84] and [LS95a] different patterns of availability are discussed for the case of parallel machine systems (parameter α_3). These are constant, zigzag, decreasing, increasing, and staircase. Let $0 = t_1 < t_2 < \cdots < t_j < \cdots < t_q$ be the points in time where the availability of a certain machine changes and let $m(t_j)$ be the number of machines being available during time interval $[t_j, t_{j+1})$ with $m(t_j) > 0$. It is assumed that the pattern is not changed infinitely often during any finite time interval. According to these cases parameter $\alpha_3 \in \{\varnothing, NC_{zz}, NC_{inc}, NC_{dec}, NC_{inczz}, NC_{deczz}, NC_{sc}, NC_{win}\}$ denotes the machine availability. NC relates to the *non-continuous* availability of the machines.

1. If all machines are continuously available ($t = 0$) then the pattern is called constant; $\alpha_3 = \varnothing$.

2. If there are only k or k–1 machines in each interval available then the pattern is called zigzag; $\alpha_3 = NC_{zz}$.

3. A pattern is called increasing (decreasing) if for all j from $I\!N$ the number of machines $m(t_j) \geq max_{1 \leq u \leq j-1}\{m(t_u)\}$ ($m(t_j) \leq min_{1 \leq u \leq j-1}\{m(t_u)\}$), i.e. the number of machines available in interval $[t_{j-1}, t_j)$ is not more (less) than this number in interval $[t_j, t_{j+1})$; $\alpha_3 = NC_{inc}$ (NC_{dec}).

4. A pattern is called increasing (decreasing) zigzag if, for all j from $I\!N$, $m(t_j) \geq max_{1 \leq u \leq j-1}\{m(t_u) - 1\}$ ($m(t_j) \leq min_{1 \leq u \leq j-1}\{m(t_u) + 1\}$); $\alpha_3 = NC_{inczz}$ (NC_{deczz}).

5. A pattern is called staircase if for all intervals the availability of machine P_i implies the availability of machine P_{i+1}; $\alpha_3 = NC_{sc}$. A staircase pattern is shown in the lower part of Figure 11.1.1; grayed areas represent intervals of non-availability. Note that patterns (1)-(4) are special cases of (5).

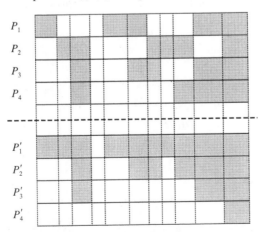

Figure 11.1.1 *Rearrangement of arbitrary patterns.*

6. A pattern is called arbitrary if none of the conditions (1)-(5) applies; $\alpha_3 = NC_{win}$. Such a pattern is shown in the upper part of Figure 11.1.1 for machines P_1, P_2, P_3, P_4; patterns defined in (1)-(5) are special cases of the one in (6).

Machine systems with arbitrary patterns of availability can always be translated to a composite machine system forming a staircase pattern [Sch84]. A composite machine is an artificial machine consisting of at most m original machines. The transformation process works in the following way. An arbitrary pattern is separated in as many time intervals as there are distinct points in time where the availability of at least one machine changes. Now in every interval periods of non-availability are moved from machines with smaller index to machines with greater index or vice versa. If there are $m(t_j)$ machines available in some interval $[t_j, t_j+1)$ then after the transformation machines $P_1, \cdots, P_{m(t_j)}$ will be available in $[t_j, t_j+1)$ and $P_{m(t_j+1)}, \cdots, P_m$ will not be available, where $0 < m(t_j) < m$. Doing this for every interval we generate composite machines. Each of them consists of at most m original machines with respect to the planning horizon.

An example for such a transformation where periods of non-availability are moved from machines with greater index to machines with smaller index, considering $m = 4$ machines, is given in Figure 11.1.1 Non-availability is represented by the grayed areas. From machines P_1, P_2, P_3, P_4 composite machines P_1', P_2', P_3', P_4' are formed. Composite machines which do not have intervals of availability can be omitted from the problem description. Then the number of composite machines in each interval is the maximum number of machines simultaneously available. The time complexity of the transformation is $O(qm)$ where q is the number of points in time, where the availability of an original machine is changing. If this number is polynomial in n or m machine scheduling problems with arbitrary patterns of non-availability can be transformed in polynomial time to a staircase pattern. This transformation is useful as, first, availability at time t is given by the number of available composite machines and, second, some results are obtained assuming this hypothesis.

The second field $\beta = \beta_1, \cdots, \beta_8$ describes task (job) and resource characteristics. We will only refer here to parameter β_1.

Parameter $\beta_1 \in \{\varnothing, t - pmtn, pmtn\}$ indicates the possibilities of preemption:
- $\beta_1 = \varnothing$: no preemption is allowed,
- $\beta_1 = t - pmtn$: tasks may be preempted, but each task must be processed by only one machine,
- $\beta_1 = pmtn$: tasks may be arbitrarily preempted.

Here we assume that not only task ($\beta_1 = t - pmtn$) but also arbitrary (task and machine) preemptions are possible ($\beta_1 = pmtn$). If there is only one machine dedicated to each task then task preemptions and arbitrary preemptions become equivalent. For single machine and shop problems this difference has not to be considered. Of course the rearrangement of an arbitrary pattern to a staircase pattern is only used when arbitrary preemption is allowed. In what follows the number of preemptions may be a criterion to appreciate the value of an algorithm. When the algorithm applies to staircase patterns, the number of preemptions for an arbitrary pattern is increased by at most mq.

The third field, γ, denotes a *single* optimality criterion (performance measure). In some recent papers *multiple* criteria scheduling models with limited machine availability are investigated, see e.g. [QBY02, LY03]. We will further investigate models with single optimality criteria.

Many of the problems considered later are solved applying simple priority rules which can be executed in $O(n \log n)$ time. The rules order the tasks in some way and then iteratively assign them to the most lightly loaded machine. The following rules as already introduced in Chapter 3 are the most prominent.

- Shortest Processing Time (*SPT*) rule. With this rule the tasks are ordered according to non-decreasing processing times.
- Longest Processing Time (*LPT*) rule. The tasks are ordered according to non-increasing processing times.
- Earliest Due Date (*EDD*) rule. Applying this rule all tasks are ordered according to non-decreasing due dates.

11.2 One Machine Problems

One machine problems are of fundamental character. They can be interpreted as building blocks for more complex problems. Such formulations may be used to represent bottleneck machines or an aggregation of a machine system. For one machine scheduling problems the only availability pattern which has to be investigated is a special case of zigzag with $k = 1$.

Let us consider first problems where preemption of tasks (jobs) is not allowed. If there is only a single interval of non-availability and $\sum C_j$ is the objective $(1, NC_{zz} \| \sum C_j)$ [ABFR89] show that the problem is NP-hard. The Shortest Processing Time (*SPT*) rule leads to a tight relative error of $R_{SPT} \le 2/7$ for this problem [LL92]. [SPR+05] presents a modified *SPT*-heuristic with an improved relative error of 3/17. He also develops a dynamic programming algorithm for the same problem capable of solving problem instances with up to 25000 tasks. It is easy to see that also problem $1, NC_{win} \| C_{max}$ is NP-hard [Lee96].

If preemption is allowed the scheduling problem becomes easier. For $1, NC_{win} | pmtn | C_{max}$, it is obvious that every schedule is optimal which starts at time zero and has no unforced idle time, that is, the machine never remains idle while some task is ready for processing. Preemption is never useful except when some task cannot be finished before an interval of non-availability occurs. This property is still true for completion time based criteria if there is no precedence constraint and no release date, as it is assumed in the rest of this section.

While the sum of completion times $(1, NC_{win} \mid pmtn \mid \sum C_j)$ is minimized by the *SPT* rule the problem of minimizing the weighted sum $(1, NC_{win} \mid pmtn \mid \sum w_j C_j)$ is NP-hard [Lee96]. Note that without availability constraints Smith's rule [Smi56] solves the problem. Maximum lateness is minimized by the Earliest Due Date (*EDD*) rule [Lee96]. If the number of tardy tasks has to be minimized $(1, NC_{win} \mid pmtn \mid \sum U_j)$ the *EDD* rule of Moore and Hodgson's algorithm [Moo68] can be modified to solve this problem also in $O(n \log n)$ time [Lee96]. Note that if we add release times or weights for the jobs the problem is NP-hard already for a continuously available machine ([LRB77] or [Kar72]). Details can be found in Chapter 4.

Lorigeon et al. [LBB02a] investigate a one-machine problem where each task has a release date r_j and a delivery duration q_j. The machine is not available for processing during a single given interval. A task may only be preempted for the duration of the non-availability interval and resumed as the machine becomes available again. The objective is to find a schedule minimizing $max_j \{C_j + q_j\}$. The problem is a generalization of a well-known NP-hard problem studied by Carlier [Car82]. Lorigeon et al. provide a branch-and-bound algorithm which solves 2133 out of 2250 problems instances with up to 50 tasks.

There are also results concerning problems where an interval of non-availability is regarded as a decision variable. Qi et al. [QCT99] study a model in which the machine has to be maintained after a maximum of δ_1 time units. Each such maintenance activity has a constant duration of δ_2 time units. The goal is to find a non-preemptive schedule which obeys the maintenance restrictions and minimizes $\sum C_j$. The problem is proved to be NP-hard in the strong sense. Qi et al. propose heuristics and a branch-and-bound algorithm.

Graves and Lee [GL99] study several variants of the same problem. Besides processing a task requires a setup operation on the machine. If a task is preempted by an interval of non-availability an additional (second) setup is required before the processing of the task can be resumed. Maintenance activities have to be carried out after a maximum of δ_1 time units. If there are at most two maintenance periods then the problem is NP-hard in the ordinary sense for the objectives C_{max}, $\sum C_j$, $\sum w_j C_j$, and L_{max}. Dynamic programming algorithms are provided to solve the problems in pseudo-polynomial time. If there is exactly one period of maintenance the problem is polynomially solvable for the objectives $\sum C_j$ (by a modification of the *SPT* rule) and L_{max} (by a modification of the EDD rule). Minimizing $\sum w_j C_j$ turns out to be NP-hard in the ordinary sense. Two pseudo-polynomial time dynamic programming algorithms are provided to solve this problem.

Lee and Leon [LL01] study a problem in which a production rate modifying activity of a given duration has to be scheduled in addition to n tasks. A task T_j processed before the activity requires p_j time units on the machine while the processing time of the same task becomes $\sigma_j p_j$ if it is scheduled after the production rate modifying activity. Preemption is not allowed. The objective is to find a starting time for the rate-modifying activity and a task sequence such that several regular functions are optimized. The problem can be solved in polynomial time for the objectives C_{max} and $\sum C_j$. For the objective $\sum w_j C_j$ the authors develop pseudo-polynomial dynamic programming algorithms. For the objective L_{max} the *EDD* rule is optimal for the practical case where the production rate is increased by the activity. The general case with arbitrary σ_j is NP-hard.

11.3 Parallel Machine Problems

In this section we cover formulations of parallel machine scheduling problems with availability constraints.

11.3.1 Minimizing the Sum of Completion Times

In case of continuous availability of the machines $(P \,||\, \sum C_j)$ the problem can be solved applying the *SPT* rule. If machines have only different beginning times B_i (this corresponds to an increasing pattern of availability) the problem can also be solved by the *SPT* rule [KM88, Lim91]. If $m = 2$ and there is only one finish time F_i^s on one machine which is finite (this corresponds to a zigzag pattern of availability) the problem becomes NP-hard [LL93]. In the same paper Lee and Liman show that for $P2, NC_{ZZ} \,||\, \sum C_j$, where machine P_2 is continuously available and machine P_1 has one finish time which is smaller than infinity, the *SPT* rule with the following modification leads to a tight relative error of $R_{SPT} < 1/2$:

Step 1:   Assign the shortest task to P_2.

Step 2:   Assign the remaining tasks in *SPT* order alternately to both machines until some time when no other task can be assigned to P_1 without violating F_1.

Step 3:   Assign the remaining tasks to P_1.

Figure 11.3.1 illustrates how that bound can be reached asymptotically (when ε tends toward 0). In both examples, the modified *SPT* rule leads to a large idle time for machine P_1. For fixed m the *SPT* rule is asymptotically optimal if there is no more than one interval of non-availability for each machine [Mos94].

In case there is only one interval of non-availability for each machine, the problem is NP-hard. In [LC00] a branch and bound algorithm based on the column generation approach is given which also solves the problem where $\sum w_j C_j$ is minimized.

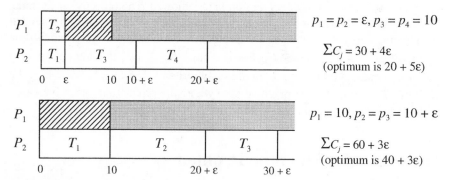

Figure 11.3.1 *Examples for the modified SPT rule.*

11.3.2   Minimizing the Makespan

Let us first investigate non-preemptive scheduling. J. D. Ullman [Ull75] analyses the complexity of the problem $P, NC_{win} \| C_{max}$. It is NP-hard in the strong sense for arbitrary m (3-partition is a special case) even if the machines are continuously available. If machines have different beginning times B_i ($P, NC_{inc} \| C_{max}$) the Longest Processing Time (*LPT*) rule leads to a relative error of $R_{LPT} < 1/2 - 1/(2m)$ or of $R_{MLPT} < 1/3$ if the rule is appropriately modified [Lee91]. The first bound is tight. The modification uses dummy tasks to simulate the different machine starting times B_i. For each machine P_i a task T_j with processing time $p_j = B_i$ is inserted. The dummy tasks are merged into the original task set and then all tasks are scheduled according to the *LPT* rule under an additional restriction that only one dummy task is assigned to each machine. After finishing the schedule, all dummy tasks are moved to the head of the machines followed by the remaining tasks assigned to each P_i. The *MLPT* rule runs in $O((n+m) \cdot log(n+m) + (n+m) \cdot m)$ time. In [LHYL97] Lee's bound of 1/3 reached by *MLPT* is improved to 1/4.

Using the bin-packing algorithm called the *MULTIFIT* it is shown in [CH98] that the bound of this algorithm is $2/7 + 2^{-k}$, where k is the selected number of the major iterations in *MULTIFIT*.

Note that the *LPT* algorithm leads to a relative error of $R_{LPT} < 1/3 - 1/(3m)$ for continuously available machines [Gra69]. H. Kellerer [Kel98] presents a dual approximation algorithm using a bin packing approach leading to a tight bound of 1/4, too.

In [LSL05] a problem with two machines and one interval of non-availability is considered. For non-resumable and resumable cases the problem is solved by enumerative techniques.

Now let us investigate results for preemptive scheduling. If all machines are only available in one and the same time interval $[B, F)$ and tasks are independent the problem is of type $P \mid pmtn \mid C_{max}$ Following [McN59] it can be shown that there exists a feasible machine preemptive schedule if and only if $max_j\{p_j\} \leq (F - B)$ and $\sum_j p_j \leq m(F - B)$. There exists an $O(n)$ algorithm which generates at most $m - 1$ preemptions to construct this schedule. If all machines are available in an arbitrary number $S = \sum_i S_i$ of time intervals $[B_i^s, F_i^s)$, $s = 1, \cdots, S_i$ and the machine system forms a staircase pattern, it is possible to generalize McNaughton's condition and show that a feasible preemptive schedule exists if and only if the following m conditions are met [Sch84]:

$$\sum_{j=1}^{k} p_j \leq \sum_{i=1}^{k} PC_i \quad \forall k = 1, \cdots, m-1, \qquad (11.3.1\text{-}k)$$

$$\sum_{j=1}^{n} p_j \leq \sum_{i=1}^{m} PC_i \qquad (11.3.1\text{-}m)$$

with $p_1 \geq p_2 \geq \cdots \geq p_n$ and $PC_1 \geq PC_2 \geq \cdots \geq PC_m$, where PC_j is the total processing capacity of machine P_i. Such a schedule can be constructed in $O(n + m \cdot \log m)$ time after the processing capacities PC_i are computed, with at most $S-1$ preemptions in case of a staircase pattern (remember that any arbitrary pattern of availability can be converted into a staircase one at the price of additional preemptions). Note that in the case of the same availability interval $[B, F)$ for all machines McNaughton's conditions are obtained from (11.3.1-1) and (11.3.1-m) alone. This remains true for zigzag patterns as then (11.3.1-2) , $\cdots$, (11.3.1-m-1) are always verified if (11.3.1-1) is true (there is one availability interval for all machines but P_m). The algorithm to solve the problem applies five rules which are explained now.

Let us consider two arbitrary processors P_k and P_l with $PC_k > PC_l$ as shown in Figure 11.3.2. Let Φ_k^a, Φ_k^b, and Φ_k^c denote the processing capacities of processor P_k in the intervals $[B_k^1, B_l^1]$, $[B_l^1, F_l^{N(l)}]$, and $[F_l^{N(l)}, F_k^{N(k)}]$, respectively. Then obviously, $PC_k = \Phi_k^a + \Phi_k^b + \Phi_k^c$.

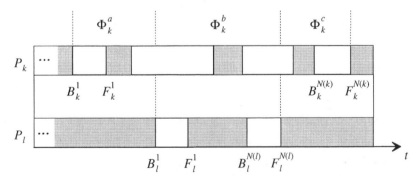

Figure 11.3.2 *Staircase pattern for two arbitrary processors.*

Assume that the tasks are ordered according to non-increasing processing times and that the processors form a staircase pattern as defined above. All tasks T_j are scheduled in the given order one by one using one of the five rules given below. Rules 1 - 4 are applied in the case where $1 \leq j < m$, $p_j > \min_i \{PC_i\}$, and if there are two processors P_k and P_l such that $PC_l = \max_i \{PC_i \mid PC_i < p_j\}$ and $PC_k = \min_i \{PC_i \mid PC_i \geq p_j\}$. Rule 5 is used if $m \leq j \leq n$ or $p_j \leq \min_i \{PC_i\}$. First we describe the rules, and after that we prove that their application always constructs a feasible schedule, if one exists. To avoid cumbersome notation we present the rules in a semi-formal way.

Rule 1. *Condition*: $p_j = PC_k$.

Schedule task T_j on processor P_k such that all the intervals $[B_k^r, F_k^r]$, $r = 1, \cdots, N(k)$, are completely filled; combine processors P_k and P_l to form a *composite processor*, denoted again by P_k, which is available in all free processing intervals of the original processor P_l, i.e. define $PC_k = PC_l$ and $PC_l = 0$.

Rule 2. *Condition*: $p_j - PC_l > \max\{\Phi_k^a, \Phi_k^c\}$ and $p_j - \Phi_k^b \geq \min\{\Phi_k^a, \Phi_k^c\}$.

Schedule task T_j on processor P_k in its free processing intervals within $[B_l^1, F_l^{N(l)}]$. If Φ_k^a (respectively Φ_k^c) is minimum use all the free processing intervals of P_k in $[B_k^1, B_l^1]$ ($[F_l^{N(l)}, F_k^{N(k)}]$) to schedule T_j, and schedule the remaining processing requirements of that task (if there is any) in the free processing intervals of P_k within $[F_l^{N(l)}, F_k^{N(k)}]$ ($[B_k^1, B_l^1]$) from left to right (right to left) such that the rth processing interval is completely filled with T_j before the $r+1$st ($r-1$st) interval is used, respectively. Combine processors P_k and P_l to a composite processor P_k which is available in the remaining free processing intervals of the original processors P_k and P_l, i.e. define $PC_k = PC_k + PC_l - p_j$ and $PC_l = 0$.

Rule 3. *Condition*: $p_j - PC_l > \max\{\Phi_k^a, \Phi_k^c\}$ and $p_j - \Phi_k^b < \min\{\Phi_k^a, \Phi_k^c\}$.

If Φ_k^a (Φ_k^c) is minimum, schedule task T_j on processor P_k such that its free processing intervals in $[B_k^1, B_l^1]$ ($[F_l^{N(l)}, F_k^{N(k)}]$) are completely filled with T_j, further fill processor P_k in the intervals $[B_l^r, F_l^r]$, $r = 1, \cdots, N(l)$, completely with T_j and use the remaining processing capacity of P_k in the interval $[B_l^1, F_l^{N(l)}]$ to schedule task T_j with its remaining processing requirement such that T_j is scheduled from left to right (right to left) where the $r+1$st ($r-1$st) interval is not used before the rth interval has been completely filled with T_j, respectively. After doing this there will be some time t in the interval $[B_l^1, F_l^{N(l)}]$ up to (after) this time task T_j is continuously scheduled on processor P_k. Time t always exists because $p_j - \min\{\Phi_k^a, \Phi_k^c\} < \Phi_k^b$. Now move T_j with its processing requirement which is scheduled after (before) t on processor P_k to processor P_l in the corresponding time intervals. Combine processors P_k and P_l to a composite processor P_k which is available in the remaining free processing intervals of the original processors P_k and P_l, i.e. define $PC_k = PC_k + PC_l - p_j$ and $PC_l = 0$.

Rule 4. *Condition*: $p_j - PC_l \leq \max\{\Phi_k^a, \Phi_k^c\}$.
Schedule task T_j on processor P_l such that all its intervals $[B_l^r, F_l^r]$, $r = 1, \cdots, N(l)$ are completely filled with T_j. If Φ_k^a (Φ_k^c) is maximum, schedule task T_j with its remaining processing requirement on processor P_k in the free processing intervals of $[B_k^1, B_l^1]$ ($[F_l^{N(l)}, F_k^{N(k)}]$) from left to right (right to left) such that the rth processing interval is completely filled with T_j before the $r+1^{\text{st}}$ ($r-1^{\text{st}}$) interval is used, respectively. Combine processors P_k and P_l to a composite processor P_k which is available in the remaining free processing intervals of the original processor P_k, i.e. define $PC_k = PC_k + PC_l - p_j$ and $PC_l = 0$.

Rule 5. *Condition*: remaining cases.
Schedule task T_j and the remaining tasks in any order in the remaining free processing intervals successively from left to right starting with processor P_k, switch to a processor P_i, $i < k$ only if the $i+1^{\text{st}}$ processor is already completely filled.

To show the optimality of rules 1 - 5 one may use the following lemma and theorem [Sch84].

Lemma 11.3.1 *After having scheduled a task T_j, $j \in \{1, \cdots, m-1\}$, on some processor P_k according to rules 1 or 2, or on P_k and P_l according to rules 3 or 4, the following observations are true*:

(1) *The remaining free processing intervals of processors P_k and P_l are disjoint.*

(2) *Combining processors P_k and P_l to a composite processor P_k results in a new staircase pattern.*

(3) *If all inequalities of (11.3.1-k), $k = 1, \cdots, m$ hold before scheduling task T_j, the remaining processing requirements and processing capacities after scheduling T_j still satisfy inequalities (11.3.1-k), $k = 1, \cdots, m$.*

(4) *The number of completely filled or completely empty intervals is $\sum\limits_{i=1}^{m} N(i) - K$ where K is the number of only partially filled intervals, $K \leq j < m$.* $\square$

We are now ready to prove the following theorem. The proof is constructive and leads to an algorithm that solves our problem.

Theorem 11.3.2 *For a system of m semi-identical processors with staircase pattern of availability and a given set T of n tasks there will always be a feasible preemptive schedule if and only if all inequalities* (11.3.1) *hold.*

Proof. We assume that $p_j > \min_i \{PC_i\}$ for $j = 1, \cdots, m-1$; otherwise the theorem is always true if and only if the inequality (11.3.1-m) holds, as can easily be seen. There always exists a feasible preemptive schedule for T_1. Now assume that the first z tasks have been scheduled feasibly according to rules 1 - 4. We show that T_{z+1} also can be scheduled feasibly:

(i) $1 < z < m$: after scheduling task T_z all inequalities (11.3.1) hold according to Lemma 11.3.1. Then $p_{z+1} \leq PC_1^z$, hence task T_{z+1} can be scheduled feasibly on processor P_1.

(ii) $m \leq z \leq n$: after scheduling the first $m-1$ tasks using rules 1-4, $m-1$ processors are completely filled with tasks. Since $PC_2^{z-1} = PC_3^{z-1} = \cdots = PC_m^{z-1} = 0$ and $PC_1^{z-1} \geq \sum_{j=z}^{n} p_j$, task T_z can be scheduled on processor P_1, and the remaining tasks can also be scheduled on this processor by means of rule 5.   □

The following algorithm makes appropriate use of the five scheduling rules.

Algorithm 11.3.3 *Algorithm by Schmidt* [Sch84] *for semi-identical processors.*

```
begin
Order the m largest tasks Tⱼ according to non-increasing processing times and
    schedule them in the given order;
for all i ∈ {1,···,m} do PCᵢ := Σ_{r=1}^{N(i)} PCᵢʳ;
repeat
    if j < m and pⱼ > min_i {PCᵢ}
    then
        begin
        Find processor Pₗ with PCₗ = max_i {PCᵢ | PCᵢ < pⱼ} and processor Pₖ with
            PCₖ = min_i {PCᵢ | PCᵢ ≥ pⱼ};
        if PCₖ = pⱼ
        then call rule 1
        else
            begin
            Calculate Φₖᵃ, Φₖᵇ, and Φₖᶜ;
            if pⱼ − PCₗ > max{Φₖᵃ, Φₖᶜ}
            then
```

```
      if  p_j-Φ_k^b ≥ min {Φ_k^a, Φ_k^c}
        then call rule 2 else call rule 3;
      else call rule 4;
        end;
    end
  else call rule 5;
until j = n;
end;
```

The number of preemptions generated by the above algorithm and its complexity are estimated by the following theorems [Sch84].

Theorem 11.3.4 *Given a system of m processors $P_1, \cdots, P_m$ of non-continuous availability, where each processor P_i is available in $N(i)$ time intervals. Then, if the processor system forms a staircase pattern and the tasks satisfy the inequalities* (11.3.1), *Algorithm* 11.3.3 *generates a feasible preemptive schedule with at most $(\sum_{i=1}^{m} N(i)) - 1$ preemptions.*   □

Theorem 11.3.5 *The time complexity of Algorithm* 11.3.3 *is $O(n + m \log m)$.*   □

Notice that if all processors are only available in a single processing interval and all these intervals have the staircase property the algorithm generates feasible schedules with at most $m - 1$ preemptions. If we further assume that $B_i = B$ and $F_i = F$ for all $i = 1, \cdots, m$ Algorithm 11.3.3 reduces to McNaughton's rule [McN59] with time complexity $O(n)$ and at most $m - 1$ preemptions.

There is a number of more general problems that can be solved by similar approaches.

(1) Consider the general problem where the intervals of m semi-identical processors are arbitrarily distributed as shown in Figure 11.3.3(a) for an example problem with $m = 3$ processors. Reordering the original intervals leads to a staircase pattern which is illustrated in Figure 11.3.3(b). Now each processor P_i', with $PC_i' > 0$ is a *composite processor* combining processors $P_i, P_{i+1}, \cdots, P_m$, and each interval $[B_i'^r, F_i'^r]$ is a *composite interval* combining intervals of availability of processors $P_i, P_{i+1}, \cdots, P_m$. The numbers in the different intervals of Figure 11.3.3(b) correspond to the number of original processors where that interval of availability is related to. After reordering the original intervals this way the problem consists of at most $Q' \leq Q = \sum_{i=1}^{m} N(i)$ intervals of availability. Using Algorithm 11.3.3, $O(m)$ preemptions are possible in each interval and thus $O(mQ)$ is an upper bound on the number of preemptions which will be generated for the original problem.

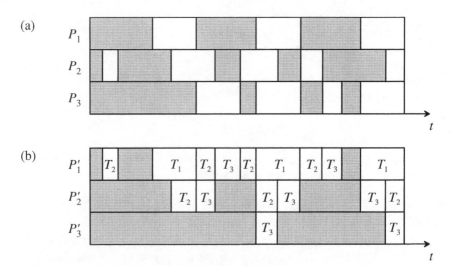

Figure 11.3.3 *Example for arbitrary processing intervals*
             (a) general intervals of availability,
             (b) corresponding staircase pattern.

(2) If there is no feasible preemptive schedule for the problem at least one of the inequalities of (11.3.1) is violated; this means that the processing capacity of at least one processor is insufficient. We now increase the processing capacity in such a way that all the tasks can be feasibly processed. An *overtime cost function* might be introduced that measures the required increase of processing capacity. Assume that an increase of one time unit of processing capacity results in an increase of one unit of cost. If some inequality (11.3.1-q) is violated we have to increase the total capacity of the first q processors by $\sum_{j=1}^{q}(p_j-PC_j)$ in case of $1 \leq q < m$; hence the processing capacity of each of the processors $P_1, \cdots, P_q$ is increased by $\frac{1}{q}\sum_{j=1}^{q}(p_j-PC_j)$. If inequality (11.3.1-$m$) is violated, the cost minimum increase of all processing capacities is achieved if the processing capacity of each processor is increased by $\frac{1}{m}(\sum_{j=1}^{n}p_j - \sum_{j=1}^{m}PC_j)$. Now Algorithm 11.3.3 can be used to construct a feasible preemptive schedule of minimum total overtime cost. Checking and adjusting the m inequalities can be done in $O(m)$ time, and then Algorithm 11.3.3 can be applied. Hence a feasible schedule of minimal overtime cost can be constructed in $O(n+m\log m)$ time.

(3) If each task T_j also has a deadline $\tilde{d}_j$ the problem is not only to meet start and finish times of all intervals but also all deadlines. The problem can be solved by using a similar approach where the staircase patterns and the given deadlines are considered. Since all the tasks may have different deadlines, the resulting time

complexity is $O(nm\log n)$. A detailed description of this procedure can be found in [Sch88]. It is also proved there that the algorithm generates at most $Q+m(s-1)-1$ preemptions if the semi-identical processor system forms a staircase pattern, and $m(Q+s-1)-1$ preemptions in the general case, where s is the number of different deadlines. We mention that this approach is not only dedicated to the deadline problem. It can also be applied to a problem where all the tasks have different ready times and the same deadline, as these two situations are of the same structure.

The corresponding optimization problem $(P,NC_{sc}\,|\,pmtn\,|\,C_{max})$ is solved by an algorithm that first computes the lower bounds LB_1, LB_2, $\cdots$, LB_m obtained from the conditions above (see Figure 11.3.4). C_{max} cannot be smaller than LB_k, $k = 1, \cdots, m-1$, obtained from (11.3.1-k). The sum of availabilities of machines $P_1, \cdots, P_k$ during time interval $[0, LB_k)$ may not be smaller than the sum of processing times of tasks $T_1, \cdots, T_k$. The sum of all machine availabilities during time interval $[0, LB_m)$ must also be larger than or equal to the sum of processing times of all tasks. In the example of Figure 11.3.4, $C_{max} = LB_3$. The number of preemptions is $S - 2$.

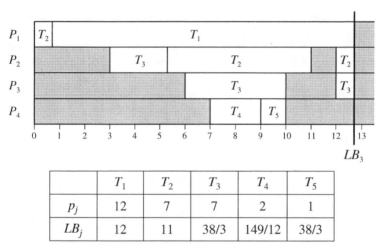

		T_1	T_2	T_3	T_4	T_5
p_j		12	7	7	2	1
LB_j		12	11	38/3	149/12	38/3

Figure 11.3.4 *Minimizing the makespan on a staircase pattern.*

When precedence constraints are added, Liu and Sanlaville [LS95a] show that problems with chains and arbitrary patterns of non-availability (i.e. $P,NC_{win}\,|\,pmtn, chains\,|\,C_{max}$) can be solved in polynomial time applying the Longest Remaining Path (*LRP*) first rule and the processor sharing procedure of [MC70]. In the same paper it is also shown that the *LRP* rule could be used to solve problems with decreasing (increasing) zigzag patterns and tasks forming an outforest (inforest) $(P,NC_{deczz}\,|\,pmtn, out\text{-}forest\,|\,C_{max}$ or $P,NC_{inczz}\,|\,pmtn, in\text{-}forest\,|\,C_{max})$. In case of only two machines and arbitrary (which means zigzag for $m = 2$) patterns of non-availability ($P2, NC_{win} \mid pmtn, prec \mid C_{max}$) this rule also solves problems with arbitrary task precedence constraints with time complexity and number of preemptions of $O(n^2)$. These results are deduced from those obtained for unit execution time scheduling by list algorithms (see Dolev and Warmuth [DW85b, DW85a]). The *LRP* algorithm is nearly on-line, as are all priority algorithms which extend list algorithms to preemption [Law82]. Indeed these algorithms first build a schedule admitting processor sharing. These schedules execute tasks of the same priority at the same speed. This property is respected when McNaughton's rule is applied. If machine availability changes unexpectedly, the property does not hold any more.

Applying the *LRP* rule results in a time complexity of $O(n \cdot log\, n + nm)$ and a number of preemptions of $O((n + m)^2 - nm)$ which both can be improved. Therefore in [BDF+00] an algorithm is given which solves problem $P, NC_{win} \mid pmtn, chains \mid C_{max}$ with $N < n$ chains in $O(N + m \cdot log\, m)$ time generating a number of preemptions which is not greater than the number of intervals of availability of all machines. If all machines are only available in one processing interval and all intervals are ordered in a staircase pattern the algorithm generates feasible schedules with at most $m - 1$ preemptions. This result is based on the observation that preemptive scheduling of chains for minimizing schedule length can be solved by applying an algorithm for the independent tasks problem. Having more than two machines in the case of arbitrary precedence constraints or an arbitrary number of machines in the case of a tree precedence structure makes the problem NP-complete [BDF+00].

When tasks require more than one processor they are called multiprocessor tasks. In [BDDM03] polynomial algorithms are given for the following cases:
- tasks have various ready times and require either one or all processors;
- sizes of the tasks are powers of 2.

11.3.3   Dealing with Due Date Involving Criteria

In [Hor74] it is shown that $P \mid pmtn, r_j, \tilde{d}_j \mid -$ can be solved in $O(n^3 \cdot min\{n^2, log\, n + log\, p_{max}\})$ time. The same flow-based approach can be coupled with a bisection search to minimize maximum lateness L_{max} (see [LLLR79], where the method is also extended to uniform machines). A slightly modified version of the algorithm still applies to the corresponding problem where the machines are not continuously available. If the number of changes of machine availabilities during any time interval is linear in the length of the interval this approach can be implemented in $O(n^3 p_{max}^3 \cdot (log\, n + log\, p_{max}))$ [San95]. When no ready times are given but due dates have to be considered, maximum lateness can be minimized for the problem ($P, NC_{win} \mid pmtn \mid L_{max}$) using the approach suggested by [Sch88] in $O(nm \cdot log\, n)$ time. The method needs to know all possible events before the next due date.

If there are not only due dates but also ready times are to be considered (problem $P, NC_{win} | r_j, pmtn | L_{max}$) Sanlaville [San95] suggests a nearly on-line priority algorithm with an absolute error of $A \leq (m - 1/m)p_{max}$ if the availability of the machines follows a constant pattern and of $A \leq p_{max}$ if machine availability refers to an increasing zigzag pattern. The priority is calculated according to the Smallest Laxity First (SLF) rule, where laxity (or slack time) is the difference between the task's due date and its remaining processing time. The SLF algorithm runs in $O(n^2 p_{max})$ time and is optimal in the case of a zigzag pattern and no release dates.

[LS95a] shows that results for C_{max} minimization in cae of in-forest precedence graphs and increasing zigzag patterns ($P, NC_{inczz} | pmtn, in\text{-}forest | C_{max}$) can be extended to L_{max}, using the SLF rule on the modified due dates. Figure 11.3.5 shows an optimal SLF schedule for the given precedence constraints.

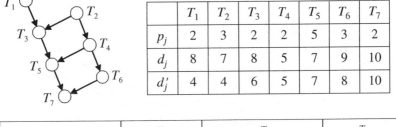

	T_1	T_2	T_3	T_4	T_5	T_6	T_7
p_j	2	3	2	2	5	3	2
d_j	8	7	8	5	7	9	10
d_j'	4	4	6	5	7	8	10

Figure 11.3.5 *Minimizing L_{max} on an increasing zigzag pattern.*

The modified due date is given by $d_j' = \min\{d_j', d_{s(j)}' + p_{s(j)}\}$ where $T_{s(j)}$ is the successor of T_j when it exists. In the same way, minimizing L_{max} on two machines with availability constraints is achieved using SLF with a different modification scheme. If there are due dates, release dates and chain precedence constraints to be considered ($P, NC_{win} | r_j, chains, pmtn | L_{max}$) the problem can be solved using a binary search procedure in combination with a linear programming formulation [BDF+00]. In case of multiprocessor tasks there exists a polynomial algorithm to minimize L_{max} if the number of processors is fixed [BDDM03].

Lawler and Martel [LM89] solved the weighted number of tardy jobs problem on two uniform machines, i.e. $Q2 | pmtn | \sum w_j U_j$. The originality of their paper comes from the fact that they show a stronger result, as the speeds of the processors may change continuously (and even be 0) during the execution. Hence, it includes as a special case availability constraints on two uniform machines. They use dynamic programming to propose pseudo-polynomial algorithms ($O(\sum w_j n^2)$, or $O(n^2 p_{max})$ to minimize the number of tardy jobs). Nothing however is said about the effort needed to compute processing capacity in one interval.

If there are more than two uniform machines to be considered and the problem is to minimize maximum lateness for jobs which have different release dates ($Q, NC_{win} | r_j, pmtn | L_{max}$) the problem can be solved in polynomial time by a combined strategy of binary search and network flow [BDF+00]. In the same paper the problem is generalized taking unrelated machines, i.e. machine speeds cannot be represented by constant factors, into account. This problem can also be solved in polynomial time applying a combination of binary search and the two-phase method given in [BEP+96].

11.4   Shop Problems

The literature on shop scheduling problems with limited machine availability is concentrated on flow shops and open shops. We are aware of only two papers dealing with the job shop. The paper of Aggoune [Agg04b] studies the two-job special case of this problem under the makespan criterion. He proposes extensions of the well known geometric algorithm by Akers and Friedman [AF55] for problems $J, NC_{win} | pmtn, n = 2 | C_{max}$ and $J, NC_{win} | n = 2 | C_{max}$. The algorithms run in polynomial time. Braun et al. [BLS05] investigate problem $J2, NC_{win} | pmtn | C_{max}$ and derive sufficient conditions for the optimality of Jackson's rule.

11.4.1   Flow Shop Problems

The flow shop scheduling problem for two machines with a constant pattern of availability minimizing C_{max} ($F2 || C_{max}$ and $F2 | pmtn | C_{max}$) can be solved in polynomial time by Johnson's rule [Joh54]. C.-Y. Lee [Lee97] has shown that this problem becomes already NP-hard if there is a single interval of non-availability on one machine only. For the case where the tasks can be resumed he also gives approximation algorithms which have relative errors of 1/2 if this interval is on machine P_1 or of 1/3 if the interval of non-availability is on machine P_2. The approximation algorithms are based on a combination of Johnson's rule and a modification of the ratio rule given in [MP93]. Lee also proposes a dynamic programming algorithm for the case with one interval only.

Improved approximation algorithms for the resumable problem with one interval are presented in [CW00], [Bre04a] and [NK04]. In the first paper a 1/3-

approximation for the case with the interval on P_1 is presented. The second paper provides a 1/4-approximation for the case with the interval occurring on P_2. The third paper finally describes a fully polynomial-time approximation scheme for the general case with one interval of non-availability, no matter on which machine. Ng and Kovalyov show that these two problems are in fact symmetrical. A polynomial-time approximation scheme for the case where general preemption is allowed (not only resumability) is presented in [Bre04b].

In [KBF+02] it is shown that the existence of approximation algorithms for flow shop scheduling problems with limited machine availability is more of an exception. It is proved that no polynomial time heuristic with a finite worst case bound can exist for $F2, NC_{win} | pmtn | C_{max}$ when at least two intervals of non-availability are allowed to occur. Furthermore it is shown that makespan minimization becomes NP-hard in the strong sense if an arbitrary number of intervals occurs on one machine only. On the other hand, there always exists an optimal schedule where the permutation of jobs scheduled between any two consecutive intervals obeys Johnson's order. However, the question which jobs to assign between which intervals remains intractable.

Due to these negative results a branch and bound algorithm is developed in [KBF+02] to solve $F2, NC_{win} | pmtn | C_{max}$. The approach uses Johnson's order property of jobs scheduled between two consecutive intervals. This property helps to reduce the number of solutions to be enumerated. Computational experiments were carried out to evaluate the performance of the branch-and-bound algorithm. In the test problem instances intervals of non-availability were allowed to occur either only on P_1, or only on P_2, or on both machines. The first result of the tests was that these instances were equally difficult to solve. The second result was that the algorithm performed very well when run on randomly generated problem instances; 1957 instances out of 2000 instances could be solved to optimality within a time limit of 1000 seconds. However, it could also be shown that there exist problem instances which are much harder to solve for the algorithm. These were instances in which the processing time of a job on the second machine was exactly twice its processing time on the first machine.

In order to speed up the solution process, a parallel implementation of the branch and bound algorithm is presented in [BFKS97]. Computations have been performed on 1, 2, 3, up to 8 processors. The experiment has been based on instances for which computational times of the sequential version of the algorithm were long. The maximum speed up gained was between 1.2 and 4.8 in comparison to the sequential version for 8 processors being involved in the computation.

Based on the above results in [BBF+01] constructive and improvement heuristics are designed for $F2, NC_{win} | pmtn | C_{max}$. They are empirically evaluated using test data from [KBF+02] and new difficult test data. It turned out that a combination of two constructive heuristics and a simulated annealing algorithm could solve 5870 out of 6000 easy problem instances and 41 out of 100 difficult instances. The experiments were run on a PC and the time limit to achieve this result was roughly 60 seconds per instance. The worst relative errors were 2.6%

and 44.4% above the optimum, respectively. The combination of two constructive heuristics could only solve 5812 out of 6000 easy instances and 13 out 100 difficult instances with an average computation time of 0.33 seconds and 3.96 seconds per instance, respectively. These results in [BBF+01] suggest that the heuristic algorithms are very good options for solving flow shop scheduling problems with limited machine availability.

In [Bra02] and [BLSS02] sufficient conditions for the optimality of Johnson's rule in the case of one or more intervals of non-availability (i.e. for $F2, NC_{win} | pmtn | C_{max}$) are derived. To find the results the technique of stability analysis is used and it is shown that in most cases Johnson's permutation remains optimal. These results are comparable to [KBF+02] but improve the running time for finding optimal solutions, such that instances with 10,000 jobs and 1,000 intervals of non-availability can be treated.

The non-preemptive case of the two-machine flow shop with limited machine availability is studied by [CW99]. In general, this problem is not approximable for the makespan criterion if at least two intervals of non-availability may occur. Cheng and Wang investigate the case where there are exactly two such intervals. One of them starts at the same time when the other one ends (consecutive intervals). They provide a 2/3-approximation algorithm for this problem.

[Lee99] studies the two-machine flow shop with one interval of non-availability under the semi-resumable scenario. He provides dynamic programming algorithms for this problem as well as approximation algorithms with worst case errors of 1 and 1/2, depending on whether the interval occurs on the first or on the second machine.

Quite a few papers exist on the two-machine no-wait flow shop. For constant machine availability and the makespan criterion this problem is polynomially solvable [GG64, HS96]. Espinouse et al. [EFP99, EFP01] study the case with one interval of non-availability. They show that the problem is NP-hard no matter if preemption is allowed or not, and not approximable if at least two intervals occur. They also provide approximation algorithms with a worst-case error of 1. Improved heuristics with worst-case errors of 1/2 are presented by [CL03a]. They also treat the case where each of the two machines has an interval of non-availability and these two intervals overlap. In the second paper [CL03b] provides a polynomial-time approximation scheme for this problem. [KS04] also study the case with one interval of non-availability. They provide a 1/2-approximation algorithm capable of handling the semi-resumable scenario and a 1/3-approximation algorithm for the resumable scenario. The non-preemptive m-machine flow shop with two intervals of non-availability on each machine and the makespan objective is studied by [Agg04a] and [AP03]. In [Agg04a] two cases are considered. In the first case, intervals of non-availability are fixed while in the second case intervals are assigned to time windows and their actual start times are decision variables. A genetic algorithm and a tabu search procedure are evaluated for test data with up to 20 jobs and 10 machines. The most important result is that flexible start times of the intervals of non-availability result in considerably shorter schedules. In [AP03] intervals of non-availability have fixed start and finish times. The proposed heuristic is based on the approach presented in [Agg04b]. The jobs in a sequence are grouped in pairs. Each pair is scheduled optimally using the algorithm in [Agg04b]. This approach is embedded into a tabu search algorithm. Experiments indicate that the heuristic is capable of finding good solutions for problem instances with up to 20 jobs.

11.4.2 Open Shop Problems

The literature on open shop scheduling problems (for a survey see also [BF97]) with limited machine availability is focused on the two-machine case and the objective of makespan minimization. The case with constant pattern of machine availability ($O2 \| C_{max}$) can be solved in linear time by an algorithm due to [GS76].

It is essential to distinguish between two kinds of preemption. The less restrictive case is investigated by [VS95]. They use a model where the processing of a job may be interrupted and later resumed on the same machine. In the interval between interruption and resumption the job may be processed on a different machine. It is shown that under this assumption the problem is polynomially solvable even for arbitrary numbers of machines and intervals of non-availability.

In the more restrictive case the processing of a job on a machine may be interrupted by the processing of other jobs or by intervals of non-availability. In the interval between the start and the end of a task, no other task of the same job may be processed. This model is similar to the open shop with no-pass constraints as introduced by Cho and Sahni [CS81].

J. Breit [Bre00] proves that this latter problem is NP-hard even for a single interval of non-availability and not approximable within a constant factor if at least three such intervals occur. For the case with one interval there exists a pseudopolynomial dynamic programming algorithm [LBB02b] as well as a linear time approximation algorithm with an error bound of 1/3 [BSS01]. The special case in which the interval occurs at the beginning of the planning horizon is solved by a linear time algorithm due to [LP93]. M. A. Kubzin et al. [KSBS02] present polynomial-time approximation schemes for the case with an arbitrary number of intervals on one machine and a continuously available second machine, as well as for the case with exactly one interval on each machine. The non-preemptive model is studied by J. Breit et al. [BSS03]. They provide a linear time 1/3-approximation algorithm and show that the problem with at least two intervals is not approximable within a constant factor.

11.5 Conclusions

We reviewed results on scheduling problems with limited machine availability.

The number of results shows that scheduling with availability constraints attracts more and more researchers, as the importance of the applications is recognized. The results presented here are of various kinds. In particular, when preemption is not authorized it will logically entail NP-hardness of the problem. If one is interested in solutions for non-preemptive problems enumerative algorithms have to be applied; otherwise approximation algorithms are a good choice. Performance bounds may often be obtained, but their quality will depend on the kind of availability patterns considered. If worst case bounds cannot be found, heuristics which can only be evaluated empirically have to be applied.

Most of the results reviewed are summarized in Table 11.5.1. The table differs for a given problem type between performance criteria entailing NP-hardness and those for which a polynomial algorithm exists.

Problem	Polynomially solvable	NP-hard
$1, NC_{win}$		$\Sigma C_j, C_{max}$
$1, NC_{win} \mid pmtn$	$\Sigma C_j, C_{max}, L_{max}, \Sigma U_j$	$\Sigma w_j C_j, \Sigma w_j U_j$ (constant availability)
P, NC_{inc}	ΣC_j	
P, NC_{zz}		ΣC_j
$P2, NC_{win} \mid pmtn, prec$	C_{max}, L_{max}	
$P, NC_{zz} \mid pmtn, tree$	C_{max}, L_{max} (in-tree)	C_{max} (for NC_{win})
$P, NC_{win} \mid pmtn, chains$	C_{max}, L_{max}	
$P, NC_{win} \mid pmtn, r_j$	C_{max}, L_{max}	
$Q, NC_{win} \mid pmtn, r_j$	C_{max}, L_{max}	
$F2, NC_{win} \mid pmtn$		C_{max} (single non-availability interval)
$O, NC_{win} \mid pmtn$	C_{max}	
$O2, NC_{win} \mid pmtn(no\text{-}pass)$		C_{max} (single non-availability interval)
$J, NC_{win} \mid n = 2$	C_{max}	
$J, NC_{win} \mid pmtn, n = 2$	C_{max}	

Table 11.5.1 *Summary of results.*

There are many interesting fields for future research.

1. As our review indicates there are many open questions in shop scheduling, e.g., for job shop models comparatively few results are available.

2. Stability analysis introduces sufficient conditions for schedules to be optimal in the case of machine availability restrictions. Extensions to open shops and job shops seem to be interesting in this field.

3. In almost all papers reviewed in this chapter machine availability restrictions are regarded as problem input. There are, however, many cases in which decision

makers have some influence on these restrictions. For example, one may think of a situation where the start time of a maintenance activity for a machine can be chosen within certain limits. In such situations machine availability restrictions become decision variables.

4. To the best of our knowledge there exist no papers dealing with limited machine availability and multiple objective functions. Such models may, however, be very interesting, especially in cases where machine availability restrictions are decision variables. For example, in a case where several machines have to undergo a maintenance activity it may be desirable to minimize the time span between start of the first and end of the last activity while a different objective function is applied for the task scheduling.

5. There are many practical cases where periods of non-availability are not known in advance. In these cases we might apply online scheduling. Some results are already available. In [AS01] it is shown that there are instances where no on-line algorithm can construct optimal makespan schedules if machines change availability at arbitrary points in time. It is also impossible for such an algorithm to guarantee that the solution is within a constant ratio c if there may be time intervals where no machine is available. Albers and Schmidt also report that things look better if the algorithm is allowed to be nearly on-line. In such a case we assume that the algorithm always knows the next point in time when the set of available machines changes. Now optimal schedules can be constructed. The algorithm presented has a running time of $O(qn + S)$, where q is the number of time instances where the set of available machines changes and S is the total number of intervals where machines are available. If at any time at least one machine is available, an on-line algorithm can construct schedules which differ by an absolute error c from an optimal schedule for any $c > 0$. This implies, that not knowing machine availabilities does not really hurt the performance of an algorithm, if arbitrary preemptions are allowed.

References

ABFR89   I. Adiri, J. Bruno, E. Prostig, and A. H. G. Rinnooy Kan. Single machine flow-time scheduling with a single breakdown. *Acta Inform.* 26, 1989, 679-696.

AF55   S. B. Akers and J. Friedman. A non-numerical approach to production scheduling Problems. *Oper. Res.* 3, 1955, 429-442.

Agg04a   R. Aggoune. Minimizing the makespan for the flow shop scheduling problem with availability constraints. *Eur. J. Oper. Res.* 153, 2004, 534-543.

Agg04b   R. Aggoune. Two-job shop scheduling problems with availability constraints, *Proceedings of the 14th International Conference on Automated Planning and Scheduling, ICAPS'04*, AAAI Press, 2004, 253-259.

AP03   R. Aggoune and M.-C. Portmann. Flow shop scheduling problem with limited machine availability: a heuristic approach, *International Conference on Industrial Engineering and Production Management - IEPM 2003*, 1, 2003, 140-149.

AS01   S. Albers and G. Schmidt. Scheduling with unexpected machine breakdowns. *Discrete Appl. Math.* 110, 2001, 85-99.

BBF+01   J. Blazewicz, J. Breit, P. Formanowicz, W. Kubiak, and G. Schmidt. Heuristic algorithms for the two-machine flowshop with limited machine availability. *Omega* 29, 2001, 599-608.

BDDM03   J. Blazewicz, P. Dell'Olmo, M. Drozdowski, and P. Maczka. Scheduling multi-pro-cessor tasks on parallel processors with limited availability. *Eur. J. Oper. Res.* 149, 2003, 377-389.

BDF+00   J. Blazewicz, M. Drozdowski, P. Formanowicz, W. Kubiak, and G. Schmidt. Scheduling preemtable tasks on parallel processors with limited availability. *Parallel Comput.* 26, 2000, 1195-1211.

BEP+96   J. Blazewicz, K. Ecker, E. Pesch, G. Schmidt, and J. Weglarz. *Scheduling Computer and Manufacturing Processes.* Springer, Berlin, 1996.

BF97   J. Blazewicz and P. Formanowicz. Scheduling jobs on open shops with limited machine availability. *RAIRO Oper. Res.* 36, 1997, 149-156.

BFKS97   J. Blazewicz, P. Formanowicz, W. Kubiak, and G. Schmidt. A note on a paral-lel branch and bound algorithm for the flow shop problem with limited ma-chine availability. *Working Paper*, Poznan University of Technology, Poznan, 1997.

BLS05   O. Braun, N. M. Leshchenko, and Y. N. Sotskov. Optimality of Jackson's per-mutations with respect to limited machine availability. *International Transac-tions in Operational Research* 13, 2006, 59-74.

BLSS02   O. Braun, T.-C- Lai, G. Schmidt, and Y. N. Sotskov. Stability of johnson's schedule with respect to limited machine availability. *Int. J. Prod. Res.* 40, 2002, 4381-4400.

Bra02   O. Braun. Scheduling problems with limited available processors and limited number of preemptions (in German). *PhD thesis*, Saarland University, 2002.

Bre00   J. Breit. Heuristic scheduling olgorithms for flow shops and open shops with limited machine availability (in German). *PhD thesis*, Saarland University, 2000.

Bre04a   J. Breit. An improved approximation algorithm for two-machine flow shop scheduling with an availability constraint. *Inform. Process. Lett.* 90, 2004, 273-278.

Bre04b   J. Breit. A polynomial-time approximation scheme for the two-machine flow shop scheduling problem with an availability constraint. *Comput. Oper. Res.* 33, 2006, 2143-2153.

BSS01   J. Breit, G. Schmidt, and V. A. Strusevich. Two-machine open shop scheduling with an availability constraint. *Oper. Res. Lett.* 29, 2001, 65-77.

BSS03   J. Breit, G. Schmidt, and V. A. Strusevich. Non-preemptive two-machine open shop scheduling with non-availability constraints. *Math. Method Oper. Res.* 57(2), 2003, 217-234.

Car82   J. Carlier. The one machine sequencing problem. *Eur. J. Oper. Res.* 11, 1982, 42-47.

CH98   S. Y. Chang and H.-C. Hwang. The worst-case analysis of the multifit algorithm for scheduling nonsimultaneous parallel machines. *Working Paper*, Dept. of Industr. Engin., 1998.

CL03a   T. C. E. Cheng and Z. Liu. 3/2-approximation for two-machine no-wait flowshop scheduling with availability constraints. *Inform. Process. Lett.* 88, 2003, 161-165.

CL03b   T. C. E. Cheng and Z. Liu. Approximability of two-machine no-wait flowshop scheduling with availability constraints. *Oper. Res. Lett.* 31, 2003, 319-322.

CS81   Y. Cho and S. Sahni. Preemptive scheduling of independent jobs with release and due dates times on open, flow and job shop. *Oper. Res.* 29, 1981, 511-522.

CW99   T. C. E. Cheng and G. Wang. Two-machine flowshop scheduling with consecutive availability constraints. *Inform. Process. Lett.* 71, 1999, 49-54.

CW00   T. C. E. Cheng and G. Wang. An improved heuristic for two-machine flowshop scheduling with an availability constraint. *Oper. Res. Lett.* 26, 2000, 223-229.

DW85a   D. Dolev and M. K. Warmuth. Profile scheduling of opposing forests and level Orders. *SIAM J. Alg. Disc. Meth.* 6, 1985, 665-687.

DW85b   D. Dolev and M. K. Warmuth. Scheduling flat graphs. *SIAM J. Comput.* 14, 1985, 638-657.

EFP99   M. L. Espinouse, P. Formanowicz, and B. Penz. Minimzing the makespan in the two-machine no-wait flow-shop with limited machine availability. *Comput. Ind. Eng.* 37, 1999, 497-500.

EFP01   M. L. Espinouse, P. Formanowicz, and B. Penz. Complexity results and approx-imation algorithms for the two machine no-wait flow-shop with limited machine availability. *J. Oper. Res. Soc.* 52, 2001, 116-121.

GG64   P. C Gilmore and R. E. Gomory. Sequencing a one-state variable machine: a solvable case of the traveling salesman problem. *Oper. Res.* 12, 1964, 655-679.

GGN00   M. Gourgand, N. Grangeon, and S. Norre. A review of the stochastic flowshop scheduling problem. *J. Decision Systems* 9, 2000, 183-213.

GJ79   M. R. Garey and D. S. Johnson. *Computers and Intractability*. Freeman, San Francisco, 1979.

GL99   G. H. Graves and C.-Y. Lee. Scheduling maintenance and semi-resumable jobs on a single machine. *Nav. Res. Log.* 46, 1999, 845-863.

Gra69   R. L. Graham. Bounds on multiprocessing timing anomalies. *SIAM J. Appl. Math.* 17, 1969, 263-269.

GS76   T. Gonzalez and S. Sahni. Open shop scheduling to minimize finish time. *J. Assoc. Comput. Mach.* 23, 1976, 665-679.

Hor74   W. A. Horn. Some simple scheduling algorithms. *Nav. Res. Log.* 21, 1974, 177-185.

HS96   N. G. Hall and C. Sriskandarajah. A survey of machine scheduling problems with blocking and no-wait in process. *Oper. Res.* 44, 1996, 510-525.

Joh54   S. M. Johnson. Optimal two- and three-stage production schedules with setup times included. *Nav. Res. Log.* 1, 1954, 61-68.

Kar72   R. M. Karp. Reducibility among combinatorial problems. In R. E. Miller, J. W. Thatcher (eds.), *Complexity of Computer Computations*, 1972, 85-103.

KBF+02   W. Kubiak, J. Blazewicz, P. Formanowicz, J. Breit, and G. Schmidt. Two-machine flow shops with limited machine availability. *Eur. J. Oper. Res.*136, 2002, 528-540.

Kel98   H. Kellerer. Algorithms for multiprocessor scheduling with machine release time. *IIE Trans.* 31, 1998, 991-999.

KM88   M. Kaspi and B. Montreuil. On the scheduling of identical parallel processes with arbitrary initial processor available time. *Research Report School of Industrial Engineering*, Purdue University, 88(12), 1988.

KS04   M. A. Kubzin and V. A. Strusevich. Two-machine flow shop no-wait scheduling with a nonavailability interval. *Nav. Res. Log.* 51, 2004, 613-631.

KSBS02   M. A. Kubzin, V. A. Strusevich, J. Breit, and G. Schmidt. Polynomial-time approximation schemes for the open shop scheduling problem with non-availability constraints. School of Computing and Mathematical Science, University of Greenwich, *Paper 02/IM/100*, 2002.

Law82   E. L. Lawler. Preemptive scheduling of precedence constrained jobs on parallel machines. In Dempster et al. (eds.), *Deterministic and Stochastic Scheduling*, Reidel, 1982, 101-123.

LBB02a   T. Lorigeon, J.-C. Billaut, and J.-L. Bouquard. Availability constraint for a single machine problem with heads and tails. *Proc. 8th Intern. Workshop on Project Management and Scheduling*, 2002, 240-243.

LBB02b   T. Lorigeon, J.-C. Billaut, and J.-L. Bouquard. A dynamic programming algorithm for scheduling jobs in a two-machine open shop with an availability constraint. *J. Oper. Res. Soc.* 53, 2002, 1239-1246.

LC00   C.-Y. Lee and Z.-L. Chen. Scheduling jobs and maintenance activities on parallel machines. *Nav. Res. Log.* 47, 2000, 145-165.

Lee91   C.-Y. Lee. Parallel machine scheduling with non-simultaneous machine available time. *Discrete Appl. Math.* 30, 1991, 53-61.

Lee96   C.-Y. Lee. Machine scheduling with an availability constraint. *J. Global Optim.* 9, 1996, 363-384.

Lee97   C.-Y. Lee. Minimizing the makespan in the two-machine flowshop scheduling problem with an availability constraint. *Oper. Res. Lett.* 20, 1997, 129-139, 1997.

Lee99   C.-Y. Lee. Two-machine flowshop scheduling with availability constraints. *Eur. J. Oper. Res.* 114, 1999, 420-429.

Lee04   C.-Y. Lee. Machine scheduling with availability constraints. In J. Y.-T. Leung (ed.) *Handbook of Scheduling*, CRC Press, 2004, 22.1-22.13.

LHYL97   G. Lin, Y. He, Y. Yao, and H. Lu. Exact bounds of the modified LPT algorithm applying to parallel machines scheduling with nonsimultaneous machine available times. *Appl. Math. J. Chinese Univ.* 12(1), 1997, 109-116.

Lim91   S. Liman. Scheduling with Capacities and Due-Dates. *PhD thesis*, University of Florida, 1991.

LL92   C.-Y. Lee and S. D. Liman. Single machine flow-time scheduling with scheduled maintenance. *Acta Inform.* 29, 1992, 375-382.

LL93   C.-Y. Lee and S. D. Liman. Capacitated two-parallel machine scheduling to minimize sum of job completion time. *Discrete Appl. Math.* 41, 1993, 211-222.

LL01   C.-Y. Lee and V. J. Leon. Machine scheduling with a rate-modifying activity. *Eur. J. Oper. Res.* 128, 2001, 119-128.

LLLR79   J. Labetoulle, E. L. Lawler, J. K. Lenstra, and A. H. G. Rinnooy Kan. Preemptive scheduling of uniform machines subject to due dates. *Technical Paper B W 99/79*, CWI, Amsterdam, 1979.

LM89   E. L. Lawler and C. U. Martel. Preemptive scheduling of two uniform machines to minimize the number of late jobs. *Oper. Res.* 37, 1989, 314-318.

LP93   L. Lu and M. E. Posner. An NP-hard open shop scheduling problem with polynomial average time complexity. *Math. Oper. Res.* 18, 1993, 12-38.

LRB77   J. K. Lenstra, A. H. G. Rinnooy Kan, and P. Brucker. Complexity of processor scheduling problems. *Ann. Disc. Math.* 1, 1977, 343-362.

LS95a   Z. Liu and E. Sanlaville. Preemptive scheduling with variable profile, precedence constraints and due dates. *Discrete Appl. Math.* 58, 1995, 253-280.

LS95b   Z. Liu and E. Sanlaville. Profile scheduling of list algorithms. In P. Chretienne et al. (eds.) *Scheduling Theory and its Applications*, Wiley, 1995, 91-110.

LS97   Z. Liu and E. Sanlaville. Stochastic scheduling with variable profile and precedence constraints. *SIAM J. Comput.* 26, 1997, 173-187.

LSL05   C.-J. Liao, D.-L. Shyur, and C.-H. Lin. Makespan minimization for two parallel machines with an availability constraint. *Eur. J. Oper. Res,* 160(2), 2005, 445-456.

LY03   C.-Y. Lee and G. Yu. Logistics scheduling under disruptions. *Working paper*, Department of Industrial Engineering and Engineering Management, The Hong Kong University of Science and Technology, Hong Kong, 2003.

MC70   R. Muntz and E. G. Coffman. Preemptive scheduling of real-time tasks on multiprocessor systems. *J. Assoc. Comput. Mach.* 17, 1970, 324-338.

McN59   R. McNaughton. Scheduling with deadlines and loss functions. *Management Sci.* 6, 1959, 1-12.

Moo68   J. M. Moore. An n job one machine sequencing algorithm for minimizing the number of late jobs. *Management Sci.* 15, 1968, 102-109.

Mos94   G. Mosheiov. Minimizing the sum of job completion times on capacitated parallel machines. *Math. Comput. Modelling* 20, 1994, 91-99.

MP93   T. E. Morton and D. W. Pentico. *Heuristic Scheduling Systems*. Wiley, New York, 1993.

NK04   C. T. Ng and M. Y. Kovalyov. An FPTAS for scheduling a two-machine flowshop with one unavailability interval. *Nav. Res. Log.* 51, 2004, 307-315.

QBY02   X. Qi, J. F. Bard, and G. Yu. Disruption management for machine scheduling: the case of SPT schedules. *Working paper*, Department of Management Science and Infomation Systems, College of Business Administration, The University of Texas, Austin, TX, 2002, 307-315.

QCT99   X. Qi, T. Chen, and F. Tu. Scheduling the maintenance on a single machine. *J. Oper. Res. Soc.* 50, 1999, 1071-1078.

SPR+05   C. Sadfi, B. Penz, C. Rapine, J. Blazewicz, P. Formanowicz. An improved approximation algorithm for the single machine total completion time scheduling problem with availability constraints. *Eur. J. Oper. Res.* 161, 2005, 3-10.

San95   E. Sanlaville. Nearly on line scheduling of preemptive independent tasks. *Discrete Appl. Math.* 57, 1995, 229-241.

Sch84   G. Schmidt. Scheduling on semi-identical processors. *Z. Oper. Res.* A28, 1984, 153-162.

Sch88   G. Schmidt. Scheduling independent tasks with deadlines on semi-identical processors. *J. Oper. Res. Soc.* 39, 1988, 271-277.

Sch00   G. Schmidt. Scheduling with limited machine availability. *Eur. J. Oper. Res.* 121, 2000, 1-15.

Smi56   W. E. Smith. Various optimizers for single-stage production. *Nav. Res. Log.* 3, 1956, 59-66.

SS98   E. Sanlaville and G. Schmidt. Machine scheduling with availability constraints. *Acta Inform.* 35, 1998, 795-811.

ST95   D. D. Sleator and R. E. Tarjan. Amortized efficiency of list Update and paging rules. *Commun. ACM* 28, 1995, 202-208.

Ull75   J. D. Ullman. Np-complete scheduling problems. *J. Comput. Syst. Sci.* 10, 1975, 384-393.

VS95   G. Vairaktarakis and S. Sahni. Dual criteria preemptive open-shop problems with minimum makespan. *Nav. Res. Log.* 42, 1995, 103-121.

12 Scheduling under Resource Constraints

The scheduling model we consider now is more complicated than the previous ones, because any task, besides processors, may require for its processing some additional scarce resources. Resources, depending on their nature, may be classified into types and categories. The classification into *types* takes into account only the functions resources fulfill: resources of the same type are assumed to fulfill the same functions. The classification into *categories* will concern two points of view. First, we differentiate three categories of resources from the viewpoint of resource constraints. We will call a resource *renewable*, if only its total usage, i.e. temporary availability at every moment, is constrained. A resource is called *non-renewable*, if only its total consumption, i.e. integral availability up to any given moment, is constrained (in other words this resource once used by some task cannot be assigned to any other task). A resource is called *doubly constrained*, if both total usage and total consumption are constrained. Secondly, we distinguish two resource categories from the viewpoint of resource divisibility: *discrete* (i.e. discretely-divisible) and *continuous* (i.e. continuously-divisible) resources. In other words, by a discrete resource we will understand a resource which can be allocated to tasks in discrete amounts from a given finite set of possible allocations, which in particular may consist of one element only. Continuous resources, on the other hand, can be allocated in arbitrary, a priori unknown, amounts from given intervals.

In the next three sections we will consider several basic sub-cases of the resource constrained scheduling problem. In Sections 12.1 and 12.2 problems with renewable, discrete resources will be considered. In Section 12.1 it will in particular be assumed that any task requires one arbitrary processor and some units of additional resources, while in Section 12.2 tasks may require more than one processor at a time (cf. also Chapter 6). Section 12.3 is devoted to an analysis of scheduling with continuous resources.

12.1 Classical Model

The resources to be considered in this section are assumed to be discrete and renewable. Thus, we may assume that s types of additional resources $R_1, R_2, \cdots, R_s$ are available in $m_1, m_2, \cdots, m_s$ units, respectively. Each task T_j requires for its processing one processor and certain fixed amounts of additional resources specified by the resource requirement vector $R(T_j) = [R_1(T_j), R_2(T_j), \cdots, R_s(T_j)]$, where $R_l(T_j)$ $(0 \le R_l(T_j) \le m_l)$, $l = 1, 2, \cdots, s$, denotes the number of units of resource R_l required for the processing of T_j. We will assume here that all required resources are granted to a task before its processing begins or resumes (in the case of pre-emptive scheduling), and they are returned by the task after its completion or in the case of its preemption. These assumptions define a very simple rule to prevent system deadlocks (see e.g. [CD73]) which is often used in practice, despite the fact that it may lead to a not very efficient use of the resources.

We see that such a model is of special value in manufacturing systems where tasks, besides processors, may require additional limited resources for their processing, such as manpower, tools, space etc. One should also not forget about computer applications, where additional resources can stand for primary memory, mass storage, channels and I/O devices. Before discussing basic results in that area we would like to introduce a missing part of the notation scheme introduced in Section 3.4 that describes additional resources. In fact, they are denoted by parameter $\beta_2 \in \{\varnothing, res\ \lambda\delta\rho\}$, where

$\beta_2 = \varnothing$: no resource constraints,

$\beta_2 = res\ \lambda\delta\rho$: there are specified resource constraints;

$\lambda, \delta, \rho \in \{\cdot, k\}$ denote respectively the number of resource types, resource limits and resource requirements. If

$\lambda, \delta, \rho = \cdot$ then the number of resource types, resource limits and resource requirements are respectively arbitrary, and if

$\lambda, \delta, \rho = k$, then, respectively, the number of resource types is equal to k, each resource is available in the system in the amount of k units and the resource requirements of each task are equal to at most k units.

At this point we would also like to present possible transformations among scheduling problems that differ only by their resource requirements (see Figure 12.1.1). In this figure six basic resource requirements are presented. All but two of these transformations are quite obvious. Transformation $\Pi(res\cdots) \propto \Pi(res1\cdots)$ has been proved for the case of saturation of machines and additional resources [GJ75] and will not be presented here. The second, $\Pi(res1\cdots) \propto \Pi(res\cdot11)$, has been proved in [BBKR86]; to sketch its proof, for a given instance of the first problem we construct a corresponding instance of the second problem by assuming the parameters all the same, except resource constraints. Then for each pair T_i, T_j such that $R_1(T_i) + R_1(T_j) > m_1$ (in the first problem), resource R_{ij} available in the amount of one unit is defined in the second problem. Tasks T_i, T_j require a unit of R_{ij}, while other tasks do not require this resource. It follows that $R_1(T_i) + R_1(T_j) \le m_1$ in the first problem if and only if $R_k(T_i) + R_k(T_j) \le 1$ for each resource R_k in the second problem.

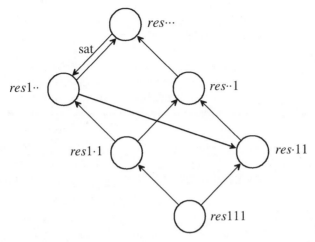

Figure 12.1.1 *Polynomial transformations among resource constrained scheduling problems.*

We will now pass to the presentation of some important results obtained for the above model of resource constrained scheduling. Space limitations prohibit us even from only quoting all these results, however, an extensive survey may be found in [BCSW86, BDM+99, Weg99]. As an example we chose the problem of scheduling tasks on parallel identical processors to minimize schedule length. Basic algorithms in this area will be presented.

Let us first consider the case of independent tasks and non-preemptive scheduling.

Problem $P2 \mid res \cdots, p_j = 1 \mid C_{max}$

The problem of scheduling unit-length tasks on two processors with arbitrary resource constraints and requirements can be solved optimally by the following algorithm.

Algorithm 12.1.1 *Algorithm by Garey and Johnson for $P2 \mid res \cdots, p_j = 1 \mid C_{max}$* [GJ75].

begin
Construct an n-node (undirected) graph G with each node labeled as a distinct
    task and with an edge joining T_i to T_j if and only if $R_l(T_i) + R_l(T_j) \le m_l$,
    $l = 1, 2, \cdots, s$;
Find a maximum matching $\mathcal{F}$ of graph G;
Put the minimal value of schedule length $C_{max}^* = n - |\mathcal{F}|$;

Process in parallel the pairs of tasks joined by the edges comprising set $\mathcal{F}$;
Process other tasks individually;
end;

Notice that the key idea here is the correspondence between maximum matching in a graph displaying resource constraints and the minimum-length schedule. The complexity of the above algorithm clearly depends on the complexity of the algorithm determining the maximum matching. There are several algorithms for finding it, the complexity of the most efficient by Kariv and Even [KE75] being $O(n^{2.5})$. An example of the application of this algorithm is given in Figure 12.1.2 where it is assumed that $n = 6$, $m = 2$, $s = 2$, $m_1 = 3$, $m_2 = 2$, $R(T_1) = [1, 2]$, $R(T_2) = [0, 2]$, $R(T_3) = [2, 0]$, $R(T_4) = [1, 1]$, $R(T_5) = [2, 1]$, and $R(T_6) = [1, 0]$.

An even faster algorithm can be found if we restrict ourselves to the one-resource case. It is not hard to see that in this case an optimal schedule will be produced by ordering tasks in non-increasing order of their resource requirements and assigning tasks in that order to the first free processor on which a given task can be processed because of resource constraints. Thus, problem $P2 \mid res1 \cdot\cdot, p_j = 1 \mid C_{max}$ can be solved in $O(n\log n)$ time.

If in the last problem tasks are allowed only for 0-1 resource requirements, the problem can be solved in $O(n)$ time even for arbitrary ready times and an arbitrary number of machines, by first assigning tasks with unit resource requirements up to m_1 in each slot, and then filling these slots with tasks having zero resource requirements [Bla78].

(a)   (b) $\mathcal{F} = \{(T_1, T_6), (T_2, T_3), (T_4, T_5)\}$

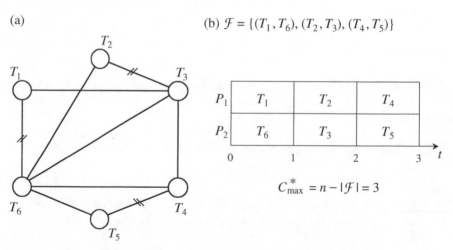

$$C_{max}^* = n - |\mathcal{F}| = 3$$

Figure 12.1.2 *An application of Algorithm 12.1.1:*
(a) graph G corresponding to the scheduling problem,
(b) an optimal schedule.

Problem $P \mid res\ sor, p_j = 1 \mid C_{max}$

When the number of resource types, resource limits and resource requirements are fixed (i.e. constrained by positive integers s, o, r, respectively), problem $P \mid res\ sor, p_j = 1 \mid C_{max}$ is still solvable in linear time, even for an arbitrary number of processors [BE83]. We describe this approach below, since it has a more general application. Depending on the resource requirement vector $[R_1(T_j), R_2(T_j), \cdots, R_s(T_j)] \in \{0, 1, \cdots, r\}^s$, the tasks can be distributed among a sufficiently large (and fixed) number of classes. For each possible resource requirement vector we define one such class. The correspondence between the resource requirement vectors and the classes will be described by a 1-1 function $f: \{0, 1, \cdots, r\}^s \rightarrow \{1, 2, \cdots, k\}$, where k is the number of different possible resource requirement vectors, i.e. $k = (r + 1)^s$. For a given instance, let n_i denote the number of tasks belonging to the ith class, $i = 1, 2, \cdots, k$. Thus all the tasks of class i have the same resource requirement $f^{-1}(i)$. Observe that most of the input information describing an instance of problem $P \mid res\ sor, p_j = 1 \mid C_{max}$ is given by the resource requirements of n given tasks (we bypass for the moment the number m of processors, the number s of additional resources and resource limits o). This input may now be replaced by the vector $v = (v_1, v_2, \cdots, v_k) \in I\!N_0^k$, where v_i is the number of tasks having resource requirements equal to $f^{-1}(i)$, $i = 1, 2, \cdots, k$. Of course, the sum of the components of this vector is equal to the number of tasks, i.e. $\sum_{i=1}^{k} v_i = n$.

We now introduce some definitions useful in the following discussion. An *elementary instance* of $P \mid res\ sor, p_j = 1 \mid C_{max}$ is defined as a sequence $R(T_1)$, $R(T_2), \cdots, R(T_u)$, where each $R(T_i) \in \{1, 2, \cdots, r\}^s - \{(0, 0, \cdots, 0)\}$, with properties $u \leq m$ and $\sum_{i=1}^{u} R(T_i) \leq (o, o, \cdots, o)$. Note that the minimal schedule length of an elementary instance is always equal to 1. An *elementary vector* is a vector $v \in I\!N_0^k$ which corresponds to an elementary instance. If we calculate the number L of different elementary instances, we see that L cannot be greater than $(o + 1)^{(r+1)^s - 1}$, however, in practice L will be much smaller than this upper bound. Denote the elementary vectors (in any order) by $b_1, b_2, \cdots, b_L$.

We observe two facts. First, any input $R(T_1), R(T_2), \cdots, R(T_n)$ can be considered as a union of elementary instances. This is because any input consisting of one task is elementary. Second, each schedule is also constructed from elementary instances, since all the tasks which are executed at the same time form an elementary instance.

Now, taking into account the fact that the minimal length of a schedule for any elementary instance is equal to one, we may formulate the original problem as that of finding a decomposition of a given instance into the minimal number of elementary instances. One may easily see that this is equivalent to finding a decomposition of the vector $v = (v_1, v_2, \cdots, v_k) \in I\!N_0^k$ into a linear combination of elementary vectors $b_1, b_2, \cdots, b_L$, for which the sum of coefficients is minimal:

Find $e_1, e_2, \cdots, e_L \in I\!N_0^k$ such that $\sum_{i=1}^{L} e_i b_i = v$ and $\sum_{i=1}^{L} e_i$ is minimal.

Thus, we have obtained a linear integer programming problem, which in the general case, would be NP-hard. Fortunately, in our case the number of variables L is fixed. It follows that we can apply a result due to Lenstra [Len83] which states that the linear programming problem with fixed number of variables can be solved in polynomial time depending on both, the number of constraints of the integer linear programming problem and $\log a$, but not on the number of variables, where a is the maximum of all the coefficients in the linear integer programming problem. Thus, the complexity of the problem is $O(2^{L^2}(k \log a)^{c^L})$, for some constant c. In our case the complexity of that algorithm is $O(2^{L^2}(k \log n)^{c^L}) < O(n)$. Since the time needed to construct the data for this integer programming problem is $O(2^s(L + \log n)) = O(\log n)$, we conclude that the problem $P \mid res\ sor, p_j = 1 \mid C_{max}$ can be solved in linear time.

Problem $Pm \mid res\ sor \mid C_{max}$

Now we generalize the above considerations for the case of non-unit processing times and tasks belonging to a fixed number k of classes only. That is, the set of tasks may be divided into k classes and all the tasks belonging to the same class have the same processing and resource requirements. If the number of processors m is fixed, then the following algorithm, based on dynamic programming, has been proposed by Błażewicz et al. [BKS89]. A schedule will be built step by step. In every step one task is assigned to a processor at a time. All these assignments obey the following rule: if task T_i is assigned after task T_j, then the starting time of T_i is not earlier than the starting time of T_j. At every moment an assignment of processors and resources to tasks is described by a *state of the assignment process*. For any state a *set of decisions* is given each of which transforms this state into another state. A *value of each decision* will reflect the length of a partial schedule defined by a given state to which this decision led. Below, this method will be described in a more detail.

The state of the assignment process is described by an $m \times k$ matrix X, and vectors Y and Z. Matrix X reflects numbers of tasks from particular classes already assigned to particular processors. Thus, the maximum number of each entry may be equal to n. Vector Y has k entries, each of which represents the number of tasks from a given class not yet assigned. Finally, vector Z has m entries and they represent classes which recently assigned tasks (to particular processors) belong to.

The initial state is that for which matrices X and Z have all entries equal to 0 and Y has entries equal to the numbers of tasks in the particular classes in a given instance.

Let S be a state defined by X, Y and Z. Then, there is a decision leading to state S' consisting of X', Y' and Z' if and only if

$$\exists\, t \in \{1, \cdots, k\} \text{ such that } Y_t > 0,$$ (12.1.1)

$$|\mathcal{M}| = 1,$$ (12.1.2)

where $\mathcal{M}$ is any subset of

$$\mathcal{F} = \{i \mid \sum_{1 \leq j \leq k} X_{ij} p_j = \min_{1 \leq g \leq m}\{\sum_{1 \leq j \leq k} X_{gj} p_j\}\},$$

and finally

$$R_l(T_t) \leq m_l - \sum_{1 \leq j \leq k} R_l(T_j)|\{g \mid Z_g = j\}|, \qquad l = 1,2,\cdots,s,$$ (12.1.3)

where this new state is defined by the following matrices

$$X'_{ij} = \begin{cases} X_{ij} + 1 & \text{if } i \in \mathcal{M} \text{ and } j = t, \\ X_{ij} & \text{otherwise}, \end{cases}$$

$$Y'_j = \begin{cases} Y_j - 1 & \text{if } j = t, \\ Y_j & \text{otherwise}, \end{cases}$$ (12.1.4)

$$Z'_i = \begin{cases} t & \text{if } i \in \mathcal{M}, \\ Z_i & \text{otherwise}. \end{cases}$$

In other words, a task from class t may be assigned to processor P_i, if this class is non-empty (inequality (12.1.1) is fulfilled), there is at least one free processor (equation (12.1.2)), and resource requirements of this task are satisfied (equation (12.1.3)).

If one (or more) conditions (12.1.1) through (12.1.3) are not satisfied, then no task can be assigned at this moment. Thus, one must simulate an assignment of an idle-time task. This is done by assuming the following new state S'':

$$X''_{ij} = \begin{cases} X_{ij} & \text{if } i \notin \mathcal{F}, \\ X_{hj} & \text{otherwise}, \end{cases}$$

$$Y'' = Y,$$ (12.1.5)

$$Z''_i = \begin{cases} Z_i & \text{if } i \notin \mathcal{F}, \\ 0 & \text{otherwise}, \end{cases}$$

where h is one of these g, $1 \leq g \leq m$, for which

$$\sum_{1 \leq j \leq k} X_{gj} p_j = \min_{\substack{1 \leq i \leq m \\ i \notin \mathcal{F}}} \left\{ \sum_{1 \leq j \leq k} X_{ij} p_j \right\} .$$

This means that the above decision leads to state S'' which repeats a pattern of assignment for processor P_h, i.e. one which will be free as the first from among those which are busy now.

A decision leading from state S to S' has its value equal to

$$\max_{1 \leq i \leq m} \left\{ \sum_{1 \leq j \leq k} X_{ij} p_j \right\} . \qquad (12.1.6)$$

This value, of course, is equal to a temporary schedule length.

The final state is that for which the matrices Y and Z have all entries equal to 0. An optimal schedule is then constructed by starting from the final state and moving back, state by state, to the initial state. If there is a number of decisions leading to a given state, then we choose the one having the least value to move back along it. More clearly, if state S follows immediately S', and S (S' respectively) consists of matrices X, Y, Z (X', Y', Z' respectively), then this decision corresponds to assigning a task from $Y - Y'$ at the time $\min_{1 \leq i \leq m} \left\{ \sum_{1 \leq j \leq k} X_{ij} p_j \right\}$.

The time complexity of this algorithm clearly depends on the product of the number of states and the maximum number of decisions which can be taken at the states of the algorithm. A careful analysis shows that this complexity can be bounded by $O(n^{k(m+1)})$, thus, for fixed numbers of task classes k and of processors m, it is polynomial in the number of tasks.

Let us note that another dynamic programming approach has been described in [BKS89] in which the number of processors is not restricted, but a fixed upper bound on task processing times p is specified. In this case the time complexity of the algorithm is $O(n^{k(p+1)})$.

Problem $P \mid res \cdots, p_j = 1 \mid C_{max}$

It follows that when we consider the non-preemptive case of scheduling of unit length tasks we have five polynomial time algorithms and this is probably as much as we can get in this area, since other problems of non-preemptive scheduling under resource constraints have been proved to be NP-hard. Let us mention the parameters that have an influence on the hardness of the problem. First, different ready times cause the strong NP-hardness of the problem even for two processors and very simple resource requirements, i.e. problem $P2 \mid res1\cdots, r_j, p_j = 1 \mid C_{max}$ is already strongly NP-hard [BBKR86] (From Figure 12.1.1 we see that problem $P2 \mid res\cdot11, r_j, p_j = 1 \mid C_{max}$ is strongly NP-hard as well). Second, an increase in the number of processors from 2 to 3 results in the strong NP-hardness of the problem. That is, problem $P3 \mid res1\cdots, r_j, p_j = 1 \mid C_{max}$ is strongly NP-hard as proved by Garey and Johnson [GJ75]. (Note that this is the famous 3-PARTITION problem, the first strongly NP-hard problem.) Again from Figure 12.1.1 we conclude that problem $P3 \mid res\,11, r_j, p_j = 1 \mid C_{max}$ is NP-hard in the strong sense. Finally, even the simplest precedence constraints result in the NP-hardness of the scheduling problem, that is, the $P2 \mid res111, chains, p_j = 1 \mid C_{max}$ is NP-hard in the strong sense [BLRK83]. Because all these problems are NP-hard, there is a need to work out approximation algorithms. We quote some of the results. Most of the algorithms considered here are list scheduling algorithms which differ from each other by the ordering of tasks on the list. We mention three approximation algorithms analyzed for the problem [1].

1. *First fit (FF)*. Each task is assigned to the earliest time slot in such a way that no resource and processor limits are violated.

2. *First fit decreasing (FFD)*. A variant of the first algorithm applied to a list ordered in non-increasing order of $R_{max}(T_j)$, where $R_{max}(T_j) = \max\{R_l(T_j)/m_l \mid 1 \le l \le s\}$.

3. *Iterated lowest fit decreasing (ILFD* - applies for $s = 1$ and $p_j = 1$ only*)*. Order tasks as in the *FFD* algorithm. Put C as a lower bound on C^*_{max}. Place T_1 in the first time slot and proceed through the list of tasks, placing T_j in a time slot for which the total resource requirement of tasks already assigned is minimum. If we ever reach a point where T_j cannot be assigned to any of C slots, we halt the iteration, increase C by 1, and start over.

Below we will present the main known bounds for the case $m < n$. In [KSS75] several bounds have been established. Let us start with the problem $P \mid res1\cdots, p_j = 1 \mid C_{max}$ for which the three above mentioned algorithms have the following bounds:

$$\frac{27}{10} - \left\lceil \frac{37}{10m} \right\rceil < R^{\infty}_{FF} < \frac{27}{10} - \frac{24}{10m},$$

$$R^{\infty}_{FFD} = 2 - \frac{2}{m},$$

$$R_{ILFD} \le 2.$$

We see that the use of an ordered list improves the bound by about 30%. Let us also mention here that problem $P \mid res\cdots, p_j = 1 \mid C_{max}$ can be solved by the approximation algorithm based on the two machine aggregation approach by Röck

---

[1] Let us note that the resource constrained scheduling for unit task processing times is equivalent to a variant of the bin packing problem in which the number of items per bin is restricted to m. On the other hand, several other approximation algorithms have been analyzed for the general bin packing problem and the interested reader is referred to [CGJ84] for an excellent survey of the results obtained in this area.

and Schmidt [RS83], as described in Section 7.3.2 in the context of flow shop scheduling. The worst case behavior of this algorithm is $R = \lceil \frac{m}{2} \rceil$.

Problem $P \mid res \cdots \mid C_{max}$

For arbitrary processing times some other bounds have been established. For problem $P \mid res \cdots \mid C_{max}$ the first fit algorithm has been analyzed by Garey and Graham [GG75]:

$$R_{FF}^{\infty} = \min\{\frac{m+1}{2}, s+2-\frac{2s+1}{m}\}\ .$$

Finally, when dependent tasks are considered, the first fit algorithm has been evaluated for problem $P \mid res \cdots, prec \mid C_{max}$ by the same authors:

$$R_{FF}^{\infty} = m\ .$$

Unfortunately, no results are reported on the probabilistic analysis of approximation algorithms for resource constrained scheduling.

Problem $P \mid pmtn, res1 \cdot 1 \mid C_{max}$

Now let us pass to preemptive scheduling. Problem $P \mid pmtn, res1 \cdot 1 \mid C_{max}$ can be solved via a modification of McNaughton's rule (Algorithm 5.1.8) by taking

$$C_{max}^{*} = \max\{\max_{j}\{p_j\}, \sum_{j=1}^{n} p_j/m, \sum_{T_j \in Z_R} p_j/m_1\}$$

as the minimum schedule length, where Z_R is the set of tasks for which $R_1(T_j) = 1$. The tasks are scheduled as in Algorithm 5.1.8, the tasks from Z_R being scheduled first. The complexity of the algorithm is obviously $O(n)$.

Problem $P2 \mid pmtn, res \cdots \mid C_{max}$

Let us consider now the problem $P2 \mid pmtn, res \cdots \mid C_{max}$. This can be solved via a transformation into the transportation problem [BLRK83].

Without loss of generality we may assume that task $T_j, j = 1, 2, \cdots, n$, spends exactly $p_j/2$ time units on each of the two processors. Let $(T_j, T_i), j \neq i$, denote a resource feasible task pair, i.e. a pair for which $R_l(T_j) + R_l(T_i) \leq m_l, l = 1, 2, \cdots, s$. Let Z be the set of all resource feasible pairs of tasks. Z also includes all pairs of the type $(T_j, T_{n+1}), j = 1, 2, \cdots, n$, where T_{n+1} is an idle time (dummy) task. Now we may construct a transportation network. Let $n+1$ sender nodes correspond to the $n+1$ tasks (including the idle time task) which are processed on processor P_1 and let $n+1$ receiver nodes correspond to the $n+1$ tasks processed on processor P_2. Stocks and requirements of nodes corresponding to T_j, $j = 1, 2, \cdots, n$, are equal to $p_j/2$, since the amount of time each task spends on each processor is equal to $p_j/2$. The stock and the requirement of two nodes corresponding to T_{n+1} are equal to $\sum_{j=1}^{n} p_j/2$, since these are the maximum amounts of time each processor may be idle. Then, we draw directed arcs (T_j, T_i) and (T_i, T_j) if and only if $(T_j, T_i) \in \mathcal{Z}$, to express the possibility of processing tasks T_j and T_i in parallel on processors P_1 and P_2. In addition we draw an arc (T_{n+1}, T_{n+1}). Then, we assign for each pair $(T_j, T_i) \in \mathcal{Z}$ a cost associated with arcs (T_j, T_i) and (T_i, T_j) equal to 1, and a cost associated with the arc (T_{n+1}, T_{n+1}) equal to 0. (This is because an interval with idle times on both processors does not lengthen the schedule). Now, it is quite clear that the solution of the corresponding transportation problem, i.e. the set of arc flows $\{x_{ji}^*\}$, is simply the set of the numbers of time units during which corresponding pairs of tasks are processed (T_j being processed on P_1 and T_i on P_2).

The complexity of the above algorithm is $O(n^4 \log \sum p_j)$ since this is the complexity of finding a minimum cost flow in a network, with the number of vertices equal to $O(n)$.

Problem $Pm \mid pmtn, res \cdots \mid C_{max}$

Now let us pass to the problem $Pm \mid pmtn, res \cdots \mid C_{max}$. This problem can still be solved in polynomial time via the linear programming approach (5.1.15) - (5.1.16) but now, instead of the processor feasible set, the notion of a *resource feasible set* is used. By the latter we mean the set of tasks which can be simultaneously processed because of resource limits (including processor limit). At this point let us also mention that problem $P \mid pmtn, res \cdot 1 \mid C_{max}$ can be solved by the generalization of the other linear programming approach presented in (5.1.24) - (5.1.27). Let us also add that the latter approach can handle different ready times and the L_{max} criterion. On the other hand, both approaches can be adapted to cover the case of the unconnected activity network in the same way as that described in Section 5.1.1.

Finally, we mention that for the problem $P \mid pmtn, res1 \cdot \cdot \mid C_{max}$, the approximation algorithms FF and FFD had been analyzed by Krause et al. [KSS75]:

$$R_{FF}^{\infty} = 3 - \frac{3}{m},$$

$$R_{FFD}^{\infty} = 3 - \frac{3}{m}.$$

Surprisingly, the use of an ordered list does not improve the bound.

12.2 Scheduling Multiprocessor Tasks

In this section we combine the model presented in Chapter 6 with the resource constrained scheduling. That is, each task is assumed to require one or more processors at a time, and possibly a number of additional resources during its execution. The tasks are scheduled preemptively on m identical processors so that schedule length is minimized.

We are given a set $\mathcal{T}$ of tasks of arbitrary processing times which are to be processed on a set $\mathcal{P} = \{P_1, \cdots, P_m\}$ of m identical processors. There are also s additional types of resources, $R_1, \cdots, R_s$, in the system, available in the amounts of $m_1, \cdots, m_s \in \mathbb{N}$ units. The task set $\mathcal{T}$ is partitioned into subsets,

$$\mathcal{T}^j = \{T_1^j, \cdots, T_{n_j}^j\}, \ j = 1, 2, \cdots, k,$$

k being a fixed integer $\leq m$, denoting a set of tasks each requiring j processors and no additional resources, and

$$\mathcal{T}^{jr} = \{T_1^{jr}, \cdots, T_{n_j^r}^{jr}\}, \ j = 1, 2, \cdots, k,$$

k being a fixed integer $\leq m$, denoting a set of tasks each requiring j processors simultaneously and at most m_l units of resource type R_l, $l = 1, \cdots, s$ (for simplicity we write superscript r to denote "resource tasks", i.e. tasks or sets of tasks requiring resources). The resource requirements of any task T_i^{jr}, $i = 1, 2, \cdots, n_j^r$, $j = 1, 2, \cdots, k$, are given by the vector $R(T_i^{jr}) \leq (m_1, m_2, \cdots, m_s)$.

We will be concerned with preemptive scheduling, i.e. each task may be preempted at any time in a schedule, and restarted later at no cost (in that case, of course, resources are also preempted). All tasks are assumed to be independent, i.e. there are no precedence constraints or mutual exclusion constraints among them. A schedule will be called feasible if, besides the usual conditions each task from $\mathcal{T}^j \cup \mathcal{T}^{jr}$ for $j = 1, 2, \cdots, k$ is processed by j processors at a time, and at each moment the number of processed $\mathcal{T}^{jr}$-tasks is such that the numbers of resources used do not exceed the resource limits. Our objective is to find a feasible schedule of minimum length. Such a schedule will be called *optimal*.

First we present a detailed discussion of the case of one resource type ($s = 1$) available in r units, unit resource requirements, i.e. resource requirement of each task is 0 or 1, and $j \in \{1, k\}$ processors per task for some $k \leq m$. So the task set is assumed to be $\mathcal{T} = \mathcal{T}^1 \cup \mathcal{T}^{1r} \cup \mathcal{T}^k \cup \mathcal{T}^{kr}$. A scheduling algorithm of complexity $O(nm)$ where n is the number of tasks in set $\mathcal{T}$, and a proof of its correctness are presented for $k = 2$. Finally, a linear programming formulation of the scheduling problem is presented for arbitrary values of s, k, and resource requirements. The complexity of the approach is bounded from above by a polynomial in the input length as long as the number of processors is fixed.

Process of Normalization

First we prove that among minimum length schedules there exists always a schedule in a special normalized form: A feasible schedule of length C for the set $\mathcal{T}^1 \cup \mathcal{T}^{1r} \cup \mathcal{T}^k \cup \mathcal{T}^{kr}$ is called *normalized* if and only if $\exists\, w \in \mathbb{N}_0$, $\exists\, L \in [0, C)$ such that the number of $\mathcal{T}^k$-, $\mathcal{T}^{kr}$-tasks executed at time $t \in [0, L)$ is $w + 1$, and the number of $\mathcal{T}^k$-,$\mathcal{T}^{kr}$-tasks executed at time $t \in [L, C)$ is w (see Figure 12.2.1). We have the following theorem [BE94].

Theorem 12.2.1 *Every feasible schedule for the set of tasks* $\mathcal{T}^1 \cup \mathcal{T}^{1r} \cup \mathcal{T}^k \cup \mathcal{T}^{kr}$ *can be transformed into a normalized schedule.*

Proof. Divide a given schedule into columns such that within each column there is no change in task assignment. Note that since the set of tasks and the number of processors are finite, we may assume that the schedule consists only of a finite number of different columns. Given two columns A and B of the schedule, suppose for the moment that they are of the same length. Let n_A^j, n_A^{jr}, n_B^j, n_B^{jr} denote the number of $\mathcal{T}^j$-, $\mathcal{T}^{jr}$-tasks in columns A and B, respectively, $j \in \{1, k\}$. Let n_A^0 and n_B^0 be the numbers of unused processors in A and B, respectively. The proof is based on the following claim.

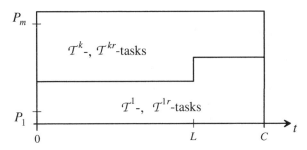

Figure 12.2.1 *A normalized form of a schedule.*

Claim 12.2.2 *Let A and B be columns as above of the same length, and $n_B^k + n_B^{kr} \geq n_A^k + n_A^{kr} + 2$. Then it is always possible to shift tasks between A and B in such a way that afterwards B contains one task of type $\mathcal{T}^k$ or $\mathcal{T}^{kr}$ less than before.* (The claim is valid for any $k \geq 2$.)

Proof. We consider two different types of task shifts, Σ_1 and Σ_2. They are presented below in an algorithmic way. Algorithm 12.2.3 tries to perform a shift of one $\mathcal{T}^k$-task from B to A, and, conversely, of some $\mathcal{T}^1$-and $\mathcal{T}^{1r}$-tasks from A to B. Algorithm 12.2.4 tries to perform a shift of some, say $j+1$ $\mathcal{T}^{kr}$-tasks from B to A, and, conversely, of j $\mathcal{T}^k$-tasks and some $\mathcal{T}^1$-, $\mathcal{T}^{1r}$-tasks from A to B.

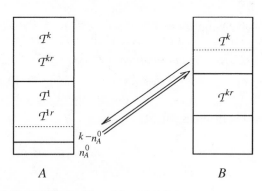

Figure 12.2.2 *Shift of tasks in Algorithm* 12.2.3.

Algorithm 12.2.3 *Shift* Σ_1.
```
begin
if  n_B^k > 0        -- i.e. B has at least one task of type T^k
then
   begin
   Shift one task of type T^k from column B to column A;
      -- i.e. remove one of the T^k-tasks from B and assign it to A
   if  n_A^0 < k
   then
      begin
      if  n_A^1 + n_A^0 ≥ k
      then  Shift k - n_A^0  T^1-tasks from A to B
      else
         if  There are at least k - n_A^0 - n_A^1 unused resources in B
         then
            begin
            Shift n_A^1  T^1-tasks from A to B;
            Shift k - n_A^0 - n_A^1  T^1r-tasks from A to B;
            end
         else  write('Σ_1 cannot be applied: resource conflict');
      end;
```

```
      end
else  write('Σ₁ cannot be applied: B has no Tᵏ-task');
end;
```

Algorithm 12.2.4 *Shift* Σ_2.

```
begin
if  n_B^{kr} > 0        -- i.e. B has at least one task of type T^{kr}
then
   begin
   if  n_A^{1r} = 0
   then
      begin
      Shift one T^{kr}-task from B to A ;
      if  n_A^0 < k
         then
            if  n_A^{kr} < r            -- i.e. no resource conflicts in A
               then  Shift k − n_A^0  T^1-tasks from A to B
               else  Write('Σ₂ cannot be applied: resource conflict');
      end
   else    -- i.e. in the case of n_A^{1r} > 0
      begin
      if  there are numbers j, λ₁, and λ₂ such that
            λ₁ + λ₂ = k − n_A^0 if n_A^0 < k, and 1 otherwise,
            0 ≤ j < λ₂,
            j ≤ n_A^k, j < n_B^{kr},
            λ₁ ≤ n_A^1, λ₂ ≤ n_A^{1r},
            n_B^{kr} + λ₂ − j − 1 ≤ r,
            n_A^{kr} + n_A^{1r} + j + 1 − λ₂ ≤ r
      then    -- perform the following shifts simultaneously
         begin
         Shift j + 1  T^{kr}-tasks from B to A ;
         Shift j  T^k-tasks from A to B ;
         Shift λ₁  T^1-tasks from A to B ;
         Shift λ₂  T^{1r}-tasks from A to B ;
         end
      else  write('Σ₂ cannot be applied');
      end;
   end
```

else write('Σ_2 cannot be applied: B has no $\mathcal{T}^{kr}$-task');
end;

Before we prove that it is always possible to change columns A and B in the proposed way by means of shifts Σ_1 and Σ_2 we formulate some assumptions and simplifications on the columns A and B (detailed proofs are left to the reader).

(a1) Without loss of generality we assume that all the tasks in A and B are pairwise independent, i.e. they are not parts of the same task.

(a2) $n_A^k + n_A^{kr} \leq n_B^k + n_B^{kr} - 2$ (condition of Claim 12.2.2). From that we get

$$n_A^{1r} + n_A^1 + n_A^0 \geq n_B^{1r} + n_B^1 + n_B^0 + 2k.$$

(a3) We restrict our considerations to the case $n_A^{1r} \geq n_B^{1r} + k$ because otherwise shift Σ_1 or Σ_2 can be applied without causing resource problems.

(a4) Next we can simplify the considerations to the case $n_B^{1r} = 0$. Following (a3) and the fact that, whatever shift we apply, at most k tasks of type $\mathcal{T}^{1r}$ are shifted from A to B (and none from B to A) we conclude that we can continue our proof without considering n_B^{1r} tasks of type $\mathcal{T}^{1r}$ in both columns.

(a5) Now we assume $n_A^0 = 0$ or $n_B^0 = 0$ as we can remove all the processors not used in both columns.

(a6) Again we can simplify our considerations by assuming $n_B^0 = 0$ and $n_B^1 = 0$. For suppose $n_B^0 > 0$ or $n_B^1 > 0$, we can remove all the idle processors and $\mathcal{T}^1$-tasks from column B and $n_B^0 + n_B^1$ idle processors or tasks of type $\mathcal{T}^1$ or $\mathcal{T}^{1r}$ from column A. This can be done because there are enough tasks $\mathcal{T}^1$ and $\mathcal{T}^{1r}$ (or idle processors) left in column A.

The two columns are now of the form shown in Figure 12.2.3.

Now we consider four cases (which exhaust all possible situations) and prove that in each of them either shift Σ_1 or Σ_2 can be applied. Let $\gamma = \min\{n_A^{1r}, \max\{k - n_A^0, 1\}\}$.

Case I: $n_B^{kr} + \gamma \leq r, n_B^k > 0$. Here Σ_1 can be applied.

Case II: $n_B^{kr} + \gamma \leq r, n_B^k = 0$. In this case Σ_2 can be applied.

Case III: $n_B^{kr} + \gamma > r, n_B^k > 0$.

If $0 \leq n_A^{1r} \leq k - n_A^0$,

    or $n_A^{1r} > k - n_A^0, k - n_A^0 \leq 0$,

    or $n_A^{1r} > k - n_A^0 > 0, n_A^0 + n_A^1 \geq k$,

or $n_A^{1r} > k - n_A^0 > 0,\ n_A^0 + n_A^1 < k,\ n_B^{kr} + k - n_A^0 - n_A^1 \leq r$,

we can always apply Σ_1. In the remaining sub-case,

$$n_A^{1r} > k - n_A^0 > 0,\ n_A^0 + n_A^1 < k,\ n_B^{kr} + k - n_A^0 - n_A^1 > r,$$

Σ_1 cannot be applied and, because of resource limits in column B, a Σ_2-shift is possible only under the additional assumption

$$n_B^{kr} + k - n_A^0 - n_A^1 - n_A^k - 1 \leq r.$$

What happens in the sub-case $n_B^{kr} + k - n_A^0 - n_A^1 - n_A^k - 1 > r$ will be discussed in a moment.

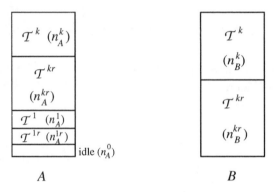

A B

Figure 12.2.3 *Restructuring columns in Claim 12.2.2.*

Case IV: $n_B^{kr} + \gamma > r,\ n_B^k = 0$.

Now, Σ_2 can be applied, except when the following conditions hold simultaneously:

$$n_A^{1r} > k - n_A^0 > 0,\ n_A^0 + n_A^1 < k - 1,\ \text{and}$$
$$n_B^{kr} + k - n_A^0 - n_A^1 - n_A^k - 1 > r.$$

We recognize that in cases III and IV under certain conditions neither of the shifts Σ_1, Σ_2 can be applied. These conditions can be put together as follows:

$$n_B^k \geq 0,\ \text{and}\ n_B^{kr} + k - n_A^0 - n_A^1 - n_A^k - 1 > r.$$

We prove that this situation can never occur: From resource limits in column A we get

$$n_B^{kr} + k - n_A^0 - n_A^1 - n_A^k - 1 > r \geq n_A^{1r} + n_A^{kr}.$$

Together with $k n_B^{kr} \leq m$ we obtain

$$(k-1)(n_A^1 + n_A^{1r} + n_A^0) - k(k-1) < 0,$$

but from (a2) we know $n_A^1 + n_A^{1r} + n_A^0 \geq 2k$, which contradicts $k > 1$.   □

Having proved Claim 12.2.2, it is not hard to prove Theorem 12.2.1. First, we observe that the number of different columns in each feasible schedule is finite. Then, applying shifts Σ_1 or Σ_2 a finite number of times we will get a normalized schedule (for pairs of columns of different lengths only a part of the longer column remains unchanged but for one such column this happens only a finite number of times).   □

Before we describe an algorithm which determines a preemptive schedule of minimum length we prove some simple properties of optimal schedules [BE94].

Lemma 12.2.5 *In a normalized schedule it is always possible to process the tasks in such a way that the boundary between $\mathcal{T}^k$-tasks and $\mathcal{T}^{kr}$-tasks contains at most k steps.*

Proof. Suppose there are more than k steps, say $k + i$, $i \geq 1$, and the schedule is of the form given in Figure 12.2.4. Suppose the step at point L lies between the first and the last step of the $\mathcal{T}^k$-, $\mathcal{T}^{kr}$-boundary.

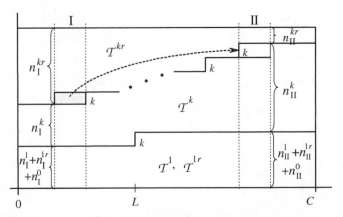

Figure 12.2.4 *k-step boundary between $\mathcal{T}^k$- and $\mathcal{T}^{kr}$-tasks.*

We try to reduce the location of the first step (or even remove this step) by exchanging parts of $\mathcal{T}^{kr}$-tasks from interval I with parts of $\mathcal{T}^k$-tasks from interval II. From resource limits we know:

$$n_{\mathrm{II}}^{1r} + n_{\mathrm{II}}^{kr} \leq r, \; n_{\mathrm{I}}^{1r} + n_{\mathrm{I}}^{kr} \leq r.$$

As there are $k + i$ steps, we have $n_{\mathrm{I}}^{kr} = n_{\mathrm{II}}^{kr} + k + i$. Consider possible sub-cases:

(*i*) If $n_{\text{II}}^{1r} + n_{\text{II}}^{kr} < r$, then exchange the $\mathcal{T}^k$- and $\mathcal{T}^{kr}$-tasks in question. This exchange is possible because in I at least one $\mathcal{T}^{kr}$-task can be found that is independent of all the tasks in II, and in II at least one $\mathcal{T}^k$-task can be found that is independent of all the tasks in I.

(*ii*) If $n_{\text{II}}^{1r} + n_{\text{II}}^{kr} = r$, then the shift described in (*i*) cannot be performed directly. However, this shift can be performed simultaneously with replacement of a $\mathcal{T}^{1r}$-task from II by a $\mathcal{T}^1$-task (or idle time) from I, as can be easily seen.

If the step at point L in Figure 12.2.4 is the leftmost or rightmost step among all steps considered so far, then the step removal works in a similar way.   □

Corollary 12.2.6 *In case $k = 2$ we may assume that the schedule has one of the forms shown in Figure 12.2.5.*   □

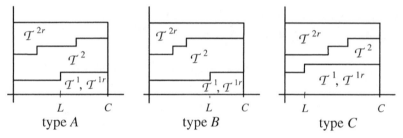

Figure 12.2.5 *Possible schedule types in Corollary 12.2.6.*

Lemma 12.2.7 *Let $k = 2$. In cases (B) and (C) of Figure 12.2.5 the schedule can be changed in such a way that one of the steps in the boundary between $\mathcal{T}^k$ and $\mathcal{T}^{kr}$ is located at point L, or it disappears.*

Proof. The same arguments as in the proof of Lemma 12.2.5 are used.   □

Corollary 12.2.8 *In case $k = 2$, every schedule can be transformed into one of the types given in Figure 12.2.6.*   □

Let us note that if in type **B1** (Figure 12.2.6) not all resources are used during interval $[L, C)$, then the schedule can be transformed into type **B2** or **C2**. If in type **C1** not all resources are used during interval $[L, C)$, then the schedule can be transformed into type **B2** or **C2**. A similar argument holds for schedules of type **A**.

The Algorithm

In this section an algorithm of scheduling preemptable tasks will be presented and its optimality will then be proved for the case $k = 2$. Now, a lower bound for the schedule length can be given. Let

$$X^j = \sum_{T_i \in \mathcal{T}^j} p_i^j, \qquad X^{jr} = \sum_{T_i^{jr} \in \mathcal{T}^{jr}} p_i^{jr}, \qquad j = 1, k.$$

It is easy to see that the following formula gives a lower bound on the schedule length,

$$C = \max\{C_{max}^r, C'\} \tag{12.2.1}$$

where

$$C_{max}^r = (X^{1r} + X^{kr})/r,$$

and C' is the optimum schedule length for all $\mathcal{T}^1$-, $\mathcal{T}^{1r}$-, $\mathcal{T}^k$-, $\mathcal{T}^{kr}$-tasks without considering resource limits (cf. Section 6.1).

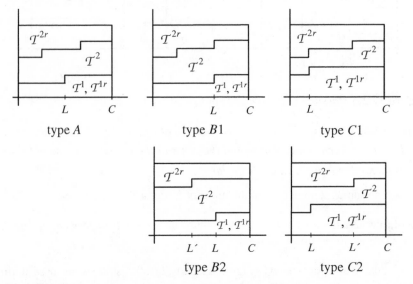

Figure 12.2.6 *Possible schedule types in Corollary* 12.2.8.

In the algorithm presented below we are trying to construct a schedule of type **B2** or **C2**. However, this may not always be possible because of resource constraints causing "resource overlapping" in certain periods. In this situation we first try to correct the schedule by exchanging some critical tasks so that resource limits are not violated, thus obtaining a schedule of type **A**, **B1** or **C1**. If this is not possible, i.e. if no feasible schedule exists, we will have to re-compute bound C in order to remove all resource overlappings.

Let L and L' be the locations of steps as shown in the schedules of type **B2** or **C2** in Figure 12.2.6. Then

$$L \equiv (X^2 + X^{2r}) \bmod C, \text{ and } L' \equiv X^{2r} \bmod C. \tag{12.2.2}$$

In order to compute resource overlapping we proceed as follows. Assign $\mathcal{T}^2$- and $\mathcal{T}^{2r}$-tasks in such a way that only one step in the boundary between these two types of tasks occurs; this is always possible because bound C was chosen properly. The schedule thus obtained is of type **B2** or **C2**. Before the $\mathcal{T}^1$- and $\mathcal{T}^{1r}$-tasks are assigned we partition the free part of the schedule into two areas, LA (left area) and RA (right area) (cf. Figure 12.2.7). Note that a task from $\mathcal{T}^1 \cup \mathcal{T}^{1r}$ fits into LA or RA only if its length does not exceed L or $C - L$, respectively. Therefore, all "long" tasks have to be cut into two pieces, and one piece is assigned to LA, and the other one to RA. We do this by assigning a piece of length $C - L$ of each long task to RA, and the remaining piece to LA (see Section 5.4.2 for detailed reasoning). The *excess* $e(T_i)$ of each such task is defined as $e(T_i) = p_i - C + L$, if $p_i > C - L$, and 0 otherwise.

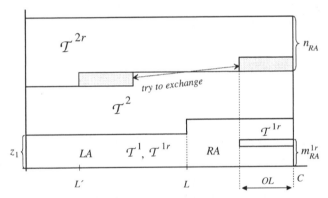

Figure 12.2.7 *Left and right areas in a normalized schedule.*

The task assignment is continued by assigning all excesses to LA, and, in addition, by putting as many as possible of the remaining $\mathcal{T}^{1r}$-tasks (so that no resource violations occur) and $\mathcal{T}^1$-tasks to LA. However, one should not forget that if there are more long tasks in $\mathcal{T}^1 \cup \mathcal{T}^{1r}$ than $z_1 + 2$ (cf. Figure 12.2.7), then each such task should be assigned according to the ratio of processing capacities of both sides LA and RA, respectively. All tasks not being assigned yet are assigned to RA. Hence only in RA resource limits may be violated. Take the sum OL of processing times of all $\mathcal{T}^{1r}$-tasks violating the resource limit. OL is calculated in the algorithm given below. Of course, OL is always less than or equal to $C - L$, and the $\mathcal{T}^{1r}$-tasks in RA can be arranged in such a way that at any time in $[L, C)$ no more than $r+1$ resources are required.

Resource overlapping (OL) of $\mathcal{T}^{1r}$- and $\mathcal{T}^{2r}$-tasks cannot be removed by exchanging $\mathcal{T}^{1r}$-tasks in RA with $\mathcal{T}^{1}$-tasks in LA, because the latter are only the excesses of long tasks. So the only possibility to remove the resource overlapping is to exchange $\mathcal{T}^{2r}$-tasks in RA with $\mathcal{T}^{2}$-tasks in LA (cf. Figure 12.2.7). Suppose that $\tau\ (\leq OL)$ is the maximal amount of $\mathcal{T}^{2}$-, $\mathcal{T}^{2r}$-tasks that can be exchanged in that way. Thus resource overlapping in RA is reduced to the amount $OL - \tau$. If $OL - \tau = 0$, then all tasks are scheduled properly and we are done. If $OL - \tau > 0$, however, a schedule of length C does not exist. In order to remove the remaining resource overlapping (which is in fact $OL - \tau$) we have to increase the schedule length again.

Let n_{RA} be the number of $\mathcal{T}^{2}$- or $\mathcal{T}^{2r}$-tasks executed at the same time during $[L, C)$. Furthermore, let z_1 be the number of processors not used by $\mathcal{T}^{2}$- or $\mathcal{T}^{2r}$-tasks at time 0, let m_{RA}^{1r} be the number of processors executing $\mathcal{T}^{1r}$-tasks in RA (cf. Figure 12.2.7), and let l_{RA}^{1} be the number of $\mathcal{T}^{1}$-tasks executed in RA and having excess in LA. The schedule length is then increased by some amount ΔC, i.e.

$$C = C + \Delta C, \text{ where } \Delta C = \min\ \{\Delta C_a, \Delta C_b, \Delta C_c\}, \tag{12.2.3}$$

and ΔC_a, ΔC_b, and ΔC_c are determined as follows.

(a) $$\Delta C_a = \frac{OL - \tau}{m_{RA}^{1r} + (m - z_1 - 2)/2 + l_{RA}^{1}}.$$

This formula considers the fact that the parts of $\mathcal{T}^{1r}$-tasks violating resource limits have to be distributed among other processors. By lengthening the schedule the following processors will contribute processing capacity:

- m_{RA}^{1r} processors executing $\mathcal{T}^{1r}$-tasks on the right hand side of the schedule,

- $(m - z_1 - 2)/2$ pairs of processors executing $\mathcal{T}^{2}$- or $\mathcal{T}^{2r}$-tasks and contributing to a decrease of L (and thus lengthening part RA),

- l_{RA}^{1} processors executing $\mathcal{T}^{1}$-tasks whose excesses are processed in LA (and thus decreasing their excesses, and hence allowing part of $\mathcal{T}^{1r}$ to be processed in LA).

(b) If the schedule length is increased by some Δ then L will be decreased by $n_{RA}\Delta$, or, as the schedule type may switch from $C2$ to $B2$ (provided L was small enough, cf. Figure 12.2.6), L would be replaced by $C + \Delta + L - n_{RA}\Delta$. In order to avoid the latter case we choose Δ in such a way that the new value of L will be 0, i.e. $\Delta C_b = L/n_{RA}$.

Notice that with the new schedule length $C + \Delta C$, $\Delta C \in \{\Delta C_a, \Delta C_b\}$, the length of the right area RA, will be increased by $\Delta C(n_{RA} + 1)$.

(c) Consider all tasks in $\mathcal{T}^1$ with non-zero excesses. All tasks in $\mathcal{T}^1$ whose excesses are less than $\Delta C(n_{RA} + 1)$ will have no excess in the new schedule. However, if there are tasks with larger excess, then the structure of a schedule of length $C + \Delta C$ will be completely different and we are not able to conclude that the new schedule will be optimal. Therefore we take the shortest task T_s of $\mathcal{T}^1$ with non-zero excess and choose the new schedule length so that T_s will fit exactly into the new RA, i.e.

$$\Delta C_c = \frac{p_s - C + L}{1 + n_{RA}} .$$

The above reasoning leads to the following algorithm [BE94].

Algorithm 12.2.9
Input: Number m of processors, number r of resource units, sets of tasks $\mathcal{T}^1$, $\mathcal{T}^{1r}, \mathcal{T}^2, \mathcal{T}^{2r}$.
Output: Schedule for $\mathcal{T}^1 \cup \mathcal{T}^{1r} \cup \mathcal{T}^2 \cup \mathcal{T}^{2r}$ of minimum length.

```
begin
    Compute bound C according to formula (12.2.1);
    repeat
    Compute L, L' according to (12.2.2), and the excesses for the tasks of
```
$\mathcal{T}^1 \cup \mathcal{T}^{1r}$,

Using bound C, find a normalized schedule for $\mathcal{T}^2$- and $\mathcal{T}^{2r}$-tasks by assigning $\mathcal{T}^{2r}$-tasks from the top of the schedule (processors $P_m, P_{m-1}, \cdots,$) and from left to right, and by assigning $\mathcal{T}^2$-tasks starting at time L, to the processors P_{z_1+1} and P_{z_1+2} from right to left (cf. Figure 12.2.8);

```
    if  the number of long 𝒯¹- and 𝒯¹ʳ-tasks is ≤ z₁+2
    then
```
Take the excesses $e(T)$ of long $\mathcal{T}^1$- and $\mathcal{T}^{1r}$-tasks, and assign them to the left area LA of the schedule in the way depicted in Figure 12.2.8
```
    else
```
Assign these tasks according to the processing capacities of both sides LA and RA of the schedule, respectively;
```
    if  LA is not completely filled
    then  Assign 𝒯¹ʳ-tasks to LA as long as resource constraints are not vio-
        lated;
    if  LA is not completely filled
```

then Assign $\mathcal{T}^1$-tasks to LA;

Fill the right area RA with the remaining tasks in the way shown in Figure 12.2.8;

if resource constraints are violated in interval $[L, C)$

then

Compute resource overlapping $OL - \tau$ and correct bound C according to (12.2.3);

until $OL - \tau = 0$;

end;

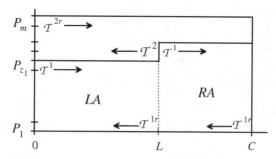

Figure 12.2.8 *Construction of an optimal schedule.*

The optimality of Algorithm 12.2.9 is proved by the following theorem [BE94].

Theorem 12.2.10 *Algorithm 12.2.9 determines a preemptive schedule of minimum length for $\mathcal{T}^1 \cup \mathcal{T}^{1r} \cup \mathcal{T}^2 \cup \mathcal{T}^{2r}$ in time $O(nm)$.* □

The following example demonstrates the use of Algorithm 12.2.9.

Example 12.2.11 Consider a processor system with $m = 8$ processors, and $r = 3$ units of resource. Let the task set contain 9 tasks, with processing requirements as given in the following table:

	T_1	T_2	T_3	T_4	T_5	T_6	T_7	T_8	T_9
processing times	10	5	5	5	10	8	2	3	7
number of processors	2	2	2	2	1	1	1	1	1
number of resource units	1	0	0	0	1	1	1	0	0

Table 12.2.1.

Then,

$$X^1 = 10, X^{1r} = 20, X^2 = 15, X^{2r} = 10,$$

$$C_{max}^r = (X^{1r} + X^{2r})/r = 10, C' = (X^1 + X^{1r} + 2X^2 + 2X^{2r})/m = 10,$$

i.e. $C = 10$ and $L = 5$. The first loop of Algorithm 12.2.9 yields the schedule shown in Figure 12.2.9. In the schedule thus obtained a resource overlapping occurs in the interval [8,10). There is no way to exchange tasks, so $\tau = 0$, and an overlapping of amount 2 remains. From equation (12.2.3) we obtain $\Delta C_a = 1/3$, $\Delta C_b = 5/2$, and $\Delta C_c = 2/3$. Hence the new schedule length will be $C = 10 + \Delta C_a = 10.33$, and $L = 4.33$, $L' = 10.0$. In the second loop the algorithm determines the schedule shown in Figure 12.2.10, which is now optimal.   □

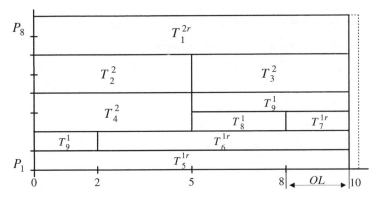

Figure 12.2.9 *Example schedule after the first loop of Algorithm* 12.2.9.

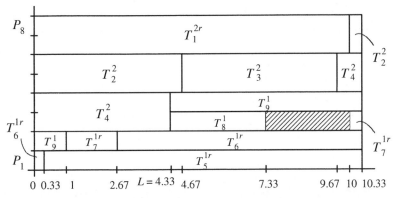

Figure 12.2.10 *Example schedule after the second loop of Algorithm* 12.2.9.

Linear Programming Approach to the General Case

In this section we will show that for a much larger class of scheduling problems one can find schedules of minimum length in polynomial time. We will consider tasks having arbitrary resource and processor requirements. That is, the task set $\mathcal{T}$ is now composed of the following subsets:

$\mathcal{T}^j, j = 1, \cdots, k$, tasks requiring j processors each and no resources, and

$\mathcal{T}^{jr}, j = 1, \cdots, k$, tasks requiring j processors each and some resources.

We present a linear programming formulation of the problem. Our approach is similar to the *LP* formulation of the project scheduling problem, cf. (5.1.15)-(5.1.16). We will need a few definitions. By a resource feasible set we mean here a subset of tasks that can be processed simultaneously because of their total resource and processor requirements. Let M be the number of different resource feasible sets. By variable x_i we denote the processing time of the ith resource feasible set, and by Q_j we denote the set of indices of those resource feasible sets that contain task $T_j \in \mathcal{T}$. Thus the following linear programming problem can be formulated:

$$\text{Minimize} \quad \sum_{i=1}^{M} x_i$$

$$\text{subject to} \quad \sum_{i \in Q_j} x_i = p_j \quad \text{for each } T_j \in \mathcal{T},$$

$$x_i \geq 0, \quad i = 1, 2, \cdots, M.$$

As a solution of the above problem we get optimal values x_i^* of interval lengths in an optimal schedule. The tasks processed in the intervals are members of the corresponding resource feasible subsets. As before, the number of constraints of the linear programming problem is equal to n, and the number of variables is $O(n^m)$. Thus, for a fixed number of processors the complexity is bounded from above by a polynomial in the number of tasks. On the other hand, a linear programming problem may be solved (using e.g. Karmarkar's algorithm [Kar84]) in time bounded from above by a polynomial in the number of variables, the number of constraints, and the sum of logarithms of all the coefficients in the *LP* problem. Thus for a fixed number of processors, our scheduling problem is solvable in polynomial time.

12.3 Scheduling with Continuous Resources

In this section we consider scheduling problems in which, apart from processors, also continuously-divisible resources are required to process tasks. Basic results will be given for problems with parallel, identical processors (Section 12.3.2) or a single processor (Sections 12.3.3, 12.3.4) and one additional type of continuous, renewable resource. This order of presentation follows from the specificity of task models used in each case.

12.3.1 Introductory Remarks

Let us start with some comments concerning the concept of a continuous resource. As we remember, this is a resource which can be allotted to a task in an arbitrary, unknown in advance amount from a given interval. We will deal with renewable resources, i.e. such for which only usage, i.e. temporary availability is constrained at any time. This "temporary" character is important, since in practice it is often ignored for some doubly constrained resources which are then treated as non-renewable. For example, this is the case of money for which usually only the consumption is considered, whereas they have also a "temporary" nature. Namely, money treated as a renewable resource mean in fact a "flow" of money, called *rate of spending* or *rate of investment*, i.e. an amount available in a given period of a fixed length (week, month). The most typical example of a (renewable) continuous resource is power (electric, hydraulic, pneumatic) which, however, is in general doubly constrained since apart from the usage, also its consumption, i.e. energy, is constrained. Other examples we get when parallel "processors" are driven by a common power source. "Processors" mean here e.g. machines with proper drives, electrolytic tanks, or pumps for refueling navy boats.

We should also stress that sometimes it is purposeful to treat a discrete (i.e. discretely-divisible) resource as a continuous one, since this assumption can simplify scheduling algorithms. Such an approach is allowed when there are many alternative amounts of (discrete) resource available for processing each task. This is, for example, the case in multiprocessor systems where a common primary memory consists of hundreds of pages (see [Weg80]). Treating primary memory as a continuous resource we obtain a scheduling problem from the class we are interested in.

In the next two sections we will study scheduling problems with continuous resources for two models of task processing characteristic (time or speed) vs. (continuous) resource amount allotted. The first model is given in the form of a continuous function: task processing speed vs. resource amount allotted at a given time (Section 12.3.2), whereas the second one is given in the form of a continuous function: task processing time vs. resource amount allotted (Section 12.3.3). The first model is more natural in majority of practical situations, since it reflects directly the "temporary" nature of renewable resources. It is also more general and allows a deep a priori analysis of properties of optimal schedules due to the form of the function describing task processing speed in relation to the allotted amount of resource. This analysis leads even to analytical results in some cases, and in general to the simplest formulations of mathematical programming problems for finding optimal schedules. However, in situations when all tasks use constant resource amounts during their execution, both models are equivalent. Then rather the second model is used as the direct generalization of the traditional, discrete model.

In Section 12.3.4 we will consider another type of problems, where task processing times are constant, but their ready times are functions of a continuous resource. This is another generalization of the traditional scheduling model which is important in some practical situations.

12.3.2 Processing Speed vs. Resource Amount Model

Assume that we have m identical, parallel processors $P_1, P_2, \cdots, P_m$, and one additional, (continuous, renewable) resource available in amount $\hat{U}$. For its processing task $T_j \in \mathcal{T}$ requires one of the processors and an amount of a continuous resource $u_j(t)$ which is arbitrary and unknown in advance within interval $(0, \hat{U}]$.

The task processing model is given in the form:

$$\dot{x}_j(t) = dx_j(t)/dt = f_j[u_j(t)], \quad x_j(0) = 0, \quad x_j(C_j) = \tilde{x}_j \tag{12.3.1}$$

where $x_j(t)$ is the state of T_j at time t, f_j is a (positive) continuous, non-decreasing function, $f_j(0) = 0$, C_j is the (unknown in advance) completion time of T_j, and $\tilde{x}_j > 0$ is the known final state, or processing demand, of T_j. Since a continuous resource is assumed to be renewable, we have

$$\sum_{j=1}^{n} u_j(t) \leq \hat{U} \quad \text{for each } t. \tag{12.3.2}$$

As we see, the above model relates task processing speed to the (continuous) resource amount allotted to this task at time t. Let us interpret the concept of a *task state*. By the state of task T_j at time t, $x_j(t)$, we mean a measure of progress of the processing of T_j up to time t or a measure of work related to this processing. This can be, for example, the number of standard instructions of a computer program already processed, the volume of a fuel bunker already refueled, the amount of a product resulting from the performance of T_j up to time t, the number of man-hours or kilowatt-hours already spent in processing T_j, etc.

Let us point out that in practical situations it is often quite easy to construct this model, i.e. to define f_j, $j = 1, 2, \cdots, n$. For example, in computer systems analyzed in [Weg80], the f_j's are progress rate functions of programs, closely related to their lifetime curves, whereas in problems in which processors use electric motors, the f_j's are functions: rotational speed vs. current density.

Let us also notice that in the case of a continuous resource changes of the resource amount allotted to a task within interval $(0, \hat{U}]$ does not mean a task preemption.

To compare formally the model (12.3.1) with the model

$$p_j = \phi_j(u_j), \quad u_j \in (0, \hat{U}] \tag{12.3.3}$$

where p_j is the processing time of T_j and ϕ_j is a (positive) continuous, non-increasing function, notice that the condition $x_j(C_j) = \tilde{x}_j$ is equivalent to

$$\int_0^{C_j} f_j[u_j(t)]dt = \tilde{x}_j. \tag{12.3.4}$$

Thus, if $u_j(t) = u_j$, i.e. is constant for $t \in (0, C_j]$, we have

$$C_j = p_j = \tilde{x}_j / f_j(u_j), \text{ i.e. } \phi_j = \tilde{x}_j / f_j(u_j). \tag{12.3.5}$$

In consequence, if T_j is processed using a constant resource amount u_j, (12.3.5) defines the relation between both models. It is worth to underline that, as we will see, on the basis of the model (12.3.1) one can easily and naturally find the conditions under which tasks are processed using constant resource amounts in an optimal schedule.

Assume now that the number n of tasks is less than or equal to the number m of machines, and that tasks are independent. The first assumption implies that in fact we deal only with the allocation of a continuous resource, since the assignment of tasks to machines is trivial. This is a "pure" (continuous) resource allocation problem, as opposed to a "mixed" (discrete-continuous) problem, when we have to deal simultaneously with scheduling on machines (considered as a discrete resource) and the allocation of a continuous resource.

If $n \leq m$ (then it is sufficient to assume $n = m$, since for $n < m$, $m - n$ machines are idle) our goal is to find a piece-wise continuous vector function $\boldsymbol{u}^*(t) = (u_1^*(t), u_2^*(t), \cdots, u_n^*(t))$, $u_j^*(t) \geq 0$, $j = 1, 2, \cdots, n$, such that (12.3.1) and (12.3.2) are satisfied, and $C_{max} = \max\{C_j\}$ reaches its minimum C_{max}^*. This problem was studied in a number of papers (see [Weg82] as a survey) under different assumptions concerning task and resource characteristics. Below we present few basic results useful in our future considerations. To this end we need some additional denotations.

Let us denote by $\mathcal{U}$ the set of *resource allocations*, i.e. all values of a vector function $\boldsymbol{u}(t)$, or all points $\boldsymbol{u} = (u_1, u_2, \cdots, u_n) \in \mathbb{R}^n$, $u_j \geq 0$ for $j = 1, 2, \cdots, n$, satisfying the relation

$$\sum_{j=1}^n u_j \leq \hat{U}.$$

Further, we will denote by $\mathcal{V}$ the set defined as follows:

$$\boldsymbol{v} = (v_1, v_2, \cdots, v_n) \in \mathcal{V} \text{ if and only if } \boldsymbol{u} \in \mathcal{U}, \\ \text{and } v_j = f_j(u_j), j = 1, 2, \cdots, n. \tag{12.3.6}$$

As the functions f_j are monotonic for $j = 1, 2, \cdots, n$, it is obvious that (12.3.5) defines a univalent mapping between $\mathcal{U}$ and $\mathcal{V}$, and thus we can call the points $\boldsymbol{v}$ *transformed resource allocations*. It is easy to prove (see, e.g. [Weg82]) that C_{max}^* as a function of final states of tasks $\tilde{\boldsymbol{x}} = (\tilde{x}_1, \tilde{x}_2, \cdots, \tilde{x}_n)$ can always be expressed as

$$C_{max}^*(\tilde{x}) = \min\{C_{max} > 0 \mid \tilde{x}/C_{max} \in co\mathcal{V}\} \qquad (12.3.7)$$

where $co\mathcal{V}$ is the convex hull of $\mathcal{V}$, i.e. the set of all convex combinations of the elements of $\mathcal{V}$. Notice that (12.3.7) gives a simple geometrical interpretation of an optimal solution of our problem. Namely, it says that C_{max}^* is always reached at the intersection point of the straight line given by the parametric equations

$$v_j = \tilde{x}_j/C_{max}, \ j = 1, 2, \cdots, n \qquad (12.3.8)$$

and the boundary of set $co\mathcal{V}$. Since, according to (12.3.6), the shape of $\mathcal{V}$, and thus $co\mathcal{V}$, depends on functions f_j, $j = 1, 2, \cdots, n$, we can study the form of optimal solutions in relation to these functions. Let us consider two special, but very important cases:

(i)   concave f_j, $j = 1, 2, \cdots, n$, and

(ii) $f_j \le c_j u_j$, $c_j = f_j(\hat{U})/\hat{U}$, $j = 1, 2, \cdots, n$.

It is easy to check that in case (i) set $\mathcal{V}$ is already convex, i.e. $co\mathcal{V} = \mathcal{V}$. Thus, the intersection point defined above is always a transformed resource allocation (see Figure 12.3.1 for $n = 2$).

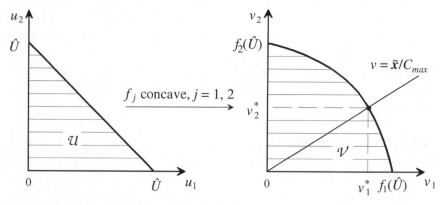

Figure 12.3.1 *The case of concave f_j, $j = 1, 2$.*

This means that in the optimal solution tasks are processed fully in parallel using constant resource amounts u_j^*, $j = 1, 2, \cdots, n$. To find these amounts let us notice that the equation of the boundary of $\mathcal{V}$ has the form $\sum_{j=1}^{n} f_j^{-1}(v_j) = \hat{U}$ (we substitute u_j from (12.3.6) for the equation of the boundary of $\mathcal{U}$, i.e. $\sum_{j=1}^{n} u_j = \hat{U}$), where f_j^{-1} is the function inverse to f_j, $j = 1, 2, \cdots, n$. Substituting v_j from (12.3.8), we get for the above equation

$$\sum_{j=1}^{n} f_j^{-1}(\tilde{x}_j/C_{max}) = \hat{U}. \tag{12.3.9}$$

For given $\tilde{x}_j$, $j = 1, 2, \cdots, n$, the (unique) positive root of this equation is equal to the minimum value C_{max}^* of C_{max}. Of course

$$u_j^* = f_j^{-1}(\tilde{x}_j/C_{max}^*), j = 1, 2, \cdots, n. \tag{12.3.10}$$

It is worth to note that equation (12.3.9) can be solved analytically for some important cases. In particular, this is the case of $f_j = c_j u_j^{1/\alpha_j}$, $c_j > 0$, $\alpha_j \in \{1, 2, 3, 4\}$, $j = 1, 2, \cdots, n$, when (12.3.9) reduces to an algebraic equation of an order ≤ 4. Furthermore, if $\alpha_j = \alpha \geq 1, j = 1, 2, \cdots, n$, we have

$$C_{max}^* = [\frac{1}{\hat{U}} \sum_{j=1}^{n} (\tilde{x}_j/c_j)^\alpha]^{1/\alpha}. \tag{12.3.11}$$

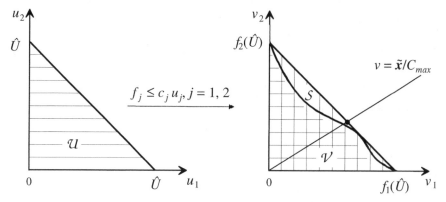

Figure 12.3.2 *The case of* $f_j \leq c_j u_j$, $c_j = f_j(\hat{U})/\hat{U}$, $j = 1, 2$.

Let us pass to the case (*ii*). It is easy to check that now set $\mathcal{V}$ lies entirely inside simplex $\mathcal{S}$ spanned on the points $(0, \cdots, 0, f_j(\hat{U}), 0, \cdots, 0)$, where $f_j(\hat{U})$ appears on the jth position, $j = 1, 2, \cdots, n$ (see Figure 12.3.2 for $n = 2$). This clearly means that $co\mathcal{V} = \mathcal{S}$, and that the intersection point of the straight line defined by (12.3.7) and the boundary of $\mathcal{S}$ most probably is not a transformed resource allocation (except for the case of linear f_j, $j = 1, 2, \cdots, n$). However, one can easily verify that the same value C_{max}^* is obtained using transformed resource allocations whose convex combination yields the intersection point just discussed. These always are, of course, the extreme points on which simplex $\mathcal{S}$ is spanned. This fact implies directly that in case (*ii*) there always exists the solution of the length $C_{max}^* = \sum_{j=1}^{n} \tilde{x}_j/f_j(\hat{U})$ in which single tasks are processed consecutively (i.e.

on a single machine) using the maximum resource amount $\hat{U}$. Of course, this solution is not unique if we assume that there is no time loss concerned with a task preemption. However, there is no reason to preempt a task if preemption does not decrease C^*_{max}.

Thus, in both cases, (i) and (ii), there exist optimal solutions in which each task is processed using a constant resource amount. Consequently, in these cases the model (12.3.1) is mathematically equivalent to the model (12.3.3).

In the general case of arbitrary functions f_j, $j = 1, 2, \cdots, n$, one must search for transformed resource allocations whose convex combination fulfills (12.3.8) and gives the minimum value of C_{max}.

Assume now that tasks are dependent, i.e. that a non-empty relation $<$ is defined on $\mathcal{T}$. To represent $<$ we will use task-on-arc digraphs, also called activity networks (see Section 3.1). In this representation we can order nodes, i.e. events in such a way that the occurrence of node i is not later than the occurrence of node j if $i < j$. As is well known, such an ordering is always possible (although not always unique) and can be found in time $O(n^2)$ (see, e.g. [Law76]). Using this ordering one can utilize the results obtained for independent tasks to solve corresponding resource allocation problems for dependent tasks. To show how it works we will need some further denotations. Denote by $\mathcal{T}_k$ the subset of tasks which can be processed in the interval between the occurrence of nodes k and $k + 1$, by $\tilde{x}_{jk} \geq 0$ a part of $T_j \in \mathcal{T}_k$ (i.e. a part of $\tilde{x}_j$) processed in the above interval, by $\Delta^*_k(\{\tilde{x}_{jk}\}_{T_j \in \mathcal{T}_k})$ the minimum length of this interval as a function of task parts $\{\tilde{x}_{jk}\}_{T_j \in \mathcal{T}_k}$, and by $\mathcal{K}_j$ the set of indices of $\mathcal{T}_k$'s such that $T_j \in \mathcal{T}_k$.

Of course, task parts $\{\tilde{x}_{jk}\}_{T_j \in \mathcal{T}_k}$ are independent for each $k = 1, 2, \cdots, K-1$; K being the total number of nodes in the network, and thus for calculating of Δ_k's as functions of these parts, we can utilize the results obtained for independent tasks. To illustrate this approach let us start with the case (ii) discussed previously. Considering the optimal solution in which task parts are processed consecutively in each interval k we see that this is equivalent to the consecutive processing of entire tasks in an order defined by relation $<$. Moreover, this result is independent on the ordering of nodes in the network. Unfortunately, the last statement is not true in general for other cases of f_j's.

Consider now the case (i) of concave f_j, $j = 1, 2, \cdots, n$, and assume that nodes are ordered in the way defined above. Thus, for calculating $\Delta^*_k(\{\tilde{x}_{jk}\}_{T_j \in \mathcal{T}_k})$, $k = 1$, $2, \cdots, K-1$, one must solve for each $\mathcal{T}_k$ an equation of type (12.3.9)

$$\sum_{T_j \in \mathcal{T}_k} f_j^{-1}(\tilde{x}_{jk}/\Delta_k) = \hat{U}. \tag{12.3.12}$$

of which Δ^*_k is the (unique) positive root for given $\{\tilde{x}_{jk}\}_{T_j \in \mathcal{T}_k}$. As already mentioned before, this equation can be solved analytically for some important cases.

The step which remains is to find a division of $\tilde{x}_j$'s into parts $\tilde{x}_{jk}^*$, $j = 1, 2, \cdots, n$; $k \in \mathcal{K}_j$ ensuring the minimum value of C_{max}. This is equivalent to the solution of the following non-linear programming problem:

$$Minimize \quad C_{max} = \sum_{k=1}^{K-1} \Delta_k^* (\{\tilde{x}_{jk}\}_{T_j \in \mathcal{T}_k}) \tag{12.3.13}$$

$$subject \ to \quad \sum_{k \in \mathcal{K}_j} \tilde{x}_{jk} = \tilde{x}_j, \qquad j = 1, 2, \cdots, n, \tag{12.3.14}$$

$$\tilde{x}_{jk} \geq 0, \qquad\qquad j = 1, 2, \cdots, n, \ k \in \mathcal{K}_j. \tag{12.3.15}$$

It can be proved (see e.g. [Weg82]) that C_{max} given by (12.3.13) is a convex function of $\tilde{x}_{jk}$'s for arbitrary f_j's, thus we have a convex programming problem with linear constraints. Its solution is the optimal solution of our problem for the preemptive case and given ordering of nodes. Using the Lagrange theorem one can verify that for $f_j = c_j u_j^{1/\alpha}$, $\alpha > 1$, when C_{max}^* is given by (12.3.11), the solution does not depend on the ordering of nodes. Of course, this is always true when the ordering of nodes is unique, i.e. for a uan (cf. Section 3.1). In general, however, in order to find a solution which is optimal over all possible orderings of nodes one must solve the corresponding convex programming problem for each of these orderings and choose a solution with the smallest value of C_{max}. Notice that it may happen in the solution thus obtained that a task is preempted because the amount of a continuous resource allotted to it is zero in some time interval. Thus, to solve the problem optimally for the non-preemptive case, one must consider all sequences of subsets of tasks such that each task appears in at least one subset and precedence constraints are obeyed. Then, for each such sequence the problem (12.3.13)-(12.3.15) has to be solved in order to end up with the best solution.

To illustrate the way of formulating the optimization problem (12.3.13)-(12.3.15) let us consider a simple example.

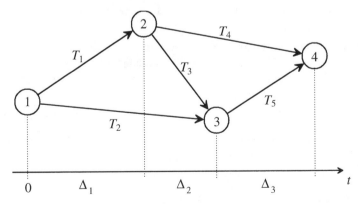

Figure 12.3.3 *Example of a uniconnected activity network.*

Example 12.3.1 Consider the uan given in Figure 12.3.3. Let $\hat{U} = 1$, $f_j = u_j$ for $j = 1, 3, 5$, and $f_j = 2u_j^{1/2}$ for $j = 2, 4$. Subsets of tasks which can be processed between the occurrence of consecutive nodes are:

$$\mathcal{T}_1 = \{T_1, T_2\}, \ \mathcal{T}_2 = \{T_2, T_3, T_4\}, \ \mathcal{T}_3 = \{T_4, T_5\}$$

Sets of indices of $\mathcal{T}_k$'s such that $T_j \in \mathcal{T}_k$ are:

$$\mathcal{K}_1 = \{1\}, \ \mathcal{K}_2 = \{1, 2\}, \ \mathcal{K}_3 = \{2\}, \ \mathcal{K}_4 = \{2, 3\}, \ \mathcal{K}_5 = \{3\}.$$

Since all the functions f_j are concave, we use equation (12.3.12) to calculate $\Delta_k^*(\{\tilde{x}_{jk}\}_{T_j \in \mathcal{T}_k})$ for $k = 1, 2, 3$. For Δ_1^* we have

$$\tilde{x}_{11}/\Delta_1^* + \tilde{x}_{21}^2/4\Delta_1^{*2} = 1,$$

and thus $\Delta_1^*(\tilde{x}_{11}, \tilde{x}_{21}) = (\tilde{x}_{11} + \sqrt{\tilde{x}_{11}^2 + \tilde{x}_{21}^2})/2$. Similarly,

$$\tilde{x}_{22}^2/4\Delta_2^{*2} + \tilde{x}_{32}/\Delta_2^* + \tilde{x}_{42}^2/4\Delta_2^{*2} = 1,$$

$$\Delta_2^*(\tilde{x}_{22}, \tilde{x}_{32}, \tilde{x}_{42}) = (\tilde{x}_{32} + \sqrt{\tilde{x}_{22}^2 + \tilde{x}_{32}^2 + \tilde{x}_{42}^2})/2$$

and

$$\tilde{x}_{43}^2/4\Delta_3^{*2} + \tilde{x}_{53}/\Delta_3^* = 1,$$

$$\Delta_3^*(\tilde{x}_{43}, \tilde{x}_{53}) = (\tilde{x}_{53} + \sqrt{\tilde{x}_{43}^2 + \tilde{x}_{53}^2})/2.$$

The problem is to minimize the sum of the above functions subject to the constraints $\tilde{x}_{11} = \tilde{x}_1$, $\tilde{x}_{21} + \tilde{x}_{22} = \tilde{x}_2$, $\tilde{x}_{32} = \tilde{x}_3$, $\tilde{x}_{42} + \tilde{x}_{43} = \tilde{x}_4$, $\tilde{x}_{53} = \tilde{x}_5$, $\tilde{x}_{jk} \geq 0$ for all j, k. Eliminating five of the variables from the above constraints, a problem with two variables remains.   □

Notice that the reasoning performed above for dependent tasks remains valid if we replace the assumption $n \leq m$ by $|\mathcal{T}_k| \leq m$, $k = 1, 2, \cdots, K-1$.

Let us now consider the case that the number of machines is less than the number of tasks which can be processed simultaneously [2]. We start with independent tasks and $n > m$. To solve the problem optimally for the preemptive case we must, in general, consider all possible assignments of machines to tasks, i.e. all m-element combinations of tasks from $\mathcal{T}$. Keeping for them denotation $\mathcal{T}_k$, $k = 1, 2, \cdots, \binom{n}{m}$, we obtain a new optimization problem of type (12.3.13)-(12.3.15).

---

[2] Recall that this assumption is not needed when in the optimal solution tasks are processed on a single machine, i.e. if $f_j \leq c_j u_j$, $c\backslash S\backslash do3(j) = f_j(\hat{U})/\hat{U}$, $j = 1, 2, ..., n$.

For the non-preemptive case we consider all maximal sequences of $\mathcal{T}_k$'s such that each task appears in at least one $\mathcal{T}_k$ and all $\mathcal{T}_k$'s containing the same task are consecutively indexed (non-preemptability!). Such sequences will be called *feasible*. It is easy to notice that a feasible sequence consists of $n - m + 1$ elements (i.e. sets $\mathcal{T}_k$). To find an optimal schedule in the general case we have to solve the problem of type (12.3.13)-(12.3.15) for each of the feasible sequences and to choose the best solution.

It is easy to see that finding an optimal schedule is computationally very difficult in general, and thus it is purposeful to construct heuristics. For the non-preemptive case the idea of a heuristic approach can be to choose one or several feasible sequences of m-tuples of tasks described above and solve a problem of type (12.3.13)-(12.3.15) for each of them. These sequences can be chosen in many different ways. A general advise is based on the following reasoning. Assume $n = 5$ and $m = 3$. Then, a feasible sequence consists of $5 - 3 + 1 = 3$ sets $\mathcal{T}_k$ of 3 elements each. Exemplary feasible sequences are: $S_1 = (\{T_1, T_2, T_3\}, \{T_2, T_3, T_4\}, \{T_3, T_4, T_5\})$, $S_2 = (\{T_1, T_2, T_3\}, \{T_1, T_2, T_4\}, \{T_1, T_2, T_5\})$.

Define now the *structure* of a sequence as the vector $(|\mathcal{K}_1|, |\mathcal{K}_2|, ..., |\mathcal{K}_n|)$ where $|\mathcal{K}_j|$ is the cardinality of the set of indices of those $\mathcal{T}_k$'s for which $T_j \in \mathcal{T}_k$. It is easy to see that the structure of S_1 is (1, 2, 3, 2, 1), whereas that of S_2 is (3, 3, 1, 1, 1). The basic idea is to study the correspondence between the structure of feasible sequences and the vector of processing demands $\tilde{x}$ of tasks in order to achieve possibly uniform workload for particular machines. If all f_i are concave and identical then we can even identify optimal sequences. For example, under the above assumptions, and $n = 5$, $m = 3$, $\tilde{x} = (10, 20, 30, 20, 10)$, sequence S_1 is optimal, whereas S_2 is optimal for $\tilde{x} = (30, 30, 10, 10, 10)$. This follows from the fact that the division of processing demands of tasks defined as $\tilde{x}_j / |\mathcal{K}_j|, j = 1, 2, ..., 5$, corresponds exactly to the uniform workload. Particular algorithms, their worst case behavior and computational results are given in [JW95].

Another idea, for an arbitrary problem type, consists of two steps:

(a) Schedule task from $\mathcal{T}$ on machines from $\mathcal{P}$ for task processing times $p_j = \tilde{x}_j / f_j(\hat{u}_j), j = 1, 2, \cdots, n$, where the $\hat{u}_j$'s are fixed resource amounts.

(b) Allocate the continuous resource among parts of tasks in the schedule obtained in step (a).

Usually in both steps we take into account the same optimization criterion (C_{max} in our case), although heuristics with different criteria can also be considered. Of course, we can solve each step optimally or heuristically. In the majority of cases step (b) can easily be solved (numbers of task parts processed in parallel are less than or equal to m; see Figure 12.3.4 for $m = 2$, $n = 4$) when, as we remember, even analytic results can be obtained for the sets $\mathcal{T}_k$. However, the complexity of step (a) is radically different for preemptive and non-preemptive scheduling. In the first case, the problem under consideration can be solved exactly in $O(n)$ time using McNaughton's algorithm, whereas in the second one it is NP-hard for any fixed value of m ([Kar72]; see also Section 5.1). In the latter case approximation algorithms as described in Section 5.1, or dynamic programming algorithms similar to that presented in Section 12.1 can be applied (here tasks are divided into classes with equal processing times).

The question remains how to define resource amounts $\hat{u}_j$, $j = 1, 2, \cdots, n$, in step (a). There are many ways to do this; some of them were described in [BCSW86] and checked experimentally in the preemptive case. Computational experiments show that solutions produced by this heuristic differ from the optimum by several percent on average. However, further investigations in this area are still needed. Notice also that we can change the amounts $\hat{u}$ when performing steps (a) and (b) iteratively.

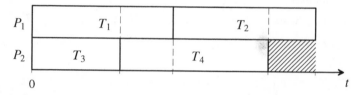

Figure 12.3.4 *Parts of tasks processed in parallel in an example schedule.*

Let us stress once again that the above two-step approach is pretty general, since it combines (discrete) scheduling problems (step (a)) with problems of continuous resource allocation among independent tasks (step (b)). Thus, in step (a) we can utilize all the algorithms presented so far in this book, as well as many others, e.g. from constrained resource project scheduling (see, e.g. [SW89]). On the other hand, in step (b) we can utilize several generalizations of the results presented in this section. We will mention some of them below, but first we say few words about dependent tasks and $|\mathcal{T}_k| > m$ for at least one k. In this case one has to combine the reasoning presented for dependent tasks and $n \leq m$, and that for independent tasks and $n > m$. This means, in particular, that in order to solve the preemptive case, each problem of type (12.3.13)-(12.3.15) must be solved for all m-elementary subsets of sets $\mathcal{T}_k$, $k = 1, 2, \cdots, K-1$.

We end this section with few remarks concerning generalizations of the results presented for continuous resource allocation problems. First of all we can deal with a doubly constrained resource, when, apart from (12.3.2), also the constraint $\sum_{j=1}^{n} \int_0^{C_j} f_j[u_j(t)]dt \leq \hat{V}$ is imposed, $\hat{V}$ being the consumption constraint [Weg81]. Second, each task may require many continuous resource types. The processing speed of task T_j is then given by $\dot{x}_j(t) = f_j[u_{j1}(t), u_{j2}(t), \cdots, u_{js}(t)]$, where $u_{jl}(t)$ is the amount of resource R_l allotted to T_j at time t, and s is the number of different resource types. Thus in general we obtain multi-objective resource allocation problems of the type formulated in [Weg91]. Third, other optimality criteria can be considered, such as $\int_{0}^{C_{max}} g[u(t)]dt$ [NZ81], $\Sigma w_j C_j$ [NZ84a, NZ84b] or L_{max} [Weg89]. Finally, sequences of sets of dependent tasks can be studied [JS88].

12.3.3 Processing Time vs. Resource Amount Model

In this section we consider problems of scheduling non-preemptable tasks on a single machine, where task processing times are linear and continuous functions of a continuous resource. The task processing model is given in the form

$$p_j = b_j - a_j u_j, \underline{u}_j \leq u_j \leq \tilde{u}_j, j = 1, 2, \cdots, n \qquad (12.3.16)$$

where $a_j > 0$, $b_j > 0$, and $\underline{u}_j$ and $\tilde{u}_j \in [0, b_j/a_j]$ are known constants. The continuous resource is available in maximal amount $\hat{U}$, i.e. $\sum\limits_{j=1}^{n} u_j \leq \hat{U}$. Although now the resource is not necessarily renewable (this is not a temporary model), we will keep denotations as introduced in Section 12.3.2. Scheduling problems using the above model were broadly studied by Janiak in a number of papers we will refer to in the sequel. Without loss of generality we can restrict our considerations to the case that lower bounds $\underline{u}_j$ of resource amounts allotted to the particular tasks are zero. This follows from the fact that in case of $\underline{u}_j > 0$ the model can be replaced by an equivalent one in the following way: replace b_j by $b_j - a_j \underline{u}_j$ and $\tilde{u}_j$ by $\tilde{u}_j - \underline{u}_j$, $j = 1, 2, \cdots, n$, and $\hat{U}$ by $\hat{U} - \sum\limits_{i=1}^{n} \underline{u}_i$, finally, set all $\underline{u}_j = 0$. Given a set of tasks $\mathcal{T} = \{T_1, \cdots, T_n\}$, let $z = [z(1), \cdots, z(n)]$ denote a permutation of task indices that defines a feasible task order for the scheduling problem, and let $\mathcal{Z}$ be the set of all such permutations. A schedule for $\mathcal{T}$ can then be characterized by a pair $(z, u) \in \mathcal{Z} \times \mathcal{U}$. The value of a schedule (z, u) with respect to the optimality criterion γ will be denoted by $\gamma(z, u)$. A schedule with an optimal value of γ will briefly be denoted by (z^*, u^*).

Let us start with the problem of minimizing C_{max} for the case of equal ready times and arbitrary precedence constraints [Jan88a]. Using a slight modification of the notation introduced in Section 3.4, we denote this type of problems by $1 \mid prec, p_j = b_j - a_j u_j, \Sigma u_j \leq \hat{U} \mid C_{max}$. It is easy to verify that an optimal solution (z^*, u^*) of the problem is obtained if we chose an arbitrary permutation $z \in \mathcal{Z}$ and allocate the continuous resource according to the following algorithm.

Algorithm 12.3.2 *for finding* u^* *for* $1 \mid prec, p_j = b_j - a_j u_j, \Sigma u_j \leq \hat{U} \mid C_{max}$
[Jan88a].
```
begin
for j := 1 to n do u_j^* := 0;
while T ≠ ∅ and Û > 0 do
   begin
```
Find $T_k \in T$ for which $a_k = \max_j \{a_j\}$;
$u_k^* := \min\{\tilde{u}_k, \hat{U}\}$;
$\hat{U} := \hat{U} - u_k^*$;
$T := T - \{T_k\}$;
```
   end;
```
$u^* := [u_1^*, \cdots, u_n^*];$ $--$ u^* is an optimal resource allocation
```
end;
```

Obviously, the time complexity of this algorithm is $O(n \log n)$.

Consider now the problem with arbitrary ready times, i.e. $1 \mid prec, r_j, p_j = b_j - a_j u_j, \Sigma u_j \leq \hat{U} \mid C_{max}$. One can easily prove that an optimal solution (z^*, u^*) of the problem is always found if we first schedule tasks according to an obvious modification of Algorithm 4.4.2 by Lawler - thus having determined z^* - and then allocate the resources according to Algorithm 12.3.3.

Algorithm 12.3.3 *for finding* u^* *for* $1 \mid prec, r_j, p_j = b_j - a_j u_j, \Sigma u_j \leq \hat{U} \mid C_{max}$
[Jan88a].
```
begin
for j := 1 to n do u_j^* := 0;
```
$S_{z^*(1)} := r_{z^*(1)};$
$l := 1;$
```
for j := 2 to n do
```
$S_{z^*(j)} := \max\{r_{z^*(j)}, S_{z^*(j-1)} + b_{z^*(j-1)}\};$
   $--$ starting times of tasks in permutation z^* for u^* have been calculated
$J := z^*;$ $--$ construct set J
```
while J ≠ ∅ and Û ≠ 0 do
   begin
```
Find the biggest index k, $l \leq k \leq n$, for which $r_{z^*(k)} = S_{z^*(k)};$
$J := \{z^*(j) \mid k \leq j \leq n, \text{ and } u_{z^*(j)}^* < \tilde{u}_{z^*(j)}\};$
Find index t for which $a_{z^*(t)} = \max\{a_{z^*(j)} \mid z^*(j) \in J\};$
$d := \min\{S_{z^*(i)} - r_{z^*(i)} \mid t < i \leq n\};$
$y := \min\{\tilde{u}_{z^*(t)}, \hat{U}, d/a_{z^*(t)}\};$
$u_{z^*(t)}^* := u_{z^*(t)}^* + y;$
$\hat{U} := \hat{U} - y;$
```
for i := t to n do
```
$S_{z^*(i)} := S_{z^*(i)} - y a_{z^*(t)};$

$l := k;$
    $--$ new resource allocation and task starting times have been calculated
  end;
$u^* := [u_1^*, \cdots, u_n^*];$ $--$ u^* is an optimal resource allocation
end;

The complexity of this algorithm is $O(n^2)$, and this is the complexity of the whole approach for finding (z^*, u^*), since Algorithm 4.4.2 is also of complexity $O(n^2)$.

Let us now pass to the problems of minimizing maximum lateness L_{max}. Since problem $1 \mid prec, p_j = b_j - a_j u_j, \Sigma u_j \le \hat{U} \mid L_{max}$ is equivalent to problem $1 \mid prec, r_j, p_j = b_j - a_j u_j, \Sigma u_j \le \hat{U} \mid C_{max}$ (as in the case without additional resources), its optimal solution can always be obtained by finding z^* according to the Algorithm 4.4.2 and u^* according to a simple modification of Algorithm 12.3.3.

It is also easy to see that problem $1 \mid r_j, p_j = b_j - a_j u_j, \Sigma u_j \le \hat{U} \mid L_{max}$ is strongly NP-hard, since the restricted version $1 \mid r_j \mid L_{max}$ is already strongly NP-hard (see Section 4.3). For the problem $1 \mid prec, r_j, p_j = b_j - a_j u_j, \Sigma u_j \le \hat{U} \mid L_{max}$ where in addition precedence constraints are given, an exact branch and bound algorithm was presented by Janiak [Jan86c].

Finally, consider problems with the optimality criteria ΣC_j and $\Sigma w_j C_j$. Problem $1 \mid prec, p_j = b_j - a_j u_j, \Sigma u_j \le \hat{U} \mid \Sigma C_j$ is NP-hard, and problem $1 \mid r_j, p_j = b_j - a_j u_j, \Sigma u_j \le \hat{U} \mid \Sigma C_j$ is strongly NP-hard, since the corresponding restricted versions $1 \mid prec \mid \Sigma C_j$ and $1 \mid r_j \mid \Sigma C_j$ are NP-hard and strongly NP-hard, respectively (see Section 4.2). The complexity status of problem $1 \mid p_j = b_j - a_j u_j, \Sigma u_j \le \hat{U} \mid \Sigma w_j C_j$ is still an open question. It is easy to verify for any given $z \in \mathcal{Z}$ the minimum value of $\Sigma w_j C_j$ in this problem is always obtained by allocating the resource according to the following algorithm of complexity $O(n \log n)$.

Algorithm 12.3.4 *for finding* u^* *for* $1 \mid p_j = b_j - a_j u_j, \Sigma u_j \le \hat{U} \mid \Sigma w_j C_j$ [Jan88a].
begin
$\mathcal{J} := z;$ $--$ construct set $\mathcal{J}$
while $\mathcal{J} \ne \emptyset$ **do**
  begin
    Find $z(k) \in \mathcal{J}$ for which $a_{z(k)} \sum_{j=k}^{n} w_{z(j)} = \max_{z(i) \in \mathcal{J}} \{ a_{z(i)} \sum_{j=i}^{n} w_{z(j)} \};$
    $u_{z(k)}^* := \min\{\tilde{u}_{z(k)}, \max\{0, \hat{U}\}\};$
    $\hat{U} := \hat{U} - u_{z(k)}^*;$
    $\mathcal{J} := \mathcal{J} - \{z(k)\};$
    end;
$u^* := [u_1^*, \cdots, u_n^*];$ $--$ u^* is an optimal resource allocation
end;

An exact algorithm of the same complexity can also be given for this problem if for any two tasks T_i, T_j either $T_i < T_j$ or $T_j < T_i$, where $T_i < T_j$ means that $b_i \le b_j$, $a_i \ge a_j$, $\tilde{u}_i \ge \tilde{u}_j$, and $w_i \ge w_j$. In this case the optimal permutation z^* is obtained by ordering the jobs according to $<$, and the algorithm of the optimal resource allocation is as follows: $u^*_{z^*(j)} = \min\{\tilde{u}_{z^*(j)}, \max\{0, \hat{U}_j\}\}$ for $j = 1, 2, \cdots, n$, where $\hat{U}_1 = \hat{U}$, $\hat{U}_{j+1} = \hat{U}_j - u^*_{z^*(j)}$, $j = 1, 2, \cdots, n-1$.

Now let us pass to the criterion which is specific to scheduling problems with additional continuous resources, namely to the criterion denoting the total resource utilization, i.e. $U = \sum_{j=1}^{n} u_j$. This criterion should be minimized subject to the constraint $\gamma < \hat{\gamma}$ where γ is a classical schedule performance measure and $\hat{\gamma}$ is a given value of γ. Of course, scheduling problems of minimizing Σu_j are closely related to corresponding problems with criterion γ. Additionally, we use the fact that for the considered problems it is easy to calculate the maximum value $\tilde{\gamma}$ of γ.

We illustrate this idea for the criterion $\gamma = C_{max}$, i.e. for problem $1 \,|\, prec, p_j = b_j - a_j u_j, C_{max} < \hat{C} \,|\, \Sigma u_j$. It is obvious that the upper bound for C_{max}, $\tilde{C}_{max} = \min_{z \in Z} \{C_{max}(z, 0)\} = C_{max}(z^*, 0)$. Thus, we have the following modification of Algorithm 12.3.2.

Algorithm 12.3.5 *for finding u^* for $1 \,|\, prec, p_j = b_j - a_j u_j, C_{max} \le \hat{C} \,|\, \Sigma u_j$* [Jan91a].
begin
for $j := 1$ **to** n **do** $u^*_j := 0$;
$U := 0$;
$C_{max} := \tilde{C}_{max}$;
while $\mathcal{T} \ne \varnothing$ **and** $C_{max} > \hat{C}$ **do**
  begin
  Find $T_k \in \mathcal{T}$ for which $a_k = \max_j\{a_j\}$;
  $u^*_k := \min\{\tilde{u}_k, \max\{0, (C_{max} - \hat{C})/a_k\}\}$;
  $U := U + u^*_k$;
  $C_{max} := C_{max} - a_k u^*_k$;
  $\mathcal{T} := \mathcal{T} - \{T_k\}$;
  end;
if $\mathcal{T} = \varnothing$ **and** $C_{max} > \hat{C}$
then no solution exists
else $u^* := [u^*_1, \cdots, u^*_n]$;   -- u^* is an optimal resource allocation
end;

Knowing how to solve a problem for criteria γ and Σu_j, one can also find the set of all Pareto-optimal (i.e. efficient or non-dominated) solutions (z^P, u^P) for *bi-*

criterion problems ([Jan91a]). As an example, consider the problem $1 \mid prec, p_j = b_j - a_j u_j \mid C_{max} \wedge \Sigma u_j$. Of course, $\underline{C}_{max} = \min_{z \in Z} \{C_{max}(z, \tilde{u})\} = \sum_{j=1}^{n} (b_j - a_j \tilde{u}_j)$ is a lower bound for C_{max}. In our problem, for each value $C_{max} \in [\underline{C}_{max}, \tilde{C}_{max}]$, any feasible permutation $z \in Z$ can be taken as Pareto-optimal permutation z^P. In order to find the set $\mathcal{U}^P$ of all Pareto-optimal resource allocations u^P, we determine the Pareto curve (which is a convex, decreasing and piece-wise linear function) from the following algorithm of time complexity $O(n \log n)$.

Algorithm 12.3.6 *for finding the Pareto curve in* $1 \mid prec, p_j = b_j - a_j u_j \mid$
$C_{max} \wedge \Sigma u_j$ [Jan91a].
begin
for $j := 1$ **to** n **do** $u^*_{z(j)} := 0$;
$i := 0$;
$C^0_{max} := \tilde{C}_{max}$;
$U^0 := 0$;
while $T \neq \varnothing$ **do**
  begin
  $i := i + 1$;
  Find $T_k \in T$ for which $a_k = \max_j \{a_j\}$;

  $u^*_k := \tilde{u}_k$;
  $C^i_{max} := C^{i-1}_{max} - a_k \tilde{u}_k$;
  $U^i := U^{i-1} + \tilde{u}_k$;
  $a^i := 1/a_k$;
  $T := T - \{T_k\}$;
  for $l := 1$ **to** n **do** $u^i_l := u^*_l$;
  end;
end;

Obtained pairs (C^0_{max}, U^0), $(C^1_{max}, U^1), \ldots, (C^n_{max}, U^n)$ are consecutive break-points of the Pareto curve; a^i is the slope of the ith segment of this curve, $i = 1, 2, \ldots, n$. The set $\mathcal{U}^P$ is the sum of n segments joining the points u^i, u^{i+1}, $i = 0, 1, 2, \ldots, n-1$, where $u^0 = 0$.

In [JK96] the problem was considered with given deadlines $\tilde{d}_j$ and minimization of the total weighted resource consumption, i.e. the problem $1 \mid p_j = b_j - a_j u_j, C_j \le \tilde{d}_j \mid \Sigma w_j u_j$. This problem is solvable in $O(n \log n)$ time for a continuously-divisible resource and is NP-hard for a discrete resource. A fully polynomial approximation scheme is presented for the last case.

The paper [CJK98] is devoted to the following machine scheduling problems with linear models of task processing times and with a discrete resource: $1 \mid p_j = b_j - a_j u_j, F_1 \leq K \mid F_2$, $1 \mid p_j = b_j - a_j u_j, F_2 \leq K \mid F_1$ and $1 \mid p_j = b_j - a_j u_j \mid F_1 \wedge F_2$, where F_1 and F_2 is a criterion of resource and completion time type, respectively. More precisely, $F_1 \in \{g_{max}, \sum u_j, \sum w_j u_j\}$ and $F_2 \in \{C_{max}, c_{max}, \sum U_j, \sum w_j U_j, \sum C_j, \sum w_j C_j\}$, where $g_{max} = \max\{g_j(u_j)\}$ ($g_j(u_j)$ is a nondecreasing resource cost function), $c_{max} = \max\{c_j(C_j)\}$ ($c_j(C_j)$ is a nondecreasing penalty cost function), and $\sum w_j U_j$ is the weighted number of tardy tasks (see Section 3.1). Computational complexities of the problems and the general scheme for the construction of Pareto sets and Pareto set ε-approximations were also presented.

In [Jan99] the model (12.3.16) was extended to one with $p_j = b_j + a_j' S_j - a_j u_j$, where S_j is a task starting time and a_j' is a task model parameter. The problems of minimization of the makespan, the total completion time and the lateness with the extended model including the constraint on the maximal resource amount $\hat{U}$, and also their inverse versions, were investigated e.g. in [IJR00].

Further generalizations concern the application of the model (12.3.16) for machine setup times [Jan99]. Single machine batch scheduling with resource dependent setup and processing time was examined in [CJK01], where polynomial time algorithms were presented to find an optimal batch sequence and resource allocations such that either the total weighted consumption $\sum w_j u_j$ is minimized subject to meeting task deadlines d_j, or the maximum task lateness is minimized subject to an upper bound on the total weighted resource consumption. Next, single machine group scheduling with resource dependent setup and processing times with continuous or discrete resource were considered in [NCJK 05, JKP05] for various criteria.

To end this section let us mention some results obtained for the processing time vs. resource amount model in case of dedicated processors. Two-machine flow shop problems with linear task models were studied by Janiak [Jan88b, Jan89a], where it was proved that the problem is NP-hard for the single criteria $\gamma = C_{max}$ and $\gamma = \sum u_j$, even for identical values of a_j on one of the machines and fixed processing times on the second machine. Approximation algorithms and an exact branch and bound algorithm were also presented in these papers. Flow shop and job shop problems with convex task models were considered in [GJ87, Jan86b, Jan88c, Jan88d, JS94, JP98, CJ00].

12.3.4 Ready Time vs. Resource Amount Model

In this section we assume that task processing times are given constants but ready times are continuously dependent on the amount of allocated continuous resource, i.e.

$$r_j = f_j(u_j),\ \underline{u}_j \le u_j \le \tilde{u}_j, j = 1, 2, \cdots, n,\qquad(12.3.17)$$

where all the lower and upper bounds of resource allocations, $\underline{u}_j$ and $\tilde{u}_j$, are known constants.

As in Section 12.3.3 tasks are assumed to be non-preemptable, and we consider single machine problems only. Problems of this type appear e.g. in the ingot preheating process in steel mills [Jan91b].

Problem $1 \mid r_j = f_j(u_j),\ \Sigma u_j \le \hat{U} \mid C_{max}$

This problem is strongly NP-hard even in the special case of linear functions f_j (see (12.3.16)) and $\underline{u}_j = 0, j = 1, 2, \cdots, n$, and is NP-hard in the case of $a_j = a, j = 1, 2, \cdots, n$ [Jan91b]. However, for identical models of r_j, i.e. for $f_j = f$, $\underline{u}_j = \underline{u}$ and $\tilde{u}_j = \tilde{u}$ for all j, the problem can be solved in polynomial time. In this case we know from [Jan86c] that an optimal solution (z^*, u^*) is obtained by scheduling tasks according to non-increasing processing times p_j (thus defining permutation z^*) and by allocating the continuous resource for z^* according to the following formulas: if

$$f(\tilde{u}_{z^*(1)}) + \sum_{j=1}^{n} p_{z^*(j)} \ge f(\underline{u}) + \sum_{j=2}^{n} p_{z^*(j)}$$

where

$$\tilde{u}_{z^*(1)} = \min\{(\hat{U} - (n-1)\underline{u}), \tilde{u}\},$$

then

$$u^*_{z^*(1)} = \tilde{u}_{z^*(1)},\ u^*_{z(j)} = \underline{u},\ j = 2, 3, \cdots, n.$$

Otherwise,

$$u^*_{z^*(j)} = f^{-1}(r - (\sum_{i=j}^{k-1} p_{z^*(i)} + d)),\ j = 1, 2, \cdots, k-1,$$

$$u^*_{z^*(k)} = f^{-1}(r - d),\ u^*_{z^*(j)} = \underline{u},\ j = k+1, k+2, \cdots, n,$$

where $r = f(\underline{u})$, and $k - 1$ is the maximal natural number such that

$$(\sum_{j=1}^{k-1} f^{-1}(r - \sum_{i=j}^{k-1} p_{z^*(i)}) + (n - (k-1))\underline{u} \le \hat{U})\ \text{and}\ (f^{-1}(r - \sum_{j=1}^{k-1} p_{z^*(j)}) \le \tilde{u}),$$

$$d = \min\{(r - \sum_{j=1}^{k-1} p_{z^*(j)} - f(\tilde{u})), d'\},$$

with d' following from the equation

$$\sum_{j=1}^{k-1} f^{-1}(r - \sum_{i=j}^{k-1} p_{z^*(i)} - d') + f^{-1}(r - d') + (n - k)\underline{u} = \hat{U}.$$

Thus, if we are able to calculate f, f^{-1} and d' in time $O(g(n))$, then (z^*, u^*) is calculated in $O(\max\{g(n), n\log n\})$ time, i.e. this time is polynomial if $g(n)$ is polynomial. For example, this is the case if f is linear. In special situations where f_j is linear and $b_j = b$ for $j = 1, 2, \cdots, n$, algorithms of time complexity $O(n\log n)$ exist. These situations are as follows:

(i) $\tilde{u}_j = \tilde{u}, \ p_j = p, \ j = 1, 2, \cdots, n$,

(ii) $a_j = a, \ p_j = p, \ j = 1, 2, \cdots, n$.

An optimal solution (z^*, u^*) is obtained by scheduling the tasks according to non-increasing values of a_j in case (i), non-increasing $\tilde{u}_j$ in case (ii), and by allocating the continuous resource using corresponding modifications of the above formulae [Jan89b].

For arbitrary linear functions f_j, Janiak [Jan89b] was able to prove that for given $z \in Z$, an optimal resource allocation u_z^* can be calculated in $O(n^2)$ time using the following algorithm.

Algorithm 12.3.7 *for finding u^* for* $1 \mid r_j = b_j - a_j u_j, \ \Sigma u_j \leq \hat{U} \mid C_{max}$ [Jan89b].
begin
for $j := 1$ **to** n **do**
  begin
  $u_{z(j)}^* := 0$;

  $C_{z(j)} := b_{z(j)} + \sum\limits_{i=j}^{n} p_{z(i)}$;

  end;
$\mathcal{J} := \{z(j) \mid j = 1, 2, \cdots, n\}$;
$l := 0$;
$C_0 := 0$;
$\mathcal{J}_0 := 0$;
while $\mathcal{J} \neq \varnothing$ **do**
  begin
  $l := l + 1$;
  Find set $\mathcal{J}_l = \{z(j) \mid z(j) \in \mathcal{J}$ and $C_{z(j)} = \min\limits_{z(i) \in \mathcal{J}} \{C_{z(i)}\}\}$;

  $\mathcal{J} = \mathcal{J} - \mathcal{J}_l$;
  end;
$Q := \mathcal{J}_l$;
while not $(\hat{U} = 0$ **or** $l = 0$ **or** $\min\limits_{j \in Q} \{\tilde{u}_j - u_j^*\} = 0)$ **do**
  begin
  $x := \min \{C_q - C_p, \ \hat{U} / \sum\limits_{j \in Q} (1/a_j), \ \min\limits_{j \in Q} \{a_j(\tilde{u}_j - u_j^*)\}\}$;

-- p and q are indices of tasks belonging to sets Q and $\mathcal{J}_{l-1}$, respectively

for $j \in Q$ **do** $u_j^* := u_j^* + x/a_j$;

$\hat{U} := \hat{U} - \sum\limits_{j \in Q} x/a_j$;

$l := l-1$;

$Q := Q \cup \mathcal{J}_l$;

end;

$u_z^* := [u_1^*, \cdots, u_n^*]$;　　　 -- u_z^* is an optimal resource allocation for permutation z

end;

In the same paper it has been shown that in the case of $a_j = a$, $\tilde{u}_j = \tilde{u}$, $p_j = p$ for $j = 1, 2, \cdots, n$, an optimal solution (z^*, u^*) is obtained when tasks are scheduled in order of non-decreasing b_j and the resource is allocated according to Algorithm 12.3.7. The same is also true for problems in which the above permutation is in accordance with the non-increasing orders of a_j, $\tilde{u}_j$ and p_j. Of course, Algorithm 12.3.7 can also be used for finding resource allocations for permutations $z \in Z$ defined heuristically. In [Jan89b] 25 such heuristics with the (best possible) worst case bound 2 were compared experimentally. The best results for "low" resource level ($\hat{U} = 0.2 \cdot \sum\limits_{j=1}^{n} \tilde{u}_j$) were produced by ordering tasks according to non-decreasing b_j, whereas for "high" resource level ($\hat{U} = 0.9 \cdot \sum\limits_{j=1}^{n} \tilde{u}_j$) sorting tasks according to non-decreasing values of $b_j - a_j \tilde{u}_j$ turned out to be most efficient.

Problem $1 \mid r_j = f_j(u_j), C_{max} \leq \hat{C} \mid \Sigma u_j$

Similarly as for $1 \mid r_j = f_j(u_j), \Sigma u_j \leq \hat{U} \mid C_{max}$ it can be proved that the considered problem is already strongly NP-hard for $f_j = b_j - a_j u_j$, $j = 1, 2, \cdots, n$, and NP-hard for $a_j = a$, $j = 1, 2, \cdots, n$ (see [Jan91b]). Also similarly to the solution of the first problem, if $f_j = f$, $u_j = \underline{u}$ for all j, the problem is solved optimally by scheduling tasks according to non-increasing p_j (thus defining permutation z^*) and by allocating the resource according to the following condition. If

$$r + \sum_{j=1}^{n} p_j - \hat{C} \leq p_{z^*(1)}, \text{ where } r = f(\underline{u}),$$

then

$$u_{z^*(1)}^* = f^{-1}(\hat{C} - \sum_{j=1}^{n} p_j), \ u_{z^*(j)}^* = \underline{u}, \ j = 2, 3, \cdots, n,$$

and otherwise

$$u_{z^*(j)}^* = f^{-1}(\hat{C} - \sum_{i=j}^{n} p_{z^*(i)}) = f^{-1}(r - \sum_{i=j}^{k-1} p_{z^*(i)} - d) \text{ for } j = 1, 2, \cdots, k-1,$$

$$u_{z^*(k)}^* = f^{-1}(\hat{C} - \sum_{i=k}^{n} p_{z^*(i)}) = f^{-1}(r - d),$$

$$u_{z^*(j)}^* = f^{-1}(r) = \underline{u}, \ j = k+1, k+2, \cdots, n,$$

where k is the maximal natural number such that

$$\sum_{i=1}^{k-1} p_{z^*(i)} \le r + \sum_{j=1}^{n} p_j - \hat{C},$$

$$d = r + \sum_{j=1}^{n} p_j - \hat{C} - \sum_{i=1}^{k-1} p_{z^*(i)} = r + \sum_{i=k}^{n} p_{z^*(i)} - \hat{C}.$$

Thus, if we are able to calculate f and f^{-1} in $O(g(n))$ time, then finding (z^*, u^*) needs $O(\max\{g(n), n\log n\})$ time.

Notice that it is generally sufficient to consider $\hat{C}$ for which $\underline{C}_{max} \le \hat{C} \le \tilde{C}_{max}$, where $\underline{C}_{max} = \min_{z \in Z} C_{max}(z, \tilde{u})$ and $\tilde{C}_{max} = \min_{z \in Z} C_{max}(z, \underline{u})$. In particular, for identical $f_j, \underline{u}_j, \tilde{u}_j, j = 1, 2, \cdots, n$, we have

$$C_{max}(z, \tilde{u}) = \underline{C}_{max} = f(\tilde{u}) + \sum_{j=1}^{n} p_j$$

and

$$C_{max}(z, \underline{u}) = \tilde{C}_{max} = f(\underline{u}) + \sum_{j=1}^{n} p_j \text{ for each } z \in Z.$$

If functions f_j are not identical and linear, then for given $z \in Z$ an optimal u_z^* is obtained in $O(n)$ time using the formula [Jan91b]

$$u_{z(j)}^* = \max\{0, (b_{z(j)} + \sum_{i=j}^{n} p_{z(i)} - \hat{C})/a_{z(j)}\}, \ j = 1, 2, \cdots, n. \quad (12.3.18)$$

This follows simply from the linear programming formulation of the problem. On the same basis it is easy to see that the cases:

(i) $b_j = b, \ \tilde{u}_j = \tilde{u}, \ p_j = p, \ j = 1, 2, \cdots, n,$

(ii) $b_j = b, \ a_j = a, \ p_j = p, \ j = 1, 2, \cdots, n,$

(iii) $a_j = a, \ \tilde{u}_j = \tilde{u}, \ p_j = p, \ j = 1, 2, \cdots, n$

are solvable in $O(n\log n)$ time by scheduling tasks according to non-increasing a_j in case (i), non-increasing $\tilde{u}_j$ in case (ii), and non-increasing b_j in case (iii), and by allocating the resource according to (12.3.18). For each of these cases z^* does not depend on $\hat{C}$, and $\underline{C}_{max} = C_{max}(z^*, \tilde{u})$, $\tilde{C}_{max} = C_{max}(z^*, 0)$.

Heuristics in which z is defined heuristically and u_z^* is calculated according to (12.3.18) were studied in [Jan91b]. The best results were obtained by scheduling tasks according to non-decreasing b_j. Unfortunately, the worst-case performance of these heuristics is not known.

On the basis of the presented results, the set of all Pareto-optimal solutions can be constructed for some bi-criterion problems of type $1 \mid r_j = f_j(u_j) \mid C_{max} \wedge \Sigma u_j$ using the ideas described in [JC94]. For linear models this set was constructed in [Jan 91b].

The problems considered in this section were generalized in [Jan 97] for the case with arbitrary precedence constraints, where it was proved that they are NP-hard even for identical linear models of r_j. When additionally all processing times are identical, the optimal solution (z^*, u^*) can be constructed in $O(n^2)$ time.

In [JL94, Jan99] the single and parallel machine scheduling problems with nonlinear function: release time vs. resource consumption, common for all tasks, with different task resource consumption rates were considered. The following criteria were minimized: the total weighted task completion time subject to a constrained maximal resource amount [JL94], the total resource utilization subject to a constrained total weighted completion time, and the bi-criteria approach [Jan99]. The borders between NP-hard and polynomially solvable cases were found.

Further generalization of the release time model was made in [Jan99], where the single machine scheduling problem with the model (12.3.1) applied to release times was considered. Due to some problem properties, the difficult dynamic resource allocation problem was reduced to a simple convex programming one. Some approximation algorithms with the worst case analysis were also presented.

References

BBKR86   J. Błażewicz, J. Barcelo, W. Kubiak, H. Röck, Scheduling tasks on two processors with deadlines and additional resources, *European J. Oper. Res.* 26, 1986, 364-370.

BCSW86   J. Błażewicz, W. Cellary, R. Słowiński, J. Węglarz, *Scheduling under Resource Constraints: Deterministic Models*, J. C. Baltzer, Basel, 1986.

BDM+99   P. Brucker, A. Drexl, R. Möhring, K. Neumann, E. Pesch, Resource-constrained project scheduling: notation, classification, models, and methods, *European J. Oper. Res.* 112, 1999, 3-41.

BE83   J. Błażewicz, K. Ecker, A linear time algorithm for restricted bin packing and scheduling problems, *Oper. Res. Lett.* 2, 1983, 80-83.

BE94   J. Błażewicz, K. Ecker, Multiprocessor task scheduling with resource requirements, *Real-Time Systems* 6, 1994, 37-54.

BKS89   J. Błażewicz, W. Kubiak, J. Szwarcfiter, Scheduling independent fixed-type tasks, in: R. Słowiński, J. Węglarz (eds.), *Advances in Project Scheduling*, Elsevier, Amsterdam, 1989, 225-236.

Bla78   J. Błażewicz, Complexity of computer scheduling algorithms under resource constraints, *Proc. I Meeting AFCET - SMF on Applied Mathematics*, Palaiseau, 1978, 169-178.

BLRK83   J. Błażewicz, J. K. Lenstra, A. H. G. Rinnooy Kan, Scheduling subject to resource constraints: classification and complexity, *Discrete Appl. Math.* 5, 1983, 11-24.

CD73   E. G. Coffman Jr., P. J. Denning, *Operating Systems Theory*, Prentice-Hall, Englewood Cliffs, N. J., 1973.

CGJ84   E. G. Coffman Jr., M. R. Garey, D. S. Johnson, Approximation algorithms for bin-packing - an updated survey, in: G. Ausiello, M. Lucertini, P. Serafini (eds.), *Algorithms Design for Computer System Design*, Springer, Vienna, 1984, 49-106.

CGJP83   E. G. Coffman Jr., M. R. Garey, D. S. Johnson, A. S. La Paugh, Scheduling file transfers in a distributed network, *Proc. 2nd ACM SIGACT-SIGOPS Symp. on Principles of Distributed Computing*, Montreal, 1983.

CJ00   T.C.E. Cheng, A. Janiak, A permutation flow-shop scheduling problem with convex models of operation processing times, *Annals of Oper. Res.* 96, 2000, 39-60.

CJK98   T.C.E. Cheng, A. Janiak , M.Y. Kovalyov, Bicriterion single machine scheduling with resource dependent processing times, *SIAM J. Optim.* 8, 1998, 617-630.

CJK01   T.C.E. Cheng, A. Janiak , M.Y. Kovalyov, Single machine batch scheduling with resource dependent setup and processing times, *European J. Oper. Res.* 135, 2001, 177-183.

GG75   M. R. Garey, R. L. Graham, Bounds for multiprocessor scheduling with resource constraints, *SIAM J. Comput.* 4, 1975, 187-200.

GJ75   M. R. Garey, D. S. Johnson, Complexity results for multiprocessor scheduling under resource constraints, *SIAM J. Comput.* 4, 1975, 397-411.

GJ87   J. Grabowski, A. Janiak, Job-shop scheduling with resource-time models of operations, *European J. Oper. Res.* 28, 1987, 58-73.

IJR00   D. Iwanowski, A. Janiak, A. Rogala, Scheduling jobs with start time and resource dependent processing times, in: K. Inderfurth, G. Schwodianer, W. Domschke, F. Juhnke, P. Kleinschmidt, G. Wascher (eds), *Oper. Res. Proc. 1999*, Springer, 2000, 389-396.

Jan86a   A. Janiak, One-machine scheduling problems with resource constraints, in: A. Prékopa, J. Szelezán, B. Strazicky (eds.), *System Modelling and Optimization*, Lecture Notes in Control and Information Sciences, Vol. 84, Springer, Berlin, 1986, 358-364.

Jan86b   A. Janiak, Flow-shop scheduling with controllable operation processing times, in: H. P. Geering, M. Mansour (eds.), *Large Scale Systems: Theory and Applications*, Pergamon Press, 1986, 602-605.

Jan86c   A. Janiak, Time-optimal control in a single machine problem with resource constraints, *Automatica* 22, 1986, 745-747.

Jan88a   A. Janiak, Single machine sequencing with linear models of jobs subject to precedence constraints, *Archiwum Aut. i Telem.* 33, 1988, 203-210.

Jan88b   A. Janiak, Permutacyjny problem przepływowy z liniowymi modelami operacji, *Zeszyty Naukowe Politechniki Śląskiej. ser. Automatyka* 94, 1988, 125-138.

Jan88c   A. Janiak, Minimization of the total resource consumption in permutation flow-shop sequencing subject to a given makespan, *J. Model. Simul. Control* 13, 1988, 1-11.

Jan88d   A. Janiak, General flow-shop scheduling with resource constraints, *Internat. J. Production Res.* 26, 1988, 1089-1103.

Jan89a   A. Janiak, Minimization of resource consumption under a given deadline in two-processor flow-shop scheduling problem, *Inform. Process. Lett.* 32, 1989, 101-112.

Jan89b   A. Janiak, Minimization of the blooming mill standstills - mathematical model. Suboptimal algorithms, *Zesz. Nauk. AGH s. Mechanika* 8, 1989, 37-49.

Jan91a   A. Janiak, Dokladne i przyblizone algorytmy szeregowania zadan i rozdzialu zasobow w dyskretnych procesach przemyslowych, Prace Naukowe Instytutu Cybernetyki Technicznej Politechniki Wroclawskiej 87, Monografie 20, Wroclaw, 1991.

Jan91b   A. Janiak, Single machine scheduling problem with a common deadline and resource dependent release dates, *European J. Oper. Res.* 53, 1991, 317-325.

Jan97   A. Janiak, Computational complexity analysis of single machine scheduling problems with job release dates dependent on resources, in: U. Zimmermann, U. Derigs, W. Gaul, R.H. Möhring, K.P. Schuster (eds.), *Oper. Res. Proc. 1996*, Springer, Berlin, 1997, 203-207.

Jan98a   A. Janiak, Single machine sequencing with linear models of release dates, *Naval Res. Logistics*, 45, 1998, 99-113.

Jan98b   A. Janiak, Minimization of the makespan in a two-machine problem under given resource constraints, *European J. Oper. Res.* 107, 1988, 325-337.

Jan99   A. Janiak, *Chosen Problems and Algorithms of Scheduling and Resource Allocation*, Akademicka Oficyna Wydawnicza PLJ, Warszawa 1999.

JC94   A. Janiak, T. C. E. Cheng, Resource optimal control in some simple-machine scheduling problems, *IEEE Trans. Aut. Control* 39, 1994, 1243-1246.

JK96   A. Janiak, M.Y. Kovalyov, Single machine scheduling subject to deadlines and resource dependent processing times, *European J. Oper. Res.* 94, 1996, 284-291.

JKP05   A. Janiak , M.Y. Kovalyov, M.-C. Portmann, Single machine group scheduling with resource dependent setups and processing times, *European J. Oper. Res.* 162, 2005, 112-121.

JL94   A. Janiak, C.-L. Li, Scheduling to minimize the total weighted completion time with a constraint on the release time resource consumption, *Math. Comput. Modelling* 20, 1994, 53-58.

JP98   A. Janiak, M.-C. Portmann, Genetic algorithm for the permutation flow-shop scheduling problem with linear models of operations, *Annals of Oper. Res.* 83, 1998, 95-114.

JS88   A. Janiak, A. Stankiewicz, On time-optimal control of a sequence of projects of activities under time-variable resource, *IEEE Trans. Aut. Control* 33, 1988, 313-316.

JS94   A. Janiak, T. Szkodny, Job-shop scheduling with convex models of operations, *Math. Comput. Modelling* 20, 1994, 59-68.

JW95   J. Józefowska, J. Węglarz, On a methodology for discrete-continuous scheduling, Research Report RA-004/95, Institute of Computing Science, Poznań University of Technology, Poznań, 1995.

Kar84   N. Karmarkar, A new polynomial-time algorithm for linear programming, *Combinatorica* 4, 1984, 373-395.

KE75   O. Kariv, S. Even, An $O(n^2)$ algorithm for maximum matching in general graphs, *Proc. 16th Annual IEEE Symp. on Foundations of Computer Science*, 1975, 100-112.

KSS75   K. L. Krause, V. Y. Shen, H. D. Schwetman, Analysis of several task-scheduling algorithms for a model of multiprogramming computer systems, *J. Assoc. Comput. Mach.* 22, 1975, 522-550. Erratum: *J. Assoc. Comput. Mach.* 24, 1977, 527.

Law76   E. L. Lawler, *Combinatorial Optimization: Networks and Matroids*, Holt, Rinehart and Winston, New York 1976.

Len83   H. W. Lenstra, Jr., Integer programming with a fixed number of variables, *Math. Oper. Res.* 8, 1983, 538-548.

McN59   R. McNaughton, Scheduling with deadlines and loss functions, *Management Sci.* 12, 1959, 1-12.

NCJK05 C.T. Ng, T.C.E. Cheng, A. Janiak, M. Y. Kovalyov, Group scheduling with controllable setup and processing times: Minimizing total weighted completion time, *Annals of Oper. Res.* 133, 2005, 163-174.

NZ81   E. Nowicki, S. Zdrzalka, Optimal control of a complex of independent operations, *Internat. J. Systems Sci.* 12, 1981, 77-93.

NZ84a   E. Nowicki, S. Zdrzalka, Optimal control policies for resource allocation in an activity network, *European J. Oper. Res.* 16, 1984, 198-214.

NZ84b   E. Nowicki, S. Zdrzalka, Scheduling jobs with controllable processing times as an optimal control problem, *Internat. J. Control* 39, 1984, 839-848.

SW89   R. Słowiński, J. Węglarz (eds.), *Advances in Project Scheduling*, Elsevier, Amsterdam, 1989.

WBCS77   J. Węglarz, J. Błażewicz, W. Cellary, R. Słowiński, An automatic revised simplex method for constrained resource network scheduling, *ACM Trans. Math. Software* 3, 295-300, 1977.

Weg80   J. Węglarz, Multiprocessor scheduling with memory allocation - a deterministic approach, *IEEE Trans. Comput.* C-29, 1980, 703-709.

Weg81   J. Węglarz, Project scheduling with continuously-divisible, doubly constrained resources, *Management Sci.* 27, 1981, 1040-1052.

Weg82   J. Węglarz, Modelling and control of dynamic resource allocation project scheduling systems, in: S. G. Tzafestas (ed.), *Optimization and Control of Dynamic Operational Research Models*, North-Holland, Amsterdam, 1982.

Weg89   J. Węglarz, Project scheduling under continuous processing speed vs. resource amount functions, 1989. in: R. Słowiński, J. Węglarz (eds.), *Advances in Project Scheduling*, Elsevier, 1989.

Weg91   J. Węglarz, Synthesis problems in allocating continuous, doubly constrained resources, in: H. E. Bradley (ed.), *Operational Research '90 - Selected Papers from the 12th IFORS International Conference*, Pergamon Press, Oxford, 1991, 715-725.

Weg99   J. Węglarz (ed.), *Project Scheduling - Recent Models, Algorithms and Applications*, Kluwer Academic Publ., 1999.

13 Constraint Programming and Disjunctive Scheduling

Constraint propagation is an elementary method for reducing the search space of combinatorial search and optimization problems which has become more and more important in the last decades. The basic idea of constraint propagation is to detect and remove inconsistent variable assignments that cannot participate in any feasible solution through the repeated analysis and evaluation of the variables, domains and constraints describing a specific problem instance.

This chapter is based on Dorndorf et al. [DPP00] and its contribution is twofold. The first contribution is a description of efficient constraint propagation methods also known as consistency tests for the disjunctive scheduling problem (DSP) which is a generalization of the classical job shop scheduling problem (JSP). By applying an elementary constraint based approach involving a limited number of search variables, we will derive consistency tests that ensure 3-b-consistency. We will further present and analyze both new and classical consistency tests which to some extent are generalizations of the aforementioned consistency tests involving a higher number of variables, but still can be implemented efficiently with a polynomial time complexity. Further, the concepts of energetic reasoning and shaving are analyzed and discussed.

The other contribution is a classification of the consistency tests derived according to the domain reduction achieved. The particular strength of using consistency tests is based on their repeated application, so that the knowledge derived is propagated, i.e. reused for acquiring additional knowledge. The deduction of this knowledge can be described as the computation of a fixed point. Since this fixed point depends upon the order of the application of the tests, we first derive a necessary condition for its uniqueness. We then develop a concept of dominance which enables the comparison of different consistency tests as well as a simple method for proving dominance. An extensive comparison of all consistency tests is given. Quite surprisingly, we will find out that some apparently stronger consistency tests are subsumed by apparently weaker ones. At the same time an open question regarding the effectiveness of energetic reasoning is answered.

13.1 Introduction

Exact solution methods for solving combinatorial search and optimizations problems generally consist of two components: (a) a search strategy which organizes the enumeration of all potential solutions and (b) a search space reduction strategy which diminishes the number of potential solutions. However, due to the exponentially growing size of the search space, even an intelligent organization of the search will eventually fail, so that only the application of efficient search space reduction mechanisms will allow the solution of more difficult problems. Consequently, as an elementary method of search space reduction, constraint propagation has become more and more important in the last decades. Constraint propagation has its origins in the popular field of constraint programming which models combinatorial search problems as special instances of the *constraint satisfaction problem* (CSP) . The basic idea of constraint propagation is to evaluate implicit constraints through the repeated analysis of the variables, domains and constraints that describe a specific problem instance. This analysis makes it possible to detect and remove *inconsistent* variable assignments that cannot participate in any solution by a merely partial problem analysis.

One of our main objectives is to present and derive efficient constraint propagation techniques also known as consistency tests for the *disjunctive scheduling problem* (DSP) which is a generalization of the classical job shop scheduling problem (JSP). The DSP constitutes a perfect object of study due to the trade-off between its computational complexity and its simple description. On the one hand, within the class of NP-hard problems the DSP has been termed to be one of the most intractable problems. This view is best supported by the notorious 10×10 problem instance of the JSP introduced by Muth and Thompson [MT63] which resisted any solution attempts for several decades and was only solved more than 25 years later by Carlier and Pinson [CP89]. On the other hand, the disjunctive model introduced by Roy and Sussman [RS64] provides an illustrative and simple representation of the DSP which is only based on two types of constraints which in scheduling are known as *precedence* and *disjunctive constraints*.

An elementary analysis of the DSP involving a limited number of search variables derives the consistency tests that ensure 3-*b*-consistency. These consistency tests can be generalized and, although their application does not establish a higher level of consistency, they enable powerful domain reductions in polynomial time. Notice, that establishing *n*-consistency for any *n* is NP-hard, thus the existence of a polynomial algorithm is not very probable. Furthermore the concepts of energetic reasoning and shaving are presented.

The other objective of this chapter is a classification of the consistency tests derived according to the domain reduction achieved. A new dominance criterion that allows a comparison of consistency tests in the aforementioned sense and simple methods for proving dominance are presented. An extensive study of all consistency tests is given. Quite surprisingly, comparing the extent of the search space reduction induced, we will find out that some apparently stronger consistency tests are subsumed by apparently weaker ones.

The remainder of this chapter is organized as follows. Section 13.2 introduces the CSP. Several concepts of consistency are proposed which may serve as a theoretical basis for constraint propagation techniques. We define consistency tests and present the aforementioned dominance criterion for comparing them.

Section 13.3 describes the DSP and examines its relation to the CSP. Section 13.4 extensively describes constraint propagation techniques for the DSP. Notice that although we focus on the basic DSP, the results of this work also apply in an unchanged manner to some important extensions of the DSP, for instance, the DSP with release times and due dates. Section 13.5 finally summarizes the results.

13.2 Constraint Satisfaction

Search and optimization problems such as the disjunctive scheduling problem are generally modelled as special subclasses of the *constraint satisfaction problem* (CSP) or the *constraint optimization problem* (COP) . We will give a short introduction to these problem classes in subsection 13.2.1. In subsection 13.2.2 we will then describe constraint propagation methods and different concepts of consistency.

13.2.1 The Constraint Satisfaction and Optimization Problem

The CSP can be roughly described as follows: "Given a domain specification, find a solution x, such that x is a member of a set of possible solutions and it satisfies the problem conditions" [Ama70]. The COP additionally requires that the solution found optimizes some objective function.

The CSP was first formalized and studied by Huffman [Huf71], Clowes [Clo71] and Waltz [Wal75] in vision research for solving line-labelling problems. Haralick and Shapiro [HS79, HS80] and Mackworth [Mac92] discuss general algorithms and applications of CSP solving. Van Hentenryck [Hen92] and Cohen [Coh90] tackle the CSP from a constraint logic programming viewpoint. Comprehensive overviews on the CSP are provided by Meseguer [Mes89] and Kumar [Kum92]. An exhaustive study of the theory of constraint satisfaction and optimization can be found in [Tsa93]. We will only present the necessary aspects and start with some basic definitions.

The *domain* of a variable is the set of all values that can be assigned to the variable. We will assume in this section that domains are finite and later allow for infinite but discrete domains. The domain associated with the variable x is denoted by $\mathcal{D}(x)$. If $\mathcal{V} = \{x_1, \cdots, x_n\}$ is a set of variables and $\mathcal{DOM} = \{ \mathcal{D}(x_1), \cdots, \mathcal{D}(x_n) \}$ the set of domains, then an *assignment* $a = \{a_1, \cdots, a_n\}$ is an element of the Cartesian product $\mathcal{D}(x_1) \times \cdots \times \mathcal{D}(x_n)$; in other words, an assignment instantiates each variable x_i with a value $a_i \in \mathcal{D}(x_i)$ from its domain.

A *constraint* c on $\mathcal{DOM}$ is a function $c \colon \mathcal{D}(x_{i_1}) \times \cdots \times \mathcal{D}(x_{i_k}) \to \{\text{true, false}\}$, where $\mathcal{V}' := \{x_{i_1}, \cdots, x_{i_k}\}$ is a non empty set of variables. The cardinality $|\mathcal{V}'|$ is also called the arity of c. If $|\mathcal{V}'| = 1$ or $|\mathcal{V}'| = 2$ then we speak of unary and binary constraints respectively. An assignment $a = \mathcal{D}(x_1) \times \cdots \times \mathcal{D}(x_n)$ satisfies c iff $c(a_{i_1}, \cdots, a_{i_k}) = $ true.

Definition 13.2.1

An instance I of the *constraint satisfaction problem* (CSP) is defined by a tuple $I = (\mathcal{V}, \mathcal{DOM}, \mathcal{CONS})$, where $\mathcal{V}$ is a finite set of variables, $\mathcal{DOM}$ the set of associated domains and $\mathcal{CONS}$ a finite set of constraints on $\mathcal{DOM}$. An assignment a is *feasible* iff it satisfies all constraints in $\mathcal{CONS}$. A feasible assignment is also called a *solution* of I. We denote with $\mathcal{F}(I)$ the set of all feasible assignments (solutions) of I.

Given an instance I of the CSP, the associated problem is to find a solution $a \in \mathcal{F}(I)$ or to prove that I has no solution.

As distinguished from the constraint satisfaction problem, the constraint optimization problem searches for a solution which optimizes a given objective function. We will only consider the case of minimization, as maximization can be handled symmetrically.

Definition 13.2.2

An instance of the *constraint optimization problem* (COP) is defined by a tuple $I = (\mathcal{V}, \mathcal{DOM}, \mathcal{CONS}, z)$, where $(\mathcal{V}, \mathcal{DOM}, \mathcal{CONS})$ is an instance of the CSP and z an objective function $z : \mathcal{D}(x_1) \times \cdots \times \mathcal{D}(x_n) \to \mathbb{R}$. Defining

$$z_{min}(I) := \begin{cases} \min\limits_{b \in \mathcal{F}(I)} z(b) & \text{if } \mathcal{F}(I) \neq \varnothing, \\ \infty & \text{otherwise,} \end{cases}$$

an assignment a is called an *optimal solution* of I iff a is feasible and $z(a) = z_{min}(I)$.

Given an instance I of the COP, the associated problem is to find an optimal solution of I and to determine $z_{min}(I)$.

It is not hard to see that the CSP and the COP are intractable and belong to the class of NP-hard problems (c.f. Section 2.2).

An instance of the CSP can be represented by means of a graph (*constraint graph*) which visualizes the interdependencies between variables that are induced by the constraints. If we restrict our attention to unary and binary constraints then the definition of a constraint graph G is quite straightforward. The vertex set of G corresponds to the set of all variables $\mathcal{V}$, while the edge set is defined as follows: two vertices $x_i, x_j \in \mathcal{V}$, $i \neq j$, are connected by an undirected edge iff there exists a constraint $c(x_i, x_j) \in \mathcal{CONS}$. This can be generalized to constraints of arbitrary arity using the notion of hypergraphs [Tsa93]. Figure 13.2.1 shows a typical CSP instance and the corresponding constraint graph.

13.2.2 Constraint Propagation

From a certain point of view, the CSP and the COP are quite simple problems. Since we assumed that the domains of a CSP instance I are finite which for most interesting problems is not a serious restriction, I can be solved by a simple generate-and-test algorithm that works as follows: enumerate all assignments $a \in \mathcal{D}(x_1) \times \cdots \times \mathcal{D}(x_n)$ and verify whether a satisfies all constraints $c \in \mathcal{CONS}$; stop if the answer is "yes". The COP can be solved by enumerating all feasible assignments and storing the one with minimal objective function value.

Unfortunately, this method is not practicable due to the size of the search space which grows exponentially with the number of variables. In the worst case, all assignments of a CSP instance have to be tested which cannot be carried out efficiently except for problem instances too small to be of any practical value. Thus, it suggests itself to examine methods which reduce the search space prior to starting (or during) the search process.

One such method of search space reduction which only makes use of simple inference mechanisms and does not rely on problem specific knowledge is known as constraint propagation. The origins of constraint propagation go back to Waltz [Wal72] who more than three decades ago developed a now well-known filtering algorithm for labelling three-dimensional line diagrams.

The basic idea of constraint propagation is to make implicit constraints more visible through the repeated analysis and evaluation of the variables, domains and constraints describing a specific problem instance. This makes it possible to detect and remove inconsistent variable assignments that cannot participate in any solution by a merely partial problem analysis.

Two complexity related problems arise when performing constraint propagation. One problem depends upon the number of variables and constraints that are examined simultaneously, while the other problem is caused by the size of the domains. These problems are usually tackled by limiting the number of variables and constraints (local consistency with respect to all subsets of k variables) and the number of domain assignments (domain- or d-consistency, bound- or b-consistency) that are considered in the examination. These different concepts will be discussed further below. We start with some simple examples, as this is the easiest way to introduce constraint propagation.

Example 13.2.3

Let $I = (\mathcal{V}, \mathcal{DOM}, \mathcal{CONS})$ be the CSP instance shown in Figure 13.2.1. A simple analysis of the constraints (i) to (vi) allows us to reduce the domains of the variables x_1, x_2 and x_3. We distinguish between the domains $\mathcal{D}(x_i)$ and the reduced domains $\delta(x_i)$. At the beginning, of course, $\delta(x_i) = \mathcal{D}(x_i)$ for $i \in \{1, 2, 3\}$.

$$\mathcal{V} = \{x_1, x_2, x_3\},$$
$$\mathcal{D}(x_1) = \{1, \cdots, 10\},$$
$$\mathcal{D}(x_2) = \{1, \cdots, 10\},$$
$$\mathcal{D}(x_3) = \{1, \cdots, 10\},$$
(i) $1 \le x_1 \le 4,$
(ii) $1 \le x_2 \le 4,$
(iii) $1 \le x_3 \le 4,$
(iv) $x_1 + x_2 = 4,$
(v) $x_1 + x_3 = 5,$
(vi) $x_2 + x_3 \ge 6.$

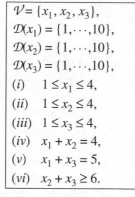

Figure 13.2.1 *Example* 13.2.3.

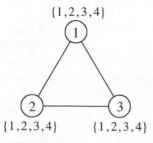

Figure 13.2.2 *Step* 1.   **Figure 13.2.3** *Steps* 2, 3 *and* 4.

1. The unary constraints (i) - (iii) yield the trivial but considerable reduction $\delta(x_1) := \delta(x_2) := \delta(x_3) := \{1,2,3,4\}$ (see Figure 13.2.2).

2. We next examine pairs of variables. Let us start with the pair (x_1, x_2) and the constraint (iv). If we choose, for instance, the assignment $a_1 = 4$ then there obviously exists no assignment $a_2 \in \delta(x_2) = \{1, \cdots, 4\}$ which satisfies (iv) $x_1 + x_2 = 4$. Hence, the value 4 can be removed from $\delta(x_1)$. The same argument is not applicable to $a_1 = 1,2,3$, so we currently can only deduce $\delta(x_1) := \{1,2,3\}$.

3. Since (iv) is symmetric in x_1 and x_2, we can as well set $\delta(x_2) := \{1,2,3\}$.

4. Consider now the pair (x_2, x_3) and constraint (vi). As $a_2 \in \{1,2,3\}$, i.e. $a_2 \le 3$, the constraint (vi), $x_2 + x_3 \ge 6$, is only satisfied for $a_3 \ge 3$. We therefore obtain $\delta(x_3) := \{3,4\}$ (see Figure 13.2.3).

5. Now let us turn to the pair (x_1, x_3) and constraint (v). Since $a_3 = 3$ or $a_3 = 4$, constraint (v), $x_1 + x_3 = 5$, yields $a_1 \ne 3$, and we can set $\delta(x_1) := \{1,2\}$.

6. Finally, studying constraint (iv) once more, we can remove $a_2 = 1$ and set

$\delta(x_2) := \{2,3\}$ (see Figure 13.2.4).

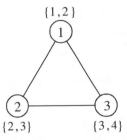

$\{1,2\}$

$\{2,3\}$ $\{3,4\}$

Figure 13.2.4 *Steps 5 and 6.*

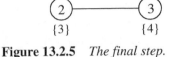

$\{1\}$

$\{3\}$ $\{4\}$

Figure 13.2.5 *The final step.*

At this point, no more values can be excluded from the current domains through the examination of pairs of variables. If we stop propagation now then the search space reduction is already of a considerable size. Prior to our simple analysis, the search space was of cardinality $|\mathcal{D}(x_1) \times \mathcal{D}(x_2) \times \mathcal{D}(x_3)| = 10 \cdot 10 \cdot 10 = 1000$, afterwards the cardinality dropped down to $|\delta(x_1) \times \delta(x_2) \times \delta(x_3)| = 2 \cdot 2 \cdot 2 = 8$.

Extending our analysis to triples of variables reduces the search space even more. Given, for instance, $a_1 = 2$, constraint (*iv*) implies $a_2 = 2$, while (*v*) implies $a_3 = 3$. Since $a_2 + a_3 = 5 < 6$, this is a contradiction to the constraint (*vi*). Reducing $\delta(x_1)$ to $\{1\}$, we can immediately deduce $\delta(x_2) = \{3\}$ and $\delta(x_3) = \{4\}$ which is shown in Figure 13.2.5. Hence, only the assignment $a = (1,3,4)$ is feasible and $\mathcal{F}(I) = \{(1,3,4)\}$ is the solution space of I. □

Example 13.2.4

Consider now the CSP instance $I = (\mathcal{V}, \mathcal{DOM}, \mathcal{CONS})$ shown in Figure 13.2.6. Here, the constraint $a \bmod b = c$ yields true, if a divided by b has a remainder of c. It is possible to show that this CSP instance has eight feasible solutions:

$$\mathcal{F}(I) = \{(4,7,5),(4,7,10),(5,6,1),(5,6,6),(9,2,5),(9,2,10),(10,1,1),$$
$$(10,1,6)\}$$

$\mathcal{V} = \{x_1, x_2, x_3\}$,
$\mathcal{D}(x_1) = \{1,\cdots,10\}$,
$\mathcal{D}(x_2) = \{1,\cdots,10\}$,
$\mathcal{D}(x_3) = \{1,\cdots,10\}$,
(*i*) $(x_1 + x_2) \bmod 10 = 1$,
(*ii*) $(x_1 \cdot x_3) \bmod 5 = 0$,
(*iii*) $(x_2 + x_3) \bmod 5 = 2$.

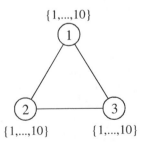

$\{1,...,10\}$

$\{1,...,10\}$ $\{1,...,10\}$

Figure 13.2.6 *Example* 13.2.4.

However, finding these solutions using only constraint propagation is not as easy as in Example 13.2.3. It is not hard to see that the corresponding current domains $\delta(x_1)$, $\delta(x_2)$ and $\delta(x_3)$ cannot be reduced by examining pairs of variables. Consider, for instance, the pair (x_1, x_2) and constraint (i): for each assignment $a_1 \in \delta(x_1)$, there exists an assignment $a_2 \in \delta(x_2)$ such that (i) is satisfied. Similar conclusions can be drawn if the roles of x_1 and x_2 are interchanged or if we study the pairs (x_2, x_3) and (x_1, x_3).

To derive further information, we have to examine pairs of assignments. We may, for instance, find out that the assignments $\{1\} \times \{1, \cdots, 9\}$ of the variables x_1 and x_2 cannot participate in any feasible solution, since they do not satisfy constraint (i). Thus given $a_1 = 1$, the only interesting assignment is $a_2 = 10$. Similar results can be obtained for $a_1 = 2$, etc. This analysis, however, increases the overhead in terms of computational complexity and storage capacity considerably, since pairs of assignments have to be dealt with, and it is not clear at all whether this additional overhead can be offset by the search space reduction achieved. □

These examples demonstrate that constraint propagation can be quite powerful, reducing the search space of a "favourable" CSP instance to a great extent after a few steps of propagation. In the worst case, however, constraint propagation does not yield a substantial reduction of the search space and even slows down the complete solution process due to the additional computations. In general, the outcome of constraint propagation lies between these two extremes: some but not all infeasible solutions can be discarded if constraint propagation is restricted to techniques which can be implemented efficiently. Thus, constraint propagation complements, but does not replace a systematic search.

After this intuitive introduction to constraint propagation, it is now necessary to provide a theoretical environment which allows us to design and assess constraint propagation techniques. We have informally described constraint propagation as "the reduction of the search space of a CSP instance through the analysis of variables, domains and constraints". The question how far this reduction should be carried out, we would readily answer "as far as possible". Remember, however, that any CSP instance is uniquely determined through its variables, domains and constraints. Thus, if we took this description literally then constraint propagation would just be a synonym to *solving* the CSP which of course is not sensible, because we initially have introduced constraint propagation in order to *simplify* the solution of the CSP. Further, we already have seen that constraint propagation is only useful up to a certain extent due to an increasing computational complexity. We therefore present different concepts of consistency which may serve as a theoretical basis for propagation techniques. Roughly speaking, a concept of consistency defines the maximal search space reduction that is possible regarding some specific criteria.

k-Consistency

The first concepts of consistency have been presented in the early seventies by Montanari [Mon74], who introduced the notions of *node-*, *arc-* and *path-consistency*. Roughly speaking, these concepts are based on the examination of constraints containing k variables, where $k = 1, 2, 3$, with their names being derived from the representation of a CSP instance as a constraint graph. Notice, that in the last section examples have been given of how to achieve node- and arc-consistency which will be seen more clearly further below. These concepts of consistency have been generalized by Freuder [Fre78] in a natural manner to the notion of *k-consistency*. For a detailed analysis of k-consistency see for instance [Tsa93]. We will only describe the basic ideas in an informal way.

In order to define k-consistency we have to introduce the notion of *k-feasibility*. Let $a = (a_1, \cdots, a_n)$ be an assignment of a given CSP instance. A partial assignment of k variables $(a_{i_1}, \cdots, a_{i_k})$ is *k-feasible*, if it satisfies all constraints which contain these variables only (or any subset of them). The motivation of the definition of k-consistency is based on the following observation: a can only be feasible, if for a given k any partial assignment $(a_{i_1}, \cdots, a_{i_k})$ is k-feasible. Inversely, any partial assignment of k variables, that is not feasible, is not interesting and hints at an inconsistent state.

In Freuder's words [Fre78] k-consistency is achieved if for any $(k-1)$-feasible assignment of $k - 1$ variables (taken from a set $\delta(x_{i_1}, \cdots, x_{i_{k-1}}) \subseteq \mathcal{D}(x_{i_1}) \times \cdots \times \mathcal{D}(x_{i_{k-1}}))$ and any choice of a kth variable, there exists an assignment of the kth variable (taken from a set $\delta(x_{i_k}) \subseteq \mathcal{D}(x_{i_k})$), such that the assignment of the k variables taken together is k-feasible.

Note that the *property* of k-consistency is always relative to the sets $\delta(x_{i_1}, \cdots, x_{i_{k-1}})$ and $\delta(x_{i_k})$. Thus, in order to *establish* k-consistency, starting from an inconsistent state, this implicitly requires a $(k-1)$-dimensional administration of these sets. At the beginning, these sets contain all assignments, that is, $\delta(x_{i_1}, \cdots, x_{i_{k-1}}) := \mathcal{D}(x_{i_1}) \times \cdots \times \mathcal{D}(x_{i_{k-1}})$ and $\delta(x_{i_k}) := \mathcal{D}(x_{i_k})$. Inconsistent assignments are then eventually discarded, until k-consistency is reached.

1-consistency is quite easy to achieve: if $x_i \in \mathcal{V}$ is a variable and $c(x_i)$ is a unary constraint then all assignments $a_i \in \delta(x_i)$ for which $c(a_i) = \text{false}$ are removed. In order to establish 2-consistency, pairs of variables $x_i, x_j \in \mathcal{V}$ and binary constraints $c(x_i, x_j)$ have to be examined: an assignment $a_i \in \delta(x_i)$ can be removed if $c(a_i, a_j) = \text{false}$ for all $a_j \in \delta(x_j)$. Analogously, 3-consistency requires the examination of triples of variables $x_i, x_j, x_k \in \mathcal{V}$ and removes pairs of assignments $(a_i, a_j) \in \delta(x_i, x_j)$, etc. As already mentioned, 1- and 2-consistency coincide with the notions of node- and arc-consistency, whereas 2- and 3-consistency taken together are equivalent to path-consistency, see e.g. [Mon74,

Mac77, MH86, Tsa93]. 1-, 2- and 3-consistency have also been summarized under the name of *lower-level* consistency as opposed to *higher-level* consistency, since only small subsets of variables, domains and constraints are evaluated simultaneously.

Efficient algorithms for establishing 1-, 2- and 3-consistency and an analysis of their complexity have been presented, among others, by Montanari [Mon74], Mackworth [Mac77], Mackworth and Freuder [MF85], Mohr and Henderson [MH86], Dechter and Pearl [DP88], Han and Lee [HL88], Cooper [Coo89] and Van Hentenryck et al. [HDT92]. Improved arc consistency algorithms AC-6 and AC-7 have been presented by Bessière [Bes94] and by Bessière et al. [BFR99]. Chen [Che99] has proposed a new arc consistency algorithm, AC-8, which requires less computation time and space than AC-6 and AC-7. Cooper developed an optimal algorithm which achieves k-consistency for arbitrary k [Coo89]. Jeavons et al.[JCC98] have identified a number of constraint classes for which some fixed level of local consistency is sufficient to ensure global consistency. They characterize all possible constraint types for which strong k-consistency guarantees global consistency, for each $k \geq 2$. Other methods for solving the CSP through the sole application of constraint propagation (*solution synthesis*) have been proposed by Freuder [Fre78], Seidel [Sei81] and Tsang and Foster [TF90]. The deductive approach proposed by Bibel [Bib88] is closely related to solution synthesis.

Domain-Consistency

Cooper's optimal algorithm [Coo89] for achieving k-consistency requires testing all subsets $\delta(x_{i_1}, \cdots, x_{i_{k-1}}) \subseteq \mathcal{D}(x_{i_1}) \times \cdots \times \mathcal{D}(x_{i_{k-1}})$ of $(k-1)$-feasible assignments which is only practicable for small values of k. We therefore describe two weaker concepts of consistency.

The first concept is based on only storing the 1-dimensional sets $\delta(x_i) \subseteq \mathcal{D}(x_i)$ for all variables $x_i \in \mathcal{V}$. For reasons near at hand, $\delta(x_i)$ is also called the *current domain* of x_i. Intuitively, we can at most discard all values $a_i \in \delta(x_i)$ for which there exist no assignments $a_j \in \delta(x_j), j \neq i$, such that $(a_1, \cdots, a_i, \cdots, a_n)$ is feasible. Alternatively, the feasibility condition can be replaced with the sufficient condition of k-feasibility which leads to a lower level of consistency. We refer to this concept of consistency as *domain-consistency* or *k-d-consistency*. Domain-consistency has been used, among others, by Nuijten [Nui94]. Formal definitions are provided below.

Definition 13.2.5

Let $I = (\mathcal{V}, \mathcal{DOM}, \mathcal{CONS})$ be an instance of the CSP. If $\delta(x_i) \subseteq \mathcal{D}(x_i)$ is the current domain of the variable $x_i \in \mathcal{V}$ then $\delta(x_i)$ is *complete* iff, for all feasible assignments $a = (a_1, \cdots, a_n)$, the value a_i is contained in $\delta(x_i)$.

Definition 13.2.6

Let $I = (\mathcal{V}, \mathcal{DOM}, \mathcal{CONS})$ be an instance of the CSP and $\Delta := \{\, \delta(x_i) \mid x_i \in \mathcal{V} \}$ be the set of current domains, so that $\delta(x_i) \subseteq \mathcal{D}(x_i)$ is complete[1].

1. Δ is *k-d-consistent* for $1 \le k \le n$ iff, for all subsets $\mathcal{V}' := \{x_{i_1}, \cdots, x_{i_{k-1}}\}$ of $k-1$ variables and any k^{th} variable $x_{i_k} \notin \mathcal{V}'$, the following condition holds:

$$\forall\, a_{i_k} \in \delta(x_{i_k}), \exists\, a_{i_1} \in \delta(x_{i_1}), \cdots, \exists\, a_{i_{k-1}} \in \delta(x_{i_{k-1}}):$$
$$(a_{i_1}, \cdots, a_{i_k}) \text{ is } k\text{-feasible.}$$

2. Δ is *strong k-d-consistent* for $1 \le k \le n$ iff Δ is k'-d-consistent for all $1 \le k' \le k$.

The following naive algorithm establishes k-d-consistency: start with $\delta(x_i) := \mathcal{D}(x_i)$ for all $x_i \in \mathcal{V}$; choose variable x_{i_k} and assignment $a_{i_k} \in \delta(x_{i_k})$; test whether there exists a subset of $k-1$ variables $\mathcal{V}' := \{x_{i_1}, \cdots, x_{i_{k-1}}\}$ which does not contain x_{i_k}, so that $(a_{i_1}, \cdots, a_{i_{k-1}}, a_{i_k})$ is not k-feasible for all $a_{i_1} \in \delta(x_{i_1}), \cdots, a_{i_{k-1}} \in \delta(x_{i_{k-1}})$; if the answer is "yes" then remove the assignment a_{i_k} from $\delta(x_{i_k})$; repeat this process with other assignments and/or variables until no more domain reductions are possible.

Example 13.2.7

Let us reconsider Example 13.2.4. After establishing n-d-consistency, the reduced domains $\delta(x_i)$ contain only assignments $a_i \in \mathcal{D}(x_i)$ for which there exists a feasible solution $(a_1, a_2, a_3) \in \mathcal{F}(I)$. Since the solution space is

$$\mathcal{F}(I) = \{(4,7,5), (4,7,10), (5,6,1), (5,6,6), (9,2,5), (9,2,10), (10,1,1), (10,1,6)\}$$

we obtain $\delta(x_1) = \{4,5,9,10\}$, $\delta(x_2) = \{1,2,6,7\}$, and $\delta(x_3) = \{1,5,6,10\}$. After the reduction, the search space is of size $|\delta(x_1) \times \delta(x_2) \times \delta(x_3)| = 4 \cdot 4 \cdot 4 = 64$ as compared to the original search space of size $|\mathcal{D}(x_1) \times \mathcal{D}(x_2) \times \mathcal{D}(x_3)| = 10 \cdot 10 \cdot 10 = 1000$ which is considerably larger. □

This gives us an indication of the maximal search space reduction that is possible if a solely domain oriented approach is chosen. Notice, however, that we did not yet discuss how to establish n-d-consistency other than to apply the naive algorithm, so an important question is whether there exists an efficient implementation after all. Before we deal with this issue, however, we will first present another concept of consistency.

---

[1] The completeness property which is usually omitted in other definitions of consistency ensures that no feasible solutions are removed. Without this property, $\Delta := \{\varnothing, \ldots, \varnothing\}$ would be n-d-consistent which obviously is not intended.

Bound-Consistency

Storing all values of the current domains $\delta(x_1), \cdots, \delta(x_n)$ still might be too costly. An interval oriented encoding of $\delta(x_i)$ provides an alternative if $\mathcal{D}(x_i)$ is totally ordered, for instance, if $\mathcal{D}(x_i) \subseteq I\!N_0$. In this case, we can identify $\delta(x_i)$ with the interval $\delta(x_i) := [l_i, r_i] := \{l_i, l_i + 1, \cdots, r_i - 1, r_i\}$, so that only the "left" and "right" bounds of $\delta(x_i)$ have to be stored. Therefore, this concept of consistency is usually referred to as *bound-consistency* or *k-b-consistency*. Bound-consistency has been discussed, among others, by Moore [Moo66], Davis [Dav87], van Beek [Bee92] and Lhomme [Lho93].

Definition 13.2.8 (*k-b-consistency*).

Let $I = (\mathcal{V}, \mathcal{DOM}, \mathcal{CONS})$ be an instance of the CSP and $\Delta := \{\ \delta(x_i) \mid x_i \in \mathcal{V}\}$ be the set of current domains, so that $\delta(x_i) \subseteq \mathcal{D}(x_i)$ is complete.

1. Δ is *k-b-consistent* for $1 \leq k \leq n$ iff, for all subsets $\mathcal{V}' := \{x_{i_1}, \cdots, x_{i_{k-1}}\}$ of $k - 1$ variables and any k^{th} variable $x_{i_k} \notin \mathcal{V}'$, the following condition holds:

$$\forall\ a_{i_k} \in \{l_{i_k}, r_{i_k}\}, \exists\ a_{i_1} \in \delta(x_{i_1}), \cdots, \exists a_{i_{k-1}} \in \delta(x_{i_{k-1}}):$$

$$(a_{i_1}, \cdots, a_{i_k}) \text{ is } k\text{-feasible.}$$

2. Δ is *strongly k-b-consistent* for $1 \leq k \leq n$ iff Δ is *k'-b*-consistent for all $1 \leq k' \leq k$.

A naive algorithm for establishing *k-b*-consistency is obtained by slightly modifying the naive *k-d*-consistency algorithm: instead of choosing $a_{i_k} \in \delta(x_{i_k})$, we may only choose (and remove) $a_{i_k} \in \{l_{i_k}, r_{i_k}\}$.

As a negative side effect, only the bounds l_i and r_i, but no intermediate value $l_i < a_i < r_i$ can be discarded, except, if due to the repeated removal of other assignments, a_i eventually becomes the left or right bound of the current domain. Thus, bound-consistency is a weaker concept than domain-consistency.

Example 13.2.9

We again examine the Examples 13.2.4 and 13.2.7. Establishing *n-b*-consistency must lead to the domain intervals $\delta(x_1) = [4,10]$, $\delta(x_2) = [1,7]$ and $\delta(x_3) = [1,10]$. Here, the size of the reduced search space is $|\delta(x_1) \times \delta(x_2) \times \delta(x_3)| = 7 \cdot 7 \cdot 10 = 490$ compared with the size of the original search space (1000) and the size of the *n-d*-consistent search space (64). □

Unfortunately, the following complexity result applies.

Theorem 13.2.10

Establishing n-b-consistency for the CSP is an NP-hard problem.

Proof. Consider an instance I of the CSP. Let $\Delta = \{\, \delta(x_i) \mid x_i \in \mathcal{V} \,\}$ be the corresponding set of current domains, such that Δ is N-b-consistent. Obviously, $\mathcal{F}(I)$ is not empty iff there exists $x_i \in \mathcal{V}$ satisfying $\delta(x_i) \neq \varnothing$. □

A similar proof shows that establishing n-d-consistency is NP-hard as well.

Consistency Tests

In general, establishing k-consistency is ruled out due to the complex data structures that are necessary for the administration of the k-feasible subsets. In the last subsection we have further seen that establishing n-d- or n-b-consistency is an NP-hard problem. Consequently, using constraint propagation in order to solve the CSP is only sensible if we content ourselves with approximations of the concepts of consistency that have been introduced.

An important problem is to derive simple rules which will lead to efficient search space reductions, but at the same time can be implemented efficiently with a low polynomial time complexity. These rules are known as *consistency tests* and are generally described through a condition-instruction pair $\mathcal{Z}$ and $\mathcal{B}$. Intuitively, the semantics of a consistency test is as follows: whenever condition $\mathcal{Z}$ is satisfied, $\mathcal{B}$ has to be executed. $\mathcal{Z}$ may be, for instance, an equation or inequality, while $\mathcal{B}$ may be a domain reduction rule. We will often use the shorthand notation $\mathcal{Z} \Rightarrow \mathcal{B}$ for consistency tests.

Example 13.2.11

Let us derive a consistency test for the CSP instance I described in Example 13.2.3. Consider the constraint (vi) $x_2 + x_3 \geq 6$. Given an assignment a_2 of x_2, we can remove a_2 from $\delta(x_2)$ if there exists no assignment $a_3 = \delta(x_3)$ satisfying (vi). However, we do not really have to test all assignments in $\delta(x_3)$, because if (vi) is not satisfied for $a_3 = \max \delta(x_3)$ then it is not satisfied for any other assignment in $\delta(x_3)$ and vice versa. Hence, for any $a_2 \in \mathcal{D}(x_2)$,

$$\gamma(a_2) : a_2 + \max \delta(x_3) < 6 \implies \delta(x_2) := \delta(x_2) \setminus \{a_2\}$$

defines a consistency test for I. □

Of course, this example is quite simple and it may not seem clear whether any advantages can be drawn from such elementary deductions. Surprisingly, however, an analogously simple analysis will allow us to derive powerful consistency tests for particular classes of constraints as will be seen in one of the subsequent sections.

One of our objectives is to compare consistency tests. This requires a condition which enables us to determine whether certain consistency tests are "at least as good" as certain others. Intuitively, this applies if the deductions implied by a set of consistency tests are "at least as good" as those implied by another set. In order to elaborate this rather vague description, we will focus on *domain consistency tests*, i.e. consistency tests which deduce domain reductions. Similar results, however, apply for other types of consistency tests.

Let us derive a formal definition of domain consistency tests. Let $\Theta := 2^{\mathcal{D}(x_1)} \times \cdots \times 2^{\mathcal{D}(x_n)}$, where $2^{\mathcal{D}(x_i)}$ denotes the set of all subsets of $\mathcal{D}(x_i)$. Given $\Delta, \Delta' \in \Theta$, that is, $\Delta = \{ \delta(x_i) \mid x_i \in \mathcal{V} \}$ and $\Delta' = \{ \delta'(x_i) \mid x_i \in \mathcal{V} \}$, we say that

1. $\Delta \subseteq \Delta'$ iff $\delta(x_i) \subseteq \delta'(x_i)$ for all $x_i \in \mathcal{V}$,

2. $\Delta \subsetneq \Delta'$ iff $\Delta \subseteq \Delta'$, and there exists $x_i \in \mathcal{V}$, such that $\delta(x_i) \subsetneq \delta'(x_i)$.

Domain consistency tests have to satisfy two conditions. First, current domains are either reduced or left unchanged. Second, only assignments $a_i \in \delta(x_i)$ are removed for which no feasible assignment $a = (a_1, \cdots, a_i, \cdots, a_n)$ exists, because otherwise solutions would be lost. Since, however, we do not need the second condition in order to derive the results of this section, only the first one is formalized.

Definition 13.2.12

A domain consistency test γ is a function $\gamma : \Theta \to \Theta$ satisfying $\gamma(\Delta) \subseteq \Delta$ for all $\Delta \in \Theta$.

Suppose now that a set of domain consistency tests is given. In order to obtain the maximal domain reduction possible, these tests have to be applied repeatedly in an iterative fashion rather than only once. The reason for this is that, after the reduction of some domains, additional domain adjustments can possibly be derived using some of the tests which have previously failed in deducing any reductions. This has been demonstrated, for instance, in Example 13.2.3. Thus, the deduction process should be carried out until no more adjustments are possible or, in other words, until the set Δ of current domains becomes a fixed point. The standard fixed point procedure is shown in Algorithm 13.2.13.

Algorithm 13.2.13 *Fixed point*

Input: Δ: set of current domains;

```
begin
  repeat
    Δ_old := Δ;
      for all (γ ∈ Γ) do Δ := γ(Δ);   -- Γ is a set of consistency tests
    until (Δ := Δ_old);
end;
```

It is important to mention that the fixed point computed does not have to be unique and usually depends upon the order of the application of the consistency tests. For this reason we will only study *monotonous* consistency tests for which the order of application does not affect the outcome of the domain reduction process. This result will be derived in the following.

Definition 13.2.14

A consistency test γ is *monotonous* iff the following condition is satisfied:

$$\forall \Delta, \Delta' \in \Theta : \Delta \subseteq \Delta' \Rightarrow \gamma(\Delta) \subseteq \gamma(\Delta') . \tag{13.2.1}$$

Let us first define the Δ-fixed-point mentioned above. Let Γ be a set of monotonous domain consistency tests. For practical reasons we will always assume that Γ is finite. Let $\gamma_\infty := (\gamma_g)_{g \in I\!N} \in \Gamma^{I\!N}$ be a series of domain consistency tests in Γ, such that

$$\forall \gamma \in \Gamma, \forall h \in I\!N, \exists g > h : \gamma_g = \gamma . \tag{13.2.2}$$

The series γ_∞ determines the order of application of the consistency tests. The last condition ensures that every consistency test in Γ is (a priori) infinitely often applied. Starting with an arbitrary set Δ of current domains, we define the series of current domain sets $(\Delta_g)_{g \in I\!N}$ induced by γ_∞ through the following recursive equation

$$\Delta_0 := \Delta ,$$
$$\Delta_g := \gamma_g(\Delta_{g-1}) .$$

Since all domains $\mathcal{D}(x_i)$ are finite and $\Delta_g \subseteq \Delta_{g-1}$ due to Definition 13.2.12, there obviously exists $g^* \in I\!N$, such that $\Delta_g = \Delta_{g^*}$ for all $g \geq g^*$. We can therefore define $\gamma_\infty(\Delta) := \Delta_{g^*}$. The next question to answer is whether $\gamma_\infty(\Delta)$ really depends on the chosen series γ_∞.

Theorem 13.2.15 *Unique fixed points.* [DPP00].

If Γ is a set of monotonous domain consistency tests and $\gamma_\infty , \gamma'_\infty \in \Gamma^{I\!N}$ are series satisfying (13.2.2) then $\gamma_\infty(\Delta) = \gamma'_\infty(\Delta)$.

Proof. For reasons of symmetry we only have to show $\gamma_\infty(\Delta) \subseteq \gamma'_\infty(\Delta)$.

Let $(\Delta_g)_{g \in I\!N}$ and $(\Delta'_g)_{g' \in I\!N}$ be the series induced by γ_∞ and γ'_∞ respectively. It is sufficient to prove that for all $g' \in I\!N$, there exists $g \in I\!N$, such that $\Delta_g \subseteq \Delta'_{g'}$. This simple proof will be carried out by induction.

The assertion is obviously true for $g' = 0$. For $g' > 0$, we have $\Delta'_{g'} = \gamma'_{g'}(\Delta'_{g'-1})$. By the induction hypothesis, there exists $h \in I\!N$, such that $\Delta_h \subseteq \Delta'_{g'-1}$. Further, (13.2.2) implies that there exists $g > h$ satisfying $\gamma_g = \gamma'_{g'}$. Since $g > h$, we know that $\Delta_{g-1} \subseteq \Delta_h$. Using the monotony property of γ_g, we can conclude

$$\Delta_g = \gamma_g(\Delta_{g-1}) \subseteq \gamma_g(\Delta_h) \subseteq \gamma_g(\Delta'_{g'-1}) = \gamma_{g'}(\Delta'_{g'-1}) = \Delta'_{g'} \,.$$

This completes the induction proof.   □

Definition 13.2.16

Let Γ be a set of monotonous domain consistency tests, Δ a set of current do-
mains and $\gamma_\infty \in \Gamma^{I\!N}$ an arbitrary series satisfying (13.2.2). We define $\Gamma(\Delta) := \gamma_\infty$
(Δ) to be the unique Δ-*fixed-point* induced by Γ and Δ.

Based on these observations, we can now propose a dominance criterion for domain consistency tests.

Definition 13.2.17

Let Γ, Γ' be sets of monotonous consistency tests.

  1. Γ *dominates* Γ' ($\Gamma \succeq \Gamma'$) iff $\Gamma(\Delta) \subseteq \Gamma'(\Delta)$ for all $\Delta \in \Theta$.

  2. Γ *strictly dominates* Γ' ($\Gamma \succ \Gamma'$) iff $\Gamma \succeq \Gamma'$, and there exists $\Delta \in \Theta$, such
     that $\Gamma(\Delta) \subsetneq \Gamma'(\Delta)$.

  3. Γ is *equivalent* to Γ' ($\Gamma \sim \Gamma'$) iff ($\Gamma \succeq \Gamma'$) and ($\Gamma' \succeq \Gamma$).

The next theorem provides a simple condition for testing dominance of domain
consistency tests. Basically, the theorem states that a set of domain consistency
tests Γ dominates another set Γ' if all domain reductions implied by the tests in Γ'
can be simulated by a finite number of tests in Γ.

Theorem 13.2.18

Let Γ, Γ' *be sets of monotonous consistency tests. If for all* $\gamma' \in \Gamma'$ *and all* $\Delta \in \Theta$,
there exist $\gamma^1, \cdots, \gamma^d \in \Gamma$, *so that*

$$(\gamma^d \circ \dots \circ \gamma^1)(\Delta) \subseteq \gamma'(\Delta) \tag{13.2.3}$$

then $\Gamma \succeq \Gamma'$.

Proof. Let γ_∞ and $\gamma'_\infty \in \Gamma^{I\!N}$ be series satisfying (13.2.2) . Let, further, $(\Delta_g)_{g \in I\!N}$
and $(\Delta'_{g'})_{g' \in I\!N}$ be the series induced by γ_∞ and γ'_∞ respectively. Again, we will
prove by induction that for all $g' \in I\!N$, there exists $g \in I\!N$, such that $\Delta_g \subseteq \Delta'_{g'}$,
since this immediately implies $\Gamma(\Delta) \subseteq \Gamma'(\Delta)$.

The assertion is obviously true for $g' = 0$. Therefore, let $g' > 0$ and $\Delta'_{g'} = \gamma'_{g'}$
$(\Delta'_{g'-1})$. By the induction hypothesis, there exists $h \in I\!N$, such that $\Delta_h \subseteq \Delta'_{g'-1}$.
Let $\gamma^1, \cdots, \gamma^d \in \Gamma$ be the sequence of consistency tests satisfying (13.2.3) for

$\gamma'_{g'}$ and Δ_h. There exist $g_d > \cdots > g_1 > h$ satisfying $\gamma_{g_1} = \gamma^1, \cdots, \gamma_{g_d} = \gamma^d$ due to (13.2.2). Without loss of generality, we assume that $g_d = h+d, \cdots, g_1 = h+1$, so that

$$\Delta_{h+d} = (\gamma_{h+d} \circ \cdots \circ \gamma_{h+1})(\Delta_h) \subseteq \gamma'_{g'}(\Delta_h) \subseteq \gamma'_{g'}(\Delta'_{g'-1}) = \Delta'_{g'}$$

which proves the induction step. This verifies the dominance relation $\Gamma \trianglerighteq \Gamma'$. $\square$

Example 13.2.19

Let us reconsider the consistency tests derived in Example 13.2.11:

$$\gamma(a_2) : a_2 + \max \delta(x_3) < 6 \; \Rightarrow \; \delta(x_2) := \delta(x_2) \setminus \{a_2\} \; .$$

Instead of defining a consistency test for each $a_2 \in \mathcal{D}(x_2)$, it is sufficient to apply a single consistency test to obtain the same effects. Observe that if a_2 can be removed then all assignments $a'_2 < a_2$ can be removed as well, so that we can replace $a_2 \in \delta(x_2)$ with $\min \delta(x_2)$. This leads to the consistency test:

$$\gamma : \min \delta(x_2) + \max \delta(x_3) < 6 \; \Rightarrow \; \delta(x_2) := \delta(x_2) \setminus \{ \min \delta(x_2) \} \; .$$

Obviously, if a_2 can be removed from $\delta(x_2)$ using $\gamma(a_2)$ then γ removes a_2 after at most $a_2 - \min \delta(x_2) + 1$ steps. Thus, $\Gamma := \{\gamma\}$ dominates $\Gamma' := \{ \gamma(a_2) \mid a_2 \in \mathcal{D}(x_2) \}$. Accordingly, Γ' dominates Γ, because $\Gamma' \supseteq \Gamma$. This proves that Γ and Γ' are equivalent. $\square$

13.3 The Disjunctive Scheduling Problem

The *disjunctive scheduling problem* (DSP) is a natural generalization of important scheduling problems like the job shop scheduling problem (JSP) which has been extensively studied in the last decades, or the open shop scheduling problem (OSP) which only in recent years has attracted more attention in scheduling research.

The *DSP* can be described as follows [Pha00]: a finite set of tasks each of which has a specific processing time, has to be scheduled with the objective of minimizing the *makespan*, i.e. the maximum of the completion times of all tasks. Preemption is not allowed which means that tasks must not be interrupted during their processing. In general, tasks cannot be processed independently from each other due to additional technological requirements or scarcity of resources. The DSP considers two kinds of constraints between pairs of tasks which model special classes of restrictions: *precedence* and *disjunctive constraints*.

- *Precedence constraints* which are also known as *temporal constraints* specify a fixed processing order between pairs of tasks. Precedence constraints cover technological requirements of the kind that some task T_i must finish before another task T_j can start, for instance, if the output of T_i is the input of T_j.

- *Disjunctive constraints* prevent the simultaneous or overlapping processing of tasks without, however, specifying the processing order. If a disjunctive constraint is defined between two tasks T_i and T_j then one of the alternatives "T_i before T_j" or "T_j before T_i" must be enforced, but which one is not predetermined. Disjunctive constraints model the resource demand of tasks in a scheduling environment with scarce resource supply. More precisely, the capacity of each resource like special machines, tools or working space is one unit per period of processing time. Tasks use at most a (constant) unit amount of each resource per processing period. Due to the limited amount of resources, two tasks requiring the same resource cannot be processed in parallel.

Note that the term disjunctive constraint, as introduced here and as commonly used in scheduling, is a special case of the general concept of disjunctive constraints.

The DSP and its subclasses have been extensively studied in academic research, since its simple formulation, on the one hand, and its intractability, on the other hand, make it a perfect candidate for the development and analysis of efficient solution techniques. Indeed, the solution techniques that have been derived for the DSP have contributed a lot to the improvement of methods for less idealized and more practice oriented problems. Extensions of the DSP generally consider sequence-dependent setup times, minimal and maximal time lags, multipurpose and parallel machines, non-unit resource supply and demand, machine breakdowns, stochastic processing times, etc.

Section 13.3.1 formulates the DSP as a constraint optimization problem with disjunctive constraints as proposed by Roy and Sussman [RS64] for the JSP. The strength of this model becomes apparent later once the common graph theoretical interpretation of the disjunctive scheduling model is presented. In Section 13.3.2, solution methods for the DSP that are based on constraint propagation are briefly discussed.

13.3.1 The Disjunctive Model

Let $B = \{1, \cdots, n\}$ be the index set of tasks to be scheduled. The processing time of task T_i, $i \in B$ is denoted with p_i. By choosing sufficiently small time units, we can always assume that the processing times are positive integer values. With each task there is associated a start time domain variable st_i with domain set $\mathcal{D}(st_i) = I\!N_0$.

If a precedence or disjunctive constraint is defined between two tasks then we say that these tasks are in *conjunction* or *disjunction* respectively. The tasks in conjunction are specified by a relation $C \subseteq B \times B$. If $(i,j) \in C$ then task T_i has to finish before task T_j can start. Instead of writing $(i,j) \in C$ we will therefore use the more suggestive $i \rightarrow j \in C$. The tasks in disjunction are specified by a symmetric relation $D \subseteq B \times B$. Whenever $(i,j) \in D$, tasks T_i and T_j cannot be processed in parallel. Since $(i,j) \in D$ implies $(j,i) \in D$, we will write $i \leftrightarrow j \in D$. Finally, let $Z = \{\, p_i \mid i \in B \,\}$ be the set of processing times.

An instance of the DSP is uniquely determined by the tuple $I = (B,C,D,Z)$. Since we want to minimize the makespan, i.e. the maximal completion time of all tasks, the objective function is $C_{max}(I) = \max_{i \in B}\{st_i + p_i\}$. The DSP can be written as follows:

$$minimize \ \{C_{max}(I)\}$$

$$st_i \in \mathcal{D}(st_i) = I\!N_0 \qquad\qquad\qquad i \in B,$$

$$(i) \quad st_i + p_i \le st_j \qquad\qquad\qquad i \to j \in C,$$

$$(ii) \quad st_i + p_i \le st_j \ \lor \ st_j + p_i \le st_i \qquad i \leftrightarrow j \in D.$$

Let us first define an assignment $ST = (st_1, \cdots, st_n) \in \mathcal{D}(st_1) \times \cdots \times \mathcal{D}(st_n)$ of all start time variables. For the sake of simplicity, we will use the same notation for variables and their assignments. An assignment ST is *feasible*, i.e. it defines a schedule (cf. Section 3.1), if it satisfies all precedence constraints (i) and all disjunctive constraints (ii). Reformulating the DSP, the problem is to find a feasible schedule with minimal objective function value $C_{max}(I)$. Obviously, for each instance of the DSP, there exists a feasible and optimal schedule.

A Graph Theoretical Approach

The significance of the disjunctive scheduling model for the development of efficient solution methods is revealed if we consider its graph theoretical interpretation. In analogy to Section 10.1, a *disjunctive graph* is a weighted graph $G = (B,C,D,W)$ with node set B, arc sets C, $D \subseteq B \times B$ where D is symmetric, and weight set W. C is called the set of precedence arcs, D the set of disjunctive arcs. Each arc $i \to j \in C \cup D$ is labelled with a weight $w_{i \to j} \in W$. Since D is symmetric, we will represent disjunctive arcs as doubly directed arcs and sometimes refer to $i \leftrightarrow j$ as a disjunctive edge. Notice that $i \leftrightarrow j \in D$ is labelled with two possibly different weights, $w_{i \to j}$ and $w_{j \to i}$.

Let $I = (B,C,D,Z)$ be an instance of the *DSP*. In order to define the associated disjunctive graph $G(I)$, we first introduce two dummy tasks *start* (0) and *end* (∗) so as to obtain a connected graph. Obviously, *start* precedes all tasks, while *end* succeeds all tasks. Further, the processing times of *start* and *end* are zero.

Definition 13.3.1

If $I = (B,C,D,Z)$ is an instance of the DSP then $G(I) := (B^*,C^*,D,W)$ is the associated disjunctive graph, where

$$B^* := B \cup \{0,*\},$$

$$C^* := C \cup \{ 0 \to i \mid i \in B \cup \{*\} \} \cup \{ i \to * \mid i \in B \cup \{0\} \},$$
$$W = \{ w_{i \to j} = p_i \mid i \to j \in C^* \cup D \} \,.$$

Example 13.3.2

Let $I = (B, C, D, Z)$ be an instance of the *DSP* with $B = \{1, \cdots, 8\}$, $C = \{1 \to 2 \to 3, 4 \to 5, 6 \to 7 \to 8\}$ and $D = \{1 \leftrightarrow 4, 1 \leftrightarrow 6, 4 \leftrightarrow 6, 2 \leftrightarrow 7, 3 \leftrightarrow 5, 3 \leftrightarrow 8, 5 \leftrightarrow 8\}$. The corresponding disjunctive graph $G = (B^*, C^*, D, W)$ is shown in Figure 13.3.1.[2]

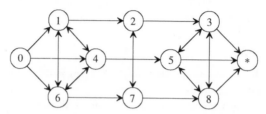

Figure 13.3.1 *A disjunctive graph.*

A disjunctive graph is transformed into a directed graph by orienting disjunctive edges.

Definition 13.3.3

Let $G = (B, C, D, W)$ be a disjunctive graph, and $S \subseteq D$.

1. S is a *partial selection* iff $i \to j \in S$ implies $j \to i \notin S$ for all $i \leftrightarrow j \in D$.

2. S is a *complete selection* iff either $i \to j \in S$ or $j \to i \in S$ for all $i \leftrightarrow j \in D$.

3. A complete selection S is *acyclic* iff the directed graph $G_S = (B, C \cup S)$ is acyclic.

Thus, we obtain a complete (partial) selection if (at most) one edge orientation is chosen from each disjunctive edge $i \leftrightarrow j \in D$. The selection is acyclic if the resulting directed graph is acyclic, ignoring any remaining undirected disjunctive edges. There is a close relationship between complete selections and schedules (let us remind that schedules are always feasible, as defined in Section 3.1). Indeed, if we are only interested in optimal schedules, then it is sufficient to search through the space of all selections which is of cardinality $2^{|D|}$ instead of the space of all schedules which is of cardinality $|I\!N_0|^n$. The DSP can thus be restated as a graph theoretical problem: find a complete and acyclic selection, such that the length of the longest path in the associated directed graph is minimal.

---

[2] We have not depicted all of the trivial edges involving the dummy operations *start* and *end*. Further, the specification of the weights has been omitted.

13.3.2 Solution Methods for the DSP

Countless is the number of solution methods proposed for the JSP which constitutes the most famous subclass of the DSP. A detailed survey is provided by Błażewicz et al. in [BDP96]. We only focus on solution methods which have incorporated constraint propagation techniques in some way or another. Particularly, constraint propagation has been used in exact solution methods most of which are based on a search space decomposition approach of the branch-and-bound kind. It seems fair to say that the advances in solving the JSP that have been made in the last decade can be attributed to a large extent to the development of efficient constraint propagation techniques. Undoubtedly, the algorithm of Carlier and Pinson presented in [CP89] marked a milestone in the JSP history, since for the first time an optimal solution for the notorious 10×10 problem instance proposed by Muth and Thompson [MT63] has been found and its optimality proven. Amazingly, due to the evolution of solution techniques and growing computational power, this formerly unsolvable instance can now be solved within several seconds. Important contributions towards this state of the art have been made among others by Applegate and Cook [AC91], Carlier and Pinson [CP90], Brucker et al. [BJS94, BJK94], Caseau and Laburthe [CL95], Baptiste and Le Pape [BL95] and Martin and Shmoys [MS96], to name only a few. In addition to using constraint propagation techniques in exact solution methods, the opinion eventually gains ground that combining constraint propagation with heuristic solution methods is most promising. Advances in this direction have been reported by Nuijten [Nui94], Pesch and Tetzlaff [PT96], Phan Huy [Pha96] and Nuijten and Le Pape [NL98].

13.4 Constraint Propagation and the DSP

In Section 13.2.2, constraint propagation has been introduced as an elementary method of search space reduction for the CSP or the COP. In this section, we examine how constraint propagation techniques can be adapted to the DSP. An important issue is the computational complexity of the techniques applied which has to be weighed against the search space reduction obtained. Recall that establishing n-, n-d- and n-b-consistency for instances of the CSP or the COP are NP-hard problems. It is not difficult to show that the same complexity result applies if we confine ourselves to the more special DSP. Thus, if constraint propagation is to be of any use in solving the DSP, we will have to content ourselves with approximations of the consistency levels mentioned above.

In the past years, two constraint propagation approaches have been studied with respect to the DSP: a time oriented and a sequence oriented approach. The time oriented approach is based on the concept of domain or bound-consistency. Each task has a current domain of possible start times. *Domain consistency tests* remove inconsistent start time assignments from current domains and, by this, reduce the set of schedules that have to be examined. In contrast to the time oriented approach, the sequence oriented approach reduces the set of complete selections by detecting sequences of tasks, i.e. selecting disjunctive edge orientations which must occur in every optimal solution. Hence, the latter approach has been often labelled *immediate selection* (see e.g. [CP89, BJK94]) or *edge-finding* (see e.g. [AC91]). We will use the term *sequence consistency test* as used in [DPP99].

Domain and sequence consistency tests are two different concepts which complement each other. Often, a situation occurs in which either only reductions of the current domains or only edge orientations are deducible. The best results, in fact, are obtained by applying both types of consistency tests, as fixing disjunctive edges may initiate additional domain reductions and vice versa.

Section 13.4.1 introduces some notation which will be used later. The subsequent sections are concerned with the definition of domain and sequence consistency tests for the *DSP*. For the sake of simplicity, precedence and disjunctive constraints will be treated separately. At first, the simple question of how to implement constraint propagation techniques for precedence constraints is discussed in Sections 13.4.2.

In Sections 13.4.3 through 13.4.8, disjunctive constraints are examined, and both already known and new consistency tests will be presented. We assume that precedence constraints are not defined and that all tasks are in disjunction which leads to the special case of a single-machine scheduling problem [Car82].

Section 13.4.3 examines which consistency tests have to be applied in order to establish *lower-level bound-consistency*, that is, strong *3-b-consistency*. Sections 13.4.4 and 13.4.5 present the well-known *input/output* and *input/output negation* consistency tests first proposed by Carlier and Pinson [CP89] and compare different time bound adjustments. Section 13.4.6 describes a class of new consistency tests which is based on the *input-or-output* conditions and is due to Dorndorf et al. [DPP99]. Section 13.4.7 takes a closer look at the concept of *energetic reasoning* proposed by Erschler et al. [ELT91] and classifies this concept with respect to the other consistency tests defined. Section 13.4.8, finally, deals with a class of consistency tests known as *shaving* which has been introduced by Carlier and Pinson [CP94] and Martin and Shmoys [MS96].

In Section 13.4.9, the results for the disjunctive constraints are summarized. Finally, Section 13.4.10 discusses how to interleave the application of the precedence and disjunctive consistency tests derived. It is worthwhile to mention that a separate analysis of precedence and disjunctive constraints leads to weaker consistency tests as compared to cases where both classes of constraints are *simultaneously* evaluated. However, it remains an open question whether simple and efficient consistency tests can be developed in this case.

13.4.1 Some Basic Definitions

For the rest of this subsection, let $I = (B, C, D, Z)$ be an instance of the DSP. Each task T_i, $i \in B$ has a current domain $\delta(st_i) \subseteq \mathcal{D}(st_i)$. In order to avoid misinterpretations between the start time variable st_i and its assignment (for which the notation st_i is used as well), we will write δ_i instead of $\delta(st_i)$. We assume that some real or hypothetical upper bound UB on the optimal makespan is known or given, so that actually $\delta_i \subseteq [0, UB - p_i]$. This is necessary, since most of the consistency tests derived only deduce domain reductions or edge orientations if the current domains are finite. In general, the tighter the upper bound, the more information can be derived.

The earliest and latest start time of task T_i are given by $est_i := \min \delta_i$ and $lst_i := \max \delta_i$. We will interpret δ_i as an *interval* of start times, i.e. $\delta_i = [est_i, lst_i]$ $= \{ est_i, est_i + 1, \cdots, lst_i - 1, lst_i \}$, although a set oriented interpretation is possible as well. We also need the earliest and latest completion time $ect_i := est_i + p_i$ and $lct_i := lst_i + p_i$ of task T_i.

Sometimes, it is important to distinguish between the earliest and latest start time *before* and *after* a domain reduction. We will then use the notation est_i^* and lst_i^* for the adjusted earliest and latest start times. We will often examine subsets $A \subseteq B$ of tasks and define $p(A) := \sum_{i \in A} p_i$, $EST_{min}(A) := \min_{i \in A} est_i$, and $LCT_{max}(A) := \max_{i \in A} lct_i$. Finally, $C_{max}(p_\delta(A))$ and $C_{max}(p_\delta^{pr}(A))$ denote the optimal makespan if all tasks in A are scheduled within their current domains without preemption or with preemption allowed.

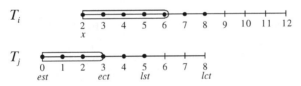

Figure 13.4.1   *Two tasks T_i, T_j with $p_i = 4$ and $p_j = 3$.*

Examples of consistency tests will be illustrated as in Figure 13.4.1 [Nui94] which shows two tasks T_i and T_j. For task T_j, the interval $[est_j, lct_j] = [0,8]$ of times at which T_j may be in process is shown as a horizontal line segment. Possible start times $[est_j, lst_j] = [0,5]$ are depicted as black circles, while the remaining times $[lst_j+1, lct_j] = [6,8]$ are marked with tick marks. A piston shaped bar of size $p_j = 3$, starting at $est_j = 0$, indicates the processing time of task T_j. The chosen representation is especially well-suited for describing the effect of domain consistency tests. If a starting time is proven to be inconsistent then the corresponding time will be marked with an x, as for instance the start time 2 on the time scale of task T_i.

13.4.2 Precedence Consistency Tests

Precedence constraints determine the order in which two specific tasks T_i and T_j have to be processed. If, for instance, task T_i has to finish before task T_j can start, then the earliest start time of T_j has to be greater than or equal to the earliest completion time of T_i. Likewise, an upper bound of the latest completion time of T_i is the latest start time of T_j. This proves the following well-known theorem.

Theorem 13.4.1 *Precedence consistency test.*

If $i, j \in B$ and $i \rightarrow j \in C$ then the following domain reduction rules apply:

$$est_j := \max\{\, est_j, est_i + p_i \,\}, \tag{13.4.1}$$

$$lst_i := \min\{\, lst_i, lst_j - p_i \,\}. \tag{13.4.2}$$

Of course, applying the consistency tests (13.4.1) and (13.4.2) until no more updates are possible is equivalent to the computation of a longest (precedence) path in the disjunctive graph, see [Chr75] for a standard algorithm. This algorithm traverses all tasks in a topological order which ensures that (13.4.1) and (13.4.2) only have to be applied *once* for each precedence arc.

13.4.3 Lower-Level Bound-Consistency

From this Section through Section 13.4.8, we will study the more interesting class of disjunctive constraints. For the sake of simplicity, we assume that B is a clique, i.e. all tasks in B are in disjunctions. We, further, assume that the set of precedence constraints is empty. We will, at first, discuss how disjunctive constraints interact with respect to some concept of consistency. For two reasons we opted for bound-consistency as the concept of consistency to work with. First of all, bound-consistency requires the least amount of storage capacity, since the current domains can be interpreted as intervals, so only the earliest and latest start times have to be memorized. Second, the most powerful consistency tests described in the following only affect/use the earliest and latest start times. Indeed, no efficient consistency tests which make use of "inner" start times are currently known.

Symbol	Description
$\gamma_{A,i}^{(h)}$	$h \le 4$: output consistency test for the couple (A,i), $h \ge 5$: input negation consistency test for the couple (A,i)
δ_i	current domain of T_i : $\delta_i \subseteq I\!N_0$
est_i	earliest start time of T_i : $est_i = \min \delta_i$
est_i^*	adjusted earliest start time of T_i
ect_i	earliest completion time of T_i : $ect_i = est_i + p_i$
lct_i	latest completion time of T_i : $lct_i = lst_i + p_i$
lst_i	latest start time of T_i : $lst_i = \max \delta_i$
lst_i^*	adjusted latest start time of T_i
$p_i(t_1,t_2)$	interval processing time of T_i in the time interval $[t_1,t_2)$
$[t_1,t_2)$	time interval: $[t_1,t_2) = \{\, t_1, t_1 + 1, \cdots, t_2 - 1 \,\}$
$[t_1,t_2]$	time interval: $[t_1,t_2] = \{\, t_1, t_1 + 1, \cdots, t_2 \,\}$
A	subset of tasks: $A \subseteq B$
$A \to i \;\; (i \to A)$	T_i has to be processed after (before) all tasks in A
$C_{max}(p_\delta(A))$	optimal makespan if all tasks in A are scheduled without preemption
$C_{max}(p_\delta^{pr}(A))$	optimal makespan if all tasks in A are scheduled with preemption allowed
$\Gamma_{\neg in}(h)$	set of input negation consistency tests
$\Gamma_{out}(h)$	set of output consistency tests
$EST_{min}(A)$	minimal earliest start time in A : $EST_{min}(A) = \min_{i \in A}\{est_i\}$
$LB_h(A)$	time bound adjustment for output consistency tests
$LB_h(A,i)$	time bound adjustment for input negation consistency tests
$LCT_{max}(A)$	maximal latest completion time in A : $LCT_{max}(A) = \max_{i \in A}\{lct_i\}$
$B_{(t_1,t_2)}$	subset of tasks which must be processed completely or partially in the time interval $[t_1,t_2)$: $B_{(t_1,t_2)} = \{\, i \in B \mid p_i(t_1,t_2) > 0 \,\}$
$p(A)$	sum of processing times in A : $p(A) = \sum_{i \in A} p_i$
$p(A,t_1,t_2)$	sum of interval processing times in A in the time interval $[t_1,t_2)$: $p(A,t_1,t_2) = \sum_{i \in A} p_i(t_1,t_2)$
$T(A)$	task set of A : $T(A) = T(EST_{min}(A), LCT_{max}(A))$
$T(t_1,t_2)$	task set: $T(t_1,t_2) = \{\, i \in B \mid t_1 \le est_i, \; lct_i \le t_2 \,\}$

Table 13.4.1: *List of symbols.*

Our goal is to examine which domain consistency tests have to be applied in order to establish strong 3-*b*-consistency which is also known as *lower-level bound-consistency*. 1-*b*-consistency is trivially established, since unary constraints are not involved, so only 2-*b*- and 3-*b*-consistency remain to be studied.

The corresponding consistency tests will be derived through an elementary and systematic evaluation of all constraints. This "bottom up" approach is quite technical, but it closes the gap that is usually left by the consistency tests which are due to the researcher's inspiration and insight into the problem's nature. As a consequence, we will rediscover most of these consistency tests which have been "derived" in a "top down" fashion in a slightly stronger version.

2-b-Consistency

In order to test for 2-*b*-consistency, pairs of different tasks have to be examined. If T_i, $i \in B$ is a task and $st_i \in \{est_i, lst_i\}$ an assignment of its start time, then st_i is (currently) consistent and cannot be removed if there exists another task T_j, $j \in B$, and an assignment $st_j \in \delta_j$, such that st_i and st_j satisfy the disjunctive constraint $i \leftrightarrow j$:

$$\exists\, st_j \in \delta_j : st_i + p_i \le st_j \ \lor \ st_j + p_j \le st_i . \tag{13.4.3}$$

Of course, if (13.4.3) is satisfied for all pairs (i,j) then 2-*b*-consistency is established. Since $\delta_j = [est_j, lst_j]$, this condition can be simplified as follows:

$$st_i + p_i \le lst_j \ \lor \ est_j + p_j \le st_i . \tag{13.4.4}$$

Suppose now that 2-*b*-consistency is not yet established. We will first show how to derive a well-known consistency test which removes an inconsistent assignment $st_i = est_i$ through a simple evaluation of (13.4.4). Similar arguments lead to a consistency test for removing the assignment $st_i = lst_i$. These consistency tests have been first proposed by Carlier and Pinson [CP89]. Obviously, if (13.4.4) is not satisfied for $st_i = est_i$ then we can remove est_i, i.e.

$$est_i + p_i > lst_j \ \land \ est_j + p_j > est_i \ \Rightarrow \ est_i = est_i + 1. \tag{13.4.5}$$

Observe that after adjusting est_i, the condition $est_i + p_i > lst_j$ on the left side of (13.4.5) is still satisfied. Therefore, we can increase est_i as long as $est_j + p_j > est_i$, i.e. until $est_j + p_j \le est_i$. This leads to the improved consistency test

$$est_i + p_i > lst_j \ \Rightarrow \ est_i = \max\{est_i, est_j + p_j\}. \tag{13.4.6}$$

Analogously, testing $st_i = lst_i$ leads to the consistency test

$$est_j + p_j > lst_i \ \Rightarrow \ lst_i = \min\{lst_i, lst_j - p_i\}. \tag{13.4.7}$$

Let Γ_2 be the set of consistency tests defined by (13.4.6) and (13.4.7) for all tasks $T_i \ne T_j$. The next lemma in combination with Theorem 13.2.15 ensures that there exists a unique fixed point $\Gamma_2(\Delta)$, i.e. applying the consistency tests in Γ_2 in an arbitrary order until no more updates are possible will always result in the same set of current domains.

Lemma 13.4.2

Γ_2 *is a set of monotonous consistency tests.*

Proof. For reasons of symmetry, it is sufficient to examine the consistency tests given by (13.4.6). Let $\Delta = \{ [est_l, lst_l] \mid l \in B \}$ and $\Delta' = \{ [est_l', lst_l'] \mid l \in B \}$. If $\Delta \subseteq \Delta'$, that is, $est_l' \le est_l$ and $lst_l \le lst_l'$ for all $l \in B$ then

$$est_i' + p_i > lst_j' \quad \Rightarrow \quad est_i + p_i > lst_j$$

$$\overset{(13.4.6)}{\Rightarrow} \quad est_i^* = \max\{ est_i, est_j + p_j \}$$

$$\Rightarrow \quad est_i^* \ge \max\{ est_i', est_j' + p_j \}$$

$$\Rightarrow \quad est_i^* \ge est_i'^*$$

As all other earliest and latest start times remain unchanged, $est_l'^* \le est_l^*$ and $lst_l^* \le lst_l'^*$ for all $l \in B$ which proves the monotony property. □

Altogether, the following theorem has been proven, see also [Nui94].

Theorem 13.4.3

For all $\Delta \in \Theta$, $\Gamma_2(\Delta)$ is 2-b-consistent. □

Example 13.4.4

Consider the situation that has been depicted in Figure 13.4.1. Since $est_i + p_i = 6 > 5 = lst_j$, we can adjust $est_i = \max\{est_i, est_j + p_j\} = \max\{2,3\} = 3$ according to (13.4.6). Note that the current domain of task T_j remains unchanged if (13.4.7) is applied. □

$\Gamma_2(\Delta)$ can be computed by repeatedly testing all pairs $i,j \in B$, $i \ne j$, until no more updates are possible. We will discuss other algorithms which subsume the tests for 2-b-consistency at a later time. As a generalization of the pair test Focacci and Nuijten [FN00] have proposed two consistency tests for shop scheduling, with sequence dependent setup times between pairs of tasks processed by the same disjunctive resource.

3-b-Consistency

In order to test for 3-b-consistency, triples of pairwise different tasks have to be examined. Again, let T_i, $i \in B$, be a task, and $st_i \in \{est_i, lst_i\}$. The start time st_i is (currently) consistent and cannot be removed if there exist $j, k \in B$, such that $i, j,$ k are indices of pairwise different tasks, and there exist assignments $st_j \in \delta_j$, $st_k \in \delta_k$, such that st_i, st_j, and st_k satisfy the disjunctive constraints $i \leftrightarrow j$, $i \leftrightarrow k$, and $j \leftrightarrow k$. Let us first consider this condition for $st_i = est_i$:

$$\exists\, st_j \in \delta_j,\, \exists\, st_k \in \delta_k : \begin{cases} (est_i + p_i \leq st_j \ \lor \ st_j + p_j \leq est_i) \ \land \\ (est_i + p_i \leq st_k \ \lor \ st_k + p_k \leq est_i) \ \land \\ (st_j + p_j \leq st_k \ \lor \ st_k + p_k \leq st_j). \end{cases} \quad (13.4.8)$$

Again, if (13.4.8) is satisfied for all triples (i,j,k) then 3-b-consistency is established. This condition is equivalent to

$$\exists\, st_j \in \delta_j,\, \exists\, st_k \in \delta_k : \begin{cases} (est_i + p_i \leq st_j \ \land \ st_j + p_j \leq st_k) \ \lor \\ (est_i + p_i \leq st_k \ \land \ st_k + p_k \leq st_j) \ \lor \\ (st_j + p_j \leq est_i \ \land \ est_i + p_i \leq st_k) \ \lor \\ (st_k + p_k \leq est_i \ \land \ est_i + p_i \leq st_j) \ \lor \\ (st_j + p_j \leq st_k \ \land \ st_k + p_k \leq est_i) \ \lor \\ (st_k + p_k \leq st_j \ \land \ st_j + p_j \leq est_i). \end{cases} \quad (13.4.9)$$

Each line of (13.4.9) represents a permutation of the tasks T_i, T_j, T_k, e.g. the first line corresponds to the sequence $i \to j \to k$. Since $\delta_j = [est_j, lst_j]$ and $\delta_k = [est_k, lst_k]$, (13.4.9) is equivalent to:

$$\exists\, st_j \in \delta_j,\, \exists\, st_k \in \delta_k : \begin{cases} (est_i + p_i \leq st_j \quad\ \land \ st_j + p_j \leq lst_k) \ \lor & (i) \\ (est_i + p_i \leq st_k \quad\ \land \ st_k + p_k \leq lst_j) \ \lor & (ii) \\ (est_j + p_j \leq est_i \ \land \ est_i + p_i \leq lst_k) \ \lor & (iii) \\ (est_k + p_k \leq est_i \ \land \ est_i + p_i \leq lst_j) \ \lor & (iv) \\ (est_j + p_j \leq st_k \quad\ \land \ st_k + p_k \leq est_i) \ \lor & (v) \\ (est_k + p_k \leq st_j \ \land \ st_j + p_j \leq est_i). & (vi) \end{cases}$$
$$(13.4.10)$$

In analogy to the case of establishing 2-b-consistency, we can increase $est_i := est_i + 1$ if (13.4.10) is not satisfied. However, in spite of the previous simplifications, testing (13.4.10) still is too costly, since the expression on the right side has to be evaluated for all $st_j \in \delta_j$ and $st_k \in \delta_k$. In the following lemmas, we therefore replace the conditions (i), (ii), (v) and (vi) which either contain st_j or st_k with simpler conditions.

Lemma 13.4.5

If Δ is 2-b-consistent and the conditions (iii) and (vi) are not satisfied then the following equivalence relations hold:

$$\exists\, st_j \in \delta_j,\, \exists\, st_k \in \delta_k : \begin{cases} (est_i + p_i \leq st_j \quad\ \land \ st_j + p_j \leq lst_k) \ \lor & (i) \\ (est_i + p_i \leq st_k \quad\ \land \ st_k + p_k \leq lst_j) & (ii) \end{cases}$$
$$(13.4.11)$$

$$\Leftrightarrow\ est_i + p_i + p_j \leq lst_k\ \vee\ est_i + p_i + p_k \leq lst_j \tag{13.4.12}$$

$$\Leftrightarrow\ \max\{\,lct_j - est_i\,,\ lct_k - est_i\,\} \geq p_i + p_j + p_k \tag{13.4.13}$$

Proof. Let us prove the first equivalence. The direction $\Rightarrow$ is obvious, so only $\Leftarrow$ has to be shown. Let (13.4.12) be satisfied. Without loss of generality, we can assume that either (a) $est_i + p_i + p_j \leq lst_k$ and $est_i + p_i + p_k > lst_j$, or that (b) $lst_k \geq lst_j$ if both, $est_i + p_i + p_j \leq lst_k$ and $est_i + p_i + p_k \leq lst_j$. Studying the two cases $est_i + p_i \geq est_j$ and $est_i + p_i < est_j$ separately, we can show that in both cases there exists $st_j \in \delta_j$, such that condition (*i*) is satisfied.

Case 1: Let $est_i + p_i \geq est_j$. If we can prove that $est_i + p_i \leq lst_j$ then choosing $st_j := est_i + p_i$ is possible, as then $st_j \in [est_j\,,\ lst_j] = \delta_j$, $est_i + p_i \leq st_j$ and $st_j + p_j = est_i + p_i + p_j \leq lst_k$. Thus, condition (*i*) is satisfied. In order to prove $est_i + p_i \leq lst_j$, we use the assumption that condition (*iii*) is not satisfied, i.e. that $est_j + p_j > est_i$ or $est_i + p_i > lst_k$. It follows from $est_i + p_i < est_i + p_i + p_j \leq lst_k$ that the second inequality cannot be satisfied, so that actually $est_j + p_j > est_i$. Thus, indeed, $est_i + p_i \leq lst_j$, as we have assumed 2-b-consistency (see (13.4.6)).

Case 2: Let $est_i + p_i \leq est_j$. If $est_j + p_j \leq lst_k$, setting $st_j := est_j \in \delta_j$ again satisfies condition (*i*). We now have to show that, in fact, $est_j + p_j \leq lst_k$. Again, we will use the assumption that 2-b-consistency is established. If $est_j + p_j > lst_k$ then (13.4.7) implies $lst_k \leq lst_j - p_k$ and $lst_k < lst_j$. Further, as $est_i + p_i + p_j \leq lst_k \leq lst_j - p_k$ we can conclude $est_i + p_i + p_k \leq lst_j$. So both inequalities of (13.4.12) are satisfied, but $lst_k < lst_j$. This is a contradiction to the assumption (b).

The second equivalence is easily proven by adding p_k and p_j, respectively, on both sides of inequalities (13.4.12) . $\qquad\square$

Lemma 13.4.6

If Δ is 2-b-consistent then the following equivalence relations hold:

$$\exists\, st_j \in \delta_j,\ \exists\, st_k \in \delta_k: \left\{ \begin{array}{ll} (est_i + p_j \leq st_k\ \wedge\ st_k + p_k \leq est_i) & \vee \quad (v) \\ (est_k + p_k \leq st_j\ \wedge\ st_j + p_j \leq est_i) & \quad (vi) \end{array} \right. \tag{13.4.14}$$

$$\Leftrightarrow \quad \begin{array}{l} est_i \geq \max\{est_j + p_j + p_k,\ est_k + p_k\}\ \vee \\ est_i \geq \max\{est_k + p_k + p_j,\ est_j + p_j\} \end{array} \tag{13.4.15}$$

$$\Leftrightarrow\quad est_i \geq \max\{\min\{est_j, est_k\} + p_j + p_k,\ est_j + p_j,\ est_k + p_k\} \tag{13.4.16}$$

Proof. We prove the first equivalence. Again, the direction $\Rightarrow$ is obvious, so we only have to show $\Leftarrow$. Let (13.4.15) be satisfied. We assume without loss of generality that $est_j \leq est_k$. This implies $\max\{est_k + p_k + p_j,\ est_j + p_j\} \geq est_k + p_k + p_j \geq \max\{est_j + p_j + p_k,\ est_k + p_k\}$, so that $est_i \geq \max\{est_j + p_j + p_k,\ est_k + p_k\}$ (*).

Case 1: Let $est_j + p_j \geq est_k$. If $est_j + p_j > lst_k$ then the 2-b-consistency (13.4.6) implies $est_j \geq est_k + p_k$ and $est_j \geq est_k$ which is a contradiction, so that actually $est_j + p_j < lst_k$. We can set $st_k := est_j + p_j \in [est_k, lst_k] = \delta_k$, and condition ($v$) is satisfied due to (∗).

Case 2: Let $est_j + p_j < est_k$. Choosing $st_k := est_k \in \delta_k$ again satisfies condition (v) due to (∗). A standard proof verifies the second equivalence.   □

Given that 2-b-consistency is established, we can therefore replace (13.4.10) with the following equivalent and much simpler condition which can be tested in constant time:

$$(\max\{lct_j - est_i, lct_k - est_i\} \geq p_i + p_j + p_k) \ \vee \qquad\qquad (i + ii)$$
$$(est_j + p_j \leq est_i \ \wedge \ est_i + p_i \leq lst_k) \ \vee \qquad\qquad\qquad (iii)$$
$$(est_k + p_k \leq est_i \ \wedge \ est_i + p_i \leq lst_j) \ \vee \qquad\qquad\qquad (iv)$$
$$(est_i \geq \max\{\min\{est_j, est_k\} + p_j + p_k, est_j + p_j, est_k + p_k\}) \,. \qquad (v + vi)$$
$$(13.4.17)$$

Resuming our previous thoughts, we can increase $est_i := est_i + 1$ if (13.4.17) is not satisfied. Observe that if ($i + ii$) is not satisfied *before* increasing est_i then it is not satisfied *after* increasing est_i. Therefore, we can proceed as follows: first, test whether ($i + ii$) holds. If this is not the case then increase est_i until one of the conditions (iii), (iv) or ($v + vi$) is satisfied. Fortunately, this incremental process can be accelerated by defining appropriate time bound adjustments.

Deriving the correct time bound adjustments requires a rather lengthy and painstaking analysis which is provided in Section 13.6 (Appendix). At the moment, we will only present an intuitive development of the results which avoids the distraction of the technical details.

Two cases have to be distinguished. In the first case, increasing est_i will never satisfy conditions ($i + ii$), (iii) and (iv). This can be interpreted as the situation in which T_i can neither be processed at the first, nor at the second position, but must be processed after T_j and T_k. We then have to increase est_i until condition ($v + vi$) is satisfied. Notice that this is always possible by choosing est_i sufficiently large, i.e. by setting

$$est_i := \max\{est_i, \min\{est_j, est_k\} + p_j + p_k, est_j + p_j, est_k + p_k\} \,.$$

However, it is possible to show that the seemingly weaker adjustment

$$est_i := \max\{est_i, \min\{est_j, est_k\} + p_j + p_k\}$$

is sufficient if it is combined with the tests for establishing 2-b-consistency or, more precisely, if after the application of this adjustment the 2-b-consistency tests are again applied. This leads to the following two consistency tests:

$$\max_{u \in \{i,j,k\},\, v \in \{j,k\},\, u \neq v} \{lct_v - est_u\} < p_i + p_j + p_k$$
$$\Rightarrow\ est_i := \max\{est_i,\, \min\{est_j, est_k\} + p_j + p_k\}, \tag{13.4.18}$$

$$est_i + p_i > \max\{lst_j, lst_k\}$$
$$\Rightarrow\ est_i := \max\{est_i,\, \min\{est_j, est_k\} + p_j + p_k\}. \tag{13.4.19}$$

It is both important to establish 2-*b*-consistency prior and after the application of these consistency tests, since the application of the latter test can lead to a 2-b-inconsistent state.

A generalization of these tests will be later described under the name input/output consistency tests. Trivial though it may seem, it should nevertheless be mentioned that the consistency tests (13.4.18) and (13.4.19) are not equivalent. Furthermore, observe that if the left side of (13.4.19) is satisfied then the consistency tests for pairs of tasks (13.4.6) can be applied to both (i,j) and (i,k), but may lead to weaker domain adjustments. We will give some examples which confirm these assertions.

Example 13.4.7

Consider the example depicted in Figure 13.4.2. Since

$$\max_{u \in \{i,j,k\},\, v \in \{j,k\},\, u \neq v} \{lct_v - est_u\} = 9 < 10 = p_i + p_j + p_k,$$

we can adjust $est_i := \max\{est_i,\, \min\{est_j, est_k\} + p_j + p_k\} = \max\{3,7\} = 7$ according to (13.4.18). By comparison, no deductions are possible using (13.4.19), as $est_i + p_i = 6 < 7 = \max\{lst_j, lst_k\}$. □

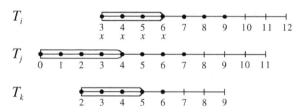

Figure 13.4.2 *Consistency test* (13.4.18).

Example 13.4.8

In Figure 13.4.3 another example is shown. Here, the consistency test (13.4.18) fails, as

$$\max_{u \in \{i,j,k\},\, v \in \{j,k\},\, u \neq v} \{lct_v - est_u\} = 9 = p_i + p_j + p_k.$$

The consistency test for pairs of tasks described in (13.4.6) can be applied to (i,j) and (i,k), but leaves est_j unchanged, since $est_j + p_j = est_k + p_k = 3 < 4 = est_i$. Only the consistency test (13.4.19) correctly adjusts $est_i := \max\{est_i,\, \min\{est_j,$

$est_k\} + p_j + p_k\} = \max\{4,6\} = 6.$ □

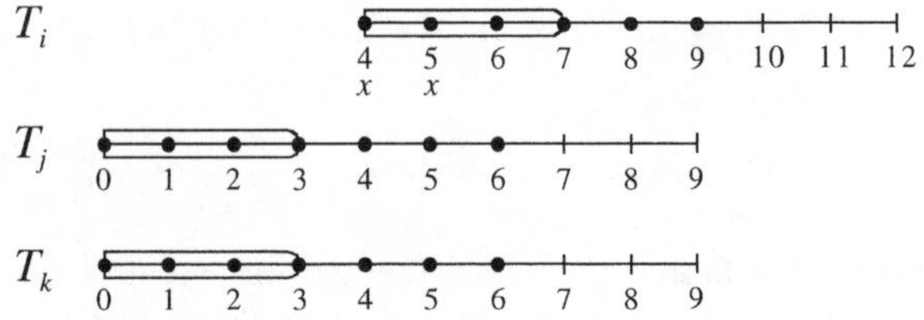

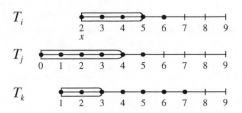

Figure 13.4.3 *Consistency test* (13.4.19).

Let us now turn to the second case in which the condition $(i + ii)$ is not satisfiable, but increasing est_i will eventually satisfy (iii) or (iv). This can be interpreted as the situation in which T_i cannot be processed first, but either $j \to i \to k$ or $k \to i \to j$ are feasible. The corresponding consistency test is as follows:

$$\max_{v \in \{j,k\}} \{lct_v - est_i\} < p_i + p_j + p_k$$
$$\Rightarrow \quad est_i := \max\{est_i, \min\{ect_j, ect_k\}\}. \tag{13.4.20}$$

A generalization of this test will be later described under the name input/output negation consistency test.

Example 13.4.9

Consider the example of Figure 13.4.4. No domain reductions are possible using the consistency tests (13.4.18) and (13.4.19). Since, however, $\max_{v \in \{j,k\}}\{lct_v - est_i\} = 7 < 9 = p_i + p_j + p_k$, we can adjust $est_i := \max\{est_i, \min\{ect_j, ect_k\}\} = \max\{2, 3\} = 3$ using the consistency test (13.4.20).   □

Figure 13.4.4 *Consistency test* (13.4.20).

The adjustments of the latest start times can be handled symmetrically. The same line of argumentation allows us to derive the following three consistency tests:

$$\max_{u \in \{j,k\}, \, v \in \{i,j,k\}, \, u \neq v} \{lct_v - est_u\} < p_i + p_j + p_k$$
$$\Rightarrow \quad lst_i := \min\{lst_i, \max\{lct_j, lct_k\} - p_j - p_k - p_i\}, \tag{13.4.21}$$

$$\min\{est_j + p_j, est_k + p_k\} > lst_i \tag{13.4.22}$$

$$\Rightarrow \quad lst_i := \min\{lst_i, \max\{lct_j, lct_k\} - p_j - p_k - p_i\},$$

$$\max_{u \in \{j,k\}} \{lct_i - est_u\} < p_i + p_j + p_k \tag{13.4.23}$$

$$\Rightarrow \quad lst_i := \min\{lst_i, \max\{lst_j, lst_k\} - p_i\}.$$

Let Γ_3 be the set of consistency tests defined in (13.4.18)-(13.4.23) for all pairwise different triples of tasks with indices $i, j, k \in B$, and let $\Gamma_{2,3} := \Gamma_2 \cup \Gamma_3$. It can be shown that all consistency tests in $\Gamma_{2,3}$ are monotonous, so $\Gamma_{2,3}(\Delta)$ is well defined. We have proven the following theorem.

Theorem 13.4.10

For all $\Delta \in \Theta$, $\Gamma_{2,3}(\Delta)$ is strongly 3-b-consistent. □

Notice that $\Gamma_3(\Gamma_2(\Delta))$ does not have to be strongly 3-b-consistent, since the application of some of the consistency tests in Γ_3 can result in current domains which are not 2-b-consistent. So, indeed, the consistency tests in Γ_2 and Γ_3 have to be applied in alternation.

Obviously, $\Gamma_{2,3}(\Delta)$ can be computed by repeatedly testing all pairwise different pairs and triples of tasks. However, as will be seen in the following sections, there exist more efficient algorithms.

13.4.4 Input/Output Consistency Tests

In the last section, domain consistency tests for pairs and triples of tasks have been described. It suggests itself to derive domain consistency tests for a greater number of tasks through a systematic evaluation of a greater number of disjunctive constraints. For the sake of simplicity, we will refrain from this rather technical approach and follow the historical courses which finally leads to the definition of these powerful consistency tests. Note, however, that we must not expect that the consistency tests derived will establish some higher level of bound-consistency, since great store has been set on an efficient implementation.

At first, we will present generalizations of the consistency tests (13.4.18) and (13.4.19). A closer look at these tests reveals that not only domain reductions but also processing orders of tasks can be deduced. It is convenient to first introduce these sequence consistency tests so as to simplify the subsequent proofs.

Sequence Consistency Tests

Given a subset of task indices $A \subsetneq B$ and an additional task T_i, $i \notin A$, Carlier and Pinson [CP89] were the first to derive conditions which imply that T_i has to be processed *before* or *after* all tasks T_j, $j \in A$. In the first case, they called i the *input* of A, in the second case, the *output* of A, and so the name *input/output conditions* seems justified.

Theorem 13.4.11 (*Input/Output Sequence Consistency Tests*).

Let $A \subsetneq B$ and $i \notin A$. If the input condition

$$\max_{u \in A,\, v \in A \cup \{i\},\, u \neq v} \{lct_v - est_u\} < p(A \cup \{i\}) \tag{13.4.24}$$

is satisfied then task T_i has to be processed before all tasks in A, for short, $i \rightarrow A$. Likewise, if the output condition

$$\max_{u \in A \cup \{i\},\, v \in A,\, u \neq v} \{lct_v - est_u\} < p(A \cup \{i\}) \tag{13.4.25}$$

is satisfied then task T_i has to be processed after all tasks in A, for short, $A \rightarrow i$.

Proof. If T_i is not processed before all tasks in A then the maximal amount of time for processing all tasks in $A \cup \{i\}$ is bounded by $\max_{u \in A,\, v \in A \cup \{i\},\, u \neq v} \{lct_v - est_u\}$. This leads to a contradiction if (13.4.24) is satisfied. Analogously, the second assertion can be shown. $\qquad\square$

The original definition of Carlier and Pinson is slightly weaker. It replaces the input condition with

$$LCT_{max}(A \cup \{i\}) - EST_{min}(A) < p(A \cup \{i\}). \tag{13.4.26}$$

Likewise, the output condition is replaced with

$$LCT_{max}(A) - EST_{min}(A \cup \{i\}) < p(A \cup \{i\}). \tag{13.4.27}$$

We will term these conditions the *modified input/output conditions.*. There are situations in which only the input/output conditions in their stricter form lead to a domain reduction. For a discussion of the computational complexity of algorithms that implement these tests see the end of Section 13.4.

Example 13.4.12

In Example 13.4.7 (see Figure 13.4.2), we have seen that

$$\max_{u \in \{i,j,k\},\, v \in \{j,k\},\, u \neq v} \{lct_v - est_u\} = 9 < 10 = p_i + p_j + p_k,$$

so that the output (13.4.25) implies $\{j,k\} \rightarrow i$. By comparison, the modified output condition is not satisfied since

$$LCT_{max}(\{j,k\}) - EST_{min}(\{i,j,k\}) = lct_j - est_j = 11 > 10 = p_i + p_j + p_k. \qquad\square$$

Domain Consistency Tests

Domain consistency tests that are based on the input/output conditions can now be simply derived. Here and later, we will only examine the adjustment of the earliest start times, since the adjustment of the latest start times can be handled analogously. Clearly, if i is the output of a subset A then T_i cannot start before all tasks of A have finished. Therefore, the earliest start time of T_i is at least $C_{max}(p_\delta(A))$, i.e. the makespan if all tasks in A are scheduled without preemption. Unfortunately, however, determining $C_{max}(p_\delta(A))$ requires the solution of the NP-hard single-machine scheduling problem [GJ79]. Thus, if the current domains are to be updated efficiently, we have to content ourselves with approximations of this bound. Some of these approximations are proposed in the next theorem which is a generalization of the consistency test (13.4.19) derived in the last subsection. This theorem is mainly due to Carlier and Pinson [CP90], Nuijten [Nui94], Caseau and Laburthe [CL95] and Martin and Shmoys [MS96]. The proof is obvious and is omitted.

Theorem 13.4.13 *(output domain consistency tests, part 1).*

If the output condition is satisfied for $A \subsetneq B$ and $i \notin A$ then the earliest start time of T_i can be adjusted to $est_i := \max\{est_i, LB_h(A)\}, h \in \{1,2,3,4\}$, where

(i) $LB_1(A) := \max_{u \in A}\{ect_u\}$,

(ii) $LB_2(A) := EST_{min}(A) + p(A)$,

(iii) $LB_3(A) := C_{max}(p_\delta^{pr}(A))$,

(iv) $LB_4(A) := C_{max}(p_\delta(A))$. $\square$

Dominance Relations

Let us compare the domain reductions that are induced by the output domain consistency tests and the different bounds. For each $h \in \{1,2,3,4\}$, we denote with $\Gamma_{out}(h) := \{ \gamma_{A,i}^{(h)} \mid A \subsetneq B, i \notin A \}$ the set of output domain consistency tests defined in Theorem 13.4.13:

$$\gamma_{A,i}^{(h)} := \max_{u \in A \cup \{i\}, v \in A, u \neq v} \{lct_v - est_u\} < p(A \cup \{i\}) \Rightarrow est_i := \max\{est_i, LB_h(A)\}.$$

Lemma 13.4.14

The following dominance relations hold:

1. $\Gamma_{out}(1) \trianglelefteq \Gamma_{out}(3) \trianglelefteq \Gamma_{out}(4)$,

2. $\Gamma_{out}(2) \trianglelefteq \Gamma_{out}(3) \trianglelefteq \Gamma_{out}(4)$.

Proof. As $LB_3(A) \le LB_4(A)$, the relation $\gamma_{A,i}^{(4)}(\Delta) \subseteq \gamma_{A,i}^{(3)}(\Delta)$ holds for all $A \subsetneq B$, $i \notin A$ and $\Delta \in \Theta$. Theorem 13.2.18 then implies that $\Gamma_{out}(3) \trianglerighteq \Gamma_{out}(4)$. Further, Carlier [Car82] has shown the following identity for the preemptive bound:

$$LB_3(A) = \max_{\emptyset \ne V \subseteq A} \{EST_{min}(V) + p(V) \}. \tag{13.4.28}$$

Since the maximum expression in (13.4.28) considers all single-elemented sets and A itself, $LB_1(A) \le LB_3(A)$ and $LB_2(A) \le LB_3(A)$. Again, using Theorem 13.2.18, we can conclude that $\Gamma_{out}(1) \trianglerighteq \Gamma_{out}(3)$ and $\Gamma_{out}(2) \trianglerighteq \Gamma_{out}(3)$. □

Intuitively, it seems natural to assume that $\Gamma_{out}(1)$ is strictly dominated by $\Gamma_{out}(3)$, while $\Gamma_{out}(3)$ is strictly dominated by $\Gamma_{out}(4)$. Indeed, this is true. Remember that, since $\Gamma_{out}(1) \trianglerighteq \Gamma_{out}(3)$ has already been shown, we only have to find an example in which $\Gamma_{out}(3)$ leads to a stronger domain reduction than $\Gamma_{out}(1)$ in order to verify $\Gamma_{out}(1) \prec \Gamma_{out}(3)$. The same naturally holds for $\Gamma_{out}(3)$ and $\Gamma_{out}(4)$.

Example 13.4.15

Consider the situation illustrated in Figure 13.4.5 with five tasks with indices i, j, k, l, m. The table in Figure 13.4.5 lists all feasible sequences and the associated schedules. Examining the start times of the feasible schedules shows that the domains $\delta_j, \delta_k, \delta_l, \delta_m$ cannot be reduced. Likewise, it can be seen that i is the output of $A = \{j, k, l, m\}$ with the earliest start time being $LB_4(A) = 10$. In fact, the output condition holds, as

$$\max_{u \in A \cup \{i\}, \, v \in A, \, u \ne v} \{lct_v - est_u\} = 10 < 11 = p(A \cup \{i\}),$$

so that we can adjust est_i using one of the bounds of Theorem 13.4.13. Apart from $LB_4(A) = 10$, it is possible to show that $LB_1(A) = 7$, $LB_2(A) = 9$ and $LB_3(A) = 9$. Obviously, $LB_1(A) < LB_3(A) < LB_4(A) = 10$. Notice that, after the adjustment of est_i, no other adjustments are possible if the same lower bound is used again, so that a fixed point is reached. This confirms the conjecture $\Gamma_{out}(1) \prec \Gamma_{out}(3) \prec \Gamma_{out}(4)$. □

It remains to classify $\Gamma_{out}(2)$. Comparing $LB_1(A)$ and $LB_2(A)$ shows that all three cases $LB_1(A) < LB_2(A)$, $LB_1(A) = LB_2(A)$ and $LB_1(A) > LB_2(A)$ can occur. Further, comparing $LB_2(A)$ and $LB_3(A)$ reveals that $LB_2(A) \le LB_3(A)$ and sometimes $LB_2(A) < LB_3(A)$. So we would presume that $\Gamma_{out}(1)$ and $\Gamma_{out}(2)$ are not comparable, while $\Gamma_{out}(2)$ is strictly dominated by $\Gamma_{out}(3)$. This time, however, our intuition fails, since in fact $\Gamma_{out}(2)$ and $\Gamma_{out}(3)$ are equivalent.

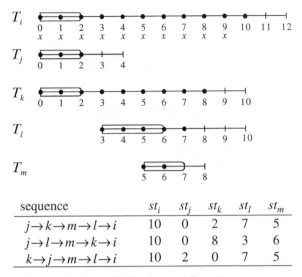

sequence	st_i	st_j	st_k	st_l	st_m
$j{\to}k{\to}m{\to}l{\to}i$	10	0	2	7	5
$j{\to}l{\to}m{\to}k{\to}i$	10	0	8	3	6
$k{\to}j{\to}m{\to}l{\to}i$	10	2	0	7	5

Figure 13.4.5 *Comparing $\Gamma_{out}(1)$, $\Gamma_{out}(3)$ and $\Gamma_{out}(4)$.*

Theorem 13.4.16 (*dominance relations for output consistency tests*). [DPP00]

$$\Gamma_{out}(1) \prec \Gamma_{out}(2) \sim \Gamma_{out}(3) \prec \Gamma_{out}(4).$$

Proof. We only have to prove $\Gamma_{out}(3) \trianglerighteq \Gamma_{out}(2)$. It is sufficient to show that for all $A \subsetneq B, i \notin A$ and all $\Delta \in \Theta$, one of the following cases applies:

(1) $\gamma_{A,i}^{(3)}(\Delta) = \gamma_{A,i}^{(2)}(\Delta)$,

(2) $\exists V \subsetneq A : \gamma_{A,i}^{(3)}(\Delta) = \gamma_{V,i}^{(2)}(\gamma_{A,i}^{(2)}(\Delta))$.

Once more, Theorem 13.2.18 will then lead to the desired result. Let us assume that the output condition (13.4.25) is satisfied for some $A \subsetneq B$ and $i \notin A$. We have to compare the bounds:

(i) $LB_2(A) = EST_{min}(A) + p(A)$,

(ii) $LB_3(A) = \max_{\varnothing \neq V \subseteq A} \{EST_{min}(V) + p(V)\}$,

If $LB_2(A) = LB_3(A)$ then $\gamma_{A,i}^{(2)}$ and $\gamma_{A,i}^{(3)}$ deduce the same domain reductions and case (1) applies. Let us therefore assume that $LB_2(A) < LB_3(A)$. Since the pre-emptive bound is determined by (13.4.28), there exists $V \subset A, V \neq \varnothing$, such that $LB_3(A) = EST_{min}(V) + p(V)$. Since $LB_2(A) < LB_3(A)$, this is equivalent to

$$EST_{min}(A) + p(A) < EST_{min}(V) + p(V). \tag{13.4.29}$$

Subtracting $p(V)$ from both sides yields

$$EST_{min}(A) + p(A - V) < EST_{min}(V) \tag{13.4.30}$$

The last inequality will be used at a later time. Assume now that est_i has been adjusted by applying $\gamma_{A,i}^{(2)}$. Note that this means that est_i is increased or remains unchanged. Thus, if the output condition is satisfied for the couple (A, i) *prior* the adjustment of est_i then it is satisfied *after* the adjustment, so that

$$\max_{u \in A \cup \{i\}, v \in A, u \neq v} \{lct_v - est_u^*\} < p(A \cup \{i\}) \tag{13.4.31}$$

still holds for $est_i^* := \max\{est_i, LB_2(A)\}$ and $est_u^* = est_u$ for all $u \neq i$. If we do not maximize over all but only a subset of values then we obtain a lower bound of the left side of this inequality and

$$\max_{u \in A, v \in V, u \neq v} \{lct_v - est_u^*\} < p(A \cup \{i\}) . \tag{13.4.32}$$

Rewriting $p(A \cup \{i\}) = p(V \cup \{i\}) + p(A - V)$ then leads to

$$\max_{u \in A, v \in V, u \neq v} \{lct_v - (est_u^* + p(A - V))\} < p(A \cup \{i\}) . \tag{13.4.33}$$

The left side of (13.4.33) can be simplified using the identity

$$\max_{u \in A, v \in V, u \neq v} \{lct_v - (est_u^* + p(A - V))\}$$
$$= \max_{v \in V} \{lct_v - (EST_{min}^*(A) + p(A - V))\}. \tag{13.4.34}$$

This is not apparent at once and requires some explanations. At first, the term on the left side of (13.4.34) seems to be less than or equal to the term on the right side, since $EST_{min}^*(A) \leq est_u^*$ for all $u \in A$. We now choose $u' \in A$ such that $est_{u'}^* = EST_{min}^*(A)$. If $u' \in V \subset A$ then $EST_{min}^*(V) = EST_{min}^*(A)$. Since the earliest start times of all tasks with indices in A did not change, this is a contradiction to (13.4.30). Thus, the left side of (13.4.34) assumes the maximal value for $u = u' \notin V$, and both terms are indeed identical. Therefore, (13.4.33) is equivalent to

$$\max_{v \in V} \{lct_v - (EST_{min}^*(A) + p(A - V))\} < p(V \cup \{i\}). \tag{13.4.35}$$

The left side of (13.4.35) can be approximated using (13.4.30) which tells us that for all $u \in V$:

$$est_u^* > EST_{min}^*(A) + p(A - V) \tag{13.4.36}$$

Likewise, we can deduce

$$est_i^* \geq LB_2(A) = EST_{min}^*(A) + p(A) > EST_{min}^*(A) + p(A - V) . \tag{13.4.37}$$

So, $EST_{min}^*(A) + p(A - V)$ in (13.4.35) can be replaced by est_u^* for all $u \in V \cup \{i\}$ which yields

$$\max_{u \in V \cup \{i\}, v \in V, u \neq v} \{lct_v - est_u^*\} < p(V \cup \{i\}) \tag{13.4.38}$$

Observe that this is nothing but the output condition for the couple (V, i).

Since $LB_2(V) = EST^*_{min}(V) + p(V) = LB_3(A)$, a subsequent application of $\gamma^{(2)}_{V,i}$ leads to the same domain reduction and the second case (2) applies. This completes our proof.   □

Sequence Consistency Tests Revisited

It has already been mentioned that applying both sequence and domain consistency tests together can lead to better search space reductions. Quite evidently, any domain reductions deduced by Theorem 13.4.13 can lead to additional edge orientations deduced by Theorem 13.4.11. We will now discuss the case in which the inverse is also true.

Imagine a situation in which $A \to i$ can be deduced for a subset of tasks, but in which the output condition does not hold for the couple (A,i). Such a situation can actually occur as has, for instance, been shown in Example 13.4.8 for the three tasks T_i, T_j, T_k: while $j \to i$ and $k \to i$ can be separately deduced without, however, implying a domain reduction, the output condition fails for the couple $(\{j,k\},i)$. This motivates the following obvious theorem as an extension of Theorem 13.4.13.

Theorem 13.4.17 (*Input/Output Domain Consistency Tests, part 2*).

Let $A \subsetneq B$ and $i \notin A$. If $A \to i$ then the earliest start time of task T_i can be adjusted to $est_i := \max\{est_i, LB_h(A)\}$, $h \in \{1,2,3,4\}$.   □

Algorithms and Implementation Issues

An important question to answer now is whether there exist efficient algorithms that implement the input/output consistency tests. There are two obstacles which have to be overcome: the computation of the domain adjustments and the detection of the couples (A,i) which satisfy the input/output conditions.

Regarding the former, computing the non-preemptive bound is ruled out due to the NP-hardness result. At the other extreme, the "earliest completion time bound" (LB_1) is a too weak approximation. Therefore, only the "sum bound" (LB_2) or the preemptive bound (LB_3) remain candidates for the domain adjustments. Recall that both bounds are equivalent with respect to the induced Δ-fixed-point. Regarding the computational complexity, however, the two bounds are quite different: on the one hand, computing LB_2 requires linear time complexity $O(|A|)$ in contrast to the $O(|A| \log |A|)$ time complexity for computing LB_3. On the other hand, establishing the Δ-fixed-point, LB_2 usually has to be computed more often than LB_3, and it is not clear which factor - complexity of bound computation or number of iterations - dominates the other.

Let us turn to the second problem. An efficient implementation of the input/output consistency tests is obviously not possible if all pairs (A,i) of subsets

$A \subsetneq B$ and tasks T_i, $i \notin A$ are to be tested separately. Fortunately, it is not necessary to do so as has been first shown by Carlier and Pinson [CP90]. They developed an $O(n^2)$ algorithm (with $n = |B|$) which deduces all edge orientations and all domain reductions that are implied by the modified input/output conditions and the preemptive bound adjustment[3]. The fundamental idea was to test the modified input/output conditions and to compute the preemptive bound adjustments *simultaneously*. Several years later, Carlier and Pinson [CP94] and Brucker et al. [BJK94] presented $O(n \log n)$ algorithms which until now have the best asymptotic performance, but require quite complex data structures.

Nuijten [Nui94], Caseau and Laburthe [CL95] and Martin and Shmoys [MS96] have chosen a solely domain oriented approach and proposed different algorithms for implementing Theorem 13.4.13 based again on the modified input/output conditions. Nuijten developed an $O(n^2)$ algorithm which as well can be applied to scheduling problems with discrete resource capacity. Caseau and Laburthe presented an $O(n^3)$ algorithm based on the concept of *task sets* which works in an incremental fashion, so that $O(n^3)$ is a seldom worst case. The algorithm introduced by Martin and Shmoys [MS96] has a time complexity of $O(n^2)$.

An $O(n^3)$ algorithm which deduces all edge orientations implied by Theorem 13.4.11 has been derived by Phan Huy [Pha00]. He also presents an $O(n^2 \log n)$ for deriving all domain adjustments implied by Theorem 13.4.17.

13.4.5 Input/Output Negation Consistency Tests

In the last subsection, conditions have been described which imply that a task has to be processed before (after) another set of tasks. In this subsection, the inverse situation that a task *cannot* be processed first (last) is studied.

Sequence Consistency Tests

The following theorem is due to Carlier and Pinson [CP89]. For reasons near at hand, we have chosen the name input/output negation for the conditions described in this theorem.

Theorem 13.4.18 (*Input/Output Negation Sequence Consistency Tests*).

Let $A \subsetneq B$ and $i \notin A$. If the input negation condition

---

[3] It is common practice to only report the time complexity for applying all consistency tests *once*. In general, the number of iterations necessary for computing the Δ-fixed-point has to be considered as well. In the worst case, this accounts for an additional factor c which depends upon the size of the current domains. In practice, however, c is a rather small constant.

$$LCT_{max}(A) - est_i < p(A \cup \{i\}) \tag{13.4.39}$$

is satisfied then task T_i cannot be processed before all tasks $T_j, j \in A$. Likewise, if the output negation condition

$$lct_i - EST_{min}(A) < p(A \cup \{i\}) \tag{13.4.40}$$

is satisfied then task T_i cannot be processed after all other tasks $T_j, j \in A$.

Proof. If T_i is processed before $T_j, j \in A$ then all tasks with indices in A have to be processed within the time interval $[est_i, LCT_{max}(A))$. This leads to a contradiction if (13.4.39) is satisfied. The second assertion can be shown analogously. $\square$

The input/output negation conditions are a relaxation of the input/output conditions and so are more often satisfied. However, the conclusions drawn in Theorem 13.4.18 are usually weaker than those drawn in Theorem 13.4.11, except for A contains a single task[4]. An important issue is therefore the development of strong domain reduction rules based on the limited information deduced.

Domain Consistency Tests

We will only study the input negation condition and the adjustments of earliest start times. Let us suppose that (13.4.39) is satisfied for $A \subsetneq B$ and $i \notin A$. Since, then, T_i cannot be processed before all tasks $T_j, j \in A$, there must be a task in A which starts and finishes before T_i, although we generally do not know which one. Thus, a lower bound of the earliest start time of T_i is

$$LB_5(A, i) = \min_{u \in A} \{ect_u\} \tag{13.4.41}$$

Caseau and Laburthe [CL95] made the following observation: if T_i cannot be processed first then, in any feasible schedule, there must exist a subset $\emptyset \neq V \subseteq A$, so that $V \rightarrow i \rightarrow A - V$. As a necessary condition, this subset V has to satisfy

$$LCT_{max}((A - V) \cup \{i\}) - EST_{min}(V) \geq p(A \cup \{i\}) . \tag{13.4.42}$$

Consequently, they proposed

$$LB_6(A, i) = \min_{\emptyset \neq V \subseteq A} \{ LB_2(V) \mid V \text{ satisifies } (13.4.42) \} \tag{13.4.43}$$

as a lower bound for the earliest start time of T_i. Notice, however, that if V satisfies (13.4.42) then the one-elemented set $V' := \{u\} \subseteq V$ with $est_u = EST_{min}(V)$ satisfies (13.4.42) as well. Further, $LB_2(V) = EST_{min}(V) + p(V) = est_u + p(V)$

---

[4] In this case, the input/output negation sequence consistency test coincides with the input/output sequence consistency test for pairs of operations.

$\geq est_u + p_u = LB_2(V')$, so that the expression in (13.4.43) is minimal for a one-element set. Therefore, setting $A_u := (A - \{u\}) \cup \{i\}$ we can rewrite

$$LB_6(A,i) = \min_{u \in A}\{ ect_u \mid LCT_{max}(A_u) - est_u \geq p(A_u \cup \{u\})\} \qquad (13.4.44)$$

This bound has a quite simple interpretation: the minimal earliest completion time is only chosen among all tasks which do not satisfy the input negation condition, because those who do, cannot start at the first position.

Up to now, est_i has been adjusted to the earliest completion time of some single task. The time bound adjustment can be improved if a condition is derived that detects a situation in which more than one task have to be processed before T_i. Observe that if for a subset $\emptyset \neq V \subseteq A$ the sequence $V \rightarrow i \rightarrow A - V$ is feasible then the following condition must hold:

$$LCT_{max}((A - V) \cup \{i\}) - est_i \geq p((A - V) \cup \{i\}) . \qquad (13.4.45)$$

This implies the lower bounds on the earliest start time:

$$LB_7(A,i) := \min_{\emptyset \neq V \subseteq A} \{ LB_2(V) \mid V \text{ satisfies } (13.4.45)\} \qquad (13.4.46)$$

$$LB_8(A,i) := \min_{\emptyset \neq V \subseteq A} \{ LB_3(V) \mid V \text{ satisfies } (13.4.45)\} \qquad (13.4.47)$$

Finally, we can try to find the exact earliest start time of task T_i by computing

$$LB_9(A,i) := \min_{\emptyset \neq V \subseteq A} \{ LB_4(V) \mid V \rightarrow i \rightarrow A - V \text{ is feasible}\} . \qquad (13.4.48)$$

The following theorem which is a generalization of the consistency test (13.4.20) summarizes the results derived above.

Theorem 13.4.19 (*Input/Output Negation Domain Consistency Tests*).

If the input negation condition is satisfied for $A \subsetneq B$ and $i \notin A$ then the earliest start time of task T_i can be adjusted to $est_i := \max\{est_i, LB_h(A,i)\}$, $h \in \{5,6,7, 8,9\}$.

Dominance Relations

For $h \in \{5,6,7,8,9\}$, let $\Gamma_{\neg in}(h) := \{ \gamma_{A,i}^{(h)} \mid A \subsetneq B, i \notin A\}$ denote the set of input negation domain consistency tests defined in Theorem 13.4.19:

$$\gamma_{A,i}^{(h)} : LCT_{max}(A) - est_i < p(A \cup \{i\}) \Rightarrow est_i := \max\{est_i, LB_h(A,i)\} .$$

Lemma 13.4.20

The following dominance relations hold:

  1. $\Gamma_{\neg in}(5) \trianglelefteq \Gamma_{\neg in}(6) \trianglelefteq \Gamma_{\neg in}(9),$

2. $\Gamma_{\neg in}(5) \preceq \Gamma_{\neg in}(7) \preceq \Gamma_{\neg in}(8) \preceq \Gamma_{\neg in}(9)$.

Lemma 13.4.21

$\Gamma_{\neg in}(5) \sim \Gamma_{\neg in}(6)$.

Proof. We only have to prove that $\Gamma_{\neg in}(6) \preceq \Gamma_{\neg in}(5)$. It is sufficient to show that
for all $A \subsetneq B$, $i \notin A$ and $\Delta \in \Theta$, there exist $A^1, \cdots, A^r \subsetneq B$ such that

$$(\gamma_{A^r,i}^{(5)} \circ \cdots \circ \gamma_{A^1,i}^{(5)})(\Delta) \subseteq \gamma_{A,i}^{(6)}(\Delta) \tag{13.4.49}$$

For the sake of simplicity, we omit an exact proof but only describe the basic
ideas. Let $U \subseteq A$ denote the index set of tasks satisfying the input negation condition, i.e. $U := \{ u \in A \mid LCT_{max}(A_u) - est_u < p(A_u \cup \{u\}) \}$ with $A_u := (A - \{u\}) \cup \{i\}$.
$\{i\}$.
 Recall that

(i) $LB_5(A,i) = \min_{u \in A} \{ect_u\}$,

(ii) $LB_6(A,i) = \min_{u \in A - U} \{ect_u\}$.

If both bounds are identical then, obviously, $\gamma_{A,i}^{(6)}(\Delta) = \gamma_{A,i}^{(5)}(\Delta)$. This identity, for
instance, holds if U is empty. Thus, in the following, we restrict our attention to
the case $|U| > 0$. If $u \in A$ is a task satisfying $ect_u = LB_5(A,i) < LB_6(A,i)$ then
$u \in U$ and

$$est_u + p(A_u \cup \{u\}) = ect_u + p(A_u) > LCT_{max}(A_u) .$$

If the earliest start time of T_i has been adjusted to $est_i^* := \max\{est_i, LB_5(A,i)\}$ by
applying $\gamma_{A,i}^{(5)}$ then we have $est_i^* \geq ect_u$, so

$$est_i^* + p(A_u) > LCT_{max}(A_u) \geq LCT_{max}(A_u - \{i\})$$

or

$$est_i^* + p((A - \{u\}) \cup \{i\}) > LCT_{max}(A - \{u\})$$

which is the input negation condition for the couple $(A - \{u\}, i)$. Therefore, est_i^*
can be adjusted once more to $LB_5(A - \{u\}, i)$. If $LB_5(A - \{u\}, i) = LB_6(A - \{u\}, i)$
then we are done, since $LB_6(A - \{u\}, i) \geq LB_6(A,i)$. Otherwise, we are in the
same situation as above which allows us to continue in the same manner. Finally,
observe that the number of adjustments is finite and bounded by $|A|$. $\square$

Example 13.4.22

Consider the example shown in Figure 13.4.6 with four tasks indexed as i, j, k, l.
A closer look at the set of feasible schedules reveals that δ_j, δ_k and δ_l cannot be
reduced. Likewise, it can be seen that i cannot be the input of $A = \{j, k, l\}$ which

is detected by the input negation condition, since $LCT_{max}(A) - est_i = 11 - 5 < 11 = p(A \cup \{i\})$. Using LB_5, no time bound adjustment is possible, since $LB_5(A,i) = 3$. However, there exists no feasible schedule in which only one task is processed before T_i. Indeed, $LB_7(A,i) = 6$ leads to a stronger time bound adjustment. After the domain reduction, a fixed point is reached, so this example and Lemma 13.4.20 prove that $\Gamma_{\neg in}(5) \prec \Gamma_{\neg in}(7)$.   □

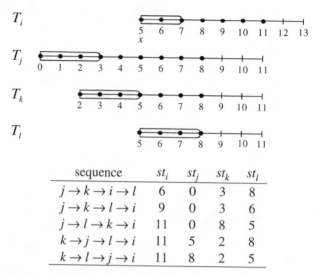

sequence	st_i	st_j	st_k	st_l
$j \to k \to i \to l$	6	0	3	8
$j \to k \to l \to i$	9	0	3	6
$j \to l \to k \to i$	11	0	8	5
$k \to j \to l \to i$	11	5	2	8
$k \to l \to j \to i$	11	8	2	5

Figure 13.4.6 *Comparing $\Gamma_{\neg in}(5)$ and $\Gamma_{\neg in}(7)$.*

Lemma 13.4.23

$\Gamma_{\neg in}(7) \sim \Gamma_{\neg in}(8)$.

Proof. Similar to Theorem 13.4.16.   □

Example 13.4.24

Consider the situation in Figure 13.4.7 with five tasks indexed as i, j, k, l, m. Again, $\delta_j, \delta_k, \delta_l$ and δ_m cannot be reduced. Further, it can be seen that i is the output of $A = \{j,k,l,m\}$ with the earliest start time being $LB_9(A,i) = 9$. However, the output condition is not satisfied for the couple (A,i). The input negation condition holds, since $LCT_{max}(A) - est_i = 11 - 1 < 11 = p(A \cup \{i\})$, but $LB_h(A,i) = 1$ for all $h \in \{5,6,7,8\}$. Thus, the current domain of T_i remains unchanged if these bound adjustments are applied, i.e. a fixed point is reached. This and Lemma 13.4.20 prove the relation $\Gamma_{\neg in}(8) \prec \Gamma_{\neg in}(9)$.   □

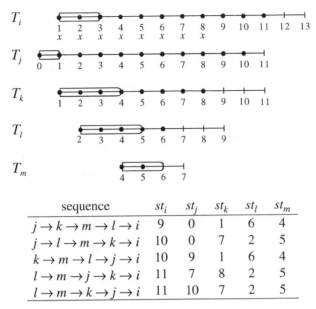

sequence	st_i	st_j	st_k	st_l	st_m
$j \to k \to m \to l \to i$	9	0	1	6	4
$j \to l \to m \to k \to i$	10	0	7	2	5
$k \to m \to l \to j \to i$	10	9	1	6	4
$l \to m \to j \to k \to i$	11	7	8	2	5
$l \to m \to k \to j \to i$	11	10	7	2	5

Figure 13.4.7 Comparing $\Gamma_{\neg in}(8)$ and $\Gamma_{\neg in}(9)$.

Altogether, we have proven the following theorem.

Theorem 13.4.25 (*dominance relations for input negation consistency tests*).

$$\Gamma_{\neg in}(5) \sim \Gamma_{\neg in}(6) \prec \Gamma_{\neg in}(7) \sim \Gamma_{\neg in}(8) \prec \Gamma_{\neg in}(9) . \qquad \square$$

Algorithms and Implementation Issues

Input negation consistency tests which use the "simple earliest completion time bound" (LB_5) as time bound adjustment and their output negation counterparts have been applied by Nuijten [Nui94], Baptiste and Le Pape [BL95] and Caseau and Laburthe [CL95]. Caseau and Laburthe have integrated the tests in their scheduling environment based on task sets in a straightforward manner which yields an algorithm with time complexity $O(n^3)$. All these algorithms only test some, but not all interesting couples (A,i). An algorithm which deduces all domain reductions with time complexity $O(n^2)$ has only been developed by Baptiste and Le Pape [BL96]. A similar implementation is proposed by Phan Huy in [Pha00]. Nuijten and Le Pape [NL98] derived several consistency tests which are similar to the input/output negation consistency tests with the time bound adjustment LB_8 and can be implemented with time complexity $O(n^2 \log n)$ and $O(n^3)$ respectively.

13.4.6 Input-or-Output Consistency Tests

In this subsection, some new consistency tests are presented which are not sub-sumed by the consistency tests presented in the previous subsections. They are based on the input-or-output conditions which have been introduced by Dorndorf et al. [DPP99].

Domain and Sequence Consistency Tests

The input-or-output conditions detect situations in which either (a) a task T_i has to be processed first or (b) a task T_j has to be processed last within a set of tasks. There exists a sequence and a domain oriented consistency test based on the in-put-or-output condition. Both tests are summarized in the next theorem.

Theorem 13.4.26 (*input-or-output consistency tests*).

Let $A \subsetneq B$ and $i, j \notin A$. If the input-or-output condition

$$\max_{u \in A \cup \{j\}, v \in A \cup \{i\}, u \neq v} \{lct_v - est_u\} < p(A \cup \{i,j\}) \qquad (13.4.50)$$

is satisfied then either task T_i has to be processed first or task T_j has to be proc-essed last within $A \cup \{i,j\}$. If $i \neq j$ then task T_i has to be processed before T_j and the domains of T_i and T_j can be adjusted as follows:

$$est_j := \max\{est_j, est_i + p_i\},$$

$$lst_j := \min\{lst_i, lst_j - p_i\}.$$

Proof. If T_i is neither processed before, nor T_j processed after all other tasks in $A \cup \{i,j\}$ then all tasks in $A \cup \{i,j\}$ have to be processed within a time interval of maximal size

$$\max_{u \in A \cup \{j\}, v \in A \cup \{i\}, u \neq v} \{lct_v - est_u\}.$$

This is a contradiction to (13.4.50).

Now, since T_i has to be processed first or T_j processed last within $A \cup \{i,j\}$, we can deduce that T_i has to be processed before T_j if $i \neq j$. This immediately implies the domain deductions described above.   □

By substituting (13.4.50) with

$$LCT_{max}((A \cup \{i\}) - EST_{min}(A \cup \{j\}) < p(A \cup \{i,j\}), \qquad (13.4.51)$$

we obtain the *modified input-or-output conditions* which can be tested more eas-ily, but are less often satisfied than the input-or-output conditions.

Example 13.4.27

In Figure 13.4.8 an example for the application of the input-or-output consistency tests with four tasks indexed as i, j, k, l is shown.
Since

$$\max_{u \in \{j,k,l\}, v \in \{i,k,l\}, u \neq v} \{lct_v - est_u\} = 6 < 7 = p(\{i,j,k,l\})$$

we can conclude that T_i has to be processed before T_j. Thus, we can adjust $est_j :=$ 4 and $lst_i := 4$. $\qquad \square$

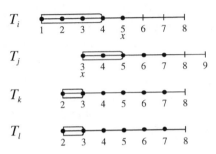

Figure 13.4.8 *The input-or-output consistency test.*

Algorithms and Implementation Issues

Deweß [Dew92] and Brucker et al. [BJK94] discuss conditions which examine all permutations of a fixed length r and which are thus called r-set conditions. Brucker et al. [BJK94] developed an $O(n^2)$ algorithm for testing all 3-set conditions which is equivalent to testing all input-or-output conditions for triples of tasks. Phan Huy [Pha00] developed an $O(n^3)$ algorithm for deriving all edge orientations implied by the modified input-or-output conditions. This algorithm can be generalized to an $O(n^4)$ algorithm which deduces all edge orientations implied by the input-or-output conditions.

13.4.7 Energetic Reasoning

The conditions described in the previous subsections for testing consistency were all founded on the principle of comparing a time interval in which a set of tasks A has to be processed with the total processing time $p(A)$ of these tasks. The time intervals chosen were defined through the earliest start and latest completion times of some of the tasks. This fundamental principle can be generalized by considering arbitrary time intervals $[t_1, t_2)$, on the one hand, and replacing simple processing time $p(A)$ with interval processing time $p(A, t_1, t_2)$, on the other hand. Erschler et al. [ELT91], see also [LEE92], were the first to introduce this idea under the name of *energetic reasoning*. Indeed, the interval processing time can be interpreted as resource energy demand which encounters a limited resource energy supply that is defined through the time interval. The original concept of Erschler et al. considered cumulative scheduling problems with discrete resource capacity. Their results have been improved by Baptiste and Le Pape [BL95] for disjunctive constraints. We will take a closer look at these results and compare them to the consistency tests described so far.

Interval Processing Time

Let us first define the interval processing time of a task T_i for a given time interval $[t_1, t_2)$, $t_1 < t_2$. The interval processing time $p_i(t_1, t_2)$ is the smallest amount of time during which T_i has to be processed within $[t_1, t_2)$. Figure 13.4.9 shows four possible situations: (1) T_i can be completely contained within the interval, (2) overlap the entire interval, (3) have a minimum processing time in the interval when started as early as possible or (4) have a minimum processing time when started as late as possible. The fifth situation not depicted applies whenever, given the current domains, T_i does not necessarily have to be processed within the given time interval. Consequently,

$$p_i(t_1, t_2) := \max\{\, 0,\, \min\{p_i, t_2 - t_1, ect_i - t_1, t_2 - lst_i\}\,\}. \qquad (13.4.52)$$

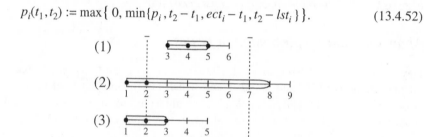

Figure 13.4.9 *Types of relations between a task and a time interval.*

The interval processing time of a subset of tasks A is given by $p(A, t_1, t_2) := \sum_{i \in A} p_i(t_1, t_2)$. Finally, let $B_{(t_1, t_2)} := \{\, i \in B \mid p_i(t_1, t_2) > 0 \,\}$ denote the set of tasks which have to be processed completely or partially within $[t_1, t_2)$.

Energetic Input/Output Consistency Tests

Baptiste and Le Pape [BL95] examined situations in which the earliest start time of a task T_i can be updated using the concept of interval processing times. Assume, for instance, that T_i finishes before t_2. The interval processing time of T_i in

$[t_1,t_2)$ would then be $p_i'(t_1,t_2) = \min\{p_i, t_2 - t_1, ect_i - t_1\}$.[5] However, if $t_2 - t_1$ $< p(B - \{i\}, t_1,t_2) + p_i'(t_1,t_2)$ then the assumption cannot be true, so that T_i has to finish after t_2. Baptiste and Le Pape showed that est_i can be then updated to

$$est_i := \max\{est_i, t_1 + p(B - \{i\}, t_1,t_2)\}. \tag{13.4.53}$$

A stronger domain reduction rule is presented in the following theorem.

Theorem 13.4.28 *Energetic output conditions.*

Let $i \in B$ and $t_1 < t_2$. If the energetic output condition

$$t_2 - t_1 < p(B - \{i\}, t_1,t_2) + \min\{p_i, t_2 - t_1, ect_i - t_1\} \tag{13.4.54}$$

is satisfied then $B_{(t_1,t_2)} - \{i\}$ is not empty, and T_i has to be processed after all tasks of $B_{(t_1,t_2)} - \{i\}$. Consequently, est_i can be adjusted to $est_i := \max\{est_i,$ $LB_h(B_{(t_1,t_2)} - \{i\})\}, h \in \{1,2,3,4\}$.

Proof. If (13.4.54) is satisfied then $p(B - \{i\}, t_1,t_2) > 0$ and $B_{(t_1,t_2)} - \{i\}$ is not empty. Furthermore, T_i must finish after t_2. By definition, all tasks in $B_{(t_1,t_2)} - \{i\}$ have positive processing times in the interval $[t_1,t_2)$ and so must start and finish before T_i. This proves $B_{(t_1,t_2)} - \{i\} \to i$ from which follows the domain reduction rule. □

Energetic input conditions can be defined in a similar way. Observe that the domain adjustment in Theorem 13.4.28 is stronger than the one defined in (13.4.53) if the "sum bound" (LB_2) or a stronger bound is used. We omit the simple proof due to the observations made in the following.

Up to now, it remained an open question which time intervals were especially suited for testing the energetic input/output conditions in order to derive strong domain reductions. We will sharpen this question and ask whether Theorem 13.4.28 really leads to stronger domain reductions at all if compared with other known consistency tests. Quite surprisingly, the answer is "no".

Theorem 13.4.29 *(comparing output and energetic output conditions).*

If the energetic output condition

$$t_2 - t_1 < p(B - \{i\}, t_1,t_2) + \min\{p_i, t_2 - t_1, ect_i - t_1\}$$

is satisfied for a task T_i, $i \in B$ and the time interval $[t_1,t_2)$ then the output condition

$$\max_{u \in A \cup \{i\}, v \in A, u \neq v} \{lct_v - est_u\} < p(A \cup \{i\})$$

---

[5] Here and later, we will assume that $p_i'(t_1, t_2) \geq 0$ which is not a serious restriction.

is satisfied for the couple $(B_{(t_1,t_2)} - \{i\}, i)$.

Proof. If the energetic output condition is satisfied then $B_{(t_1,t_2)} - \{i\}$ is not empty, and there exists a task T_v with $v \in B_{(t_1,t_2)} - \{i\}$. Let us first consider the case $u \in B_{(t_1,t_2)} - \{i\}$, $u \neq v$. We can approximate the right side of (13.4.54) and obtain

$$t_2 - t_1 < p(B - \{i\}, t_1, t_2) + p_i$$

$$= p(B - \{i, u, v\}, t_1, t_2) + p_u(t_1, t_2) + p_v(t_1, t_2) + p_i. \tag{13.4.55}$$

Since $u, v \in B_{(t_1,t_2)}$, we know from (13.4.52) that $t_2 - lst_v \geq p_v(t_1, t_2)$ and $ect_u - t_1 \geq p_u(t_1, t_2)$, and we can approximate

$$t_2 - t_1 < p(B - \{i, u, v\}, t_1, t_2) + ect_u - t_1 + t_2 - lst_v + p_i \tag{13.4.56}$$

which is equivalent to

$$lst_v - ect_u < p(B - \{i, u, v\}, t_1, t_2) + p_i. \tag{13.4.57}$$

Note that $p(B - \{i, u, v\}, t_1, t_2) \leq p(B_{(t_1,t_2)} - \{i, u, v\})$, so we arrive at

$$lst_v - ect_u < p(B_{(t_1,t_2)} - \{u, v\}). \tag{13.4.58}$$

or, equivalently,

$$lct_v - est_u < p(B_{(t_1,t_2)}). \tag{13.4.59}$$

Now, consider the case $u = i \neq v$. Using (13.4.54), we have

$$t_2 - t_1 < p(B - \{i\}, t_1, t_2) + ect_i - t_1$$

$$= p(B - \{i, v\}, t_1, t_2) + p_v(t_1, t_2) + ect_i - t_1. \tag{13.4.60}$$

We can, again, substitute $p_v(t_1, t_2)$ with $t_2 - lst_v$ and obtain

$$lst_v - ect_i < p(B - \{i, v\}, t_1, t_2). \tag{13.4.61}$$

A similar line of argumentation as above leads to

$$lct_v - est_i < p(B_{(t_1,t_2)}). \tag{13.4.62}$$

Finally, combining (13.4.59) and (13.4.62) leads to the output condition for the couple $(B_{(t_1,t_2)} - \{i\}, i)$ which proves our assertion. □

A similar result applies for the energetic input condition. Inversely, a quite simple proof which is omitted shows that the input/output conditions are subsumed by the energetic generalizations, so that both concepts are in fact equivalent.

Other Energetic Consistency Tests

It is possible to derive input/output negation conditions and input-or-output conditions that are based on energetic reasoning. However, as in the case of the input/output conditions, they do not imply additional domain reductions which are not also deduced by the corresponding non-energetic conditions. We therefore omit a detailed presentation of these conditions.

The results of this subsection have an important implication. They tell us that for the disjunctive scheduling problem, all known consistency tests that are based on energetic reasoning are not more powerful than their non-energetic counterparts. It is not clear whether this holds for arbitrary consistency tests, although we strongly assume this. A step towards proving this conjecture has been made in [DPP99] where it has been shown that, regardless of the chosen consistency tests, the interval processing times $p(A, t_1, t_2)$ can always be replaced by the simple processing times $p(A)$.

13.4.8 Shaving

All consistency tests presented so far share the common idea that a possible start time st_i of a task T_i can be removed from its current domain δ_i if there exists no feasible schedule in which T_i actually starts at that time. In this context, the consistency tests that have been introduced in the Sections 13.4.3 through 13.4.7 can be interpreted as sufficient conditions for proving that no feasible schedule can exist which involve a specific start time assignment st_i. In Section 13.4.3, for instance, we have tested the sufficient condition whether there exists a 2- or 3-feasible start time assignment.

This general approach has been summarized by Martin and Shmoys under the name *shaving* [MS96]. They proposed additional shaving variants. *Exact one-machine shave* verifies whether a non-preemptive schedule exists by solving an instance of the one-machine scheduling problem in which the start time st_i ∈ $\{est_i, lst_i\}$ is fixed. Quite obviously, exact one-machine shave is NP-hard and equivalent to establishing *n-b*-consistency. *One-machine shave* relaxes the non-preemption requirement and searches for a (possibly) preemptive schedule.

Carlier and Pinson [CP94] and Martin and Shmoys [MS96] independently proposed the computation of Δ-fixed-points as a method for proving the non-existence of a feasible schedule. Given a set of consistency tests Γ and a set of current domains, say Δ', a feasible schedule cannot exist if a current domain in Γ(Δ') is empty. Carlier and Pinson, and Martin and Shmoys who coined the name *C-P shave* have chosen the modified input/output domain consistency tests and the precedence consistency tests as underlying set of consistency tests. Martin and Shmoys have further proposed *double shave* which applies C-P shave for detecting inconsistencies. Torres and Lopez [TL00] review possible extensions of shaving techniques that have been proposed for job shop scheduling. Dorndorf et al. [DPP01] very successfully apply shaving techniques to the open shop scheduling problem (OSP), which is a special case of the DSP (cf. Chapter 9).

13.4.9   A Comparison of Disjunctive Consistency Tests

Let us summarize the results derived so far. In Figure 13.4.10, the dominance relations between different levels of bound-consistency and classes of consistency tests are shown[6]. A strict dominance is represented by an arc $\rightarrow$, while $\leftrightarrow$ stands for an equivalence relation. An encircled "+" means that the corresponding classes of consistency tests taken together imply a dominance relation. Since the dominance relation is transitive, we do not display all relations explicitly.

Let us start with the upper half of the figure. Obviously, n-b-consistency and exact one-machine shave are equivalent and strictly dominate all other consistency tests. On the left side, n-b-consistency, of course, subsumes all levels of k-b-consistency for $k \leq n$.

In the center of the figure, the consistency tests with an input/output component in their names are shown. As has been proven in Section 13.4.7, the energetic consistency tests are equivalent to the non-energetic ones. In Example 13.4.12, we have verified that the input/output consistency tests dominate the modified input/output consistency tests. The same dominance relation holds for the input-or-output tests when compared to the modified tests. In Section 13.4.3 we have shown that the input/output and input/output negation consistency tests taken together establish strong 3-b-consistency if for the former the "sum bound" (LB_2) and for the latter the "simple earliest completion time bound" (LB_5) are applied for adjusting the current domains. The input/output and input/output negation tests usually imply more than 3-b-consistency as can be seen in Example 13.4.15. However, if only pairs and triples of tasks are considered then the equivalence relation holds. Further, it has been shown in Section 13.4.3 that applying the input/output consistency tests for pairs of tasks is equivalent to establishing 2-b-consistency if the "earliest completion time bound" (LB_1) is used as time bound adjustment.

Let us now turn to the right side of the figure. It is not hard to show that double shave strictly dominates C-P shave which in turn strictly dominates one-machine shave. Apart from this, there exists no particular relationship between double shave and C-P shave and the other consistency tests. However, double shave and C-P shave usually lead to significantly stronger domain reductions as has been verified empirically. Finally, Martin and Shmoys [MS96] have shown that one-machine shave is equivalent to the modified input/output domain consistency tests.

---

[6] Although the dominance relation has only been defined for sets of consistency tests, it can be extended in a straightforward manner to the levels of bound-consistency.

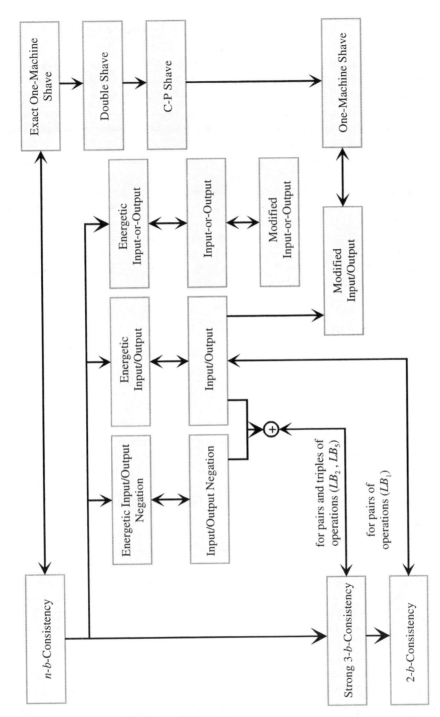

Figure 13.4.10 *Dominance relations.*

13.4.10   Precedence vs. Disjunctive Consistency Tests

The consistency tests which have been developed for the disjunctive constraints can be applied to an instance of the DSP by decomposing this instance into (preferably maximal) cliques. Since all consistency tests presented are monotonous, they can be applied in an arbitrary order and always result in the same Δ-fixed-point. However, the runtime behaviour differs extremely depending on the order of application that has been chosen.

An ordering rule which has been proven to be quite effective is to perform the sequence consistency tests that are likely to deduce more edge orientations and have a lower time complexity in the beginning. A particular set of consistency tests is only triggered if all "preceding" consistency tests do not imply any deductions any more. This ensures that the more costly consistency tests are only seldomly applied and contribute less in the overall computational costs.

Finally, Nuijten and Sourd [NS00] have recently described consistency checking techniques for the DSP that are based on the simultaneous consideration of precedence constraints and disjunctive constraints.

13.5   Conclusions

Constraint propagation is an elementary method which reduces the search space of a search or optimization problem by analyzing the interdependencies between the variables, domains and constraints that define the set of feasible solutions. Instead of achieving full consistency with respect to some concept of consistency, we generally have to content ourselves with approximations due to reasons of complexity. In this context, we have evaluated classical and new consistency tests for the DSP which are simple rules that reduce the domains of variables (domain consistency tests) or derive knowledge in a different form, e.g. by determining the processing sequences of a set of tasks (sequence consistency tests).

The particular strength of this approach is based on the repeated application of the consistency tests, so that the knowledge derived is propagated, i.e. reused for acquiring additional knowledge. The deduction of this knowledge can be described as the computation of a fixed point. Since this fixed point depends upon the order of the application of the consistency tests, Dorndorf et al. [DPP00] at first have derived a necessary condition for its uniqueness and have developed a concept of dominance which enables to compare different consistency tests. With respect to this dominance relation, they have examined the relationship between several concepts of consistency (bound-consistency, energetic reasoning and shaving) and the most powerful consistency tests known as the input/output, input/output negation and input-or-output consistency tests. They have been able to improve the well-known result that the input/output consistency tests for pairs of tasks imply 2-b-consistency by deriving the tests which establish strong 3-b-

consistency. These consistency tests are slightly stronger than the famous ones derived by Carlier and Pinson [CP89, CP90]. Dorndorf et al. [DPP00] have analyzed the input/output, input/output negation and input-or-output consistency tests and have classified different lower bounds which are used for the reduction of domains. They have shown that apparently weaker bounds still induce the same fixed point. Finally, an open question regarding the concept of energetic reasoning has been answered. In contrast to scheduling problems with discrete resource supply, they have shown that the known consistency tests based on energetic reasoning are equivalent to the tests based on simple processing times.

13.6 Appendix: Bound Consistency Revisited

In this section, we derive the time bound adjustments for establishing 3-*b*-consistency as has been announced in Section 13.4.3. Let us assume that the following condition

$$(\max\{lct_j - est_i, lct_k - est_i\} \geq p_i + p_j + p_k) \quad \vee \qquad\qquad (i+ii)$$

$$(est_j + p_j \leq est_i \ \wedge \ est_i + p_i \leq lst_k) \quad \vee \qquad\qquad (iii)$$

$$(est_k + p_k \leq est_i \ \wedge \ est_i + p_i \leq lst_j) \quad \vee \qquad\qquad (iv)$$

$$(est_i \geq \max\{\min\{est_j, est_k + p_j + p_k, est_j + p_j, est_k + p_k\}) \qquad\qquad (v+vi)$$

$$\qquad\qquad\qquad\qquad\qquad\qquad\qquad\qquad\qquad\qquad\qquad\qquad (13.6.1)$$

is not satisfied given the current earliest and latest start times. As already mentioned, there exist two cases. In the first case, increasing est_i will never satisfy conditions $(i + ii)$, (iii) and (iv). Therefore, we have to adjust est_i so as to satisfy condition $(v+vi)$. In the second case, condition $(i + ii)$ is not satisfiable, but increasing est_i eventually satisfies (iii), (iv) or $(v+vi)$. Here, the minimal earliest start time for which (iii) or (iv) holds is not greater than the minimal earliest start time for which $(v+vi)$ holds. This will be proven in the remainder of this subsection.

We will first deal with the problem of how to distinguish between the two cases. The corresponding time bound adjustments will then be derived at a later time. In Lemma 13.6.1, a necessary and sufficient condition for the existence of $est_i^* \geq est_i$ satisfying condition (iii) is described.

Lemma 13.6.1 *(condition (iii)).*

There exists $est_i^* \geq est_i$ such that condition (iii) is satisfied iff

$$\max\{est_j + p_j + p_i, est_i + p_i\} \leq lst_k . \qquad\qquad (13.6.2)$$

The smallest start time which then satisfies (iii) is $est_i^* = \max\{est_i, est_j + p_j\}$.

Proof. If condition (*iii*) is satisfied for $est_i^* \geq est_i$ then $est_j + p_j \leq est_i^*$ and $est_i^* + p_i$ $\leq lst_k$, so that $\max\{est_j + p_j + p_i, est_i + p_i\} \leq lst_k$. This proves the direction $\Rightarrow$. In order to show $\Leftarrow$, let $\max\{est_j + p_j + p_i, est_i + p_i\} \leq lst_k$. If $est_i < est_j + p_j$ then $est_i^* = est_j + p_j$ is the smallest value which satisfies (*iii*). Otherwise, if $est_i \geq est_j + p_j$ then $est_i^* = est_i$ is the smallest value which satisfies (*iii*). $\square$

Changing the roles of j and k in Lemma 13.6.1 leads to a similar result for condition (*iv*).

Corollary 13.6.2 *(conditions (iii) and (iv)).*

There exists $est_i^* \geq est_i$ *which satisfies (iii) or (iv) iff*

$$(\max\{est_j + p_j + p_i, est_i + p_i\} \leq lst_k) \ \vee$$
$$(\max\{est_k + p_k + p_i, est_i + p_i\} \leq lst_j) \tag{13.6.3}$$

If Δ is 2-b-consistent then (13.6.3) is equivalent to

$$(est_j + p_j + p_i \leq lst_k \ \vee \ est_k + p_k + p_i \leq lst_j) \ \wedge$$
$$(est_i + p_i \leq lst_k \ \vee \ est_i + p_i \leq lst_j) \tag{13.6.4}$$

Proof. The first assertion follows directly from Lemma 13.6.1. Let us show the second equivalence and assume that 2-b-consistency is established. Obviously, (13.6.3) immediately implies (13.6.4). The other direction, however, is not apparent at once.

Hence, let (13.6.4) be satisfied. It is sufficient to study the case $est_j + p_j + p_i$ $\leq lst_k$, since $est_k + p_k + p_i \leq lst_j$ leads to a similar conclusion. Given (13.6.4), we can deduce that $est_i + p_i \leq lst_k$ or $est_i + p_i \leq lst_j$ ($*$).

Now, if $est_i + p_i \leq lst_k$ then the first condition $\max\{est_j + p_j + p_i, est_i + p_i\} \leq lst_k$ of (13.6.3) is satisfied. If, however, $est_i + p_i > lst_k$ then 2-b-consistency implies $est_k + p_k \leq est_i$. Further, $est_i + p_i \leq lst_j$ due to ($*$). Therefore, $est_k + p_k + p_i$ $\leq lst_j$, and the second condition $\max\{est_k + p_k + p_i, est_i + p_i\} \leq lst_j$ of (13.6.3) is satisfied. $\square$

Given these results, it is now quite easy to describe the adjustments of the earliest start times.

Lemma 13.6.3 *(adjusting earliest start times, part 1).*

Let Δ be 2-b-consistent. If

$$\max_{u \in \{j,k\}, v \in \{i,j,k\}, u \neq v} \{lct_v - est_u\} < p_i + p_j + p_k \tag{13.6.5}$$

or

$$est_i + p_i > \max\{lst_j, lst_k\} \tag{13.6.6}$$

then $(i+ii)$, (iii), (iv) are not satisfiable for any $est_i^* \geq est_i$. The minimal earliest start time $est_i^* \geq est_i$ satisfying $(v+vi)$ is then defined by

$$est_i^* := \max\{est_i, \min\{est_j, est_k\} + p_j + p_k, est_j + p_j, est_k + p_k\}. \qquad (13.6.7)$$

Proof. We have shown in Lemma 13.4.5 that there exists no $est_i^* \geq est_i$ satisfying condition $(i + ii)$ iff

$$\max_{v \in \{j,k\}} \{lct_v - est_i\} < p_i + p_j + p_k. \qquad (13.6.8)$$

Likewise, we have shown in Lemma 13.6.1 that there exists no $est_i^* \geq est_i$ satisfying condition (iii) or (iv) iff (13.6.4) is not satisfied, i.e. iff

$$\begin{aligned}
&(est_j + p_j + p_i > lst_k \ \wedge \ est_k + p_k + p_i > lst_j) \ \vee \\
&(est_i + p_i > lst_k \ \wedge \ est_i + p_i > lst_j)
\end{aligned} \qquad (13.6.9)$$

which is equivalent to

$$\begin{aligned}
&(lct_k - est_j < p_i + p_j + p_k \ \wedge \ lct_j - est_k < p_i + p_j + p_k) \ \vee \\
&est_i + p_i > \max\{lst_j, lst_k\}.
\end{aligned} \qquad (13.6.10)$$

(13.6.8) and (13.6.10) together imply that $(i+ii)$, (iii) and (iv) are not satisfiable, so we have to choose the minimal earliest start time est_i^* satisfying condition $(v+vi)$ which leads to

$$est_i^* := \max\{est_i, \min\{est_j, est_k\} + p_j + p_k, est_j + p_j, est_k + p_k\}. \qquad (13.6.11)$$

It remains to combine (13.6.8) and (13.6.10) to one single condition. Making use of the fact that $est_i + p_i > \max\{lst_j, lst_k\}$ already implies (13.6.8), we can deduce that these two conditions are equivalent to:

$$(\max_{u \in \{j,k\}, v \in \{i,j,k\}, u \neq v} \{lct_v - est_u\} < p_i + p_j + p_k) \ \vee \ (est_i + p_i > \max\{lst_j, lst_k\}).$$

This completes the proof. $\qquad\qquad\qquad\qquad\qquad\qquad\qquad\qquad\qquad\qquad\qquad\qquad$ □

Lemma 13.6.4 *(adjusting earliest start times, part 2).*

Let Δ be 2-b-consistent. If (13.6.5) and (13.6.6) are not satisfied but

$$\max_{u \in \{j,k\}} \{lct_i - est_u\} < p_i + p_j + p_k \qquad (13.6.12)$$

then $(i+ii)$ is not satisfiable for any $est_i^* \geq est_i$. The minimal earliest start time $est_i^* \geq est_i$ satisfying (iii), (iv) or $(v+vi)$ is then defined through

$$est_i^* := \max\{est_i, \min\{v_j, v_k\}\}, \qquad (13.6.13)$$

where

$$v_j := \begin{cases} est_j + p_j & \text{if } \max\{est_j + p_j + p_i, est_i + p_i\} \leq lst_k, \\ est_k + p_k & \text{otherwise}, \end{cases}$$

$$v_k := \left\{ \begin{array}{ll} est_k + p_k & \text{if } \max\{est_k + p_k + p_i, est_i + p_i\} \le lst_j, \\ est_j + p_j & \text{otherwise.} \end{array} \right.$$

Proof. The assumptions imply that $(i + ii)$ is not satisfiable. From Lemma 13.6.1, we know that $est_i^* := \max\{est_i, \min\{v_1, v_2\}\}$ is the minimal earliest start time which satisfies (iii) or (iv). Further, Lemma 13.6.3 implies that there exists no smaller est_i^* satisfying $(v + vi)$, so indeed est_i^* is the correct adjustment.   □

Lemma 13.6.3 leads to the consistency tests

$$\max_{u \in \{i,j,k\}, v \in \{j,k\}, u \neq v} \{lct_v - est_u\} < p_i + p_j + p_k \Rightarrow$$

$$est_i := \max\{est_i, \min\{est_j, est_k\} + p_j + p_k, est_j + p_j, est_k + p_k\}, \quad (13.6.14)$$

$$est_i + p_i > \max\{lst_j, lst_k\} \Rightarrow$$

$$est_i := \max\{est_i, \min\{est_j, est_k\} + p_j + p_k, est_j + p_j, est_k + p_k\}. \quad (13.6.15)$$

which correspond with the two different versions of the output domain consistency tests for triples of tasks (see Theorems 13.4.13 and 13.4.17). Observe that

$$LB_3(\{j,k\}) = \max\{\min\{est_j, est_k\} + p_j + p_k, est_j + p_j, est_k + p_k\}$$

is the optimal makespan if the tasks T_j and T_k are scheduled with preemption allowed. From Theorem 13.4.16, we know that the time bound adjustment $LB_3(\{j,k\})$ can be replaced with $LB_2(\{j,k\}) = \min\{est_j, est_k\} + p_j + p_k$, so that instead of (13.6.14) the following consistency test can be applied:

$$\max_{u \in \{i,j,k\}, v \in \{j,k\}, u \neq v} \{lct_v - est_u\} < p_i + p_j + p_k \Rightarrow$$

$$est_i := \max\{est_i, \min\{est_j, est_k\} + p_j + p_k\}. \quad (13.6.16)$$

Likewise, we can replace (13.6.15) with the equivalent consistency test

$$est_i + p_i > \max\{lst_j, lst_k\} \Rightarrow$$

$$est_i := \max\{est_i, \min\{est_j, est_k\} + p_j + p_k\}. \quad (13.6.17)$$

This follows from the fact that the 2-*b*-consistency tests already ensure

$$est_i \ge \max\{est_j + p_j, est_k + p_k\} \text{ if } est_i + p_i > \max\{lst_j, lst_k\}.$$

Lemma 13.6.4 derives the consistency test

$$\max_{u \in \{j,k\}} \{lct_v - est_i\} < p_i + p_j + p_k \Rightarrow est_i := \max\{est_i, \min\{v_j, v_k\}\} \quad (13.6.18)$$

which corresponds to the input negation domain consistency test for triples of tasks (see Theorem 13.4.19). Again, we can replace the time bound adjustment $LB_6(\{j,k\}) = \min\{v_j, v_k\}$ with $LB_5(\{j,k\}) = \min\{ect_j, ect_k\}$ due to Lemma

13.4.21 which leads to the equivalent consistency test

$$\max_{u \in \{j,k\}} \{lct_v - est_i\} < p_i + p_j + p_k \Rightarrow est_i := \max\{est_i, \min\{ect_j, ect_k\}\} \quad (13.6.19)$$

This proves the assertions made in Section 13.4.3.

References

AC91   D. Applegate, W. Cook, A computational study of the job shop scheduling problem, *ORSA J. Comput.* 3, 1991, 149-156.

Ama70   S. Amarel, On the representation of problems and goal-directed procedures for computers, in R. Banerji and M. Mesarovic, editors, *Theoretical Approaches to Non-Numerical Problem Solving*, Springer, Heidelberg, 1970, 179-244.

BDP96   J. Błażewicz, W. Domschke, E. Pesch, The job shop scheduling problem: Conventional and new solution techniques, *European J. Oper. Res.* 93, 1996, 1-33.

Bee92   P. van Beek, Reasoning about qualitative temporal information, *Artificial Intelligence* 58, 1992, 297-326.

Bes94   C. Bessiere, Arc-consistency and arc-consistency again, *Artificial Intelligence* 65, 1994, 179-190.

BFR99   C. Bessiere, E. C. Freuder, J.-C. Regin, Using constraint metaknowledge to reduce arc consistency computation, *Artificial Intelligence* 107, 1999, 125-148.

Bib88   W. Bibel, Constraint satisfaction from a deductive viewpoint, *Artificial Intelligence* 35, 1988, 401-413.

BJK94   P. Brucker, B. Jurisch, Z. Krämer, The job shop problem and immediate selection, *Ann. Oper. Res.* 50, 1994, 73-114.

BJS94   P. Brucker, B. Jurisch, B. Sievers, A fast branch and bound algorithm for the job shop scheduling problem, *Discrete Appl. Math.* 49, 1994, 107-127.

BL95   P. Baptiste, C. LePape, A theoretical and experimental comparison of constraint propagation techniques for disjunctive scheduling, in *Proceedings of the 14th International Joint Conference on Artificial Intelligence*, Montreal, 1995, 136-140.

BL96   P. Baptiste, C. LePape. Edge-finding constraint propagation algorithms for disjunctive and cumulative scheduling, in *Proceedings of the 15th Workshop of the U. K. Planning Special Interest Group*, Liverpool, 1996.

Car82   J. Carlier, The one machine sequencing problem, *European J. Oper. Res.* 11, 1982, 42-47.

Che99   Y. Chen, Arc consistency revisited, *Information Processing Letters* 70, 1999, 175-184.

Chr75   N. Christofides, *Graph Theory: An Algorithmic Approach*, Academic Press, London, 1975.

CL95   Y. Caseau, F. Laburthe, Disjunctive scheduling with task intervals, Technical Report 95-25, Laboratoire d'Informatique de l'Ecole Normale Superieure, Paris, 1995.

Clo71   M. B. Clowes, On seeing things, *Artificial Intelligence* 2, 1971, 179-185.

Coh90   J. Cohen, Constraint logic programming languages, *Communications of the ACM* 33, 1990, 52-68.

Coo89   M. C. Cooper, An optimal k -consistency algorithm, *Artificial Intelligence* 41, 1989, 89-95.

CP89   J. Carlier, E. Pinson, An algorithm for solving the job shop problem, *Management Sci.* 35, 1989, 164-176.

CP90   J. Carlier, E. Pinson, A practical use of Jackson's preemptive schedule for solving the job shop problem, *Ann. Oper. Res.* 26, 1990, 269-287.

CP94   J. Carlier, E. Pinson, Adjustments of heads and tails for the job shop problem, *European J. Oper. Res.* 78, 1994, 146-161.

Dav87   E. Davis, Constraint propagation with interval labels, *Artificial Intelligence* 32, 1987, 281-331.

Dew92   G. Deweß, An existence theorem for packing problems with implications for the computation of optimal machine schedules, *Optimization* 25, 1992, 261-269.

DP88   R. Dechter, J. Pearl, Network-based heuristics for constraint satisfaction problems, *Artificial Intelligence* 34, 1988, 1-38.

DPP99   U. Dorndorf, T. Phan-Huy, E. Pesch, A survey of interval capacity consistency tests for time and resource constrained scheduling, in J. Weglarz, editor, *Project Scheduling - Recent Models, Algorithms and Applications*, Kluwer Academic Publishers, Boston, 1999, 213-238.

DPP00   U. Dorndorf, E. Pesch, T. Phan-Huy, Constraint propagation techniques for disjunctive scheduling problems, *Artificial Intelligence* 122, 2000, 189-240.

DPP01   U. Dorndorf, E. Pesch, T. Phan-Huy, Solving the open shop scheduling problem, *Journal of Scheduling* 4, 2001, 157-174.

ELT91   J. Erschler, P. Lopez, C. Thuriot, Raisonnement temporel sous contraintes de ressource et problèmes d'ordonnancement, *Revue d'Intelligence Artificielle* 5, 1991, 7-32.

FN00   F. Focacci, W. Nuijten, A constraint propagation algorithm for scheduling with sequence dependent setup times, in U. Junker, S.E. Karisch, S. Tschöke, editors, *Proceedings of the 2nd International Workshop on the Integration of AI and OR Techniques in Constraint Programming for Combinatorial Optimization Problems*, 2000, 53-55.

Fre78   E. C. Freuder, Synthesizing constraint expressions, *Journal of the ACM* 21, 1978, 958-966.

GJ79   M. R. Garey, D. S. Johnson, *Computers and Intractability: A Guide to the Theory of NP-Completeness*, Freeman, San Francisco, 1979.

HDT92   P. van Hentenryck, Y. Deville, C.-M. Teng, A generic arc consistency algorithm and its specializations, *Artificial Intelligence* 57, 1992, 291-321.

Hen92   P. van Hentenryck, *Constraint Satisfaction in Logic Programming*, MIT Press, Cambridge, 1992.

HL88   C.-C. Han, C.-H. Lee, Comments on Mohr and Henderson's path consistency algorithm, *Artificial Intelligence* 36, 1988, 125-130.

HS79   R. M. Haralick, L. G. Shapiro, The consistent labelling problem: Part I, *IEEE Transactions PAMI* 1, 1979,173-184.

HS80   R. M. Haralick, L. G. Shapiro, The consistent labelling problem: Part II, *IEEE Transactions PAMI* 2, 1980, 193-203.

Huf71   D. Z. Huffman, Impossible objects as nonsense sentences, *Machine Intelligence* 6, 1971, 295-323.

JCC98   P. Jeavons, D. Cohen, M.C. Cooper, Constraints, consistency and closure, *Artificial Intelligence* 101, 1998, 251-265.

Kum92   V. Kumar, Algorithms for constraint satisfaction problems, *AI Magazine* 13, 1992, 32-44.

LEE92   P. Lopez, J. Erschler, P. Esquirol, Ordonnancement de tâches sous contraintes: une approche énergétique, *RAIRO Automatique, Productique, Informatique Industrielle* 26, 1992, 453-481.

Lho93   O. Lhomme, Consistency techniques for numeric CSPs, in *Proceedings of the 13th International Joint Conference on Artificial Intelligence*, Chambery, France, 1993, 232-238.

Mac77   Z. K. Mackworth, Consistency in networks of relations, *Articial Intelligence* 8, 1977, 99-118.

Mac92   Z. K. Mackworth, The logic of constraint satisfaction, *Artificial Intelligence* 58, 1992, 3-20.

Mes89   P. Meseguer, Constraint satisfaction problems: An overview, *AI Communications*, 2, 1989, 3-17.

MF85   Z. K. Mackworth, E. C. Freuder, The complexity of some polynomial network consistency algorithms for constraint satisfaction problems, *Artificial Intelligence* 25, 1985, 65-74.

MH86   R. Mohr, T. C. Henderson, Arc and path consistency revisited, *Artificial Intelligence* 28, 1986, 225-233.

Mon74   U. Montanari, Networks of constraints: Fundamental properties and applications to picture processing, *Information Sciences* 7, 1974, 95-132.

Moo66   R. E. Moore, *Interval Analysis*, Prentice Hall, Englewood Cliffs, 1966.

MS96   P. Martin, D. B. Shmoys, A new approach to computing optimal schedules for the job shop scheduling problem, in *Proceedings of the 5th International IPCO Conference*, 1996.

MT63   J. F. Muth, G. L. Thompson (eds.), *Industrial Scheduling*, Prentice Hall, Englewood Cliffs, 1963.

NL98   W. P. M. Nuijten, C. Le Pape. Constraint-based job shop scheduling with ILOG scheduler, *Journal of Heuristics* 3, 1998, 271-286.

NS00   W. Nuijten, F. Sourd, New time bound adjustment techniques for shop schedul-
            ing, in P. Brucker, S. Heitmann, J. Hurink, S. Knust, editors, *Proceedings of the
            7th International Workshop on Project Management and Scheduling*, 2000, 224-
            226.

Nui94   W. P. M. Nuijten, *Time and Resource Constrained Scheduling:* A *Constraint
            Satisfaction Approach,* PhD thesis, Eindhoven University of Technology, 1994.

Pha96   T. Phan-Huy, *Wissensbasierte Methoden zur Optimierung von Produktions-
            abläufen*, Master's thesis, University of Bonn, Bonn, 1996.

Pha00   T. Phan-Huy, *Constraint Propagation in Flexible Manufacturing*, Springer,
            2000.

PT96   E. Pesch, U. Tetzlaff, Constraint propagation based scheduling of job shops,
            INFORMS J. Comput. 8, 1996, 144-157.

RS64   B. Roy, B. Sussman, *Les problèmes d`ordonnancement avec contraintes dis-
            jonctives*, Note D. S. 9, SEMA, Paris, 1964.

Sei81   R. Seidel, A new method for solving constraint satisfaction problems, in *Pro-
            ceedings of the 7th International Joint Conference on AI*, 1981, 338-342.

TF90   E. P. K. Tsang, N. Foster, Solution synthesis in the constraint satisfaction prob-
            lem, Technical report csm-142, Department of Computer Sciences, University of
            Essex, Essex, 1990.

TL00   P. Torres, P. Lopez, Overview and possible extensions of shaving techniques for
            job-shop problems, in U. Junker, S. E. Karisch, S. Tschöke, editors, *Proceedings
            of the 2nd International Workshop on the Integration of AI and OR Techniques
            in Constraint Programming for Combinatorial Optimization Problems*, 2000,
            181-186.

Tsa93   E. Tsang, *Foundations of Constraint Satisfaction*, Academic Press, Essex, 1993.

Wal72   D. L. Waltz, Generating semantic descriptions from drawings of scenes with
            shadows, Technical report AI-TR-271, M.I.T., 1972.

Wal75   D. L. Waltz. Understanding line drawings of scenes with shadows, in P. H.
            Winston, editor, *The Psychology of Computer Vision*, 19-91. McGraw-Hill,
            1975.

14 Scheduling in Flexible Manufacturing Systems

14.1 Introductory Remarks

An important application area for machine scheduling theory comes from Flexible Manufacturing Systems (*FMS*s). This relatively new technology was introduced to improve the efficiency of a job shop while retaining its flexibility. An FMS can be defined as an integrated manufacturing system consisting of flexible machines equipped with tool magazines and linked by a material handling system, where all system components are under computer control [BY86a]. Existing FMSs mainly differ by the installed hardware concerning machine types, tool changing devices and material handling systems. Instances of machine types are dedicated machines or parallel multi-purpose ones. Tool changing devices can be designed to render automatic online tool transportation and assignment to the machines' magazines while the system is running. In other cases tool changes are only possible if the operations of the system are stopped. Most of the existing FMSs have automatic part transportation capabilities.

Different problems have to be solved in such an environment which comprise design, planning and scheduling. The vital factors influencing the solutions for the latter two are the FMS-hardware and especially the existing machine types and tool changing devices. In earlier (but still existing) FMSs NC-machines are used with limited versatility; several different machines are needed to process a part. Moreover the machines are not very reliable. For such systems shop scheduling models are applicable; in classical, static formulations they have been considered in Chapters 8 through 10. Recent developments in FMS-technology show that the machines become more versatile and reliable. Some FMSs already are implemented using mainly only one machine type. These general purpose machine tools make it possible to process a part from the beginning to the end using only one machine [Jai86]. A prerequisite to achieve this advantage without or with negligible setup times is a *tool changing system* that can transfer tools between the machines' tool magazines and the central tool storage area while all machines of the system are in operation. Some FMSs already fulfill this assumption and thus incorporate a high degree of flexibility. Results from queuing theory using closed queuing networks show that the expected production rate is maximized under a configuration which incorporates only general purpose machines [BY86b, SS85, SM85].

With the notation of machine scheduling theory this kind of FMS design can be represented by parallel machine models, and thus they were treated relatively broadly in Chapter 5. The most appropriate type of these models depends on the particular scheduling situation. All the machines might be identical or they have to be regarded as uniform or unrelated. Parts (i.e. jobs) might have due dates or deadlines, release times, or weights indicating their relative importance. The possibilities of part processing might be restricted by certain precedence constraints, or each operation (i.e. task) can be carried out independently of the others which are necessary for part completion. Objectives might consist of minimizing schedule length, mean flow time or due date involving criteria. All these problem characteristics are well known from traditional machine scheduling theory, and had been discussed earlier.

Most of the FMS-scheduling problems have to take into account these problem formulations in a quite general framework and hence are NP-hard. Thus, with the view from today, they are computationally intractable for greater problem instances.

The difficulties in solving these problem types are sometimes overcome by considering the possibility of preempting part processing. As shown in former chapters, quite a lot of intractable problems are solvable in their preemptive versions in polynomial time. In the context of FMSs one has to differ between two kinds of preemptions. One occurs if the operation of a part is preempted and later resumed on the same machine (part-preemption). The other one appears if the operation of a part is preempted and then resumed at the same point of time or at a later time on another machine (part-machine-preemption). The consequences of these two kinds of preemption are different. Part-preemption can be carried out without inducing a change of machines and thus it does not need the use of the FMS material handling system. Part-machine-preemption requires its usage for part transportation from one machine to another. A second consequence comes from the buffer requirements. In either case of preemption storage capacity is needed for preempted and not yet finished parts. If it is possible to restrict the number and kind of preemptions to a desirable level, this approach is appealing. Some computationally intractable problem types are now efficiently solvable and for most measures of performance the quality of an optimal preemptive schedule is never worse than non-preemptive one. To consider certain inspection, repair or maintenance requirements of the machine tools, processing availability restrictions have to be taken into account. The algorithmic background of these formulations can be found in [Sch84, Sch88] and were already discussed in Chapter 5. Some non-deterministic aspects of these issues will be studied in Chapter 15.

In the context of FMSs another model of scheduling problems is also of considerable importance. In many cases tools are a very expensive equipment and under such an assumption it is unlikely that each tool type is available in an unrestricted amount. If the supply of tools of some type is restricted, this situation leads to parallel machine models with resource constraints. In an FMS-environment the number of resource types will correspond to the number of tool types, the resource limits correspond to the number of available tools of each type and the resource requirements correspond to the number of tools of each type which are necessary to perform the operation under consideration. Models of this kind are extensively treated in [BCSW86], and some recent results had been given in Section 12.1.

There is another aspect which has to be considered in such an environment. In many cases of FMS production scheduling it is desired to minimize part movements inside the system to avoid congestion and unnecessary repositioning of parts which would occur if a part is processed by more than one machine or if it is preempted on the same machine. FMSs which consist mainly of general purpose machine tools have the prerequisite to achieve good results according to the above objectives. In the best case repositioning and machine changeovers can be avoided by assigning each part to only one machine where it is processed from the beginning to the end without preemption. A modeling approach to represent this requirement would result in a formulation where all operations which have to be performed at one part would be summed up resulting in one super-operation having different resource requirements at discrete points of time. From this treatment models for project scheduling would gain some importance [SW89]. A relaxed version of this approach to avoid unnecessary part transportation and repositioning has to consider the minimization of the number of preemptions in a given schedule.

Let us also mention that any FMS scheduling problem can be decomposed into single machine problems, as it was suggested in [RRT89]. Then the ideas and algorithms presented in Chapter 4 can be utilized.

From the above issues, we can conclude that traditional machine and project scheduling theory has a great impact on advanced FMS-environments. Besides this, different problems are raised by the new technology which require different or modified models and corresponding solution approaches. There are already many results from machine and project scheduling theory available which can also be used to support the scheduling of operations of an FMS efficiently, while some others still have to be developed (see [RS89] as a survey). Various modeling approaches are investigated in [Sch89], some more recent, selected models are investigated in the following three sections. We stress the scheduling point of view in making this selection, due to the character of this book, and, on the other hand, the prospectivity of the subject. Each of the three models selected opens some new directions for further investigations. The first one deals with dynamic job shops (i.e. such in which some events, particularly job arrivals, occur at unknown times) and with the approach solving a static scheduling problem at each time of the occurrence of such an event, and then implementing the solution on a rolling horizon basis. The second considers simultaneous task scheduling and vehicle routing in a class of FMS, which was motivated by an application in a factory producing helicopter parts. Last but not least, a practical implementation of the FMS model in acrylic glass production will be presented.

14.2   Scheduling Dynamic Job Shops

14.2.1   Introductory Remarks

In this section we consider *dynamic job shops*, i.e. such in which job arrival times are unknown in advance, and we allow for the occurrence of other non-deterministic events such as machine breakdowns. The scheduling objective will be mean job tardiness which is important in many manufacturing systems, especially those that produce to specific customer orders. In low to medium volume of discrete manufacturing, typified by traditional job shops and more recently by flexible manufacturing systems, this objective was usually operationalized through the use of priority rules. A number of such rules were proposed in the literature, and a number of investigations were performed dealing with the relative effectiveness of various rules, e.g. in [Con65, BB82, KH82, BK83, VM87]. Some deeper tactical aspects of the interaction between priority rules and the methods of assigning due-dates were studied in [Bak84].

Below we will present a different approach to the problem, proposed recently by Raman, Talbot and Rachamadugu [RTR89a], and Raman and Talbot [RT92]. This approach decomposes the dynamic problem into a series of static problems. A static problem is generated at each occurrence of a non-deterministic event in the system, then solved entirely, and the solution is implemented on a rolling horizon basis. In this procedure the entire system is considered at each instance of the static problem, in contrast to priority rules which consider only one machine at a time. Of course, when compared with priority rules, this approach requires greater computational effort, but also leads to significantly better system performance. Taking into account the computing power available today, this cost seems to be worth to pay. Moreover, the idea of the approach is pretty general and can be implemented for other dynamic scheduling problems. Let us remind that the approach was originally used by Raman, Rachamadugu and Talbot [RRT89] for a single machine.

The static problem mentioned above can be solved in an exact or a heuristic way. An example of an exact method is a modification of the depth-first search branch and bound algorithm developed by Talbot [Tal82] for minimizing schedule length in a *project scheduling* problem. We will not describe this modification which is presented in [RT92] and used for benchmarking a heuristic method proposed in the same paper. This heuristic is especially interesting for practical applications, and thus will be described in more detail. It is based on decomposing the multiple machine problem, and constructing the schedule for the entire system around the bottleneck machine. For this purpose relative job priorities are established using task due dates (TDDs). However, in comparison with the traditional usage of task milestones, in this approach TDDs are derived by taking into account other jobs in the system, and TDDs assignment is combined with task scheduling. In the next two sections the heuristic algorithm will be described and results of computational experiments will be presented.

14.2.2 Heuristic Algorithm for the Static Problem

In papers dealing with priority rules applied to our scheduling problem it has been shown the superiority of decomposing job due dates into task due dates, and using TDDs for setting priorities. In particular, Baker [Bak84] found that the *Modified Task Due Date (MTD) rule* performs well across a range of due date tightness. It selects the task with the minimum MTD, where the MTD of task T_{ij} is calculated as

$$MTD_{ij} = \max\,(\tau + p_{ij}, d_{ij}),\tag{14.2.1}$$

and where τ is the time when the scheduling decision needs to be made and d_{ij} is the TDD of T_{ij}. Raman, Talbot and Rachamadugu [RTR89b] proved that for a given set of TDDs the total tardiness incurred by two adjacent tasks in a non-delay schedule on any given machine does not increase if they are re-sequenced according to the *MTD* rule. It means that if TDDs are set optimally, the *MTD* rule guarantees local optimality between adjacent tasks at any machine for a non-delay schedule. Most existing implementations of the *MTD* rule set TDDs by decomposing the total flow $d_j - p_j$ of job J_j where $p_j = \sum_{i=1}^{n_j} p_{ij}$, into individual task flows in a heuristic way. Vepsalainen and Morton [VM87] proposed to estimate each TDD by netting the lead time for the remaining tasks from the job due date. In this way the interactions of all jobs in the system are taken into account. The heuristic by Raman and Talbot also takes explicitly into account this interactions, and, moreover, considers TDD assignment and task scheduling simultaneously. Of course, the best set of TDDs is one which yields the best set of priorities, and thus the goodness of a given set of TDDs can be determined only when the system is scheduled simultaneously. In consequence, the heuristic is not a single pass method, but it considers global impact of each TDD assignment within a schedule improvement procedure. The initial solution is generated by the *MTD* rule with TDDs at the maximum values that they can assume without delaying the corresponding jobs. Machines are then considered one by one and an attempt is made to revise the schedule of tasks on a particular machine by modifying their TDDs. Jobs processed on all machines are ranked in the non-increasing order of their tardiness. For any task in a given job with positive tardiness, first the interval for searching for the TDD is determined and for each possible value in this interval the entire system is rescheduled. The value which yields the minimum total tardiness is taken as the TDD for that task. This step is repeated for all other tasks of that job processed on the machine under consideration, for all other tardy jobs on that machine following their rank order, and for all machines in the system. The relative workload of a given machine is used to determine its criticality;

the algorithm ranks all the machines from the most heavily loaded (number 1) and considers them in this order. Since the relative ranking of machines does not change, in the sequel they are numbered according to their rank.

In order to present the algorithm we need two additional denotations. The ordered sequence of jobs processed on P_k, $k = 1, 2, \cdots, m$ will be denoted by $\mathcal{J}_k$, the set of tasks of $J_j \in \mathcal{J}_k$ on P_k by $\mathcal{T}_{kj}$, and the number of tasks in $\mathcal{T}_{kj}$ by n_{kj}.

Algorithm 14.2.1 *Heuristic for the static job shop to minimize mean tardiness* [RT92].
begin   -- initialization
for each task T_{ij} **do** $d_{ij} := d_j - t_{ij} + p_{ij}$;
      -- a set of new task due dates has been assigned, taking into account
      -- the cumulative processing time t_{ij} of J_j up to and including task T_{ij}
call *MTD rule*;
      -- the initial sequence has been constructed
Order and number all the machines in non-increasing order of their total workloads $\sum\limits_{J_j \in \mathcal{J}_k} \sum\limits_{T_{ij} \in \mathcal{T}_{kj}} p_{ij}$;

$r := 1$; $z(0) := \infty$; $z(1) := \sum\limits_{j=1}^{n} D_j$;
      -- initial values of counters are set up
while $z(r) < z(r-1)$ **do**
  begin
  $\mathcal{P}_1 := \mathcal{P}$;
      -- the set of unscanned machines $\mathcal{P}_1$ is initially equal to the set of all machines $\mathcal{P}$

  while $\mathcal{P}_1 \neq \varnothing$ **do**
    begin
    Find $k^* := \min\{k \mid P_k \in \mathcal{P}_1\}$;   -- machine P_{k^*} is selected (scanned)
    while $\mathcal{J}_{k^*} \neq \varnothing$ **do**
      begin
      Select J_{j^*} as the job with the largest tardiness among jobs
        belonging to $\mathcal{J}_{k^*}$;
      for $l = 1$ **to** $n_{k^* j^*}$ **do**
        begin   -- schedule revision
        Determine interval $[a_l, b_l]$ of possible values for the TDD value d_{lj^*}
          of task T_{lj^*};   -- this will be described separately
        for $x = a_l$ **to** b_l **do**
          begin
          Generate the due dates of other tasks of J_{j^*};
              -- this will be described separately
          call *MTD* rule;

```
                    -- all machines are rescheduled
        Record total tardiness D(x) = Σ Dⱼ;
                                      j
            end;
        Find x such that D(x) is minimum;
        d_{lj*} := x;
        Reassign due dates of other tasks of J_{j*} accordingly;
                        -- task due dates are chosen so that the value of the
                        -- total tardiness is minimized; this will be described separately
            end;
        for j = 1 to n do
        Calculate Dⱼ;          -- new tardiness values are calculated
        𝒥_{k*} := 𝒥_{k*} - {J_{j*}};
                    -- the list of unscanned jobs on P_{k*} is updated
        end;
        𝒫₁ = 𝒫₁ - {P_{k*}};          -- the list of unscanned machines is updated
        end;
    r := r + 1;
    z(r) := Σ Dⱼ;
            j
        end;
end;
```

We now discuss in more details the schedule revision loop, the major part of the algorithm, which is illustrated in Figure 14.2.1. As we see, the solution tree is similar to a branch and bound search tree with the difference that each node represents a complete solution.

Given the initial solution, we start with machine P_1 (which has the maximum workload), and job J_j (say) with the maximum tardiness among all jobs in $\mathcal{J}_1$. Consider task T_{1j} whose initial TDD is d_{1j}. The algorithm changes now this TDD to integer values in the interval $[L_1, U_1]$, where $L_1 = \sum_{l=1}^{l_1} p_{lj}$, $U_1 = d_j$. It follows from (14.2.1) that for any $d_{1j} < L_1$, the relative priority of T_{1j} remains unchanged, since L_1 is the earliest time by which T_{1j} can be completed.

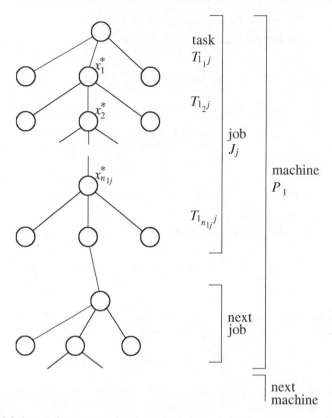

Figure 14.2.1   *Solution tree for the scheduling algorithm.*

Now, a descendant node is generated for each integer x in this interval. For a given x, the TDDs of other tasks of J_j are generated as follows

$$d_{ij} = d_{i-1j} + (x - p_{1_1 j})p_{ij}/t_{1_1-1j}, \; i = 1, 2, \cdots, 1_1 - 1$$

and

$$d_{ij} = d_{i-1j} + (d_j - x)p_{ij}/(p_j - t_{1_1 j}), \; i = 1_1 + 1, 1_1 + 2, \cdots, n_j,$$

where $t_{ij} = \sum_{l=1}^{i} p_{lj}$. Thus, we split J_j into three "sub-jobs" J_{j_1}, J_{j_2}, J_{j_3}, where J_{j_1} consists of all tasks prior to $T_{1_1 j}$, J_{j_2} contains only $T_{1_1 j}$, and J_{j_3} comprises all tasks following $T_{1_1 j}$. Due dates of all tasks within a sub-job are set independently of other sub-jobs. They are derived from the due date of the corresponding sub-job by assigning them flows proportional to their processing times, due dates of J_{j_1}, J_{j_2} and J_{j_3} being $x - p_{1_1 j}$, x, and d_j, respectively. TDDs of tasks of other jobs remain unchanged. The solution value for the descendant is determined by re-scheduling all jobs at all machines for the revised set of TDDs using the *MTD* rule. The branch corresponding to the node with the minimum total tardiness is selected, and the TDD of T_{1_1j} is fixed at the corresponding value of x, say x_1^*. TDDs of all tasks of J_j preceding T_{1_1j} are updated as follows

$$d_{ij} = d_{i-1j} + (x_1^* - p_{1_1j})t_{ij}/t_{1_1-1j}, \ i = 1, 2, \cdots, 1_1 - 1.$$

Next, the due date of task T_{1_2} is assigned. The interval scanned for d_{1_2j} is $[L_2, U_2]$

where $L_2 = \sum\limits_{i=1_1+1}^{1_2} t_{ij} + x_1^*$, $U_2 = d_j$. In the algorithm it is assumed $a_l = \lceil L_2 \rceil$ and $b_l = \lfloor U_2 \rfloor$. For a given value of x for the TDD of T_{1_2j}, the due dates of tasks of J_{j_1}, excluding T_{1_1j}, T_{1_2j} and those which precede T_{1_1j}, are generated as follows

$$d_{ij} = d_{i-1j} + (x - x_1^* - p_{1_2j})p_{ij}/(t_{1_2-1j} - t_{1_1j}), \ i = 1_1 + 1, \ 1_1 + 2, \cdots, 1_2 - 1$$

and

$$d_{ij} = d_{i-1j} + (d_j - x)p_{ij}/(p_j - t_{1_2j}), \ i = 1_2 + 1, \ 1_2 + 2, \cdots, n_j.$$

TDDs of tasks preceding and including T_{1_1j} remain unchanged.

In the general step, assume we are considering TDD reassignment of task T_{ij} at P_k (we omit index k for simplicity). Assume further that after investigating P_1 through P_{k-1} and all tasks of J_j prior to T_{ij} on P_k, we have fixed the due dates of tasks $T_{\overline{1}j}, T_{\overline{2}j}, \cdots, T_{\overline{z}j}$. Let T_{ij} be processed between $T_{\overline{l}j}$ and $T_{\overline{l+1}j}$ with fixed due dates of x_l^* and x_{l+1}^*, respectively, i.e. the ordered sequence of tasks of J_j is $(T_{1j}, T_{2j}, \cdots, T_{\overline{1}j}, \cdots, T_{\overline{2}j}, \cdots, T_{\overline{l}j}, \cdots, T_{ij}, \cdots, T_{\overline{l+1}j}, \cdots, T_{\overline{z}j}, \cdots, T_{n_jj})$. Then, for assigning the TDD of T_{ij}, we need to consider only the interval $\left[\sum\limits_{r=\overline{l+1}}^{i} p_{rj} + x_l^*, x_{l+1}^* - p_{\overline{l+1}j} \right]$. Moreover, while reassigning the TDD of T_{ij}, TDDs need to be generated for only those tasks which are processed between $T_{\overline{l}j}$ and $T_{\overline{l+1}j}$.

Of course, as the algorithm runs, the search interval becomes smaller. However, near the top of the tree, it can be quite wide. Thus, Raman and Talbot propose the following improvement of the search procedure. In general, while searching for the value x for a given task at any machine, we need theoretically consider the appropriate interval $[L, U]$ in unit steps. However, a task can only occupy a given number of positions α in any sequence. For a permutation schedule on a single machine we have $\alpha = n$, whereas in a job shop $\alpha > n$ because of the forced idle times on different machines. Nonetheless, it is usually much smaller than the number of different values of x. In consequence, TDDs, and thus total tardiness remains unchanged for many subintervals within $[L, U]$.

The procedure is a modification of the binary search method. Assume that we search in the interval $[L_0, U_0]$ (see Figure 14.2.2). First, we compute total tardiness $D(L_0)$ and $D(U_0)$ of all jobs for $x = L_0$ and $x = U_0$, respectively. Next, we divide the interval into two equal parts, compute the total tardiness in the midpoint of each half-interval, and so on. Within any generated interval, scanning for the next half-interval is initially done to the left, i.e. $U_i = \dfrac{L_0 + U_{i-1}}{2}$, $i = 1$, 2, 3, and terminates when a half-interval is fathomed, i.e. when the total tardiness at end-points and the midpoint of this interval are the same (e.g. $[L_0, U_3]$ in Figure 14.2.2).

Notice that this search procedure may not always find the best value of x. This is because it ignores (rather unlikely in real problems) changes in total tardiness within an interval, if the same tardiness is realized at both its end points and its midpoint.

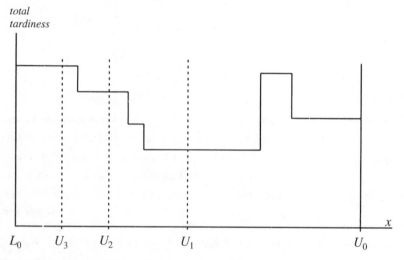

Figure 14.2.2   *A modification binary search procedure.*

After finishing of left-scanning, the procedure evaluates the most recently generated and unfathomed interval to its right. If the total tardiness at both end-points and the midpoint of that interval are not the same, another half-interval is generated and left scanning is resumed. The procedure stops when all half-intervals are fathomed.

Note that it is desirable to increase the upper limit of the search interval for the initial tasks of $\mathcal{J}_j$ from d_j to some arbitrarily large value (for example, the length of the initial solution). This follows from the fact that it can happen that the position of any task of a given job which corresponds to the minimum total tardiness results in that job itself being late. The presented search procedure reduces the computational complexity of each iteration from $O(m^2 N^3 \sum_{j=1}^{n} p_j)$ to $O(m^2 N^4)$, where $N = \sum_{j=1}^{n} n_j$.

14.2.3 Computational Experiments

Raman and Talbot [RT92] conducted extensive computational experiments to evaluate their algorithm (denoted *GSP - Global Scheduling Procedure*) for both static and dynamic problems.

For the static problem two sets of experiments were performed. The first compared *GSP* with the following priority rules known from literature: *SPT* (Shortest Processing Time), *EDD* (Earliest Due Date), *CRIT* (Critical Ratio), *MDD* (Modified Job Due Date), *MTD*, and *HYB* (Hybrid - see [RTR89a], uses *MTD* for scheduling jobs on non-bottleneck machines, and *MDD* on bottleneck machines), and with the exact algorithm running with a time trap of 14 sec. *GSP* provided the best results yielding an average improvement of 12.7% over the next best rule.

In the second set of experiments 100 problems (in four scenarios of 25 problems) were solved optimally as well as by the *GSP* algorithm. In majority of cases in each scenario *GSP* found the optimal solution. The performance of *GSP* relative to the optimum depends upon parameter ρ which determines the range of job due dates. *GSP* solutions are quite close to the optimum for large ρ, and the difference increases for smaller ρ.

Experiments for the dynamic problem were performed to study the effectiveness of implementing *GSP* solutions of the static problem on a rolling horizon basis. As we mentioned, in a dynamic environment a static problem is generated at each occurrence of a non-deterministic event in the system, such as an arrival of a new job. At such point in time a tree shown in Figure 14.2.1 is generated taking into account the tasks already in process. Of course, at that time some machines can be busy - they are blocked out for the period of commitment since we deal with the non-preemptive scheduling. The solution found by *GSP* is implemented until the next event occurs, as in the so-called reactive scheduling which will be discussed in more details in Chapter 15.

In the experiment job arrivals followed a Poisson process. Each job has assigned a due date that provided it a flow proportional to its processing time. Each job had a random routing through the system of 5 machines. Task processing times on each machine were sampled from a uniform distribution which varied to obtain two levels of relative machine workloads. The obtained machine utilizations ranged from 78% to 82% for the case of balanced workloads, and from 66% to 93% for unbalanced workloads. In both cases, the average shop utilization was about 80%. *GSP* was implemented with a time trap of 1.0 sec. per static problem, and compared with the same priority rules as in the case of the static problem experiment. The computational results are shown in Table 14.2.1. Since among the priority rules *MTD* performed the best for all scenarios, it has been used as a benchmark. Also given in the table is the corresponding level of significance α for one-tailed tests concerning paired differences between *MTD* and *GSP*. As we see, *GSP* retains its effectiveness for all flows, and for both levels of workload balance, and this holds for $\alpha = 0.15$ or less.

Flow	Balanced workloads			Unbalanced workloads		
	MTD	*GSP*	α	*MTD*	*GSP*	α
2	268	252	0.15	396	357	0.01
3	151	139	0.04	252	231	0.07
4	84	68	0.09	154	143	0.06
5	39	28	0.11	111	81	0.09

Table 14.2.1 *Experimental results for the dynamic problem.*

14.3 Simultaneous Scheduling and Routing in some FMS

14.3.1 Problem Formulation

In FMS scheduling literature majority of papers deal with either part and machine scheduling or with Automated Guided Vehicle (AGV) routing separately. In this section both issues are considered together, and the objective is to construct a schedule of minimum length [BEF+91].

The FMS under consideration has been implemented by one of the manufacturers producing parts for helicopters. A schematic view of the system is presented in Figure 14.3.1 and its description is as follows.

Pieces of raw material from which the parts are machined are stored in the automated storage area AS (1). Whenever necessary, an appropriate piece of material is taken from the storage and loaded onto the pallet and vehicle at the stand (2). This task is performed automatically by computer controlled robots. Then, the piece is transported by an AGV (7) to the desired machine (6) where it is automatically unloaded at (8). Every machine in the system is capable of processing any machining task. This versatility is achieved by a large number of tools and fixtures that may be used by the machines. The tool magazines (4) of every machine have a capacity of up to 130 tools which are used for the various machining operations. The tools of the magazines are arranged in two layers so that the longer tools can occupy two vertical positions. The tools are changed automatically. Fixtures are changed manually. It should be noted that a large variety of almost 100 quite different parts can be produced by each of these machines in this particular FMS. Simpler part types require about 30 operations (and tools) and the most complicated parts need about 80 operations. Therefore, the tool magazines have sufficient capacity to stock the tools for one to several consecutive parts in a production schedule. In addition, the tools are loaded from a large automated central tool storage area (3) which is located closely to the machines. No tool competition is observed, since the storage area contains more than 2000 tools (including many multiple tools) and there are 4 NC-machines. The delivered raw material is mounted onto the appropriate fixture and processed by the tools which are changed according to a desired plan. The tool technology of this particular system allows the changing of the tools during execution of the jobs. This is used to eliminate the setup times of the tools required for the next job and occasionally a transfer of a tool to another machine (to validate completely the no-resource competition). The only (negligible) transition time in the FMS that could be observed was in fact the adjustment in size of the spindle that holds the tool whenever the next tool is exchanged with the previous one. After the completion the finished part exchanges its position with the raw material of the next job that is waiting for its processing. It is then automatically transported by an AGV to the inspection section (9). Parts which passed the inspection are transported and unloaded at the storage area (10).

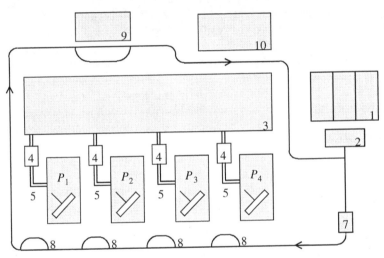

Figure 14.3.1   *An example FMS.*

We see that the above system is very versatile and this feature is gained by the usage of many tools and large tool magazines. As it was pointed out in Section 14.1, it is a common tendency of modern flexible manufacturing systems to become so versatile that most of the processes on a part can be accomplished by just one or at most two machine types. As a result many systems consist of identical parallel machines. On the other hand, the existence of a large number of tools in the system allows one not to consider resource (tool) competition. Hence, our problem here reduces in fact to that of simultaneous scheduling and routing of parts among parallel machines. The inspection stage can be postponed in that analysis, since it is performed separately on the first-come-first-served basis.

Following the above observations, we can model the considered FMS using elements described below. Given a set of n independent single-task jobs (parts) $J_1, J_2, \cdots, J_n$ with processing times $p_j, j = 1, 2, \cdots, n$, that are to be processed without preemptions on a set of m parallel identical machines $P_1, P_2, \cdots, P_m$, m not being a very large number. Here parallelism means that every machine is capable of processing any task. Setup times connected with changing tools are assumed to be zero since the latter can be changed on-line during the execution of tasks. Setup times resulting from changing part fixtures are included in the processing times.

As mentioned above, machines are identical except for their locations and thus they require different *delivery times*. Hence, we may assume that k ($k < m$) AGVs $V_1, V_2, \cdots, V_k$, are to deliver pieces of raw material from the storage area to specified machines and the time associated with the delivery is equal to $\tau_i, i = 1, 2, \cdots, m$. The delivery time includes loading time at the storage area and unloading time at the required machine, their sum being equal to a. During each trip exactly one piece of raw material is delivered; this is due to the dimension of parts to be machined. After delivery of a piece of raw material the vehicle takes a pallet with a processed part (maybe from another machine), delivers it to the inspection stage and returns to the storage area (1). The round trip takes A units of time, including two loading and two unloading times. It is apparent that the most efficient usage of vehicles in the sense of a throughput rate for pieces delivered is achieved when the vehicles are operating at a cyclic mode with cycle time equal to A. In order to avoid traffic congestion we assume that starting moments of consecutive vehicles at the storage area are delayed by a time units.

The problem is now to construct a schedule for machines and vehicles such that the whole job set is processed in a minimum time.

It is obvious that the general problem stated above is NP-hard, as it is already NP-hard for the non-preemptive scheduling of two machines (see Section 5.1). In the following we will consider two variants of the problem. In the first, the *production schedule* (i.e. the assignment of jobs to machines) is assumed to be known, and the objective is to find a feasible schedule for vehicles. This problem can be solved in polynomial time. The second consists of finding a composite schedule, i.e. one taking into account simultaneous assignment of vehicles and machines to jobs.

14.3.2 Vehicle Scheduling for a Fixed Production Schedule

In this section we consider the problem of vehicle scheduling given a production schedule. Suppose an (optimal) non-preemptive assignment of jobs to machines is given (cf. Figure 14.3.2). This assignment imposes certain deadlines d_j^i on delivery of pieces of raw material to particular machines, where d_j^i denotes the latest moment by which raw material for part J_j should be delivered to machine P_i.

The lateness in delivery could result in exceeding the planned schedule length C. Below we describe an approach that allows us to check whether it is possible to deliver all the required pieces of raw material to their destinations (given some production schedule), and if so, a vehicle schedule will be constructed. Without loss of generality we may assume that at time 0 at every machine there is already a piece of material to produce the first part; otherwise one should appropriately delay starting times on consecutive machines (cf. Figure 14.3.3).

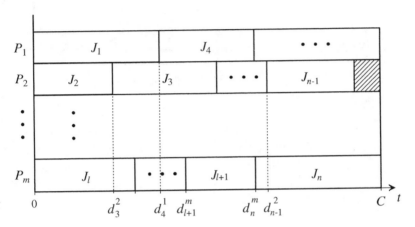

Figure 14.3.2   *An example production schedule.*

Our vehicle scheduling problem may now be formulated as follows. Given a set of deadlines d_j^i, $j = 1, 2, \cdots, n$, and delivery times from the storage area to particular machines τ_i, $i = 1, 2, \cdots, m$, is that possible to deliver all the required pieces of raw material on time, i.e. before the respective deadlines. If the answer is positive, a feasible vehicle schedule should be constructed. In general, this is equivalent to determining a feasible solution to a *Vehicle Routing with Time Windows* (see e.g., [DLSS88]). Let J_0 and J_{n+1} be two dummy jobs representing the first departure and the last arrival of every vehicle, respectively. Also define two dummy machines P_0 and P_{m+1} on which J_0 and J_{n+1} are executed, respectively, and let $\tau_0 = 0$, $\tau_{m+1} = M$ where M is an arbitrary large number. Denote by $i(j)$ the index of the machine on which J_j is executed. For any two jobs J_j, $J_{j'}$, let $c_{jj'}$ be the travel time taken by a vehicle to make its delivery for job $J_{j'}$ immediately after its delivery for J_j.

$$c_{jj'} = \begin{cases} \tau_{i(j')} - \tau_{i(j)} & \text{if } \tau_{i(j')} \geq \tau_{i(j)} \\ A - \tau_{i(j')} - \tau_{i(j)} & \text{if } \tau_{i(j')} < \tau_{i(j)} \end{cases} \quad j' = 0, \ldots, n+1, \ j \neq j'.$$

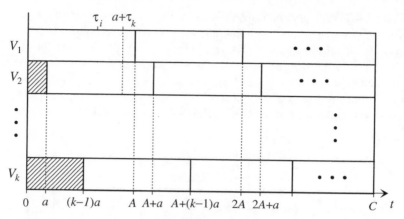

Figure 14.3.3   *An example vehicle schedule.*

If $\tau_j + c_{jj'} \le \tau_{j'}$, define a binary variable $x_{jj'}$ equal to 1 if and only if a vehicle makes its delivery for $J_{j'}$ immediately after its delivery for J_j. Also, let u_j be a non-negative variable denoting the latest possible delivery time of raw material for job J_j, $j = 1, \cdots, n$. The problem then consists of determining whether there exist values of the variables satisfying

$$\sum_{j'=1}^{n} x_{0j'} = \sum_{j=1}^{n} x_{j\,n+1} = k, \tag{14.3.1}$$

$$\sum_{j=0,\,j\ne l}^{n+1} x_{jl} = \sum_{j'=0,\,j'\ne l}^{n+1} x_{j'l} = 1, \qquad l = 1,\cdots,n, \tag{14.3.2}$$

$$u_j - u_{j'} + M x_{jj'} \le M - c_{jj'}, \qquad j,j' = 1,\cdots,n, j\ne j', \tag{14.3.3}$$

$$0 \le u_j \le d_j^i. \tag{14.3.4}$$

In this formulation, constraint (14.3.1) specifies that k vehicles are used, while constraints (14.3.2) associate every operation with exactly one vehicle. Constraints (14.3.3) and (14.3.4) guarantee that the vehicle schedule will satisfy time feasibility constraints. They are imposed only if $x_{jj'}$ is defined. This feasibility problem is in general NP-complete [Sav85]. However, for our particular problem, it can be solved in polynomial time because we can use the cyclic property of the schedule for relatively easily checking of the feasibility condition of the vehicle schedule for a given production schedule. The first schedule does not need to be constructed. When checking this feasibility condition one uses the job latest transportation starting times (using the assumption given at the beginning of this section) defined as follows

$$s_j = d_j^i - \tau_i, \quad j = m+1, m+2, \cdots, n.$$

The feasibility checking is given in Lemma 14.3.1.

Lemma 14.3.1 *For a given ordered set of latest transportation starting times* s_j, $s_j \leq s_{j+1}, j = m+1, m+2, \cdots, n,$ *one can construct a feasible transportation schedule for k vehicles if and only if*

$$s_j \geq (\lceil \tfrac{j-m}{k} \rceil - 1)A + [j - m - (\lceil \tfrac{j-m}{k} \rceil - 1)k - 1]a$$

for all $j = m+1, m+2, \cdots, n,$ *where* $\lceil \tfrac{j}{k} \rceil$ *denotes the smallest integer not smaller than j/k.*

Proof. It is not hard to prove the correctness of the above formula taking into account that its two components reflect, respectively, the time necessary for an integer number of cycles and the delay of an appropriate vehicle in a cycle needed for a transportation of the jth job in order. □

The conditions given in Lemma 14.3.1 can be checked in $O(n\log n)$ time in the worst case. If one wants to construct a feasible schedule, the following polynomial time algorithm will find it, whenever one exists. The basic idea behind the algorithm is to choose for transportation a job whose deadline, less corresponding delivery time, is minimum - i.e., the most urgent delivery at this moment. This approach is summarized by the following algorithm.

Algorithm 14.3.2 *for finding a feasible vehicle schedule given a production schedule with m machines* [BEF+91].
```
begin
t := 0; l := 0;
for j = m+1 to n do
Calculate job's J_j latest transportation starting time;    -- initial values are set up
Sort all the jobs in non-decreasing values of their latest transportation starting
    times and renumber them in this order;
for j = m+1 to n do
  begin
  Calculate slack time of the remaining jobs; sl_j := s_j - t;
  If any slack time is negative then stop;   -- no feasible vehicle schedule exists
  Load job J_j onto an available vehicle;
  l := l+1;
  if l ≤ k-1 then t := t+a
  else
    begin
    t := t - (k-1)a + A;
    l := 0;
    end;
  end;   -- all jobs are loaded onto the vehicles
end;
```

A basic property of Algorithm 14.3.2 is proved in the following theorem.

Theorem 14.3.3 *Algorithm* 14.3.2 *finds a feasible transportation schedule whenever one exists.*

Proof. Suppose that Algorithm 14.3.2 fails to find a feasible transportation schedule while such a schedule S exists. In this case there must exist in S two jobs J_i and J_j such that $sl_i < sl_j$ and J_j has been transported first. It is not hard to see that exchanging these two jobs, i.e., J_i being transported first, we do not cause the unfeasibility of the schedule. Now we can repeat the above pattern as long as such a pair of jobs violating the earliest slack time rule exists. After a finite number of such changes one gets a feasible schedule constructed according to the algorithm, which is a contradiction.   □

Let us now calculate the complexity of Algorithm 14.3.2 considering the off-line performance of the algorithm. Then its most complex function is the ordering of jobs in non-decreasing order of their slack times. Thus, the overall complexity would be $O(n\log n)$. However, if one performs the algorithm in the on-line mode, then the selection of a job to be transported next requires only linear time, provided that an unordered sequence is used. In both cases a low order polynomial time algorithm is obtained. We see that the easiness of the problem depends mainly on its regular structure following the cyclic property of the vehicle schedule.

Example 14.3.4 To illustrate the use of the algorithm, consider the following example. Let m the number of machines, n the number of jobs, and k the number of vehicles be equal to 3, 9 and 2, respectively. Transportation times for respective machines are $\tau_1 = 1$, $\tau_2 = 1.5$, $\tau_3 = 2$, and cycle and loading and unloading times are $A = 3$, $a = 0.5$, respectively. A production schedule is given in Figure 14.3.4(a). Thus the deadlines are $d_5^1 = 3$, $d_7^1 = 7$, $d_6^2 = 6$, $d_8^2 = 7$, $d_4^3 = 2$, $d_9^3 = 8$. They result in the latest transportation starting times $s_4 = 0$, $s_5 = 2$, $s_6 = 4.5$, $s_7 = 6$, $s_8 = 5.5$, $s_9 = 6$. The corresponding vehicle schedule generated by Algorithm 14.3.2 is shown in Figure 14.3.4(b). Job J_9 is delivered too late and no feasible transportation schedule for the given production plan can be constructed.   □

The obvious question is now what to do if there is no feasible transportation schedule. The first approach consists of finding jobs in the transportation schedule that can be delayed without lengthening the schedule. If such an operation is found, other jobs that cannot be delayed are transported first. In our example (Figure 14.3.4(a)) job J_7 can be started later and instead J_9 can be assigned first to vehicle V_1. Such an exchange will not lengthen the schedule. However, it may also be the case that the production schedule reflects deadlines which cannot be exceeded, and therefore the jobs cannot be shifted. In such a situation, one may use an alternative production schedule, if one exists. As pointed out in [Sch89], it is often the case at the FMS planning stage that several such plans may be constructed, and the operator chooses one of them. If none can be realized because of a non-feasible transportation schedule, the operator may decide to construct optimal production and vehicle schedules at the same time. One such approach based on dynamic programming is described in the next section.

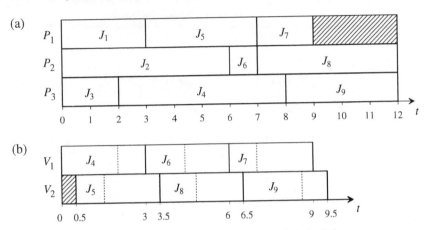

Figure 14.3.4 *Production and non-feasible vehicle schedules*
   (a) *production schedule,*
   (b) *vehicle schedule: J_9 is delivered too late.*

14.3.3 Simultaneous Job and Vehicle Scheduling

In this section, the problem of simultaneous construction of production and vehicle schedules is discussed. As mentioned above, this problem is NP-hard, although not strongly NP-hard. Thus, a pseudopolynomial time algorithm based on dynamic programming can be constructed for its solution.
Assume that jobs are ordered in non-increasing order of their processing times, i.e. $p_1 \geq \cdots \geq p_{n-1} \geq p_n$. Such an ordering implies that longer jobs will be processed first and processing can take place on machines further from the storage area, which is a convenient fact from the viewpoint of vehicle scheduling.
Now let us formulate a dynamic programming algorithm using the ideas presented in [GLL+79]. Define

$$x_j(t_1, t_2, \cdots, t_m) = \begin{cases} \textbf{true} & \text{if jobs } J_1, J_2, \cdots, J_j \text{ can be scheduled on} \\ & \text{machines } P_1, P_2, \cdots, P_m \text{ in such a way that} \\ & P_i \text{ is busy in time interval } [0, t_i], i = 1, \\ & 2, \cdots, m \text{ (excluding possible } idle\ time \text{ fol-} \\ & \text{lowing from vehicle scheduling), and the} \\ & \text{vehicle schedule is feasible} \\ \textbf{false} & \text{otherwise} \end{cases}$$

where

$$x_0(t_1, t_2, \cdots, t_m) = \begin{cases} \textbf{true} & \text{if } t_i = 0, i = 1, 2, \cdots, m \\ \textbf{false} & \text{otherwise.} \end{cases}$$

Using these variables, the recursive equation can be written in the following form

$$x_j(t_1, t_2, \cdots, t_m) =$$

$$\bigvee_{i=1}^{m} [x_{j-1}(t_1, t_2, \cdots, t_{i-1}, t_i - p_i, t_{i+1}, \cdots, t_m) \wedge Z_{ij}(t_1, t_2, \cdots, t_{i-1}, t_i, t_{i+1}, \cdots, t_m)]$$

where

$$Z_{ij}(t_1, t_2, \cdots, t_{i-1}, t_i, t_{i+1}, \cdots, t_m) =$$

$$\begin{cases} \textbf{true} & \textbf{if } t_i - p_j - \tau_i \geq \\ & (\lceil \frac{i-m}{k} \rceil - 1)A + [j - m - (\lceil \frac{i-m}{k} \rceil - 1)k - 1]a \\ & \textbf{or } j \leq m \\ \textbf{false} & \text{otherwise} \end{cases}$$

is the condition of vehicle schedule feasibility, given in Lemma 14.3.1.

Values of $x_j(\cdot)$ are computed for $t_i = 0, 1, \cdots, C, i = 1, 2, \cdots, m$, where C is an upper bound on the minimum schedule length C_{max}^*. Finally, C_{max}^* is determined as

$$C_{max}^* = \min\{\max\{t_1, t_2, \cdots, t_m\} \mid x_n(t_1, t_2, \cdots, t_m) = \textbf{true}\}.$$

The above algorithm solves our problem in $O(nC^m)$ time. Thus, for fixed m, it is a pseudopolynomial time algorithm, and can be used in practice, taking into account that m is rather small. To complete our discussion, let us consider once more the example from Section 14.3.2. The above dynamic programming approach yields schedules presented in Figure 14.3.5. We see that it is possible to complete all the jobs in 11 units and deliver them to machines in 8 units.

To end this section let us notice that various extensions of the model are possible and worth considering. Among them are those including different routes for particular vehicles, an inspection phase as the second stage machine, resource competition, and different criteria (e.g., maximum lateness). These issues are currently under investigation.

Further extensions of the described FMS model have been presented in [BBFW94] and [KL95].

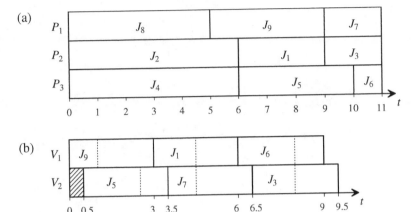

Figure 14.3.5 *Optimal production and vehicle schedule*
      (a) *production schedule,*
      (b) *vehicle schedule.*

14.4 Batch Scheduling in Flexible Flow Shops under Resource Constraints

The cast-plate-method of manufacturing acrylic-glass gives raise to a batch scheduling problem on parallel processing units under resource constraints. This chapter introduces the real world problem as well as an appropriate mathematical programming formulation. The problem finally is solved heuristically.

14.4.1 Introduction - Statement of the Problem

The cast-plate-method for manufacturing acrylic-glass essentially consists of the preparation of a viscous chemical solution, pouring it in a mould i.e. between two plates of mineral glass (like a sandwich) and polymerizing the syrup to solid sheets. Sandwiches of the same product are collected on storage racks. Figure 14.4.1 shows a manufacturing plant and its production facilities.

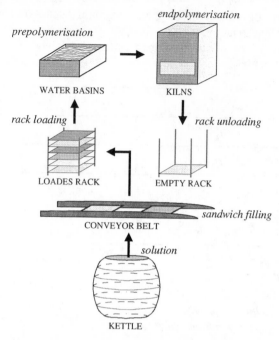

Figure 14.4.1 *Procedure of manufacturing acrylic-glass.*

Depending on the included product, each rack has to be successively put into one to four water basins holding particular temperatures for pre-polymerization. Subsequently, every rack has to be designed to a kiln where end-polymerization takes place, using a particular kiln temperature (temper cycle). As soon as polymerization is concluded the racks are unloaded, i.e. the mineral glass plates are removed and the hardened acrylic-glass plates may be taken to the quality control section. The production cycle for the empty rack starts again, using a "new" solution, i.e. loading the rack with next sheets, see [FKPS91].

The goal of the optimization process was to find optimal (at least "good") production schedules for the weekly varying manufacturing program in order to improve the plant's polymerization capacity utilization.

14.4.2 Mathematical Formulation

For a given manufacturing program (say: product mix) and a fixed time horizon we examine the polymerization area:

- *Pre-polymerization* in (water-)basins:
  There is the choice between a given number of basins different in size and with different temperatures. Every basin can be heated to any temperature. Racks may be put into or taken out of the basins at any time.
- *End-polymerisation* in kilns:

The present kilns solely differ in size. Every kiln can handle any temper cycle (i.e. heating up, processing, and cooling down). Kilns must not be opened while such a cycle is still in progress.

The production units before and after the above mentioned polymerization area such as conveyer belts, kettles and quality control are linked through buffers and hence remain uncritical. Thus, we do not need to include those areas in our considerations about optimal schedules and consequently define the polymerization area to be the optimization field. However, we have to consider the fact, that the number of storage racks for every type of rack is limited at any time (cf. global rack restrictions).

The interval $[0, H]$ is the given planning period ($H = 10080$ in minutes, i.e. a week); $t \in [0, H]$ denotes entry times, H is the planning horizon. For technical reasons we allow t to be an integer, but only decision variables with $t \in [0, \cdots, H]$ in a feasible solution are of importance.

The customers' orders, as part of a given product mix, can be divided according to their characteristic attributes such as type, size, and diameter. A *job* *(product)* J_i, $i = 1, \cdots, n$, denotes all orders of the same type, size, and diameter.

Size and diameter determine the type of rack to be used for job J_i, whereas the number τ_i of racks needed can be figured out according to the racks' holding capacity and the total requirement of sheets for that job.

Each basin and each kiln is large enough in order to hold at least one rack regardless of its type. Furthermore the breadth of all basins and kilns is almost the same. They only differ in their length. Racks may only placed one after another into basins or kilns. Even putting racks of the smallest breadth next to each other is impossible. Hence to satisfy capacity constraints we only need to consider the production units' length. So, a real number κ_i according to the particular rack's size is assigned to every job, where κ_i equals the rack's breadth if the (rack's breadth $\le$) rack's length is smaller or equal the unit's breadth, and where κ_i equals the rack's length if the (rack's breadth $\le$) unit's breadth is smaller than the rack's length. The κ_i should be chosen such that shunter distances are taken into consideration.

A 1-*basin job* is a job that needs to pre-polymerize in one basin only. A 2-*basin job* (3-basin, 4-basin) *job* is a job the sandwiches of which need to pass through two (three, four) basins with different temperatures. $\mathcal{J}$ is the set of all jobs and $I = \{1, \cdots, n\}$ is the set of indices for all jobs $J_i \in \mathcal{J}$. I_υ, $\upsilon = 1, \cdots, 4$, is the set of indices for all υ-basin jobs. Thus, $I = I_1 \cup I_2 \cup I_3 \cup I_4$; $I_k \cap I_j = \varnothing$ for all $k, j \in \{1, \cdots, 4\}$ and $k \ne j$.

Each job has its unique *work schedule* to be applied to each rack of that job. Different jobs require different work schedules. A work schedule is a table of the following non-negative real numbers (considering multiple-basin jobs):

- Maximum allowed waiting time in front of the basin area;
- Basin 1: temperature and duration of stay,

- Basin 2: temperature and duration of stay,
- Basin 3: temperature and duration of stay,
- Basin 4: temperature and duration of stay;
- Maximum allowed waiting time between basin- and kiln area;
- Kiln: temperature and duration of stay.

A rack, holding filled sandwiches, is the smallest unit to be scheduled. For racks holding sandwiches of the same job J_i we use the index l, $l = 1, \cdots, \tau_i$.

There are K different types of (empty) racks; a_k gives the number of present racks for every type k, $k = 1, \cdots, K$.

A basin q, $q = 1, \cdots, \Omega$, is characterized by its temperature u_q and its length v_q. Ω_μ ($\mu = 1, \cdots, 4$) gives the number of basins of size μ and $\Omega = \Omega_1 + \Omega_2 + \Omega_3 + \Omega_4$ is the total number of basins in the considered plant. Besides the four distinct basin sizes we consider two distinct kiln sizes.

A kiln r, $r = 1, \cdots, N$, is characterized by its length v_r. N_1 (N_2) gives the number of large (small) kilns in the plant, $N = N_1 + N_2$ is the total number of kilns.

The Model

The following definitions and parameters are used to describe a mathematical model of the problem.
Definitions and parameters:

A *fictitious kiln* r, $r = 1, \cdots, m$, is one of the kilns supposed to run with a particular temperature u_r.

m	total number of fictitious kilns ($m = Nrt$, where rt is the number of different temperatures required)
r_1 (r_2)	number of fictitious large (small) kilns ($r_1 = N_1 rt$, $r_2 = N_2 rt$)
$\{1, \cdots, r_1\}$	set of indices concerning fictitious large kilns
$\{r_1 + 1, \cdots, m\}$	set of indices concerning fictitious small kilns

A *fictitious basin* q, $q = 1, \cdots, Q$, is one of the basins heated to a particular temperature.

Q	total number of fictitious basins ($Q = \Omega qt$, where qt is the number of different temperatures needed)
q_1 (q_2, q_3, q_4)	number of fictitious small (medium-size, large, extra-large) basins ($q_\upsilon = \Omega_\upsilon qt$)
$\{1, \cdots, q_1\}$	set of indices concerning small basins
$\{q_1 + 1, \cdots, q_1 + q_2\}$	set of indices concerning medium-size basins
$\{q_1 + q_2 + 1, \cdots, q_1 + q_2 + q_3\}$	set of indices concerning large basins
$\{q_1 + q_2 + q_3 + 1, \cdots, Q\}$	set of indices concerning extra-large basins

Each rack l of job J_i with $l > \tau_i$ is called a *fictitious rack* of job J_i. Fictitious racks are auxiliary tools. Let us illustrate their purpose:

Consider rack l, where $1 \leq l \leq \tau_i$, containing a 3-basin job. This rack has to pre-polymerize in three successive basins. We define fictitious racks as copies of the present rack in accordance to the steps of pre-polymerisation. A fictitious rack $l + \tau_i$ is created to describe the original rack l in its second basin. Another fictitious rack $l + 2\tau_i$ arises in accordance to its third step of pre-polymerisation. Thus, the rack index corresponds with the sequence of basins the existing rack passes through. Obviously, fictitious racks occur for multiple-basin jobs only, i.e. for $i \in I_2 \cup I_3 \cup I_4$.

Let $\bar{\tau}_i$ be the number of (real and fictitious) racks for job J_i, then $\bar{\tau}_i := \upsilon \tau_i$ if $i \in I_\upsilon$ ($\upsilon = 1, \cdots, 4$).

We call jobs J_i and J_j *compatible* if their work schedules contain identical temperatures and durations for end-polymerization, meaning that racks holding sandwiches of those jobs may be assigned to the same kiln at one time (regard, a kiln cannot be opened during polymerization). Hence we define $A = (a_{ij})$, $i, j = 1, \cdots, n$, as the *job-compatibility matrix*, that characterizes the jobs' compatibility with respect to its mere chemical features (they determine the temperatures required in a kiln), its temper-time, and possible preferences. The matrix is defined as

$$a_{ij} := \begin{cases} 1 & \text{if jobs } J_i \text{ and } J_j \text{ must not join the same kiln} \\ 0 & \text{else.} \end{cases}$$

Finally we introduce some time parameters:

α_{il} time, when the lth rack of job J_i leaves the basin area

σ_{ilr} (duration of) stay of the lth rack of job J_i in kiln r

σ_{ilq} (duration of) stay of the lth rack of job J_i in basin q

ω_i maximum allowed waiting time for a rack of job J_i between basin- and kiln area

H_{max} maximum allowed waiting time before entering the basin area

Z_{max} maximum allowed waiting time for a rack of a multiple-basin job between two basins

where $1 \leq i \leq n$; $1 \leq l \leq \tau_i$, $\bar{\tau}_i$, respectively; $1 \leq r \leq m$; $1 \leq q \leq Q$.

Infeasible or undesirable assignments of job J_i to basin q (or kiln r) can be prevented by fixing $\sigma_{ilq} = \infty$ (or $\sigma_{ilr} = \infty$) for all $l = 1, \cdots, \tau_i$. We may assume that σ_{ilq} includes any kind of necessary setup times. Analogously, σ_{ilr} includes setup times as well as heating up and cooling down times that occur in the temper cycle.

Remark: We assumed, that at time $t = 0$ every rack of the given production plan potentially stands by at the beginning of the basin area. (Of course, later on we have to make sure, that the global rack restrictions are satisfied.) Granting exception to this assumption, the program can easily be remodeled by introducing times when rack l of job J_i enters and leaves the basin area.

Our decision variables are

$$x_{ilrt} \in \{0, 1\} \quad \text{for } i = 1, \cdots, n, l = 1, \cdots, \tau_i, r = 1, \cdots, m, t = 0, \cdots, H,$$

$$x_{ilqt} \in \{0, 1\} \quad \text{for } i = 1, \cdots, n, l = 1, \cdots, \overline{\tau}_i, q = 1, \cdots, Q, t = 0, \cdots, H,$$

where

$$x_{ilrt} := \begin{cases} 1 & \text{if the } l^{\text{th}} \text{ rack of job } J_i \text{ enters kiln } r \text{ at time } t, \\ 0 & \text{else.} \end{cases}$$

$$x_{ilqt} := \begin{cases} 1 & \text{if the } l^{\text{th}} \text{ rack of job } J_i \text{ enters basin } q \text{ at time } t, \\ 0 & \text{else.} \end{cases}$$

For technical reasons we define auxiliary variables $x_{ilqt} \in \{0, 1\}$ for $t < 0$; $i = 1, \cdots, n; l = 1, \cdots, \overline{\tau}_i; q = 1, \cdots, Q$ analogously.

An entire allocation scheme for the basin area is denoted by $\overline{x}$ whereas $\hat{x}$ denotes an entire allocation scheme for the kiln area. According to the definition of x_{ilqt} and x_{ilrt}, both $\overline{x}$ and $\hat{x}$ represent a binary four-dimensional matrix. An allocation scheme for the entire polymerization area is denoted by $(\overline{x}, \hat{x})$.

A mathematical programming model for the *basin area* is given below.

Objective function:

$$\text{Minimize } \varphi_1(\overline{x}) = \sum_{i \in I_1} \sum_{l=1}^{\tau_i} \sum_{q=1}^{Q} \sum_{t=0}^{H} (t + \sigma_{ilq}) x_{ilqt} +$$

$$\sum_{i \in I_2} \sum_{l=1}^{\tau_i} \sum_{q=1}^{Q} \sum_{t=0}^{H} (t + \sigma_{i(\tau_i+l)q}) x_{i(\tau_i+l)qt} +$$

$$\sum_{i \in I_3} \sum_{l=1}^{\tau_i} \sum_{q=1}^{Q} \sum_{t=0}^{H} (t + \sigma_{i(2\tau_i+l)q}) x_{i(2\tau_i+l)qt} +$$

$$\sum_{i \in I_4} \sum_{l=1}^{\tau_i} \sum_{q=1}^{Q} \sum_{t=0}^{H} (t + \sigma_{i(3\tau_i+l)q}) x_{i(3\tau_i+l)qt}$$

$$\text{Minimize } \varphi_2(\overline{x}) = \max_{\substack{i \in I \\ l=1,\dots,\overline{\tau}_i \\ q=1,\dots,Q \\ t=0,\dots,H}} \{(t + \sigma_{ilq}) x_{ilqt}\}$$

subject to

"waiting time constraints":   (14.4.1)

$$tx_{ilqt} \leq H_{max} \qquad \text{for } i \in I; \; l = 1, \cdots, \tau_i ;$$

$$q = 1, \cdots, Q; \; t = 0, \cdots, H \qquad (14.4.1a)$$

$$tx_{i(l+\tau_i)\mu t} - (\tau + \sigma_{ilq})x_{ilq\tau} \leq Z_{max}$$

$$i \in I_2 \cup I_3 \cup I_4; \; l = 1, \cdots, \tau_i ;$$

$$\mu, q = 1, \cdots, Q; \; \tau, t = 0, \cdots, H \qquad (14.4.1b)$$

$$t \cdot x_{i(l+2\tau_i)\mu t} - (\tau + \sigma_{i(l+\tau_i)q}) \cdot x_{i(l+\tau_i)q\tau} \leq Z_{max}$$

$$i \in I_3 \cup I_4; \; l = 1, \cdots, \tau_i ;$$

$$\mu, q = 1, \cdots, Q; \; \tau, t = 0, \cdots, H \qquad (14.4.1c)$$

$$t \cdot x_{i(l+3\tau_i)\mu t} - (\tau + \sigma_{i(l+2\tau_i)q})x_{i(l+2\tau_i)q\tau} \leq Z_{max}$$

$$i \in I_4; \; l = 1, \cdots, \tau_i ;$$

$$\mu, q = 1, \cdots, Q; \; \tau, t = 0, \cdots, H \qquad (14.4.1d)$$

"*basin capacity constraints*": $\qquad\qquad\qquad\qquad\qquad\qquad (14.4.2)$

$$\sum_{i=1}^{n} \sum_{l=1}^{\overline{\tau}_i} \kappa_i [x_{ilqt} + x_{ilq(t-1)} + \cdots + x_{ilq(t-\sigma_{ilq}+1)}] \leq \upsilon_q$$

$$\text{for } q = 1, \cdots, Q; \; t = 0, \cdots, H$$

"*basin coordination constraints*": $\qquad\qquad\qquad\qquad (14.4.3)$

Using $\pi_1 := q_1 + 1$, $\pi_2 := q_1 + q_2$, $\pi_3 := q_1 + q_2 + 1$, $\pi_4 := q_1 + q_2 + q_3$ and $\pi_5 := q_1 + q_2 + q_3 + 1$ and the signum function sign: $I\!R \to I\!R_{>0}$ defined as

$$\text{sign}(x) = \begin{cases} 1 & \text{if } x > 0 \\ 0 & \text{if } x = 0 \\ -1 & \text{if } x < 0 \end{cases}$$

we have the following constraints:

$$\sum_{q=1}^{q_1} \text{sign}\Big[\sum_{i=1}^{n} \sum_{l=1}^{\overline{\tau}_i} (x_{ilqt} + x_{ilq(t-1)} + \cdots + x_{ilq(t-\sigma_{ilq}+1)})\Big] \leq \Omega_1 \qquad \text{for } t = 0, \cdots, H;$$

$$\sum_{q=\pi_1}^{\pi_2} \text{sign}\Big[\sum_{i=1}^{n} \sum_{l=1}^{\overline{\tau}_i} (x_{ilqt} + x_{ilq(t-1)} + \cdots + x_{ilq(t-\sigma_{ilq}+1)})\Big] \leq \Omega_2 \qquad \text{for } t = 0, \cdots, H;$$

$$\sum_{q=\pi_3}^{\pi_4} \text{sign}\Big[\sum_{i=1}^{n} \sum_{l=1}^{\overline{\tau}_i} (x_{ilqt} + x_{ilq(t-1)} + \cdots + x_{ilq(t-\sigma_{ilq}+1)})\Big] \leq \Omega_3 \qquad \text{for } t = 0, \cdots, H;$$

$$\sum_{q=\pi_5}^{Q} \text{sign}\Big[\sum_{i=1}^{n} \sum_{l=1}^{\overline{\tau}_i}(x_{ilqt} + x_{ilq(t-1)} + \cdots + x_{ilq(t-\sigma_{ilq}+1)})\Big] \leq \Omega_4 \qquad \text{for } t = 0, \cdots, H;$$

"throughput constraints": (14.4.4)

$$\sum_{q=1}^{Q} \sum_{t=0}^{H} x_{ilqt} = 1 \qquad \text{for } i = 1, \cdots, n; \ l = 1, \cdots, \overline{\tau}_i$$

"basin-sequence constraints": (14.4.5)

$$x_{ilqt} + x_{i(l+\tau_i)\mu\tau} \leq 1$$

$$\text{for } i \in I_2; \ l = 1, \cdots, \tau_i; \ \mu, q = 1, \cdots, Q;$$

$$t = 0, \cdots, H; \tau = 0, \cdots, t + \sigma_{ilq} \qquad (14.4.5a)$$

$$x_{ilqt} + x_{i(l+\tau_i)\mu\tau} \leq 1$$

$$\text{for } i \in I_3; \ l = 1, \cdots, 2\tau_i; \ \mu, q = 1, \cdots, Q;$$

$$t = 0, \cdots, H; \tau = 0, \cdots, t + \sigma_{ilq} \qquad (14.4.5b)$$

$$x_{ilqt} + x_{i(l+\tau_i)\mu\tau} \leq 1$$

$$\text{for } i \in I_4; \ l = 1, \cdots, 3\tau_i; \ \mu, q = 1, \cdots, Q;$$

$$t = 0, \cdots, H; \tau = 0, \cdots, t + \sigma_{ilq} \qquad (14.4.5c)$$

"binary constraints": (14.4.6)

$$x_{ilqt} \in \{0, 1\} \qquad \text{for } i = 1, \cdots, n; \ l = 1, \cdots, \overline{\tau}_i; \ q = 1, \cdots, Q; \ t = 0, \cdots, H$$

"prohibiting infeasible assignments": (14.4.7)

$$x_{ilqt} = 0 \qquad \text{for } l = 1, \cdots, \overline{\tau}_i; \ t = 0, \cdots, H$$

and for all $(i, q) \in I \times \{1, \cdots, Q\}$
such that job J_i is not to be designed to basin q

"technical constraints": (14.4.8)

$$x_{ilqt} = 0 \qquad \text{for } t < 0; \ i = 1, \cdots, n; l = 1, \cdots, \overline{\tau}_i; \ q = 1, \cdots, Q.$$

The objective function $\varphi_1(\overline{x})$ minimizes the sum of flow times of the racks in the basins area in order to gain an optimal throughput with respect to the given product mix. The objective function $\varphi_2(\overline{x})$ gives the instant when the very last rack (as part of the given product mix) leaves the basin area. Minimizing $\varphi_2(\overline{x})$ corresponds to finishing processing of the given product mix in the basin area as soon as possible (i.e. minimizing the makespan).

The waiting time constraints (14.4.1) assure that the maximum allowed waiting time is observed by every rack before entering the first basin for all $i \in I$ (14.4.1a), between the first and the second basin for all $i \in I_2 \cup I_3 \cup I_4$ (14.4.1b), between the second and third basin for all $i \in I_3 \cup I_4$ (14.4.1c), and before entering the fourth basin for all $i \in I_4$ (14.4.1d). The basin capacity constraints (14.4.2) guarantee that each basin's capacity is never exceeded. The basin coordination constraints (14.4.3) state that the number of fictitious basins to be chosen at any time is limited by the number of physically present basins. This holds for basins of all different sizes. Throughput constraints (14.4.4) require that every fictitious rack is assigned to exactly one basin during the regarded planning period. The basin-sequence constraints (14.4.5) assure, that pre-polymerization for racks of multiple-basin jobs proceeds in due succession with respect to the particular work schedule. Inequalities (14.4.5a) guarantee that, considering 2-basin jobs, pre polymerization in the first basin has to be finished before continuing in a second basin. (14.4.5b) enforce the appropriate order of basins 1, 2, 3 for all 3-basin jobs, for 4-basin jobs (14.4.5c) perform analogously. For every fictitious rack l of a job J_i a binary constraint (14.4.6) characterizes whether the rack is assigned to basin q at time t or not. The set of equations (14.4.7) are tools for precluding infeasible or objectionable assignments. Negative time subscripts t may occur in constraints (14.4.3). Thus, in (14.4.8) we also define decision variables x_{ilqt} for $t < 0$, though they are of no importance for the problem in practice.

A mathematical programming model for the *kiln area* is formulated as follows:

Objective function:

$$\text{Minimize } \psi_1(\hat{x}) = \sum_{i=1}^{n} \sum_{l=1}^{\tau_i} \sum_{r=1}^{m} \sum_{t=0}^{H} [(t - \alpha_{il}) + \sigma_{ilr}] x_{ilrt}$$

$$\text{Minimize } \psi_2(\hat{x}) = \max_{\substack{i=1,\ldots,n \\ l=1,\ldots,\tau \\ r=1,\ldots,m \\ t=0,\ldots,H}} \{(t + \sigma_{ilr}) x_{ilrt}\}$$

subject to

"availability constraints": (14.4.9)

$$t x_{ilrt} \geq \alpha_{il} \qquad \text{for } i = 1, \cdots, n ; \ l = 1, \cdots, \tau_i ; \ r = 1, \cdots, m ; \ t = 0, \cdots, H$$

"waiting time constraints": (14.4.10)

$$(t - \alpha_{il}) x_{ilrt} \leq \omega_i \ \text{ for } i = 1, \cdots, n ; \ l = 1, \cdots, \tau_i ; \ r = 1, \cdots, m ; \ t = 0, \cdots, H$$

"kiln capacity constraints": (14.4.11)

$$\sum_{i=1}^{n} \sum_{l=1}^{\tau_i} x_{ilrt}\kappa_i \le v_r \qquad\qquad \text{for } r = 1, .., m \ ; \ t = 0, \cdots, H$$

"*kiln coordination constraints*": $\qquad\qquad\qquad\qquad\qquad\qquad$ (14.4.12)

$$\sum_{r=1}^{r_1} \text{sign}\Big[\sum_{i=1}^{n} \sum_{l=1}^{\tau_i} (x_{ilrt} + x_{ilr(t-1)} + \cdots + x_{ilr(t-\sigma_{ilr}+1)})\Big] \le N_1$$

$$\text{for } t = 0, \cdots, H \qquad\qquad (14.4.12a)$$

$$\sum_{r=r_1+1}^{m} \text{sign}\Big[\sum_{i=1}^{n} \sum_{l=1}^{\tau_i} (x_{ilrt} + x_{ilr(t-1)} + \cdots + x_{ilr(t-\sigma_{ilr}+1)})\Big] \le N_2$$

$$\text{for } t = 0, \cdots, H \qquad\qquad (14.4.12b)$$

"*product compatibility constraints*": $\qquad\qquad\qquad\qquad\qquad$ (14.4.13)

$$a_{ij}(x_{ilrt} + x_{jkrt}) \le 1$$

$$\text{for } i, j = 1, \cdots, n \ ; \ l = 1, \cdots, \tau_i \ ; \ k = 1, \cdots, \tau_j \ ; \ r = 1, \cdots, m; \ t = 0, \cdots, H;$$

"*kiln closed constraints*": $\qquad\qquad\qquad\qquad\qquad\qquad\qquad$ (14.4.14)

$$(x_{ilr\tau} + x_{jkrt}) \le 1 \qquad\quad \text{for } i, j = 1, \cdots, n \ ; \ l = 1, \cdots, \tau_i \ ; \ k = 1, \cdots, \tau_j \ ;$$

$$r = 1, \cdots, m \ ; \ \tau = 0, \cdots, H \ ; \ t = \tau+1, \cdots, \tau+\sigma_{ilr}$$

"*throughput constraints*": $\qquad\qquad\qquad\qquad\qquad\qquad\qquad$ (14.4.15)

$$\sum_{r=1}^{m} \sum_{t=0}^{H} x_{ilrt} = 1 \qquad\quad \text{for } i = 1, \cdots, n; \ l = 1, \cdots, \tau_i$$

"*binary constraints*": $\qquad\qquad\qquad\qquad\qquad\qquad\qquad\qquad$ (14.4.16)

$$x_{ilrt} \in \{0, 1\} \qquad \text{for } i = 1, \cdots, n \ ; \ l = 1, \cdots, \tau_i \ ; \ r = 1, \cdots, m \ ; \ t = 0, \cdots, H$$

"*job completion constraints*": $\qquad\qquad\qquad\qquad\qquad\qquad$ (14.4.17)

$$(t+\sigma_{ilr})\dot{x}_{ilrt} \le H \quad \text{for } i = 1, \cdots, n \ ; \ l = 1, \cdots, \tau_i \ ; \ r = 1, \cdots, m \ ; \ t = 0, \cdots, H$$

"*prohibiting infeasible assignments*": $\qquad\qquad\qquad\qquad\quad$ (14.4.18)

$$x_{ilrt} = 0 \qquad\qquad\qquad \text{for } l = 1, \cdots, \tau_i \ ; \ t = 0, \cdots, H$$

$$\text{for all } (i, r) \in I \times \{1, \cdots, m\} \text{ such that job } J_i \text{ is not to}$$
$$\text{be assigned to a fictitious kiln } r$$

The objective function $\psi_1(\hat{x})$ sums up the times that the racks of the given product mix spend in the kiln area in accordance to the basin exit times α_{il} given as a function of a basin schedule $\bar{x}$. We minimize $\psi_1(\hat{x})$ in order to gain an optimal throughput with respect to the kiln area. The objective function $\psi_2(\hat{x})$ determines the time, when the very last rack of the given product mix leaves the kiln area. Minimizing $\psi_2(\hat{x})$ corresponds with finishing polymerisation of all given racks in the kiln area as soon as possible, i.e. minimizing makespan.

The availability constraints (14.4.9) assure, that no rack is assigned to a kiln prior to termination of its pre-polymerization in the basin area. Constraints (14.4.10) prevent that the maximum waiting time between release in the basin area and assignment to a kiln is exceeded for any rack. Constraints (14.4.11) make sure, that the capacity of each kiln is observed at any time. Constraints (14.4.12) state, that the number of fictitious kilns chosen at any time is limited by the number of physically present kilns. Constraints (14.4.13) require, that racks which are assigned to the same kiln at a time need to hold compatible jobs. The kiln closed constraints (14.4.14) state, that there is neither recharging nor untimely removal of racks while polymerization is still in progress. Throughput constraints (14.4.15) require that every fictitious rack is assigned to exactly one kiln during the planning period. For every fictitious rack a binary constraint of (14.4.16) characterizes whether the rack is assigned to a kiln r at time t. The job completion constraints (14.4.17) state that all racks have to leave the kiln area during the planning period, i.e. until the planning horizon H. Constraints (14.4.18) exclude infeasible or objectionable assignments like priorities of particular customer orders.

The holding time of each rack consists of the period of time spent in the optimization area and the time needed for filling (mounting) and dismounting. Assume that t_{il} is an empirical upper bound for this period of time for rack l of job J_i. Let $\bar{\sigma}_i$ be an upper bound for the flow time of job J_i in the basin area. It consists of waiting times before and between the basins as well as of the pre-polymerization times spent in the basins. Hence we obtain $\bar{\sigma}_i \leq \max_q \{\sigma_{ilq}\} + H_{max}$ for all $l = 1, \cdots, \tau_i$ and all 1-basin jobs J_i, $i \in I_1$. The flow time for multiple-basin jobs is given by the sum of flow time in the respective basins and the waiting times spent between them. For a rack l of a 3-basin job J_i, $i \in I_3$, for instance, we get

$$\bar{\sigma}_i \leq \max_q \{\sigma_{ilq}\} + \max_q \{\sigma_{i(\tau_i+l)q}\} + \max_q \{\sigma_{i(2\tau_i+l)q}\} + H_{max} + 2Z_{max}.$$

Furthermore, we define $\mathcal{J}_k := \{i \in I \mid \text{job } J_i \text{ requires racks of type } k\}$ for every $k \in \{1, \cdots, K\}$. Of course, $I = \bigcup_{k=1}^{K} \mathcal{J}_k$.

Using these notations, the global rack constraints can be formulated as follows:

"*global rack constraints*":   (14.4.19)

$$\sum_{i \in \mathcal{I}_k} \sum_{l=1}^{\tau_i} \sum_{q=1}^{Q} (x_{ilqt} + \cdots + x_{ilq(t-t_{il}+1)}) \leq a_k \qquad \text{for } k = 1, \cdots, \ K; t = 0, \cdots, H$$

$$\sum_{i \in \mathcal{I}_k} \sum_{l=1}^{\tau_i} \sum_{r=1}^{m} (x_{ilr(t+\omega_i+\bar{\sigma}_i)} + \cdots + x_{ilr(t+\omega_i+\bar{\sigma}_i-t_{il}+1)}) \leq a_k$$

$$\text{for } k = 1, \cdots, \ K; t = 0, \cdots, H.$$

The number of empty racks of any size is limited. Therefore the set of above constraints guarantees, that, with respect to each particular type of rack, no racks are scheduled unless/until a previously used (empty) storage rack falls vacant.

Remark: The t_{il} are functions of $(\bar{x}, \hat{x})$. Thus, it is impossible to determine the values of t_{il} exactly a priori. The actual choice of t_{il} determines whether the global rack constraints are too restrictive with respect to the real world problem or whether they give a relaxation for the problem of short rack capacities. Efficient algorithms should dynamically adapt the t_{il}. The $\bar{\sigma}_i$ depend upon the quality of a presupposed basin schedule $\bar{x}$. The above remarks on t_{il} apply to $\bar{\sigma}_i$ analogously.

14.4.3 Heuristic Solution Approach

Several exact solution methods for 0-1 programming problems are proposed in literature (cf. [Bal67, Sch86, NW88]), however all of these are applicable for small problem sizes only. Hence, as our problem may include up to about 10^8 binary variables the only suitable solution methods are heuristics.

On the first glance it seems to be appropriate to develop solution methods for the basin area and for the kiln area independently. However a final combination of two independently derived schedules could be impossible without exceeding the feasible buffers between the two areas, so that the waiting time constraints might be violated. Thus, the only reasonable line of attack is an integrated optimization. During a *forward computation* feasible schedules for the basin area yield also feasible schedules for the kiln area, while a *backward computation* derives feasible schedules for the basin area from ones of the kiln area.

In our practical problem we decided to use backward computation. This decision was based on the observation that the kilns' capacity already has been noticed to be a bottleneck for particular product mixes while the basins' capacity appears to be less critical. Hence forward computation more often results in infeasible solutions.

One kind of backward computation, called *simple backward computation*, first generates a feasible schedule for the kiln area and tries to adapt this schedule to the basin area such that none of the constraints becomes violated. The other kind, called *simultaneous backward computation* works rack after rack. First, one rack is assigned to some kiln and to some basin. In step i the ith rack is tried to fit into the partial schedule such that none of the constraints will be violated. The procedure terminates when all racks have been considered. Before we are going into details, we provide a short description of the heuristics used for generation of feasible kiln schedules.

Generation of Feasible Kiln Schedules

We are going to introduce three greedy heuristics as well as two regret heuristics capable to generate feasible solutions for the kiln area. Both kinds of heuristics are based on priority rules.

In order to get good feasible solutions intuition tells us that it seems to be reasonable to consider compatibility properties of the jobs. In order to avoid a waste of the kiln's length racks of less compatible jobs should more likely be put into small kilns than racks of jobs of high compatibility. Hence we first need a measure for the job incompatibility. A useful measure will be the percentage of racks to which a job is incompatible, i.e. the values

$$u_i^1 := \frac{1}{\tau}\left(\sum_{j=1}^n \tau_i a_{ij}\right) \text{ where } \tau := \sum_{j=1}^n \tau_i.$$

Value τ is the sum of all racks in use and a_{ij} are coefficients of our compatibility matrix A as defined in the previous section.

Similarly, we consider the rack size required for job J_i, and also the kiln size. Thus, let $u_i^2 := \frac{\kappa_i}{\max_i\{\kappa_i\}}$ for all $J_i \in I$, and $u_r^3 := \frac{\kappa_r}{\max_r\{\kappa_r\}}$ for all $r = 1, \cdots, N$. If we use $\overline{u}_i^j := 1 - u_i^j$ for $j := 1, 2$ and $\overline{u}_r^3 := 1 - u_r^3$ then we are able to define a characteristic (job × kiln)-matrix $S(\alpha,\beta,\gamma) = (s(\alpha,\beta,\gamma)_{ir})$ where α, β, $\gamma \in [0,1]$ and

$$s(\alpha,\beta,\gamma)_{ir} := [(1-\alpha)u_i^1 + \alpha\overline{u}_i^1][(1-\beta)u_i^2 + \beta\overline{u}_i^2][(1-\gamma)u_r^3 + \gamma\overline{u}_r^3].$$

The triple (α,β,γ) is called a *strategy* and determines the "*measure of quality*" of assignment "rack of job J_i to kiln r". Among all possible matrices especially the entries of $S(0,0,1)$ and $S(1,1,0)$ correspond to our intuition.

First we determine a feasible schedule for the kilns and then calculate a feasible schedule for the basins. Hence it might be better to prefer multiple-basin jobs while filling the kilns. According to our strategy this implies that almost all basins are available for multiple-basin jobs, so that unfeasibilities are prevented. Thus delays of the product mix completion time are reduced. Furthermore, the number of racks to be produced of a particular job and their processing times in a kiln as well as in the basin area should be considered. This is motivated by the observation that jobs that will be in process for a long time probably determine the completion time of the schedule. Thus let

$$u_i := \sum_{l=1}^{\bar{\tau}_i} \max_q \{\sigma_{ilq}\} + \sum_{l=1}^{\tau_i} \max_r \{\sigma_{ilr}\}, \quad u_i^4 := \frac{u_i}{\max_i \{u_i\}}, \quad \text{and} \quad \bar{u}_i^4 := 1 - u_i^4.$$

Moreover the basin size is included by $u_q^5 := \dfrac{\upsilon_q}{\max_q \{\upsilon_q\}}$ and $\bar{u}_q^5 := 1 - u_q^5$, for $q = 1, \cdots, \Omega$. Taking these values into account for the matrix entries of S yields another characteristic matrix $S(\alpha,\beta,\gamma,\delta,\varepsilon) = (s(\alpha,\beta,\gamma,\delta,\varepsilon)_{ir})$ where δ, $\varepsilon \in [0, 1]$ and $s(\alpha,\beta,\gamma,\delta,\varepsilon)_{ir} := s(\alpha,\beta,\gamma)_{ir}[(1-\delta)u_i^4 + \delta\bar{u}_i^4][(1-\varepsilon)u_q^5 + \varepsilon\bar{u}_q^5]$. The tuple $(\alpha,\beta,\gamma,\delta,\varepsilon)$ will also be called a *strategy*.

All subsequent heuristics should be applied several times in order to create good schedules. To increase the variety of solutions we finally add some random elements to the above mentioned strategies. Let z be a random variable uniformly distributed in $[0,1]$. Then $S'(\alpha,\beta,\gamma) = (s'(\alpha,\beta,\gamma)_{ir})$ and $S(\alpha,\beta,\gamma,\delta,\varepsilon) = (s'(\alpha,\beta,\gamma,\delta,\varepsilon)_{ir})$ are defined as $s'(\alpha,\beta,\gamma)_{ir} := s(\alpha,\beta,\gamma)_{ir}z$ and $s'(\alpha,\beta,\gamma,\delta,\varepsilon)_{ir} := s(\alpha,\beta,\gamma,\delta,\varepsilon)_{ir}z$, respectively. For convenience we only speak of matrix $S = (s_{ir})$, however, always keeping in mind that any of the four above mentioned matrices might be used.

Now we provide three greedy procedures $GREEDY_1$, $GREEDY_2$, and $GREEDY_3$. $GREEDY_1$ first chooses a kiln and then its jobs with respect to matrix S. $GREEDY_2$ proceeds just the other way round whereas $GREEDY_3$ searches for the maximum entry in S among all remaining feasible "job to kiln" assignments.

Algorithm 14.4.1 $GREEDY_1$
```
begin
repeat
    J := set of all jobs of which still racks have to be polymerized;
    if there are empty kilns
    then
        begin
        Choose an empty kiln r;
        repeat
            Choose a job J_i ∈ J such that s_{ir} is maximum;
            if a_{ij} = 0 for a rack of job J_j already in kiln r
            then fill kiln r as far as possible with racks of job J_i;
            J := J - {J_i};
        until J = ∅ or r is full;      -- run kiln r
        end
    else wait for the next time when a kiln will be unloaded;
until all racks have been in polymerization;
end;
```

Algorithm 14.4.2 *GREEDY*$_2$
```
begin
J := set of all jobs;
repeat
    K := set of all kilns not in process;        -- i.e. those not completely loaded
    if  K ≠ ∅
    then
      begin
      Choose a job Jᵢ ∈ J of which the most racks are not polymerized;
      repeat
          Choose a kiln r ∈ K such that sᵢᵣ is maximum;
          if  aᵢⱼ = 0 for a rack of job Jⱼ already in kiln r
          then  fill kiln r as far as possible with racks of job Jᵢ;
          K := K- {r};
      until  K = ∅  or  there are no more racks of job Jᵢ;
      J := J- {Jᵢ};        -- if J is empty then run all newly loaded kilns
      end
    else
    begin
    Wait for the next time when a kiln will be unloaded;
    J := set of all jobs of which still racks have to be polymerized;
    end;
until  all racks have been in polymerization;
end;
```

Algorithm 14.4.3 *GREEDY*$_3$
```
begin
repeat
    K := set of all kilns not in process;        -- i.e. those not completely loaded
    J := set of all jobs of which still racks have to polymerize;
    SS := S;         -- SS = (ssᵢᵣ)
    if  K ≠ ∅  and  J ≠ ∅
    then
      repeat
        for all  r ∈ K do
            for all  Jᵢ ∈ J do  choose (i, r) such that ssᵢᵣ is maximum;
        if  aᵢⱼ = 0  for racks of a job Jⱼ already in kiln r
        then  fill kiln r as far as possible with racks of job Jᵢ;
        ssᵢᵣ := 0;
      until  SS = 0;
    else
```

```
   if 𝒦 = ∅
   then wait for the next time when a kiln will be unloaded;
until  all racks have been in polymerization;
end;
```

The *regret heuristics* (as special greedy heuristics) are also based upon some matrix S. They are not only greedily grasping for the highest value in rows or columns of S but consider the differences of the entries. So assume that for each $i = 1, \cdots, n$ we have a descending ordering of the values $s_{i r_1} \geq \cdots \geq s_{i r_N}$. Let also $s_{i_1 r} \geq \cdots \geq s_{i_m r}$ be descending orderings for all $r = 1, \cdots, N$. We define regrets $\xi_i := s_{i r_1} - s_{i r_2}$, $i = 1, \cdots, n$, and $\zeta_r := s_{i_1 r} - s_{i_2 r}$, $r = 1, \cdots, N$, and first try assignments where the above differences are largest. Heuristics $REGRET_1$ and $REGRET_2$ correspond almost completely to $GREEDY_1$ and $GREEDY_2$, respectively; it is sufficient to point to the slight distinction. In $REGRET_1$ the empty kilns are chosen according to the descending list of ζ_r, i.e. kiln r where ζ_r is maximum comes first. Similarly, $REGRET_2$ chooses the jobs J_i according to the descending ordering of ξ_i, i.e. job J_i where ξ_i is maximum comes first. Especially for the regret heuristics, variety of solutions increases if random elements influence the matrix entries. Otherwise many of the regrets ξ_i and ζ_r will become zero.

We resigned new computation of the regrets, whenever a "waiting" job or a "waiting" kiln "disappears", because of the insignificant solution improvement compared to the raise of computational complexity.

Generation of Feasible Schedules

Each time a kiln is in process it runs according to a special temper cycle. It is heated up to a particular temperature and later on cooled down. While the kiln is in process it must not be opened. This restriction does not apply to the basins. The basin temperatures are much lower than the kiln temperatures and racks may be put into or removed from basins at any time. Hence the basin temperature should be kept constant or temperature changes should be reduced to a minimum. The initial assignment of a particular temperature to each basin is done according to the number and size of the racks which have to pre-polymerize in this particular temperature as well as to the basin size. Procedures as used for job to kiln assignments should somewhat be adjusted.

When the initial basin heating is completed backward computation may start. We first give an outline of the simple backward computation algorithm. Let t_{ilq} and t_{ilr} be the time when rack l ($l = 1, \cdots, \bar{\tau}_i$ or τ_i) of job J_i ($i = 1, \cdots, n$) enters basin q ($q = 1, \cdots, \Omega$), and kiln r ($r = 1, \cdots, N$), respectively. Furthermore, let random(z) be a random number generator that initializes z with a random number from interval [0, 1]. Consider the sums of (at most 4) subsequent pre-polymerization times in basins of a rack of job J_i. Let σ_i be sums' mean value, $i = 1, \cdots, n$.

Algorithm 14.4.4 *Simple Backward Computation*
Input: The algorithm starts with a feasible solution for the kiln area

```
begin
for all i ∈ I do
    for l := 1 to τ_i do
        begin -- compute possible starting times for prepolymerization
        random(z);
```
$$t_{ilq} := t_{ilr} - \sigma_i - z\omega_i;$$
```
        if i ∈ I_2 ∪ I_3 ∪ I_4 then
```
$t_{i(\tau_i+l)q} := t_{ilq} + \sigma_{ilq};$
```
        if i ∈ I_3 ∪ I_4 then
```
$t_{i(2\tau_i+l)q} := t_{i(\tau_i+l)q} + \sigma_{i(\tau_i+l)q};$
```
        if i ∈ I_4 then
```
$t_{i(3\tau_i+l)q} := t_{i(2\tau_i+l)q} + \sigma_{i(2\tau_i+l)q};$
```
        end
repeat
```
Take the earliest possible starting time t_{ilq};
```
    if rack l of job J_i may be put into some basin q
    then pre-polymerize in q
    else
        if there is no basin for rack l available
        then
            if there is an empty basin
            then adapt its temperature for job J_i
            else pre-polymerization is impossible;
until all racks have been considered or pre-polymerization is impossible;
end;
```

If pre-polymerization is impossible for some rack we can start the above procedure once again and compute new starting times for pre-polymerization of all or some racks. Another possibility would be to generate a new schedule $\hat{x}$ for the kilns.

The simultaneous backward computation is a simple extension of the heuristics mentioned above for the kiln area. Whenever a rack is chosen (in any of these heuristics) to be assigned to some kiln at time t try to assign this rack to some basin at its possible pre-polymerization starting time (that is computed as in simple backward computation). If this assignment is possible it will also be realized and the next rack will be considered according to the heuristic in use. If this assignment is not possible the possible pre-polymerization starting time may be changed randomly or the procedure starts again with the next time step $t+1$ for a kiln assignment of the considered rack.

Several advises should be given in order to achieve some acceleration. Very long basin processing times that may occur, only for jobs J_i, $i \notin I_4$, should be split into two processing times. Hence 1-basin jobs become 2-basin jobs and so on. During pre-polymerization basin changes are allowed and provide more flexibility. Gaps of time during which a basin is not completely filled may be split when a new rack is put into. It should be observed that splittings of small gaps are preferred to avoid unnecessary splittings of long basin processing times.

The Main Algorithm

Up to now we only described the way a feasible schedule is computed. According to this schedule and its fitness (= value of its objective function) efforts were done for improvement. Slight changes of the parameters α, β, γ, δ, ε according to the objective function values lead to new solutions. A brief outline will describe the main idea. Five strategies $(\alpha,\beta,\gamma,\delta,\varepsilon)$ and related schedules (see the preceding section) are randomly generated. Later on take the last five schedules and subdivide the interval $[0, 1]$ with respect to the fitness, i.e. the best schedule gets the largest part and the worst schedule the smallest one. Generate five random numbers in $[0, 1]$. Sum up the values of α whereby the random numbers belong to the subparts in $[0, 1]$ corresponding to the schedules with matrix parameters α. The new value α is this sum divided by 5. Do the same procedure for the remaining parameters and generate the new schedule. This kind of search may be considered as a variant of tabu search.

For convenience we use in our algorithmic description below α_1^j, α_2^j, α_3^j, α_4^j, α_5^j instead of α, β, γ, δ, ε, respectively, in the jth solution.

Algorithm 14.4.5 *Schedule Improvement*
```
begin
for j := 1 to 5 do
```
Generate tuple $\xi_j := (\alpha_1^j, \alpha_2^j, \alpha_3^j, \alpha_4^j, \alpha_5^j)$ at random and generate a
    schedule $(\bar{x}, \hat{x})$ and its fitness $f_j := \varphi_1(\bar{x}) + \psi_1(\hat{x})$;
```
j := 5;
repeat
    j := j + 1;
    ξ_j := 0;
```

$$\textbf{for } s := j-5 \textbf{ to } j-1 \textbf{ do} \quad z_s := \cfrac{1}{f_s\left(\cfrac{1}{f_{j-5}} + \cfrac{1}{f_{j-4}} + \cfrac{1}{f_{j-3}} + \cfrac{1}{f_{j-2}} + \cfrac{1}{f_{j-1}}\right)};$$

Subdivide the interval $[0, 1]$ in 5 parts, each of lengths z_{j-5}, z_{j-4}, z_{j-3}, z_{j-2}, z_{j-1};
```
    for r := 1 to 5 do
        for i := 1 to 5 do
            begin
```

random(z);

if z belongs to the subinterval of [0, 1] corresponding to the sth solution $(j-5 \leq s \leq j-1)$

then $\alpha_r^j := \alpha_r^j + \alpha_r^s;$

end;

$\xi_j := \frac{1}{5}(\alpha_1^j, \alpha_2^j, \alpha_3^j, \alpha_4^j, \alpha_5^j);$

Generate a new schedule $(\overline{x}, \hat{x})$ according to ξ_j and its fitness

$f_j := \varphi_1(\overline{x}) + \psi_1(\hat{x});$

until j is sufficiently large **or** some other stopping criteria are satisfied;
end;

This algorithm always looks for a better parameter constellation incorporating some random elements. The algorithm may also be applied to the alternative objective functions.

14.4.4 Implementation and Computational Experiment

The system ComPlex is an interactive computer implementation of the provided algorithms written in TURBO PASCAL. Input and output data are memorized in dBASE files and can be edited in dBASE III+.

After the user has logged in he gets a main menu as shown in Table 14.4.1. While choices "0" and "4" are self-explaining selection of "1" displays the users system settings as in Table 14.4.2. "Start time" determines the day (of the year) and time (of this day) when kilns may start processing. Similarly when "Finish time" is reached computation and processing has to stop. Whenever a kiln is unloaded and becomes available again the greedy heuristics start assigning racks. Instead of supervising finish times of kilns in process (increment = 0) it is sufficient to check availability at specified intervals, i.e. every Increment minutes. Jobs may become urgent if due date comes close. Regard, basins or kilns may be blocked for production (entry 1 in the string of zeros). The last five entries of Table 14.4.2 are the randomly chosen or predetermined parameters $\alpha, \beta, \gamma, \delta, \varepsilon$.

ComPlex Menu	
0	Esc
1	System setup
2	Current schedule
3	Data presentation
4	Help

Table 14.4.1 *Main menu.*

ComPlex - System setups		
Start time	[Day]	10-00:00
Finish time	[Day]	00-00:00
Increment	[Min]	20
Level of urgency	[Min]	600
Heat change duration	[Min]	30
Additional cooling	[Min]	180
Week programs	[No]	1
Iteration	[No]	500
Display		less
Reserved basins	[No]	0000000000000000000000000000
Reserved kilns	[No]	100000100000
Distance in basins	[cm]	10
Distance in kilns	[cm]	50
Compatibility α	[0..1]	0.6648
Rack size β	[0..1]	0.6456
Kiln size γ	[0..1]	0.4892
Processing time δ	[0..1]	0.8394
Basin size ε	[0..1]	0.5763

Table 14.4.2 *System setup.*

ComPlex - Computation			
res. kilns	2	current week	1/1
res. basins	0	current iteration	3/500
		number of jobs	107
		time of kiln loading	14-03:00
		number of racks	330/330
first assignment to kiln		**best iterations**	
job	5	no.	finish time
kiln	1	2	14-14:30
rack	10	1	14-19:00
basin	2		
adding racks to fill up kiln		**parameters**	
job	...	α	0.6632
rack	...	β	0.7642
basin	...	γ	0.3464
		δ	0.2453
		ε	0.8338

Table 14.4.3 *Computation.*

Selection "2" in the main menu displays data during execution as in Table 14.4.3. Via "Data presentation" results will be displayed as in Table 14.4.4, for instance, in case of rack 14 of job 82.

rack 14 of job 82				key: 2293/1/15/315*213/15.0			
	no.	type	cm×cm	from	for	until	temp.
rack	25	6	320×320	08-08:35	4835	11-17:10	
filling				08-08:35	25	08-09:00	
waiting				08-09:00	0	08-09:00	
basin1	3	1	850×370	08-09:00	730	08-21:10	35
basin2	21	4	780×270	08-21:10	590	09-07:00	30
basin3	6	1	850×370	09-07:00	2190	10-19:30	25
basin4	1	1	850×370	10-19:30	270	11-00:00	40
waiting				11-00:00	120	11-02:00	
kiln	1	1	900×565	11-02:00	720	11-14:00	90
heating up				11-02:00	120	11-04:00	
polymerization				11-04:00	240	11-08:00	90
cooling1				11-08:00	90	11-09:30	80
cooling 2				11-09:30	120	11-11:30	65
cooling 3				11-11:30	150	11-14:00	20
cleaning				11-17:00	10	11-17:10	

Table 14.4.4 *Schedule of a rack.*

References

Bak75   K. R. Baker, A comparative study of flow shop algorithms, *Oper. Res. 23*, 1975, 62-73.

Bak84   K. R. Baker, Sequencing rules and due date assignments in a job shop, *Management Sci.* 30, 1984, 1093-1104.

Bal67   E. Balas, Discrete programming by the filter method, *Oper. Res.* 15, 1967, 915-957.

BB82   K. R. Baker, J. M. W. Bertrand, A dynamic priority rule for sequencing against due dates, *J. Oper. Management* 3, 1982, 37-42.

BBFW94   J. Błażewicz, Burkard, G. Finke, G. J. Woeginger, Vehicle scheduling in two-cycle flexible manufacturing system, *Math. Comput. Modelling* 20, 1994, 19-31.

BCSW86   J. Błażewicz, W. Cellary, R. Słowiński, J. Węglarz, *Scheduling Under Resource Constraints - Deterministic Models*, J. C. Baltzer, Basel, 1986.

BEF+91   J. Błażewicz, H. Eiselt, G. Finke, G. Laporte, J. Węglarz, Scheduling tasks and vehicles in a flexible manufacturing system, *Internat. J. FMS* 4, 1991, 5-16.

BFR75   P. Bratley, M. Florian, P. Robillard, Scheduling with earliest start and due date
            constraints on multiple machines, *Naval Res. Logist. Quart.* 22, 1975, 165-173.

BH91   S. A. Brah, J. L. Hunsucker, Branch and bound algorithm for the flow shop
            with multiple processors, *European J. Oper. Res.* 51, 1991, 88-99.

BK83   K. R. Baker, J. J. Kanet, Job shop scheduling with modified due dates, *J. Oper.
            Management* 4, 1983, 11-22.

Bra88   S. A. Brah, Scheduling in a flow shop with multiple processors, Ph.D. thesis,
            University of Houston, Houston, TX., 1988.

BY86a   J. A. Buzacott, D. D. Yao, Flexible manufacturing systems: a review of ana-
            lytical models, *Management Sci.* 32, 1986, 890-905.

BY86b   J. A. Buzacott, D. D. Yao, On queuing network models for flexible manufac-
            turing systems, *Queuing Systems* 1, 1986, 5-27.

Con65   R. W. Conway, Priority dispatching and job lateness in a job shop, *J. Industrial
            Engineering* 16, 1965, 123-130.

DLSS89   M. Desrochers, J. K. Lenstra, M. W. P. Savelsbergh, F. Soumis, Vehicle rout-
            ing with time windows, in: B. L. Golden, A. A. Assad (eds.), *Vehicle Routing:
            Methods and Studies*, North-Holland, Amsterdam, 1988, 65-84.

Fre82   S. French, Sequencing and Scheduling: *An Introduction to the Mathematics of
            Job-Shop*, J. Wiley, New York, 1982.

FKPS91   H. Friedrich, J. Keßler, E. Pesch, B. Schildt, Batch scheduling on parallel units
            in acrylic-glass production, *ZOR* 35, 1991, 321-345.

GLL+79   R. L. Graham, E. L. Lawler, J. K. Lenstra, A. H. G. Rinnooy Kan, Optimiza-
            tion and approximation in deterministic sequencing and scheduling theory: A
            survey, *Ann. Discrete Math.* 5, 1979, 287-326.

Gra66   R. L. Graham, Bounds for certain multiprocessing anomalies, *Bell System
            Technical J.* 54, 1966, 1563-1581.

Gup70   J. N. D. Gupta, M-stage flowshop scheduling by branch and bound, *Opsearch*
            7, 1970, 37-43.

Jai86   R. Jaikumar, Postindustrial manufacturing, *Harvard Buss. Rev.* Nov./Dec.,
            1986, 69-76.

KH82   J. J. Kanet, J. C. Hayya, Priority dispatching with operation due dates in a job
            shop, *J. Oper. Management* 2, 1982, 155-163.

KL95   V. Kats, E. Levner, The contrained cyclic robotic flowshop problem: a solvable
            case, Proc.WISOR-95, 1995.

KM87   S. Kochbar, R. J. T. Morris, Heuristic methods for flexible flow line schedul-
            ing, *J. Manuf. Systems* 6, 1987, 299-314.

Lan87   M. A. Langston, Improved LPT scheduling identical processor systems,
            RAIRO Technique et Sci. Inform. 1, 1982, 69-75.

MB67   G. B. McMahon, P. G. Burton, Flow shop scheduling with the branch and
            bound method, *Oper. Res.* 15, 1967, 473-481.

NW88   G. L. Nemhauser, L. A. Wolsey, *Integer and combinatorial optimization*, Wiley, New York, 1988.

RRT89   N. Raman, R. V. Rachamadugu, F. B. Talbot, Real time scheduling of an automated manufacturing center, *European J. Oper. Res.* 40, 1989, 222-242.

RS89   R. Rachamadugu, K. Stecke, Classification and review of FMS scheduling procedures, Working Paper No 481C, The University of Michigan, School of Business Administration, Ann Arbor MI, 1989.

RT92   N. Raman, F. B. Talbot, The job shop tardiness problem: a decomposition approach, *European J. Oper. Res.*, 1992.

RTR89a N. Raman, F. B. Talbot, R. V. Rachamadugu, Due date based scheduling in a general flexible manufacturing system, *J. Oper. Management* 8, 1989, 115-132.

RTR89b N. Raman, F. B. Talbot, R. V. Rachamadugu, Scheduling a general flexible manufacturing system to minimize tardiness related costs, Working Paper # 89-1548, Bureau of Economic and Business Research, University of Illinois at Urbana - Champaign, Champaign, IL, 1989.

Sal73   M. S. Salvador, A solution of a special class of flowshop scheduling problems, *Proceedings of the Symposium on the theory of Scheduling and its Applications*, Springer, Berlin, 1975, 83-91.

Sav85   M. W. P. Savelsbergh, Local search for routing problems with time windows, *Annals Oper. Res.* 4, 1985, 285-305.

Sch84   G. Schmidt, Scheduling on semi-identical processors, *ZOR* 28, 1984, 153-162.

Sch86   A. Schrijver, *Theory of linear and integer programming*, J. Wiley, New York, 1986.

Sch88   G. Schmidt, Scheduling independent tasks on semi-identical processors with deadlines, *J. Oper. Res. Soc.* 39, 1988, 271-277.

Sch89   G. Schmidt, *CAM: Algorithmen und Decision Support für die Fertigungssteuerung*, Springer, Berlin, 1989.

SM85   K. E. Stecke, T. L. Morin, The optimality of balancing workloads in certain types of flexible manufacturing systems, *European J. Oper. Res.* 20, 1985, 68-82.

SS85   K. E. Stecke, J. J. Solberg, The optimality of unbalancing both workloads and machine group sizes in closed queuing networks for multiserver queues, *Oper. Res.* 33, 1985, 882-910.

SS89   C. Sriskandarajah, S. P. Sethi, Scheduling algorithms for flexible flowshops: worst and average case performance, *European J. Oper. Res.* 43, 1989, 143-160.

SW89   R. Słoviński, J. Węglarz (eds.), *Advances in Project Scheduling*, Elsevier, Amsterdam, 1989.

Tal82   F. B. Talbot, Resource constrained project scheduling with time resource tradeoffs: the nonpreemptive case, *Management Sci.* 28, 1982, 1197-1210.

VM87   A. P. J. Vepsalainen, T. E. Morton, Priority rules for job shops with weighted tardiness costs, *Management Sci.* 33, 1987, 1035-1047.

Wit85   R. J. Wittrock, Scheduling algorithms for flexible flow lines, *IBM J. Res. Develop.* 29, 1985, 401-412.

Wit88   R. J. Wittrock, An adaptable scheduling algorithms for flexible flow lines, *Oper. Res.* 33, 1988, 445-453.

15 Computer Integrated Production Scheduling

Within all activities of production management, *production scheduling* is a major part covering planning and control functions. By *production management* we mean all activities which are necessary to carry out production. The two main decisions to be taken in this field are production *planning* and production *control*. Production scheduling is a common activity of these two areas because scheduling is needed not only on the planning level as mainly treated in the preceding chapters but also on the control level. From the different aspects of production scheduling problems we can distinguish *predictive production scheduling* or *offline-planning* (*OFP*) and *reactive production scheduling* or *online-control* (*ONC*). Predictive production scheduling serves to provide guidance in achieving global coherence in the process of local decision making. Reactive production scheduling is concerned with revising predictive schedules when unexpected events force changes. OFP generates the requirements for ONC, and ONC creates feedback to OFP.

Problems of production scheduling can be modeled on the basis of distributed planning and control loops, where data from the actual manufacturing process are used. A further analysis of the problem shows that job release to, job traversing inside the manufacturing system and sequencing in front of the machines are the main issues, not only for production control but also for short term production planning.

In practice, scheduling problems arising in manufacturing systems are of discrete, distributed, dynamic and stochastic nature and turn out to be very complex. So, for the majority of practical scheduling purposes simple and rigid algorithms are not applicable, and the manufacturing staff has to play the role of the flexible problem solver. On the other hand, some kind of Decision Support Systems (*DSS*) has been developed to support solving these scheduling problems. There are different names for such systems among which "Graphical Gantt Chart System" and "Leitstand" are the most popular. Such a DSS is considered to be a shop floor scheduling system which can be regarded as a control post mainly designed for short term production scheduling. Many support systems of this type are commercially available today. A framework for this type of systems can be found in [EGS97].

Most of the existing shop floor production scheduling systems, however, have two major drawbacks. First, they do not have an integrated architecture for the solution process covering planning and control decisions, and second, they do not take sufficient advantage from the results of manufacturing scheduling theory. In the following, we concentrate on designing a system that tries to avoid these drawbacks, i.e. we will introduce intelligence to the modeling and to the solution process of practical scheduling problems.

Later in this chapter we suggest a special DSS designed for *short term production scheduling* that works on the planning and on the control level. It makes appropriate use of scheduling theory, knowledge-based and simulation techniques. The DSS introduced will also be called "*Intelligent Production Scheduling System*" or IPS later.

This chapter is organized as follows. First we give a short idea about the environment of production scheduling from the perspective of problem solving in computer integrated manufacturing (Section 15.1). Based on this we suggest a reference model of production scheduling for enterprises (Section 15.2). Considering the requirements of a DSS for production scheduling we introduce an architecture for scheduling manufacturing processes (Section 15.3). It can be used either for an open interactive (Section 15.3.1) or a closed loop solution approach (Section 15.3.2). Based on all this we use an example of a flexible manufacturing cell to show how knowledge-based approaches and ideas relying on traditional scheduling theory can be integrated within an interactive approach (Section 15.3.3). Note that, in analogy, the discussion of all these issues can be applied to other scheduling areas than manufacturing.

15.1 Scheduling in Computer Integrated Manufacturing

The concept of *Computer Integrated Manufacturing* (CIM) is based on the idea of combining information flow from technical and business areas of a production company [Har73]. All steps of activities, ranging from customer orders to product and process design, master production planning, detailed capacity planning, predictive and reactive scheduling, manufacturing and, finally, delivery and service contribute to the overall information flow. Hence a sophisticated common database support is essential for the effectiveness of the CIM system. Usually, the database will be distributed among the components of CIM. To integrate all functions and data a powerful communication network is required. Examples of network architectures are hierarchical, client server, and loosely connected computer systems. Concepts of CIM are discussed in detail by e.g. Ranky [Ran86] and Scheer [Sch91].

We repeat briefly the main structure of CIM systems. The more technically oriented components are *Computer Aided Design* (CAD) and *Computer Aided Process Planning* (CAP), often comprised within *Computer Aided Engineering* (CAE), *Computer Aided Manufacturing* (CAM), and *Computer Aided Quality Control*(CAQ). More businesslike components are the *Production Planning System* (PPS) and the already mentioned IPS. The concept of CIM is depicted in Figure 15.1.1 where edges represent data flows in either directions. In CAD, development and design of products is supported. This includes technical or physical calculations and drafting. CAP supports the preparation for manufacturing through process planning and the generation of programs for numeric controlled machines. Manufacturing and assembly of products are supported by CAM which is responsible for material and part transport, control of machines and transport systems, and for supervising the manufacturing process. Requirements for product quality and generation of quality review plans are delivered by CAQ. The objective of PPS is to take over all planning steps for customer orders in the sense of material requirements and resource planning. Within CIM, the IPS organizes the execution of all job- or task-oriented activities derived from customer orders.

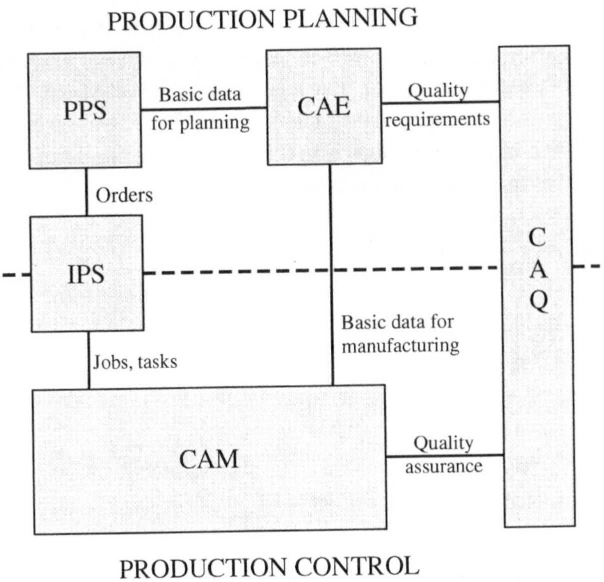

PRODUCTION PLANNING

PRODUCTION CONTROL

Figure 15.1.1 *The concept of CIM.*

Problems in production planning and control could theoretically be represented in a single model and then solved simultaneously. But even if all input data would be available and reliable this approach would not be applicable in general because of prohibitive computing times for finding a solution. Therefore a practical approach is to solve the problems of production planning and control sequentially using a hierarchical scheme. The closer the investigated problems are to the bottom of the hierarchy the shorter will be the time scale under consideration and the more detailed the needed information. Problems on the top of the hierarchy incorporate more aggregated data in connection with longer time scales. Decisions on higher levels serve as constraints on lower levels. Solutions for problems on lower levels give feedback to problem solutions on higher levels. The relationship between PPS, IPS and CAM can serve as an example for a hierarchy which incorporates three levels of problem solving. It is of course obvious that a hierarchical solution approach cannot guarantee optimality. The number of levels to be introduced in the hierarchy depends on the problem under consideration, but for the type of applications discussed here a model with separated tactical (PPS), operational (IPS), and physical level (CAM) seems appropriate.

In production planning the material and resource requirements of the customer orders are analyzed, and production data such as ready times, due dates or deadlines, and resource assignments are determined. In this way, a midterm or tactical production plan based on a list of customer orders to be released for the next manufacturing period is generated. This list also shows the actual production requirements. The production plan for short term scheduling is the output of the production scheduling system IPS on an operational level. IPS is responsible for the assignment of jobs or tasks to machines, to transport facilities, and for the provision of additional resources needed in manufacturing, and thus organizes job and task release for execution. On a physical level CAM is responsible for the real time execution of the output of IPS. In that way, IPS represents an interface between PPS and CAM as shown in the survey presented in Figure 15.1.2. In detail, there are four major areas the IPS is responsible for [Sch89a].

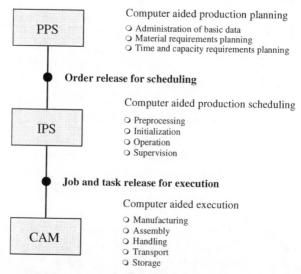

Figure 15.1.2   *Production planning, scheduling and execution.*

(1) *Preprocessing*: Examination of production prerequisites; the customer orders will only be released for manufacturing if all needed resources such as materials, tools, machines, pallets, and NC-programs are available.

(2) *System Initialization*: The manufacturing system or parts thereof have to be set up such that processing of released orders can be started. Depending on the

type of job, NC-programs have to be loaded, tools have to be mounted, and materials and equipment have to be made available at specific locations.

(3) *System Operation*: The main function of short term production scheduling is to decide about releasing jobs for entering the manufacturing system, how to traverse jobs inside the system, and how to sequence them in front of the machines in accordance with business objectives and production requirements.

(4) *System Supervision and Monitoring*: The current process data allow to check the progress of work continuously. The actual state of the system should always be observed, in order to be able to react quickly if deviations from a planned state are diagnosed.

Offline planning (OFP) is concerned with preprocessing, system initialization and system operation on a predictive level, while online control (ONC) is focused mainly on system operation on a reactive level and on system supervision and monitoring. Despite the fact that all these functions have to be performed by the IPS, following the purpose of this chapter we mainly concentrate on short term production scheduling on the predictive and the reactive level.

One of the basic necessities of CIM is an integrated database system. Although data are distributed among the various components of a CIM system, they should be logically centralized so that the whole system is virtually based on a single database. The advantage of such a concept would be redundancy avoidance which allows for easier maintenance of data and hence provides ways to assure consistency of data. This is a major requirement of the *database management system* (DBMS). The idea of an integrated data management within CIM is shown in Figure 15.1.3.

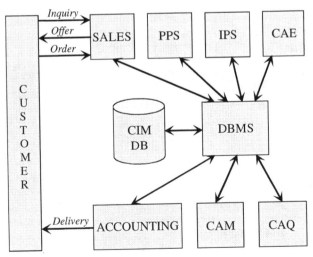

Figure 15.1.3 *CIM and the database.*

The computer architecture for CIM follows the hierarchical approach of problem solving which has already been discussed earlier in this section. The hierarchy can be represented as a tree structure that covers the following decision oriented levels of an enterprise: strategic planning, tactical planning, operational scheduling, and physical manufacturing. At each level a host computer is coordinating one or more computers on the next lower level; actions at each level are carried out independently, as long as the requirements coming from the supervising level are not violated. The output of each subordinated level meets the requirements for correspondingly higher levels and provides feedback to the host. The deeper the level of the tree is, the more detailed are the processed data and the shorter has to be the computing time; in higher levels, on the other hand, the data are more aggregated. Figure 15.1.4 shows a distributed computer architecture, where the boxes assigned to the three levels PPS, IPS and CAM represent computers or computer networks. The leaves of the tree represent physical manufacturing and are not further investigated.

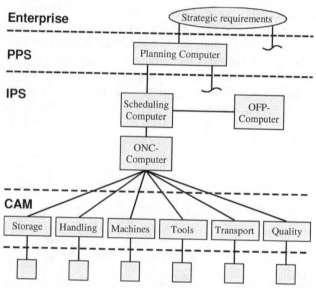

Figure 15.1.4 *Computer system in manufacturing.*

Apart from a vertical information flow, a horizontal exchange of data on the same level between different computers must be provided, especially in case of a distributed and global environment for production scheduling. Generally, different network architectures to meet these requirements may be thought of. Standard protocols and interfaces should be utilized to allow for the communication between computers from different vendors.

15.2 A Reference Model for Production Scheduling

In order to implement the solution approaches presented in the previous chapters within a framework of an IPS we need a basic description of the scheduling system. Here we introduce a modeling approach integrating declarative representation and algorithmic solution [Sch96]. Problem representation and problem solution are strongly interconnected, i.e. data structures and solution methods have to be designed interdependently [Wir76]. We will suggest a reference model for production scheduling and show how problem description and problem solution can be integrated. To achieve this we follow the object-oriented modeling paradigm.

Object-oriented modeling attempts to overcome the disadvantage of modeling data, functions, and communication, separately. The different phases of the modeling process are analysis, design, and programming. Analysis serves as the main representation formalism to characterize the requirements from the viewpoint of the application domain; design uses the results of analysis to obtain an implementation-oriented representation, and programming means translating this representation using some programming language into code. Comparing object-oriented modeling with traditional techniques its advantages lie in data abstraction, reusability and extensibility of the model, better software maintenance, and direct compatibility of the models of different phases of the software development process. Often it is also claimed that this approach is harmonizing the decentralization of organizations and their support by information systems. We will now develop an open object-oriented analysis model for production scheduling. In comparison to other models of this kind (see e.g. [RM93]) the model presented here is a one to one mapping of the classification scheme of deterministic scheduling theory introduced in Chapter 3 to models of information systems for production scheduling. Following this approach we hope to achieve a better transformation of theoretical results to practical applications.

A model built by object-oriented analysis consists of a set of objects communicating via messages which represent dynamic relations of pairs of them. Each object consists of attributes and methods here also called algorithms. Methods are invoked by messages and methods can also create messages themselves. Objects of the same type are classified using the concept of classes; with this concept inheritance of objects can be represented. The main static relations between pairs of objects are generalization/specialization and aggregation/decomposition.

Different methods for generating object-oriented models exist [DTLZ93], [WBJ90]. From a practical point of view the method should make it easy to develop and maintain a system; it should assist project management by defining deliverables and effective tool support should be available. Without loss of generality the object model for production scheduling which will be introduced here is based on the modeling approach called Object-Oriented Analysis or OOA suggested by [CY91]. It is easy to use, easy to understand, and fulfils most of the above mentioned criteria.

In Figure 15.2.1 the main classes and objects for production scheduling are represented using OOA notation. Relationships between classes or objects are represented by arcs and edges; edges with a semi-circle represent generalization/specialization relations, edges with triangles represent aggregation/decomposition, and arcs between objects represent communications by message passing. The arc direction indicates a transmitter/receiver relationship. The introduced classes, objects, attributes, methods, and relations are complete in the sense that applying the proposed model a production schedule can be generated; nevertheless it is easy to enlarge the model to represent additional business requirements.

Each customer order is translated into a manufacturing order, also called job, using process plans and bill of materials. Without loss of generality we want to assume that a job refers always to the manufacturing of one part where different tasks have to be carried out using different resources.

While in Figure 15.2.1 a graphical notation related to OOA is used, we will apply in the following a textual notation. We will denote the names of classes and objects by capital letters, the names of attributes by dashes, and the names of methods by brackets. In OOA notation relationships between classes or objects will be represented by arcs and edges; edges with a semi-circle represent generalization/specialization relations, edges with triangles represent aggregation, and arcs represent communications between objects by message passing. The direction of the arc indicates a transmitter-receiver relationship. The introduced classes, objects, attributes, methods, and relations are complete in the sense that applying the proposed model a production schedule can be generated; nevertheless it is easy to enlarge the model to represent additional business requirements.

The main classes of production scheduling are JOB, BOM (BILL_OF_MATERIALS), PP (PROCESS_PLAN), TASK, RESOURCE, and SCHEDULE. Additional classes are ORDER specialized to PURCHASING_ORDER and DISPATCH_ORDER and PLANNING specialized to STRATEGIC_P, TACTICAL_P, and OPERATIONAL_P. The class RESOURCE is a generalization of MACHINE, TOOL, and STAFF. Without loss of generality we concentrate the investigation here only on one type of resources which is MACHINE; all other types of resources could be modeled in the same manner. In order to find the attributes of the different classes and objects we use the classification scheme introduced in Chapter 3.

The objects of class BOM generate all components or parts to be produced for a customer order. With this the objects of class JOB will be generated. Each object of this class communicates with the corresponding objects of class PP which includes a list of the technological requirements to carry out some job. According to these requirements all objects of class TASK will be generated, which are necessary to process all jobs.

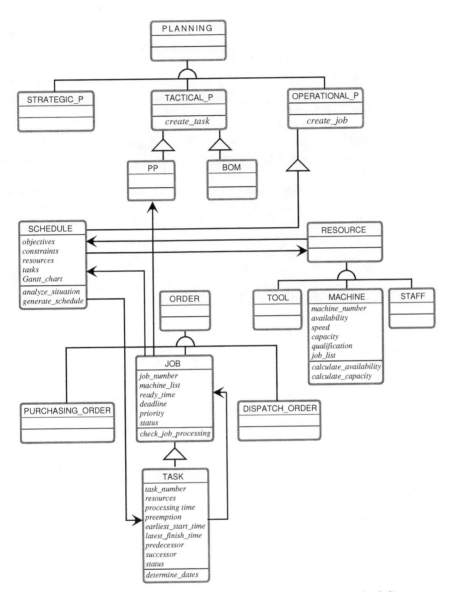

Figure 15.2.1 *Object-oriented analysis model for production scheduling.*

An object of class JOB is characterized by the attributes "job_number", "machines", "machine_list", "ready_time", "deadline", "completion_time", "flow_time", "priority", and "status". Some values of the attributes concerning time and priority considerations are determined by the earlier mentioned Production Planning System (PPS). The value of the attribute "machines" refers to these machines which have the qualification to carry out the corresponding job; after generating the final production schedule the value of "machine_list" refers to the ordered number of these machines to which the job is assigned. The value of the attribute "status" gives an answer to the question if the job is open, scheduled, or finished. The method used by JOB is here <check_job_processing> which has the objective to supervise the progress of processing the job. Communication between JOB and SCHEDULE results in determining the values of "machine_list", "completion_time", "flow_time" and "status".

Each object of class TASK contains structural attributes like "task_number", "resources", "processing_time", "completion_time", "finish_time", "preemption", "earliest_start_time", "latest_finish_time" and additional attributes like "predecessor", "successor", and "status". The values of the two attributes referring to earliest start and latest finish time are determined by the object-owned method <determine_dates>. The parameters for this method are acquired by communication with objects of the class JOB. Again the attribute "status" is required for analyzing the current state of processing of the task under consideration.

Objects of class MACHINE are described by the attributes "machine_number", "availability", "speed", "capacity", "qualification", and "job_list". The value of "qualification" is the set of tasks which can be carried out by the machine. The value of "job_list" is unknown at the beginning; after generating the schedule the value refers to the set of jobs and corresponding tasks to be processed by this machine. In the same sense the values of "availability" and "capacity" will be altered using the methods <calculate_availability> and <calculate_capacity>.

The task of the object SCHEDULE is to generate the final production schedule. In order to do this the actual manufacturing situation has to be analyzed in terms of objective function and constraints to be considered. This leads to the determination of the values for the attributes "objectives" and "constraints" using the method <analyze_situation>. The method <generate_schedule> is constructing the desired schedule. Calling this method the communication links to the objects of classes RESOURCE, JOB, and TASK respectively, are activated. To the attributes "resources" and "tasks" the input values for <generate_schedule> are assigned. The result of the method is a depiction of the production schedule which is assigned to the attribute "Gantt_chart". The required data concerning tasks and resources like machines, availability, speed, processing times etc. are available through the communication links to the objects of classes TASK and RESOURCE.

Example 15.2.1 The following example shows how an object-oriented model for production scheduling can be generated. When we refer to the objects of a particular class the first time we declare the name of the corresponding object, its attributes, and the value of the attributes. Later, we only declare the name of the object and the value of the attributes. All entries are abbreviated.

```
JOB1            "j_no"        J₁;
                "machines"    P₁, P₂;
                "mach_list"   open;
                "ready"       0;
                "deadline"    open;
                "prio"        none;
                "stat"        open;
JOB2  (J₂;  P₂;  open;  0;  open;  none;  open)
JOB3  (J₃;  P₁;  open;  2;  open;  none;  open)
```

There are three jobs which have to be processed. No given sequence for J_1 exists but J_2 can only be processed on P_2 and J_3 can only be processed on P_1. The jobs can start to be processed at times 0 and 2; there is no deadline which has to be obeyed, all jobs have the same priority. The machine list and status of the jobs are open at the beginning; later they will assume the values of the permutation of the machines and scheduled, in_process, or finished, respectively.

```
TASK11          "t_no"        T₁₁;
                "res"         P₁, P₂;
                "p_time"      3;
                "preempt"     no;
                "e_s_t"       0;
                "l_f_t"       open;
                "pre"         ∅;
                "suc"         T₁₂, T₁₃;
                "stat"        open;
TASK12  (T₁₂;  P₁, P₂;  13;  no;  3;  open;  T₁₁;  ∅;  open)
TASK13  (T₁₃;  P₁, P₂;   2;  no;  3;  open;  T₁₁;  ∅;  open)
TASK20  (T₂₀;  P₂;        4;  no;  0;  open;  ∅;  ∅;  open)
TASK31  (T₃₁;  P₁;        2;  no;  2;  open;  ∅;  T₃₂, T₃₃, T₃₄;  open)
TASK32  (T₃₂;  P₁;        4;  no;  4;  open;  T₃₁;  ∅;  open)
TASK33  (T₃₃;  P₁;        4;  no;  4;  open;  T₃₁;  ∅;  open)
TASK34  (T₃₄;  P₁;        2;  no;  4;  open;  T₃₁;  ∅;  open)
```

The three jobs consist of eight tasks; all tasks of job J_1 can processed on all machines, all other tasks are only allowed to be processed on machine P_2 or only on machine P_1. Processing times, precedence constraints and ready times are known, preemption is not allowed, and again deadlines do not exist. The status of the tasks is open at the beginning; later it will also assume the values scheduled, in_process, or finished.

```
MACHINE1        "m_no"        P₁;
                "avail"       [0,∞);
                "speed"       1;
```

                              "capac" PC_1;
                              "qualif" T_{11}, T_{12}, T_{13}, T_{31}, T_{32}, T_{33}, T_{34};
                              "j_list"   open;
            MACHINE2 (P_2; [0, ∞); 1; PC_2; T_{11}, T_{12}, T_{13}, T_{20}; open)

There are two machines available for processing. Both machines have the same
speed. They are available throughout the planning horizon, capacity and qualification are known. The job list, i.e. the sequence the jobs are processed by the
machines is not yet determined.

        SCHEDULE   "object"   makespan;
                        "constr"   open;
                        "res" P_1, P_2;
                        "tasks" T_{11}, T_{12}, T_{13}, T_{31}, T_{32}, T_{33}, T_{34};
                        "Gantt_chart"   open;

The objective here is to minimize the makespan, i.e. to find a schedule where
$\max\{C_i\}$ is minimized. Besides task and machine related constraints no other
constraints have to be taken into account. All input data to generate the desired
production schedule is given, the schedule itself is not yet known. Calling the
method <generate_schedule> will result in a time oriented assignment of tasks to
machines. Doing this the attributes will assume the following values.

JOB1 (J_1; P_1, P_2; 0; 17; none; scheduled)
JOB2 (J_2; P_2; 0; 4; none; scheduled)
JOB3 (J_3; P_1; 3; 15; none; scheduled)

TASK11 (T_{11}; P_1; 3; no; 0; 3; ∅; T_{12}, T_{13}; scheduled)
TASK12 (T_{12}; P_2; 13; no; 4; 17; T_{11}; ∅; scheduled)
TASK13 (T_{13}; P_1; 2; no; 15; 17; T_{11}; ∅; scheduled)
TASK20 (T_{20}; P_2; 4; no; 0; 4; ∅; ∅; scheduled)
TASK31 (T_{31}; P_1; 2; no; 3; 5; ∅; T_{32}, T_{33}, T_{34}; scheduled)
TASK32 (T_{32}; P_1; 4; no; 5; 9; T$_{31}$; ∅; scheduled)
TASK33 (T_{33}; P_1; 4; no; 9; 13; T$_{31}$; ∅; scheduled)
TASK34 (T_{34}; P_1; 2; no; 13; 15; T_{31}; ∅; scheduled)

All jobs and the corresponding tasks are now scheduled; job J_1 will be processed
on machines P_1 and P_2 within the time interval [0,17], job J_2 on machine P_2 in
the interval [0,4] and job J_3 on machine P_1 in the interval [3,15].

```
MACHINE1   (P₁;  [17,∞);  1;  PC₁; T₁₁, T₁₂, T₁₃, T₃₁, T₃₂, T₃₃, T₃₄;
                                   T₁₁, T₃₁, T₃₂, T₃₃, T₃₄, T₁₃)
MACHINE2   (P₂;  [17,∞);  1;  PC₂; T₁₁, T₁₂, T₁₃, T₂₀; T₂₀, T₁₂)
```

The availability of machines P_1 and P_2 has now been changed. Machine P_1 is processing tasks T_{11}, T_{13}, and all tasks of job J_3, machine P_2 is processing tasks T_{20} and T_{12}. The processing sequence is also given.

```
SCHEDULE        "object"        makespan;
                "constr"        open;
                "res"           P₁, P₂;
                "tasks"         T₁₁, T₁₂, T₁₃, T₃₁, T₃₂, T₃₃, T₃₄;
                "Gantt_chart"   generated;
```

The schedule has now been generated and is depicted by a Gantt chart shown in Figure 15.2.2. adaptation   □

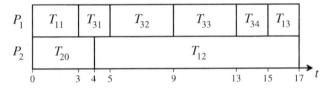

Figure 15.2.2 *Gantt chart for the example problem.*

Example 15.2.2 We now want to use the classical job shop scheduling problem as an example to show how the approach can be applied to dedicated models. Here we will concentrate especially on the interaction between problem representation and problem solution. The general job shop problem is treated in Chapter 8. The object model is characterized by the classes JOB, TASK, MACHINE and SCHEDULE. Investigating attributes of the objects we only concentrate on some selection of them. The class JOB can be described as follows.

```
JOB             "j_no"          Jⱼ;
                "machines "     Permutation over Pᵢ;
                "mach_list"     open;
                "ready"         0;
                "deadline"      open;
                "prio"          none;
                "stat"          open;
```

As we are investigating a simple job shop problem each job is assigned to all machines following some pre-specified sequence, ready times for all jobs are zero; deadlines and priorities have not to be considered.

Each job consists of different tasks which are characterized by the machine where the task has to be processed and the corresponding processing time; preemption is not allowed. Each task can be described by its predecessor or successor task. Input data for the algorithm are the values of the attributes "res", "p_time", "pre" and "suc". The values of "e_s_t" are not obligatory because they can be derived from the values of the attributes "pre" and "suc". With this the class TASK can be described as follows.

TASK	`"t_no"`	T_{ij};
	`"res"`	P_i;
	`"p_time"`	p_{ij};
	`"preempt"`	no;
	`"e_s_t"`	r_{ij};
	`"l_f_t"`	open;
	`"pre"`	T_{kj};
	`"suc"`	T_{lj};
	`"stat"`	open;
MACHINE	`"m_no"`	P_i;
	`"avail"`	$[0, \infty)$;
	`"speed"`	1;
	`"capac"`	PC_i;
	`"qualif"`	T_{ij};
	`"j_list"`	open;

All machines are continuously available in the planning period under consideration. The value of the attribute "capac" is not necessary to apply the algorithm, it is only introduced for completeness reasons.

SCHEDULE	`"object"`	makespan;
	`"constr"`	open;
	`"res"`	$P_1, \ldots, P_m$;
	`"tasks"`	T_{ij};
	`"Gantt_chart"`	open;
	`<generate_schedule>`	simulated annealing; □

The objective is again to find a production schedule which minimizes the maximum completion time. Additional information for describing the scheduling situation is not available. The input data for the algorithm are the available machines, the processing times of all jobs on all machines and the corresponding sequence of task assignment. After the application of an appropriate algorithm (compare to Chapter 8) the corresponding values describing the solution of the scheduling problem are assigned to the attributes and the Gantt chart will be generated.

We have shown using some examples that the object-oriented model can be used for representing scheduling problems which correspond to those investigated in the theory of scheduling. It is quite obvious that the model can be specified to various individual problem settings. Thus we can use it as some reference for developing production scheduling systems.

15.3 IPS: An Intelligent Production Scheduling System

The problems of short term production scheduling are highly complex. This is not only caused by the inherent combinatorial complexity of the scheduling problem but also by the fact that input data are dynamic and rapidly changing. For example, new customer orders arrive, others are cancelled, or the availability of resources may change suddenly. This lack of stability requires permanent revisions, and previous solutions are due to continuous adaptations. Scheduling models for manufacturing processes must have the ability to partially predict the behavior of the entire shop, and, if necessary, to react quickly by revising the current schedule. Solution approaches to be applied in such an environment must have especially short computing times, i.e. time- and resource-consuming models and methods are not appropriate on an operational level of production scheduling.

All models and methods for these purposes so far developed and partially reviewed in the preceding chapters are either of descriptive or of constructive nature. Descriptive models give an answer to the question "*what happens if ...?*", whereas constructive models try to answer the question "*what has to happen so that ...?*". Constructive models are used to find best possible or at least feasible solutions; descriptive models are used to evaluate decision alternatives or solution proposals, and thus help to get a deeper insight into the problem characteristics. Examples of descriptive models for production scheduling are queuing networks on an analytical and discrete simulation on an empirical basis; constructive models might use combinatorial optimization techniques or knowledge of human domain experts.

For production scheduling problems one advantage of descriptive models is the possibility to understand more about the dynamics of the manufacturing system and its processes, whereas constructive models can be used to find solutions directly. Coupling both model types the advantages of each would be combined. The quality of a solution generated by constructive models could then be evaluated by descriptive ones. Using the results, the constructive models could be revised until an acceptable schedule is found. In many cases there is not enough knowledge available about the manufacturing system to build a constructive model from the scratch. In such situations descriptive models can be used to get a better understanding of the relevant problem parameters.

From another perspective there also exist approaches trying to change the system in order to fit into the scheduling model, others simplify the model in order to permit the use of a particular solution method. In the meantime more model realism is postulated. Information technology should be used to model the problem without distortion and destruction. In particular it can be assumed that in practical settings there exists not only one scheduling problem all the time and there is not only one solution approach to each problem, but there are different problems at different points in time. On the other hand the analysis of the computational complexity of scheduling problems gives also hints how to simplify a manufacturing process if alternatives for processing exist.

Short term production scheduling is supported by shop floor information systems. Using data from an aggregated production plan a detailed decision is made in which sequence the jobs are released to the manufacturing system, how they traverse inside the system, and how they are sequenced in front of the machines. The level of shop floor scheduling is the last step in which action can be taken on business needs for manufacturing on a predictive and a reactive level.

One main difference between these two scheduling levels is the liability of the input data. For predictive scheduling input data are mainly based on expectations and assumptions. Unforeseen circumstances like rush orders, machine breakdowns, or absence of employees can only be considered statistically, if at all. This situation is different in reactive scheduling where actual data are available. If they are not in coincidence with the estimated data, situation-based revisions of previous decisions have to be made. Predictive scheduling has to go hand in hand with reactive scheduling.

Shop floor information systems available commercially today are predominately data administration systems. Moreover, they collect and monitor data available from machines and the shop floor. Mainly routine operations are carried out by the shop floor system; the production manager is supported by offering the preliminary tools necessary for the development of a paperless planning and control process. Additionally, some systems are also offering various scheduling strategies but with limited performance and without advice when to apply them. It can be concluded that the current shop floor information systems are good at data administration, but for the effective solution of production scheduling problems they are of very little help [MS92a, MS92b].

An intuitive job processing schedule, based solely upon the experience of skilled production managers, does not take advantage of the potential strengths of an integrated IPS. Thus, the development of an intelligent system which integrates planning and control within scheduling for the entire operation and supports effectively the shop floor management, becomes necessary. Such a system could perform all of the functions of the current shop floor scheduling systems and would also be able to generate good proposals for production schedules, which also take deviations from the normal routine into consideration. With the help of such concepts the problems involved in initializing and operating a manufacturing system should be resolved.

Practical approaches to production scheduling on the planning and control level must take also into account the dynamic and unpredictable environment of the shop floor. Due to business and technical considerations, most decisions must be made before all the necessary information has been gathered. Production scheduling must be organized in advance. Predictive scheduling is the task of production planning and the basis for production control; where reactive scheduling has to be able to handle unexpected events. In such a situation, one attempt is to adapt to future developments using a chronological and functional hierarchy within the decision making steps of production scheduling. This helps to create a representation of the problem that considers all available information [Sch89a].

The chronological hierarchy leads to the separation of offline planning (OFP) and online control (ONC). Problems involved in production scheduling are further separated on a conceptual and a specific level in order to produce a functional hierarchy, too. The purpose of the chronological approach to prioritization is to be able to come to a decision through aggregated and detailed modeling, even if future information is unspecific or unavailable. Aside from fulfilling the functional needs of the organization, the basic concept behind the functional hierarchy is to get a better handle on the combinatorial difficulties that emerge from the attempt of simultaneously solving all problems arising in a manufacturing environment. The IPS should follow hierarchical concepts in both, the chronological and the functional aspect. The advantage of such a procedure consists not only in getting a problem-specific approach for investigation of the actual decision problem, but also in the representation of the decision making process within the manufacturing organization.

Models and methods for the hierarchically structured scheduling of production with its planning and control parts have been developed over the years and are highly advanced; see e.g. [KSW86, Kus86, Ste85, LGW86]. However, they lack integration in the sense of providing a concept, which encompasses the entire planning and control process of scheduling. With our proposal for an IPS we try to bring these methods and models one step closer to practical application. The rudimentary techniques of solving predictive scheduling problems presented here work on a closed Analysis-Construction-Evaluation loop (ACE loop). This loop has a feedback mechanism creating an IPS on the levels of OFP and ONC [Sch92]. An overview over the system is shown in Figure 15.3.1.

The OFP module consists of an analysis, a construction and an evaluation component. First, the problem instance is analyzed (A) in terms of objectives, constraints and further characteristics. In order to do this the first step for (A) is to describe the manufacturing environment with the scheduling situation as detailed as necessary. In a second step from this description a specific model has to be chosen from a set of scheduling models in the library of the system. The analysis component (A) can be based upon knowledge-based approaches, such as those used for problems like classification.

The problem analysis defines the parameters for the construction (C) phase. From the basic model obtained in (A), a solution for the scheduling problem is generated by (C) using some generic or specific algorithms. The result is a complete schedule that has then to be evaluated by (E). Here the question has to be answered if the solution can be implemented in the sense that manufacturing according to the proposed solution meets business objectives and fulfils all constraints coming from the application. If the evaluation is satisfactory to the user, the proposed solution will be implemented. If not, the process will repeat itself until the proposed solution delivers a desirable outcome or no more improvements appear to be possible in reasonable time.

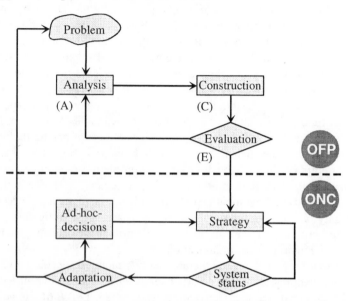

Figure 15.3.1 *Intelligent problem solving in manufacturing.*

The construction component (C) of the ACE loop generates solutions for OFP. It bases its solution upon exact and heuristic problem solving methods. Unfortunately, with this approach we only can solve static representations of quite general problems. The dynamics of the production process can at best be only approximately represented. In order to obtain the necessary answers for a dynamic process, the evaluation component (E) builds up descriptive models in the form of queuing networks at aggregated levels [BY86] or simulation on a specific level [Bul82, Ca86]. With these models one can evaluate the various outcomes and from this if necessary new requirements for problem solution are set up.

Having generated a feasible and satisfactory predictive schedule the ONC module will be called. This module takes the OFP schedule and translates its requirements to an ONC strategy, which will be followed as long as the scheduling problem on the shop floor remains within the setting investigated in the analysis phase of OFP. If temporary disturbances occur, a time dependent strategy in the form of an ad-hoc decision must be devised. If the interruption continues for such a long time that a new schedule needs to be generated, the system will return to the OFP module and seek for an alternative strategy on the basis of a new problem instance with new requirements and possibly different objectives within the ACE loop. Again a new ONC strategy has to be found which will then be followed until again major disturbances occur.

As already mentioned, production scheduling problems are changing over time; a major activity of the problem analysis is to characterize the problem setting such that one or more scheduling problems can be modeled and the right method or a combination of methods for constructing a solution can be chosen from a library of scheduling methods or from knowledge sources coming from different disciplines. With this there are three things to be done; first the manufacturing situation has to be described, second the underlying problem has to be modeled and third an appropriate solution approach has to be chosen. From this point of view one approach is using expert knowledge to formulate and model the problem using the reference model presented in the preceding section, and then using "deep"-knowledge from the library to solve it.

The function of OFP is providing flexibility in the development and implementation of desirable production schedules. OFP applies algorithms which can either be selected from the library or may also be developed interactively on the basis of simulation runs using all components of the ACE loop. The main activity of the interaction of the three components of the loop is the resolution of conflicts between the suggested solution and the requirements coming from the decision maker. Whenever the evaluation of some schedule generated by (C) is not satisfactory then there exists at least some conflict between the requirements or business objectives of a problem solution and the schedule generated so far. Methods to detect and resolve these conflicts are discussed in the next section.

The search for a suitable strategy within ONC should not be limited to routine situations, rather it should also consider e.g. breakdowns and their predictable consequences. ONC takes into consideration the scheduling requirements coming from OFP and the current state of the manufacturing system. To that end, it makes the short term adjustments, which are necessary to handle failures in elements of the system, the introduction of new requirements for manufacturing like rush orders or the cancellation of jobs. An algorithmic reaction on this level of problem solving based on sophisticated combinatorial considerations is generally not possible because of prohibitive computing times of such an approach. Therefore, the competence of human problem solvers in reaching quality, real-time decisions is extremely important.

OFP and ONC require suitable diagnostic experience for high quality decision making. Schedules generated in the past should be recorded and evaluated, for the purpose of using this experience to find solutions for actual problems to be solved. Knowledge-based systems, which could be able to achieve the quality of "self-learning" in the sense of case-based reasoning [Sch98], can make a significant contribution along these lines.

Solution approaches for scheduling problems mainly come from the fields of Operations Research (*OR*) and Artificial Intelligence (*AI*). In contrast to OR-approaches to scheduling, which are focused on *optimization* and which were mainly covered in the preceding chapters, AI relies on *satisfaction*, i.e. it is sufficient to generate solutions which are accepted by the decision maker. Disregarding the different paradigm of either disciplines the complexity status of the scheduling problems remains the same, as it can be shown that the decision variant of a problem is not easier than the corresponding optimization problem (see Section 2.2). Although the OR- and AI-based solution approaches are different, many efforts of either disciplines for investigating scheduling problems are similar; examples are the development of priority rules, the investigation of bottleneck resources and constraint-based scheduling. With priority scheduling as a job- or task-oriented approach, and with bottleneck scheduling as a resource-oriented one, two extremes for rule-based schedule generation exist.

Most of the solution techniques can be applied not only for predictive but also for reactive scheduling. Especially for the latter case priority rules concerning job release to the system and job traversing inside the system are very often used [BPH82, PI77]. Unfortunately, for most problem instances these rules do not deliver best possible solutions because they belong to the wide field of *heuristics*. Heuristics are trying to take advantage from special knowledge about the characteristics of the domain environment or problem description respectively and sometimes from analyzing the structure of known good solutions. Many AI-based approaches exist which use domain knowledge to solve predictive and reactive scheduling problems, especially when modeled as constraint-based scheduling.

OR approaches are built on numerical constraints, the AI approach is considering also non-numerical constraints distinguishing between *soft* and *hard* *constraints*. In this sense scheduling problems also can be considered as *constraint satisfaction problems* with respect to hard and soft constraints. Speaking of hard constraints we mean constraints which represent necessary conditions that must be obeyed. Among hard constraints are given precedence relations, routing conditions, resource availability, ready times, and setup times. In contrast to these, soft constraints such as desirable precedence constraints, due dates, work-in-process inventory, resource utilization, and the number of tool changes, represent rather *preferences* the decision maker wants to be considered. From an OR point of view they represent the aspect of optimization with respect to an objective function. Formulating these preferences as constraints too, will convert the optimization problem under consideration into a feasibility or a decision problem. In practical cases it turns out very often that it is less time consuming to decide on the feasibility of a solution than to give an answer to an optimization problem.

The *constraint satisfaction problem* (*CSP*) deals with the question of finding values for the variables of a set $X = \{x_1, \cdots, x_n\}$ such that a given collection C of constraints $c_1, \cdots, c_m$ is satisfied. Each variable x_i is assigned a domain z_i which defines the set of values x_i may assume. Each constraint is a subset of the Cartesian product $z_1 \times z_2 \times \cdots \times z_n$ that specifies conditions on the values of the variables $x_1, \cdots, x_n$. A subset $\mathcal{Y} \subseteq z_1 \times z_2 \times \cdots \times z_n$ is called a *feasible solution* of the

constraint satisfaction problem if $\mathcal{Y}$ meets all constraints of C, i.e. if $\mathcal{Y} \subseteq \bigcap\limits_{j=1}^{n} c_j$.

The analysis of a constraint satisfaction problem either leads to feasible solutions or to the result that for a given constraint set no such solution exists. In the latter case *conflict resolution* techniques have to be applied. The question induced by a constraint satisfaction problem is an NP-complete problem [GJ79] and one of the traditional approaches to solve it is backtracking. In order to detect *unfeasibility* it is sometimes possible to avoid this computationally expensive approach by carrying out some preprocessing steps where conflicts between constraints are detected in advance.

Example 15.3.1 For illustration purposes consider the following example problem with $X = \{x_1, x_2, x_3\}$, $z_1 = z_2 = z_3 = \{0, 1\}$, and $C = \{c_1, c_2, c_3\}$ representing the constraints

$$x_1 + x_2 = 1 \qquad\qquad\qquad\qquad\qquad\qquad (15.3.1)$$

$$x_2 + x_3 = 1 \qquad\qquad\qquad\qquad\qquad\qquad (15.3.2)$$

$$x_1 + x_3 = y \text{ for } y \in \{0, 2\} . \qquad\qquad\qquad\qquad (15.3.3)$$

Feasible solutions for this example constraint satisfaction problem are given by $\mathcal{Y}_{11} = \{(0, 1, 0)\}$ and $\mathcal{Y}_{12} = \{(1, 0, 1)\}$. If a fourth constraint represented by

$$x_2 + x_3 = 0 \qquad\qquad\qquad\qquad\qquad\qquad (15.3.4)$$

is added to C, conflicts arise between (15.3.2) and (15.3.4) and between (15.3.1), (15.3.3), and (15.3.4). From these we see that no feasible solution exists. Notice that no backtracking approach was needed to arrive at this result.   □

To solve constraint satisfaction problems most AI scheduling systems construct a search tree and apply some search technique to find a feasible solution. A common technique to find feasible solutions quickly is constraint directed search. The fundamental philosophy uses a priori *consistency checking techniques* [DP88, Fre78, Mac77, Mon74]. The basic concept is to prune the search space before unfeasible combinations of variable values are generated. This technique is also known as *constraint propagation*.

Apart from the discussed focus on constraints, AI emphasizes the role of domain specific knowledge in decomposing the initial problem according to several perspectives like bottleneck resources, hierarchies of constraints, conflicting subsets of constraints, while ignoring less important details. Existing AI-based scheduling systems differentiate between *knowledge representation* (models) and *scheduling methodology* (algorithms). They focus rather on a particular application than on general problems. The scheduling knowledge refers to the manufacturing system itself, to constraints and to objectives or preferences. Possible representation techniques are semantic networks (declarative knowledge), predicate logic (especially for constraints), production rules (procedural knowledge) and frames (all of it). Scheduling methodology used in AI is mainly based on production rules (operators), heuristic search (guides the application of operators), opportunistic reasoning (different views of problem solving, e.g. resource-based or job-based), hierarchical decomposition (sub-problem solution, abstraction and distributed problem solving), pattern matching (e.g. using the status of the manufacturing system and given objectives for the application of priority rules), constraint propagation, reinforcement or relaxation techniques.

In the next three sections we describe two approaches which use AI-based solution techniques to give answers to production scheduling problems. In Section 15.3.1 we demonstrate open loop interactive scheduling and in Section 15.3.2 we discuss some closed loop approaches using expert knowledge in the solution process of scheduling problems. In Section 15.3.3 we present an example for integrated problem solving combining OR- and AI-based solution approaches.

15.3.1 Interactive Scheduling

We now want to describe how a constraint-based approach can be used within the ACE-loop to solve predictive scheduling problems interactively. Following Schmidt [Sch89b], decomposable problems can be solved via a heuristic solution procedure based on a hierarchical "relax and enrich" strategy (*REST*) with look ahead capabilities. Using *REST* we start with a solution of some relaxed feasibility problem considering hard constraints only. Then we enrich the problem formulation step by step by introducing preferences from the decision maker. These preferences can be regarded as soft constraints. We can, however, not expect in general that these additional constraints can be met simultaneously, due to possible conflicts with hard constraints or with other preferences. In this case we have to analyze all the preferences by some *conflict detection* procedure. Having discovered conflicting preferences we must decide which of them should be omitted in order to *resolve contradictions*. This way a feasible and acceptable solution can be generated.

REST appears to be appealing in a production scheduling environment for several reasons. The separation of hard constraints from preferences increases scheduling flexibility. Especially, preferences very often change over time so that plan revisions are necessary. If *relaxation* and *enrichment* techniques are applied, only some preferences have to be altered locally while very often major parts of the present schedule satisfying hard constraints can be kept unchanged. A similar argument applies for acceptable partial schedules which may be conserved and the solution procedure can concentrate on the unsatisfactory parts of the schedule only.

This problem treatment can be incorporated into the earlier mentioned DSS framework for production scheduling which then includes an *algorithmic* module to solve the problem under the set of hard constraints, and a *knowledge-based* module to take over the part of conflict detection and implementation of consistent preferences. Without loss of generality and for demonstration purposes only we want to assume in the following that the acceptability of a solution is the greater the more preferences are incorporated into the final schedule. For simplicity reasons it is assumed that all preferences are of equal importance.

In this section we describe the basic ideas of *REST* quite generally and demonstrate its application using an example from precedence constrained scheduling. We start with a short discussion of the types of constraints we want to consider. Then we give an overview on how to detect conflicts between constraints and how to resolve them. Finally, we give a simple example and present the working features of the scheduling system based on *REST*.

Analyzing Conflicts

Given a set of tasks $\mathcal{T} = \{T_1, \cdots, T_n\}$, let us assume that preferences concern the order in which tasks are processed. Hence the set of preferences $\mathcal{PR}$ is defined as a subset of the Cartesian product, $\mathcal{T} \times \mathcal{T}$. *Conflicts* occur among contradictory constraints. We assume that the given hard constraints are not contradictory among themselves, and hence that and thus a feasible schedule that obeys all the hard constraints always exists. Obviously, conflicts can only be induced by the preferences. Then, two kinds of contradictions have to be taken into account: conflicts between the preferences and the hard constraints, and conflicts among preferences themselves. Following the strategy of *REST* we will not extract all of these conflicts in advance. We rather start with a feasible schedule and aim to add as many preferences as possible to the system.

The conflicting preferences are mainly originated from desired task orderings, time restrictions and limited resource availabilities. Consequently, we distinguish between logically conflicting preferences, time conflicting preferences, and resource conflicting preferences.

Logical conflicts between preferences occur if a set of preferred task orderings contains incompatible preferences. Logical conflicts can easily be detected by investigating the directed graph $G = (\mathcal{T}, \mathcal{PR})$. This analysis can be carried out by representing the set of preferences as a directed graph $G = (\mathcal{T}, \mathcal{LC})$ where $\mathcal{T}$ is the set of tasks and $\mathcal{LC} \subseteq \mathcal{T} \times \mathcal{T}$ represents the preferred processing orders among them.

Example 15.3.2 To illustrate the approach we investigate an example problem where a set $\mathcal{T} = \{T_1, T_2, T_3, T_4\}$ of four tasks has to be scheduled. Let the preferred task orderings be given by $\mathcal{PR} = \{PR_1, PR_2, PR_3, PR_4, PR_5\}$ with $PR_1 = (T_1, T_2)$, $PR_2 = (T_2, T_3)$, $PR_3 = (T_3, T_2)$, $PR_4 = (T_3, T_4)$, and $PR_5 = (T_4, T_1)$.

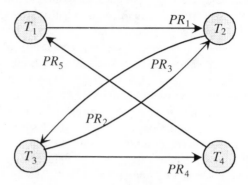

Figure 15.3.2 $G = (\mathcal{T}, \mathcal{PR})$ *representing preferences in Example* 15.3.2.

Logical conflicts in $G = (\mathcal{T}, \mathcal{PR})$ can be detected by finding all cycles of G (see Figure 15.3.2). From this we get two sets of conflicts, $\mathcal{LC}_1 = \{PR_2, PR_3\}$ and $\mathcal{LC}_2 = \{PR_1, PR_2, PR_4, PR_5\}$.   □

Time conflicts occur if a set of preferences is not consistent with time restrictions following from the initial solution obtained on the basis of the hard constraints. To detect time conflicts we must explicitly check all time conditions between the tasks. Hard constraints implying earliest beginning times EB_j, latest beginning times LB_j and processing times p_j restrict the preferences that can be realized. So, if

$$EB_u + p_u > LB_v \tag{15.3.5}$$

for tasks T_u and T_v, the preference (T_u, T_v), would violate the time restrictions. More generally, suppose that for some $k \in I\!N$ and for tasks $T_{u_1}, \cdots, T_{u_k}$ and T_u there are preferences (T_{u_1}, T_{u_2}), $(T_{u_2}, T_{u_3}), \cdots, (T_{u_{k-1}}, T_{u_k})$ and (T_{u_k}, T_v). These preferences imply that the tasks should be processed in order $(T_{u_1}, \cdots, T_{u_k}, T_v)$. However, if this task sequence has the property

$$Z_{u_k} + p_{u_k} > LB_v \tag{15.3.6}$$

where

$$Z_{u_k} = \max \{EB_{u_k}, \max_l \{EB_{u_l} + \sum_{j=l}^{k-1} p_j\}\}$$

then obviously the given set of preferences is conflicting. If (15.3.6) is true the time constraint coming from the last task of the chain will be violated.

Example 15.3.2 - continued - To determine time conflicts, assume that each task T_j has a given earliest beginning time EB_j and a latest beginning time LB_j as specified together with processing times p_j in the following Table 15.3.1.

T_j	EB_j	LB_j	p_j
T_1	7	7	5
T_2	3	12	4
T_3	13	15	2
T_4	12	15	0

Table 15.3.1 *Time parameters for Example 15.3.2.*

To check time conflicts we have to investigate time compatibility of the preferences PR_i, $i = 1, \cdots, 5$. Following (15.3.5), a first consistency investigation shows that each of the preferences, PR_3 and PR_5, is in conflict with the time constraints. The remaining preferences, PR_1, PR_2, and PR_4 would suggest execution of the tasks in order (T_1, T_2, T_3, T_4). To verify feasibility of this sequence we have to check all its subsequences against (15.3.6). The subsequences of length 2 are time compatible because the only time conflicting sequences would be (T_3, T_2) and (T_4, T_1). For the total sequence (T_1, T_2, T_3, T_4) we get $Z_3 = \max \{EB_3, EB_1 + p_1 + p_2, EB_2 + p_2\} = 15$ and $Z_3 + p_3 > LB_4$, thus the subset $\{PR_1, PR_2, PR_4\}$ of preferences creates a time conflict. Similarly the two subsequences of length 3 are tested: the result is that sequence (T_1, T_2, T_3) realized by preferences PR_1 and PR_2 establishes a time conflict, whereas (T_2, T_3, T_4) does not. So we end up with four time conflicting sets of preferences, $TC_1 = \{PR_3\}$, $TC_2 = \{PR_5\}$, $TC_3 = \{PR_1, PR_2\}$, and $TC_4 = \{PR_1, PR_2, PR_4\}$. □

If the implementation of some preference causes a resource demand at some time t such that it exceeds resource limits at this time, i.e.

$$\sum_{T_i \in T_t} R_k(T_j) > m_k, \; k = 1, \cdots, s, \tag{15.3.7}$$

then a *resource conflict* occurs. Here T_t denotes the set of tasks being processed at time t, $R_k(T_j)$ the requirement of resource of type R_k of task T_j, and m_k the corresponding resource maximum supply.

Example 15.3.2 - continued - As to the *resource conflicts*, assume that $s = 1$, $m_1 = 1$, and $R_1(T_j) = 1$ for all $j = 1, \cdots, 4$. Taking the given time constraints into account, we detect a conflict for PR_1 from (15.3.7) since T_2 cannot be processed in parallel with tasks T_3 and T_4. Thus an additional conflicting set $RC_1 = \{PR_1\}$ has to be introduced. □

Coping with Conflicts

Let there be given a set T of tasks, and a set PR of preferences concerning the processing order of tasks. Assume that logical conflicting sets $LC_1, \cdots, LC_\lambda$, time conflicting sets $TC_1, \cdots, TC_\tau$, and resource conflicting sets $RC_1, \cdots, RC_\rho$ have been detected. We then want to find out if there is a solution schedule that meets all the restrictions coming from these conflicting sets. This means that we need to find a subset PR' of PR of maximal cardinality such that none of the conflicting sets LC_i, TC_j, RC_k contradicts PR', i.e. is contained in PR'.

Let $LC := \{LC_1, \cdots, LC_\lambda\}$ be the set of all logically conflicting sets; the set TC of time conflicting sets and the set RC of resource conflicting sets are defined analogously. Define $C := LC \cup TC \cup RC$, i.e. C contains all the conflicting sets of the system. The pair $IH := (PR, C)$ represents a *hypergraph* with *vertices* PR and *hyperedges* C. Since IH describes all conflicts arising in the system we refer to IH as the *conflict hypergraph*.

Our aim is to find a suitable subset PR', i.e. one that does not contain any of the hyperedges. We notice that if $H_1 \subseteq H_2$ for hyperedges H_1 and H_2, we need not to consider H_2 since H_1 represents the more restrictive conflicting set. Observing this we can simplify the hypergraph by eliminating all hyperedges that are supersets of other hyperedges. The hypergraph then obtained is referred to as the *reduced conflict hypergraph*.

According to our *measure of acceptability* we are interested in the maximum number of preferences that can be accepted without loosing feasibility. This is justified if all preferences are of equal importance. If the preferences have different weights we might be interested in a subset of preferences of maximum total weight. All these practical questions result in NP-hard problems [GJ79].

To summarize the discussion we have to perform three steps to solve the problem.

Step 1: Detect all the logically, time, and resource conflicting sets.

Step 2: Build the reduced conflict hypergraph.

Step 3: Apply some conflict resolution algorithm.

Algorithm 15.3.3 *frame* $(IH = (PR, C))$;

```
begin
S := ∅;        -- initialization of the solution set
while PR ≠ ∅ do
   begin
   Reduce hypergraph (PR, C);
   Following some underlying heuristic, choose preference PR ∈ PR;    (15.3.8)
```

$\mathcal{PR} := \mathcal{PR} - \{PR\};$

if $C \nsubseteq S \cup \{PR\}$ **for all** $C \in \mathcal{C}$ **then** $S := S \cup \{PR\};$

$--$ $\mathcal{PR}$ is accepted if the temporal solution set does not contain any conflicting preferences

    for all $C \in \mathcal{C}$ **do** $C := C \cap (\mathcal{PR} \times \mathcal{PR});$

        $--$ the hypergraph is restricted to the new (i.e. smaller) set of vertices

   end;

end;

The algorithm is called *frame* because it has to be put in concrete form by intro-ducing some specific heuristics in line (15.3.8). Based on the conflict hypergraph $I\!H = (\mathcal{PR}, \mathcal{C})$ heuristic strategies can easily be defined. We also mention that if preferences are of different importance their weight should be considered in the definition of the heuristics in (15.3.8).

In the following we give a simple example of how priority driven heuristics can be defined. Each time (15.3.8) is performed, the algorithm chooses a prefer-ence of highest priority. In order to gain better adaptability we allow that priori-ties are re-computed before the next choice is taken. This kind of dynamics is important in cases where the priority values are computed from the hypergraph structure, because as the hypergraph gets smaller step by step its structure changes during the execution of the algorithm, too.

Heuristic *DELTA-decreasing* (δ_{dec}): Let $\delta: \mathcal{PR} \rightarrow I\!N^0$ be the *degree* that as-signs - in analogy to the notion of degree in graphs - each vertex $PR \in \mathcal{PR}$ the number of incident hyperedges, i.e. the number of hyperedges containing vertex PR. The heuristic δ_{dec} then arranges the preferences in order of non-increasing degree. This strategy follows the observation that the larger the degree of a pref-erence is, the more subsets of conflicting preferences exist; thus such a prefer-ence has less chance to occur in the solution set. To increase this chance we give such preference a higher priority.

Heuristic *DELTA-increasing* (δ_{inc}): Define $\delta_{inc} := -\delta_{dec}$. This way preferences of lower degree get higher priority. This heuristic was chosen for comparison against the δ_{dec} strategy.

Heuristic *GAMMA-increasing* (γ_{inc}): Define $\gamma: \mathcal{PR} \rightarrow I\!N^0$ as follows: For $PR \in \mathcal{PR}$, let $\gamma(PR)$ be the number of vertices that do not have any common hyper-edge with PR. The heuristic γ_{inc} then arranges the preferences in order of non-decreasing cardinalities. The idea behind this strategy is that a preference with small γ-value has less chance to be selected to the solution set. To increase this chance we give such preference a higher priority.

Heuristic *GAMMA-decreasing* (γ_{dec}): Define $\gamma_{dec} := -\gamma_{inc}$. This heuristic was chosen for comparison against the γ_{inc} strategy.

The above heuristics have been compared by empirical evaluation [ES93]. There it turned out that *DELTA-decreasing*, *GAMMA-increasing* behave considerably better than *DELTA-increasing* and *GAMMA-decreasing*.

Example 15.3.2 - continued - Summarizing all conflicts we get the conflict hypergraph $IH := (\mathcal{PR}, C)$ where the set C contains the hyperedges

> $\{PR_2, PR_3\}$
> $\{PR_1, PR_2, PR_4, PR_5\}$ (logically conflicting sets)
>
> $\{PR_3\}$
> $\{PR_5\}$
> $\{PR_1, PR_2\}$ (time conflicting sets)
> $\{PR_1, PR_2, PR_4\}$
>
> $\{PR_1\}$ (resource conflicting set).

Figure 15.3.3 shows the hypergraph where encircled vertices are hyperedges.

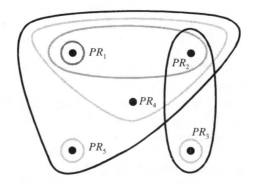

Figure 15.3.3 $IH = (\mathcal{PR}, C)$ *representing conflicts of the example problem.*

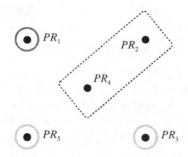

Figure 15.3.4   *Reduced hypergraph representing conflicts of the example problem.*

Since each of the hyperedges of cardinality > 1 contains a conflicting set of cardinality one, the reduced hypergraph has only the three hyperedges $\{PR_1\}$, $\{PR_3\}$ and $\{PR_5\}$, see Figure 15.3.4. A subset of maximal cardinality that is not in conflict with any of the conflicting sets is $\{PR_2, PR_4\}$. Each of the above algorithms finds this solution as can easily be verified. ☐

Below we present a more complex example where not only the heuristics can be nontrivially applied; the example also demonstrates the main idea behind an interactive schedule generation.

Working Features of an Interactive Scheduling System

The above described approach of *REST* with conflict detection mechanisms can be integrated into a DSS [EGS97]. Its general outline is shown in Figure 15.3.5.

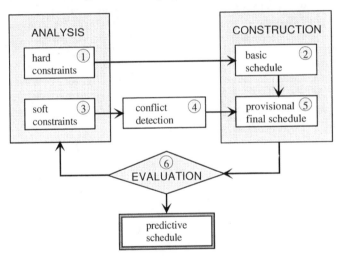

Figure 15.3.5   *A DSS for the REST-approach.*

The DSS consists of four major modules: problem analysis, schedule generation, conflict detection and evaluation. Their working features can be organized by incorporating six phases. The first phase starts with some problem analysis investigating the hard constraints which have to be taken into account for any problem solution. Then, in the second phase a first feasible solution (basic schedule) is generated by applying some scheduling algorithm. The third phase takes over the part of analyzing the set of preferences of task constraints. In the fourth phase their interaction with the results of the basic schedule is clarified via the conflict detection module. In the fifth phase a compatible subset of soft constraints according to the objectives of the decision maker is determined, from which a revised schedule is generated. In the last phase the revised evaluated. If the evaluation is satisfactory a solution for the predictive scheduling problem is found; if not, the schedule has to be revised by considering new constraints from the decision maker. The loop stops as soon as a satisfactory solution has been found.

The DSS can be extended to handle a dynamic environment. Whenever hard constraints have to be revised or the set of preferences is changing we can apply this approach on a rolling basis.

Example 15.3.4 To demonstrate the working feature of the scheduling system consider an extended example. Let there be given a set of tasks $\mathcal{T} = \{T_1, T_2, T_3, T_4, T_5, T_6, T_7, T_8\}$, and hard constraints as shown in Figure 15.3.6(a). Processing times and earliest and latest beginning times are given as triples (p_j, EB_j, LB_j) next to the task nodes. In addition, concurrent task execution is restricted by two types of resources and resource requirements of the tasks are $R(T_1) = [2, 0]$, $R(T_2) = [2, 4]$, $R(T_3) = [0, 1]$, $R(T_4) = [4, 2]$, $R(T_5) = [1, 0]$, $R(T_6) = [2, 5]$, $R(T_7) = [3, 0]$, $R(T_8) = [0, 1]$. The total resource supply is $m = [5, 5]$.

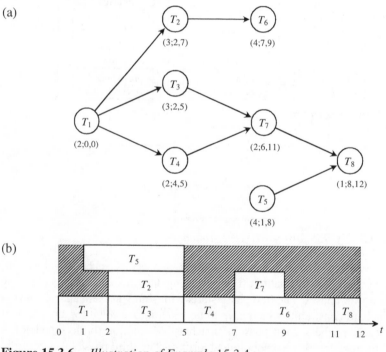

Figure 15.3.6   *Illustration of Example* 15.3.4 :
            (a) *hard constraints,*
            (b) *a basic schedule.*

Having analyzed the hard constraints we generate a feasible basic schedule by applying some scheduling algorithm. The result is shown in Figure 15.3.6(b).

Feasibility of the schedule is gained by assigning a starting time s_j to each task such that $EB_j \le s_j \le LB_j$ and the resource constraints are met.

Describing the problem in terms of the constraint satisfaction problem, the variables refer to the starting times of the tasks, their domains to the intervals of corresponding earliest and latest beginning times and the constraints to the set of preferences. Let the set of preferences be given by $\mathcal{PR} = \{PR_1, \cdots, PR_7\}$ with $PR_1 = (T_3, T_4)$, $PR_2 = (T_2, T_3)$, $PR_3 = (T_4, T_3)$, $PR_4 = (T_7, T_5)$, $PR_5 = (T_5, T_2)$, $PR_6 = (T_5, T_6)$, and $PR_7 = (T_4, T_5)$ (see Figure 15.3.7). Notice that the basic schedule of Figure 15.3.6(b) realizes just two of the preferences.

Analyzing conflicts we start with the detection of logical conflicts. From the cycles of the graph in Figure 15.3.7 we get the logically conflicting sets $\mathcal{LC}_1 = \{PR_1, PR_3\}$ and $\mathcal{LC}_2 = \{PR_1, PR_2, PR_5, PR_7\}$.

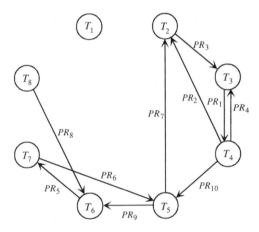

Figure 15.3.7 $G = (\mathcal{T}, \mathcal{PR})$ representing preferences in Example 15.3.4.

Task sequence	Time conflicting set of preferences
(T_4, T_3)	$\mathcal{TC}_1 = \{PR_3\}$
(T_2, T_3, T_4)	$\mathcal{TC}_2 = \{PR_1, PR_2\}$
(T_7, T_5, T_6)	$\mathcal{TC}_3 = \{PR_4, PR_6\}$
(T_2, T_3, T_4, T_5)	$\mathcal{TC}_4 = \{PR_1, PR_2, PR_7\}$
(T_3, T_4, T_5, T_2)	$\mathcal{TC}_5 = \{PR_1, PR_5, PR_7\}$
(T_4, T_5, T_2, T_3)	$\mathcal{TC}_6 = \{PR_2, PR_5, PR_7\}$
(T_5, T_2, T_3, T_4)	$\mathcal{TC}_7 = \{PR_1, PR_2, PR_5\}$

Table 15.3.2 *Subsets of preferences being in time conflict.*

For the analysis of time constraints we start with task sequences of length 2. We see that there is only one such conflict, $\mathcal{TC}_1$. Next, task sequences of length greater than 2 are checked. Table 15.3.2 summarizes non-feasible task sequences and their corresponding time conflicting subsets of preferences.

So far we found 2 logically conflicting sets and 7 time conflicting sets of preferences. In order to get the reduced hypergraph, all sets that already contain a conflicting set must be eliminated. Hence there remain 5 conflicting sets of preferences, $\{PR_3\}$ $\{PR_1, PR_2\}$, $\{PR_4, PR_6\}$, $\{PR_1, PR_5, PR_7\}$ and $\{PR_2, PR_5, PR_7\}$. The corresponding hypergraph is sketched in Figure 15.3.8.

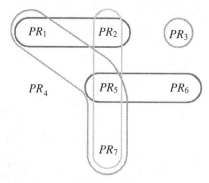

Figure 15.3.8 $G_c = (\mathcal{PR}, \mathcal{E})$ representing logical and time conflicts of Example 15.3.4.

We did, however, not consider the resource constraints so far. To detect resource conflicts we had to find all combinations of tasks which cannot be scheduled simultaneously because of resource conflicts. Since in general the number of these sets increases exponentially with the number of tasks, we follow another strategy: First create a schedule without considering resource conflicts, then check for resource conflicts and introduce additional precedence constraints between tasks being in resource conflict. In this manner we proceed until a feasible solution is found.

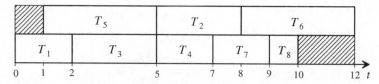

Figure 15.3.9   Schedule for Example 15.3.4 without considering resource conflicts.

To construct a first schedule, we aim to find a set of non-conflicting preferences of maximum cardinality, i.e. a maximum set that does not contain any of the hyperedges of the above hypergraph. For complexity reasons we content ourselves with an approximate solution and apply algorithm *frame*. Heuristics GAMMA-increasing, for example, selects the subset $\{PR_1, PR_5, PR_6\}$ and we result in the schedule presented in Figure 15.3.9. Remember that we assumed for simplicity reasons that all preferences are equally weighted.

The schedule of Figure 15.3.9 shows two resource conflicts, for T_2, T_4 and for T_6, T_8. Hence T_2 and T_4 (and analogously T_6 and T_8) cannot be processed simultaneously, and we have to choose an order for these tasks. This way we end up with two additional hard constraints in the precedence graph shown in figure 15.3.10(a). Continuing the analysis of constraints we result in a schedule that realizes the preferences PR_5 and PR_6 (Figure 15.3.10(b)).   □

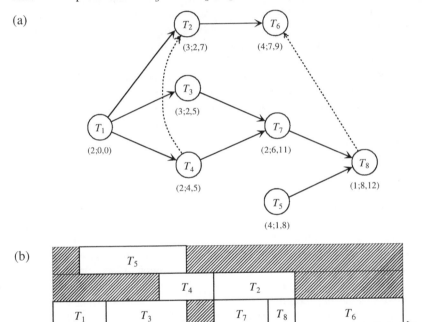

Figure 15.3.10 *Final schedule for Example* 15.3.4 :
    (a) *precedence graph of Figure* 15.3.6(a) *with additional hard constraints* (T_4, T_2) *and* (T_8, T_6),
    (b) *a corresponding schedule.*

We can now summarize our approach by the following algorithm:

Algorithm 15.3.5 *for interactive scheduling.*
begin
Initialize 'Basic Schedule';
Collect preferences;

Detect conflicts;
while conflicts exist **do** Apply *frame*;
Generate final schedule;
end;

Reactive Scheduling

Now assume that the predictive schedule has been implemented and some unforeseen disturbance occurs. In this case within the ACE loop (compare Figure
15.3.1) reactive scheduling is concerned with revising predictive schedules as
unexpected events force changes. Now we want to present some ideas how to
interactively model parts of the reactive scheduling process using fuzzy logic.
The idea is to apply this approach for monitoring and diagnosing purposes only.
Based on the corresponding observations detailed reactive scheduling actions can
be taken by the decision maker [Sch94].

An algorithmic reaction on the reactive level of problem solving based on
sophisticated combinatorial considerations is generally not possible because of
prohibitive computing times; therefore, the competence of human problem
solvers in reaching quality, real-time decisions is extremely important. The human problem solver should be supported by advanced modeling techniques. In
order to achieve this we suggest the application of fuzzy logic because it allows
to represent the vague, qualitative view of the human scheduler most conveniently. The underlying theory of fuzzy sets [Zad65] concentrates on modeling
vagueness due to common sense interpretation, data aggregation and vague relations. Examples for common sense interpretation in terms of a human production
scheduler are e.g. 'long queues of jobs' or 'high machine processing speed'. Data
aggregation is used in expressions like 'skilled worker' or 'difficult situation' and
vague relations are represented by terms like 'not much more urgent than' or
'rather equal'.

Reactive scheduling requires suitable diagnostic support for quick decision
making. This is intended with our approach modeling reactive scheduling by
fuzzy logic. The two main components of the model we use are (1) linguistic
variables [Zad73] and (2) fuzzy rules or better decision tables [Kan86]. A linguistic variable L can be represented by the tuple $L = (X, U, f)$ where set X represents the feasible values of L, set U represents the domain of L, and f is the membership function of L which assigns to each element $x \in X$ a fuzzy set
$A(x) = \{u, f_x(u)\}$ where $f_x(u) \in [0, 1]$.

A decision table (DT) consists of a set of conditions (if-part) and a set of actions (then-part). In case of multi conditions or multi actions conditions or actions respectively have to be connected by operators. If all conditions have only
one precise value we speak of *deterministic* DT. In case we use fuzzy variables
for representing conditions or actions we also can build non-deterministic DT.
These tables are very much alike of how humans think. In order to represent the

interaction of linguistic variables in DT we have to introduce set-theoretic operations to find the resulting membership function. The most common operations are union, intersection and complement. In case of an union we have $f_C(u) = \max\{f_A(u), f_B(u)\}$, in case of an intersection $f_D(u) = \min\{f_A(u), f_B(u)\}$, and in case of the complement A° of A we have $f_{A^\circ}(u) = 1 - f_A(u)$. To understand the approach of modeling reactive scheduling by fuzzy logic better consider the following scenario.

There are queues of jobs in front of machines on the shop floor. For each job J_j the number of jobs N_j waiting ahead in the queue, its due date d_j and its slack time $s_j = d_j - t$ are known where t is the current time. Processing times of the jobs are subject to disturbances. Due date and machine assignment of the jobs are determined by predictive scheduling. The objective is to diagnose critical jobs, i.e. jobs which are about to miss their due dates in order to reschedule them. N_j and s_j are the linguistic input variables and "becomes critical" is the output variable of the DT. Membership functions for the individual values of the variables are determined by a knowledge acquisition procedure which will not be described here. The following DT shown in Table 15.3.3 represents the fuzzy rule system.

AND	Small	Medium	Great
Few	soon	later	not to see
Some	now	later	not to see
Many	now	soon	not to see
Very	now	soon	later

Table 15.3.3 *Decision table for fuzzy rule system.*

The rows represent the values of the variable N_j and the columns represent the values of the variable s_j. Both variables are connected by an AND-operator in any rule. With the above Table 15.3.3 twelve rules are represented. Each element of the table gives one value of the output variable "becomes critical" depending on the rule. To find these results the membership functions of the input variables are merged by intersection operations to a new membership function describing the output variable of each rule. The resulting fuzzy sets of all rules are then combined by operations which are applied for the union of sets. This procedure was tested to be most favorable from an empirical point of view.

As a result the decision maker on the shop floor gets the information which jobs have to be rescheduled now, soon, later, or probably not at all. From this two possibilities arise; either a complete new predictive schedule has to be generated or local ad-hoc decisions can be taken on the reactive scheduling level. Control decisions based on this fuzzy modeling approach and their consequences should be recorded and evaluated, for the purpose of using these past decisions to find better solutions to current problems. Fuzzy case-based reasoning systems, which should be able to achieve the quality of a self-learning system, could make a significant contribution along these lines.

We have implemented our approach of modeling reactive scheduling by fuzzy logic as a demonstration prototype. Two screens serve as the user interface. On the first screen the jobs waiting in a machine queue and the slack time of the job under investigation are shown. An example problem is shown in Figure 15.3.11.

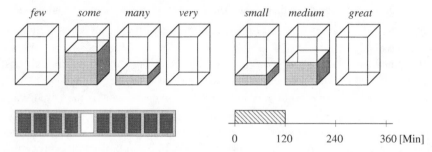

few   some   many   very   small   medium   great

0   120   240   360 [Min]

Figure 15.3.11   *Interface of fuzzy scheduler.*

There are ten jobs waiting in a queue to be processed by some machine P_i, the job under consideration J_j is shown by a white rectangle. Above the queue the different fuzzy sets concerning the linguistic variable N_j and the values of the corresponding membership functions are represented. The slack time s_j of job J_j is currently 130 minutes; again fuzzy sets and membership functions of this linguistic variable are represented above the scale.

Applying the rules of the DT results in a representation which is shown in Figure 15.3.12. The result of the first part of inference shows that for job J_j the output variable "becomes critical" is related to some positive values for "soon" and for "later" shown by white segments. From this it is concluded by the second part of inference that this job has to be checked again before it is about to be re-scheduled.

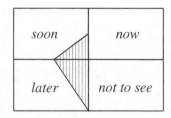

| soon | now |
| later | not to see |

Figure 15.3.12   *Result of inference.*

15.3.2   Knowledge-based Scheduling

Expert Systems are special kinds of knowledge-based systems. Expert systems are designed to support problem modeling and solving with the intention to simulate the capabilities of domain experts, e.g. problem understanding, problem solving, explaining the solution, knowledge acquisition, and knowledge restructuring. Most expert systems use two types of knowledge: *descriptive knowledge* or *facts*, and *procedural knowledge* or knowledge about the *semantics* behind facts. The architecture of expert systems is mainly based on a *closed loop solution approach*. This consists of the four components *storing knowledge, knowledge acquisition, explanation of results*, and *problem solution*. In the following we will concentrate on such a closed loop problem solution processes.

There is an important difference between expert systems and conventional problem solving systems. In most expert systems the model of the problem description and basic elements of problem solving are stored in a knowledge base. The complete solution process is carried out by some inference module interacting with the knowledge base. Conventional systems do not have this kind of separated structure; they are rather a mixture of both parts in one program.

In order to implement an expert system one needs three types of *models*: a model of the domain, a model of the elementary steps to be taken, and a model of inference that defines the sequence of elementary steps in the process of problem solution. The domain is represented using descriptive knowledge about objects, their attributes and their relations as introduced in Section 15.2. In production scheduling for example, objects are machines, jobs, tasks or tools, attributes are machine states, job and task characteristics or tool setup times, and relations could be the subsumption of machines to machine types or tasks to jobs. The model of elementary steps uses production rules or other representations of procedural knowledge. For if-then rules there exists a unique input-output description. The model of inference uses combinations or sets of elementary steps to represent the solution process where a given start state is transformed to a desired goal state. This latter type of knowledge can also be knowledge of domain experts or domain independent knowledge. The goal of the expert system approach is mainly to improve the modeling part of the solution process to get closer to reality.

To give a better understanding of this view we refer to an example given by Kanet and Adelsberger [KA87]: "... consider a simple scheduling situation in which there is a single machine to process jobs that arrive at different points in time within the planning period. The objective might be to find a schedule which minimizes mean tardiness. An algorithmic approach might entertain simplifying the formulation by first assuming all jobs to be immediately available for processing. This simplified problem would then be solved and perhaps some heuristic used to alter the solution so that the original assumption of dynamic arrivals is back in tack. The approach looks at reformulation as a means to 'divide et impera'. On the other hand a reformulative approach may ... seek to find a 'richer' problem formulation. For example the question might be asked 'is working over-time a viable alternative?', or 'does there exist another machine that can accomplish this task?', or 'is there a subset of orders that are less critical than others?', and so on."

On the other hand systems for production scheduling should not only replicate the expert's schedule but extend the capabilities by doing more problem solving. In order to achieve this AI systems separate the scheduling model from a general solution procedure. In [Fox90] the shop floor scheduling model described uses terms from AI. It is considered to be time based planning where tasks or jobs must be selected, sequenced, and assigned to resources and time intervals for execution. Another view is that of a multi agent planning problem, where each task or job represents a separate agent for which a schedule is to be created; the agents are uncooperative, i.e. each is attempting to maximize its own goals. It is also claimed that expert systems appear inappropriate for the purpose of problem solution especially for two reasons: (1) problems like production scheduling tend to be so complex that they are beyond the cognitive capabilities of the human scheduler, and (2) even if the problem is relatively easy, factory environments change often enough so that any expertise built up over time becomes obsolete very quickly.

We believe that it is nevertheless possible to apply an expert system approach for the solution of production scheduling problems but with a different perspective on problem solving. Though, as already stated, expert systems are not appropriate for solving combinatorial search problems, they are quite reasonable for the *analysis* of models and their solutions. In this way expert systems can be used for building or selecting models for scheduling problems. An appropriate solution procedure can be selected for the model, and then the expert system can again support the evaluation of the solution.

The scheduling systems reviewed next are not expert systems in their purest sense and thus we will use the more general term *knowledge-based system*. ISIS [SFO86, Fox87, FS84], OPIS [SPP+90] and CORTES [FS90] are a family of systems with the goal of modeling knowledge of the manufacturing environment using mainly constraints to support *constraint guided search*; knowledge about constraints is used in the attempt to decrease the underlying search space. The systems are designed for both, predictive and reactive scheduling.

ISIS-1 uses pure constraint guided search, but was not very successful in solving practical scheduling problems. ISIS-2 uses a more sophisticated search technique. Search is divided into the four phases job selection, time analysis, resource analysis, and resource assignment. Each phase consists in turn of the three sub-phases pre-search analysis (model construction), search (construction of the solution), and post-search analysis (evaluation of the solution). In the job selection phase a priority rule is applied to select the next job from the given set of available jobs. This job is passed to the second phase. Here earliest start and latest finish times for each task of the job are calculated without taking the resource requirements into account. In phases three and four the assignment of resources and the calculation of the final start and finish times of all tasks of the job under consideration is carried out. The search is organized by some *beam search* method. Each solution is evaluated within a rule-based post-search analysis. ISIS-3 tries to schedule each job using more information from the shop floor, especially about bottleneck-resources. With this information the job-centered scheduling approach as it is realized in ISIS-2 was complemented by a resource-centered scheduler.

As the architecture of ISIS is inflexible as far as modifications of given schedules are concerned, a new scheduling system called OPIS-1 was developed. It uses a *blackboard approach* for the communication of the two knowledge sources analysis and decision. These use the blackboard as shared memory to post messages, partial results and any further information needed for the problem solution. The blackboard is the exclusive medium of communication. Within OPIS-1 the "analyzer" constructs a rough schedule using some *balancing heuristic* and then determines the bottlenecks. Decision is then taken by the resource and the job scheduler already implemented in ISIS-3. Search is centrally controlled. OPIS-1 is also capable to deal with reactive scheduling problems, because all events can be communicated through the blackboard. In OPIS-2 this event management is supported by two additional knowledge sources which are a "right shifter" and a "demand swapper". The first one is responsible for pushing jobs forward in the schedule, and the second for exchanging jobs. Within the OPIS systems it seems that the most difficult operation is to decide which knowledge source has to be activated.

The third system of the family we want to introduce briefly is CORTES. Whereas the ISIS systems are primarily job-based and OPIS switches between job-based and resource-based considerations, CORTES takes a task-oriented point of view, which provides more flexibility at the cost of greater search effort. Within a five step heuristic procedure a task is assigned to some resource over some time interval.

Knowledge-based systems using an expert system approach should concentrate on finding good models for the problem domain and the description of elementary steps to be taken during the solution process. The solution process itself may be implemented by a different approach. One example for model development considering knowledge about the domain and elementary steps to be taken can be found in [SS90]. Here a reactive scheduling problem is solved along the same line as OPIS works using the following problem categorization: (1) machine breakdown, (2) rush jobs, (3) new batch of jobs, (4) material shortage, (5) labor absenteeism, (6) job completion at a machine, and (7) change in shift. Knowledge is modularized into independent knowledge sources, each of them designed to solve a specific problem. If a new event occurs it is passed to some meta-analyzer and then to the appropriate knowledge source to give a solution to the analyzed scheduling problem. For instance, the shortage of some specific raw material may result in the requirement of rearranging the jobs assigned to a particular machine. This could be achieved by using the human scheduler's heuristic or by an appropriate algorithm to determine some action to be taken.

As a representative for many other knowledge-based scheduling systems - see [Ata91] for a survey - we want to describe *SONIA* which integrates both predictive and reactive scheduling on the basis of hard and soft constraints [CPP88]. The scheduling system is designed to detect and react to inconsistencies (conflicts) between a predictive schedule and the actual events on the shop floor. SONIA consists of two analyzing components, a capacity analyzer and an analyzer of conflicts, and further more a predictive and a reactive component, each containing a set of heuristics, and a component for managing schedule descriptions.

For representing a schedule in SONIA the resources needed for processing jobs are described at various levels of detail. Individual resources like machines are elements of resource groups called work areas. Resource reservation constraints are associated with resources. To give an example for such a constraint, (*res*; t_1, t_2, n; *list-of-motives*) means that n resources from resource group *res* are not available during the time interval (t_1, t_2) for the reasons given in the *list-of-motives*.

Each job is characterized by a ready time, a due date, precedence constraints, and by a set of tasks, each having resource requirements. To describe the progress of work the notions of an actual status and a schedule status are introduced. The *actual status* is of either kind "completed", "in-process", "not started", and the *schedule status* can be "scheduled", "selected" or (deliberately) "ignored". There may also be temporal constraints for tasks. For example, such a constraint can be described by the expression (*time* $\leq t_1t_2$, k) where t_1 and t_2 are points in time which respectively correspond to the start and the finish time of processing a task, and k represents the number of time units; if there have to be at least t time units between processing of tasks T_j and T_{j+1}, the corresponding expression would be (*time* $\leq$ (*end* T_j)(*start* T_{j+1}), t). To represent actual time values, the origin of time and the current time have to be known.

SONIA uses constraint propagation which enables the detection of inconsistencies or conflicts between predictive decisions and events happening on the shop floor. Let us assume that as a result of the predictive schedule it is known that task T_j could precede task T_{j+1} while the actual situation in the workshop is such that T_j is in schedule status "ignored" and T_{j+1} is in actual status "in process". From this we get an inconsistency between these temporal constraints describing the predictive schedule and the ones which come from the actual situation. The detection of conflicts through constraint propagation is carried out using propagation axioms which indicate how constraints and logic expressions can be combined and new constraints or conflicts can be derived. The axioms are utilized by an interpreter.

SONIA distinguishes between the three major kinds of conflicts: delays, capacity conflicts and breakdowns. The class of delays contains all conflicts which result from unexpected delays. There are four subclasses to be considered, "Task Delay" if the expected finish time of a task cannot be respected, "Due-Date Delay" if the due date of a manufacturing job cannot be met, "Interruption Delay" if some task cannot be performed in a work shift determined by the predictive schedule, and "Global Tardiness Conflict" if it is not possible to process all of the selected tasks by the end of the current shift. The class of capacity conflicts refers to all conflicts that come from reservation constraints. There are three subclasses to be considered. If reservations for tasks have to be cancelled because of break-downs we speak of "Breakdown Capacity Conflicts". In case a resource is assigned to a task during a work shift where this resource is not available, an "Out-Of-Shift Conflict" occurs. A capacity conflict is an "Overload" if the number of tasks assigned to a resource during a given interval of time is greater than the available capacity. The third class consists of breakdowns which contains all subclasses from delays and capacity conflicts caused only by machine break-downs. In the following we give a short overview of the main *components* of the SONIA system and its *control architecture*.

(*i*) *Predictive Components* The predictive components are responsible for generating an off-line schedule and consist of a selection and an ordering component. First a set of tasks is selected and resources are assigned to them. The selection depends on other already selected tasks, shop status, open work shifts and jobs to be completed. Whenever a task is selected its schedule status is "selected" and the resulting constraints are created by the schedule management system. The ordering component then uses an iterative constraint satisfaction process utilizing heuristic rules. If conflicts arise during schedule generation, backtracking is carried out, i.e. actions coming from certain rules are withdrawn. If no feasible schedule can be found for all the selected tasks a choice is made for the tasks that have to be rejected. Their schedule status is set to "ignored" and the corresponding constraints are deleted.

(*ii*) *Reactive Components.* For reactive scheduling three approaches to resolve conflicts between the predictive schedule and the current situation on the shop floor are possible: Predictive components can generate a complete new schedule, the current schedule is modified globally forward from the current date, or local changes are made. The first approach is the case of predictive scheduling which already been described above. The easiest reaction to modify the current schedule is to reject tasks, setting their scheduling status to "ignored" and deleting all related constraints. Of course, the rejected task should be that one causing the conflicts. If several rejections are possible the problem gets far more difficult and applicable strategies have still to be developed. Re-scheduling forward from the current date is the third possibility of reaction considered here. In this case very often due dates or ends of work shifts have to be modified. An easy reaction would simply by a right shift of all tasks without modifying their ordering and the resource assignments. In a more sophisticated approach some heuristics are applied to change the order of tasks.

(*iii*) *Analysis Components*. The purpose of the analyzers is to determine which of the available predictive and reactive components should be applied for schedule generation and how they should be used. Currently, there are two analysis components implemented, a capacity analyzer and a conflict analyzer. The capacity analyzer has to detect bottleneck and under-loaded resources. These detections lead to the application of scheduling heuristics, e.g. of the kind that the most critical resources have to be scheduled first; in the same sense, under-loaded resources lead to the selection of additional tasks which can exploit the resources. The conflict analyzer chooses those available reactive components which are most efficient in terms of conflict resolution.

(*iv*) *Control Architecture*. Problem solving and evaluating knowledge have to be integrated and adjusted to the problem solving context. A blackboard architecture is used for these purposes. Each component can be considered as an independent knowledge source which offers its services as soon as predetermined conditions are satisfied. The blackboard architecture makes it possible to have a flexible system when new strategies and new components have to be added and integrated. The domain blackboard contains capacity of the resources determined by the capacity analyzer, conflicts which are updated by the schedule management, and results given by predictive and reactive components. The control blackboard contains the scheduling problem, the sub-problems to be solved, strategies like heuristic rules or meta-rules, an agenda where all the pending actions are listed, policies to choose the next pending action and a survey of actions which are currently processed.

SONIA is a knowledge-based scheduling system which relies on constraint satisfaction where the constraints come from the problem description and are then further propagated. It has a very flexible architecture, generates predictive and reactive schedules and integrates both solution approaches. A deficiency is that nothing can be said from an ex-ante point of view about the quality of the solutions generated by the conflict resolution techniques. Unfortunately also a judgement from an ex-post point of view is not possible because there is no empirical data available up to now which gives reference to some quality measure of the schedule. Also nothing is known about computing times. As far as we know, this lack of evaluation holds for many knowledge-based scheduling systems developed until today.

15.3.3   Integrated Problem Solving

In this last section we first want to give an example to demonstrate the approach of integrating algorithms and knowledge within an interactive approach for OFP and ONC relying on the ACE loop. For clarity purposes, the example is very simple. Let us assume, we have to operate a flexible manufacturing cell that consists of identically tooled machines all processing with the same speed. These kinds of cells are also called pools of machines. From the production planning system we know the set of jobs that have to be processed during the next period of time e.g. in the next shift. As we have identical machines we will now speak of tasks instead of jobs which have to be processed. The business need is that all tasks have to be finished at the end of the next eight hour shift. With this the problem is roughly stated.

Using further expert knowledge from scheduling theory for the analysis of the problem we get some insights using the following knowledge sources (see Chapter 5 for details):

(1) The schedule length is influenced mainly by the sequence the tasks enter the system, by the decision to which machine an entering task is assigned next, and by the position an assigned task is then given in the corresponding machine queue.

(2) As all machines are identically tooled each task can be processed by all machines and with this also preemption of tasks between machines might be possible.

(3) The business need of processing all tasks within the next eight hour shift can be translated in some objective which says that we want to minimize schedule length or makespan.

(4) It is well known that for identical machines, independent tasks and the objective of minimizing makespan, schedules with preemptions of tasks exist which are never worse than schedules where task preemption is not allowed.

From the above knowledge sources (1)-(4) we conclude within the problem analysis phase to choose McNaughton's rule [McN59] to construct a first basic schedule. From an evaluation of the generated schedule it turns out that all tasks could be processed within the next shift. Another observation is that there is still enough idle time to process additional tasks in the same shift. To evaluate the dynamics of the above manufacturing environment we simulate the schedule taking also transportation times of the preempted tasks to the different machines into account. From the results of simulation runs we now get a better understanding of the problem. It turns out that considering transportation of tasks the schedule constructed by McNaughton's rule is not feasible, i.e. in conflict according to the restriction to finish all tasks within the coming shift. The transport times which were neglected during static schedule generation have a major impact on the schedule length.

From this we must analyze the problem again and with the results from the evaluation process we derive the fact that the implemented schedule should not have machine change-overs of any task, to avoid transport times between machines.

Based on this new constraint and further knowledge from scheduling theory we decide now to use the longest processing time heuristic to schedule all tasks. It is shown in Chapter 5 that LPT gives good performance guarantees concerning schedule length and problem settings with identical machines. Transport times between machines do not have to be considered any more as each task is only assigned to one machine. Let us assume the evaluation of the LPT-schedule is satisfactory.

Now, we use earliest start and latest finish times for each task as constraints for ONC. These time intervals can be determined using the generated OFP-schedule. Moreover we translate the LPT rule into a more operational scheduling rule which says: release all the tasks in a non-increasing order of processing times to the flexible manufacturing cell and always assign a task to the queue of a machine which has least actual total work to process. The machine itself selects tasks from its own queue according to a first-come-first-served (FCFS) strategy.

As long as the flexible manufacturing cell has no disturbances ONC can stick to the given translation of the LPT-strategy. Now, assume a machine breaks down and that the tasks waiting in the queue have to be assigned to queues of the remaining machines. Let us further assume that under the new constraints not all the tasks can be finished in the current shift. From this a new objective occurs for reactive scheduling which says that as many tasks as possible should be finished. Now, FCFS would not be the appropriate scheduling strategy any longer; a suitable ad-hoc decision for local repair of the schedule has to be made. Finding this decision on the ONC-level means again to apply some problem analysis also in the sense of diagnosis and therapy, i.e. also ad-hoc decisions follow some analysis-construction sequence. If there is enough time available also some simulation runs could be applied, but in general this is not possible. To show a way how the problem can be resolved similar rules as these from Table 15.3.4 could be used.

For the changed situation, the shortest processing time (SPT) rule would now be applied. The SPT rule is proposed due to the expectation that this rule helps to finish as many tasks as possible within the current shift. In case of further disturbances that cause major deviations from the current system status, OFP has to be reactivated for a global repair of the schedule.

At the end of this section we want to discuss shortly the relationship of our approach to solve production scheduling problems and the requirements of integrated problem solving within computer integrated manufacturing. The IPS has to be connected to existing information systems of an enterprise. It has interfaces to the production planning systems on a tactical level of decision making and the real-time oriented CAM-systems. It represents this part of the production scheduling system which carries out the feedback loop between planning and execution. The vertical decision flow is supplemented by a horizontal information flow from CAE and CAQ. The position of the IPS within CIM is shown in Figure 15.3.13.

We gave a short introduction to an IPS which uses an interactive scheduling approach based on the ACE loop. Analysis and evaluation are carried out mainly by the decision maker, construction is mainly supported by the system. To that end a number of models and methods for analysis and construction have been devised, from which an appropriate selection should be possible. The modular and open architecture of the system offers the possibility of a step by step implementation which can be continuously adapted to changing requirements.

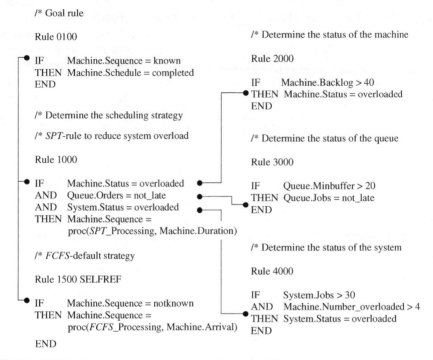

/* Goal rule

Rule 0100

```
IF      Machine.Sequence = known
THEN  Machine.Schedule = completed
END
```

/* Determine the scheduling strategy

/* *SPT*-rule to reduce system overload

Rule 1000

```
IF      Machine.Status = overloaded
AND   Queue.Orders = not_late
AND   System.Status = overloaded
THEN  Machine.Sequence =
         proc(SPT_Processing, Machine.Duration)
```

/* *FCFS*-default strategy

Rule 1500 SELFREF

```
IF      Machine.Sequence = notknown
THEN  Machine.Sequence =
         proc(FCFS_Processing, Machine.Arrival)

END
```

/* Determine the status of the machine

Rule 2000

```
IF      Machine.Backlog > 40
THEN  Machine.Status = overloaded
END
```

/* Determine the status of the queue

Rule 3000

```
IF      Queue.Minbuffer > 20
THEN  Queue.Jobs = not_late
END
```

/* Determine the status of the system

Rule 4000

```
IF      System.Jobs > 30
AND   Machine.Number_overloaded > 4
THEN  System.Status = overloaded
END
```

Table 15.3.4 *Example problem for reactive scheduling.*

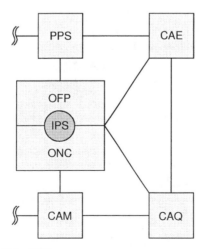

Figure 15.3.13 *IPS within CIM.*

A further application of the system lies in a distributed production scheduling environment. The considered manufacturing system has to be modeled and appropriately decomposed into subsystems. For the manufacturing system and each of its subsystems corresponding IPS apply, which are implemented on different computers connected by an appropriate communication network. The IPS on the top level of the production system serves as a coordinator of the subsystem IPS. Each IPS on the subsystem level works independently fulfilling the requirements from the master level and communicating also with the other IPS on this level. Only if major decisions which requires central coordination the master IPS is also involved.

References

Ata91   H. Atabakhsh, A survey for constraint based scheduling systems using an artificial intelligence approach, *Artif. Intell. Eng.* 6, 1991, 58-73.

BPH82   J. H. Blackstone, D. T. Phillips, G. L. Hogg, A state-of-the-art survey of dispatching rules for manufacturing job shop operations, *Int. J. Prod. Res.* 20, 1982, 27-45.

Bul82   W. Bulgren, *Discrete System Simulation*, Prentice-Hall, 1982.

BY86   J. A. Buzacott, D. D. Yao, FMS: a review of analytical models, *Management Sci.* 32, 1986, 890-905.

Car86   A. S. Carrie, The role of simulation in FMS, in: A. Kusiak (ed.), *Flexible Manufacturing Systems: Methods and Studies*, Elsevier, 1986, 191-208.

CPP88   A. Collinot, C. Le Pape, G. Pinoteau, SONIA: a knowledge-based scheduling system, *Artif. Intell. Eng.* 3, 1988, 86-94.

CY91   P. Coad, E. Yourdon, *Object-Oriented Analysis*, Prentice-Hall, 1991.

DP88   R. Dechter, J. Pearl, Network-based heuristics for constraint-satisfaction problems, *Artif. Intell.* 34, 1988, 1-38.

DTLZ93   J. Drake, W. T. Tsai, H. J. Lee, Object-oriented analysis: criteria and case study, *Int. J. of Software Engineering and Knowledge Engineering* 3, 1993, 319-350.

EGS97   K. Ecker, J. N. D. Gupta, G. Schmidt, A framework for decision support systems for scheduling problems, *European J. Oper. Res.* 101, 1997, 452-462.

ES93   K. Ecker, G. Schmidt, Conflict resolution algorithms for scheduling problems, in: K. Ecker, R. Hirschberg (eds.), *Workshop on Parallel Processing*, TU Clausthal, 1993, 81-90.

Fox87   M. S. Fox, *Constraint Directed Search: A Case Study of Job-Shop Scheduling*, Morgan Kaufmann, 1987.

Fox90   M. S. Fox, Constraint-guided scheduling - a short history of research at CMU, *Computers in Industry* 14, 1990, 79-88

Fre78   E. C. Freuder, Synthesizing constraint expressions, *Comm. ACM* 11, 1978, 958-966.

FS84   M. S. Fox, S. F. Smith, ISIS - a knowledge-based system for factory scheduling, *Expert Systems* 1, 1984, 25-49.

FS90   M. S. Fox, K. Sycara, Overview of CORTES: a constraint based approach to production planning, scheduling and control, *Proc. 4th Int. Conf. Expert Systems in Production and Operations Management*, 1990, 1-15

GJ79   M. R. Garey, D. S. Johnson, *Computers and Intractability: A Guide to the Theory of NP-Completeness.* W. H. Freeman, 1979.

Har73   J. Harrington, *Computer Integrated Manufacturing*, Industrial Press, 1973.

KA87   J. J. Kanet, H. H. Adelsberger, Expert systems in production scheduling, *European J. Oper. Res.* 29, 1987, 51-59.

Kan86   Kandel, A., *Fuzzy Mathematical Techniques with Applications*, Addison-Wesley, 1986.

Kus86   A. Kusiak, Application of operational research models and techniques in flexible manufacturing systems, *European J. Oper. Res.* 24, 1986, 336-345.

KSW86   M. V. Kalkunte, S. C. Sarin, W. E. Wilhelm, Flexible Manufacturing Systems: A review of modelling approaches for design, justification and operation, in: A. Kusiak (ed.), *Flexible Manufacturing Systems: Methods and Studies*, Elsevier, 1986, 3-28.

LGW86   A. J. Van Looveren, L. F. Gelders, N. L. Van Wassenhove, A review of FMS planning models, in: A. Kusiak (ed.), *Modelling and Design of Flexible Manufacturing Systems*, Elsevier, 1986, 3-32.

Mac77   A. K. Mackworth, Consistency in networks of relations, *Artif. Intell.* 8, 1977, 99-118.

McN59   R. McNaughton, Scheduling with deadlines and loss functions, *Management Sci.* 12, 1959, 1-12.

Mon74   U. Montanari, Networks of constraints: Fundamental properties and applications to picture processing, *Inform. Sci.* 7, 1974, 95-132.

MS92a   K. Mertins, G. Schmidt (Hrsg.), *Fertigungsleitsysteme 92*, IPK Eigenverlag, 1992

MS92b   W. Mai, G. Schmidt, Was Leitstandsysteme heute leisten, *CIM Management* 3, 1992, 26-32.

NS91   S. Noronha, V. Sarma, Knowledge-based approaches for scheduling problems, *IEEE Trans. Knowledge and Data Engineering* 3(2), 1991, 160-171.

PI77   S. S. Panwalkar, W. Iskander, A survey of scheduling rules, *Oper. Res.* 25, 1977, 45-61.

RM93   A. Ramudhin, P. Marrier, An object-oriented logistic tool-kit for schedule modeling and representation, *Int. Conf. on Ind. Eng. and Prod. Man.*, Mons, 1993, 707-714.

Ran86   P. G. Ranky, *Computer Integrated Manufacturing*, Prentice-Hall, 1986.

Sch89a   G. Schmidt, *CAM: Algorithmen und Decision Support für die Fertigungssteuerung*, Springer, 1989.

Sch89b   G. Schmidt, Constraint satisfaction problems in project scheduling, in: R. Słowiński, J. Węglarz (eds.), *Advances in Project Scheduling*, Elsevier, 1989, 135-150.

Sch91   A.-W. Scheer, *CIM - Towards the Factory of the Future*, Springer, 1991.

Sch92   G. Schmidt, A decision support system for production scheduling, *J. Decision Syst.* 1(2-3), 1992, 243-260.

Sch94   G. Schmidt, How to apply fuzzy logic to reactive scheduling, in: E. Szelke, R. Kerr (eds.), *Knowledge Based Reactive Scheduling*, North-Holland, 1994, 57-57.

Sch96   G. Schmidt, Modelling Production Scheduling Systems, *Int. J. Production Economics* 46-47, 1996, 106-118.

Sch98   G. Schmidt, Case-based reasoning for production scheduling, *Int. J. Production Economics* 56-57, 1998, 537-546.

SFO86   S. F. Smith, M. S. Fox, P. S: Ow, Constructing and maintaining detailed production plans: investigations into the development of knowledge-based factory scheduling systems, *AI Magazine* 7, 1986, 45-61.

Smi92   S. F. Smith, Knowledge-based production management: approaches, results and prospects, *Production Planning and Control* 3(4), 1992, 350-380.

SPP+90   S. F. Smith, S. O. Peng, J.-Y. Potvin, N. Muscettola, D. C. Matthys, An integrated framework for generating and revising factory schedules, *J. Oper. Res. Soc.* 41, 1990, 539-552

SS90   S. C. Sarin, R. R. Salgame, Development of a knowledge-based system for dynamic scheduling, *Int. J. Prod. Res.* 28, 1990, 1499-1512.

Ste85   K. E. Stecke, Design, planning, scheduling and control problems of flexible manufacturing systems, *Ann. Oper. Res.* 3, 1985, 3-121.

WBJ90   J. R. Wilfs-Brock, R. E. Johnson, Surveying current research in object oriented design, *Comm. ACM* 33, 1990, 104-124.

Wir76   N. Wirth, *Algorithms + Data Structures = Programs*, Prentice-Hall, 1976.

Zad65   L. A. Zadeh, Fuzzy sets, *Information and Control* 8, 1965, 338-353.

Zad73   L. A. Zadeh, The concept of linguistic variables and its application to approximate reasoning, Memorandum ERL-M411, UC Berkeley, 1973.

Index

– S –